Elsevier Academic Press
30 Corporate Drive, Suite 400, Burlington, MA 01803, USA
525 B Street, Suite 1900, San Diego, California 92101-4495, USA
84 Theobald's Road, London WC1X 8RR, UK

This book is printed on acid-free paper.

Cover designer: Neringa Stjernman

Library of Congress Cataloging-in-Publication Data
Application submitted

British Library Cataloguing in Publication Data
A catalogue record for this book is available from the British Library

ISBN: 0-12-437552-9
For all information on all Elsevier Academic Press publications
visit our Web site at www.books.elsevier.com

Printed in the United States of America
05   06   07   08   09   10   9   8   7   6   5   4   3   2   1

Contents

Preface   vii

Chapter 1.   **Introduction**   1

1.1   Biomedical Signal Processing: Objectives and Contexts . . .   2

    1.1.1   Objectives   2

    1.1.2   Contexts   4

1.2   Basics of Bioelectrical Signals   6

    1.2.1   On the Cellular Level   6

    1.2.2   On the Body Surface   9

    1.2.3   Bioelectrical Signals   11

1.3   Signal Acquisition and Analysis   14

1.4   Performance Evaluation   16

    1.4.1   Databases   17

    1.4.2   Simulation   21

Bibliography   22

Chapter 2.   **The Electroencephalogram—A Brief Background**   25

2.1   The Nervous System   27

    2.1.1   Neurons   28

    2.1.2   The Cerebral Cortex   30

2.2   The EEG—Electrical Activity Measured on the Scalp   31

    2.2.1   EEG Rhythms and Waveforms   33

    2.2.2   Categorization of EEG Activity   36

2.3   Recording Techniques   37

2.4   EEG Applications   39

    2.4.1   Epilepsy   40

    2.4.2   Sleep Disorders   44

    2.4.3   Brain–Computer Interface   47

Bibliography   50

Chapter 3.   EEG Signal Processing   **55**
  3.1 Modeling the EEG Signal 55
     3.1.1 Deterministic and Stochastic Signal Properties 55
     3.1.2 Stochastic Properties 56
     3.1.3 Linear, Stochastic Models 63
     3.1.4 Nonlinear Modeling of the EEG 67
  3.2 Artifacts in the EEG 71
     3.2.1 Artifact Characteristics 73
     3.2.2 Artifact Processing 76
     3.2.3 Artifact Reduction Using Linear Filtering 77
     3.2.4 Artifact Cancellation Using Linearly Combined
           Reference Signals 78
     3.2.5 Adaptive Artifact Cancellation Using Linearly
           Combined Reference Signals 83
     3.2.6 Artifact Cancellation Using Filtered Reference Signals 87
  3.3 Nonparametric Spectral Analysis 91
     3.3.1 Fourier-based Power Spectrum Analysis 92
     3.3.2 Spectral Parameters 97
  3.4 Model-based Spectral Analysis 103
     3.4.1 The Autocorrelation/Covariance Methods 106
     3.4.2 The Modified Covariance Method 110
     3.4.3 Burg's Method 112
     3.4.4 Performance and Design Parameters 116
     3.4.5 Spectral Parameters 119
  3.5 EEG Segmentation 125
     3.5.1 Spectral Error Measure—The Periodogram Approach 128
     3.5.2 Spectral Error Measure—The Whitening Approach . . 131
     3.5.3 Segmentation Performance and Beyond 134
  3.6 Joint Time–Frequency Analysis 135
     3.6.1 The Short-Time Fourier Transform 137
     3.6.2 The Ambiguity Function 142
     3.6.3 The Wigner–Ville Distribution 147
     3.6.4 General Time–Frequency Distributions—Cohen's Class 153
     3.6.5 Model-based Analysis of Slowly Varying Signals 158
  Bibliography 162
  Problems 175

Chapter 4.   Evoked Potentials   **181**
  4.1 Evoked Potential Modalities 183
     4.1.1 Auditory Evoked Potentials 185
     4.1.2 Somatosensory Evoked Potentials 187
     4.1.3 Visual Evoked Potentials 188

4.1.4   Evoked Potentials and Cognition 189
4.2   Noise Characteristics 190
4.3   Noise Reduction by Ensemble Averaging 192
4.3.1   Averaging of Homogeneous Ensembles 193
4.3.2   Ensemble Averaging Interpreted as Linear Filtering . 200
4.3.3   Exponential Averaging 202
4.3.4   Averaging of Inhomogeneous Ensembles 207
4.3.5   Spike Artifacts and Robust Averaging 219
4.3.6   The Effect of Latency Shifts 225
4.3.7   Estimation of Latency Shifts 229
4.3.8   Weighting of Averaged Evoked Potentials Using
Ensemble Correlation 236
4.4   Noise Reduction by Linear Filtering 241
4.4.1   Time-Invariant, A Posteriori Filtering 243
4.4.2   Limitations with Time-Invariant, A Posteriori
Filtering 252
4.5   Single-Trial Analysis Using Basis Functions 253
4.5.1   Orthogonal Expansions 254
4.5.2   Sets of Basis Functions 260
4.5.3   Karhunen–Loève Expansion—Optimal Basis Functions 264
4.5.4   Interpretation as Linear, Time-Variant Filtering . . . 271
4.5.5   Modeling with Damped Sinusoids 272
4.6   Adaptive Analysis Using Basis Functions 278
4.6.1   The Instantaneous LMS Algorithm 279
4.6.2   The Block LMS Algorithm 285
4.7   Wavelets 286
4.7.1   The Wavelet Transform 288
4.7.2   Multiresolution Signal Analysis 292
4.7.3   Multiresolution Signal Analysis—A Classical Example 297
4.7.4   Implementation of the Discrete Wavelet Transform
Using Filter Banks 300
4.7.5   Wavelet Examples 306
4.7.6   Denoising 312
Bibliography 318
Problems 329

Chapter 5.   The Electromyogram   **337**
5.1   The Electrical Activity of Muscles 338
5.1.1   Action Potentials and Motor Units 338
5.1.2   Recording of Myoelectric Signals 343
5.1.3   EMG Applications 345
5.2   Amplitude Estimation in the Surface EMG 347

5.2.1   Signal Model and ML Estimation 347
5.2.2   Modifications of the ML Amplitude Estimator 352
5.2.3   Multiple Electrode Sites 356
5.3   Spectral Analysis of the Surface EMG 361
5.4   Conduction Velocity Estimation 365
5.4.1   Two-Channel Time Delay Estimation 367
5.4.2   Multichannel Time Delay Estimation 369
5.5   Modeling the Intramuscular EMG 371
5.5.1   A Signal Model of the MUAP Train 372
5.5.2   MUAP Train Amplitude 377
5.5.3   MUAP Train Power Spectrum 378
5.6   Intramuscular EMG Signal Decomposition 383
5.6.1   MUAP Feature Extraction and Clustering 387
5.6.2   Resolution of Superimposed MUAP Waveforms 391
Bibliography 398
Problems 406

Chapter 6.   The Electrocardiogram—A Brief Background   411
6.1   Electrical Activity of the Heart 412
6.2   Generation and Recording of an ECG 415
6.2.1   Depolarization and Repolarization 415
6.2.2   ECG Recording Techniques 419
6.2.3   ECG Waves and Time Intervals 426
6.3   Heart Rhythms 430
6.3.1   Sinus Rhythm 431
6.3.2   Premature Beats 433
6.3.3   Atrial Arrhythmias 434
6.3.4   Ventricular Arrhythmias 436
6.3.5   Conduction Blocks 438
6.4   Heartbeat Morphologies 438
6.4.1   Myocardial Ischemia 438
6.4.2   Myocardial Infarction 439
6.5   Noise and Artifacts 440
6.6   Clinical Applications 443
6.6.1   Resting ECG 443
6.6.2   Intensive Care Monitoring 444
6.6.3   Ambulatory Monitoring 444
6.6.4   Stress Test 445
6.6.5   High-Resolution ECG 447
Bibliography 449

Chapter 7. ECG Signal Processing 453

  7.1 Baseline Wander 457
      7.1.1 Linear, Time-Invariant Filtering 458
      7.1.2 Linear, Time-Variant Filtering 467
      7.1.3 Polynomial Fitting 470
  7.2 Powerline Interference (50/60 Hz) 473
      7.2.1 Linear Filtering 473
      7.2.2 Nonlinear Filtering 476
      7.2.3 Estimation–Subtraction 479
  7.3 Muscle Noise Filtering 484
  7.4 QRS Detection 485
      7.4.1 Signal and Noise Problems 487
      7.4.2 QRS Detection as an Estimation Problem 488
      7.4.3 Detector Preprocessing 497
      7.4.4 Decision Rules 504
      7.4.5 Performance Evaluation 507
  7.5 Wave Delineation 510
  7.6 Data Compression 514
      7.6.1 Lossless Compression 517
      7.6.2 Lossy Compression—Direct Methods 519
      7.6.3 Lossy Compression—Transform-based Methods 526
      7.6.4 Interbeat Redundancy 533
      7.6.5 Interlead Redundancy 536
      7.6.6 Quantization and Coding 539
      7.6.7 Performance Evaluation 541
  Bibliography 545
  Problems 557

Chapter 8. ECG Signal Processing: Heart Rate Variability 567

  8.1 Acquisition and RR Interval Conditioning 568
  8.2 Time Domain Measures 570
  8.3 Heart Rhythm Representations 573
      8.3.1 The Integral Pulse Frequency Modulation Model . . . 574
      8.3.2 Interval Series Representions 578
      8.3.3 Event Series Representation 582
      8.3.4 Heart Timing Representation 585
  8.4 Spectral Analysis of Heart Rate Variability 589
      8.4.1 Direct Estimation from Unevenly Spaced Samples . . 592
      8.4.2 The Spectrum of Counts 593
      8.4.3 Lomb's Periodogram 597
  8.5 Clustering of Beat Morphologies 603
  8.6 Dealing with Ectopic Beats 605

8.6.1   Correlation-based Correction 608

8.6.2   Interpolation-based Correction 609

8.6.3   The Heart Timing Signal and Ectopic Beats 611

8.7   Interaction with Other Physiological Signals 614

Bibliography 621

Problems 628

Appendix A. Review of Important Concepts   633

A.1   Matrix Fundamentals 633

A.1.1   Definitions 633

A.1.2   Matrix Decomposition 638

A.1.3   Matrix Optimization 640

A.1.4   Linear Equations 641

A.2   Discrete-Time Stochastic Processes 642

A.2.1   Definitions 642

A.2.2   Stationarity 644

A.2.3   Ergodicity 645

A.2.4   Bias and Consistency 645

A.2.5   Power Spectrum 646

A.2.6   White Noise 646

A.2.7   Filtering of Stochastic Processes 647

Bibliography 648

Appendix B. Symbols and Abbreviations   649

B.1   Mathematical Symbols 649

B.2   Abbreviations 658

Index   661

Preface

Bioelectrical signals have been recorded and analyzed for several decades but still continue to excite physicians and engineers. Novel signal processing techniques have helped uncover information which completely changed the way various diseases previously were diagnosed. In fact, it is today difficult to imagine a situation when diseases related to the heart, the brain, or the muscles are diagnosed without also including certain information derived from bioelectrical signals. Such information is essential to therapeutic devices in cardiac, neurological, and neuromuscular applications, and will in the future, when systematically fused with other types of biomedical signals, continue to improve the quality of life of many patients. Monitoring of home-based patients is becoming increasingly popular in health care, frequently involving bioelectrical signals which can be safely and comfortably recorded using noninvasive techniques.

The aim of this book is to present a comprehensive overview of techniques with particular relevance to the processing of bioelectrical signals. The presentation is problem-driven and deals with issues having received considerable attention from both a scientific viewpoint, i.e., in the form of publications and conference presentations, and a viewpoint of product development. Since biomedical signal processing has been largely synonymous with the processing of ECG, EEG, EMG, and evoked potentials, we have focused the presentation on issues related to these four types of bioelectrical signals. It is yet our conviction that the reader is fully capable of transferring the way of thinking developed herein for bioelectrical signals, as well as to transfer the developed methods, when later dealing with other types of biomedical signals.

Choosing a problem-driven presentation means, in this book, that different methods are described within the context of a certain bioelectrical signal. For example, power spectral analysis is described within the context of EEG signal processing though such analysis is certainly well-established in other biomedical applications as well. While some may feel that the realm of a method's usefulness is depreciated with this kind of presentation, we hope nonetheless that the power in connecting a particular type of signal to a particular method outweighs the disadvantages. On occasion, the problem-driven presentation also means the display of a smorgasbord of methods developed to solve a certain problem such as the cancellation of powerline interference. We hope that this way of dealing with a problem would serve the reader well by offering an idea of the diversity with which a problem can be solved. Not all methods considered in this textbook are directly applicable in clinical practice but may require one or several heuristic add-ons before their performance become satisfactory; the exact definition of such add-ons is rarely disclosed in the original publication of a method but needs to be developed by those interested in the method's pursuit.

With the display of different methods for solving a particular problem comes the natural wish of knowing which method offers the best performance. We have, however, abstained from making such comparisons due to the many pitfalls associated with choosing performance measure, data set, and so forth. We would instead like to challenge the reader to delve into this important aspect.

Biomedical signal processing has today reached certain maturity as an academic subject and is now supported by the availability of a handful of textbooks. Being an interdisciplinary subject by nature, biomedical signal processing has to be taught quite differently depending on the educational program. For students in biomedical engineering a course in physiology is part of the curriculum, whereas students in electrical engineering and computer science usually lack such a course. In order to maintain the interdisciplinary nature when teaching the latter group of students, we have included chapters or sections with brief, self-contained introductions to the underlying electrophysiology, recording techniques, and some important clinical applications. Without any prior knowledge of these aspects, a course in biomedical signal processing runs the risk of losing its very essence.

It is evident that a course on biomedical signal processing may embrace widely different contents—an observation which not only applies to the choice of biomedical signals but also to the choice of methodologies. Rather than yield to the temptation to include as much as possible, we have deliberately avoided to cover certain important techniques including pattern recognition, artificial neural networks, higher-order statistics, and nonlinear dynamics. Though important in biomedical applications, the fundamentals of these techniques are well-covered by a number of textbooks.

This book is intended for final year undergraduate students and graduate students in biomedical engineering, electrical engineering, and computer science. It is suitable for a one-quarter or one-semester course depending on the content covered and the amount of emphasis put on problem solving and project work. A necessary prerequisite is the fundamentals of digital signal processing as presented in textbooks such as [1, 2]. Since many recent methods used in biomedical applications are based on concepts from statistical modeling and signal processing, a basic course in probability theory and stochastic processes is another important prerequisite. It is also desirable that the reader has certain familiarity with linear algebra so that common matrix operations (summarized in Appendix A) can be performed. Readers who want to achieve a deeper understanding of statistical signal processing are referred to a number of highly recommended textbooks on this topic [3–5]. Adaptive filtering is a topic which is just briefly touched upon in this book; comprehensive coverage of such filters and their properties can be found in [6].

This book may also be used as a comprehensive reference for practicing engineers, physicians, researchers and, of course, anyone interested in finding out what information can be derived from bioelectrical signals. For practicing engineers, we have used selected parts of the book for a short course on biomedical signal processing (i.e., 2–3 days); in such cases, the main emphasis should be put on the significance of different methods rather than on mathematical details.

Contents Overview

Chapter 1 puts biomedical signal processing in context, and gives a brief description of bioelectricity and its manifestation on the body surface as signals. General aspects on signal acquisition and performance evaluation are briefly considered.

Chapter 2 provides the reader with the basics of the brain, serving as a background to the following chapter on EEG signal processing. Some common EEG patterns are described and their relationships to cerebral pathology are pointed out. An understanding of EEG signal characteristics, as well as the purposes for which the characteristics can be exploited, is essential information when assimilating the contents of Chapter 3. The main themes in Chapter 3 on EEG signal processing are related to artifact rejection and spectral analysis; two techniques of critical importance to EEG interpretation. Special attention is given to the multitude of spectral analysis techniques and a section on time–frequency analysis is included.

Chapter 4 provides a comprehensive overview of noise reduction techniques for use with event-related signals, here treated within the context of evoked potentials. Similar to EEG signals and spectral analysis, evoked potentials and signal averaging became "partners" at a very early stage in the history of biomedical signal processing, thus motivating the emphasis on this partnership. The overview of noise reduction techniques covers both ensemble averaging (and its spawn) and more advanced approaches where the signal is modeled and filtered using a basis function expansion; wavelets represent one such popular approach.

Chapter 5 deals with myoelectric activity and the related EMG signal recorded either by noninvasive or invasive techniques (and makes a minor departure from the general framework of dealing with signals recorded on the body surface). Of the many developed methods for EMG signal analysis, we cover some central ones related to muscle force and conduction velocity where signal modeling and statistical estimation techniques are involved.

Chapter 6 contains a background to the electrophysiology of the heart, describes the main characteristics of the ECG signal in terms of morphology and rhythm, and prepares the way for Chapters 7 and 8 by mentioning the most important ECG applications. Chapter 7 describes a suite of methods, essential to any system which performs ECG signal analysis, developed for the purpose of noise reduction, heartbeat detection and delineation, and data compression. Chapter 8 is completely devoted to the analysis of heart rate variability—an area of considerable clinical and technical interest in recent years—and describes techniques for representing and characterizing such variability in the time and frequency domain.

We have included an extensive, but not exhaustive, number of references which give the interested reader rich possibilities to further explore the original presentations of the methods. References are almost exclusively made to journal publications since these are easily retrieved from libraries. As a result, the very first publication of a method, often appearing in a conference proceeding, is not acknowledged for which we apologize.

A collection of problems has been developed in order to illustrate the presented methods and their applications. While some problems are straightforward to solve, others require considerable effort and background knowledge and are intended as "appetizers" for students interested in pursuing research in biomedical signal processing. An accompanying manual with detailed solutions to all problems is available at the publisher's web site

www.books.elsevier.com/0124375529

to instructors who adopt the book.

Any course on biomedical signal processing must include one or several projects which give the student an opportunity to process signals and to learn the pros and cons of a method. We have developed a companion web site where several project descriptions are listed and signals available for download; its location is

www.biosignal.lth.se

An important goal with this web site is to allow the inclusion of new projects so that projects can be submitted by anyone interested in teaching the contents of this book (submission instructions are available at the web site).

Those interested in using this book for a one-quarter course may want to omit the sections on time–frequency analysis (Section 3.6), basis functions and related adaptive analysis (Sections 4.5 and 4.6), wavelets (Section 4.7), and certain parts of Chapter 8 dealing with heart rate variability; the mathematical level is relatively advanced in all these parts. A shorter course may to a lesser extent deal with problem solving, however, we strongly encourage the inclusion of at least one project since it provides the student with experiences essential to the understanding of biomedical signal processing.

Acknowledgments

The making of an interdisciplinary textbook is only possible with the help of many people. We are fortunate to have colleagues and friends who, most generously, have supported us with their expertise and helpful advices.

We would like to express our deep gratitude to Juan Pablo Martínez (Zaragoza), Olivier Meste (Nice), Salvador Olmos (Zaragoza), and Olle Pahlm (Lund), who willingly reviewed the material in different parts of the manuscript and provided us with numerous suggestions and corrections. Their profound understanding of biomedical signals has considerably contributed to improve the manuscript.

We are grateful for the input given by Jan Anders Ahnlide (Lund), Magnus Lindgren (Oslo), Miguel Angel Mañanas (Barcelona), and Roberto Merletti (Torino); their expertise in the areas of neurophysiology and electromyography has been invaluable when preparing the parts of the book concerning these aspects. We are also grateful to Vaidotas Marozas (Kaunas), Nitish Thakor (Baltimore), and Patrik Wahlberg (Lund); they read parts of the manuscript and provided us with helpful comments and unique perspectives on biomedical signal processing.

We would like to acknowledge the helpful comments and feedback from our doctoral students; special thanks goes to Joakim Axmon, Javier Mateo, Frida Nilsson, Kristian Solem, Martin Stridh, and Magnus Åström, for proofreading, designing problems and projects, and providing us with data. Since this course was introduced some five years ago at our universities, we have received several comments from our students to whom we would like to express our appreciation; in particular, we thank Åsa Robertsson for help with Matlab programming.

We also wish to thank Rute Almeida (Porto), Ed Berbari (Indianapolis), Per Ola Börjesson (Lund), Christina Grossmann (Lund), Bo Hjorth (Stockholm), Arturas Janušauskas (Kaunas), Algimantas Krisçiukaitis (Kaunas), Gerhard Kristensson (Lund), Arunas Lukoseviçius (Kaunas), Jonas Pettersson (Lund), Ana Paula Rocha (Porto), Helen Sheppard (Lund), Owe Svensson (Lund), and Viktor Öwall (Lund). Their different talents contributed significantly to improve the content of this book.

We are grateful to Kerstin Brauer (Lund) for skillfully helping us with the illustrative figures, and to Lena Eliasson (Lund) for preparing Figure 1.1. A number of figures are based on signals which generously were made available to us; Stefan Sauermann (Vienna) provided the signals for Figure 1.4(a) and (c), Jan Anders Ahnlide (Lund) for Figure 1.4(b), and José Luis Martínez de Juan (Valencia) for Figure 1.5(b). We are also grateful to Fabrice Wendling (Rennes) who made Figure 3.6 available, Nils Löfgren (Borås) for Figure 3.31, Patrick Celka (Neuchâtel) for Figure 3.38, Rodrigo Quian Quiroga (Leicester) for Figure 4.49, Maria Hansson (Lund) for Figure 6.13, Franc Jager (Ljubljana) for Figure 8.1, and Luca Mainardi (Milano) for Figure 8.24.

Special thanks are due to Christine Minihane and the staff at Elsevier for their cooperation and continuing encouragement during the authoring and production process.

The Spanish government kindly provided financial support through the grants SAB2003-0130 and TXT1996-1793. Support was also received from the European Community through the SOCRATES/ERASMUS Teaching Staff Mobility program.

We are infinitely grateful to our respective families for their support and encouragement throughout this project. Above all, we admire their endless patience when we got lost in yet another discussion on "the book." Parts of the manuscript were conceived in a picturesque village of the Spanish Pyrenees; we would like to express our sincere thanks to relatives and friends living there who made these occasions productive and highly enjoyable.

We are deeply indebted to artist Neringa Stjernman who enriched the book cover by her interpretation of bioelectrical signals; it was a great surprise, and a pure joy, to find out about her unique relationship to electrophysiology. It would be incomplete not to acknowledge the influence of Palle Mikkelborg whose music has been a vital source of inspiration in the making of this book. His trumpet paints a spectrum full of colors—blue in green, orange with electric red, anything but grey—which makes one sing a song.

Leif Sörnmo
Pablo Laguna

Bibliography

[1] J. G. Proakis and D. G. Manolakis, *Digital Signal Processing. Principles, Algorithms, and Applications.* New Jersey: Prentice-Hall, 3rd ed., 1996.

[2] A. V. Oppenheim, R. W. Schafer, and J. R. Buck, *Discrete-Time Signal Processing.* New Jersey: Prentice-Hall, 2nd ed., 1999.

[3] C. W. Therrien, *Discrete Random Signals and Statistical Signal Processing.* New Jersey: Prentice-Hall, 1992.

[4] S. M. Kay, *Fundamentals of Statistical Signal Processing. Estimation Theory.* New Jersey: Prentice-Hall, 1993.

[5] M. Hayes, *Statistical Digital Signal Proccessing and Modeling.* New York: John Wiley & Sons, 1996.

[6] S. Haykin, *Adaptive Filter Theory.* New Jersey: Prentice-Hall, 4th ed., 2002.

Chapter 1

Introduction

The function of the human body is frequently associated with signals of electrical, chemical, or acoustic origin. Such signals convey information which may not be immediately perceived but which is hidden in the signal's structure. This information has to be "decoded" or extracted in some way before the signals can be given meaningful interpretations. The signals reflect properties of their associated underlying biological systems, and their decoding has been found very helpful in explaining and identifying various pathological conditions. The decoding process is sometimes straightforward and may only involve very limited, manual effort such as visual inspection of the signal on a paper print-out or computer screen. However, the complexity of a signal is often quite considerable, and, therefore, biomedical signal processing has become an indispensable tool for extracting clinically significant information hidden in the signal.

Biomedical signal processing represents an interdisciplinary topic. Knowledge of the physiology of the human body is crucial to avoid the risk of designing an analysis method which distorts, or even removes, significant information. It is also valuable to have a sound knowledge of other topics such as anatomy, linear algebra, calculus, statistics, and circuit design.

Biomedical signal processing has, by some, been viewed as a stepping-stone for developing diagnostic systems which offer fully automated analysis. Some decades ago when computers first arrived in the area of medicine, automation was the overriding goal. However, this goal has been considerably modified over the years, not only because of the inherent difficulties in developing such systems, but equally so because the physician must be ultimately responsible for the diagnostic decisions taken. While fully automated analysis may be warranted in a few situations, today's goal is rather to develop computer systems which offer advanced aid to the physician in making well-founded decisions. In these systems biomedical signal processing has come to play a very important role.

Research in biomedical signal processing has so far mainly been concerned with the analysis of one particular signal type at a time ("unimodal signal analysis"); a fact, which to a large extent, influences the content of the present textbook. However, the emerging interest in multimodal signal analysis will definitely help to explain, in more detail, how different physiological subsystems interact with each other, such as the interaction between blood pressure and heart rate in the cardiovascular system. By exploring the mutual information contained in different signals, more qualified diagnostic decisions can be made. The increased algorithmic complexity associated with multimodal analysis is not a serious limitation since it will be met by the rapid advancement of computer technology and the ever-increasing computational speed.

1.1 Biomedical Signal Processing: Objectives and Contexts

1.1.1 Objectives

Biomedical signal processing has many objectives, and some of the most important ones are presented below. We also describe the main contexts in which biomedical signal processing is applied. Other challenging objectives and contexts can certainly be defined by those interested in pursuing a career in this fascinating, interdisciplinary field.

Historically, biomedical signals have often been assessed visually, and manual ruler-based procedures were developed to make sure that measurements could be obtained in a standardized manner. However, it is well-known that there is relatively poor concordance between manually obtained measurements, and this may lead to unreliable diagnostic conclusions. A fundamental objective of biomedical signal processing is therefore to *reduce the subjectivity* of manual measurements. The introduction of computer-based methods for the purpose of objectively quantifying different signal characteristics is the result of a desire to improve measurement accuracy as well as reproducibility.

In addition to reducing measurement subjectivity, biomedical signal processing is used in its own right for developing methods that *extract features* to help characterize and understand the information contained in a signal. Such feature extraction methods can be designed to mimic manual measurements, but are equally often designed to extract information which is not readily available from the signal through visual assessment. For example, small variations in heart rate that cannot be perceived by the human eye have been found to contain very valuable clinical information when quantified in detail using a suitable signal processing technique; see Chapter 8 for more details on this particular topic. Although it is certainly desirable to extract features that have an intuitive meaning to the physician, it is not necessarily those features which yield the best performance in clinical terms.

In many situations, the recorded signal is corrupted by different types of noise and interference, sometimes originating from another physiological process of the body. For example, situations may arise when ocular activity interferes with the desired brain signal, when electrodes are poorly attached to the body surface, or when an external source such as the sinusoidal 50/60 Hz powerline interferes with the signal. Hence, *noise reduction* represents a crucial objective of biomedical signal processing so as to mitigate the technical deficiencies of a recording, as well as to separate the desired physiological process from interfering processes. In fact, the desired signal is in certain situations so dramatically masked by noise that its very presence can only be revealed once appropriate signal processing has been applied. This is particularly evident for certain types of transient, very low-amplitude activity such as evoked potentials, which are part of brain signals, and late potentials, which are part of heart signals.

Certain diagnostic procedures require that a signal be recorded on a long timescale, sometimes lasting for several days. Such recordings are, for example, routinely done for the purpose of analyzing abnormal sleep patterns or to identify intermittently occurring disturbances in the heart rhythm. The resulting recording, which often involves many channels, amounts to huge data sizes, which quickly fill up hard disk storage space once a number of patients have been examined. Transmission of biomedical signals across public telephone networks is another, increasingly important application in which large amounts of data are involved. For both these situations, *data compression* of the digitized signal is essential and, consequently, another objective of biomedical signal processing. General-purpose methods of data compression, such as those used for sending documents over the internet, do not perform particularly well since the inherent characteristics of the biomedical signal are not at all exploited. Better performance can be obtained by applying tailored algorithms for data compression of biomedical signals. Data compression can also be understood in a wider sense as the process in which clinical information from a long-term recording is condensed into a smaller data set that is more manageable for the person analyzing the data. In this latter sense, it is highly desirable to develop signal processing algorithms which are able to determine and delimit clinically significant episodes.

Mathematical *signal modeling* and *simulation* constitute other important objectives in biomedical signal processing which can help to attain a better understanding of physiological processes. With suitably defined model equations it is possible to simulate signals which resemble those recorded on the cellular level or on the body surface, thereby offering insight into the relationship between the model parameters and the characteristics of the observed signal. Examples of bioelectrical models include models of the head and brain for localizing sources of neural activity and models of the thorax and the heart for simulating different cardiac rhythms. Signal modeling is also central to the branch of signal processing called "model-based signal processing," where algorithm development is based on the optimization of an appropriately selected performance criterion. In employing the model-based approach, the suggested signal model is fitted to the observed signal by selecting those values of the model parameters which optimize the performance criterion. While model-based biomedical signal processing represents a systematic approach to the design of algorithms—to be frequently adopted in the present textbook—it does not always lead to superior performance; heuristic approaches may actually perform just as well and sometimes even better. It is a well-known fact that many commercial, medical devices rely on the implementation of ad hoc techniques in order to achieve satisfactory performance.

The complexity of a signal model depends on the problem to be solved. In most signal processing contexts, it is fortunately not necessary to develop a multilevel model which accounts for cellular mechanisms, current propagation in tissue, and other biological properties. Rather, it is often sufficient to develop a "phenomenological" model which only accounts for phenomena which are relevant to the specific problem at hand.

1.1.2   Contexts

The other purpose of this section is to point out the three major clinical contexts in which algorithms for biomedical signal processing are designed, namely, the contexts of

- *diagnosis*,

- *therapy*, and

- *monitoring*.

In the *diagnostic context*, medical conditions are identified from the examination of signal information, reflecting the function of an organ such as the brain or the heart, in combination with other symptoms and clinical signs. A signal is often acquired by a noninvasive procedure which makes the examination less taxing on the patient. Most of these procedures are also associated with inexpensive technology for acquisition and analysis, thus increasing the likelihood that the technology can be disseminated to countries with less developed economies. A diagnostic decision rarely requires immediate availability of the results from signal analysis, but it is usually acceptable to wait a few minutes for the analysis to be completed. Hence, signal analysis can be done off-line on a personal computer, thus relying on standardized hardware and operating system, possibly supplemented with a digital signal processor (DSP) board for accelerating certain bottleneck computations. Algorithms for biomedical signal processing do not define the entire diagnostic computer system, but their scope ranges from performing a simple filtering operation to forming a more substantial part of the clinical decision-making.

Therapy generally signifies the treatment of disease and often involves drug therapy or surgery. With regard to biomedical signal processing, therapy may imply a narrower outlook in the sense that an algorithm is used to directly modify the behavior of a certain physiological process, for example, as the algorithms of a pacemaker do with respect to cardiac activity. In a therapeutic context, an algorithm is commonly designed for implementation in an implantable device like a heart defibrillator, and, therefore, it must, unlike an algorithm operating in a diagnostic context, strictly comply with the demands of on-line, real-time analysis. Such demands pose some serious constraints on algorithmic complexity as well as on the maximal acceptable time delay before a suitable action needs to be taken. Low power consumption is another critical factor to be considered in connection with devices that are implanted through a surgical procedure; for example, the battery of an implantable device is expected to last up to ten years. Hence, algorithms which involve computationally demanding signal processing techniques are less suitable for use in a therapeutic context.

Biomedical signal processing algorithms form an important part of real-time systems for *monitoring* of patients who suffer from a life-threatening condition. Such systems are usually designed to detect changes in cardiac or neurological function and to predict the outcome of a patient admitted to the intensive care unit (ICU). Since such changes may be reversible with early intervention, irreversible damage can sometimes be prevented. Similar to therapeutic contexts, the signal is processed during monitoring in an essentially sequential fashion such that past samples constitute the main basis for a decision, while just a few seconds of the future samples may also be considered—a property which usually stands in sharp contrast to signal processing for diagnostic purposes, where the signal is acquired in its entirety prior to analysis. Thus, a noncausal approach to signal analysis can only be adopted in the diagnostic context which mimics that of a human reader who interprets a signal by making use of both past and future properties. Constraints need to be imposed on the algorithmic design in terms of maximal delay time because the occurrence of a life-threatening event must be notified to the ICU staff within a few seconds. Another important issue to be considered is the implications of a clinical event that is missed by the algorithm or the implications of a nonevent that is falsely detected causing the staff to be notified.

1.2   Basics of Bioelectrical Signals

Although the scope of the present textbook is to present signal processing techniques useful for the analysis of electrical signals recorded on the *body surface*, it may still be well-motivated to consider the genesis of bioelectrical signals from a cellular perspective. Bioelectrical signals are related to ionic processes which arise as a result of electrochemical activity of a special group of cells having the property of *excitability*. The mechanisms which govern the activity of such cells are similar, regardless of whether the cells are part of the brain, the heart, or the muscles. In particular, the electrical force of attraction has central importance for the processing and transmission of information in the nervous system, as well as for sustaining the mechanical work done by the heart and the muscles. Since the origin of these voltages is only briefly described below, the interested reader is referred to textbooks on human physiology which offer a much more detailed description of the cellular aspects [1, 2]. The basic concepts introduced for mathematical modeling of bioelectrical phenomena are described in [3], while more comprehensive reading is found in [4–6].

1.2.1   On the Cellular Level

A cell is bounded by a plasma membrane which basically consists of lipid layers with poor ability to conduct an electrical current. The membrane possesses permeability properties which allow certain substances to pass from the inside of the cell to the outside through different channels, defined by body fluids, while other substances remain blocked. Intracellular and extracellular fluids mainly consist of water, which is electrically neutral; however, the fluids become electrically conductive since they contain several types of ions. The dominant ions in a nerve cell (neuron) are sodium (Na^+), potassium (K^+), and chloride (Cl^-). Other ions such as calcium (Ca^{2+}) are also present but play roles of varying importance depending on where the excitable cell is located; the calcium ion is much more important in the cells of the heart than in the nerves, for example.

Under resting conditions, the inside of a cell is negatively charged with respect to the outside, and, therefore, a negative transmembrane potential results since the outside is assumed to have zero voltage. The difference in charge is due to the fact that the concentration of negatively charged ions inside the cell is higher than on the outside, whereas the opposite relation applies to the concentration of positive ions. In addition to the difference in ion concentration, the actual magnitude of the resting transmembrane potential is also determined by the permeability of the membrane to the different ions.

A potential arises when membrane channels open so that a certain ion may diffuse across the membrane. This process can be illustrated by the simplified situation in which potassium ions are assumed to be inside the cell and sodium ions outside and when the initial transmembrane potential is equal to zero. When the potassium channels are opened, an increase in positive electrical charge outside the cell is created as a result of the diffusion process; at the same time, the inside of the cell becomes increasingly negative and a potential arises across the membrane. This electrical potential constitutes the other force which causes ions to move across the membrane. As the outside of the cell becomes increasingly positive, the resulting potential will increasingly influence the outbound movement of potassium ions. The ion movement ceases when the concentration force balances the electrical force; an equilibrium potential is then said to have been reached. It should be noted that some other active transport mechanisms, not considered here, also come into play when a potential is created.

The resting transmembrane potential of a cell is determined by the equilibrium potentials of the different ions involved and is thus not equal to any of the equilibrium potentials of an individual type of ion. For the situation considered above with open potassium channels, the equilibrium potential for potassium in a nerve cell is found to be about -90 mV, while the equilibrium potential for sodium—assuming instead open sodium channels—is about $+60$ mV. The resting transmembrane potential is within the range of -60 to -100 mV, depending on the type of cell.

When a cell is stimulated by a current, rapid alterations in membrane ion permeability take place which give rise to a change in the membrane potential and generate a signal referred to as an *action potential*. The propagation of action potentials is the very mechanism which makes the heart contract and the nervous system communicate over short and long distances. The stimulus current must exceed a certain threshold level in order to elicit an action potential, otherwise the cell will remain at its resting potential. An excited cell exhibits nonlinear behavior: once a stimulus intensity exceeds the threshold level the resulting action potential is identical and independent of intensity—the *all-or-nothing principle*. An action potential consists mainly of two phases: *depolarization* during which the membrane potential changes toward zero so that the inside of the cell becomes less negative, and ultimately reverses to become positive, and *repolarization* during which the potential returns to its resting level so that the inside again becomes more negative.

The membrane potential remains at its resting level until it is perturbed by some external stimulus, such as a current propagating from neighboring cells. Depolarization is then initiated, and the membrane permeability changes so that sodium channels are opened and the sodium ions can rush into the cell. At the same time, potassium ions try to exit since these are concentrated on the inside, but cannot, thereby causing the charge inside the cell to become increasingly positive, and eventually the membrane potential reverses polarity. Once the rush of sodium ions into the cell has stopped and the membrane potential approaches the sodium equilibrium potential, the peak amplitude of an action potential is reached. During repolarization, sodium channels close and potassium channels open so that the membrane potential can return to its resting, negative potential. The activity of a potassium channel is illustrated in Figure 1.1.

The duration of an action potential varies much more than its amplitude: the repolarization phase of a cardiac cell is much longer than the depolarization phase and lasts from 200 to 300 milliseconds, while for a neuron the two phases combined only last for about one millisecond with both phases having about the same duration. Figure 1.2 shows the action potentials for cells of the brain (motor neuron), the skeletal muscle, and the heart. From these waveforms, it can be observed that the cardiac action potential differs considerably from the others in its lack of an immediate repolarization phase. Instead, there is a plateau in the action potential because the membrane channels of the different ions open and close at different speeds.

Once an action potential has been elicited, the membrane cannot immediately respond to a new stimulus but remains in a "refractory" state for a certain period of time. The refractory period is related to changes that take place in sodium and potassium permeability of the membrane. Obviously, the refractory period imposes an upper limit on the frequency at which action potentials can be communicated through the nervous system or the heart can beat.

The propagation of an action potential exhibits special behavior since it travels a distance through the triggering of new action potentials rather than by traveling itself along the membrane. The current created by the initial membrane depolarization triggers an adjacent membrane so that a new action potential results, and so on. This process repeats itself until the membrane ends and delivers an action potential which is identical to the initial action potential. Due to the refractory period, the action potential travels away from membranes which recently have been excited and continues to do so until it reaches a point on the membrane where the voltage is insufficient for further stimulation.

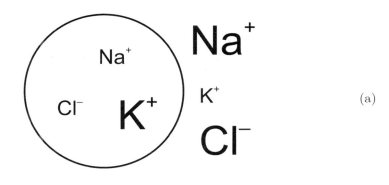

(a)

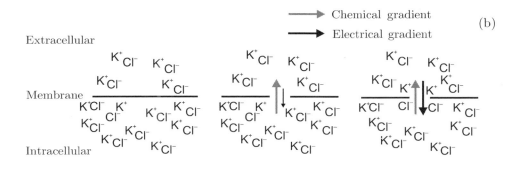

(b)

Figure 1.1: Cellular activity of potassium channels (which is similar for sodium but the reverse). (a) Concentration distribution of potassium (K^+), sodium (Na^+), and chloride (Cl^-) ions inside and outside a cell. (b) The relationship between chemical gradient and electrical gradient for K^+ ions and K^+ channels.

1.2.2 On the Body Surface

The ability of excitable cell membranes to generate action potentials causes a current to flow in the tissue that surrounds the cells. With the tissue being a conducting medium, commonly referred to as a volume conductor, the collective electrical activity of many cells can be measured noninvasively on the body surface [4–6]. The recording of a bioelectrical signal in clinical practice is done by attaching at least two electrodes to the body surface. In its simplest form, a signal is recorded by making use of two electrodes: the "exploring" electrode, placed close to the electrical source, and the "indifferent" electrode, placed elsewhere on the body surface [7]. Multiple electrode configurations are commonly used in clinical practice to obtain a spatial de-

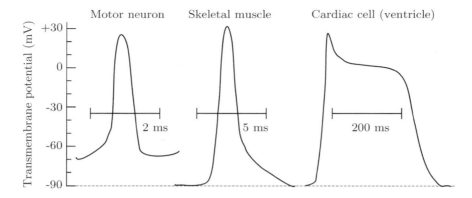

Figure 1.2: Examples of action potentials with shapes that range from the spike-like waveform of a nerve cell (left) to the much more extended waveform of a cardiac cell (right). The transmembrane potential difference was measured by placing one microelectrode inside the cell and another outside. It should be noted that the timescale differs from waveform to waveform.

scription of the bioelectrical phenomenon. Since the activity of excitable cells is viewed from a distance by the electrodes, with different tissues in between, such as blood, skeletal muscles, fat, and bone, it is impossible to noninvasively determine detailed information about cellular properties and propagation patterns. Nonetheless, significant empirical knowledge has over the years been acquired from analyzing the patterns of signals recorded on the body surface, which have been found crucial for clinical decision-making; this observation constitutes an important motivation for the writing of the present textbook.

The problem of characterizing the electrical source by noninvasive measurements has, in spite of the above-mentioned limitations, been the subject of considerable research due to the far-reaching clinical implications of its potential solution. In order to arrive at a meaningful solution, it is necessary to introduce a mathematical model in which the collective electrical cellular activity is treated as a volume source, i.e., it is defined by a fixed dipole, a multiple dipole, or some other source model. Furthermore, by introducing a model for the volume conductor which accounts for essential properties of the human body, such as geometry and resistivity, the electrical field measured on the body surface can be modeled. The important *inverse problem* consists of determining the electrical source from measurements on the body surface under the assumption that the geometry and electrical properties of the volume conductor are known [5].

1.2.3 Bioelectrical Signals

The present textbook deals with the processing of electrical signals that describe the activity of the brain, the heart, and the muscles. Some of these signals reflect spontaneous, ongoing activity, while others only occur as the result of external stimulation. The properties of these signals call for widely different processing techniques; an individual waveform can in some signals be directly linked to a specific clinical diagnosis, while in other signals the composite of many waveforms must be analyzed before a meaningful interpretation can be made.

The **electroencephalogram** (EEG) reflects the electrical activity of the brain as recorded by placing several electrodes on the scalp, see Figure 1.3(a). The EEG is widely used for diagnostic evaluation of various brain disorders such as determining the type and location of the activity observed during an epileptic seizure or for studying sleep disorders. The brain activity may also be recorded during surgery by attaching the electrodes directly to the uncovered brain surface; the resulting invasive recording is named an *electrocorticogram* (ECoG). The background to EEG signals is presented in Chapter 2 and is then followed by Chapter 3 where different EEG signal processing techniques are described.

Evoked potentials (EPs) constitute a form of brain activity which usually is evoked by a sensory stimulus such as one of visual or acoustic origin. Their clinical use includes the diagnosis of disorders related to the visual pathways and the brainstem. An EP, also referred to as an event-related potential, is a transient signal which consists of waves of such tiny amplitudes that its presence in the "background EEG" is typically invisible to the human eye, see Figure 1.4(a). Evoked potentials are recorded using an electrode configuration similar to that of an EEG. Chapter 4 contains an overview of methods developed for "revealing" EPs and for analyzing the resulting signal waveform.

The **electrocardiogram** (ECG) reflects the electrical activity of the heart and is obtained by placing electrodes on the chest, arms, and legs, see Figure 1.3(b). With every heartbeat, an impulse travels through the heart which determines its rhythm and rate and which causes the heart muscle to contract and pump blood. The ECG represents a standard clinical procedure for the investigation of heart diseases such as myocardial infarction. The *electrogram* (EG) is an intracardiac recording where the electrodes have been placed directly within the heart; the EG signal is used in implantable devices such as pacemakers and defibrillators. The background to ECG signals is presented in Chapter 6, while Chapters 7 and 8 present different ECG signal processing techniques.

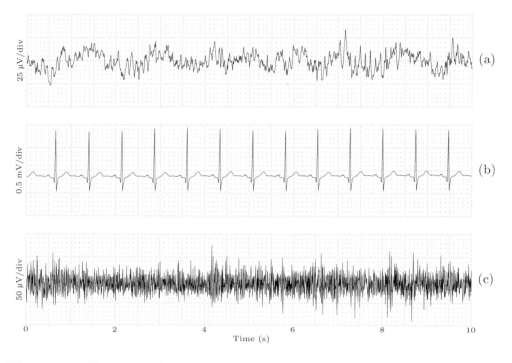

Figure 1.3: Examples of the three major bioelectrical signals recorded from the body surface: (a) an electroencephalogram (EEG) containing alpha activity, (b) an electrocardiogram (ECG) during sinus rhythm, and (c) an electromyogram (EMG) obtained from the chin in the waking state. All three signals were obtained from different normal subjects.

The **electromyogram** (EMG) records the electrical activity of skeletal muscles which produce an electrical current, usually proportional to the level of activity, see Figure 1.3(c). The EMG is used to detect abnormal muscular activity which occurs in many diseases such as muscular dystrophy, inflammation of muscles, and injury to nerves in arms and legs. Recording the surface EMG involves placing the electrodes on the skin overlying the muscle, whereas the intramuscular EMG involves inserting needle electrodes through the skin into the muscle to be examined. Chapter 5 presents an overview of EMG signal processing techniques.

Some other types of bioelectrical signals also deserve mentioning although their related signal analysis will not be further considered in the present textbook.

The *electroneurogram* (ENG) results from the stimulation of a peripheral nerve with an electric shock such that the response along the nerve can be measured. The ENG, usually acquired with needle electrodes, is used

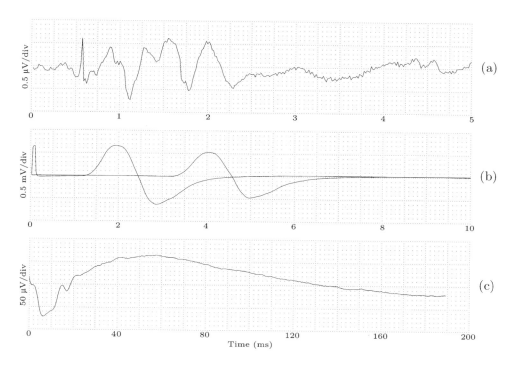

Figure 1.4: Examples of bioelectrical signals resulting from stimulation. (a) An evoked potential (EP) resulting from auditory stimulation (the brainstem response). The displayed signal is actually the result of averaging several responses in order to reduce the high noise level of the original signal; see Section 4.3 for details on noise reduction. (b) An electroneurogram (ENG) recorded at two electrode locations, where the delay between the two signals is used to estimate nerve conduction velocity. (c) An electroretinogram (ERG) obtained during stimulation with a flash of light.

to determine the conduction velocity of the nerve, thereby assisting in the diagnosis of nerve injury. By stimulating a nerve at two different sites separated by a well-defined distance, it is possible to estimate the conduction velocity from the distance by which the resulting two signal waveforms are separated, see the example in Figure 1.4(b). The ENG can be measured both invasively and noninvasively.

An *electroretinogram* (ERG) is used for studying the electrical potentials generated by the retina of the eye during light stimulation [8, 9], see Figure 1.4(c). The ERG is recorded by placing the exploring electrode, encapsulated in a contact lens, on the cornea. The ERG has been found useful for assessing the electrical response of the rods and cones, i.e., the visual cells at the back of the retina. A normal ERG shows appropriate responses with increased light intensity, while an abnormal ERG is obtained in conditions such as arteriosclerosis of the retina or detachment of the retina. The algorithms described in Chapter 4 for signal processing of EPs are, by and large, also applicable to the analysis of ERGs.

The *electrooculogram* (EOG) is the recording of the steady corneal–retinal potential which is proportional to vertical and horizontal movements of the eye, thus offering an objective way to quantify the direction of the gaze [5, 10], see Figure 1.5(a). The EOG is of particular interest in patients who suffer from sleep disorders, where the presence of rapid eye movement (REM) is important for determining certain sleep stages. The EOG is recorded when studying nystagmus, i.e., a rapid, involuntary oscillation of the eyeballs, for example, in patients suffering from vertigo and dizziness. The EOG is also useful in virtual reality environments where a device for eye-tracking may be needed. The EOG is briefly touched upon in Chapter 3 in connection with EEG signal processing since the electrical activity caused by eye movements often interferes with the EEG and, therefore, needs to be cancelled.

The *electrogastrogram* (EGG) is a recording of the impulses which propagate through the muscles of the stomach and which control their contractions [11], see Figure 1.5(b). The EGG is studied when the muscles of the stomach or the nerves controlling the muscles are not working normally, for example, when the stomach does not empty food normally. The EGG is recorded by attaching a number of electrodes over the stomach during fasting and subsequent to a meal. In normal individuals a regular "rhythmic" signal is generated by the muscles of the stomach, having an amplitude which increases after a meal; the normal frequency of the gastric rhythm is approximately 3 cycles/minute. However, in symptomatic patients the rhythm is often irregular and sometimes without the increase in amplitude that follows a meal. A small selection of papers describing technical means of analyzing the EGG signal can be found in [12–16].

1.3   Signal Acquisition and Analysis

The acquisition of bioelectrical signals is today accomplished by means of relatively low-cost equipment which appropriately amplifies and digitizes the signal. As a result, several clinical procedures based on bioelectrical signals are in widespread use in hospitals around the world. In many situations, PC-based systems can be utilized as an efficient and cost-effective solution for signal analysis, especially considering the availability of expansion cards for data acquisition. Such a system includes one or several sensors, external hardware for patient insulation and signal amplification, an acquisition

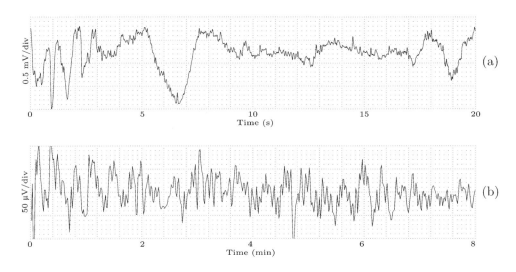

Figure 1.5: Recordings which exemplify (a) an electrooculogram (EOG) of the right eye and (b) an electrogastrogram (EGG). Note that the two timescales differ.

card with analog/digital (A/D) conversion, and software for signal analysis (Figure 1.6) [17]. In situations where the analysis is performed in an implantable device, the system design involves additional considerations, e.g., those related to the design of application-specific integrated circuitry and the selection of appropriate battery technology.

In the digitization process, it is usually sufficient to use 12–14 bits for amplitude quantization in order to cover the dynamic range of a signal; it is presumed that very slow, large-amplitude drift in the direct current (DC) level has been removed prior to quantization without modifying the physiological content of the signal. The amplitude of individual bioelectrical waveforms ranges from 0.1 μV, observed in certain types of EPs once subjected to noise reduction, to several millivolts, as observed in the ENG, ECG, and EOG.

Most bioelectrical signals recorded on the body surface have a spectral content confined to the interval well below 1 kHz, and thus the sampling rate—chosen to be at least the Nyquist rate—rarely exceeds a few kilohertz. However, since signals measured on the body surface are subjected to lowpass filtering caused by the intermediate tissue, invasively recorded signals, such as those on action potentials, generally exhibit a much higher frequency content.

In a PC-based system, signal analysis is often done locally by relying either on the internal CPU or an expansion digital signal processor (DSP) card.

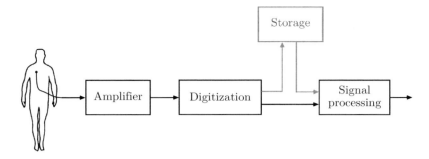

Figure 1.6: Block diagram describing the main steps in biomedical signal analysis. The signal is often processed at the time of its acquisition, but may also be stored on a local hard disk or a server on the web for later retrieval and processing.

However, with today's availability of web-based resources, it is no longer necessary to perform the entire signal analysis locally. It is equally possible to acquire the signal at one physical location, using the PC-based system, and then to process it at another location, i.e., relying on a client/server solution [18]. Since the acquired signal in most cases is stored in a database that resides on a server, it can be advantageous to also process the signal on the server since it may offer more computational power.

1.4   Performance Evaluation

Performance evaluation is an important and challenging part of biomedical signal processing required before any algorithm can be implemented in a clinical context. Unlike many other engineering applications where the information in the signal source is known a priori, the message "sent" by a bioelectrical source is unknown and has to be unmasked in some manual way in order to render performance evaluation possible. For example, the evaluation of an algorithm for detecting heartbeats is relatively straightforward since it is an easy task for a physician to determine the times of occurrence of the heartbeats; the performance figures would then be designed to reflect how well the output of the algorithm agrees with the manually obtained times of occurrence. The performance evaluation becomes much more complicated when the goal is to develop an algorithm that computes a parameter set which accurately discriminates signals obtained from healthy subjects and patients who suffer from a particular disease. In such cases, an assessment of the output of the algorithm cannot be carried out simply because the "truth" cannot be retrieved from the observed signal. Instead, the performance may be evaluated in terms of its ability to correctly discriminate between the two

Table 1.1: Definitions of the performance measures sensitivity, specificity, positive predictive value, and negative predictive value.

Performance measure	Definition	Interpretation
Sensitivity	$\dfrac{N_{TP}}{N_{TP} + N_{FN}}$	The probability of a positive result for the diseased subjects
Specificity	$\dfrac{N_{TN}}{N_{FP} + N_{TN}}$	The probability of a negative result for the healthy subjects
Positive predictive value	$\dfrac{N_{TP}}{N_{TP} + N_{FP}}$	The probability of disease when the result is positive
Negative predictive value	$\dfrac{N_{TN}}{N_{FN} + N_{TN}}$	The probability of health when the result is negative

N_{TP} = the number of diseased subjects with a positive result (True Positive)

N_{TN} = the number of healthy subjects with a negative result (True Negative)

N_{FN} = the number of diseased subjects with a negative result (False Negative)

N_{FP} = the number of healthy subjects with a positive result (False Positive)

groups of healthy and diseased subjects. The most commonly used performance measures for describing such discrimination are those of sensitivity, specificity, positive predictive value, and negative predictive value, whose definitions are given in Table 1.1.

It has been pointed out that "while new analytic technologies seem very promising when they are first applied, the initial glitter often fades when the method is systematically evaluated" [19]. This statement not only underlines the importance of performance evaluation, but also that a great deal of effort must be devoted to algorithm development before satisfactory performance can be achieved.

1.4.1 Databases

The availability of signal databases is of vital importance for both development and evaluation of signal processing algorithms. The immense diversity of waveform patterns which exists among subjects necessitates evaluation of the algorithm on a database of considerable size before its performance can be judged as satisfactory for use in a clinical setting. Needless to say, one part of a database must be used for algorithm development, while the remaining part is kept for performance evaluation in order to assure that no learning of the evaluation data has taken place.

The word "database" is here interpreted as a collection of signals that has been obtained using the same recording protocol from suitably selected groups of healthy subjects and patients. A database often includes signals of one particular type, such as EEGs or ECGs, but may just as well include other types of concurrently recorded signals. Annotations are another important type of database information which define the time instants at which certain events occur in the signal, such as the presence of heartbeats or epileptic seizures. The annotations may also account for more complex signal properties as well as for nonphysiological information such as the presence of noise episodes and technical deficiencies due to poorly attached electrodes (Figure 1.7). The annotations are determined manually by one or several physicians who must carefully scrutinize the signal with respect to the properties to be annotated. The inclusion of several annotators generally implies that more reliable annotations are obtained. However, it is inevitable that discrepancies arise among the annotators which must be resolved by consensus, thus adding further labor to an already laborious process.

In addition to the signal and its annotation, the database may include additional information on subjects such as gender, race, age, weight, medication, and data from other clinical procedures which may be valuable when evaluating performance.

A substantial number of databases have been collected over the years for the purpose of addressing various clinical issues. The MIT–BIH arrhythmia database is the most widely used database for evaluation of methods designed for detecting abnormalities in cardiac rhythms and is almost certainly also the most popular database overall in biomedical signal processing [21, 22]. The MIT–BIH arrhythmia database contains ECG signals which have been recorded during ambulatory conditions such as working and eating. Another widely used ECG database is the AHA database, which was developed for evaluation of ventricular arrhythmia detectors [23]. More recent additions to the list of databases include the European ST–T and LTST databases, which were collected for the purpose of investigating the occurrence of insufficient blood supply to the cardiac muscle cells (myocardial ischemia) [20, 24]. An interesting adjunct to the MIT–BIH arrhythmia database is the MIT–BIH noise stress test database which contains several recordings of noise typically encountered in ambulatory conditions: by adding a calibrated amount of noise to a "clean" ECG signal, the noise immunity of an algorithm can be tested with this database [25].

Multimodal databases have also been collected and may include signals that reflect brain activity, heart activity, muscle activity, blood pressure, respiration, as well as other types of activity, see Figure 1.8. Some examples are the MIMIC database [26], the IMPROVE database [27], and the IBIS database [28, 29], which all contain continuously recorded data obtained

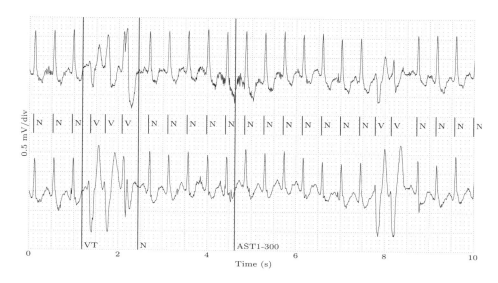

Figure 1.7: Example of a manually annotated, two-channel ECG from a patient with myocardial ischemia. The sequence of short, vertical bars shows the times of occurrence of the heartbeats, and the related labels "N" and "V" indicate whether the beat is of normal or ventricular origin. The three longer bars indicate the onset of a new rhythm (VT: ventricular tachycardia, N: sinus rhythm, and AST1-300: maximum ST depression of –300 μV). The signal was taken from the European ST–T database [20].

from intensive care monitoring, while other databases have been collected for investigating sleep disorders [30, 31]. Most databases described in the literature are publicly available, either at no cost or at a charge, while some remain the private property of those who collected the data. Databases of biomedical signals have proven to be equally valuable for researchers and instrument manufacturers.

The increasing availability of databases certainly makes it more convenient and less time-consuming to pursue projects on algorithm development. Because of the easy access to databases, now available on different sites on the World Wide Web, it is possible to develop and evaluate signal processing algorithms without having to deal with the cumbersome and often labor-intensive task of data collection. The *PhysioNet* (`www.physionet.org`) is a website which constitutes a tremendous leap forward, being a resource where various types of physiological signals are freely available for download [32]. The PhysioNet maintains different classes of databases, ranging from those which are carefully scrutinized and thoroughly annotated to those which are unannotated and sometimes not yet completely acquired.

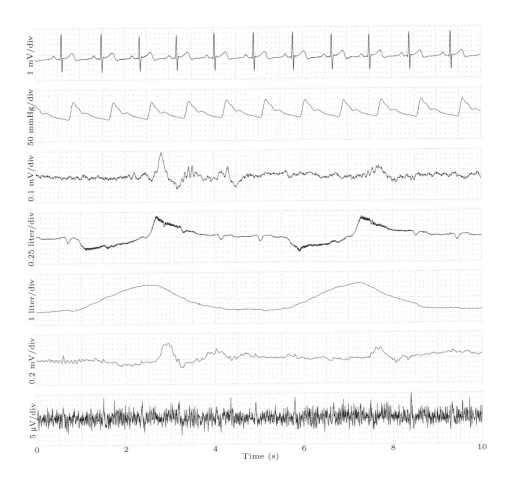

Figure 1.8: Concurrently recorded signals from a multimodal database, from top to bottom: ECG, blood pressure, EEG, nasal respiration, abdominal respiration, EOG, and EMG. This type of recording is used for studying sleep disorders, see Section 2.4.2. The signals were taken from the MIT–BIH polysomnographic database [30].

With the easy availability of databases comes also the potential risk of omitting medical expertise from projects since hospital-based activities are no longer needed. If the worst comes to the worst, a lack of clinically experienced eyes may lead to the introduction of clinically unacceptable distortion into the signal via the algorithm, rather than improving its interpretation. Hence, it is always important for the project's outcome to establish a viable liaison between engineers and physicians. Another potential risk when downloading a database is that its original clinical purpose is tweaked into answering questions for which the database was never intended. In cases when no suitable database is available, it is necessary to develop an appropriate recording protocol for data collection of one's own and then, of course, to perform the actual signal acquisition. Anyone embarking on a project in biomedical signal processing is, in addition to considering the use of available databases, strongly encouraged to also deal with the details of collecting signals.

1.4.2 Simulation

A simulation describes quantitatively some physiological behavior by mathematical equations and is used to replicate signals which are generated in the body. An advantage of simulations is the possibility to investigate conditions which are difficult to deal with experimentally. Another advantage, of particular relevance for algorithmic performance evaluation, is that the properties of a simulated signal can be exactly controlled by a set of parameters. As a result, the agreement between the "true" parameter values of the simulated signal and those determined by an estimation method can be quantitatively assessed and expressed in terms of a suitable performance measure. The exact definition of such a measure depends on the case at hand and may involve rates of missed events and false events in detection problems and bias and variance in parameter estimation problems.

Signal modeling and simulation are intimately linked together because a simulation is based on an existing model. Models producing highly realistic-looking signals are often associated with high complexity and do not easily lend themselves to parameter estimation. Simpler models, which can only account for a partial phenomenon of the signal, are still very useful for algorithm development and have, in fact, often been considered.

In biomedical signal processing, a model of the physiological, "clean" signal is often accompanied by a model of the noise sources, and the combination of the two models makes it possible to simulate signals observed on the body surface. The term "noise" is here used in a wide sense which includes physiological activities other than the one under study which may interfere with the desired signal. Simulated signals with different signal-to-noise ratios (SNRs) can be easily produced using this approach. While performance evaluation is mostly concerned with accuracy, i.e., the difference between the true value and the estimated value, it is also important to study the reproducibility of an algorithm. Reproducibility is the ability of an algorithm to produce repeated measurements which cohere, obviously under the assumption that the same signal conditions apply to all measurements. Although reproducibility is best investigated by sequentially repeating an experiment on the same patient, simulated signals represent a powerful and much more manageable means of evaluating the reproducibility of an algorithm. The performance is then evaluated by using the algorithm to process a series of simulated signals, each time by adding a different noise realization to the clean signal.

In addition to simulations based on mathematical models for both signal and noise, it may in certain cases be appropriate to evaluate the performance by employing simulated signals to which "real world" noise is instead added. The reverse situation with "real world" signals and simulated noise may sometimes also be of interest.

We conclude this section by noting that the simulation approach represents a useful step in algorithm development, provided of course that the signal model is adequate. However, databases consisting of collected signals must constitute the lion's share of the evaluation work so that the clinical utility of an algorithm can be thoroughly established.

Bibliography

[1] A. C. Guyton and J. E. Hall, *Textbook of Medical Physiology.* Philadelphia: W. B. Saunders, 10th ed., 2000.

[2] A. J. Vander, J. H. Sherman, and D. S. Luciano, *Human Physiology. The Mechanisms of Body Function.* New York: McGraw–Hill, 5th ed., 1990.

[3] J. Enderle, "Bioelectric phenomena," in *Introduction to Biomedical Engineering* (J. Enderle, S. Blanchard, and J. Bronzino, eds.), ch. 3, pp. 79–138, San Diego: Academic Press, 2000.

[4] R. Plonsey and R. C. Barr, *Bioelectric Phenomena: A Quantitative Approach.* New York: Plenum, 1988.

[5] J. Malmivuo and R. Plonsey, *Bioelectromagnetism.* Oxford: Oxford University Press, 1995.

[6] R. M. Gulrajani, *Bioelectricity and Biomagnetism.* Montreal: John Wiley & Sons, 1998.

[7] M. R. Neuman, "Biopotential electrodes," in *Medical Instrumentation. Application and Design* (J. G. Webster, ed.), ch. 5, pp. 183–232, New York: John Wiley & Sons, 1998.

[8] J. C. Armington, *The Electroretinogram.* New York: Academic Press, 1974.

[9] J. R. Heckenlively and G. B. Arden (eds.), *Principles and Practices of Clinical Electrophysiology of Vision.* St. Louis, MO: Mosby Year Book, 1991.

[10] R. H. S. Carpenter, *Movements of the Eyes.* London: Pion, 2nd ed., 1988.

[11] W. C. Alvarez, "The electrogastrogram and what it shows," *JAMA*, vol. 28, pp. 1116–1118, 1922.

[12] D. A. Linkens and S. P. Datardina, "Estimation of frequencies of gastrointestinal electrical rhythms using autoregressive modelling," *Med. Biol. Eng. & Comput.*, vol. 16, pp. 262–268, 1978.

[13] R. H. Smallwood, D. A. Linkens, H. L. Kwok, and C. J. Stoddard, "Use of autoregressive-modelling techniques for the analysis of colonic myoelectrical activity in man," *Med. Biol. Eng. & Comput.*, vol. 18, pp. 591–600, 1980.

[14] N. Mirizzi and U. Scafoglieri, "Optimal direction of the electrogastrographic signal in man," *Med. Biol. Eng. & Comput.*, vol. 21, pp. 385–389, 1983.

[15] J. D. Z. Chen, W. R. Stewart Jr., and R. W. McCallum, "Spectral analysis of episodic rhythmic variations in the cutaneous electrogastrogram," *IEEE Trans. Biomed. Eng.*, vol. 40, pp. 128–135, 1993.

[16] Z. M. Zhou, Z. Hui, R. Shaw, and F. S. Barnes, "Real-time multichannel computerized electrogastrograph," *IEEE Trans. Biomed. Eng.*, vol. 44, pp. 1228–1236, 1997.

[17] J. G. Webster, *Medical Instrumentation. Application and Design.* Boston: John Wiley & Sons, 1998.

[18] N. H. Lovell, F. Magrabi, B. G. Celler, K. Huynh, and H. Garsden, "Web-based acquisition, storage, and retrieval of biomedical signals," *IEEE Eng. Med. Biol. Mag.*, vol. 20, pp. 38–44, 2001.

[19] A. S. Gevins and A. Rémond, *Handbook of Electroencephalography and Clinical Neurophysiology: Methods of Analysis of Brain Electrical and Magnetic Signals*, vol. 1. Amsterdam/New York: Elsevier, 1987.

[20] A. Taddei, G. Distante, M. Emdin, P. Pisani, G. B. Moody, C. Zeelenberg, and C. Marchesi, "The European ST-T database: Standards for evaluating systems for the analysis of ST-T changes in ambulatory electrocardiography," *Eur. Heart J.*, vol. 13, pp. 1164–1172, 1992.

[21] R. G. Mark, P. S. Schluter, G. B. Moody, P. H. Devlin, and D. Chernoff, "An annotated ECG database for evaluating arrhythmia detectors," in *Proc. IEEE Frontiers Eng. Health Care*, pp. 205–210, 1982.

[22] G. B. Moody and R. G. Mark, "The impact of the MIT-BIH arrhythmia database. History, lessons learned, and its influence on current and future databases," *IEEE Eng. Med. Biol. Mag.*, vol. 20, pp. 45–50, 2001.

[23] R. E. Hermes, D. B. Geselowitz, and G. C. Oliver, "Development, distribution, and use of the American Heart Association database for ventricular arrhythmia detector evaluation," in *Proc. Computers in Cardiology*, pp. 263–266, IEEE Computer Society Press, 1980.

[24] F. J. Jager, A. Taddei, G. B. Moody, M. Emdin, G. Antolič, R. Dorn, A. Smrdel, C. Marchesi, and R. G. Mark, "Long-term ST database: A reference for the development and evaluation of automated ischaemia detectors and for the study of the dynamics of myocardial ischaemia," *Med. Biol. Eng. & Comput.*, vol. 41, pp. 172–183, 2003.

[25] G. B. Moody, W. K. Muldrow, and R. G. Mark, "A noise stress test for arrhythmia detectors," in *Proc. Computers in Cardiology*, pp. 381–384, IEEE Computer Society Press, 1984.

[26] G. B. Moody and R. G. Mark, "A database to support development and evaluation of intelligent intensive care monitoring," in *Proc. Computers in Cardiology*, pp. 657–660, IEEE Press, 1996.

[27] I. Korhonen, J. Ojaniemi, K. Nieminen, M. van Gils, A. Heikelä, and A. Kari, "Building the IMPROVE data library," *IEEE Eng. Med. Biol. Mag.*, vol. 16, pp. 25–32, 1997.

[28] J. Gade, I. Korhonen, M. van Gils, P. Weller, and L. Pesu, "Technical description of the IBIS data library," *Comp. Meth. Prog. Biomed.*, vol. 3, pp. 175–186, 2000.

[29] S. M. Jakob, K. Nieminen, J. Karhu, and J. Takala, "IBIS data library: Clinical description of the Finnish database," *Comp. Meth. Prog. Biomed.*, vol. 3, pp. 161–166, 2000.

[30] Y. Ichimaru and G. B. Moody, "Development of the polysomnographic database on CD-ROM," *Psychiatry Clin. Neurosci.*, vol. 53, pp. 175–177, 1999.

[31] G. Klösch, B. Kemp, T. Penzel, A. Schlögl, P. Rappelsberger, E. Trenker, G. Gruber, J. Zeitlhofer, B. Saletu, W. M. Herrmann, S. L. Himanen, D. Kunz, M. L. Barbanoj, J. Röschke, A. Värri, and G. Dorffner, "The SIESTA project polygraphic and clinical database," *IEEE Eng. Med. Biol. Mag.*, vol. 20, pp. 51–57, 2001.

[32] G. B. Moody, R. G. Mark, and A. L. Goldberger, "PhysioNet: A web-based resource for study of physiologic signals," *IEEE Eng. Med. Biol. Mag.*, vol. 20, pp. 70–75, 2001.

Chapter 2

The Electroencephalogram— A Brief Background

The human brain is the most complex organic matter known to mankind and has, not surprisingly, been the subject of extended research. Its complexity has spurred multifaceted research in which brain functionality is explored from low-level chemical and molecular properties in individual neurons to high-level aspects such as memory and learning. An early discovery established that the brain is associated with the generation of electrical activity. Richard Caton had demonstrated already in 1875 that electrical signals in the microvolt range can be recorded on the cerebral cortex of rabbits and dogs. Several years later, Hans Berger recorded for the first time electrical "brain waves" by attaching electrodes to the human scalp; these waves displayed a time-varying, oscillating behavior that differed in shape from location to location on the scalp [1]. Berger made the interesting observation that brain waves differed not only between healthy subjects and subjects with certain neurological pathologies, but that the waves were equally dependent on the general mental state of the subject, e.g., whether the subject was in a state of attention, relaxation, or sleep.

The experiments conducted by Berger became the foundation of *electroencephalography*, later to become an important noninvasive clinical tool in better understanding the human brain and for diagnosing various functional brain disturbances. The clinical interpretation of the EEG has evolved into a discipline in its own right, where the human reader is challenged to draw conclusions based on the frequency, amplitude, morphology, and spatial distribution of the brain waves. So far, no single biological or mathematical model has been put forward which fully explains the diversity of EEG patterns, and, accordingly, EEG interpretation largely remains a phenomenological clinical discipline [2].

Visual scrutiny was for many years the sole approach to EEG interpretation but has today been supplemented by the capabilities offered by modern, powerful computers. The interpretation is significantly facilitated, although not even close to being fully automated, by an array of digital signal processing methods designed for a variety of purposes, e.g., improvement of SNR, quantification of various signal characteristics, and extraction of new information not readily available by visual inspection [3–5]. Signal processing methods can be divided into two general categories: methods developed for the analysis of spontaneous brain activity (the "background EEG"[1]) and brain potentials which are evoked by various sensory and cognitive stimuli (evoked potentials, EPs). While the former category of methods certainly has helped to gain a better understanding of the EEG, the analysis of EPs is critically dependent on the availability of signal processing techniques.

In recent years, the study of brain function has been revolutionized by the introduction of various imaging modalities: positron emission tomography (PET), single photon emission computed tomography (SPECT), and magnetic resonance imaging (MRI), which can produce two- or three-dimensional images with good spatial resolution. These modalities extend the information inferred from an electrophysiological investigation by providing detailed information on, e.g., anatomy and blood flow in different regions of the brain. As a result, the EEG has today lost part of its dominance in clinical routine; however, it remains a very powerful tool in the diagnosis of many diseases such as epilepsy, sleep disorders, and dementia. Furthermore, the EEG signal is important for real-time monitoring in the operating theatre and in the intensive care unit, e.g., when monitoring the progress of patients in a coma or with encephalopathies. In monitoring applications, the fraction-of-a-second temporal resolution of the EEG is unsurpassed compared to the above-mentioned imaging modalities. Another aspect in favor of the EEG is that the total cost associated with recording instrumentation, and technicians required to manage the equipment, is dramatically lower than that associated with neuroimaging. The technical demands on equipment for recording EEGs are relatively modest and are, for a basic recording setup, restricted to a set of electrodes, a signal amplifier, and a PC for data storage, signal analysis, and graphical presentation.

The magnetoencephalogram (MEG) is yet another noninvasive technique which quantifies the weak magnetic field of mass neural activity by using an extremely sensitive magnetic field sensor—the SQUID. The main advantage of the MEG technique is that the magnetic field is less distorted by the skull than is the electrical potential. While the MEG originally was believed to

---

[1]The term "background EEG" is here used in a wider sense than the clinical convention and also includes abnormal brain activity such as epilepsy.

provide information which is independent of the EEG [6], it has recently
been shown that the EEG and MEG signals have strong interdependence [7,
8]. Since these two types of recording technology, as well as the imaging
techniques mentioned above, exhibit both strengths and weaknesses, they
should ultimately be used to complement each other [9, 10].

This chapter first presents a brief description of the nervous system and
the electrical activity of the brain (Section 2.1); for further details, the interested reader is referred to the multitude of textbooks which contain comprehensive descriptions of the human brain. A variety of common EEG patterns
and waveforms are presented in Section 2.2 which are of special interest in
the subsequent chapter on EEG signal processing methods. Section 2.3 describes the standard technique used for recording an EEG in clinical routine.
Finally, Section 2.4 provides a brief overview of some important EEG applications.

2.1   The Nervous System

The nervous system gathers, communicates, and processes information from
various parts of the body and assures that both internal and external changes
are handled rapidly and accurately. The nervous system is commonly divided
into the central nervous system (CNS), consisting of the brain and the spinal
cord, and the peripheral nervous system (PNS), connecting the brain and
the spinal cord to the body organs and sensory systems. The two systems
are closely integrated because sensory input from the PNS is processed by
the CNS, and responses are sent by the PNS to the organs of the body. The
nerves transmitting signals to the CNS are called *afferent* or, alternatively,
sensory nerves. The nerves transmitting signals from the CNS are called
efferent or, alternatively, *motor* nerves since these signals may elicit muscle
contraction.

Another important division of the nervous system is based on its functionality: the *somatic* nervous system and the *autonomic* nervous system.
The somatic system includes those nerves which control muscle activity in
response to conscious commands. This system also relays the physical sensa-
tions. The autonomic nervous system regulates the bodily activities which
are beyond conscious control, e.g., cardiac activity and muscle activity in
internal organs such as the bladder and uterus. The autonomic nervous system actually consists of two subsystems which operate against each other:
the *sympathetic* nervous system, which dominates when physical activity is
called for, and the *parasympathetic* nervous system, which dominates during
relaxation. Both these subsystems innervate the same organs and act so as to
maintain the correct balance of the internal organ environment. For example,

during physical exercise or when a subject experiences fear, the sympathetic system causes the heart rate to increase while the parasympathetic system decreases the rate. Heart rate variability as a result of the antagonistic effect between the two subsystems has been the subject of considerable research in order to better understand the relation between neurological diseases and dysfunction of the autonomic nervous system. Chapter 8 describes methods developed for quantification of heart rate variability.

2.1.1 Neurons

The basic functional unit of the nervous system is the nerve cell—the *neuron*—which communicates information to and from the brain. All nerve cells are collectively referred to as neurons although their size, shape, and functionality may differ widely. Neurons can be classified with reference to morphology or functionality. Using the latter classification scheme, three types of neurons can be defined: *sensory neurons*, connected to sensory receptors, *motor neurons*, connected to muscles, and *interneurons*, connected to other neurons.

The archetypal neuron consists of a cell body, the *soma*, from which two types of structures extend: the *dendrites* and the *axon*, see Figure 2.1(a). Dendrites can consist of as many as several thousands of branches, with each branch receiving a signal from another neuron. The axon is usually a single branch which transmits the output signal of the neuron to various parts of the nervous system. The length of an axon ranges from less than 1 mm to longer than 1 m; the longer axons are those which run from the spinal cord to the feet. Dendrites are rarely longer than 2 mm.

The transmission of information from one neuron to another takes place at the *synapse*, a junction where the terminal part of the axon contacts another neuron. The signal, initiated in the soma, propagates through the axon encoded as a short, pulse-shaped waveform, i.e., the action potential. Although this signal is initially electrical, it is converted in the presynaptic neuron to a chemical signal ("neurotransmitter") which diffuses across the synaptic gap and is subsequently reconverted to an electrical signal in the postsynaptic neuron, see Figure 2.1(b).

Summation of the many signals received from the synaptic inputs is performed in the postsynaptic neuron. The amplitude of the summed signal depends on the total number of input signals and how closely these signals occur in time; the amplitude decreases when the signals become increasingly dispersed in time. The amplitude of the summed signal must exceed a certain threshold in order to make the neuron fire an action potential. Not all neurons contribute, however, to the excitation of the postsynaptic neuron; inhibitory effects can also take place due to a particular chemical structure

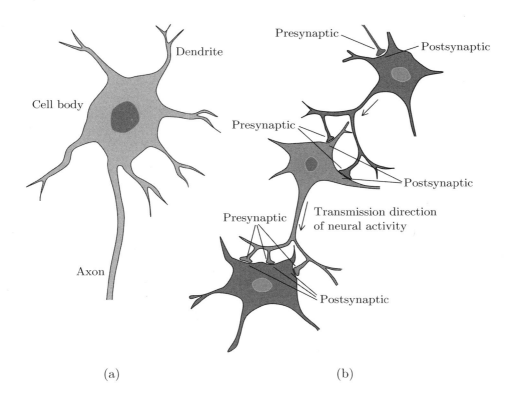

(a) (b)

Figure 2.1: (a) An archetypal neuron and (b) three interconnected neurons. A presynaptic neuron transmits the signal toward a synapse, whereas a postsynaptic neuron transmits the signal away from the synapse.

associated with certain neurons. A postsynaptic neuron thus receives signals which are both excitatory and inhibitory, and its output depends on how the input signals are summed together. This input/output operation is said to represent one neural computation and is performed repeatedly in billions of neurons.

In contrast to the electrical activity measured on the scalp, electrical activity propagating along the axon is manifested as a series of action potentials, all waveforms having identical amplitudes. This remarkable feature is explained by the "on/off" property of the neuron which states that an action potential is either elicited with a fixed amplitude or does not occur at all. The intensity of the input signals is instead modulated by the firing rate of the action potentials. For example, this signal property implies that a high firing rate in sensory neurons is associated with considerable pain or, in motor neurons, with a powerful muscle contraction. Furthermore, it is fascinating to realize that this modulation system is particularly well-suited for transmission of information over long distances and is tolerant to local failures. The upper bound of the firing rate is related to the refractory period of the neuron, i.e., the time interval during which the neuron is electrically insensitive.

Neurons are, of course, not working in splendid isolation, but are interconnected into different circuits ("neural networks"), and each circuit is tailored to process a specific type of information. A well-known example of a neural circuit is the knee-jerk reflex. This particular circuit is activated by muscle receptors which, by a hammer tap, initiate a signal that travels along an afferent pathway. The received sensory information stimulates motor neurons through synaptic contacts, and a new signal is generated which travels peripherally back, giving rise to muscle contraction and the associated knee-jerk response.

2.1.2   The Cerebral Cortex

The cerebral cortex is the most important part of the CNS, and the different regions of cortex are responsible for processing vital functions such as sensation, learning, voluntary movement, speech, and perception. The cortex is the outermost layer of the cerebrum and has a thickness of 2–3 mm. The cortical surface is highly convoluted by ridges and valleys of varying sizes and thus increases the neuronal area; the total area is as large as 2.5 m^2 and includes more than 10 billion neurons. The cortex consists of two symmetrical *hemispheres*—left and right—which are separated by the deep sagittal fissure (the central sulcus). Each hemisphere is divided into four different lobes: the *frontal, temporal, parietal,* and *occipital lobes,* see Figure 2.2.

Voluntary movement is primarily controlled by the area of the frontal lobe just anterior to the central sulcus—the motor cortex. Tasks requiring considerable muscle control, e.g., speech, certain facial expressions, and finger movements, are associated with the largest subarea of the motor cortex. Sensory information is processed in various parts of the lobes: the auditory cortex is located in the superior part of the temporal lobe, the visual cortex is located at the posterior part of the occipital lobes, and the somatic sensory cortex is located just posterior to the central sulcus of the parietal lobe.

The above-mentioned cortical areas are referred to as primary areas since these neurons are specialized for a particular purpose. The primary areas are relatively small in size, but are supplemented with larger, surrounding areas which are essential for the mental abilities that are characteristic of human beings. The neurons of a secondary area analyze, for example, visual information in further detail with respect to shape, color, and size of an object. These neurons also provide associative references to other senses and

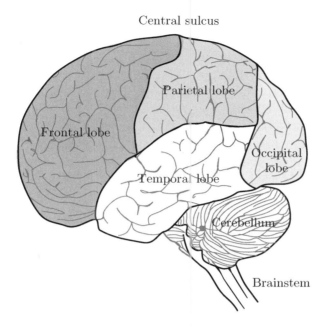

Figure 2.2: The cerebral cortex and the four lobes.

will, ultimately, integrate the present information with earlier experiences and knowledge.

2.2 The EEG—Electrical Activity Measured on the Scalp

The collective electrical activity of the cerebral cortex is usually referred to as a *rhythm* because the measured signals often exhibit oscillatory, repetitive behavior. The activity of a single cortical neuron cannot be measured on the scalp due to thick layers of tissue (fluids, bones, and skin) which attenuate the electrical signal when it propagates toward the electrode.[2] However, the joint activity of millions of cortical neurons, at a depth down to several millimeters, produces an electrical field which is sufficiently strong to be measured on the scalp; this depth depends on the "strength" of the neural source. The electrical field is mainly generated by currents that flow during synaptic excitation of the dendrites, the excitatory postsynaptic potentials.

---

[2]It is possible to invasively investigate the electrical behavior of only a few neurons by the use of microelectrodes. The specific properties of signals acquired by such intracerebral electrodes will, however, not be given further consideration in the present text.

The diversity of EEG rhythms is enormous and depends, among many other things, on the mental state of the subject, such as the degree of attentiveness, waking, and sleeping. Figure 2.3 illustrates a number of EEG rhythms observed during different states. The rhythms are conventionally characterized by their frequency range and relative amplitude.

The amplitude of the EEG signal is related to the degree of synchrony with which the cortical neurons interact. Synchronous excitation of a group of neurons produces a large-amplitude signal on the scalp because the signals originating from individual neurons will add up in a time-coherent fashion. Repetition of the synchronous excitation results in a rhythmic EEG signal, consisting of large-amplitude waveforms occurring at a certain repetition rate. On the other hand, asynchronous excitation of the neurons results in an irregular-looking EEG with low-amplitude waveforms. In both cases, the excitation may very well involve an identical number of neurons, but, depending on the time dispersion of the neuronal input, different amplitudes of the EEG result.

The frequency, or the oscillatory rate, of an EEG rhythm is partially sustained by input activity from the thalamus. This part of the brain consists of neurons which possess pacemaker properties, i.e., they have the intrinsic ability to generate a self-sustained, rhythmic firing pattern. Another reason to the rhythmic behavior is coordinated interactions arising between cortical neurons themselves in a specific region of the cortex. In the latter case, no pacemaker function is involved, but the rhythm is rather an expression of a feedback mechanism that may occur in a neuronal circuit [11].

High-frequency/low-amplitude rhythms reflect an active brain associated with alertness or dream sleep, while low-frequency/large-amplitude rhythms are associated with drowsiness and nondreaming sleep states. "This relationship is logical because when the cortex is most actively engaged in processing information, whether generated by sensory input or by some internal process, the activity level of cortical neurons is relatively high but also relatively unsynchronized. In other words, each neuron, or very small group of neurons, is vigorously involved in a slightly different aspect of a complex cognitive task; it fires rapidly, but not quite simultaneously with most of its neighbors. This leads to low synchrony, so the EEG amplitude is low. By contrast, during deep sleep, cortical neurons are not engaged in information processing, and large numbers of them are phasically excited by a common, rhythmic input. In this case synchrony is high, so the EEG amplitude is large" [11].

The meaning of different brain rhythms largely remains unexplained, although several hypotheses have been put forward. Despite this gap in understanding, quantification of EEG rhythm characteristics has nevertheless

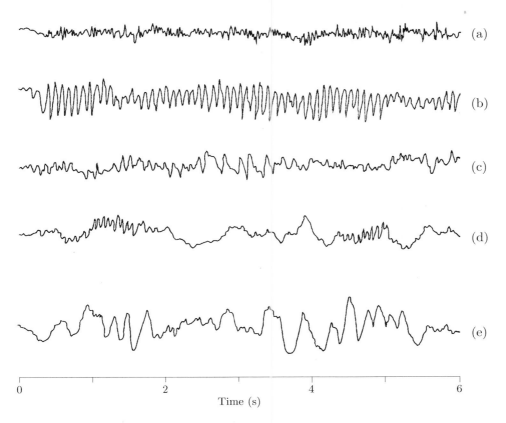

Figure 2.3: Electroencephalographic rhythms observed during various states from wakefulness to sleep: (a) excited, (b) relaxed, (c) drowsy, (d) asleep, and (e) deeply asleep. This example is classical and was originally presented by the famous EEG pioneer H.H. Jasper [12].

proved to be an extremely useful clinical approach in studying functional states of the brain.

2.2.1 EEG Rhythms and Waveforms

The characteristics of the most frequently occurring rhythms and waveforms will now be briefly summarized. Signals recorded from the scalp have, in general, amplitudes ranging from a few microvolts to approximately 100 μV and a frequency content ranging from 0.5 to 30–40 Hz. Electroencephalographic rhythms, also referred to as background rhythms, are conventionally classified into five different frequency bands. The interpretation of these bands in terms of "normal" or "abnormal" is relative and depends on the age and mental state of the subject. For example, the EEG of a newborn is drastically different from that of an adult and has, in general, a considerably higher frequency content. The frequency bands indicated below are somewhat coarse, but nevertheless provide a clinically useful categorization of different rhythms (the band definitions below follow those presented in [13]).

Delta rhythm, <4 Hz. The delta rhythm is typically encountered during deep sleep and has a large amplitude. It is usually not observed in the awake, normal adult, but is indicative of, e.g., cerebral damage or brain disease (encephalopathy).

Theta rhythm, 4–7 Hz. The theta rhythm occurs during drowsiness and in certain stages of sleep.

Alpha rhythm, 8–13 Hz. This rhythm is most prominent in normal subjects who are relaxed and awake with eyes closed; the activity is suppressed when the eyes are open. The amplitude of the alpha rhythm is largest in the occipital regions.

Beta rhythm, 14–30 Hz. This is a fast rhythm with low amplitude, associated with an activated cortex and which can be observed, e.g., during certain sleep stages. The beta rhythm is mainly observed in the frontal and central regions of the scalp.

Gamma rhythm, >30 Hz. The gamma rhythm is related to a state of active information processing of the cortex. Using an electrode located over the sensorimotor area and connected to a high-sensitivity recording technique, the gamma rhythm can be observed during finger movements [14].

Most of the above rhythms may persist up to several minutes, while others occur only for a few seconds, such as the gamma rhythm. It is important to realize that one rhythm is not present at all times, but an irregular, "arrhythmic"-looking signal may prevail during long time intervals.

Spikes and sharp waves. Spikes and sharp waves (SSWs) are transient waveforms that stand out from the background EEG with an irregular, unpredictable temporal pattern (paroxysmal activity). Their presence indicates a deviant neuronal behavior often found in patients suffering from epileptic seizures [15]. Because of their relation to seizures, SSWs are often referred to as interictal since they occur between ictal events, i.e., epileptic seizures.

The clinical definition of SSWs is somewhat ambiguous, but both types of waveforms are generally characterized by a steep initial upstroke. A spike is differentiated from a sharp wave by its duration: a spike has a duration in the range 20–70 ms, while a sharp wave is 70–200 ms long. Although the waveform morphology is essentially monophasic, it is not uncommon to observe both bi- and triphasic waveforms. The waveform morphology is naturally dependent on where the electrode is located on the scalp.

Spikes may occur as isolated events or in various types of runs. Such runs are collectively referred to as *spike-wave complexes*, and each complex consists of a spike followed by a slow wave [15]. Spike-wave complexes occur at repetition rates which range from less than 3 to 6 Hz; the repetition rate may correlate with different clinical interpretations. An example of spike-wave complexes is presented in Figure 2.4.

Certain artifacts in a normal EEG can occasionally be mistaken for SSWs. For example, cardiac activity may interfere with the EEG to such a degree that a heartbeat (particularly the waves of the QRS complex) masquerades as a spike.

Sleep rhythms.   The brain has three essential functional states: awake, sleep without *rapid eye movement* (REM), and sleep with REM. The two sleep states, commonly referred to as non-REM and REM sleep, are passed through several times during one night. Non-REM sleep is an "idle" state associated with resting of the brain and the bodily functions. Slow, large-amplitude EEG rhythms during non-REM sleep indicate a high degree of synchrony of the underlying cortical neurons. This sleep state can be further subdivided into four distinct stages related to the degree of sleep depth, see Table 2.1.

A number of transient waveforms usually occur which are characteristic of the different sleep stages: *vertex waves*, *sleep spindles*, and *K complexes*, see Table 2.1 and Figure 2.5. Vertex waves occur during the earlier sleep stages and constitute responses to external stimuli such as sounds. Sleep spindles are bursts of alpha-like activity with a duration of 0.5–1 s. The K complexes can be viewed as the fusion of sleeps spindles and vertex waves.

Rapid eye movement sleep corresponds to an active brain, probably occupied with dreaming. It is therefore not surprising that the EEG closely resembles that of the waking brain and that beta rhythms are present. A prominent feature of the REM sleep state is that the eyes, with the lids closed, move rapidly back and forth in an irregular pattern. These eye movements produce a sawtooth pattern in the EEG when the electrodes are attached close to the eyes.

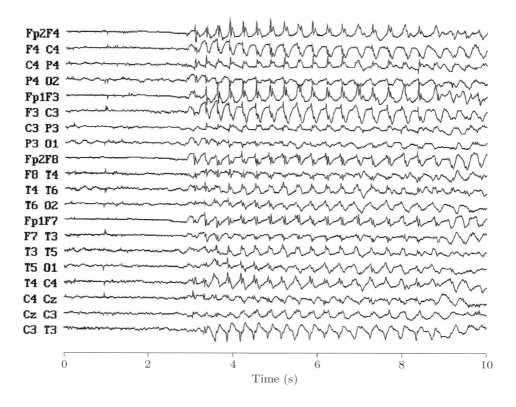

Figure 2.4: A multichannel EEG with spike-wave complexes occurring at a 3-Hz repetition rate. Each channel results from two electrodes placed at locations defined by the codes displayed next to the signal; the definitions of electrode placements are given in Figure 2.7. (Reprinted from Wong [16] with permission.)

Ictal EEG. During an epileptic seizure the EEG is referred to as an ictal EEG, manifested by an abnormal rhythm with a sudden increase in amplitude, as illustrated in Figure 2.6. The onset of a seizure is also associated with a sudden change in frequency content which often evolves into a rhythm with a spiky wave pattern. The ictal EEG may exhibit considerable variability from seizure to seizure, making its detection, whether approached manually or automatically, difficult.

2.2.2   Categorization of EEG Activity

The above-mentioned activities can be roughly categorized into the following four groups with respect to their degree of nonstationarity. The categoriza-

Table 2.1: Essential characteristics of the four non-REM sleep stages and REM sleep [17].

Sleep stage	Sleep depth	Waveforms
1	Drowsiness	From alpha dropouts to vertex waves
2	Light sleep	Vertex waves, spindles, K complexes
3	Deep sleep	Much slowing, K complexes, some spindles
4	Very deep sleep	Much slowing, some K complexes
REM	REM sleep	Desynchronization with faster frequencies

tion was originally presented in [18], and the categories were defined with special reference to their suitability for spectral analysis.

Activity without major temporal changes. Normal, spontaneous waking activity at rest, e.g., with open or closed eyes; various kinds of alpha, beta, and theta rhythms.

Slowly time-varying activity. Sleep background activity, postictal background activity, lengthy seizure discharges.

Intermittent activity. Intermittent, slow rhythm, sleep spindles, i.e., activity with stable patterns over intervals of several seconds.

Paroxysmal activity. Spikes, sharp waves, spike-wave complexes, 3-Hz spike-wave formations, K complexes, and vertex waves observed during sleep, i.e., different types of transient activity.

2.3 Recording Techniques

The clinical EEG is commonly recorded using the International 10/20 system, which is a standardized system for electrode placement [19]. This particular recording system (electrode montage) employs 21 electrodes attached to the surface of the scalp at locations defined by certain anatomical reference points; the numbers 10 and 20 are percentages signifying relative distances between different electrode locations on the skull perimeter, see Figure 2.7. Bipolar as well as so-called unipolar electrodes are used in clinical routine, with the latter type requiring a reference electrode either positioned distantly or taken as the average of all electrodes.

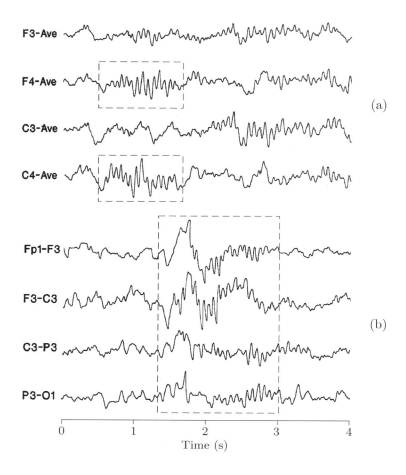

Figure 2.5: Electroencephalographic signals recorded during sleep with (a) sleep spindles and (b) K complexes; note that each K complex is the fusion of a vertex wave and a sleep spindle. (Reprinted from Wong [16] with permission.)

The spacing of electrodes with the 10/20 system is relatively sparse: the interelectrode distance is approximately 4.5 cm on a typical adult head. Improved spatial resolution may be required when brain mapping is of interest. Mapping constitutes a spatial analysis technique in which the EEG activity is represented as a topographic map projected onto the scalp [20]. Using too few electrodes may result in aliasing in the spatial domain, and, consequently, the electrical activity will be inaccurately represented. Studies have indicated that the total number of electrodes used in brain mapping applications should be 64 or higher in order to provide sufficient detail [21].

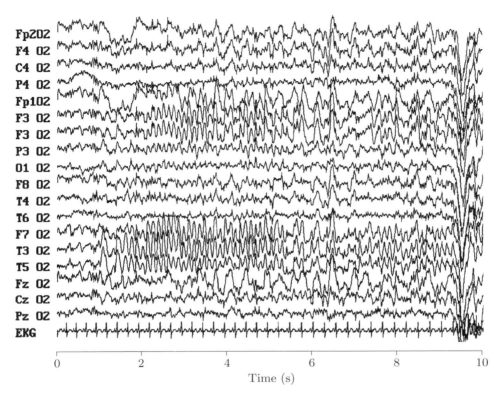

Figure 2.6: A multichannel EEG showing the onset of an epileptic seizure, occurring after the first second. The onset is characterized by an increase in amplitude and a change in spectral content. The seizure is particularly pronounced in certain channels. Note that the ECG is displayed at the bottom (the abbreviations EKG and ECG are synonymous). (Reprinted from Wong [16] with permission.)

The sampling rate for EEG signal acquisition is usually selected to be at least 200 Hz, when taking the frequency ranges of the rhythmic activities previously given into account. A more detailed analysis of transient, evoked waveforms may, however, necessitate a considerably higher sampling rate; see Chapter 4 which describes the analysis of EPs.

2.4 EEG Applications

This section considers two of the most important clinical applications of the EEG, namely, the study of epilepsy and sleep disorders. The design of a brain–computer interface is another EEG application which is considered; so far, this has primarily been studied from a research-oriented perspective.

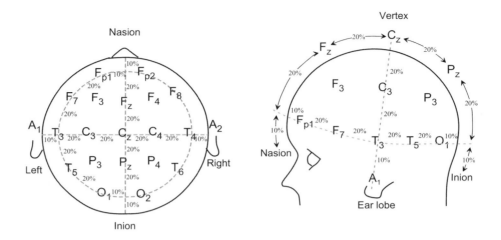

Figure 2.7: The International 10/20 system for recording of clinical EEGs. The anatomical reference points are defined as the top of the nose (nasion) and the back of the skull (inion). The letters F, P, C, T, O, and A denote frontal, parietal, central, temporal, occipital, and auricle, respectively. Note that odd-numbered electrodes are on the left side, even-numbered electrodes are on the right side, and z (zero) is along the midline.

While the descriptions by necessity are kept brief, it is nevertheless hoped that they will help illustrate the importance of signal processing in various EEG applications. The interested reader is referred to the specialist literature for more information on these topics.

2.4.1 Epilepsy

A person with epilepsy suffers from seizures during which sudden bursts of uncontrolled electrical activity occur in a group of neurons of the cerebral cortex. Epileptic seizures are manifested in many different ways depending on where the origin (focus) of the electrical activity is located and how different areas of the brain become successively recruited during a seizure. For example, a seizure which begins in the sensory areas of the cortex is usually manifested by some visual or auditive sensation. The epileptic focus is defined by a group of neurons whose functionality is impaired, whereas the other areas involved in a seizure are often normal.

The interplay between excitatory signals, which increase the electrical activity of the brain by causing nerve cells to fire, and inhibitory signals, which decrease the activity by preventing nerve cells from firing, is well-balanced during normal conditions. However, an imbalance between the two activities is believed to be an important cause of epilepsy. In particular, the neurotransmitters that chemically convey the signals in the synapse are central to causing such an imbalance; if the excitatory neurotransmitters are too active or the inhibitory ones are not active enough, the likelihood of a seizure increases. As a result, bursts of uncontrolled electrical activity will occur. Recently developed antiepileptic drugs are aimed at changing this impaired balance of the neurotransmitters by either decreasing the excitatory activity or increasing the inhibitory activity.

Some seizures are difficult to observe and only result in minor mental confusion, while others cause loss of consciousness, although rarely leading to permanent injury or death. Seizures are typically recurrent events at a highly variable rate, ranging from a few seizures during a lifetime to a few dozen during a single day. The duration of each seizure ranges from a few seconds to a few minutes. Since the manifestations of epileptic seizures differ widely, a scheme for classifying seizures into different groups has been established based on the characteristics of the EEG [22]. The two main groups are defined by the location at which the seizure starts: *partial seizures* start in a restricted (focal) area of the brain, while *primary generalized seizures* involve the entire brain from their onset (Figure 2.8). The seizures belonging to the former group are related to a single epileptic focus, while this does not apply to the latter group. As a result, certain partial seizures may be cured by a surgical procedure in which a small part of the cortex is removed. The procedure must be preceded by a series of extremely thorough investigations in order to assure that the location of the epileptic focus is accurately delimited. In some cases, a partial seizure may evolve to other parts of the brain and is then referred to as a partial seizure with secondary generalization. Figure 2.9 displays an EEG which was recorded during the onset of a primary generalized seizure.

Epilepsy is caused by several pathological conditions such as brain injury, stroke, brain tumors, infections, and genetic factors. The largest group of epileptic patients has, however, an unknown etiology.

The EEG is the principal test for diagnosing epilepsy and gathering information about the type and location of seizures. For subjects with suspected epilepsy, an EEG is recorded in clinical routine for half an hour in a relatively dark and quiet room. During this period, the subject is asked to open and close his/her eyes to study changes in the EEG related to light (recall the presence or absence of alpha activity mentioned on page 34). At the end of the investigation, two "activation" methods are commonly used to provoke waveforms which are associated with epilepsy. In one activation method, the subject is instructed to breath rapidly and deeply (hyperventilation), and in the other method to face a strobe light flashing at a rate of 1–25 Hz

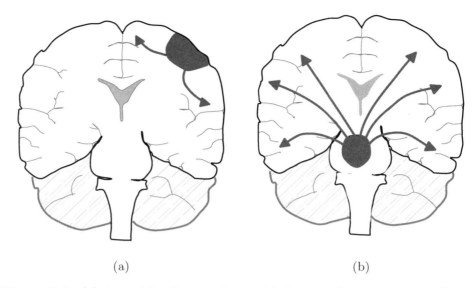

(a)   (b)

Figure 2.8: (a) A partial epileptic seizure with focus in the motor cortex. Related symptoms are muscle twitches in one of the arms, but the subject is fully conscious. (b) A primary generalized seizure which spreads across the entire brain. Symptoms are spasms and unconsciousness. In both figures, a vertical cross-section of the brain is viewed from the front.

(photic stimulation). Sleep deprivation represents another type of activation method which may also be considered.

Although the EEG obtained in clinical routine is often recorded between seizures, i.e., the interictal EEG, the signal waveforms may nevertheless indicate a tendency toward seizures. Examples of interictal waveforms have already been presented in Figure 2.4. The occurrence of SSWs in a local area of the brain, such as in the left temporal lobe, suggests that partial seizures are initiated in that particular area. Spike-wave complexes which are widespread over both hemispheres of the brain suggest, on the other hand, the occurrence of primary generalized seizures. Unfortunately, the absence of interictal waveforms does not rule out the possibility of a seizure.

In order to record an EEG during a seizure, it is often necessary to record the EEG during prolonged periods. Such recordings are often done while video filming the patient, allowing the neurologist to correlate EEG findings to visual findings in order to improve seizure assessment. This type of recording is referred to as "video EEG" and is usually done in the hospital over a period of several days. Another, more convenient, less expensive method is to record the EEG during normal, everyday conditions by a small-sized, digital recording device attached to a belt around the patient's waist [23]. This

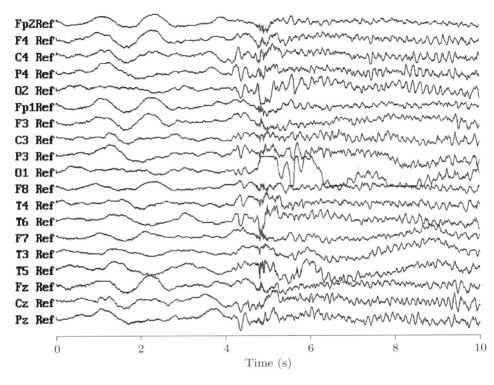

Figure 2.9: A multichannel EEG showing the onset of a primary generalized seizure about halfway into the recording. (Reprinted from Wong [16] with permission.)

type of recording, referred to as "ambulatory EEG", is done in the home for a period of 24 hours or more and therefore includes both waking and sleeping cycles. Similar to video EEG recordings, several electrodes must be attached to the scalp for long periods of time which sometimes cause itching. Scratching the head introduces noise into the EEG recording which may occasionally resemble waveforms of physiological origin. A variety of other noise types, such as those caused by blinking and frowning, can also appear, which may make the interpretation of an ambulatory recording difficult.

Whether performed in the hospital or under ambulatory conditions, long-term EEG monitoring produces large amounts of data which would be very time-consuming to scrutinize. Automatic spike and seizure detection is therefore an important means of reducing the amount of data and improving the efficiency of EEG interpretation. The design of such detection algorithms involves several signal processing considerations regarding the mathematical characterization of interictal waves and epileptic seizures [24–26]. Algorithms for noise and artifact rejection are another important part of such programs. An algorithm for seizure prediction/warning may help a patient wearing an ambulatory recording device take appropriate safety measures prior to a seizure.

A number of therapeutic devices have been developed for epileptic patients which trigger an action that prevents a seizure before it begins. Of these devices, the vagus nerve stimulator is the most well-known and is programmed to regularly stimulate the vagus nerve[3] with a series of intermittent electrical pulses [27]. As the pulses reach the brain, an antiepileptic effect has been observed in some patients, although the mechanisms behind this effect so far remain poorly understood. The vagal nerve stimulator is surgically implanted, similar to a cardiac pacemaker. The stimulating electrode is wrapped around the vagus nerve in the neck, see Figure 2.10. Current stimulators operate blindly by eliciting a preset pattern of stimulation pulses; no attempt is made to predict seizures and modify the therapy accordingly. However, the development of more intelligent vagal stimulators is underway and will involve signal processing algorithms for the prediction of seizures [28, 29].

2.4.2   Sleep Disorders

Sleep disorders, which are frequent in our society, may be caused by several conditions of medical and/or psychological origin. A commonly used scheme for classification of sleep disorders defines the following four groups [30].

Insomnia. Disorders in initiating or maintaining sleep. Most people have at some point in their lives suffered from insomnia due to an agonizing event or approaching examination; this condition is normally transient and is not treated. Depression is associated with poor sleep and causes a substantial reduction in deep sleep, i.e., stages 3 and 4, which makes the patient tired during the daytime. Alcohol and drug abuse are other factors that cause insomnia.

Hypersomnia. Disorders causing excessive sleep and somnolence. Narcolepsy is one example of hypersomnia characterized by uncontrollable daytime sleep attacks while night-time sleep is still fairly normal. Sleep apnea is another condition which indirectly causes hypersomnia. During night-time sleep, the patient suffers from frequent, prolonged suspensions of breathing (>10 s) which cause the patient to wake due to

---

[3]The vagus nerve is one of the 12 pairs of cranial nerves which emanate from the brain; it branches out into the chest and abdomen. The name "vagus" means "wandering" since this nerve is found in many different places. The vagus nerve is also of central importance for controlling heart rate, see Section 8.

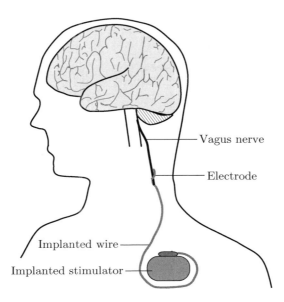

Figure 2.10: The vagus nerve stimulator can prevent epileptic seizures by electrical excitation.

snoring. As a result, a patient with sleep apnea has interrupted deep sleep and is very tired during the daytime (sleep apnea may also be classified as insomnia).

Circadian rhythm disorders. Disorders in the sleep–wake schedule. The most well-known example of such disorders results from flying across several time zones ("jet lag"). Fortunately, the difficulties related with adapting to the new sleep–wake schedule are typically temporary and disappear within a week. A more serious condition arises in subjects whose diurnal rhythm is slightly longer than 24 hours. These subjects sleep later by up to half an hour every day and progressively move into daytime sleep and then back into night-time sleep. As a result, it is difficult to maintain a normal work–rest schedule.

Parasomnia. Deviations in the normal sleep pattern. These sleep disorders are related to deviations from the normal well-being during sleep, although not necessarily leading to awakening. The nightmare is the most common type of parasomnia, being a dream which contains a threatening situation; the nightmare is related to increased autonomic activity as reflected by a drastic increase in heart rate. Sleep terror is a more serious condition, unrelated to dreams, which is characterized by piercing screams or cries; this condition is mostly seen in children

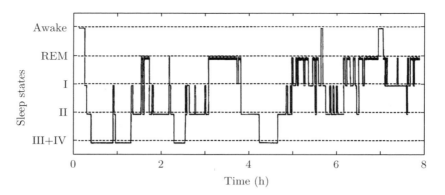

Figure 2.11: Variations in sleep stages observed during one night's sleep. Sleep stages III and IV have been combined. Dream sleep occurs during the REM stage.

and disappears with age. Sleepwalking is another condition, similar to sleep terror, which occurs during the deep stages of sleep.

Each of the different types of sleep disorder exhibits certain manifestations in the EEG. To properly diagnose each disorder, it is therefore important to quantitatively determine how the pattern of sleep stages changes over time, see Figure 2.11. This information is commonly acquired by having the patient stay overnight in a sleep laboratory with electrodes attached to the scalp. Since the manual effort associated with sleep staging is enormous, it is highly desirable to develop and implement a system that automatically performs the sleep staging described in Table 2.1. A fundamental task of such a system is obviously to detect the individual waves that characterize the different stages (i.e., vertex waves, sleep spindles, and the K complexes) and the different rhythms such as delta, theta, alpha, and beta. In order to mimic the method by which a neurologist interprets an EEG, it is important to develop a system that considers contextual information on how individual waves are distributed spatially across channels as well as temporally within each channel [31, 32].

Sleep analysis is commonly based on a polygraphic recording, i.e., a recording that involves several types of physiological signals, not only an EEG; the resulting recording is therefore referred to as *polysomnographic*. Such a recording was exemplified in Figure 1.8 in the Introduction where a number of signals such as the EEG, ECG, EMG, EOG, blood pressure, and nasal and abdominal respiration were included. Polysomnography may also include video filming as a record of the patient's behavior during sleep as expressed by sounds and body movements, see Figure 2.12. Since a polysomnographic recording contains many signals of different origins, its analysis may

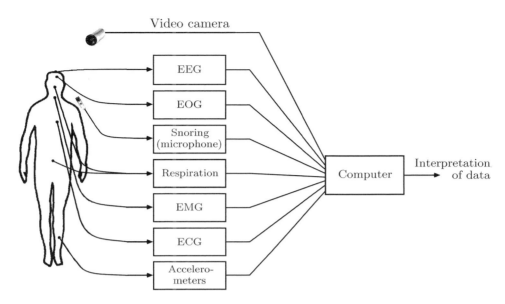

Figure 2.12: Setup for acquisition of a polysomnographic recording in the sleep laboratory. Some of the acquired signals are often multichannel. Additional physiological measures such as blood pressure and blood oxygen level may also be monitored.

be quite complicated. Computer-based analysis of such recordings makes it possible to quantify correlations that may exist between different types of signals. Similar to systems for automated recognition of epileptic seizures, noise and artifact rejection are important parts of a system for automated sleep analysis [33, 34].

2.4.3 Brain–Computer Interface

A brain–computer interface (BCI) enables a subject to communicate with and control the external world without using the brain's normal output through peripheral nerves and muscles [35–37]. Messages are conveyed by spontaneous or evoked EEG activity rather than by muscle contractions which are otherwise used for communication through speech and writing. Subjects with severe neuromuscular disorders, or sometimes those who are completely paralyzed (the "locked-in" syndrome), benefit greatly from a BCI which offers them basic communication capabilities through which they can express themselves, for example, by controlling a spelling program or operating a neuroprosthesis. Although the BCI was first conceived in the early 1970s [38], it was not until the 1990s that its development took a great leap forward [39, 40], owing to more detailed knowledge of the EEG signal and the rapid progress in computer technology.

The following two closely interrelated steps are fundamental to the design and use of a BCI:

- The mental process of the user which encodes commands in the EEG signal; and

- the BCI which, by employing sophisticated signal processing techniques, translates the EEG signal characteristics into commands which control a device.

The imagination of different simple hand and feet movements is associated with different EEG signal characteristics which can be used to encode a set of commands [35, 40, 41]. The related mental process, usually referred to as *motor imagery*, is identical to the process that results in an actual physical movement, except that the motor (muscle) activity is blocked. In order for the BCI to learn the meaning of different EEG signal characteristics, the subject is instructed to imagine one of several actions. For each imagined action, a set of descriptive parameters ("features") is extracted from the EEG signal and submitted to a classifier. By repeating the imagined actions several times, the classifier can be trained to determine which action the subject is imagining. Subsequent to the learning phase, the BCI relies on the classifier to translate the subject's motor imagery into device commands, such as the selection of a letter in a spelling program. The block diagram in Figure 2.13 presents the basic components of a BCI. Since BCIs must operate in real time, it is important that the signal processing does not introduce unacceptable time delays.

The learning phase of a BCI is unfortunately not a one-off procedure resulting in a fixed-parameter classifier, but must be repeated on a regular basis. Since the EEG exhibits considerable variability due to factors such as time of day, hormonal level, and fatigue, it is necessary to adjust the classifier in order to maintain an acceptable performance. In addition, the overall success of the BCI depends on how well the two adaptive "controllers"— the user's brain and the BCI system—are able to interact with each other. The user must develop and maintain good correlation between his/her intent and the signal features used in the BCI. The BCI system must extract signal features that the user can control and translate those features into commands correctly [36].

The most common technique for extracting features from an EEG signal is to analyze spectral power in different frequency bands [42–46]. Spectral analysis of a single channel may be useful although multichannel analysis is preferable since it accounts for spatial variations associated with different

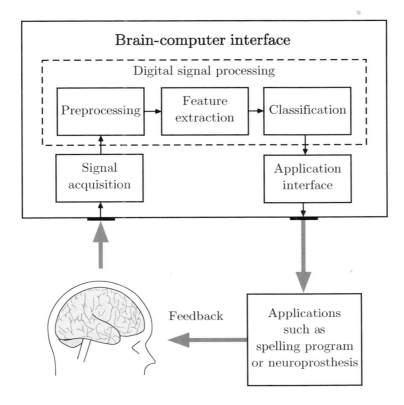

Figure 2.13: Block diagram of a brain–computer interface.

types of motor imagery; for example, differences between the hemispheres can be exploited by multichannel analysis [47]. The frequency bands are selected so that they reflect the EEG rhythms of interest: the mu rhythm[4] and the beta rhythm have been found particularly useful in a BCI. These two rhythms originate from the sensorimotor cortex, i.e., the area which is primarily responsible for the control of hand and foot movements. The extraction of spectral features is further considered in Chapter 3 where a number of spectral estimation techniques that have been implemented in BCIs are presented.

The performance of a BCI may be measured in terms of information transfer rate and is defined in bits per minute. The performance depends on the accuracy with which the different imaginative states are classified. At present, a sophisticated BCI is not able to decipher more than 10–25

---

[4]The mu rhythm has a spectral content similar to the alpha rhythm. While the alpha rhythm is related to idling activity in the visual cortex, the mu rhythm is related to idling activity in the motor cortex.

bits/minute—an information transfer rate which would enable a completely paralyzed subject to write approximately two words per minute. However, these rates are much too slow for the control of complex movements or the interaction with a neuroprosthesis. The information rate can be increased through surgical implantation of microelectrodes which record the activity of more localized populations of neurons.

The above-mentioned approach to BCI design is based on analysis of the spontaneous EEG. However, a BCI can also be based on the use of EPs which result from sensory stimulation [36, 38, 48–50].

Bibliography

[1] H. Berger, "Über das elektroenkephalogramm des menschen," *Arch. Psychiat. Nerv-enkr.*, vol. 87, pp. 527–570, 1929.

[2] J. S. Barlow, *The Electroencephalogram. Its Patterns and Origins.* Cambridge, Mass.: The MIT Press, 1993.

[3] J. S. Barlow, "Computerized clinical electroencephalography in perspective," *IEEE Trans. Biomed. Eng.*, vol. 26, pp. 377–391, 1979.

[4] A. S. Gevins, "Analysis of the electromagnetic signals of the human brain: Milestones, obstacles and goals," *IEEE Trans. Biomed. Eng.*, vol. 31, no. 12, pp. 833–850, 1984.

[5] E. Niedermeyer and F. Lopes da Silva, *Electroencephalography. Basic Principles, Clinical Applications and Related Fields.* Baltimore: Williams & Wilkins, 4th ed., 1998.

[6] R. Plonsey, "Capabilities and limitations of electrocardiography and magnetocardiography," *IEEE Trans. Biomed. Eng.*, vol. 19, pp. 239–244, 1972.

[7] J. Malmivuo, V. Suikho, and H. Eskola, "Sensitivity distributions of EEG and MEG measurements," *IEEE Trans. Biomed. Eng.*, vol. 44, pp. 196–208, 1997.

[8] J. Malmivuo, "Theoretical limits of the EEG method are not yet reached," *Int. J. Bioelectromagn.*, vol. 1, pp. 2–3, 1999. (http://www.tut.fi/ijbem/).

[9] J. P. Wikswo Jr., A. Gevins, and S. J. Williams, "The future of the EEG and the MEG," *Electroencephal. Clin. Neurophysiol.*, vol. 87, pp. 1–9, 1993.

[10] A. M. Dale and M. J. Sereno, "Improved localization of cortical activity by combining EEG and MEG with MRI cortical surface reconstruction," *J. Cogn. Neurosci.*, vol. 5, pp. 162–176, 1993.

[11] M. F. Bear, B. W. Connors, and M. A. Paradiso, *Neuroscience. Exploring the Brain.* Baltimore: Williams & Wilkins, 1996.

[12] H. H. Jasper, "Electroencephalography," in *Epilepsy and Cerebral Localization* (W. G. Penfield and T. C. Erickson, eds.), Springfield, IL.: C. C. Thomas, 1941.

[13] E. Niedermayer, "The normal EEG of the waking adult," in *Electroencephalography. Basic Principles, Clinical Applications and Related Fields* (E. Niedermayer and F. Lopes da Silva, eds.), ch. 13, pp. 149–173, Baltimore: Williams & Wilkins, 1999.

[14] C. Andrew and G. Pfurtscheller, "Event-related coherence as a tool for dynamic interaction of brain regions," *Electroencephal. Clin. Neurophysiol.*, vol. 98, pp. 144–148, 1996.

[15] E. Niedermayer, "Abnormal EEG patterns: Epileptic and paroxysmal," in *Electroencephalography. Basic Principles, Clinical Applications and Related Fields* (E. Niedermayer and F. Lopes da Silva, eds.), ch. 13, pp. 235–260, Baltimore: Williams & Wilkins, 1999.

[16] P. K. H. Wong, *Digital EEG in Clinical Practice*. Philadelphia: Lippincott - Raven, 1996.

[17] E. Niedermayer, "Sleep and EEG," in *Electroencephalography. Basic Principles, Clinical Applications and Related Fields* (E. Niedermayer and F. Lopes da Silva, eds.), ch. 10, pp. 174–188, Baltimore: Williams & Wilkins, 1999.

[18] G. Dumermuth and L. Molinari, "Spectral analysis of EEG background activity," in *Handbook of Electroencephalography and Clinical Neurophysiology: Methods of Analysis of Brain Electrical and Magnetic Signals* (A. S. Gevins and A. Rémond, eds.), ch. 4, pp. 85–130, Amsterdam/New York: Elsevier, 1987.

[19] H. Jasper, "The ten-twenty electrode system of the International Federation," *Electroencephal. Clin. Neurophysiol.*, vol. 10, pp. 371–375, 1958.

[20] F. H. Lopes da Silva, "Computer-assisted EEG diagnosis: Pattern recognition and brain mapping," in *Electroencephalography. Basic Principles, Clinical Applications and Related Fields* (E. Niedermayer and F. Lopes da Silva, eds.), pp. 1164–1189, Baltimore: Williams & Wilkins, 1999.

[21] A. S. Gevins and S. L. Bressler, "Functional topography of the human brain," in *Functional Brain Mapping* (G. Pfurtscheller and F. Lopes da Silva, eds.), pp. 99–116, Toronto: Hans Huber Publishers, 1988.

[22] E. Niedermayer, "Epileptic seizure disorders," in *Electroencephalography. Basic Principles, Clinical Applications and Related Fields* (E. Niedermayer and F. Lopes da Silva, eds.), ch. 27, pp. 476–585, Baltimore: Williams & Wilkins, 1999.

[23] E. Waterhouse, "New horizons in ambulatory electroencephalography," *IEEE Eng. Med. Biol. Mag.*, vol. 22, pp. 74–80, 2003.

[24] J. Gotman, "Automatic recognition of epileptic seizures in the EEG," *Electroencephal. Clin. Neurophysiol.*, vol. 54, pp. 530–540, 1981.

[25] F. Pauri, F. Pierelli, G. E. Chatrian, and W. W. Erdly, "Long-term EEG-video-audio monitoring: Computer detection of focal EEG seizure patterns," *Electroencephal. Clin. Neurophysiol.*, vol. 82, pp. 1–9, 1992.

[26] Y. U. Khan and J. Gotman, "Wavelet based automatic seizure detection in intracerebral electroencephalogram," *Clin. Neurophysiol.*, vol. 114, pp. 898–908, 2003.

[27] R. S. Fisher, G. L. Krauss, E. Ramsay, K. Laxer, and J. Gates, "Assessment of vagus nerve stimulation for epilepsy: Report of the therapeutics and technology assessment subcommittee of the American Academy of Neurology," *Neurology*, vol. 49, pp. 293–297, 1997.

[28] M. D'Alessandro, R. Esteller, G. Vachtsevanos, A. Hinson, J. Echauz, and B. Litt, "Epileptic seizure prediction using hybrid feature selection over multiple intracranial EEG electrode contacts: A report of four patients," *IEEE Trans. Biomed. Eng.*, vol. 50, pp. 603–615, 2003.

[29] B. Litt and J. Echauz, "Prediction of epileptic seizures," *Lancet Neurology*, vol. 1, pp. 22–30, 2002.

[30] R. Broughton, "Polysomnography: Principles and applications in sleep and arousal disorders," in *Electroencephalography. Basic Principles, Clinical Applications and Related Fields* (E. Niedermayer and F. Lopes da Silva, eds.), pp. 858–895, Baltimore: Williams & Wilkins, 1999.

[31] B. H. Jansen and B. M. Dewant, "Knowledge-based approach to sleep EEG analysis—A feasibility study," *IEEE Trans. Biomed. Eng.*, vol. 36, pp. 510–518, 1989.

[32] J. C. Principe, S. K. Gala, and T. G. Chang, "Sleep staging automaton based on the theory of evidence," *IEEE Trans. Biomed. Eng.*, vol. 36, pp. 503–509, 1989.

[33] T. Penzel and R. Conradt, "Computer based sleep recording and analysis," *Sleep Med. Rev.*, vol. 4, pp. 131–148, 2000.

[34] P. Anderer, S. J. Roberts, A. Schlögl, G. Gruber, G. Klosch, W. Herrmann, P. Rappelsberger, O. Filz, M. J. Barbanoj, G. Dorffner, and B. Saletu, "Artifact processing in computerized analysis of sleep EEG—A review," *Neuropsychobiology*, vol. 40, pp. 150–157, 1999.

[35] G. Pfurtscheller and C. Neuper, "Motor imagery and direct brain–computer communication," *Proc. IEEE*, vol. 89, pp. 1123–1134, 2001.

[36] J. R. Wolpow, N. Birbaumer, D. J. McFarland, G. Pfurtscheller, and T. M. Vaughan, "Brain–computer interfaces for communication and control," *Clin. Neurophysiol.*, vol. 113, pp. 761–791, 2002.

[37] T. Ebrahimi, J.-M. Vesin, and G. García, "Brain–computer interface in multimedia communication," *IEEE Signal Proc. Mag.*, vol. 20, pp. 14–24, 2003.

[38] J. J. Vidal, "Towards direct brain–computer communication," *Ann. Rev. Biophys. Bioeng.*, vol. 2, pp. 157–180, 1973.

[39] Z. A. Keirn and J. I. Aunon, "A new mode of communication between men and his surroundings," *IEEE Trans. Biomed. Eng.*, vol. 37, pp. 1209–1214, 1990.

[40] G. Pfurtscheller, D. Flotzinger, and J. Kalcher, "Brain–computer interface—A new communication device for handicapped persons," *J. Microcomput. Appl.*, vol. 16, pp. 293–299, 1993.

[41] G. Pfurtscheller, C. Neuper, C. Guger, W. Harkam, H. Ramoser, A. Schlögl, B. Obermaier, and M. Pregenzer, "Current trends in Graz brain–computer interface (BCI) research," *IEEE Trans. Rehab. Eng.*, vol. 8, pp. 216–219, 2000.

[42] D. J. McFarland, A. T. Lefkowicz, and J. R. Wolpaw, "Design and operation of an EEG-based brain–computer interface with digital signal processing technology," *Behav. Res. Meth. Instrum. Comput.*, vol. 29, pp. 337–345, 1997.

[43] C. W. Anderson, E. A. Stoltz, and S. Shamsunder, "Multivariate autoregressive models for classification of spontaneous electroencephalographic signals during mental tasks," *IEEE Trans. Biomed. Eng.*, vol. 45, pp. 277–286, 1998.

[44] W. D. Penny, S. J. Roberts, E. A. Curran, and M. J. Stokes, "EEG-based communication: A pattern recognition approach," *IEEE Trans. Rehab. Eng.*, vol. 8, pp. 214–215, 2000.

[45] G. Pfurtscheller, C. Neuper, A. Schlögl, and K. Lugger, "Separability of EEG signals recorded during right and left motor imagery using adaptive autoregressive parameters," *IEEE Trans. Rehab. Eng.*, vol. 6, pp. 316–325, 1998.

[46] M. Pregenzer and G. Pfurtscheller, "Frequency component selection for an EEG-based brain to computer interface," *IEEE Trans. Rehab. Eng.*, vol. 7, pp. 413–417, 1999.

[47] M. Osaka, "Peak alpha frequency of EEG during a mental task; Task difficulty and hemispheric differences," *Psychophysiology*, vol. 21, pp. 101–105, 1984.

[48] L. A. Farwell and E. Donchin, "Talking off the top of your head: Toward a mental prosthesis utilizing brain potentials," *Electroencephal. Clin. Neurophysiol.*, vol. 70, pp. 510–523, 1988.

[49] E. Donchin, K. M. Spencer, and R. Wijesinghe, "The mental prosthesis: Assessing the speed of a P300-based brain–computer interface," *IEEE Trans. Rehab. Eng.*, vol. 8, pp. 174–179, 2000.

[50] N. Birbaumer and A. Kübler, "The thought translation device (TTD) for completely paralyzed patients," *IEEE Trans. Rehab. Eng.*, vol. 8, pp. 190–193, 2000.

Chapter 3

EEG Signal Processing

The various waveforms of the EEG signal convey clinically valuable information. A waveform may represent an isolated event, or several waveforms may constitute a composite signal pattern. In both cases, it is important to develop methods for the detection and objective quantification of signal characteristics to facilitate visual interpretation. The extraction of relevant signal features is particularly crucial when the aim is to design a system for EEG classification. The cancellation of noise and artifacts is another important issue in EEG signal processing and a prerequisite for the subsequent signal analysis to be reliable.

This chapter begins by presenting an overview of signal models which, to various extents, have been adopted in EEG signal processing (Section 3.1). The models, being primarily statistical in nature, serve as a starting point for a more detailed description of analysis methods in later sections. Section 3.2 describes the characteristics of the most frequently occurring EEG artifacts and reviews methods for their reduction or cancellation. The flourishing number of approaches to spectral analysis of stationary and nonstationary signals are considered in Sections 3.3–3.4 and 3.6, respectively. Section 3.5 presents methods with which nonstationary signals can be decomposed into a series of consecutive segments having "quasistationary" properties.

3.1 Modeling the EEG Signal

3.1.1 Deterministic and Stochastic Signal Properties

A fundamental question is whether the EEG should be viewed as a deterministic or stochastic signal. Attempts to answer this question may provide some insight into the mechanisms of EEG generation, but may also have implications on the methods considered suitable for signal analysis.

In general, one cannot predict the exact characteristics of the EEG signal in terms of amplitude, duration, or morphology of the individual waves, and, therefore, it seems quite natural to view the EEG signal as a realization of a stochastic process. This view gains further strength if one makes the observation that a "pure" EEG signal which reflects only cerebral activity cannot be acquired. In fact, there is always corrupting random noise, introduced, for example, by internal noise in the amplifier equipment or the digitization process, which, even if the "pure" EEG had had deterministic properties, in the end makes it reasonable to consider the EEG as a stochastic process [1].

While we will primarily adopt a stochastic approach to signal analysis in the remaining part of this chapter, it is important to be aware that the deterministic/stochastic issue is far from being settled in current research literature. Considerable effort has recently been directed toward finding a quantitative answer to the deterministic/stochastic issue, most notably by hypothesizing that the EEG is generated by a nonlinear dynamic system. The output of such a nonlinear system is characterized by a deterministic process which may exhibit "chaotic" behavior, resembling that of a stochastic process.[1]

No strong evidence has been presented showing that the EEG is better modeled as a chaotic deterministic process, except under certain conditions such as before and during an epileptic seizure [2]. In order to achieve accurate signal modeling, the characteristics of the deterministic process have to be so complex that it cannot be easily distinguished from a stochastic process, thus implying that a stochastic description may be equally suitable [3]. Since it is beyond the scope of this book to describe techniques developed for characterizing chaotic behavior in the EEG, the interested reader may wish to consult the review papers in [4, 5].

3.1.2   Stochastic Properties

When using a stochastic signal description as the primary approach to EEG modeling and analysis, one of the first questions that arises is which type of probability density function (PDF) will provide adequate statistical characterization of the signal. Assuming that the samples $x(0), x(1), \ldots, x(N-1)$ representing the EEG signal are modeled as a real-valued stochastic process, it is of interest to determine the joint PDF that completely characterizes

---

[1]Briefly, methods of distinguishing deterministic from stochastic processes rely on the fact that a deterministic system always evolves in the same way from a given starting point. A signal can be tested for determinism by selecting a "test" state, searching the signal for a nearby state, and comparing their time evolution. A deterministic system has an error that either remains small or increases exponentially with time (the chaotic system), while a stochastic system has a randomly distributed error; the error is taken as the difference between the time evolution of the test state and the nearby state.

this process,

$$p(\mathbf{x}; \boldsymbol{\theta}) = p(x(0), x(1), \ldots, x(N-1); \boldsymbol{\theta}), \tag{3.1}$$

where the column vector $\mathbf{x}$ contains the EEG samples in the observation interval $[0, N-1]$,

$$\mathbf{x} = \begin{bmatrix} x(0) \\ x(1) \\ \vdots \\ x(N-1) \end{bmatrix}. \tag{3.2}$$

The probability density function $p(\mathbf{x}; \boldsymbol{\theta})$ in (3.1) embraces a parameter vector $\boldsymbol{\theta}$ whose elements define the specific shape of the function; the elements are assumed to be deterministic but their values are unknown. Ultimately, this parameter vector provides us with quantitative information on various signal properties. For example, assuming that $p(\mathbf{x}; \boldsymbol{\theta})$ is uniformly distributed, the vector $\boldsymbol{\theta}$ contains the definition of the amplitude intervals over which the samples in $\mathbf{x}$ are uniform. Another example is the Gaussian PDF where $\boldsymbol{\theta}$ includes the mean value and the correlation properties of $\mathbf{x}$, thus defining the location and spread of the N-dimensional, bell-shaped Gaussian PDF.

A nonparametric approach to determine $p(\mathbf{x}; \boldsymbol{\theta})$ is to first compute the amplitude histogram of the observed EEG samples and then to hypothesize about the particular structure of $p(\mathbf{x}; \boldsymbol{\theta})$. Another approach is to assume that the structure of the PDF is known a priori, i.e., based on certain neurophysiological insight, and instead focus on how to estimate the parameter vector $\boldsymbol{\theta}$. Irrespective of the approach that we decide to pursue, the PDF issue has no straightforward answer: the ever-changing properties of the EEG require a highly complex PDF structure in order to accurately model signals corresponding to various brain states.

From an engineering point of view it may be inviting to assume that the EEG signal is characterized by a multivariate Gaussian PDF. This assumption is quite plausible, considering that the EEG, as recorded on the scalp, can be viewed as the summation of signals from a very large number of individual neural generators ("oscillators"). The well-known central limit theorem states that the sum of independent random variables in the limit has a Gaussian PDF as the number of summed variables increases; each random variable is not required to be Gaussian for this theorem to hold [6]. The assumption that individual neural generators act independently of one another can, however, be questioned, considering that neural oscillators are organized in groups with substantial internal dependence (the group organization is required to produce the synchronous activity that is visible in the

EEG). On the other hand, groups of neural generators may still be independent of one another, thus suggesting that the requirements of the central limit theorem may still be valid [7].

In the early days of EEG signal analysis, several experimental studies were performed for the purpose of establishing how accurately the amplitude of the EEG samples could be described by a Gaussian PDF, see, e.g., [8–11]. The EEG was investigated under a variety of conditions, and results ranged from the EEG being considered as a Gaussian process to the EEG being considered as a highly non-Gaussian process. For example, the EEG was found to exhibit Gaussian behavior during synchronized activity, such as during the presence of alpha rhythm, whereas it was found to deviate from a Gaussian distribution during REM sleep [8]. In another study, the resting EEG was found to be Gaussian during 66% of the time, while the percentage dropped to 32% when the patients were asked to perform a mental arithmetic task [10]. In general, the amplitude distribution became increasingly non-Gaussian as the measurement interval was increased. In one study it was found that more than 90% of all 1-s intervals could be considered Gaussian, whereas less than 50% were Gaussian when the interval length was increased to 8 s [12]. It should be pointed out that the statistical procedure used for testing Gaussianity is, in itself, rather difficult to apply since one must assume certain properties of the samples to be tested, for example, that the samples constitute a set of statistically independent random variables.

Despite the mixed opinions on the role of Gaussianity in EEG signals, we will still often consider the PDF to be Gaussian since spectral analysis—one of the most popular tools in EEG signal analysis—has a natural connection to this particular distribution. In its most general form, the multivariate Gaussian PDF of a stochastic process $x(n)$ is completely characterized by its mean value

$$m_x(n) = E\left[x(n)\right] \tag{3.3}$$

and the correlation function (also called the autocorrelation function)

$$r_x(n_1, n_2) = E\left[x(n_1)x(n_2)\right], \tag{3.4}$$

which reflects the dependence between the two samples $x(n_1)$ and $x(n_2)$. It should be noted that the correlation function $r_x(n_1, n_2)$ is a symmetric function, i.e., $r_x(n_1, n_2) = r_x(n_2, n_1)$. In general, the process $x(n)$ is non-stationary since the moments defined by (3.3) and (3.4) are time-dependent functions and therefore can differ from sample to sample.

Using vector and matrix notations, the mean vector and the correlation matrix can be compactly defined by

$$\mathbf{m}_x = E\left[\mathbf{x}\right], \tag{3.5}$$

and

$$
\begin{aligned}
\mathbf{R}_x &= E\left[\mathbf{x}\mathbf{x}^T\right] \\
&= \begin{bmatrix}
r_x(0,0) & r_x(0,1) & \cdots & r_x(0,N-1) \\
r_x(1,0) & r_x(1,1) & \cdots & r_x(1,N-1) \\
\vdots & \vdots & & \vdots \\
r_x(N-1,0) & r_x(N-1,1) & \cdots & r_x(N-1,N-1)
\end{bmatrix},
\end{aligned}
\tag{3.6}
$$

respectively. These two quantities characterize the multivariate Gaussian PDF

$$
p(\mathbf{x}) = \frac{1}{(2\pi)^{\frac{N}{2}}|\mathbf{C}_x|^{\frac{1}{2}}} \exp\left[-\frac{1}{2}(\mathbf{x}-\mathbf{m}_x)^T \mathbf{C}_x^{-1}(\mathbf{x}-\mathbf{m}_x)\right],
\tag{3.7}
$$

where the matrix $\mathbf{C}_x$ describes the covariance and is related to the correlation matrix $\mathbf{R}_x$ in the following way,

$$
\mathbf{C}_x = E\left[(\mathbf{x}-\mathbf{m}_x)(\mathbf{x}-\mathbf{m}_x)^T\right] = \mathbf{R}_x - \mathbf{m}_x\mathbf{m}_x^T.
\tag{3.8}
$$

It is immediately obvious from this relation that $\mathbf{C}_x$ and $\mathbf{R}_x$ are identical when $x(n)$ is a zero-mean process, i.e., when $\mathbf{m}_x = \mathbf{0}$.[2]

For the Gaussian model above, the essential information on signal properties is contained in the correlation matrix $\mathbf{R}_x$, and, accordingly, $\mathbf{R}_x$ plays an important role in the analysis of EEG signals. Ultimately, it is our expectation that $\mathbf{R}_x$ will convey useful physiological information on, for example, a particular brain state. Unfortunately, the matrix $\mathbf{R}_x$, given in its most general form in (3.6), is difficult to reliably estimate from a single realization of data because the estimation of each entry $r_x(n_1, n_2)$ is based on only two samples, i.e., $x(n_1)$ and $x(n_2)$. Thus, the resulting estimate will have an unacceptably large variance; the variance can, of course, be reduced if several realizations are available which can be used for ensemble averaging. In many situations, however, it is reasonable to assume that the EEG signal possesses certain "restrictive" properties, for example, by viewing the signal as a stationary process, as a process with slowly changing correlation properties, or as the output of a linear, time-invariant system driven by random noise. The introduction of such restrictions implies not only that $\mathbf{R}_x$ becomes more structured, and thus easier to estimate, but, perhaps more importantly, that the information contained in $\mathbf{R}_x$ can be given an intuitive interpretation.

---

[2]The matrices $\mathbf{C}_x$ and $\mathbf{R}_x$ are sometimes referred to as the sample covariance and sample correlation matrix, respectively. These matrices are of particular interest when repetitive signals are studied and are further discussed in Chapter 4 on the analysis of evoked potentials.

Stationarity. An important restriction of a stochastic process is when its statistical properties are assumed to be time-invariant; such a process is said to be stationary. Of particular interest are processes which are *wide-sense stationary*, since then we are only required to consider the first two moments of the process, i.e., those which define the Gaussian PDF in (3.7). A process $x(n)$ is said to be wide-sense stationary if its mean function $m_x(n)$ is equal to a constant m_x for all time instants n,

$$m_x(n) = m_x, \tag{3.9}$$

and its correlation function $r_x(n_1, n_2)$ is a function only of the time lag $k = n_1 - n_2$ between the samples $x(n_1)$ and $x(n_2)$, i.e., $r_x(n, n - k) = r_x(k)$. The lag-dependent correlation function is denoted

$$r_x(k) = E\left[x(n)x(n - k)\right]. \tag{3.10}$$

The corresponding correlation matrix is defined by

$$\mathbf{R}_x = \begin{bmatrix} r_x(0) & r_x(-1) & r_x(-2) & \cdots & r_x(-N+1) \\ r_x(1) & r_x(0) & r_x(-1) & \cdots & r_x(-N+2) \\ r_x(2) & r_x(1) & r_x(0) & \cdots & r_x(-N+3) \\ \vdots & \vdots & \vdots & \ddots & \vdots \\ r_x(N-1) & r_x(N-2) & r_x(N-3) & \cdots & r_x(0) \end{bmatrix}. \tag{3.11}$$

Apart from being symmetric—recall that $r_x(k) = r_x(-k)$ for a real-valued process—the correlation matrix in (3.11) is Toeplitz since all elements in a given diagonal are identical, and equal to the correlation function at a certain time lag. It should be pointed out that the Toeplitz property applies to the correlation matrix of any stationary process; this matrix structure is of central importance in the development of computationally efficient methods for spectral estimation. In the following, the term "stationary" is used to signify that the process is wide-sense stationary.

Stationarity is particularly attractive when considering that the properties of the stochastic process can be interpreted in spectral terms. This aspect is valuable since it provides us with a more intuitive characterization of the signal than does the correlation function (although correlation analysis was popular in EEG signal processing at one point in time [13]). Spectral analysis is intimately related to the Gaussian distribution since the power spectral density or, more briefly, the power spectrum is defined as the discrete-time Fourier transform (DTFT) of the correlation function $r_x(k)$,

$$S_x(e^{j\omega}) = \sum_{k=-\infty}^{\infty} r_x(k)e^{-j\omega k}. \tag{3.12}$$

The inverse relation is

$$r_x(k) = \frac{1}{2\pi} \int_{-\pi}^{\pi} S_x(e^{j\omega})e^{j\omega k}d\omega. \tag{3.13}$$

In EEG analysis, the validity of the stationarity assumption is, of course, dependent on the type of signal to be analyzed. Normal spontaneous activity (category 1, page 37) is essentially stationary and is therefore commonly subjected to power spectral analysis. However, similar to our earlier considerations on Gaussianity and its behavior at increasing interval length, normal spontaneous activity is reasonably well-modeled by a stationary process only over relatively short intervals.

Fourier-based, nonparametric techniques for estimating the power spectrum $S_x(e^{j\omega})$ in (3.12) are described in Section 3.3, and some of the inherent limitations associated with these techniques are discussed. The extraction of features characterizing different spectral components is another topic considered in that section.

Nonstationarity. When considering long time periods, several factors make it necessary to treat the EEG signal as a nonstationary, stochastic process, i.e., a process whose mean, correlation function, and higher-order moments are time-varying. For example, the degree of wakefulness of the subject may vary slowly over time, causing the properties of the alpha rhythm to change slowly. Another contributing factor is due to the opening and closing of the eyes which cause an abrupt change in rhythmic activity. Yet another factor is the intermittent occurrence of transient waveforms such as epileptic spikes or pulse-shaped artifacts.

The above-mentioned three factors serve as good examples of major EEG nonstationarities. Specific algorithmic approaches have been presented to process each of these types.

- Slowly time-varying signal properties can be tracked by repeatedly applying the analysis of a method, originally developed for stationary signals, to consecutive, overlapping intervals. The short-time Fourier transform is a well-known technique which performs a "sliding" spectral analysis of the signal [14], see Section 3.6.1. Another approach is to design a time–frequency method which is inherently capable of characterizing time-varying spectral properties; the Wigner–Ville distribution is one such method which, together with some of its "relatives", will be described in Sections 3.6.2–3.6.4.

  Yet another approach is to develop a parametric model of the EEG signal and to estimate its parameter values recursively using an adaptive algorithm (Section 3.6.5). In any of these approaches, the output is made up of consecutive spectra reflecting slow changes in the rhythmic behavior.

- Abruptly changing activity, possibly suggesting transitions occurring between different behavioral states, can be analyzed by first decomposing the EEG signal into a series of variable-length, quasistationary segments [15]. Each individual segment can then be characterized by means of its power spectrum and related spectral parameters. The segmentation approach requires an algorithm which can determine the appropriate interval boundaries; the design details of such boundary detectors are presented in Section 3.5.

- Transient activity in the form of spikes and sharp waves represents one type of nonstationarity which calls for event detection [16–28]. The detector output provides the basis for grouping together transient waveforms with similar shapes and for characterizing their temporal occurrence pattern. In contrast to the above types of nonstationarities, transient waveforms do not lend themselves to spectral analysis, but are better described by various waveform parameters, such as amplitude and duration. Wavelet analysis is a powerful technique for extracting such parameters and will be described in the context of EPs in Section 4.7.

  Since transient waveforms are superimposed on the background EEG, the output of the event detector can instead be used to help exclude segments clearly unsuitable for spectral analysis.

Non-Gaussian signals. It may be necessary to develop analysis methods of the EEG which go beyond second-order moment analysis and which take non-Gaussian properties into account [29]. The first step in the development of such methods has been to study higher-order moments of the univariate amplitude distribution of the EEG, i.e., to obtain an estimate of

$$E\left[(x(n) - m_x)^k\right], \quad k = 3, 4, \ldots. \tag{3.14}$$

The third-order moment is proportional to the *skewness* which describes the degree of deviation from the symmetry of a Gaussian PDF, whereas the fourth-order moment is, apart from a constant, proportional to the *kurtosis* which describes the peakedness of the PDF around the mean value m_x [30].

The method of moments is simple to use when applied to univariate distributions, but does not produce any information on temporal relationships. The extension of the method to the multivariate case is, in general, too complicated to be considered from a technical, as well as from a conceptual, point of view. However, one of the most widespread techniques for studying non-Gaussian properties of the EEG takes its starting point in the third-order cumulant $\kappa_x(k_1, k_2)$ which, assuming that $x(n)$ is a zero-mean, stationary process, reflects the joint variation of three samples separated by lags defined by k_1 and k_2,

$$\kappa_x(k_1, k_2) = E\left[x(n)x(n - k_1)x(n - k_2)\right]. \tag{3.15}$$

Similar to the Fourier transform, which relates the correlation function $r_x(k)$ to the power spectrum $S_x(e^{j\omega})$, the two-dimensional Fourier transform relates $\kappa_x(k_1, k_2)$ to the *bispectrum* [31]. The bispectrum displays the spectral power as a function of two frequencies and is therefore useful for detecting relations between different frequencies. The bispectrum may be valuable in indicating the degree to which a signal follows a Gaussian distribution; the bispectrum is zero if the signal is purely Gaussian. Further information on bispectral analysis and its application to EEG signals can be found in a number of studies [32–36].

3.1.3 Linear, Stochastic Models

We will now give a brief overview of the most popular signal models in EEG analysis—the linear stochastic models. This class of mathematical models is entirely phenomenological in nature since they do not incorporate any specific anatomical or physiological information. Instead, the models are designed to account for certain landmark features of the observed signal, for example, that the signal is composed of different narrowband components. The common aim of these models is to derive clinically useful model parameters, rather than to develop a model that explains the underlying mechanisms of EEG generation.

The EEG signal is modeled as the output of a linear system driven by stationary, *white noise*[3] which is usually assumed to be Gaussian, see Figure 3.1. The parameter values defining the system are estimated by fitting the linear model to the EEG signal using a suitable error criterion, frequently taken as the mean-square error criterion. Section 3.4 describes a number of such parameter estimation methods in further detail. Several important reasons can be listed for pursuing the model-based approach [37, 38].

- It can produce a power spectrum which is more accurate than that obtained by Fourier-based analysis, especially when short-duration signals are being analyzed. Needless to say, high accuracy can *only* be achieved as long as the observed signal is in good agreement with those produced by the model.

---

[3]White noise consists of a sequence of zero-mean, uncorrelated random variables characterized by a suitable PDF.

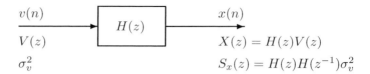

Figure 3.1: Modeling of the EEG by linear filtering of white noise $v(n)$ with the variance σ_v^2. The linear filter $H(z)$, which is characterized by a set of parameters, spectrally shapes the noise and produces the output signal $x(n)$. The complex power spectrum of the output signal $x(n)$ is denoted $S_x(z)$.

- The parametric spectral description consists of a compact set of features which may be useful for the detection and classification of various EEG rhythms.

- The development of other, nonspectral algorithms is facilitated. For example, detection of epileptiform activity and certain types of nonstationary events can be based on the deviation between a model signal and the observed signal.

- The estimated parameter values can serve as a basis for designing an EEG simulator, primarily useful for algorithmic testing and performance evaluation.

In addition, it should be pointed out that the popularity of linear models is partially explained by the existence of computationally attractive methods for parameter estimation.

The autoregressive, moving average (ARMA) model represents a general form of the linear, stochastic models, defined by the following difference equation [39, 40]

$$x(n) = -a_1 x(n-1) - \cdots - a_p x(n-p) + b_0 v(n) + \cdots + b_q v(n-q), \quad (3.16)$$

where the model parameters $a_1, \ldots, a_p, b_0, \ldots, b_q$ are fixed, and the input $v(n)$ is white noise with the variance given by

$$E\left[v^2(n)\right] = \sigma_v^2. \quad (3.17)$$

Hence, the output sample $x(n)$ is modeled as a linear combination of the p past output samples $x(n-1), \ldots, x(n-p)$, the q past input samples $v(n-1), \ldots, v(n-q)$, and the present input sample $v(n)$.

Using the z-transform, the ARMA model can be equally well-described by the rational transfer function $H(z)$,

$$H(z) = \frac{B(z)}{A(z)} = \frac{b_0 + b_1 z^{-1} + \cdots + b_q z^{-q}}{1 + a_1 z^{-1} + \cdots + a_p z^{-p}}. \tag{3.18}$$

Having estimated the model parameters $b_0, \ldots, b_p$, $a_1, \ldots, a_q$, and σ_v^2 from the observed signal, the complex power spectrum of the ARMA model can be calculated from

$$S_x(z) = H(z)H(z^{-1})\sigma_v^2, \tag{3.19}$$

or the power spectrum can be calculated by evaluating z on the unit circle, i.e., $z = e^{j\omega}$,

$$S_x(e^{j\omega}) = |H(e^{j\omega})|^2 \sigma_v^2, \tag{3.20}$$

see Figure 3.1. The ARMA power spectrum is then obtained from

$$S_x(e^{j\omega}) = \left| \frac{b_0 + b_1 e^{-j\omega} + \cdots + b_q e^{-j\omega q}}{1 + a_1 e^{-j\omega} + \cdots + a_p e^{-j\omega p}} \right|^2 \sigma_v^2. \tag{3.21}$$

The main characteristics of the power spectrum $S_x(e^{j\omega})$ in (3.21) are determined by the locations of the roots of the polynomials $B(z)$ and $A(z)$. The zeros, given by $B(z)$, are associated with spectral valleys, and the poles, given by $A(z)$, are associated with spectral peaks.

AR modeling. In the autoregressive (AR) model, i.e., for $q = 0$ and $b_0 = 1$ in (3.16), the number of parameters is restricted so the present sample $x(n)$ is assumed to be a regression of the p past output samples and the input noise $v(n)$,

$$x(n) = -a_1 x(n-1) - \cdots - a_p x(n-p) + v(n). \tag{3.22}$$

This model has come into widespread use in EEG signal processing because it provides a compact parametric description of many different EEG rhythms [39, 41–43]. The AR model parameters contain essential spectral information on a rhythm and can be used to derive information on the spectral power and dominant frequency of the rhythm (Section 3.4.5). The number of peaks that can be present in the AR power spectrum is determined by the model order p; each additional spectral peak requires an increase in order by two.

Figure 3.2 exemplifies AR modeling by presenting a measured EEG with a strong alpha rhythm and a simulated signal. The simulated signal is obtained by filtering white noise with an all-pole filter whose parameters have

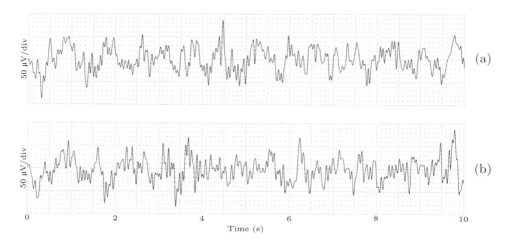

Figure 3.2: (a) An EEG signal with a prominent alpha rhythm, and (b) a simulated signal produced by an AR model whose parameters were estimated from the signal displayed in (a).

first been estimated from the EEG signal. Thus, the AR power spectra of the two signals in Figure 3.2 are identical.

As we will see in Section 3.4, the estimation of AR parameters is typically synonymous with the solution of a linear matrix equation. By exploiting the special structure of the involved correlation matrix, the AR parameter estimates can be determined in a very efficient way. This property stands in contrast to estimation procedures associated with the other two commonly used linear models, the ARMA and the moving average (MA) models, which require significantly more computations in order to determine the model parameters.

Time-varying AR modeling. The issue of nonstationarity, which we touched upon earlier, may be handled within the context of AR modeling by replacing the fixed model parameters $a_1, \ldots, a_p$ in (3.22) with their time-varying counterparts, i.e.,

$$x(n) = -a_1(n)x(n-1) - \cdots - a_p(n)x(n-p) + v(n). \qquad (3.23)$$

In general, the usefulness of this model extension is limited since the exact temporal behavior of $a_i(n)$ is unknown. However, in situations where slowly changing spectral properties are expected, it is possible to follow such parameter changes by making use of an adaptive estimation algorithm, see Section 3.6.5. Another approach is to constrain the temporal evolution of the model parameters so that it is characterized by a linear combination of a set of smooth basis functions [44–46].

Multivariate AR modeling.   Another generalization of the linear model is the multivariate model which opens up the possibility to study the spatial interaction between different regions of the brain. For the AR model in (3.22), we have the following multivariate difference equation,

$$\mathbf{x}(n) = -\mathbf{A}_1\mathbf{x}(n-1) - \cdots - \mathbf{A}_p\mathbf{x}(n-p) + \mathbf{v}(n), \tag{3.24}$$

where $\mathbf{x}(n)$ is an M-dimensional column vector that contains the samples at time n of the M channels, and $\mathbf{A}_1, \ldots, \mathbf{A}_p$ are $M \times M$ matrices that together describe temporal as well as spatial correlation properties across the scalp. The off-diagonal elements in the matrices $\mathbf{A}_i$ reflect the degree to which different electrode sites are correlated on the scalp. The statistical properties of the M-dimensional input noise vector $\mathbf{v}(n)$ are usually assumed to be zero-mean and with variances $\sigma_{v_1}^2, \sigma_{v_2}^2, \ldots, \sigma_{v_M}^2$ which differ from channel to channel.

The multivariate AR model has been considered in a large number of EEG studies [47–51]. More recently, multivariate AR parameters have been used as classification features in a brain–computer interface (Section 2.4.3) to help a disabled person unable to communicate with physical methods but with full mental control [52]. The basic idea is to detect EEG patterns which are specific to different mental tasks so that a "task alphabet" can be defined with which, for example, a paralyzed person can control the operation of a wheelchair.

AR modeling with impulse input.   As a final example of linear modeling, it may be of interest to point out that the input to the linear system $H(z)$ does not necessarily have to be defined by white noise only, but the white noise can be mixed with a train of isolated impulses at random occurrence times. Such an approach has been suggested for the modeling of transient events, e.g., K complexes and vertex waves, observed in sleep recordings [53]; the output of such a signal model is illustrated in Figure 3.3.

3.1.4   Nonlinear Modeling of the EEG

Although the above linear, filtered-noise models, with either time-invariant or time-varying parameters, have been successfully used in many EEG applications, these models are far from adequate in representing all types of signal patterns, nor do they provide deeper insight into the mechanisms of EEG generation. As a result, nonlinear simulation models have been developed in order to better understand the underlying generation process. Such

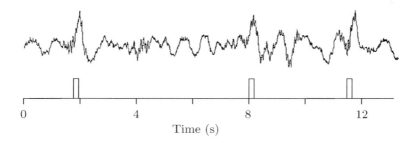

Figure 3.3: Simulation of a "sleep EEG" which contains transient events. The signal is generated by a linear system driven by both white noise, representing the background activity, and randomly occurring impulses (indicated by bars), accounting for the presence of vertex waves and K complexes. (Reprinted from da Rosa et al. [53] with permission.)

models are based on certain neurophysiological facts and may reflect how different neuron populations interact with one another as described by a set of nonlinear differential equations [2]. The usefulness of nonlinear EEG models for the design of signal processing methods has, however, yet to be demonstrated.

A nonlinear model of one cortical neuron population was originally proposed in the early 1970s for the study of rhythmic EEG activities and, in particular, the alpha rhythm [54, 55]. This model was later extended to account for multiple coupled neuron populations for the purpose of simulating signals before and during epileptic seizures [56–58]. Another important extension of the nonlinear model deals with the mechanisms behind the generation of EPs in the visual cortex [59, 60]. Below, a brief description is given of the basic model for one neuron population; the interested reader is referred to the original papers for details on the more advanced models and their mathematical analysis.

In simplified terms, the neuron population may be modeled as two interacting subpopulations of neurons—the main (pyramidal) cells and the interneurons—which are connected to each other through positive or negative feedback, see Figure 3.4 (an interneuron is any neuron that is not a sensory or motor neuron). The *first* subpopulation is composed of pyramidal cells which receive excitatory and inhibitory feedback from the interneurons (cf. page 29) as well as excitatory input from neighboring and distant populations. The *second* subpopulation is composed of interneurons and only receives excitatory input from the pyramidal cells.

The first subpopulation is modeled by two different blocks: a linear, time-invariant system (dynamic) and a nonlinear transformation (static).

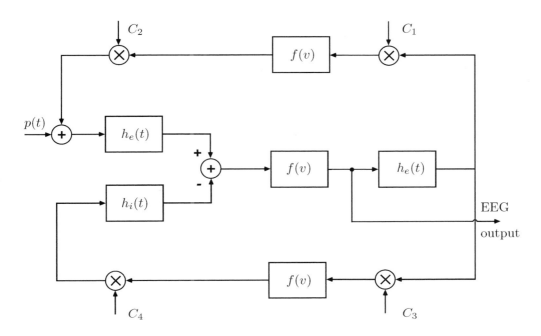

Figure 3.4: A model of a cortical neuron population consisting of main cells and interneurons. The linear systems $h_e(t)$ and $h_i(t)$ describe the excitatory and inhibitory postsynaptic potentials, respectively, of the main cells, while the interneurons are described by $h_e(t)$ alone. The parameters $C_1, C_2, C_3,$ and C_4 account for interactions between the two subpopulations.

By means of a filtering operation, the first block converts the average pulse density of action potentials arriving at the subpopulation into an average postsynaptic potential which is either excitatory or inhibitory. The word "average" here denotes the overall behavior of the neurons within the specific subpopulation. The first block is composed of two linear, time-invariant systems, defined by the impulse responses $h_e(t)$ and $h_i(t)$, which model the shapes of realistic excitatory and inhibitory postsynaptic potentials, respectively. From experimental studies, it has been found that a realistic choice of impulse responses is given by the following two expressions:[4]

$$h_e(t) = Aate^{-at}u(t), \tag{3.25}$$

and

$$h_i(t) = Bbte^{-bt}u(t), \tag{3.26}$$

---

[4]A continuous-time model representation is adopted here since this is the one commonly employed in the literature.

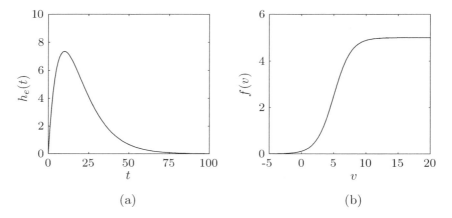

(a)   (b)

Figure 3.5: (a) The impulse response $h_e(t)$ that describes the shape of an excitatory postsynaptic potential ($A = 20, a = 0.1$). The impulse response $h_i(t)$ has the same shape as $h_e(t)$, but the amplitude and timescale differ. (b) The sigmoid function $f(v)$ that transforms the average postsynaptic potential into a pulse density of action potentials ($r = 0.75, e_0 = 2.5, v_0 = 5$).

where $u(t)$ denotes the unit step function. An example of the impulse response is presented in Figure 3.5(a). The parameters A and B determine the maximum amplitude of the postsynaptic potentials, and a and b reflect various properties associated with the dendrites, such as their average time delay.[5] The cell somas of the first subpopulation are modeled by a summation unit where excitatory and inhibitory postsynaptic potentials are summed with positive and negative signs.

The second block transforms the average postsynaptic potential of a population into a pulse density of action potentials fired by the neurons. The voltage-to-pulse transformation is accomplished by a "threshold function" $f(v)$ which essentially has two different levels: the output firing rate is equal to zero for low input potentials, whereas it is equal to a fixed, nonzero value once a certain threshold value is exceeded (cf. the on/off property of neuronal communication mentioned on page 29). A function which exhibits such a characteristic is the sigmoid function, defined by

$$f(v) = \frac{2e_0}{1 + e^{r(v_0 - v)}}. \tag{3.27}$$

With this particular function, a postsynaptic potential of $v = v_0$ corresponds to a firing rate of e_0. The parameter r defines the steepness of the function

---

[5]A model including a parameter that represents a number of different physiological characteristics is referred to as a *lumped-parameter model*.

between the two levels, see Figure 3.5(b). The maximum firing rate is equal to $2e_0$.

The influence from neighboring or distant neuron populations is modeled by the excitatory input pulse density $p(t)$ which is fed to the first subpopulation. This input may be defined by a stochastic process such as white noise and characterized by a certain type of PDF, for example, Gaussian or uniform. Furthermore, the first subpopulation of neurons receives excitatory and inhibitory feedback from the second subpopulation.

The second subpopulation is modeled in the same way as the first subpopulation but only involves one linear, time-invariant system $h_e(t)$ since the input to interneurons is primarily excitatory.

Interactions between the two subpopulations are described by four connectivity parameters C_1, C_2, C_3, and C_4, which account for the average number of synaptic contacts established between the subpopulations. Depending on the values of these four parameters, signals can be generated which resemble alpha rhythms and certain waveforms observed during epileptic states; the latter type of waveforms is obtained with values of C_1, C_2, C_3, and C_4 that cause the model to operate close to instability [55].

To generate EEG signals, the output of the model is normally taken after summation of the excitatory and inhibitory postsynaptic potentials. The electrical field generated at this point is considered to be the main source of activity observed on the scalp.

Figure 3.6 presents a number of examples obtained with a model based on multiple coupled neuron populations of the type presented in Figure 3.4 (details on the simulation can be found in [57]). The simulated signals in Figures 3.6(a)–(d) are paired with real signals in Figures 3.6(e)–(h), which were recorded before and during an epileptic seizure. In this particular case, the signals were recorded with intracerebral electrodes which reflect the spontaneous electrical activity much closer to the neurons than surface EEG electrodes. It is evident from Figure 3.6 that nonstationary signals can be generated with the model which exhibit good overall agreement with recorded signals.

3.2 Artifacts in the EEG

One of the crucial aspects in biomedical signal processing is to acquire knowledge about noise and artifacts which are present in the signal so that their influence can be minimized. Artifact processing therefore constitutes one of the cornerstones in computer-based analysis of biomedical signals and is equally important whether the signals originate from the brain, the heart, or from any other electrical source in the human body. In EEG recordings, a

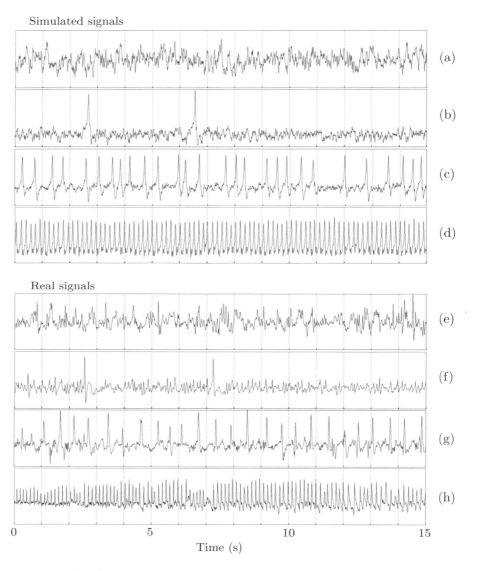

Figure 3.6: (a)–(d) Simulated EEG signals produced by a model which involves multiple, coupled neuron population models such as the one described in the block diagram in Figure 3.4. The resulting signals display a progression from (a) "normal" to (d) a rhythmic discharge of spikes as observed during an epileptic seizure. (e)–(h) Real signals recorded by intracerebral electrodes (e), (f) before and (g), (h) during an epileptic seizure. (Reprinted from Wendling et al. [57] with permission.)

wide variety of artifacts can occur, some of which can be easily identified by a simple algorithm, while others may closely mimic the brain activity and are, as a result, exceedingly difficult to distinguish, even with the eye of a well-trained electroencephalographer.

One useful categorization of artifacts is based on their origin, i.e., those of *physiological* or *technical* origin. This categorization is also applicable to other bioelectrical signals; a description of artifacts that may occur in the ECG signal can be found on page 487. While the influence of artifacts of technical origin can be reduced to a large degree by paying extra attention to the attachment of electrodes to the body surface, it is impossible to avoid the influence of artifacts of physiological origin. Accordingly, a majority of algorithms developed for EEG artifact processing are intended for the reduction of physiological artifacts.

Below, we first review the most common types of artifacts and then describe the signal processing aspects involved with various methods of artifact reduction.

3.2.1 Artifact Characteristics

The first three types of noncerebral artifacts which we will describe are of physiological origin (eye movement and blinks, cardiac activity, and muscle activity), while the fourth type (electrodes and equipment) is of technical origin. It should be emphasized that multiple types of artifacts can, of course, be present in the EEG at the same time. Moreover, certain types of artifacts are more frequently encountered in certain recording situations, for example, during sleep. The interested reader is referred to [61–63] where additional details can be found on various EEG artifacts; these references also contain descriptions of other, less common artifacts related to respiration, tongue movements, tremor, and skin potentials.

Eye movement and blinks. Eye movement produces electrical activity—the electrooculogram (EOG)—which is strong enough to be clearly visible in the EEG. The EOG reflects the potential difference between the cornea and the retina which changes during eye movement. The measured voltage is almost proportional to the angle of gaze [64]. The strength of the interfering EOG signal depends primarily on the proximity of the electrode to the eye and the direction in which the eye is moving, i.e., whether a horizontal or vertical eye movement takes place. The waveforms produced by repeated eye movement are exemplified in Figure 3.7(a). Although the resulting artifact can be easily discerned in this figure due to its repetitive character, the EOG artifact can sometimes be confused with slow EEG activity, e.g., theta and delta activities. Eye movement is not only present during the waking state,

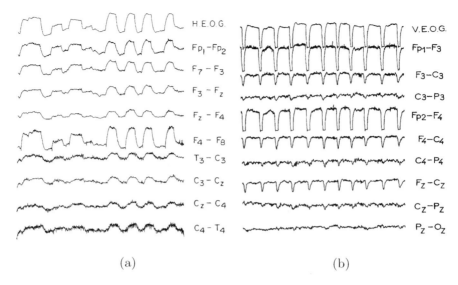

$$\text{(a)}\qquad\qquad\qquad\qquad\text{(b)}$$

Figure 3.7: Artifacts in the EEG caused by (a) eye movement and (b) repetitive, voluntary blinking. The signal at the top of each column shows the horizontal and vertical EOG, respectively. (Reprinted from Barlow [63] with permission.)

but may also interfere when rapid movements occur during sleep (REM sleep).

Another common artifact is caused by eyelid movement ("blinks") which also influences the corneal–retinal potential difference. The blinking artifact usually produces a more abruptly changing waveform than eye movement, and, accordingly, the blinking artifact contains more high-frequency components. This particular signal characteristic is exemplified in Figure 3.7(b), where the waveform produced by repetitive blinking resembles a square wave. From Figure 3.7(b) it can be observed that the amplitude of blinking artifacts in the frontal electrodes is substantially larger than that of the background EEG.

From an artifact processing viewpoint, it is highly practical if a "pure" EOG signal can be acquired by means of two reference electrodes positioned near the eye, cf. the upper traces in Figure 3.7 which do not contain any EEG activity. The availability of such *reference signals* is valuable since these are correlated with the EOG in the EEG and, accordingly, are useful for artifact cancellation purposes (see below).

Muscle activity.   Another common artifact is caused by electrical activity of contracting muscles, measured on the body surface by the EMG [65]. This type of artifact is primarily encountered when the patient is awake and

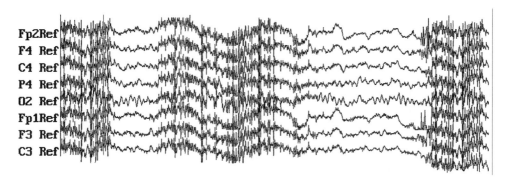

Figure 3.8: A 5-s, multichannel EEG recording contaminated with intermittent episodes of electromyographic artifacts. (Reprinted from Wong [66] with permission.)

occurs during swallowing, grimacing, frowning, chewing, talking, sucking, and hiccupping [63]. The overall shape of the EMG signal depends on the degree of muscle contraction: a weak contraction produces a train of low-amplitude spikes, while an increase in contraction strength decreases the interspike distance so that the EMG more closely exhibits the properties of a continuously varying signal ("colored noise"), see Figure 3.8. The muscle artifact is considerably reduced during relaxation and sleep.

In terms of artifact processing, the spectral properties of the EMG are much less favorable than those associated with eye movement, because they considerably overlap with beta activity in the 15–30 Hz range. Unfortunately, this disadvantage is further exacerbated by the fact that it is impossible to acquire a reference signal containing only EMG activity which would be useful for artifact cancellation.

Cardiac activity. The electrical activity of the heart, as reflected by the ECG, can interfere with the EEG. Although the amplitude of the cardiac activity is usually low on the scalp in comparison to the EEG amplitude (1–2 and 20–100 μV, respectively), it can hamper the EEG considerably at certain electrode positions and for certain body shapes, e.g., short, stout subjects with short, thick necks [62]. The repetitive, regularly occurring waveform pattern which characterizes the normal heartbeats fortunately helps to reveal the presence of this artifact. However, the spike-shaped ECG waveforms can sometimes be mistaken for epileptiform activity when the ECG is barely visible in the EEG. This situation may be further complicated during the presence of certain cardiac arrhythmias, which can exhibit considerable variability in the interbeat interval.

Similar to the eye-related artifacts mentioned above, the ECG can be acquired independently by one or several electrodes for use in canceling the ECG activity that may be superimposed on the EEG.

Electrodes and equipment. Movement of electrodes causes changes in the DC contact potential at the electrode–skin interface which produce an artifact commonly referred to as the "electrode-pop" artifact. This type of technical artifact is not unique to the EEG signal, but may occur in any bioelectric signal measured on the body surface [67, 68]. The electrode-pop artifact is usually manifested as an abrupt change in the baseline level, followed by a slow, gradual return to the original baseline level. On occasion, the electrode-pop artifact can be misinterpreted as spikes or sharp waves.

The electrode wire which connects the electrode to the acquisition equipment is another possible source of artifact. Insufficient shielding of the electrode wire makes it susceptible to electromagnetic fields caused by currents flowing in nearby powerlines or electrical devices. As a result, 50/60 Hz powerline interference is picked up by the electrodes and contaminates the EEG signal.

Finally, equipment-related artifacts include those produced by internal amplifier noise and amplitude clipping caused by an analog-to-digital converter with too narrow a dynamic range.

3.2.2   Artifact Processing

The scope of artifact processing ranges from artifact rejection, in which a simple marker is created to identify the artifact, to complete cancellation of the artifact from the EEG signal. Artifact rejection is the crudest approach since its goal is to simply reject segments of poor quality from further analysis [69–71]. Although rejection is today the main alternative in handling segments which contain excessive EMG interference, it is, as a rule, desirable to retain the data as much as possible; this is especially important when only short segments of data are available for analysis.

The demands on artifact cancellation depend on the context in which the algorithm is to be used. Artifact cancellation in EEGs for visual reading and interpretation requires that extreme measures be taken to assure that no clinical information is lost and that no new artifacts are introduced as a by-product of the cancellation procedure. These demands can be relaxed when artifact cancellation constitutes an intermediate processing step, e.g., for the purpose of designing an epileptic spike detector. In any case, it is essential that the development of algorithms for artifact cancellation is accompanied by visual assessment to assure that the performance is acceptable.

In this section, artifact processing is synonymous with a preprocessing stage which conditions the EEG signal for subsequent analysis. It should be noted, however, that artifact processing can also be part of one of the later stages in the analysis. For example, a method for spectral estimation of the EEG can be made more robust to the presence of impulsive noise. Another example is the method of weighted averaging which is employed for noise reduction of evoked potentials (see Chapter 4).

The common approach to artifact processing is to first estimate the noise $v(n)$, either from a signal measured on the scalp or from available reference signals, and then to subtract the estimate from the observed signal $x(n)$. This approach assumes implicitly that $x(n)$ can be divided into a sum of cerebral activity $s(n)$ and noise $v(n)$,

$$x(n) = s(n) + v(n). \tag{3.28}$$

Once such an assumption has been acknowledged, it obviously makes sense to estimate $s(n)$ by subtracting a noise estimate $\hat{v}(n)$ from $x(n)$. This type of model is associated not only with subtraction methods, but it also underlies noise reduction achieved through linear filtering of $x(n)$.

The great popularity of the additive model in (3.28) is explained by its simplicity and the wide range of methods developed for optimal estimation of $s(n)$. Nevertheless, it should be pointed out that this type of model is not necessarily the most appropriate, but another type of model may be preferred which assumes that the signal and noise interact in a multiplicative way,

$$x(n) = s(n)v(n). \tag{3.29}$$

Although techniques for separating multiplicative noise from $s(n)$ were developed in the 1960s ("homomorphic signal processing", [72, 73]), these techniques have only received marginal attention in the area of EEG signal processing [74, 75].

3.2.3 Artifact Reduction Using Linear Filtering

Linear, time-invariant filtering has been considered for the reduction of EMG artifacts and 50/60 Hz powerline interference. This technique mitigates the influence of such artifacts by spectrally shaping the observed signal. Unfortunately, its applicability is limited because the spectra of the EEG and the artifacts overlap each other considerably.

Lowpass filtering is useful in reducing the influence of EMG activity when the analysis of slower EEG rhythms is of particular interest [76, 77]. However, lowpass filtering must be used with caution since waveforms of cerebral origin with sharp edges become distorted. Also, bursts of EMG spikes can be smoothed to such a degree that they mimic alpha or beta rhythms [63]. Various nonlinear filter structures have been suggested to overcome these performance limitations, but such filters have not come into widespread use [77–79]. While the reduction of EMG artifacts largely remains an unsolved problem in EEG signal processing [80], it is nevertheless important to detect episodes of EMG activity and to account for such information when interpreting the EEG [81].

Removal of 50/60 Hz powerline interference can be done with a linear, time-invariant notch filter. A poorly designed notch filter can, however, introduce spurious activity, resembling the beta rhythm, due to the ringing which is associated with narrowband filtering, and can influence the shape of epileptiform spikes. These problems are particularly pronounced for filters with nonlinear phase characteristics, since different frequency components will be delayed differently. Different notch filtering techniques for cancellation of powerline interference in ECG signals are considered in detail in Section 7.2.

3.2.4   Artifact Cancellation Using Linearly Combined Reference Signals

Since artifacts due to eye movement and blinks are very common, most efforts have been directed toward developing cancellation methods for those artifacts. We will describe the most popular approach in which an estimate of the EOG artifact is first computed, based on one or several reference signals ("EOG-only" signals), and then subtracted from the EEG signal measured on the scalp [82–87]. The electrodes used to record the EOG are positioned around the eye so that horizontal and vertical movements are well-reflected, see Figure 3.9.

We assume that the EEG signal is composed of cerebral activity $s(n)$ which is additively disturbed by the EOG artifact $v_0(n)$,

$$x(n) = s(n) + v_0(n). \tag{3.30}$$

Another assumption in this approach is that the EOG reference signals $v_1(n), \ldots, v_M(n)$ are linearly transferred into the EEG signal. Hence, it seems reasonable to produce an artifact-cancelled signal $\hat{s}(n)$ by subtracting a linear combination of the reference signals from the EEG, using the weights $w_1, \ldots, w_M$,

$$\hat{s}(n) = x(n) - \sum_{i=1}^{M} w_i v_i(n) = s(n) + \left(v_0(n) - \mathbf{w}^T \mathbf{v}(n)\right), \tag{3.31}$$

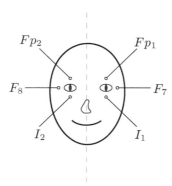

Figure 3.9: Electrode positions for the recording of EOG signals which reflect horizontal ($F_7 - F_8$) and vertical ($Fp_2 - I_2$ or $Fp_1 - I_1$) eye movement. Note that two other electrodes, I_1 and I_2, are used in addition to the electrodes of the 10/20 system shown in Figure 2.7.

where the vectors are defined as

$$\mathbf{v}(n) = \begin{bmatrix} v_1(n) \\ v_2(n) \\ \vdots \\ v_M(n) \end{bmatrix} \tag{3.32}$$

and

$$\mathbf{w} = \begin{bmatrix} w_1 \\ w_2 \\ \vdots \\ w_M \end{bmatrix}. \tag{3.33}$$

The estimate of the EOG artifact is thus obtained by $\hat{v}_0(n) = \mathbf{w}^T \mathbf{v}(n)$. In the following, it is assumed that all signals are random in nature, with zero-mean, and that $s(n)$ is uncorrelated with the EOG signals $\mathbf{v}(n)$ at each time n,

$$E[s(n)v_i(n)] = 0, \quad i = 0, \ldots, M. \tag{3.34}$$

The method is summarized by the block diagram in Figure 3.10.

Before the cancellation method can be used, it is necessary to determine the values of the different weights. One way is to minimize the *mean-square error* (MSE) $\mathcal{E}_{\mathbf{w}}$ between $x(n)$ and the linearly combined reference signals with respect to $\mathbf{w}$ [85],

$$\mathcal{E}_{\mathbf{w}} = E\left[\left(x(n) - \mathbf{w}^T \mathbf{v}(n) \right)^2 \right]. \tag{3.35}$$

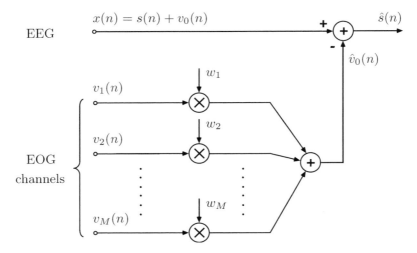

Figure 3.10: Cancellation of eye movement artifacts based on a linear combination of EOG signals, using a fixed set of weights $w_1, \ldots, w_M$. The EOG signal estimate is denoted $\hat{v}_0(n)$.

Since $s(n)$ is assumed to be uncorrelated with the EOG artifacts $v_i(n)$, the MSE $\mathcal{E}_{\mathbf{w}}$ can alternatively be expressed as

$$\mathcal{E}_{\mathbf{w}} = E\left[s^2(n)\right] + E\left[\left(v_0(n) - \mathbf{w}^T\mathbf{v}(n)\right)^2\right], \tag{3.36}$$

which makes it clear that the weights $\mathbf{w}$ should be chosen such that the error between $v_0(n)$ and $\mathbf{w}^T\mathbf{v}(n)$ is minimized. The term $E\left[s^2(n)\right]$ is independent of $\mathbf{w}$ and does not affect the outcome of the minimization.

Differentiation of $\mathcal{E}_{\mathbf{w}}$ in (3.35) with respect to the coefficient vector $\mathbf{w}$ yields

$$\begin{aligned}
\nabla_{\mathbf{w}}\mathcal{E}_{\mathbf{w}} &= \nabla_{\mathbf{w}}\left(E\left[x^2(n)\right] + \mathbf{w}^T\mathbf{R}_v(n)\mathbf{w} - 2\mathbf{w}^T\mathbf{r}_{xv}(n)\right) \\
&= 2\mathbf{R}_v(n)\mathbf{w} - 2\mathbf{r}_{xv}(n).
\end{aligned} \tag{3.37}$$

The correlation matrix $\mathbf{R}_v(n)$ of the reference signals describes the *spatial correlation* between the different channels at each time n and is defined by

$$\begin{aligned}
\mathbf{R}_v(n) &= E\left[\mathbf{v}(n)\mathbf{v}^T(n)\right] \\
&= \begin{bmatrix}
r_{v_1 v_1}(n) & r_{v_1 v_2}(n) & \cdots & r_{v_1 v_M}(n) \\
r_{v_2 v_1}(n) & r_{v_2 v_2}(n) & \cdots & r_{v_2 v_M}(n) \\
\vdots & \vdots & & \vdots \\
r_{v_M v_1}(n) & r_{v_M v_2}(n) & \cdots & r_{v_M v_M}(n)
\end{bmatrix},
\end{aligned} \tag{3.38}$$

where

$$r_{v_i v_j}(n) = E\left[v_i(n)v_j(n)\right]. \tag{3.39}$$

The cross-correlation vector $\mathbf{r}_{xv}(n)$ between $x(n)$ and $\mathbf{v}(n)$ is defined by

$$\mathbf{r}_{xv}(n) = E\left[x(n)\mathbf{v}(n)\right] = \begin{bmatrix} r_{xv_1}(n) \\ r_{xv_2}(n) \\ \vdots \\ r_{xv_M}(n) \end{bmatrix}, \tag{3.40}$$

where

$$r_{xv_i}(n) = E\left[x(n)v_i(n)\right]. \tag{3.41}$$

Although the correlation quantities $\mathbf{R}_v(n)$ and $\mathbf{r}_{xv}(n)$ change over time, we will for now assume that these quantities remain fixed over the observation interval of interest,

$$\mathbf{R}_v(n) \equiv \mathbf{R}_v, \tag{3.42}$$
$$\mathbf{r}_{xv}(n) \equiv \mathbf{r}_{xv}, \tag{3.43}$$

for $n = 0, 1, \ldots, N-1$. It should be noted that $\mathbf{R}_v$ is not, in general, a Toeplitz matrix since, for example, the signal power $r_{v_i v_i}$ typically varies from channel to channel.

Setting the gradient $\nabla_{\mathbf{w}} \mathcal{E}_{\mathbf{w}}$ in (3.37) equal to zero, we obtain the following system of linear equations,

$$\mathbf{R}_v \mathbf{w}^{\mathrm{o}} = \mathbf{r}_{xv}, \tag{3.44}$$

whose solution yields the optimal weight vector $\mathbf{w}^{\mathrm{o}}$. The corresponding minimum MSE is easily found by insertion of (3.44) in (3.35),

$$\mathcal{E}_{\min} = E\left[x^2(n)\right] - (\mathbf{w}^{\mathrm{o}})^T \mathbf{R}_v \mathbf{w}^{\mathrm{o}}. \tag{3.45}$$

In practice, the spatial correlations $r_{v_i v_j}$ need to be estimated from the measured EOG signals prior to computation of $\mathbf{w}$. Since $\mathbf{R}_v$ is considered to be fixed in time, it is estimated by simply replacing $E\left[v_i(n)v_j(n)\right]$ by the corresponding time average,

$$\hat{r}_{v_i v_j} = \frac{1}{N} \sum_{n=0}^{N-1} v_i(n)v_j(n). \tag{3.46}$$

The cross-correlation vector $\mathbf{r}_{xv}$ can be estimated in the same way. The procedure to find the values of the optimal weight vector $\mathbf{w}^o$ is then repeated

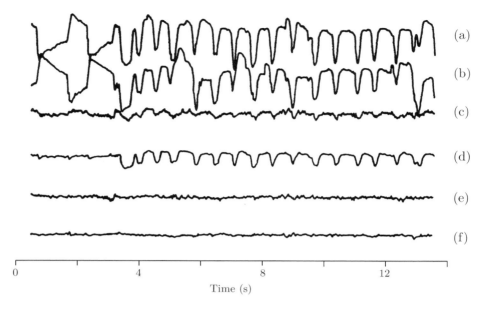

Figure 3.11: An example of ocular artifact cancellation in EEG signals. Electrooculographic signals measured for (a) the right and (b) the left eye; EEG signals from two different electrodes (c), (d) before and (e), (f) after artifact cancellation has taken place. Cancellation was based on a linear combination of the EOG signals displayed in (a) and (b); for further details, see [88–90]. (Reprinted from Ifeachor et al. [90] with permission.)

for each of the available EEG channels in order to produce channel-specific weights. The performance of the EOG cancellation method is illustrated in Figure 3.11, where artifact cancellation is based on two reference signals, i.e., $M = 2$.

Certain issues related to the use of the above cancellation method need to be addressed. Determination of $\mathbf{w}$ is usually accomplished by asking the subject to perform horizontal and vertical eye movements as well as blinking at the onset of the investigation. The weight vector estimate that results from such a learning phase is then applied in subsequent signal analysis.

Another design issue is the number of reference signals, M, required for adequate EOG cancellation. In the literature, this number ranges from only one [84] up to several signals. However, in order to account for horizontal as well as vertical eye movements, at least four reference signals should be included [85].

A major concern associated with this approach is ensuring that only EOG activity is cancelled while cerebral activity remains unaltered. Although minimization of the MSE criterion in (3.35) aims to ensure this, spurious activity may occasionally be introduced into the artifact-cancelled EEG through the EOG. For example, the EOG electrodes may pick up large-amplitude, slow activity originating from a frontal lobe focus near the EOG electrodes [63].

We conclude this subsection by mentioning that cancellation of ECG artifacts in the EEG can be performed in a way similar to the above EOG technique [85, 91–94]. Electrocardiographic cancellation, in contrast to EOG cancellation, offers the additional possibility of exploiting the fact that heartbeats are recurrent. Since the amplitude and morphology of normal heartbeats are relatively stable over time, a representative, noise-reduced beat can be obtained from time-synchronized ensemble averaging of several successive beats; Section 4.3.1 provides further details on how to compute averaged signals. The resulting averaged beats, preferably obtained from several electrode locations, are linearly combined into one signal used for subtraction in (3.31).

3.2.5 Adaptive Artifact Cancellation Using Linearly Combined Reference Signals

Electrooculographic cancellation based on a fixed set of linear weights is less appropriate when time-varying changes occur in the way the EOG interferes with the EEG. To handle such situations, it is desirable to modify the previously described algorithm for artifact cancellation so that it can track slow changes in EOG influence [95, 96]. Below, we will briefly present an adaptive algorithm—the *least mean-square (LMS) algorithm*—which makes use of linearly combined reference signals for artifact cancellation. Although this algorithm is the most commonly used, it is only one of many algorithms developed for the purpose of adaptive filtering and noise cancellation; an in-depth presentation of various adaptive algorithms and their performance in terms of stability and convergence can be found in [97, 98].

The estimate of the EOG artifact $v_0(n)$ in (3.31), as produced by $\mathbf{w}^T\mathbf{v}(n)$, will now be modified so that the weight vector $\mathbf{w}$ becomes a function of time,

$$\mathbf{w}^T\mathbf{v}(n) \rightarrow \mathbf{w}^T(n)\mathbf{v}(n).$$

Consequently, the mean-square error criterion becomes

$$\mathcal{E}_{\mathbf{w}}(n) = E\left[\left(x(n) - \mathbf{w}^T(n)\mathbf{v}(n)\right)^2\right] \tag{3.47}$$

and should be minimized with respect to $\mathbf{w}(n)$. We found earlier that the error $\mathcal{E}_{\mathbf{w}}$ in (3.35) has a unique minimum since it is a quadratic function of $\mathbf{w}$. Figure 3.12 exemplifies the bowl-like shape of $\mathcal{E}_{\mathbf{w}}(n)$ when it is plotted as a

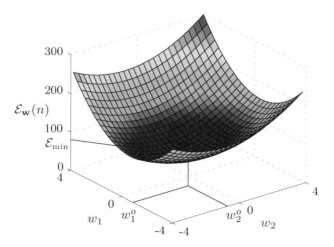

Figure 3.12: The quadratic error surface $\mathcal{E}_{\mathbf{w}}(n)$ plotted as a function of the weights w_1 and w_2. The optimal weights w_1^o and w_2^o correspond to the minimum error $\mathcal{E}_{\min}$.

function of two different weights w_1 and w_2. For time-varying characteristics of $\mathbf{v}(n)$, however, the optimal solution of $\mathcal{E}_{\mathbf{w}}(n)$ changes with time, and, accordingly, the bottom of the bowl must be searched for at every new sample n.

A common approach to the minimization of $\mathcal{E}_{\mathbf{w}}(n)$ is to search for the optimum values using the *method of steepest descent*—a classical, iterative procedure for finding extrema of nonlinear functions. The underlying idea of this method is to update the current weight estimate $\mathbf{w}(n)$ by an additive correction term which brings the next estimate $\mathbf{w}(n+1)$ closer to the desired solution. The correction of $\mathbf{w}(n)$ is achieved by taking a step in the direction of the steepest descent of the quadratic error surface. This direction is given by the *negative error gradient vector*, i.e., the vector of partial derivatives of $\mathcal{E}_{\mathbf{w}}(n)$ with respect to the weights $w_i(n)$, see (A.46) in Appendix A. Thus, $\mathbf{w}(n)$ is updated by the following equation,

$$\mathbf{w}(n+1) = \mathbf{w}(n) - \frac{1}{2}\mu\nabla_{\mathbf{w}}\mathcal{E}_{\mathbf{w}}(n), \qquad (3.48)$$

where the step size μ is a positive-valued scalar which determines the speed of adaptation. Large values of μ yield faster convergence to the optimal solution, but at the expense of a noisier estimate of $\mathbf{w}(n)$. Too large a value of μ will cause the algorithm to become unstable since the algorithm constitutes a feedback system. On the other hand, when μ is small, the algorithm will approach the optimum solution more slowly, but will provide a less noisy estimate of $\mathbf{w}(n)$.

Calculating the gradient vector of the error $\mathcal{E}_{\mathbf{w}}(n)$ with respect to $\mathbf{w}$, we obtain

$$\nabla_{\mathbf{w}}\mathcal{E}(n) = -2E[e(n)\mathbf{v}(n)], \qquad (3.49)$$

where the error $e(n)$ is

$$e(n) = x(n) - \mathbf{w}^T(n)\mathbf{v}(n). \qquad (3.50)$$

The weight update equation becomes

$$\mathbf{w}(n+1) = \mathbf{w}(n) + \mu E[e(n)\mathbf{v}(n)]. \qquad (3.51)$$

The expected value $E[e(n)\mathbf{v}(n)]$ is generally unknown and must therefore be replaced by an estimate before the algorithm can be used in practice. In the LMS algorithm, the expected value is replaced by simply taking its instantaneous estimate at time n,

$$E[e(n)\mathbf{v}(n)] \approx e(n)\mathbf{v}(n), \qquad (3.52)$$

and thus

$$\mathbf{w}(n+1) = \mathbf{w}(n) + \mu e(n)\mathbf{v}(n). \qquad (3.53)$$

Typically, the LMS algorithm is initialized by setting all weights equal to zero, i.e.,

$$\mathbf{w}(0) = \mathbf{0}, \qquad (3.54)$$

where $\mathbf{0}$ denotes a column vector whose entries are zero. The block diagram in Figure 3.13 illustrates the LMS-based artifact cancellation technique which makes use of a linear combination of reference signals $v_1(n), \ldots, v_M(n)$. While the derivation of the LMS algorithm is straightforward, the related performance analysis in terms of convergence properties and filter stability is relatively complicated and is therefore not considered here. Instead, we will restrict ourselves to summarizing some important results which are valid for stationary signals, see [98]. For the *steady-state* situation, the LMS algorithm converges in the mean to the optimal solution previously given in (3.44),

$$\lim_{n\to\infty} E\left[\mathbf{w}(n)\right] = \mathbf{w}^o = \mathbf{R}_v^{-1}\mathbf{r}_{xv}, \qquad (3.55)$$

provided that

$$0 < \mu < \frac{2}{\lambda_{\max}}, \qquad (3.56)$$

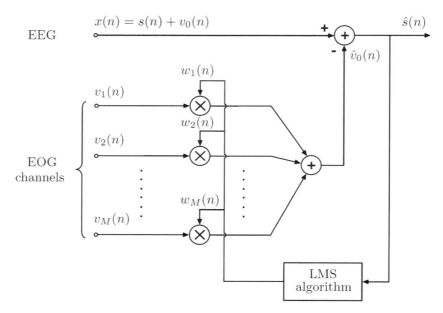

Figure 3.13: Cancellation of eye movement artifacts based on a linear combination of EOG signals, using an adaptively updated set of weights $w_1(n), \ldots, w_M(n)$.

where $\lambda_{\max}$ denotes the largest eigenvalue of $\mathbf{R}_v$. A tighter upper bound in (3.56) can be obtained by making use of the fact that $\mathrm{tr}(\mathbf{R}_v) \geq \lambda_{\max}$ so $\mu < 2/\mathrm{tr}(\mathbf{R}_v)$, where the trace of $\mathbf{R}_v$ equals the total power of the M different reference signals (cf. (A.39) in Appendix A). This bound is used more often since the signal power is much more easily estimated than is $\mathbf{R}_v$. Another quantity of interest is the time τ taken for the LMS algorithm to approach the MSE solution. An approximate expression is given by

$$\tau \approx \frac{1}{2\mu\lambda_{\min}}, \tag{3.57}$$

where τ is expressed as the number of samples and $\lambda_{\min}$ denotes the smallest eigenvalue of $\mathbf{R}_v$.

Although the weight vector $\mathbf{w}(n)$ produced by the LMS algorithm will converge in the mean for steady-state conditions, each weight will fluctuate around its optimum value because the correction is based on the noisy gradient vector given in (3.52). Therefore, we can express the weight vector $\mathbf{w}(n)$ in terms of the optimal solution $\mathbf{w}^o$ and a time-varying weight error vector $\Delta\mathbf{w}(n)$,

$$\mathbf{w}(n) = \mathbf{w}^o + \Delta\mathbf{w}(n). \tag{3.58}$$

The fluctuations reflected by $\Delta\mathbf{w}(n)$ cause the minimum MSE $\mathcal{E}_{\min}$ in (3.45) to increase by an amount commonly referred to as the *excess mean-square error*,

$$\mathcal{E}(n) = \mathcal{E}_{\min} + \mathcal{E}_{\mathrm{ex}}(n). \tag{3.59}$$

It is, in general, difficult to derive an exact expression for $\mathcal{E}_{\mathrm{ex}}(n)$; however, by assuming that the input signal $x(n)$ and the weights $\mathbf{w}(n)$ are statistically independent, it can be shown that once the adaptation has been completed, i.e., for $n = \infty$, $\mathcal{E}_{\mathrm{ex}}(n)$ is given by [98]

$$\mathcal{E}_{\mathrm{ex}}(\infty) = \mathcal{E}_{\min} \frac{\displaystyle\sum_{i=1}^{M} \frac{\mu\lambda_i}{2 - \mu\lambda_i}}{1 - \displaystyle\sum_{i=1}^{M} \frac{\mu\lambda_i}{2 - \mu\lambda_i}}, \tag{3.60}$$

where $\lambda_1, \ldots, \lambda_M$ are the eigenvalues of $\mathbf{R}_v$.

3.2.6 Artifact Cancellation Using Filtered Reference Signals

An important generalization of the above cancellation method is motivated by the interesting observation that EOG potentials exhibit frequency-dependent behavior when transferred through the intervening tissues to the EEG electrode locations on the scalp [99–101]. It has also been found that eye movement and blinks exhibit different spectral properties when transferred to the EEG. Eye movement is transferred at lower frequencies, extending up to 6 or 7 Hz, while blinks are generally transferred at higher frequencies ranging up to the alpha band (8–13 Hz) [100].

Improved artifact cancellation can therefore be expected when each of the weights w_i in (3.31) is replaced by a transfer function—here represented by the impulse response $h_i(n)$ of a linear, time-invariant system—which models the frequency-dependent transmission of the EOG activity into the EEG [102], see Figure 3.14. Thus, the improved estimate of the EOG activity is based on both spatial and temporal information.

In terms of modeling, it is again assumed that the observed signal is composed of cerebral activity $s(n)$ which is additively disturbed by stationary noise $v_0(n)$. An estimate of $v_0(n)$ can be derived from the EOG reference signals $v_1(n), \ldots, v_M(n)$ since all are assumed to be correlated with $v_0(n)$ and, accordingly, contribute to improving the estimate of $v_0(n)$. However, since each of the reference signals has been modified spectrally when propagating from the EOG electrode to the EEG electrode, the estimate $\hat{v}_0(n)$

is obtained by filtering each $v_i(n)$ with $h_i(n)$, followed by summation of the M filter outputs,

$$\hat{v}_0(n) = \sum_{l=0}^{L-1} h_1(l) v_1(n-l) + \cdots + \sum_{l=0}^{L-1} h_M(l) v_M(n-l). \qquad (3.61)$$

Each filter is defined by a finite impulse response (FIR) with length L (the special case when the filter lengths depend on electrode site is not considered). In the following, it is convenient to use the more concise vector notation in order to express the "filtered reference signal" estimator,

$$\hat{v}_0(n) = \sum_{i=1}^{M} \mathbf{h}_i^T \tilde{\mathbf{v}}_i(n), \qquad (3.62)$$

where $\tilde{\mathbf{v}}_i(n)$ indicates that the samples contained in $\mathbf{v}_i(n)$ have been reversed in time,

$$\mathbf{v}_i(n) = \begin{bmatrix} v_i(n-L+1) \\ \vdots \\ v_i(n-1) \\ v_i(n) \end{bmatrix}, \qquad \tilde{\mathbf{v}}_i(n) = \begin{bmatrix} v_i(n) \\ v_i(n-1) \\ \vdots \\ v_i(n-L+1) \end{bmatrix}. \qquad (3.63)$$

The impulse response $\mathbf{h}_i$ is defined as

$$\mathbf{h}_i = \begin{bmatrix} h_i(0) \\ h_i(1) \\ \vdots \\ h_i(L-1) \end{bmatrix}. \qquad (3.64)$$

It should be pointed out that the time-indexed vector notations in (3.63) and (3.32) are identical but with different meanings: the former vector contains *several samples of one channel*, whereas the latter contains the *samples of several channels at one particular instant in time*. It should be obvious from the context which one of the two notations is intended.

We will again consider the MSE criterion which was introduced in (3.35) to find the optimal FIR filters $\mathbf{h}_i$. The MSE is now defined by

$$\mathcal{E}_{\mathbf{h}} = E\left[\left(x(n) - \sum_{i=1}^{M} \mathbf{h}_i^T \tilde{\mathbf{v}}_i(n) \right)^2 \right]. \qquad (3.65)$$

The minimization of $\mathcal{E}_{\mathbf{h}}$ is, however, more complicated than in the case of linearly combined reference signals in (3.35). In order to proceed we assume

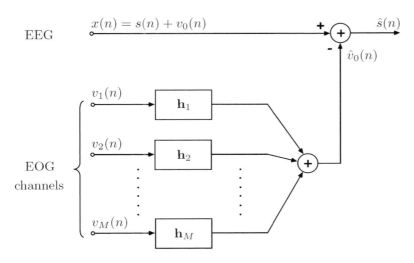

Figure 3.14: Cancellation of eye movement artifacts using an estimate based on linear FIR filtering of M different EOG channels; the impulse response of each filter is denoted $\mathbf{h}_i$.

that the EOG reference signals are modeled as stationary processes with known second-order characteristics (or which at least can be estimated). The cross-correlation between two reference signals is defined by

$$r_{v_i v_j}(k, n) = E\left[v_i(n)v_j(n - k)\right], \quad i, j = 1, \ldots, M, \tag{3.66}$$

which in the general case is a function of both lag k and time n. The cross-correlation between the EOG-contaminated signal $x(n)$ and the reference signal $v_i(n)$ is defined by

$$r_{xv_i}(k, n) = E\left[x(n)v_i(n - k)\right], \quad i = 1, \ldots, M. \tag{3.67}$$

Similar to the previous method, which made use of linearly combined reference signals, we assume that the correlation information remains fixed over time in all channels,

$$r_{v_i v_j}(k, n) \equiv r_{v_i v_j}(k), \tag{3.68}$$

$$r_{xv_i}(k; n) \equiv r_{xv_i}(k). \tag{3.69}$$

The EEG and the M reference signals are, somewhat idealized, assumed to be uncorrelated with each other,

$$E\left[s(n)v_i(n)\right] = 0, \quad i = 1, \ldots, M. \tag{3.70}$$

The optimal filter impulse responses are derived by differentiation of the error $\mathcal{E}_{\mathbf{h}}$, i.e.,

$$\nabla_{\mathbf{h}_j} \mathcal{E}_{\mathbf{h}} = 0, \quad j = 1, \ldots, M, \tag{3.71}$$

which results in

$$E\left[\tilde{\mathbf{v}}_j(n)\left(x(n) - \sum_{i=1}^{M} \mathbf{h}_i^T \tilde{\mathbf{v}}_i(n)\right)\right] = 0, \quad j = 1, \ldots, M. \tag{3.72}$$

Making use of the fact that

$$\sum_{i=1}^{M} \mathbf{h}_i^T \tilde{\mathbf{v}}_i(n) = \sum_{i=1}^{M} \tilde{\mathbf{v}}_i^T(n)\mathbf{h}_i,$$

and evaluating the expected value in (3.72), the following block matrix equation results:

$$\begin{bmatrix} \mathbf{R}_{\mathbf{v}_1\mathbf{v}_1} & \mathbf{R}_{\mathbf{v}_1\mathbf{v}_2} & \cdots & \mathbf{R}_{\mathbf{v}_1\mathbf{v}_M} \\ \mathbf{R}_{\mathbf{v}_2\mathbf{v}_1} & \mathbf{R}_{\mathbf{v}_2\mathbf{v}_2} & \cdots & \mathbf{R}_{\mathbf{v}_2\mathbf{v}_M} \\ \vdots & \vdots & & \vdots \\ \mathbf{R}_{\mathbf{v}_M\mathbf{v}_1} & \mathbf{R}_{\mathbf{v}_M\mathbf{v}_2} & \cdots & \mathbf{R}_{\mathbf{v}_M\mathbf{v}_M} \end{bmatrix} \begin{bmatrix} \mathbf{h}_1 \\ \mathbf{h}_2 \\ \vdots \\ \mathbf{h}_M \end{bmatrix} = \begin{bmatrix} \mathbf{r}_{xv_1} \\ \mathbf{r}_{xv_2} \\ \vdots \\ \mathbf{r}_{xv_M} \end{bmatrix}, \tag{3.73}$$

where

$$\mathbf{R}_{\mathbf{v}_i\mathbf{v}_j} = E\left[\tilde{\mathbf{v}}_i\tilde{\mathbf{v}}_j{}^T\right] = \begin{bmatrix} r_{v_iv_j}(0) & r_{v_iv_j}(1) & \cdots & r_{v_iv_j}(L-1) \\ r_{v_iv_j}(1) & r_{v_iv_j}(0) & \cdots & r_{v_iv_j}(L-2) \\ \vdots & \vdots & \ddots & \vdots \\ r_{v_iv_j}(L-1) & r_{v_iv_j}(L-2) & \cdots & r_{v_iv_j}(0) \end{bmatrix} \tag{3.74}$$

and

$$\mathbf{r}_{xv_i} = \begin{bmatrix} r_{xv_1}(0) \\ r_{xv_1}(1) \\ \vdots \\ r_{xv_1}(L-1) \end{bmatrix}. \tag{3.75}$$

Since the reference signals $v_i(n)$ are assumed to be stationary processes, then $\mathbf{R}_{\mathbf{v}_i\mathbf{v}_j} = \mathbf{R}_{\mathbf{v}_j\mathbf{v}_i}$. The matrix equation in (3.73), defining the impulse responses of the cancellation filters, is well-known from the area of noise cancellation based on Wiener filtering [103].

A potential limitation of the MSE criterion in (3.65) is that it does not accommodate any prior information that may be available on the properties of the impulse responses $\mathbf{h}_i$. For example, it can for various reasons be important to assure that $\mathbf{h}_i$ is reasonably close to an a priori known impulse response $\overline{\mathbf{h}}_i$ representative of a certain group of subjects [102]. A modified MSE can be introduced to account for such information, defined by

$$\mathcal{E}'_{\mathbf{h}} = \mathcal{E}_{\mathbf{h}} + \nu \sum_{i=1}^{M} (\mathbf{h}_i - \overline{\mathbf{h}}_i)^T (\mathbf{h}_i - \overline{\mathbf{h}}_i). \qquad (3.76)$$

In this definition, the second term "biases" the solution of (3.65) by also considering the squared error between $\mathbf{h}_i$ and $\overline{\mathbf{h}}_i$ in the minimization process; the parameter ν determines how much overall weight should be assigned to the prior knowledge. The minimization of $\mathcal{E}'_{\mathbf{h}}$ with respect to $\mathbf{h}_i$ results in a matrix equation given by

$$\left(\begin{bmatrix} \mathbf{R}_{\mathbf{v}_1 \mathbf{v}_1} & \cdots & \mathbf{R}_{\mathbf{v}_1 \mathbf{v}_M} \\ \vdots & & \vdots \\ \mathbf{R}_{\mathbf{v}_M \mathbf{v}_1} & \cdots & \mathbf{R}_{\mathbf{v}_M \mathbf{v}_M} \end{bmatrix} - 2\nu \mathbf{I} \right) \begin{bmatrix} \mathbf{h}_1 \\ \vdots \\ \mathbf{h}_M \end{bmatrix} = \begin{bmatrix} \mathbf{r}_{xv_1} - 2\nu \overline{\mathbf{h}}_1 \\ \vdots \\ \mathbf{r}_{xv_M} - 2\nu \overline{\mathbf{h}}_M \end{bmatrix}, \qquad (3.77)$$

where the dimension of the identity matrix $\mathbf{I}$ matches that of the block correlation matrix on the left-hand side in (3.77).

Finally, it should be mentioned that an adaptive version of the filtered reference signal method has been developed for on-line EOG cancellation [102]. Such an algorithm is able to track slow changes in the properties of the eye-to-electrode transfer function. The a priori impulse response $\overline{\mathbf{h}}_i$ was computed from a calibration stage carried out just before the actual EEG recording and during which the subject was instructed to make predefined eye movements.

3.3 Nonparametric Spectral Analysis

Spectral analysis is a powerful technique for characterization of a wide range of biomedical signals. This technique was introduced at an early stage to provide a more detailed characterization of EEG background activity than that which could be achieved by simple analysis techniques such as those relying on one-dimensional histograms of the samples. Considering the oscillatory behavior of many EEG rhythms, signal decomposition in terms of sine and cosine functions was found useful as well as feasible from a computational point of view. Spectral analysis based on the Fourier transform essentially correlates the signal with sines and cosines of various frequencies and produces a set of coefficients that defines the power spectrum. The power of a particular frequency band is readily obtained from the spectrum and can, among many other things, be used to determine whether an alpha rhythm is present or not. Fourier-based spectral analysis is commonly referred to as nonparametric spectral analysis since no parametric modeling assumptions regarding the signal are incorporated [104].

It was pointed out in Section 3.1.2 that the power spectrum $S_x(e^{j\omega})$ is a natural quantity for characterizing a stationary, Gaussian signal since $S_x(e^{j\omega})$ is defined as the Fourier transform of $r_x(k)$. The power spectrum can also be a useful measure for stationary, non-Gaussian signals, although it no longer provides complete characterization. Statistical procedures have been suggested to test whether a signal is stationary or not [12]; however, such procedures are rarely used in practice. Instead, spectral analysis is, in general, considered applicable to signals recorded during normal, spontaneous waking activity at rest (i.e., part of the first category, "EEG activity without major temporal changes", listed on page 37), having relatively short durations of about 10 s, and which are free of artifacts [105]. Further details on the EEG stationarity issue can be found in [12, 106–109].

From a historical perspective, it may be interesting to mention that Fourier-based spectral analysis of the EEG was investigated long before the advent of digital computers and the revolutionary Fast Fourier Transform (FFT), which was developed in the 1960s. Already in the early 1930s, the coefficients of the Fourier series were calculated manually for different EEG recordings [110, 111]. An alternative approach to the decomposition of the EEG into different frequency components was explored by designing a bank of analog bandpass filters—at that time implemented by analog electronics. The filter bank quantified rhythmic activities by using bandpass filters with suitably selected center and cut-off frequencies [112]. The bandpass filtering technique and Fourier-based spectral analysis are, however, intimately related to each other since the latter technique can be interpreted as a bank of narrowband filters where the output of each filter provides a measure of the power of each frequency [103].

3.3.1   Fourier-based Power Spectrum Analysis

In this section, the key points of nonparametric power spectrum estimation based on the discrete-time Fourier transform are summarized. We recall from Section 3.1 that the power spectrum of the stationary signal $x(n)$ is defined by

$$S_x(e^{j\omega}) = \sum_{k=-\infty}^{\infty} r_x(k)e^{-j\omega k},$$

where $r_x(k)$ denotes the correlation function that characterizes the samples $x(0), \ldots, x(N-1)$. Before $S_x(e^{j\omega})$ can be calculated, $r_x(k)$ has to be estimated from $x(n)$ since it is unknown in practice. Assuming that the observed signal is correlation ergodic (see Appendix A on page 645), the estimation is commonly accomplished by use of the following time average estimator,

$$\hat{r}_x(k) = \frac{1}{N} \sum_{n=0}^{N-1-k} x(n+k)x(n), \quad k = 0, \ldots, N-1, \qquad (3.78)$$

where negative lags are obtained from the symmetry property, i.e., $\hat{r}_x(k) = \hat{r}_x(-k)$. Combining the power spectrum definition with the correlation function estimate in (3.78), an estimate of the power spectrum can be obtained,

$$\hat{S}_x(e^{\jmath\omega}) = \sum_{k=-N+1}^{N-1} \hat{r}_x(k)e^{-\jmath\omega k}, \qquad (3.79)$$

also known as the *periodogram*.

While the signal $x(n)$ only appears implicitly in (3.79), it is straightforward to derive an expression which explicitly shows how $\hat{S}_x(e^{\jmath\omega})$ is related to $x(n)$. This is done by first noting that $\hat{r}_x(k)$ in (3.78) can be expressed as a convolution of $x(n)$ and its time-reversed counterpart,

$$\hat{r}_x(k) = \frac{1}{N}x(k) * x(-k), \qquad (3.80)$$

where $x(n)$ is assumed to be zero outside the interval $[0, N-1]$, and $\hat{r}_x(k)$ is symmetric, i.e., $\hat{r}_x(k) = \hat{r}_x(-k)$. Taking the Fourier transform of the convolution, we have

$$\hat{S}_x(e^{\jmath\omega}) = \frac{1}{N}|X(e^{\jmath\omega})|^2 = \frac{1}{N}\left|\sum_{n=0}^{N-1} x(n)e^{-\jmath\omega n}\right|^2. \qquad (3.81)$$

Thus, the power spectrum is estimated by simply computing the squared magnitude of the N-point DTFT of $x(n)$ and efficiently implemented by the FFT algorithm.

In order to better understand the properties of the periodogram, it is instructive to study estimator performance in terms of its mean and variance. It can be shown that the mean of the periodogram is equal to (see Problem 3.3),

$$E\left[\hat{S}_x(e^{\jmath\omega})\right] = \sum_{k=-N+1}^{N-1} E\left[\hat{r}_x(k)\right]e^{-\jmath\omega k}$$

$$= \sum_{k=-N+1}^{N-1} r_x(k)w_B(k)e^{-\jmath\omega k}, \qquad (3.82)$$

where

$$w_B(k) = \begin{cases} 1 - \dfrac{|k|}{N}, & -N \le k \le N; \\ 0, & |k| > N \end{cases} \tag{3.83}$$

is a triangular window, also known as the Bartlett window. Alternatively, the product of $r_x(k)$ and $w_B(k)$ in (3.82) can be expressed in the frequency domain as a convolution of their respective Fourier transforms,

$$E\left[\hat{S}_x(e^{\jmath\omega})\right] = \frac{1}{2\pi} \int_{-\pi}^{\pi} S_x(e^{\jmath\theta}) W_B(e^{\jmath(\omega-\theta)}) d\theta$$

$$= \frac{1}{2\pi} S_x(e^{\jmath\omega}) * W_B(e^{\jmath\omega}), \tag{3.84}$$

where $W_B(e^{\jmath\omega})$ is given by

$$W_B(e^{\jmath\omega}) = \frac{1}{N} \frac{\sin^2(\omega N/2)}{\sin^2(\omega/2)}. \tag{3.85}$$

Figure 3.15 displays the magnitude function $10 \cdot \log(W_B(e^{\jmath\omega}))$ for $N = 16$ and 64 and illustrates the principal features of $W_B(e^{\jmath\omega})$, namely, the presence of a main lobe at $\omega = 0$ and several sidelobes. It is obvious from (3.84) that the periodogram is a biased estimator since the mean $E[\hat{S}_x(e^{\jmath\omega})]$ is not equal to $S_x(e^{\jmath\omega})$ but rather to a version modified by $W_B(e^{\jmath\omega})$ through convolution. The main lobe acts as a smearer of the estimated spectrum and sets a limit on the degree to which the details of the spectrum can be resolved: frequencies closer than $f = 1/N$ cannot be separated in the periodogram. The sidelobes are associated with an undesirable effect since these lobes leak power from the main frequency band into bands with less power or even without power. The effects related to the main lobe and the sidelobes are commonly referred to as *smearing* and *leakage*, respectively. Although the periodogram is asymptotically unbiased because $W_B(e^{\jmath\omega})$ approaches a Dirac impulse as N approaches infinity, this property is of less interest in practice when the number of samples is limited for various reasons.

The leakage effect is particularly pronounced for narrowband signals and causes measurements on spectral power at a fixed frequency to be less reliable. Within the context of EEG signal analysis, this effect implies that a highly synchronized brain rhythm is likely to be better described by the power contained in a frequency band rather than by measuring the power at its spectral peak.

The variance of the periodogram is another interesting quantity which unfortunately is not as easily derived as its mean value in (3.84). However, under the assumption that the observed signal $x(n)$ is modeled by a Gaussian

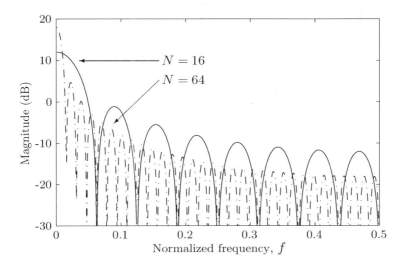

Figure 3.15: The Fourier transform of the Bartlett window $w_B(k)$ for $N = 16$ and 64. The width of the main lobe and the magnitude of the sidelobes both decrease as the total number of samples N increases.

stochastic process, it can be shown that the variance of $\hat{S}_x(e^{j\omega})$ is given by [113],

$$V\left[\hat{S}_x(e^{j\omega})\right] \approx S_x^2(e^{j\omega})\left[1 + \left(\frac{\sin\omega N}{N\sin\omega}\right)^2\right], \qquad (3.86)$$

which at large values of N is proportional to the square of $S_x(e^{j\omega})$. The significance of this result is that the variance of the periodogram does not approach zero as the number of samples increases, i.e., the periodogram does not produce a consistent estimate of the power spectrum. This disappointing result has, in combination with the effects of leakage and smearing, spurred the development of several approaches to improving the performance of the periodogram. The most important modifications of the periodogram revolve around the use of *windowing* and *averaging* [114] techniques. These two techniques aim at reducing the leakage effect and the variance of the periodogram and can be used separately as well as in combination.

Windowing is an operation which is implicitly performed when computing $\hat{S}_x(e^{j\omega})$ in (3.81): a rectangular window $w(n)$ is in a sense applied to extract the segment $x(0), \ldots, x(N-1)$ from a signal that may extend over a longer interval. It may be advantageous to replace the rectangular window with another which has smaller sidelobes. Several windows have been designed with the common purpose of reducing the amplitude of the side-lobes; Hanning, Hamming, and Blackman windows are perhaps the most well-known [103]. Windowing achieves the reduction of sidelobes at the expense of a wider main lobe. Thus, windowing provides a trade-off between leakage and spectral resolution of the power spectrum estimate. The variance of the periodogram remains unaffected by windowing. In the time domain, windowing reduces the influence of abrupt discontinuities at the interval end points since most windows deemphasize end point samples.

Variance reduction relies on first partitioning $x(n)$ into K nonoverlapping segments of length L $(N = KL)$,

$$x_i(n) = x(n + iL), \quad n = 0, \ldots, L - 1; \; i = 0, \ldots, K - 1, \tag{3.87}$$

and then averaging the periodograms $\hat{S}_{x_i}(e^{j\omega})$ resulting from each of the segments $x_i(n)$. Assuming that the segments are uncorrelated, averaging produces a consistent power spectrum estimate because the variance tends toward zero as K approaches infinity,

$$V\left[\hat{S}_x(e^{j\omega})\right] = \frac{1}{K}V\left[\hat{S}_{x_i}(e^{j\omega})\right]$$

$$\approx \frac{1}{K}S_x^2(e^{j\omega})\left[1 + \left(\frac{\sin\omega L}{L\sin\omega}\right)^2\right]. \tag{3.88}$$

Averaging of periodograms reduces the signal length from N to L samples, and, as a result, the spectral resolution is reduced by a factor K. Again, we have to deal with a trade-off between different properties of the power spectrum estimate, this time between variance and spectral resolution.

We conclude the section on nonparametric spectrum estimation by presenting an estimator which combines the above-mentioned techniques of windowing and averaging,

$$\hat{S}_x(e^{j\omega}) = \frac{1}{KLU}\sum_{i=0}^{K-1}\left|\sum_{n=0}^{L-1}x_i(n)w(n)e^{-j\omega n}\right|^2. \tag{3.89}$$

Here, U is a normalization factor related to the characteristics of the window $w(n)$,

$$U = \frac{1}{L}\sum_{n=0}^{L-1}w^2(n). \tag{3.90}$$

A variation of the method expressed in (3.89) is to let consecutive signal segments overlap to a certain degree, often a 50% overlap. This variation is

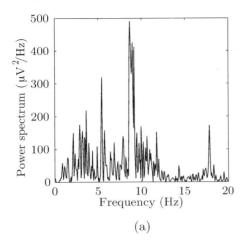 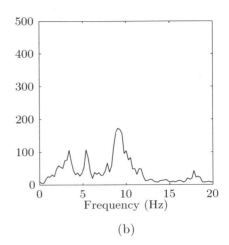

(a) (b)

Figure 3.16: Spectral analysis of an EEG with alpha rhythm. (a) The power spectrum obtained without segmentation ($N = 1024$) and (b) with segmentation using $N = 256$ and a segment overlap of 128 samples. The spectral peak related to the alpha rhythm is more easily discerned in (b). The narrow peaks above 10 Hz in (a) more or less disappear when segmentation is introduced. The analyzed EEG signal is the one displayed in Figure 3.2(a).

known as *Welch's method* and has the property of reducing the variance of $\hat{S}_x(e^{j\omega})$ in (3.89) [103]. The performance of Welch's method can be further improved by instead using the multiple window method in which special attention is paid to the design of windows in order to obtain a lower variance [115–117].

Figure 3.16 illustrates the Fourier-based spectral analysis of an EEG signal containing a strong alpha rhythm. Using Bartlett windowing, the power spectrum is computed either without segment averaging or with averaging of overlapping segments. Although the power spectrum in Figure 3.16(a) has a larger variance than that in Figure 3.16(b), its better spectral resolution is sometimes preferable when the aim is to resolve closely spaced peaks; this is not possible with the power spectrum obtained from averaging. Obviously, none of the two power spectral estimates can be singled out as being superior.

3.3.2 Spectral Parameters

Nonparametric spectrum analysis is the backbone of many systems for EEG analysis. The resulting power spectrum is, however, not readily interpreted, but must often be condensed into a compact set of representative parameters more suitable for quantitative investigations, such as classification and statistical postprocessing. The feature extraction process becomes even more important when considering the large amount of data contained in a multichannel EEG recording. This process produces a set of parameters which describe prominent features of the spectrum, such as peak amplitudes and their respective frequencies. The most frequently used spectral parameters are presented below.

Spectral parameters have been used in many different EEG applications, including the development of normative data for healthy subjects; analysis of data from patients suffering from sleep disorders, cerebral ischemia, renal failure; and for real-time monitoring purposes during surgery, see, e.g., [118, 119].

The first step in the development of a spectral parameter is to assess the properties of the estimated power spectrum using a suitable graphical presentation format. Although this step may seem self-evident, it is nevertheless important to stress the fact that visual assessment is extremely valuable in judging how representative a parameter is in describing a certain spectral property. Such assessment is also useful in understanding the way in which artifacts distort the power spectrum.

A basic decision is whether the spectral power should be presented and analyzed on a linear or logarithmic scale. In general, this decision is guided by the scope of the analysis: a logarithmic scale may be preferable when unsynchronized EEG rhythms with low amplitude are of primary interest. Figure 3.17 illustrates the use of both linear and logarithmic scales for graphical presentation; in fact, the spectral parameters presented below make use of both scales.

Power in frequency bands. The absolute power can be computed in frequency bands whose limits are determined either by clinical convention (i.e., the frequency bands of the alpha, beta, delta, and theta rhythms, described on page 34) or by a statistical technique that indicates the most important bands.

Alternatively, it may be more appropriate to compute relative power, defined as the ratio of the power in a single frequency band to either the total power $r_x(0)$ or the power contained in certain bands, see Figure 3.17(a). Power ratios of different frequency bands may be designed to reflect the relation between slow and fast EEG activity in order to characterize the degree of EEG abnormality [120]. In addition, relative power measurements may be preferable since absolute power is influenced by nonphysiological factors such as skull thickness.

A disadvantage of power measurements in fixed frequency bands is that these measurements become less representative when a peak is located at a

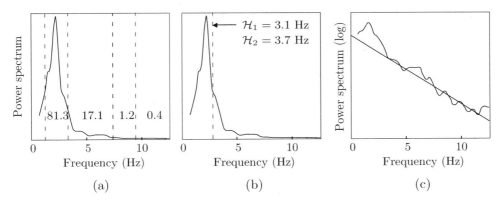

Figure 3.17: The power spectrum of an EEG and related parameters: (a) relative power (percentages) in four frequency bands reflecting delta, theta, and alpha activity (split into two bands); (b) the Hjorth parameters mobility $\mathcal{H}_1$ and complexity $\mathcal{H}_2$; and (c) the spectral slope estimated from the logarithmic power spectrum. The EEG was recorded from a child with a brainstem tumor and is dominated by slow rhythmic activity as reflected by the value of $\mathcal{H}_1$ of 3.1 Hz. (Adapted from Matthis et al. [121]).

boundary. This problem can be alleviated by instead identifying the peaks of the spectrum.

Peak frequency. Power spectral feature extraction can be accomplished by identifying the most prominent peaks. Each spectral peak is then characterized by certain parameters such as its frequency, amplitude, and width. Peak identification typically involves ad hoc criteria for judging if a maximum in the spectrum is sufficiently distinguished from its surroundings to be considered as a peak rather than belonging to the background signal [122]. Since the variance of the periodogram is large, it may be necessary to find a more robust estimate of the peak frequency than simply using the location of the maximum; the mean or median frequency of each individual peak component may be used as such estimates. The mean frequency of a spectrum is considered below in connection with the so-called Hjorth descriptors and time–frequency analysis.

Spectral slope. Based on experimental observations, it has been suggested that the power spectrum of spontaneous EEG activity is roughly composed of two components,

$$S_x(e^{j\omega}) = S_x^r(e^{j\omega})S_x^a(e^{j\omega}), \tag{3.91}$$

where $S_x^r(e^{j\omega})$ represents rhythmic activity and $S_x^a(e^{j\omega})$ represents unstructured, irregular-looking activity [123]. Since the irregular EEG component has often been found to decay exponentially as the frequency increases, the logarithm of the power spectrum $S_x(e^{j\omega})$ can be approximated by the following simple model [3],

$$\log S_x(e^{j\omega}) \approx \log S_x^r(e^{j\omega}) + b|\omega|, \tag{3.92}$$

where b is a negative-valued parameter. One approach to estimate the spectral slope parameter b is to find the particular value of b that minimizes the least-square error $\mathcal{J}(b)$ between $\log S_x(e^{j\omega})$ and the slope $b|\omega|$,

$$\mathcal{J}(b) = \int_{-\pi}^{\pi} \left(\log S_x(e^{j\omega}) - b|\omega| \right)^2 d\omega. \tag{3.93}$$

The calculation of this integral is in practice based on the discrete Fourier transform in which the frequency variable ω has been quantized; integration is replaced by summation over the discrete frequencies.

The resulting parameter estimate $\hat{b}$ has been suggested as a clinical parameter for quantifying irregular EEG activity and the decay rate of high-frequency components, see Figure 3.17(c) [121]. The spectral slope has also been suggested as a means of preconditioning $S_x(e^{j\omega})$ [123], where an estimate of the rhythmic power spectrum is obtained after subtraction of the slope from $S_x(e^{j\omega})$,

$$\hat{S}_x^r(e^{j\omega}) = 10^{(\log S_x(e^{j\omega}) - \hat{b}|\omega|)}. \tag{3.94}$$

Hjorth descriptors.   Another approach to spectral feature extraction is to calculate the moments of the power spectrum $S_x(e^{j\omega})$, originally suggested by Hjorth [124, 125]. The nth-order spectral moment $\overline{\omega}_n$ is defined by the following integral,

$$\overline{\omega}_n = \int_{-\pi}^{\pi} \omega^n S_x(e^{j\omega}) d\omega. \tag{3.95}$$

The odd-numbered moments are all identical to zero because the power spectrum is a symmetric function, i.e., $S_x(e^{j\omega}) = S_x(e^{-j\omega})$. The *Hjorth descriptors* are closely related to the even-numbered spectral moments. The first descriptor is defined by the total signal power,

$$\mathcal{H}_0 = \overline{\omega}_0 = 2\pi r_x(0), \tag{3.96}$$

also referred to as *activity*. The second descriptor, called *mobility*, reflects the dominant frequency of $x(n)$ and is defined by the square root of the normalized second-order moment,

$$\mathcal{H}_1 = \sqrt{\frac{\overline{\omega}_2}{\overline{\omega}_0}}. \tag{3.97}$$

Finally, the fourth-order moment $\overline{\omega}_4$ is used to define a measure related to the bandwidth of $x(n)$ which is termed *complexity*,

$$\mathcal{H}_2 = \sqrt{\frac{\overline{\omega}_4}{\overline{\omega}_2} - \frac{\overline{\omega}_2}{\overline{\omega}_0}}. \tag{3.98}$$

The descriptors $\mathcal{H}_1$ and $\mathcal{H}_2$ produce estimates of the dominant frequency and half the bandwidth, respectively, as long as the rhythm is characterized by a unimodal power spectrum, i.e., with only one dominant peak. For multimodal spectra the dominant frequency can, of course, still be computed, but it no longer has an intuitive relationship to spectral landmarks.

An attractive property of the above three descriptors is that they can be efficiently computed in the time domain without having to compute the Fourier transform $S_x(e^{j\omega})$ and the moment integral in (3.95) [124, 126]. This property is rather easily shown by assuming that the sampled signal $x(n)$ results from sampling of a continuous-time signal $x_c(t)$ using the sampling period T_s,

$$x(n) = x_c(nT_s), \quad n = 0, 1, \ldots, N-1. \tag{3.99}$$

The spectral zero-, second-, and fourth-order moments can then be expressed in terms of the mean power of $x_c(t)$, and its first and second derivatives, respectively,

$$\overline{\omega}_0 = 2\pi E\left[x_c^2(t)\right], \tag{3.100}$$

$$\overline{\omega}_2 = 2\pi T_s^2 E\left[\left(\frac{dx_c(t)}{dt}\right)^2\right], \tag{3.101}$$

$$\overline{\omega}_4 = 2\pi T_s^4 E\left[\left(\frac{d^2 x_c(t)}{dt^2}\right)^2\right]. \tag{3.102}$$

For a sampled signal, the first and second derivatives can be approximated by the following two difference equations, respectively,

$$x^{(1)}(n) = x(n) - x(n-1),$$

$$x^{(2)}(n) = x(n+1) - 2x(n) + x(n-1),$$

where

$$\frac{d^i x_c(t)}{dt^i} \approx \frac{x^{(i)}(n)}{T_s^i}. \tag{3.103}$$

We have tacitly assumed that the two samples $x(-1)$ and $x(N)$ outside the observation interval $[0, N - 1]$ are available. Accordingly, estimates of the spectral moments can be determined using the following time domain average,

$$\hat{\bar{\omega}}_i \approx \frac{2\pi}{N} \sum_{n=0}^{N-1} \left(x^{(i/2)}(n) \right)^2, \quad i = 0, 2, 4. \tag{3.104}$$

Since the Hjorth descriptors were originally developed for on-line EEG analysis, these were implemented as averages successively updated in time—"running averages". For example, the activity parameter $\mathcal{H}_0$ becomes a function of time, i.e., $\mathcal{H}_0(n)$, by using the samples in a sliding window of length L,

$$\mathcal{H}_0(n) = \bar{\omega}_0(n) = \frac{2\pi}{L} \sum_{k=n-L+1}^{n} x^2(k). \tag{3.105}$$

It is well-known that the computation of the derivatives in (3.101) and (3.102) is sensitive to noise, and, therefore, it is advisable to limit the bandwidth of the signal prior to this computation. Despite this disadvantage, the Hjorth descriptors have been found to be useful in various EEG applications, for example, in sleep staging. The descriptors serve as an example of the understanding that a clinically useful technique is not necessarily synonymous with a technique requiring very complex calculations.

A one-sided spectral moment definition which includes an integration interval that ranges from zero to half the sampling rate has also been considered for the purpose of deriving spectral parameters [127], see also Section 5.3. The calculation of the one-sided moment was found to be more robust than that associated with (3.101) and (3.102); however, both types of moment definitions provide similar information.

Spectral purity index. The spectral purity index (SPI) is a heuristic parameter pursued by Barlow in the study of certain EEG signals [3, 128]. The parameter is designed to reflect signal bandwidth and is related to the Hjorth complexity descriptor in (3.98). It is defined as the ratio between the squared, running second-order moment and the running total power and fourth-order moment,

$$\Gamma_{\text{SPI}}(n) = \frac{\bar{\omega}_2^2(n)}{\bar{\omega}_0(n)\bar{\omega}_4(n)}. \tag{3.106}$$

The term "purity" refers to how well the analyzed signal is described by a single frequency: the SPI is equal to unity for a noise-free, sinusoidal signal

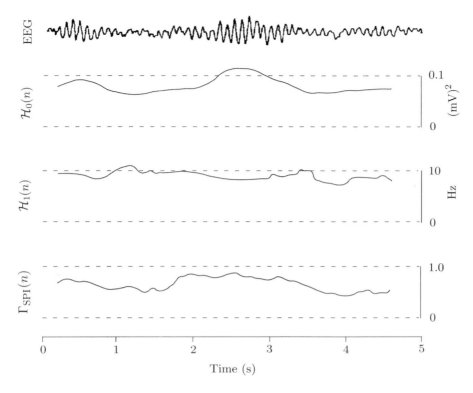

Figure 3.18: An EEG showing an alpha rhythm. The signal is characterized by three different trends obtained as running averages of the activity $\mathcal{H}_0(n)$, the mobility $\mathcal{H}_1(n)$, and the spectral purity index $\Gamma_{\mathrm{SPI}}(n)$. (Reprinted from Goncharova and Barlow [128] with permission.)

and decreases to zero as the bandwidth of the signal increases. The SPI is illustrated by the example in Figure 3.18.

The noise susceptibility aspect of the three Hjorth descriptors is equally valid for the SPI and is related to the fact that both approaches are based on signal derivatives. The SPI is actually more susceptible to noise since its definition involves the ratio between the first and the second derivatives.

3.4 Model-based Spectral Analysis

Linear stochastic modeling has frequently been considered in EEG signal analysis, especially in obtaining a parametric description of the spectral properties. As pointed out in Section 3.1, autoregressive modeling has, by far, received the most attention of the linear models, and, therefore, we

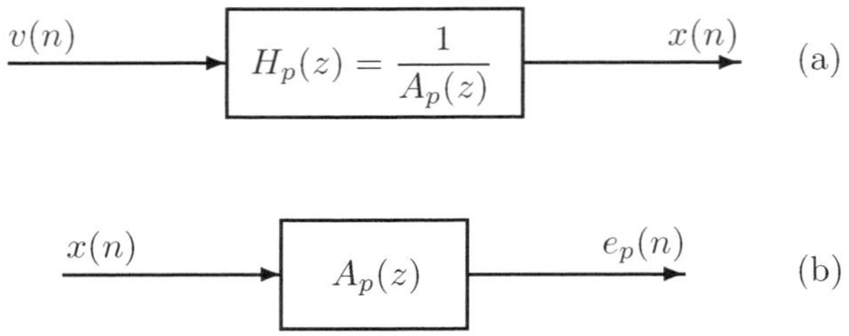

Figure 3.19: (a) Autoregressive modeling and (b) the related linear prediction error filter.

will limit our presentation to a number of methods of estimating the AR parameters $a_1, \ldots, a_p$ in

$$x(n) = -a_1 x(n-1) - \cdots - a_p x(n-p) + v(n) \qquad (3.107)$$

and the variance σ_v^2 of the input noise $v(n)$.[6] In the AR model, the output signal is generated by feeding the noise to a filter of order p with the transfer function

$$H_p(z) = \frac{1}{A_p(z)} = \frac{1}{1 + a_1 z^{-1} + \cdots + a_p z^{-p}}, \qquad (3.108)$$

see Figure 3.19(a). Apart from zeros located at $z = 0$, $H(z)$ is completely defined by its poles, and, consequently, the AR model is also referred to as an *all-pole model*.

Autoregressive modeling is closely related to the linear prediction problem in which the current sample $x(n)$ is predicted from the p previous samples $x(n-1), \ldots, x(n-p)$ using an FIR filter structure of the predictor. This relation becomes plausible when noting that the AR model in (3.107) implies a technique with which $x(n)$ can be predicted from a linear combination of the p preceding samples,

$$\hat{x}_p(n) = -a_1 x(n-1) - \cdots - a_p x(n-p). \qquad (3.109)$$

The input noise $v(n)$ of the AR model can be omitted from the FIR predictor since the noise is assumed to be stationary and white, and, accordingly, it does not contribute to improving the prediction of $x(n)$. The prediction

---

[6]The literature on linear modeling and spectral analysis is immense, and the interested reader is referred to the excellent textbooks by Therrien [129] and Kay [130] for an in-depth description of AR, as well as ARMA and MA modeling.

error $e_p(n)$ is defined as

$$e_p(n) = x(n) - \hat{x}_p(n)$$

$$= x(n) + \sum_{k=1}^{p} a_k x(n-k), \tag{3.110}$$

which, after insertion of the AR modeling equation in (3.107), becomes

$$e_p(n) = -\sum_{k=1}^{p} a_k x(n-k) + v(n) + \sum_{k=1}^{p} a_k x(n-k) = v(n) \tag{3.111}$$

and thus establishes the close relationship between AR modeling and linear prediction.

The prediction error $e_p(n)$ in (3.110) can be viewed as the output of a linear, time-invariant FIR filter which is completely characterized by its transfer function

$$A_p(z) = 1 + a_1 z^{-1} + \cdots + a_p z^{-p}, \tag{3.112}$$

where the parameter a_0 is equal to one. The filter $A_p(z)$ is commonly referred to as a *prediction error filter*. When minimizing the prediction error variance

$$\sigma_e^2 = E\left[e_p^2(n)\right], \tag{3.113}$$

estimates of the model parameters $a_1, \ldots, a_p$ can be determined.

Based on the observation that the transfer function of the AR model equals the inverse of the transfer function of prediction error filter, see Figure 3.19, the idea is to first solve the linear prediction problem and then to substitute the resulting parameter estimates into the AR model. In the same way, the variance of the input noise σ_v^2 is estimated as the variance of the prediction errors:

$$\sigma_v^2 = \sigma_e^2. \tag{3.114}$$

It is essential to bear in mind that the AR parameter estimates correspond to the parameters of the prediction error filter only as long as the orders of the AR model and the predictor are identical (or, possibly, when the predictor order is higher). Based on the resulting parameter estimates, the AR power spectrum $S_x(e^{j\omega})$ can be computed by evaluating $A_p(z)$ in (3.112) for $z = e^{j\omega}$, giving

$$S_x(e^{j\omega}) = \frac{\sigma_v^2}{|A_p(e^{j\omega})|^2} = \frac{\sigma_v^2}{\left|1 + \displaystyle\sum_{k=1}^{p} a_k e^{-j\omega k}\right|^2}. \tag{3.115}$$

We will now address the question of how to find estimates of the model parameters. Among the many methods that have been presented for this purpose, we will consider the following three in more detail.[7]

- *The autocorrelation/covariance methods* result from straightforward minimization of the prediction error variance in (3.113) (Section 3.4.1). These two methods will later be modified to handle signals with slowly time-varying spectral properties by making use of the adaptive LMS algorithm (Section 3.6.5).

- *The modified covariance method* extends the autocorrelation/covariance methods by considering both forward and backward prediction errors in the error criterion (Section 3.4.2).

- *Burg's method* is based on the same criterion as the modified covariance method but assumes that the FIR predictor has been replaced by an FIR lattice structure (Section 3.4.3). Burg's method can be extended to the slowly time-varying case, resulting in the gradient adaptive lattice algorithm (Section 3.6.5).

The choice of various parameters, such as model order and sampling rate, is discussed in Section 3.4.4, and the problem of extracting features from an AR power spectrum is treated in Section 3.4.5.

3.4.1   The Autocorrelation/Covariance Methods

We will now consider the minimization of the error variance $E\left[e_p^2(n)\right]$ in (3.113). This is an operation which can be performed by straightforward differentiation of $E\left[e_p^2(n)\right]$ with respect to a_k, thereby resulting in a system of linear equations with p unknown variables. Instead of working out the details of such an equation system, we will start our derivation by introducing certain useful vector notations. From (3.110) we know that the prediction error is given by

$$e_p(n) = \sum_{k=0}^{p} a_k x(n-k)$$
$$= \mathbf{a}_p^T \tilde{\mathbf{x}}_p(n), \qquad\qquad (3.116)$$

---

[7]Autoregressive modeling is here treated within a statistical framework. It is interesting to note, however, that a deterministic approach to AR parameter estimation based on minimization of the least-squares error criterion

$$\epsilon^2 = \sum_n e_p^2(n)$$

leads to a method identical to that resulting from the statistical criterion in (3.113) [129].

where the observed signal is contained in the time-reversed vector

$$\tilde{\mathbf{x}}_p(n) = \begin{bmatrix} x(n) \\ x(n-1) \\ \vdots \\ x(n-p) \end{bmatrix}, \tag{3.117}$$

and the parameters of the prediction error filter in the vector

$$\mathbf{a}_p = \begin{bmatrix} 1 \\ a_1 \\ \vdots \\ a_p \end{bmatrix}. \tag{3.118}$$

Using the above notations, the error variance can be expressed as

$$\begin{aligned} \sigma_e^2 &= E\left[e_p^2(n)\right] \\ &= E\left[\mathbf{a}_p^T \tilde{\mathbf{x}}_p(n)\tilde{\mathbf{x}}_p^T(n)\mathbf{a}_p\right] \\ &= \mathbf{a}_p^T \tilde{\mathbf{R}}_x \mathbf{a}_p, \end{aligned} \tag{3.119}$$

where $\tilde{\mathbf{R}}_x$ denotes the reversal of the correlation matrix for $x(n)$, defined above in (3.11). Before minimizing σ_e^2, it is necessary to introduce a constraint which assures that the first element of $\mathbf{a}_p$ in (3.118) is identical to one,

$$\mathbf{a}_p^T \mathbf{i} = 1, \tag{3.120}$$

where

$$\mathbf{i} = \begin{bmatrix} 1 \\ 0 \\ \vdots \\ 0 \end{bmatrix}. \tag{3.121}$$

The use of Lagrange multipliers is very powerful in optimization problems including one or several linear constraints (see Appendix A); in our problem the following Lagrangian $\mathcal{L}$ is to be minimized with respect to $\mathbf{a}_p$,

$$\mathcal{L} = \frac{1}{2}\mathbf{a}_p^T \tilde{\mathbf{R}}_x \mathbf{a}_p + \lambda(1 - \mathbf{a}_p^T \mathbf{i}), \tag{3.122}$$

where λ is the Lagrange multiplier (the factor $\frac{1}{2}$ is included for convenience and does not affect the end result of the minimization). The gradient of $\mathcal{L}$ is given by

$$\nabla_{\mathbf{a}_p}\mathcal{L} = \tilde{\mathbf{R}}_x \mathbf{a}_p - \lambda\mathbf{i} = 0, \tag{3.123}$$

which, after multiplication by $\mathbf{a}_p^T$, yields

$$\mathbf{a}_p^T \tilde{\mathbf{R}}_x \mathbf{a}_p - \lambda = 0, \tag{3.124}$$

or, using (3.119),

$$\lambda = \mathbf{a}_p^T \tilde{\mathbf{R}}_x \mathbf{a}_p = \sigma_e^2. \tag{3.125}$$

By combining (3.123) and (3.125), we obtain a matrix equation whose solution yields the desired parameter values,

$$\tilde{\mathbf{R}}_x \mathbf{a}_p = \sigma_e^2 \mathbf{i}. \tag{3.126}$$

This particular type of matrix equation is commonly referred to as the *normal equations* of linear prediction. For a stationary stochastic process, the correlation matrix is a symmetric, Toeplitz matrix so $\tilde{\mathbf{R}}_x = \mathbf{R}_x$.[8]

In order to indicate more clearly how the model parameters $a_1, \ldots, a_p$ and σ_e^2 are computed, it may be instructive to rewrite (3.126) as a two-step computation. The first step produces estimates of $a_1, \ldots, a_p$ from

$$\begin{bmatrix} a_1 \\ a_2 \\ \vdots \\ a_p \end{bmatrix} = \begin{bmatrix} r_x(0) & r_x(1) & \cdots & r_x(p-1) \\ r_x(1) & r_x(0) & \cdots & r_x(p-2) \\ \vdots & \vdots & \ddots & \vdots \\ r_x(p-1) & r_x(p-2) & \cdots & r_x(0) \end{bmatrix}^{-1} \begin{bmatrix} -r_x(1) \\ -r_x(2) \\ \vdots \\ -r_x(p) \end{bmatrix}, \tag{3.127}$$

followed by the second step, which yields the variance σ_e^2,

$$\sigma_e^2 = r_x(0) + \sum_{i=1}^{p} a_i r_x(i). \tag{3.128}$$

The solution of the normal equations in (3.126) turns out to be computationally rather demanding. The *Levinson–Durbin recursion* is a fast and efficient method which solves these equations by exploiting the symmetry and Toeplitz properties of the correlation matrix $\mathbf{R}_x$. This recursion avoids not only the matrix inversion in (3.127), but, equally important, it provides a new, fresh perspective on linear prediction by introducing the lattice filter in a natural way; the lattice structure forms part of the AR parameter estimation presented in Section 3.4.3. The Levinson–Durbin recursion is defined

---

[8]These normal equations are equivalent to the well-known Yule–Walker equations of an AR model (not described here) and establish the identity between the linear prediction problem and AR modeling.

by [129, p. 422ff.]

$$\gamma_j = \frac{\mathbf{r}_{j-1}^T \tilde{\mathbf{a}}_{j-1}}{\sigma_{e_{j-1}}^2}, \tag{3.129}$$

$$\mathbf{a}_j = \begin{bmatrix} \mathbf{a}_{j-1} \\ 0 \end{bmatrix} - \gamma_j \begin{bmatrix} 0 \\ \tilde{\mathbf{a}}_{j-1} \end{bmatrix}, \tag{3.130}$$

$$\sigma_{e_j}^2 = (1 - \gamma_j^2)\sigma_{e_{j-1}}^2, \tag{3.131}$$

for $j = 1, \ldots, p$, where

$$\mathbf{r}_j = \begin{bmatrix} r_x(1) \\ r_x(2) \\ \vdots \\ r_x(j+1) \end{bmatrix}. \tag{3.132}$$

The recursion is initialized by

$$\mathbf{a}_0 = 1, \tag{3.133}$$

$$\mathbf{r}_0 = r_x(1), \tag{3.134}$$

$$\sigma_{e_0}^2 = r_x(0). \tag{3.135}$$

The index j has been attached to the error variance σ_e^2 to indicate model order. It should be noted that the recursion yields the desired parameter vector $\mathbf{a}_p$ *as well as* all parameter vectors $\mathbf{a}_j$ of lower order ($j < p$). This "order-recursive" property, being highly attractive in many situations, has no parallel in the computationally much more demanding direct solution of the normal equations in (3.126).

The correlation matrix $\tilde{\mathbf{R}}_x$ has, so far, been considered to be known a priori; however, in practice we have to estimate $\tilde{\mathbf{R}}_x$ from the observed signal. The problem of estimating $r_x(k)$ has already been touched upon for the purpose of nonparametric spectral analysis, cf. the estimator in (3.78). However, instead of first estimating $r_x(k)$ and then constructing $\tilde{\mathbf{R}}_x$, the estimation may be based on a data matrix $\mathbf{X}_p$.

In the literature, the *covariance method*[9] is synonymous with that of a data matrix defined by

$$\mathbf{X}_p = \begin{bmatrix} x(p) & x(p-1) & \cdots & x(0) \\ x(p+1) & x(p) & \cdots & x(1) \\ \vdots & \vdots & & \vdots \\ x(N-2) & x(N-3) & \cdots & x(N-p-2) \\ x(N-1) & x(N-2) & \cdots & x(N-p-1) \end{bmatrix} \tag{3.136}$$

[9]The name "covariance method" is actually a misnomer since there is no relationship to the statistical term; however, the name has today become well-established in the literature.

which is used to estimate the reversed correlation matrix,

$$\hat{\tilde{\mathbf{R}}}_{\mathbf{x}} = \frac{1}{N-p}\tilde{\mathbf{X}}_p^T\tilde{\mathbf{X}}_p, \tag{3.137}$$

where

$$\tilde{\mathbf{X}}_p = \begin{bmatrix} x(N-p-1) & \cdots & x(N-2) & x(N-1) \\ x(N-p-2) & \cdots & x(N-3) & x(N-2) \\ \vdots & & \vdots & \vdots \\ x(1) & \cdots & x(p) & x(p+1) \\ x(0) & \cdots & x(p-1) & x(p) \end{bmatrix}. \tag{3.138}$$

Although the correlation matrix estimate in (3.137) is not Toeplitz, it is nonetheless frequently used since only measured data is included and no zero-padding is required to take into account nonexistent samples [129]. Note that the Levinson–Durbin recursion does not apply to this correlation matrix estimate.

The *autocorrelation method* is defined by using another definition of the data matrix $\mathbf{X}_p$ where the beginning and end of the signal are padded with zeros,

$$\mathbf{X}_p = \begin{bmatrix} x(0) & 0 & \cdots & 0 \\ x(1) & x(0) & \cdots & 0 \\ \vdots & \vdots & & \vdots \\ x(p) & x(p-1) & \cdots & x(0) \\ x(p+1) & x(p) & \cdots & x(1) \\ \vdots & \vdots & & \vdots \\ x(N-1) & x(N-2) & \cdots & x(N-p-1) \\ 0 & x(N-1) & \cdots & x(N-p-2) \\ \vdots & \vdots & & \vdots \\ 0 & 0 & \cdots & x(N-1) \end{bmatrix}. \tag{3.139}$$

3.4.2   The Modified Covariance Method

Solving the linear prediction problem is the key to finding the AR model parameter estimates. While it is natural to formulate this problem in terms of minimizing the forward prediction errors, improved parameter estimates can actually be obtained by also taking the backward prediction errors into account. In backward prediction, the aim is to "predict" $x(n-p)$ from the p following samples $x(n-p+1), \ldots, x(n)$, again using an FIR predictor structure. The data set used for producing backward prediction errors is identical to that used for producing forward prediction errors. Below, the AR parameter estimation is based on a criterion which combines both forward and backward prediction error variances,

$$\sigma_e^2 = E\left[|e_p^+(n)|^2\right] + E\left[|e_p^-(n)|^2\right]. \tag{3.140}$$

The forward prediction errors, earlier defined in (3.116), are now denoted

$$e_p^+(n) \equiv e_p(n) \tag{3.141}$$

in order to distinguish these errors from the backward prediction errors which are defined as

$$e_p^-(n) = x(n-p) - \hat{x}(n-p). \tag{3.142}$$

It should be observed that this definition denotes the error at $n-p$ with the index n. The backward prediction $\hat{x}(n-p)$ results from a linear combination of the p most recent observations,

$$\hat{x}(n-p) = -b_1 x(n-p+1) - \cdots - b_p x(n). \tag{3.143}$$

The introduction of another set of predictor parameters $b_1, \ldots, b_p$ may initially appear troublesome since it is not obvious how these are related to $a_1, \ldots, a_p$. Before dealing with the error criterion in (3.140), we therefore briefly consider the "isolated" backward prediction problem based on the criterion

$$\sigma_{e-}^2 = E\left[|e_p^-(n)|^2\right], \tag{3.144}$$

which helps us in establishing a relationship between the parameters of the forward and backward predictors. Repeating the minimization procedure of the preceding subsection on forward prediction, the following normal equations result,

$$\mathbf{R}_x \mathbf{b}_p = \sigma_{e-}^2 \mathbf{i}. \tag{3.145}$$

Since $\tilde{\mathbf{R}}_x = \mathbf{R}_x$, a comparison of (3.145) with the normal equations of the forward prediction problem in (3.126) implies that

$$\mathbf{b}_p = \mathbf{a}_p, \tag{3.146}$$
$$\sigma_{e-}^2 = \sigma_e^2. \tag{3.147}$$

The result that the parameters of the forward and backward predictors are identical is not entirely surprising since the analyzed signal is assumed to be a stationary process.

Using the results in (3.146), the backward prediction errors can be obtained from

$$e_p^-(n) = \sum_{k=0}^{p} b_k x(n - p + k)$$
$$= \mathbf{a}_p^T \mathbf{x}_p(n), \tag{3.148}$$

which allows us to rewrite (3.140) as

$$\sigma_e^2 = E\left[\mathbf{a}_p^T \tilde{\mathbf{x}}_p(n)\tilde{\mathbf{x}}_p^T(n)\mathbf{a}_p\right] + E\left[\mathbf{a}_p^T \mathbf{x}_p(n)\mathbf{x}_p^T(n)\mathbf{a}_p\right]$$
$$= \mathbf{a}_p^T(\tilde{\mathbf{R}}_x + \mathbf{R}_x)\mathbf{a}_p. \tag{3.149}$$

The minimization of (3.149), which again is based on a Lagrangian,

$$\mathcal{L} = \mathbf{a}_p^T(\tilde{\mathbf{R}}_x + \mathbf{R}_x)\mathbf{a}_p + \lambda\left(1 - \mathbf{a}_p^T \mathbf{i}\right), \tag{3.150}$$

leads to the following set of normal equations,

$$(\tilde{\mathbf{R}}_x + \mathbf{R}_x)\mathbf{a}_p = \sigma_e^2 \mathbf{i}. \tag{3.151}$$

At first glance, this result appears to be a major setback since $\mathbf{R}_x + \tilde{\mathbf{R}}_x = 2\mathbf{R}_x$, and, consequently, the parameter estimates of (3.151) are identical to those associated with the forward prediction criterion in (3.113). However, using the correlation matrix estimate based on the data matrix in (3.136), the left-hand side of (3.151) becomes a sum of two symmetric but non-Toeplitz matrices which, in general, differ from each other. Hence, the modified covariance method is defined by the following set of equations,

$$\frac{1}{N - p}(\tilde{\mathbf{X}}_p^T \tilde{\mathbf{X}}_p + \mathbf{X}_p^T \mathbf{X}_p)\mathbf{a}_p = \sigma_e^2 \mathbf{i}. \tag{3.152}$$

The modified covariance method has been found to exhibit better performance than the covariance method [131].

3.4.3   Burg's Method

The third, and final, method considered for estimation of the AR parameters is also based on joint minimization of the forward and backward prediction error variances. In contrast to the modified covariance method, however, we will now explicitly make use of the Levinson–Durbin recursion in the minimization process in order to arrive at an efficient estimation method. The resulting method, named after its inventor John Burg, estimates the parameters of a prediction error filter with lattice structure. The parameters are then transformed into the desired, direct form FIR predictor parameters so that the power spectrum can be calculated.

We start our derivation by recalling the definitions of the forward and backward prediction errors,

$$e_p^+(n) = \mathbf{a}_p^T \tilde{\mathbf{x}}_p(n),$$
$$e_p^-(n) = \mathbf{a}_p^T \mathbf{x}_p(n).$$

Combining these two relations with the Levinson–Durbin recursion given in (3.130), i.e.,

$$\mathbf{a}_p = \begin{bmatrix} \mathbf{a}_{p-1} \\ 0 \end{bmatrix} - \gamma_p \begin{bmatrix} 0 \\ \tilde{\mathbf{a}}_{p-1} \end{bmatrix},$$

it is possible to derive recursions for $e_p^+(n)$ and $e_p^-(n)$. With appropriate partitioning of $\tilde{\mathbf{x}}_p(n)$, the forward prediction error can be computed recursively using

$$e_p^+(n) = \begin{bmatrix} \mathbf{a}_{p-1}^T & 0 \end{bmatrix} \begin{bmatrix} \tilde{\mathbf{x}}_{p-1}(n) \\ x(n-p) \end{bmatrix} - \gamma_p \begin{bmatrix} 0 & \tilde{\mathbf{a}}_{p-1}^T \end{bmatrix} \begin{bmatrix} x(n) \\ \tilde{\mathbf{x}}_{p-1}(n-1) \end{bmatrix}$$
$$= e_{p-1}^+(n) - \gamma_p e_{p-1}^-(n-1). \tag{3.153}$$

In the same way, a recursion can be established for the backward prediction error,

$$e_p^-(n) = \begin{bmatrix} \mathbf{a}_{p-1}^T & 0 \end{bmatrix} \begin{bmatrix} \mathbf{x}_{p-1}(n-1) \\ x(n) \end{bmatrix} - \gamma_p \begin{bmatrix} 0 & \tilde{\mathbf{a}}_{p-1}^T \end{bmatrix} \begin{bmatrix} x(n-p) \\ \mathbf{x}_{p-1}(n) \end{bmatrix}$$
$$= e_{p-1}^-(n-1) - \gamma_p e_{p-1}^+(n). \tag{3.154}$$

Both recursions are initialized by the input signal $x(n)$,

$$e_0^+(n) = e_0^-(n) = x(n). \tag{3.155}$$

A remarkable property of the recursive equations of $e_p^+(n)$ and $e_p^-(n)$ is that they together define one stage in a lattice filter; the entire prediction error filter is formed by cascading p different lattice stages, see Figure 3.20. The lattice realization of the prediction error filter is completely defined by the parameters $\gamma_1, \ldots, \gamma_p$, which were initially introduced as "help variables" in the Levinson–Durbin recursion, but which can now be interpreted as the parameters of a lattice filter or, as they are more commonly called, *reflection coefficients*.

Subject to the constraint that the prediction errors are generated by the above lattice equations, we minimize the error

$$\sigma_e^2 = E\left[|e_p^+(n)|^2\right] + E\left[|e_p^-(n)|^2\right] \tag{3.156}$$

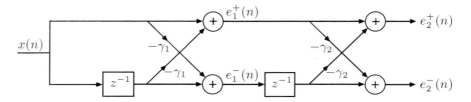

Figure 3.20: A second-order prediction error filter with FIR lattice structure ($p = 2$). The forward and backward prediction errors are available at the output of the upper and lower branch, respectively.

with respect to γ_p. This error can, after substitution of the lattice recursions in (3.153) and (3.154), be written as

$$\sigma_e^2 = \left(1 + \gamma_p^2\right) \left(E\left[|e_{p-1}^+(n)|^2\right] + E\left[|e_{p-1}^-(n-1)|^2\right] \right) \\ - 4\gamma_p E\left[e_{p-1}^+(n)e_{p-1}^-(n-1)\right], \tag{3.157}$$

which is minimized by choosing γ_p such that

$$\frac{\partial \sigma_e^2}{\partial \gamma_p} = 2\gamma_p \left(E\left[|e_{p-1}^+(n)|^2\right] + E\left[|e_{p-1}^-(n-1)|^2\right] \right) \\ - 4E\left[e_{p-1}^+(n)e_{p-1}^-(n-1)\right] = 0. \tag{3.158}$$

The optimal p^{th} reflection coefficient is given by

$$\gamma_p = \frac{2E\left[e_{p-1}^+(n)e_{p-1}^-(n-1)\right]}{E\left[|e_{p-1}^+(n)|^2\right] + E\left[|e_{p-1}^-(n-1)|^2\right])}. \tag{3.159}$$

While the reflection coefficients are essential for computing the prediction errors, our primary interest is in determining the FIR predictor parameters $a_1, \ldots, a_p$ for use in spectral analysis. These parameters can be easily computed by inserting γ_p into the Levinson–Durbin recursion in (3.130).[10]

---

[10]The reflection coefficients possess several interesting properties, perhaps most notably that the corresponding AR model is assured to be stable when $|\gamma_j| < 1$ for $j = 1, \ldots, p$ [129].

The reflection coefficient γ_j is estimated by replacing the expected values in (3.159) with their corresponding time averages

$$\hat{\gamma}_p = \frac{2\displaystyle\sum_{n=p}^{N-1} e_{p-1}^+(n)e_{p-1}^-(n-1)}{\displaystyle\sum_{n=p}^{N-1}(|e_{p-1}^+(n)|^2 + |e_{p-1}^-(n-1)|^2)}, \tag{3.160}$$

where both sums start at $n = p$ in order to assure that only existing samples are used in the computation.

From a conceptual point of view, Burg's method is more involved than the two other estimation methods presented above. It may therefore be instructive to summarize the steps required in computing estimates of the reflection coefficients $\gamma_1, \gamma_2, \ldots, \gamma_p$.

1. The procedure is initialized by setting the "zero-order" forward and backward prediction errors equal to $x(n)$ for $n = 0, \ldots, N-1$, cf. (3.155). Thus, an estimate of the first reflection coefficient γ_1 is obtained from

$$\hat{\gamma}_1 = \frac{2\displaystyle\sum_{n=1}^{N-1} x(n)x(n-1)}{\displaystyle\sum_{n=1}^{N-1}\left(x^2(n) + x^2(n-1)\right)}. \tag{3.161}$$

2. The "first-order" forward and backward prediction errors are generated as the output of the lattice predictor defined by $\hat{\gamma}_1$,

$$e_1^+(n) = x(n) - \hat{\gamma}_1 x(n-1), \tag{3.162}$$
$$e_1^-(n) = x(n-1) - \hat{\gamma}_1 x(n), \tag{3.163}$$

for $n = 1, \ldots, N-1$.

3. The reflection coefficient γ_2 of the second stage in the lattice predictor is given by

$$\hat{\gamma}_2 = \frac{2\displaystyle\sum_{n=2}^{N-1} e_1^+(n)e_1^-(n-1)}{\displaystyle\sum_{n=2}^{N-1}\left((e_1^+(n))^2 + (e_1^-(n-1))^2\right)}. \tag{3.164}$$

4. The "second-order" forward and backward prediction errors are generated by filtering $e_1^+(n)$ and $e_1^-(n)$ with the second lattice stage defined by $\hat{\gamma}_2$,

$$e_2^+(n) = e_1^+(n) - \hat{\gamma}_2 e_1^-(n-1), \qquad (3.165)$$
$$e_2^-(n) = e_1^-(n-1) - \hat{\gamma}_2 e_1^+(n), \qquad (3.166)$$

for $n = 2, \ldots, N-1$.

5. The computations are repeated by incrementing p by one until the desired model order is reached.

3.4.4   Performance and Design Parameters

At this point, it seems appropriate to consider which of the above methods of AR parameter estimation is preferred for EEG spectral analysis. It is also important to address the problem of selecting the model order p, which, so far, has been considered to be known a priori. The sampling rate is yet another design parameter that is briefly discussed.

Choosing method.   The performance of model-based methods for spectral estimation can be assessed using simulated signals as generated by the AR model in (3.107). With this approach, one can compare the parameter estimates to the true values, either directly or in power spectral terms. Based on such simulations, the covariance method, the modified covariance method, and Burg's method have been found to yield more accurate spectral estimates than does the autocorrelation method [103]. Ranking of these three methods is difficult, although the two methods based on minimization of forward and backward prediction error variances may be preferred over the covariance method.

In contrast to Burg's method, the modified covariance method does not guarantee that the parameter estimates correspond to a stable AR model, i.e., all poles lie inside the unit circle [129]. The stability issue has fortunately no significance in spectral analysis since identical power spectra will result from a model with a pair of complex-conjugated poles at $z_{1,2} = re^{\pm j\phi}$ and another model with a pair at $z_{1,2} = \frac{1}{r}e^{\pm j\phi}$ exhibiting inverse symmetry with respect to the unit circle (the radius and the angle are such that $0 < r < 1$ and $-\pi < \phi < \pi$, respectively). This property can be illustrated by the following two second-order AR models:

$$H_2^I(z) = \frac{1}{1 - z^{-1} + 0.5z^{-2}} \qquad (3.167)$$

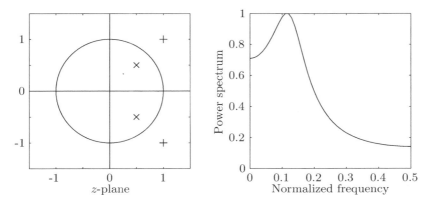

Figure 3.21: Two different second-order AR models with identical power spectra. Each AR model is defined by a pair of complex-conjugated poles, located either inside ("x") or outside ("+") the unit circle.

and

$$H_2^{II}(z) = \frac{1}{1 - 2z^{-1} + 2z^{-2}}, \tag{3.168}$$

whose poles are defined by $z_{1,2}^I = \frac{1}{\sqrt{2}}e^{\pm j0.25\pi}$ and $z_{1,2}^{II} = \sqrt{2}e^{\pm j0.25\pi}$, respectively. Figure 3.21 shows the locations of the two pole pairs and the corresponding power spectra which are identical.

The stability issue must, however, be taken into account when designing an EEG simulator that relies on AR modeling [132, 133] or when designing an EEG classifier which uses the model parameters $a_1, \ldots, a_p$ for making decisions. When using the modified covariance method, one must assure that poles located outside the unit circle are mirrored to the inside. Such mirroring requires additional calculations since the roots of the polynomial $A_p(z)$ must first be determined. Using Burg's method, it is guaranteed that all poles will be located inside the unit circle.

While simulation results provide valuable insight on method performance, it is absolutely essential to assess the performance in clinical terms before a method can be finally used. As an example, the performance of the autocorrelation method and Burg's method was studied in terms of classifying short EEG intervals with respect to their spectral characteristics [134]. A set of EEGs was divided into consecutive, 1-s intervals after which an electroencephalographer assigned each interval to one of the following seven classes: 1. artifactual, 2. artifactual, low-voltage, 3. low-voltage, 4. slow, 5. slow, low-voltage, 6. normal alpha, and 7. normal alpha, low-voltage. Classification schemes were then developed, either based directly on the AR parameters or on certain features derived from the power spectrum, such as peak frequency. For both type of schemes, the AR parameter estimates of Burg's method were found to yield much better agreement with the manual classification than did those of the autocorrelation method. In that study, classification performance was also investigated for parameters derived from nonparametric spectral analysis, and the results were found to be approximately the same as those obtained with Burg's method. Small shifts in peak frequency were more easily observed in the AR power spectrum than in the nonparametric spectrum; however, this property had no significant effect on classification performance. It should be noted that Burg's method has recently been considered in many other EEG-related studies [135–137].

Model order.   The model order p must be estimated before the AR power spectrum can be computed. The model order significantly influences the shape of the estimated power spectrum: too low a value of p results in an overly smooth spectrum with insufficient resolution, whereas too large an order introduces spurious spectral peaks.

Several criteria have been developed for estimating the model order p. A common idea to the criteria is to design a function $\mathcal{M}(p)$ that incorporates the prediction error variance $\sigma_{e_p}^2$, which is a decreasing function of p, and a penalty function, which avoids overparametrization of the AR model. The appropriate model order is given by the value of p that minimizes $\mathcal{M}(p)$. The Akaike information criterion (AIC) [129, 138] and the minimum description length (MDL) [129, 139] are two well-known criteria for choosing AR model order, defined by

$$\mathcal{M}_{\text{AIC}}(p) = N \ln \sigma_{e_p}^2 + 2p, \qquad (3.169)$$

and

$$\mathcal{M}_{\text{MDL}}(p) = N \ln \sigma_{e_p}^2 + p \log N. \qquad (3.170)$$

Both these criteria are related to the prediction error $\sigma_{e_p}^2$ but have different expressions for the penalty: the MDL increases the penalty as more samples become available, whereas the penalty of AIC is independent of the number of samples N.

Various criteria for selecting model order have been used with varying degrees of success in EEG signal processing [52, 134], which is probably explained by the fact that an AR model is fitted to non-AR data. For example, the AIC was found to estimate too low a model order, causing the power spectrum to lack certain spectral peaks which were clearly discernible in the periodogram [140]. Instead, the overwhelming majority of studies have assumed a fixed model order, where the actual order is dependent on the purpose of the analysis. Lower model orders provide sufficient spectral detail for classification of EEG spectra and assure a certain robustness against noise and artifacts [141]; a fifth- or sixth-order AR model has been successfully used in several studies [49, 52, 134, 142–144]. A higher model order should be considered when a more detailed representation of the power spectrum is required.

In deciding the model order, it is valuable to know how many spectral peaks are of interest since this number offers a lower limit of p (each additional spectral peak increases the model order by two). An upper limit of the model order is provided by the rule of thumb that $p < N/3$ [131], although the selected order is typically much lower than $N/3$.

Sampling rate. The sampling rate is a design parameter which influences the AR parameter estimates and the model order [145]. This influence becomes obvious when considering how $r_x(k)$ changes with sampling rate. It is obvious that an increased sampling rate results in a higher resolution of $r_x(k)$, since additional lags can be computed. In order to maintain the information contained in $r_x(k)$, it is necessary to use all lags of the upsampled correlation function. Thus, doubling the sampling rate means that the model order must be increased by a factor two in order to assure that all correlation lags of the original $r_x(k)$ are included when solving the normal equations. However, it is usually advisable to choose a sampling rate that is relatively close to the Nyquist rate in order to make sure that the degrees of freedom of the model are spent on describing relevant signal components rather than modeling noise that may be present in the high-frequency band.

The effect on the power spectrum when an EEG signal is analyzed at different sampling rates (128 and 128/3 Hz) with a fixed model order is illustrated in Figure 3.22. Some of the spectral details are lost when the signal with the higher sampling rate is analyzed since the same number of poles has to account for information in a much larger spectral interval (ranging to 128/2 Hz instead of 128/6 Hz), see Figure 3.22(b). Using the lower sampling rate, two spectral peaks can be discerned in the interval between 10 and 12 Hz, which are also discernible in the nonparametric power spectrum, see Figure 3.22(c).

3.4.5 Spectral Parameters

The extraction of spectral features is, of course, not only relevant to nonparametric spectral analysis, as described in Section 3.3.2, but to parametric (AR-based) analysis as well. As a rule, spectral parameters employed in the nonparametric case cannot be directly applied to a parametric power spectrum. For example, it may be tempting to use the amplitude of a spectral

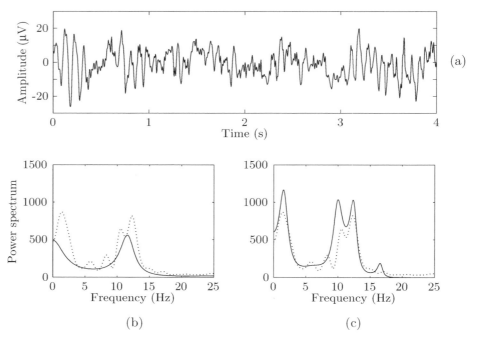

Figure 3.22: (a) A 4-s EEG signal acquired at a sampling rate of 128 Hz. Assuming a model order of $p = 10$, the parameters of the corresponding AR power spectrum are estimated by Burg's method from (b) the original signal and (c) the signal that results from three times decimation, i.e., a sampling rate of 128/3 Hz. For comparison, the power spectrum obtained by Welch's method is displayed (dotted line).

peak as a power measure of a certain EEG rhythm. While such a measure is meaningful for a nonparametric power spectrum, it is inaccurate in characterizing a parametric spectrum [131].[11] Another important reason for developing spectral parameters is that properties inherent to AR modeling can be exploited.

For an AR model, the definition of spectral parameters has its starting point in the complex power spectrum $S_x(z)$ for the stochastic process $x(n)$ given by

$$S_x(z) = \frac{\sigma_v^2}{A(z)A(z^{-1})}. \tag{3.171}$$

---

[11]It can be shown that the peak value is proportional to the square of the power when the modeled signal is sinusoidal, whereas the area under the peak component is linearly proportional to the power of the sinusoidal [131, p. 202].

This expression follows from the fact that the noise-shaping filter is $H(z) = 1/A(z)$ and that the power spectrum of the output signal is given by $S_x(z) = \sigma_v^2 H(z) H(z^{-1})$, cf. (3.19). Since $A(z)$ is a rational function of z, $S_x(z)$ can be expressed in terms of its poles d_j as follows:

$$S_x(z) = \frac{\sigma_v^2}{\displaystyle\prod_{j=1}^{p}(1 - d_j z^{-1})(1 - d_j^* z)}. \tag{3.172}$$

The polynomial $A(z)$ includes the poles d_j located inside the unit circle, i.e., with $|d_j| < 1$ for all j, assuring stability of the filter $H(z)$, and $A(z^{-1})$ includes those which are outside. In order to assure that a power spectrum is an even function, the poles of $S_x(z)$ must exhibit symmetry with respect to the real axis of the z-plane: a pole at $re^{j\phi}$ is always paired with its complex-conjugate at $re^{-j\phi}$, where $0 \leq r \leq 1$ and $-\pi \leq \phi \leq \pi$ [129]. We restrict ourselves to only considering models with even-valued orders p and with distinct poles, i.e., they are all of order one.[12] Consequently, the poles are always complex-conjugated pairs,

$$d_{2i} = d_{2i-1}^*, \quad i = 1, \ldots, p/2. \tag{3.173}$$

One approach to characterize the power spectrum $S_x(z)$ in terms of its "intrinsic" spectral components is to obtain the *partial fraction expansion* of $H(z)$ with which $H(z)$ can be expressed as a sum of first-order terms [113],

$$H(z) = \frac{1}{A(z)} = \sum_{j=1}^{p} \frac{c_j}{1 - d_j z^{-1}}. \tag{3.174}$$

The coefficients c_j are found from

$$c_j = (1 - d_j z^{-1})H(z)\big|_{z=d_j}, \quad j = 1, \ldots, p. \tag{3.175}$$

Considering that the AR model is defined by complex-conjugated pole pairs, we can alternatively express the partial fraction expansion as a sum of second-order transfer functions $H_i(z)$,

$$H(z) = \sum_{i=1}^{p/2} H_i(z). \tag{3.176}$$

---

[12]The special case of an odd-valued model order p is not considered here, but can be easily handled if necessary, see [113, p. 191]; the same applies to the presence of multiple-order (indistinct) poles.

Each $H_i(z)$ defines a second-order ARMA model since the numerator of $H_i(z)$ will include a zero:

$$
\begin{aligned}
H_i(z) &= \frac{c_{2i-1}}{1 - d_{2i-1}z^{-1}} + \frac{c_{2i}}{1 - d_{2i}z^{-1}} \\
&= \frac{(c_{2i-1} + c_{2i}) - (c_{2i-1}d_{2i} + c_{2i}d_{2i-1})z^{-1}}{(1 - d_{2i-1}z^{-1})(1 - d_{2i}z^{-1})} \\
&= \frac{2\Re(c_{2i}) - 2\Re(c_{2i}d_{2i}^*)z^{-1}}{(1 - d_{2i-1}z^{-1})(1 - d_{2i}z^{-1})},
\end{aligned}
\tag{3.177}
$$

where $i = 1, \ldots, p/2$. The last expression in (3.177) results from the fact that complex-conjugate poles, such as those in (3.173), always result in complex-conjugate coefficients in the partial fraction expansion, i.e., $c_{2i} = c_{2i-1}^*$. Since the first-order numerator polynomial of $H_i(z)$ has real-valued coefficients, its zero is located on the real axis.

In the following, we will assume that the second-order transfer functions $H_i(z)$ of the partial fraction expansion that characterize $S_x(z)$ are essentially nonoverlapping,

$$
H_i(z)H_j(z^{-1}) \approx 0, \quad i \neq j.
\tag{3.178}
$$

From a practical viewpoint, this assumption is not very restrictive since our interest is in characterizing different EEG rhythms that are spectrally well-separated from each other. If the components exhibit considerable overlap, it no longer makes sense to determine spectral parameters such as peak frequency and bandwidth since these are ill-defined. Using (3.178), the complex power spectrum can be expressed as

$$
\begin{aligned}
S_x(z) &= H(z)H(z^{-1})\sigma_v^2 \\
&= \sigma_v^2 \sum_{i=1}^{p/2} \sum_{j=1}^{p/2} H_i(z)H_j(z^{-1}) \\
&\approx \sigma_v^2 \sum_{i=1}^{p/2} H_i(z)H_i(z^{-1}) = \sum_{i=1}^{p/2} S_{x_i}(z).
\end{aligned}
\tag{3.179}
$$

Figure 3.23 illustrates the decomposition of a two-component power spectrum using the above partial fraction expansion approach; each component is characterized by a second-order transfer function $H_i(z)$. From Figure 3.23(a) it is evident that the difference between the original power spectrum $S_x(e^{j\omega})$ and the power spectrum that results from summation of the two components,

$$
S_{x_1}(e^{j\omega}) + S_{x_2}(e^{j\omega}) = |H_1(e^{j\omega})|^2 \sigma_v^2 + |H_2(e^{j\omega})|^2 \sigma_v^2,
$$

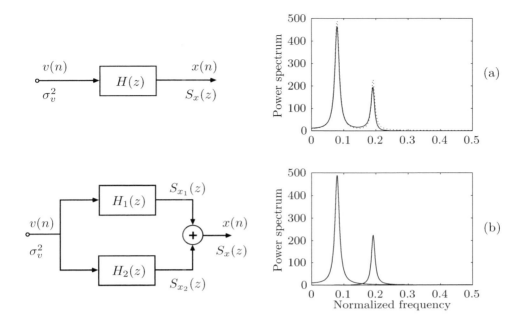

Figure 3.23: Decomposition of an AR power spectrum using partial fraction expansion. (a) The AR model and the corresponding power spectrum $S_x(e^{j\omega})$ (solid line). The spectrum has two dominant frequencies defined by two pairs of complex-conjugate poles. (b) Decomposition of the model into two second-order systems connected in parallel and their corresponding power spectra. For comparison, the power spectrum that results from summation of the two components, i.e., $|H_1(e^{j\omega})|^2\sigma_v^2 + |H_2(e^{j\omega})|^2\sigma_v^2$, has been included in the diagram of (a) as a dotted line.

is negligible. Thus, the spectral error arising from the assumption of nonoverlapping components can be neglected.

We will now present the calculation of the parameters that describe the power, the peak frequency, and the bandwidth for each of the spectral components,

$$S_{x_i}(z) = H_i(z)H_i(z^{-1})\sigma_v^2, \quad i = 1, \ldots, p/2, \tag{3.180}$$

related to the expansion in (3.179). The power $r_{x_i}(0)$ can be calculated by the z-transform integral inversion formula which relates $r_{x_i}(k)$ to $S_{x_i}(z)$ [129, p. 162],

$$r_{x_i}(k) = \frac{1}{2\pi J} \oint_C S_{x_i}(z)z^{k-1}dz, \tag{3.181}$$

where the contour of integration C is assumed to be within the region of convergence (here taken as the unit circle). Hence, the power is obtained from

$$r_{x_i}(0) = \frac{1}{2\pi \jmath} \oint_C S_{x_i}(z) z^{-1} dz$$

$$= \sum_{l=0}^{1} \text{Res} \left[S_{x_i}(z) z^{-1}, d_{2i-l} \right]. \tag{3.182}$$

The term $\text{Res} \left[S_{x_i}(z) z^{-1}, d_{2i-l} \right]$ denotes the residue of $S_{x_i}(z) z^{-1}$ at $z = d_{2i-l}$ and is determined by

$$\text{Res} \left[S_{x_i}(z) z^{-1}, d_{2i-l} \right] = \lim_{z \to d_{2i-l}} (z - d_{2i-l}) S_{x_i}(z) z^{-1}. \tag{3.183}$$

Evaluating (3.182) for the poles of $S_{x_i}(z)$ that are located within the unit circle, it can be shown that the power P_i is given by (see Problem 3.13)

$$P_i = r_{x_i}(0)$$

$$= \frac{8\sigma_v^2}{1 - |d_{2i}|^2} \left[\Re \left(\frac{\Re^2(c_{2i}) + \Re^2(c_{2i}d_{2i}^*) - \Re(c_{2i})\Re(c_{2i}d_{2i}^*)(d_{2i} + d_{2i}^{-1})}{1 + |d_{2i}|^2 - d_{2i}^2 - d_{2i}^* d_{2i}^{-1}} \right) \right]. \tag{3.184}$$

In certain situations, it may be more appropriate to use a normalized power measure,

$$P_i' = \frac{P_i}{r_x(0)}, \tag{3.185}$$

where $r_x(0)$ denotes the total power of $S_x(z)$.

The peak frequency can be determined in a number of ways, of which perhaps the most straightforward is to search for the peak location of $S_{x_i}(e^{\jmath\omega})$. A disadvantage of this approach is that $S_{x_i}(e^{\jmath\omega})$ must be computed at a sufficiently fine resolution before a meaningful peak search can be carried out. A simpler approach is to estimate the peak frequency by calculating the phase angle of the pole at d_i,

$$\omega_i = \arctan \left(\frac{\Im(d_i)}{\Re(d_i)} \right). \tag{3.186}$$

This is a useful estimate as long as the pole is relatively close to the unit circle.

In order to find the exact location of the peak frequency, however, it is necessary to determine the particular ω_i that maximizes the expression

$$S_{x_i}(e^{\jmath\omega}) = \left| \frac{2\Re(c_{2i}) - 2\Re(c_{2i}r_i e^{-\jmath\phi_i})e^{-\jmath\omega}}{(1 - r_i e^{-\jmath\phi_i} e^{-\jmath\omega})(1 - r_i e^{\jmath\phi_i} e^{-\jmath\omega})} \right|^2 \sigma_v^2. \tag{3.187}$$

Since the zero is real-valued and the pole pair of interest is close to the unit circle, it can be shown that this expression has its maximum at the minimum of its denominator, which is

$$\omega_i = \arccos\left(\frac{1 + r_i^2}{2r_i} \cos\phi_i\right), \tag{3.188}$$

see Problem 3.14. It is obvious that ω_i in (3.188) approaches ϕ_i as r_i approaches one. The frequency ω_i is undefined for poles located at the origin, i.e., for $r_i = 0$, because it corresponds to a flat spectrum; a case which is highly unlikely to occur in practice.

Finally, the 3-dB bandwidth $\Delta\omega_i$ of $S_{x_i}(e^{j\omega})$ can, for values of r_i close to one, be approximated by the following simple expression [113, p. 342],

$$\Delta\omega_i \approx 2(1 - r_i). \tag{3.189}$$

Spectral decomposition based on partial fraction expansion of $S_x(z)$ was originally developed for EEG analysis by Zetterberg and coworkers [39, 40, 146] and later used by others [137, 147]; this technique is often referred to as *spectral parameter analysis* (SPA). The SPA technique is exemplified here by findings from an animal study in which the EEG activity was monitored during brain injury, caused by progressively reduced oxygenation of the blood (hypoxia), leading to complete lack of oxygen (asphyxia) [137]. Figure 3.24(a) shows the AR power spectrum obtained from a 3.3-s interval prior to the induction of hypoxia. Figure 3.24(b) presents the trends of the three most dominant frequencies during the first minute of asphyxia, whereas Figure 3.24(c) presents their individual power relative to the value estimated prior to oxygen reduction. It is evident from these trends that the initial phase of asphyxia is accompanied by a considerable increase in relative power of the two upper frequency components of the EEG. At about 50 s, the relative power in all three spectral components decrease to zero as one would expect at asphyxia.

3.5   EEG Segmentation

Spectral analysis of EEG signals, using any of the earlier described techniques, is normally preceded by manual intervention in order to assure that only segments without "nonstationary events" are analyzed. This procedure is not really feasible for the analysis of long recordings since these frequently contain several different types of activities. It would therefore be helpful to develop a technique that can divide the EEG into segments in such a way that segments with similar spectral characteristics can be grouped together.

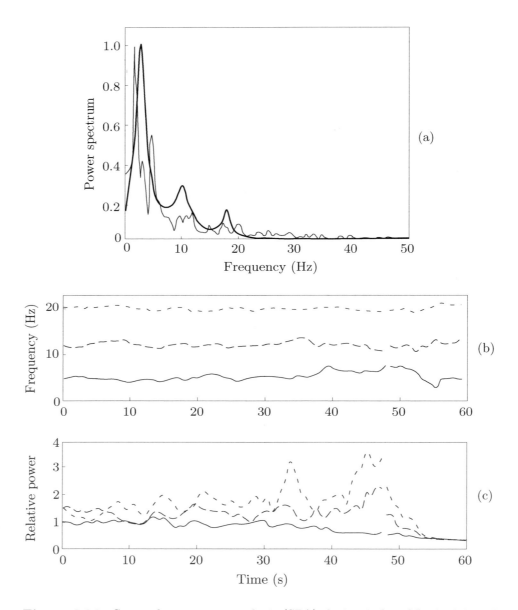

Figure 3.24: Spectral parameter analysis (SPA) during induced brain injury in piglets. (a) The AR power spectrum (heavy line) is computed by Burg's method, and the periodogram is computed by using a Hanning window. (b) Trends displaying the peak frequencies of the three most dominant spectral components during the first minute of asphyxia, and (c) trends displaying the power relative to the value before oxygen reduction. (Reprinted from Goel et al. [137] with permission.)

The output of the method would provide a quantitative and compact description of the whole recording and would serve as a basis for subsequent EEG assessment by the neurophysiologist. In particular, segmentation may facilitate the discovery of brief, diagnostically significant episodes in long-term EEG recordings, which otherwise may have been overlooked during visual scrutiny.

A popular approach to solving the segmentation problem is based on the assumption that the EEG is composed of a series of consecutive, variable-length segments, each with stationary signal properties. This assumption suggests that methods of detecting changes in the EEG should exploit the second-order statistics (i.e., spectral properties) of the signal, either in terms of a nonparametric or model-based description. The detection of changes requires that two time windows be defined: the *reference window* and the *test window*. An estimate of the signal statistics is obtained from the reference window and compared to the statistics of the sliding test window by means of a *dissimilarity measure*, here denoted $\Delta(n)$. The reference window can either be of fixed length or of a length that increases as long as no change is detected, whereas the sliding test window typically has a fixed length, see Figure 3.25. The measure $\Delta(n)$ reflects changes in signal statistics between the reference and the test window and is designed so that it essentially remains at a constant level until a change occurs, after which $\Delta(n)$ rapidly increases. A segment boundary is detected at a time $n = n_1$ when $\Delta(n)$ exceeds the threshold level η,

$$\Delta(n_1) > \eta. \qquad (3.190)$$

Once a boundary has been detected, the segmentation procedure is restarted by redefining the test window as the reference window, and a new test window is defined after the reference window. The criterion in (3.190) is often combined with another test in which $\Delta(n)$ is required to exceed the threshold η for a certain duration before a boundary is declared as detected [148].

In designing an algorithm for detecting changes in EEG signal properties, several aspects should be considered.

- The activity should remain stationary for at least a second in order to allow for accurate parameter estimation. Transient waveforms occurring in the test window should be eliminated before such estimation can take place.

- A change should be sufficiently abrupt in order to be detected. Even so, the change is detected with a certain time delay; therefore, a correction, for example, implemented as a "backtracking" procedure in $\Delta(n)$, may be required to find the exact time at which the change occurred.

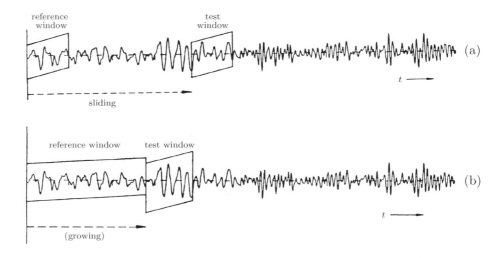

Figure 3.25: Principles for EEG segmentation based on (a) fixed-size reference and test windows and (b) a growing-size reference window and fixed-sized test window. A difference in spectral properties of the two windows constitutes the basis for change detection. (Reprinted from Appel and van Brandt [149] with permission.)

- Detector performance should be studied in theoretical terms as well as by means of simulated signals. However, the ultimate performance evaluation must be related to answering the following questions. How well does the method agree with the neurophysiologist's interpretation? Does the segmentation approach make sense from a clinical point of view? Unfortunately, these two issues are not easily integrated into the development of signal processing methodology.

Below we will present two methods for EEG segmentation which involve tests on spectral changes. Further insight into the problem of detecting abrupt signal changes can be obtained from the book by Basseville and Nikiforov which provides a theoretical basis to the signal segmentation problem [150], see also [151]. From a historical viewpoint, it is interesting to remember that EEG analysis was one of the very first applications dealing with signal segmentation, prompting subsequent theoretical development.

3.5.1   Spectral Error Measure—The Periodogram Approach

The definition of the periodogram given earlier in (3.79) is now slightly modified to include a running time index n, which indicates the interval during which the spectrum is estimated. The running periodogram,

$$S_x(e^{\jmath\omega}, n) = \sum_{k=-\infty}^{\infty} r_x(k, n) e^{-\jmath\omega k}, \tag{3.191}$$

is based on samples contained in the interval $[n, n + N - 1]$ used to estimate the autocorrelation function by

$$\hat{r}_x(k, n) = \begin{cases} \dfrac{1}{N} \displaystyle\sum_{l=0}^{N-1-k} x(l + n + k)x(l + n), & k = 0, \ldots, N - 1; \\ 0, & k = N, N + 1, \ldots. \end{cases} \tag{3.192}$$

The spectrum of the reference window is denoted $S_x(e^{\jmath\omega}, 0)$ and the spectrum of the sliding test window is denoted $S_x(e^{\jmath\omega}, n)$, i.e, the running periodogram in (3.191). The parameters N_r and N_t denote the length of the reference and test window, respectively. It is assumed that the reference window has a fixed length, although it is simple to modify the method to use a reference window with increasing length. The running periodogram in (3.191) can be viewed as a precursor of the methods for joint time–frequency analysis described in Section 3.6.

The mainstream engineering approach to defining a dissimilarity measure $\Delta(n)$ would probably be to study a spectral error measure defined as the integral of the squared error between the difference of the spectra of the test and reference windows,

$$\frac{1}{2\pi} \int_{-\pi}^{\pi} \left(S_x(e^{\jmath\omega}, n) - S_x(e^{\jmath\omega}, 0) \right)^2 d\omega.$$

A major disadvantage of such a measure, however, is its asymmetry with respect to the detection of signals having either increasing or decreasing power. This property can be illustrated by an example where the change in signal power is proportionally the same while the shape of the power spectrum is held fixed,

$$S_x(e^{\jmath\omega}, n) = \alpha S_x(e^{\jmath\omega}, 0)$$

and

$$S_x(e^{\jmath\omega}, n) = \frac{1}{\alpha} S_x(e^{\jmath\omega}, 0),$$

where α is a positive-valued constant. Inserting these two spectra into the error measure above, it becomes immediately evident that an increase in power is rewarded more than a decrease and is consequently easier to detect.

A simple way to remedy this deficiency is to normalize the spectral error measure with respect to the signal power of both the test and the reference windows such that

$$\Delta_1(n) = \frac{\dfrac{1}{2\pi}\displaystyle\int_{-\pi}^{\pi}(S_x(e^{j\omega},n) - S_x(e^{j\omega},0))^2 d\omega}{\dfrac{1}{4\pi^2}\displaystyle\int_{-\pi}^{\pi}S_x(e^{j\omega},n)d\omega\int_{-\pi}^{\pi}S_x(e^{j\omega},0)d\omega}. \qquad (3.193)$$

It is easily verified that the normalized definition of the spectral error measure in (3.193) handles the above example with changing signal power in a symmetric way.

In EEG analysis, the implementation of (3.193) has been based on its time domain counterpart [152],

$$\Delta_1(n) = \frac{\displaystyle\sum_{k=-\infty}^{\infty}(r_x(k,n) - r_x(k,0))^2}{r_x(0,n)r_x(0,0)}, \qquad (3.194)$$

replacing the autocorrelation function estimate in (3.192) with $r_x(k)$. The numerator of (3.194) is obtained using Parseval's theorem, which states that

$$\sum_{n=-\infty}^{\infty}x^2(n) = \frac{1}{2\pi}\int_{-\pi}^{\pi}|X(e^{j\omega})|^2 d\omega. \qquad (3.195)$$

The expression in (3.194) lends itself more easily to an efficient, recursive update of the test window statistics $r_x(k,n)$ than does $S_x(e^{j\omega},n)$. Recalling from (3.78) that $r_x(k,n)$ is estimated through (although now modified by the running time n)

$$\hat{r}_x(k,n) = \frac{1}{N_t}\sum_{l=0}^{N_t-1-k}x(l+n+k)x(l+n), \quad k = 0,\ldots,N_t-1, \qquad (3.196)$$

it is easily shown that $\hat{r}_x(k,n)$ can be computed from $\hat{r}_x(k,n-1)$ by correcting for the contributions of the oldest and the most recent samples,

$$\hat{r}_x(k,n) = \hat{r}_x(k,n-1)$$
$$+ \frac{1}{N_t}(x(n-1+N_t)x(n-1+N_t-k) - x(n-1+k)x(n-1)). \qquad (3.197)$$

A corresponding recursive relationship between $S_x(e^{j\omega},n)$ and $S_x(e^{j\omega},n-1)$ is, on the other hand, not readily available. The computation of $\Delta_1(n)$ can be

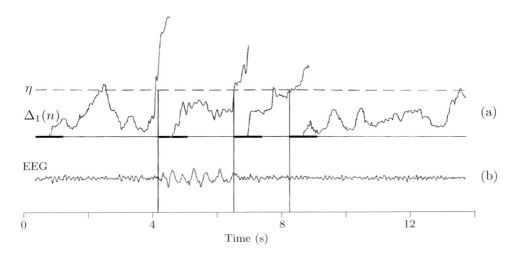

Figure 3.26: Segmentation of the EEG based on the spectral error measure $\Delta_1(n)$. (a) $\Delta_1(n)$, the detection threshold η (dashed line), and the detected segment boundaries (vertical lines), which in this example is three. The heavy bars below $\Delta_1(n)$ indicate the location of the reference windows immediately following the detected boundaries. (b) The EEG signal containing a paroxysmal event. (Reprinted from Bodenstein et al. [152] with permission.)

further simplified by truncating the sum in (3.194) so as to include only the N' shortest—and the most reliably estimated—correlation lags ($N' < N$). Equally important, truncation makes sense from a clinical viewpoint since inclusion of too many correlation lags has been found to "increase the sensitivity to insignificant signal fluctuations and lead to hypersegmentation" [152]. Correlation-based segmentation was first introduced in [153] using a dissimilarity measure closely related to that in (3.194), see also [154].

Electroencephalographic segmentation resulting from the use of $\Delta_1(n)$ is demonstrated by the example in Figure 3.26. In this case, a boundary is detected whenever $\Delta_1(n)$ exceeds the threshold η for at least 400 ms. The definitive position of the segment boundary is determined empirically by searching backwards from the threshold crossing time for the onset of the slope in $\Delta_1(n)$. Of the three detected boundaries, the last one was judged to be incorrect.

3.5.2 Spectral Error Measure—The Whitening Approach

In the previous section, segmentation was based on nonparametric estimation of the signal statistics. However, it can also be based on parametric estimation in which AR power spectra are inserted into the spectral error

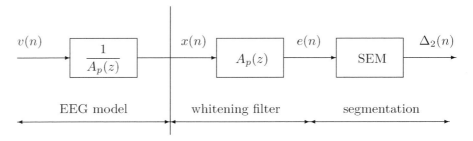

Figure 3.27: Segmentation of the EEG based on AR modeling and linear prediction. The parameters of $A_p(z)$ are estimated from $x(n)$ and are used for whitening of $x(n)$. The segmentation is based on the spectral error measure (SEM) $\Delta_2(n)$ which tests the whiteness of the prediction errors $e(n)$.

measure in (3.193). Such a straightforward approach to model-based segmentation has rarely been considered, but the validity of an AR model, estimated from the reference window, is rather tested by linear prediction in the sliding test window; the intimate relationship between AR modeling and linear prediction has already been pointed out in Section 3.4. This technique relies on the fact that a pth-order linear prediction error filter $A_p(z)$ decorrelates ("whitens") the observed signal $x(n)$ as long as it is described by the pth-order AR model for which the predictor was designed. Once a spectral change occurs in $x(n)$, the output of the linear prediction error filter no longer remains a white process, see Figure 3.27. From this observation, it seems well-motivated to develop a segmentation method based on a dissimilarity measure that reflects deviations from the whiteness of the prediction errors $e(n)$. In fact, several such methods have been presented in the literature, and below we will describe the method first developed for EEG segmentation [15].

Assuming that a suitable model order p has been chosen to describe the signal in the reference window, the prediction error variance σ_e^2 can be estimated by any of the AR methods presented in Section 3.4. The power spectrum of $e(n)$ is flat with a height given by σ_e^2,

$$S_e(e^{\jmath\omega}, 0) = r_e(0, 0) = \sigma_e^2. \tag{3.198}$$

Thus, this spectrum serves as a reference for subsequent spectra $S_e(e^{\jmath\omega}, n)$ determined from the output of the prediction error filter $A_p(z)$. In analogy with the definition of the spectral error measure $\Delta_1(n)$ in (3.193), we define a quadratic spectral error measure $\Delta_2(n)$ in order to quantify deviations between $S_e(e^{j\omega}, n)$ and $S_e(e^{j\omega}, 0)$ [15, 155, 156],

$$\Delta_2(n) = \frac{\dfrac{1}{2\pi} \displaystyle\int_{-\pi}^{\pi} \left(S_e(e^{j\omega}, n) - \sigma_e^2\right)^2 d\omega}{\left[\dfrac{1}{2\pi} \displaystyle\int_{-\pi}^{\pi} S_e(e^{j\omega}, n) d\omega\right]^2}. \tag{3.199}$$

Note that the normalization factor in (3.199) differs from that of $\Delta_1(n)$ but represents the factor suggested in the original work [15]. Again, it is advantageous to develop a time domain expression for $\Delta_2(n)$ which is better suited for implementation. In order to arrive at such an expression, we make use of the fact that any power spectrum, being an even function, can be expressed as a sum of cosine functions,

$$S_e(e^{j\omega}, n) = \sum_{k=-\infty}^{\infty} r_e(k, n) e^{-j\omega k}$$

$$= r_e(0, n) + 2 \sum_{k=1}^{\infty} r_e(k, n) \cos(\omega k). \tag{3.200}$$

The integral for symmetric functions over one period is equal to zero and thus

$$\int_{-\pi}^{\pi} r_e(k, n) \cos(\omega k) d\omega = 0. \tag{3.201}$$

Furthermore, since the cosine functions constitute an orthogonal set of functions,

$$\frac{1}{2\pi} \int_{-\pi}^{\pi} \cos(\omega k) \cos(\omega l) d\omega = \begin{cases} \dfrac{1}{2}, & k = l; \\ 0, & k \neq l, \end{cases} \tag{3.202}$$

we can express $\Delta_2(n)$ in terms of $r_e(k, n)$,

$$\Delta_2(n) = \left(\frac{r_e(0,0)}{r_e(0,n)} - 1\right)^2 + \frac{2}{r_e^2(0,n)} \sum_{k=1}^{\infty} r_e^2(k, n). \tag{3.203}$$

It is obvious that prediction errors $e(n)$ deviating from a white process with variance $r_e(0,0)$ will be penalized in (3.203). The first term is close to zero as long as the variance $r_e(0, n)$ of the test window is close to that of the reference window $r_e(0,0)$. The second term remains close to zero only as long as $e(n)$ remains white as $r_e(k, n) = \sigma_e^2 \delta(k)$, where $\delta(k)$ is the unit impulse function.

An even simpler test for the detection of a change in spectral characteristics is obtained by neglecting the second term [150, 157, 158] and by modifying the first term in (3.203) so that only the power of the prediction errors is monitored,

$$\Delta_3(n) = \frac{r_e(0,n)}{r_e(0,0)} - 1, \tag{3.204}$$

which, after insertion of the correlation function estimate of the prediction errors, is given by

$$\Delta_3(n) = \frac{1}{N_t} \sum_{k=0}^{N_t-1} \left(\frac{e^2(n+k)}{r_e(0,0)} - 1 \right). \tag{3.205}$$

This dissimilarity measure can be derived within a more rigorous statistical framework [157]; unfortunately, the performance of (3.205) suffers from a serious deficiency in that several AR models may exist with a prediction error variance identical to $r_e(0,0)$. Furthermore, the measure $\Delta_3(n)$ represents an asymmetric detection function since it is more sensitive to abrupt changes manifested by an increase in signal power than by a decrease.

The above two error measures based on model-based segmentation incorporate knowledge of one single signal model, i.e., the AR model is estimated from the reference window and then used for prediction in the test window. Neither of the two measures is symmetric with respect to the detection of signals with increasing or decreasing power. An ad hoc approach to modifying $\Delta_3(n)$ so that it becomes a symmetric test is to extend (3.205) to incorporate a "reverse" test, i.e., to also compute the prediction errors in the reference window using the AR parameters estimated from the test window [159]. Thus, a modified error measure $\Delta_4(n)$ involving knowledge on two AR models can be defined as in (3.205) but by adding a second term,

$$\Delta_4(n) = \frac{1}{N_t} \sum_{k=0}^{N_t-1} \left(\frac{e_r^2(n+k)}{r_{e_r}(0,0)} - 1 \right) + \frac{1}{N_r} \sum_{k=0}^{N_r-1} \left(\frac{e_t^2(k)}{r_{e_t}(0,n)} - 1 \right), \tag{3.206}$$

where $e_t(n)$ and $e_r(n)$ denote the prediction errors obtained by AR parameter estimates from the test and the reference windows, respectively. The corresponding prediction error variances are denoted $r_{e_t}(0,n)$ and $r_{e_r}(0,0)$. Simulations have indicated a better, more symmetric behavior for $\Delta_4(n)$ than for $\Delta_3(n)$ [159]; however, the clinical value of $\Delta_4(n)$ for EEG segmentation remains to be established.

3.5.3   Segmentation Performance and Beyond

The performance of a correlation-based method [153], closely related to the one given in (3.194), and that of the prediction-based method in (3.203) were compared using different types of simulated signals [149]. The signals were generated by switching between two AR models with different power spectra, both defined by a pair of complex-conjugated poles. Since the positions in time of the segment boundaries were known a priori, it was possible to determine the detection rate, the false alarm rate, and the accuracy in positioning the segment boundaries.

The simulation results showed that the overall performance of the two methods was quite similar, although the prediction-based method was associated with a lower false alarm rate [149]. It is interesting to contrast these findings on false alarm rate with those reported by Bodenstein and Praetorius, who originally developed the prediction-based method but later retracted from it because it too often led to misplaced segment boundaries and hypersegmentation. They replaced the prediction-based method by the much simpler correlation-based method using $\Delta_1(n)$, since it was found to be "better adapted to the practical needs of the clinical neurophysiologist" [148, 152, 160].

The discrepancy in performance results obtained using simulated signals and EEG signals can, perhaps, serve as an example of the difficulty in assessing the performance of methods by studying only simulation signals. Despite this difficulty, the problem of segmenting EEG signals continues to receive attention using spectral features and dissimilarity measures other than those described above [161–164].

Once a method of EEG segmentation with an acceptable performance has been devised, it will be of considerable interest to group multiple EEG patterns based on the outcome of the automated segmentation procedure. Pattern classification must make use of a set of features, for example, the N' shortest correlation lags [160], various spectral features [158] (the power spectrum being estimated with either a nonparametric or parametric technique), or the estimated AR model parameters. Classification produces a number of pattern classes which reflect different types of activities in the EEG recording and which ultimately may be presented to the clinician in the form of a summary sheet.

3.6 Joint Time–Frequency Analysis

A major limitation of Fourier-based spectral analysis is its inability to provide information on *when* in time different frequencies of a signal occur. The Fourier transform only reflects *which* frequencies exist during the total observation interval, because the Fourier transform integrates frequency components over the total observation interval. While such spectral analysis is adequate for stationary signals whose frequencies, on average, are equally spread in time, it is inadequate for nonstationary signals with time-dependent spectral content. Similar observations apply to parametric spectral analysis. Consequently, there is strong motivation for the development of methods that analyze signals with regard to both time and frequency so that the frequencies present at each instant in time can be displayed. Joint time–frequency information has been found extremely valuable not only in the interpretation of EEG signals, but equally so for many other types of biomedical signals exhibiting nonstationary characteristics.

Methods that produce signal representations in the time–frequency domain may be divided into the following three main categories.

1. *Linear, nonparametric* methods have in common that their time–frequency representations can be obtained from a linear filtering operation. Among these methods, the short-time Fourier transform is the classical one, which is described and exemplified in Section 3.6.1. The wavelet transform is another popular method which belongs to the category of linear methods which, due to its suitability for characterizing transient signals, will be described in Chapter 4.

2. *Quadratic, nonparametric* methods offer improved time–frequency resolution. The Wigner–Ville distribution (WVD) is the most well-known and is, together with a number of modifications introduced to address its limitations, described in Sections 3.6.2–3.6.4. Similar to the methods of the first category, these methods do not involve any particular assumptions regarding the signal.

3. *Parametric* methods produce time–frequency representations based on the assumption that the observed signal derives from a statistical model with time-varying parameters, usually with the AR model as the starting point. The methods previously described for AR parameter estimation are modified in Section 3.6.5 so that slow changes in the parameters can be tracked. The resulting parameter estimates are used for computing successive power spectra.

While the purpose of this section is restricted to presenting some introductory concepts in time–frequency analysis, the reader should also be aware that dozens of methods have been presented in this area, with each method having its own particular merits. The interested reader is therefore referred to the textbooks available on this topic which more profoundly deal with the theoretical aspects and contain a variety of applications [165–170].

When presenting the nonparametric time–frequency methods below, we have temporarily abandoned the discrete-time context since the continuous-time description of the WVD is more easily comprehended. Accordingly,

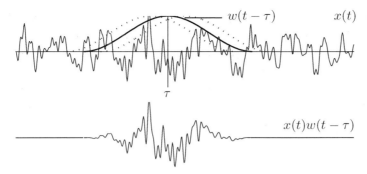

Figure 3.28: In the short-time Fourier transform, a sliding time window $w(\tau - t)$ is used for excerpting successive parts of the signal $x(t)$.

details related to digital implementation of different time–frequency algorithms are omitted, see instead [165, 168]. While it is straightforward to translate the continuous-time, short-time Fourier transform to its discrete-time counterpart, the WVD requires some further considerations. Another difference in the presentation below is that signals are considered deterministic rather than stochastic, which otherwise is our common assumption. A stochastic framework for the WVD has been presented [171], although it is far less often adopted in the literature.

3.6.1 The Short-Time Fourier Transform

In its simplest form, time–frequency analysis can be carried out by dividing the signal $x(t)$ into short, consecutive, possibly overlapping, segments which are subjected to spectral analysis, see Figure 3.28. The resulting series of spectra reflect the time-varying nature of the signal. The most well-known approach to time–frequency analysis makes use of nonparametric, Fourier-based spectral analysis (cf. Section 3.3) applied to each of the short segments—an operation referred to as the *short-time Fourier transform* (STFT). In this approach, the definition of the Fourier transform is modified so that a sliding time window $w(t)$ is included that defines each time segment to be analyzed, thus resulting in a two-dimensional function $X(t, \Omega)$ defined by

$$X(t, \Omega) = \int_{-\infty}^{\infty} x(\tau) w(\tau - t) e^{-j\Omega\tau} d\tau, \qquad (3.207)$$

where Ω denotes analog frequency. The STFT is said to produce a time–frequency representation of $x(t)$ or, equivalently, a time–frequency distribution of $x(t)$ (where the word "distribution" is not used in the probabilistic sense). The time window $w(t)$ is a positive-valued function which may have a rectangular shape, although it is often chosen from among the shapes previously described in nonparametric spectral analysis on page 96. The window length determines the resolution in time and frequency such that a short window yields good time resolution but poor frequency resolution, and the opposite when a long window is used. The actual computation of the STFT in (3.207) can be implemented by means of a linear filtering operation (Problem 3.19).

Analogous to the computation of the periodogram in (3.81), which was obtained as the squared magnitude of the Fourier transform of the signal, the *spectrogram* of $x(t)$ is obtained by computing the squared magnitude of the STFT in (3.207),

$$S_x(t, \Omega) = |X(t, \Omega)|^2. \tag{3.208}$$

The spectrogram is a real-valued, nonnegative distribution which provides a signal representation in the time–frequency domain. The stochastic (and discrete-time) version of the spectrogram has, in fact, already been introduced for the purpose of segmenting nonstationary signals; it was then called the running periodogram and was defined in (3.191).

Spectral analysis based on the Fourier transform can never achieve perfect resolution in both time and frequency. It is possible to derive an uncertainty principle which states that the product of a signal's time duration Δ_t and its bandwidth Δ_Ω must never be below a certain lower bound. In order to express this principle in mathematical terms, we introduce the following definitions of Δ_t and Δ_Ω,

$$\Delta_t = \left(\frac{\int_{-\infty}^{\infty} (t - \bar{t})^2 |x(t)|^2 dt}{\int_{-\infty}^{\infty} |x(t)|^2 dt} \right)^{\frac{1}{2}}, \tag{3.209}$$

$$\Delta_\Omega = \left(\frac{\int_{0}^{\infty} (\Omega - \overline{\Omega})^2 |X(\Omega)|^2 d\Omega}{\int_{0}^{\infty} |X(\Omega)|^2 d\Omega} \right)^{\frac{1}{2}}, \tag{3.210}$$

where $X(\Omega)$ denotes the Fourier transform of $x(t)$. Both definitions are analogous to the definition of the standard deviation of a random variable, and thus Δ_t and Δ_Ω represent measures of width in time and frequency, respectively. The parameters $\bar{t}$ and $\overline{\Omega}$ define the center point ("center-of-gravity") of $x(t)$ and $X(\Omega)$, respectively, and are obtained in a way analogous to the mean value of a random variable,

$$\bar{t} = \frac{\displaystyle\int_{-\infty}^{\infty} t|x(t)|^2 dt}{\displaystyle\int_{-\infty}^{\infty} |x(t)|^2 dt}, \tag{3.211}$$

$$\overline{\Omega} = \frac{\displaystyle\int_{0}^{\infty} \Omega|X(\Omega)|^2 d\Omega}{\displaystyle\int_{0}^{\infty} |X(\Omega)|^2 d\Omega}. \tag{3.212}$$

With the above definitions of Δ_t and Δ_Ω, it can be shown that if $x(t)$ decays to zero such that [168][13]

$$\lim_{|t|\to\infty} \sqrt{t}x(t) = 0, \tag{3.213}$$

then the *uncertainty principle* states that

$$\Delta_t \Delta_\Omega \geq \frac{1}{2}. \tag{3.214}$$

Equality is only achieved when $x(t)$ is a Gaussian signal,

$$x(t) = ce^{-(t-t_0)^2/\sigma}, \tag{3.215}$$

where σ denotes a width parameter, and c and t_0 are constants. It should be emphasized that the above uncertainty principle, with its particular width definitions, only applies to Fourier-based spectral analysis, while other bounds apply to quadratic time–frequency representations.

The spectrogram has been used in a wide range of EEG applications thanks to its capability of lucidly displaying changes that occur in the rhythmic activities of the EEG which otherwise would be difficult to perceive from the time domain signal. For example, patients with suspected epilepsy undergoing a photic stimulation test may or may not demonstrate an EEG response to such stimulation (see page 42). By looking at the spectrogram it becomes immediately clear whether such a response is present or not. Figure 3.29 shows a spectrogram obtained during photic stimulation which, for this particular subject, contains a well-defined spectral peak at each of the investigated stimulation rates. The display format in Figure 3.29 is often referred to as a *compressed spectral array* (CSA) since successive spectra, one in front of the other, are compactly presented [172]. As we will see shortly, other useful display formats are available.

---

[13]This requirement is not very restrictive since we are dealing with deterministic signals assumed to have finite energy. In fact, most signals with finite energy fulfil this assumption.

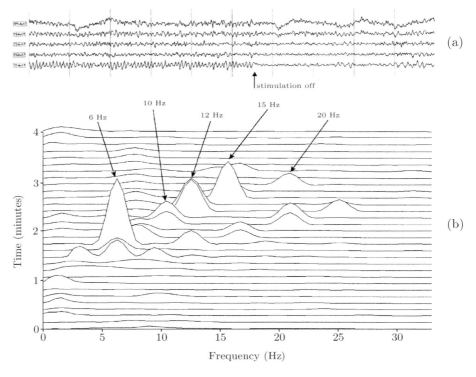

Figure 3.29: (a) Ten-second excerpt of a four-channel EEG recorded during photic stimulation at 12 Hz; the EEG signals recorded during the other stimulation rates (6, 10, 15, and 20 Hz) have been omitted. (b) The spectrogram, resulting from the five rates, exhibits a marked peak at each rate. The power spectra are computed from consecutive, 8-s intervals. (Reprinted from Scheuer [173] with permission.)

Another example which illustrates the usefulness of the spectrogram is taken from heart surgery of an infant during which brain activity was monitored, see Figure 3.30(a) [174]. Drastic spectral changes can be observed in the infant's EEG at the time when a decrease in blood pressure occurred; this decrease can be established from the simultaneously recorded measurements of diastolic pressure[14] shown in Figure 3.30(c). The changes in the EEG are primarily manifested by a reduction in the power in the interval above 7–8 Hz (Figure 3.30(b)). This reduction is probably caused by a lack of oxygenated blood which normally perfuses the brain.

While the EEG changes shown in Figures 3.29 and 3.30 took place over a time span of one or several minutes, spectral changes may evolve much more

---

[14]The diastolic blood pressure represents the pressure in the arteries when the heart is at rest.

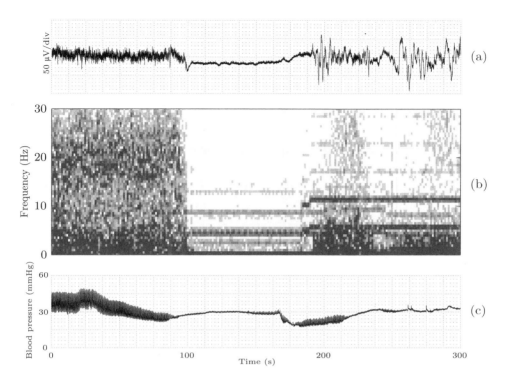

Figure 3.30: (a) The EEG recorded during heart surgery of an infant. (b) The corresponding spectrogram displays a drastic reduction in high-frequency content after 100 s, partially reverting at about 200 s. (c) The blood pressure measurements show that a reduction in diastolic blood pressure precedes the reduction in high-frequency content.

rapidly on a second-to-second basis. The latter situation is illustrated by the 12-s EEG shown in Figure 3.31(a), recorded during an epileptic seizure. Visually, one can easily establish that the temporary increase in amplitude is coupled to a more sinusoidal behavior of the signal. However, it is almost impossible for the human eye to perceive the slowing down in frequency which occurs in the interval from 4 to 8 s. By analyzing the corresponding spectrogram in Figure 3.31(b), it becomes immediately evident that the dominant frequency component, initially located around 9 Hz at 4 s, gradually declines to 6 Hz during the following 4 s. Finally, the dominant frequency dissolves into a pattern which resembles the pattern observed prior to the seizure episode. Although the physiological significance of this particular behavior of "frequency slowing" is not clear, the example illustrates nonetheless the fact that time–frequency analysis is a powerful technique for revealing nonstationary signal properties which would otherwise have remained hidden in classical spectral analysis.[15]

Figure 3.31(b)–(d) presents the spectrograms that result from using different lengths of the time window when analyzing the EEG in Figure 3.31(a). The spectrogram in Figure 3.31(c) is obtained with the longest time window and therefore exhibits the poorest time resolution of the three lengths; this property is reflected by a ridge which extends longer in time than does the ridge in Figure 3.31(d). On the other hand, the best frequency resolution is found in Figure 3.31(c) due to the longer time window, while the frequency resolution in Figure 3.31(d) is worse. It is clear from this example that one is always faced with a trade-off with respect to resolution in time and frequency.

3.6.2   The Ambiguity Function

While the spectrogram has been found to be very useful in many biomedical applications, its relatively poor time–frequency resolution has prompted the development of other techniques which better describe the time-varying nature of a signal. The WVD is probably the most well-known technique that addresses this issue and is not constrained by the uncertainty principle in (3.214) related to the Fourier transform [178–182]. The original definition of the WVD, which actually appeared already in the 1930s [178], has more recently been followed by several approaches with the common aim to mitigate its drawbacks [167, 168].

Rather than proceeding directly to the definition of the WVD, we will first introduce the *ambiguity function* $A_x(\nu, \tau)$ which is central to the WVD definition and which constitutes an important concept in time–frequency analysis. This function is designed to reflect uncertainty in both time τ and frequency ν associated with a signal $x(t)$. In order to account for such uncertainty, we introduce two versions of $x(t)$ which are shifted in time and frequency,

$$x(t; \nu, \tau) = x(t - \frac{\tau}{2})e^{-j\nu t/2}, \tag{3.216}$$

$$x(t; -\nu, -\tau) = x(t + \frac{\tau}{2})e^{j\nu t/2}. \tag{3.217}$$

A frequency shift of $x(t)$ is introduced by modulating $x(t)$ with the complex exponential $e^{j\nu t}$. The ambiguity function is a two-dimensional function

---

[15]During epileptic seizures, a monotonic frequency slowing behavior, similar to the one observed in Figure 3.31, has been mentioned in several studies which are based on observations from time–frequency analysis [94, 175–177]. The time–frequency distribution may be used as a basis for developing a detector which finds "frequency ridges" characteristic of certain EEG patterns. For example, it has been hypothesized that frequency ridges present during a seizure can be modeled by a piecewise linear function [94].

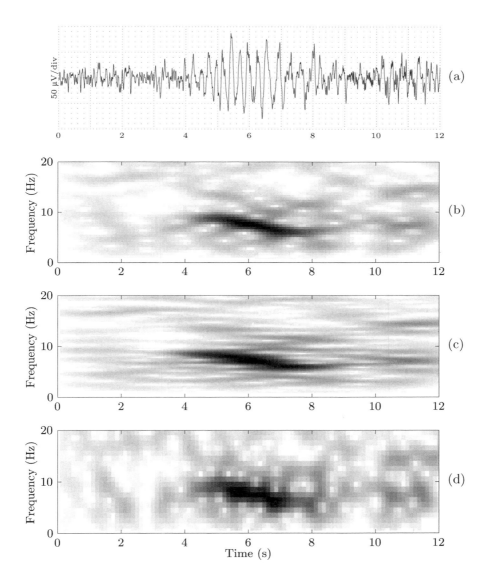

Figure 3.31: (a) The EEG at the onset of an epileptic seizure. The corresponding spectrogram is computed using a Hamming window with a length of (b) 1 s, (c) 2 s, and (d) 0.5 s.

defined as the correlation between the time and frequency shifted signal versions of which one is conjugated [183],

$$A_x(\tau, \nu) = \int_{-\infty}^{\infty} x^*(t; \nu, \tau) x(t; -\nu, -\tau) dt$$

$$= \int_{-\infty}^{\infty} x^*(t - \frac{\tau}{2}) x(t + \frac{\tau}{2}) e^{j\nu t} dt. \tag{3.218}$$

The ambiguity function can be understood as the Fourier transform in the variable t of the product $x(t - \frac{\tau}{2}) x^*(t + \frac{\tau}{2})$ describing the deterministic, instantaneous correlation of two values, separated by the time lag τ.

The definition of $A_x(\tau, \nu)$ exhibits several interesting properties. For example, the Fourier transform of $A_x(\tau, 0)$ with respect to τ yields the energy density spectrum $S_x(\Omega)$,

$$S_x(\Omega) = \int_{-\infty}^{\infty} A_x(\tau, 0) e^{-j\Omega\tau} d\tau$$

$$= \int_{-\infty}^{\infty} \left(\int_{-\infty}^{\infty} x^*(t) x(t + \tau) dt \right) e^{-j\Omega\tau} d\tau, \tag{3.219}$$

where the expression within parenthesis defines a correlation function of $x(t)$. The spectrum $S_x(\Omega)$ in (3.219) can be viewed as the deterministic counterpart to the power spectrum of a random signal. Another interesting observation is that the Fourier transform of $A_x(\tau, \nu)$ yields the instantaneous, time-dependent correlation,

$$\frac{1}{2\pi} \int_{-\infty}^{\infty} A_x(\tau, \nu) e^{-j\nu t} d\nu = x^*(t - \frac{\tau}{2}) x(t + \frac{\tau}{2}). \tag{3.220}$$

An important property of the ambiguity function is that its maximum is always located at the origin $(0, 0)$ of the τ-ν domain and is equal to the energy of $x(t)$ defined as

$$A_x(0, 0) = \int_{-\infty}^{\infty} |x(t)|^2 dt. \tag{3.221}$$

It can also be shown that $A(\tau, \nu)$ remains concentrated to the origin of the τ-ν domain, although $x(t)$ has been subjected to a shift in time and frequency.

In order to investigate the concentration property, it is helpful to first consider a signal composed of a lowpass envelope modulated by a cosine at frequency Ω_1,

$$x(t) = s(t) \cos(\Omega_1 t). \tag{3.222}$$

This signal can be decomposed into the components $x_1(t)$ and $x_2(t)$, representing positive and negative frequency, respectively,

$$x(t) = x_1(t) + x_2(t), \qquad (3.223)$$

where

$$x_1(t) = \frac{1}{2}s(t)e^{j\Omega_1 t}, \qquad (3.224)$$

$$x_2(t) = \frac{1}{2}s(t)e^{-j\Omega_1 t}. \qquad (3.225)$$

The corresponding ambiguity function becomes

$$A_x(\tau, \nu) = \int_{-\infty}^{\infty} \left(x_1^*(t - \frac{\tau}{2}) + x_2^*(t - \frac{\tau}{2}) \right) \left(x_1(t + \frac{\tau}{2}) + x_2(t + \frac{\tau}{2}) \right) e^{j\nu t} dt$$

$$= \frac{1}{2}A_s(\tau, \nu)\cos(\Omega_1 \tau) + \frac{1}{4}A_s(\tau, \nu - 2\Omega_1) + \frac{1}{4}A_s(\tau, \nu + 2\Omega_1),$$

$$(3.226)$$

where

$$A_s(\tau, \nu) = \int_{-\infty}^{\infty} s^*(t - \frac{\tau}{2})s(t + \frac{\tau}{2})e^{j\nu t} dt. \qquad (3.227)$$

Apart from the "auto-term" $\frac{1}{2}A_s(\tau, \nu)\cos(\Omega_1 \tau)$, being concentrated to the origin, the ambiguity function also includes two undesirable terms with identical shape but translated $\pm 2\Omega_1$, reflecting an undesired cross-correlation between positive and negative frequencies.

Fortunately, it is possible to remove such cross-correlation without sacrificing signal information. Since real signals have symmetric frequency components, of which one is redundant, we only need to consider positive frequencies of the spectrum. This part can be isolated by using the *analytic signal* $x_A(t)$ of $x(t)$, which in the frequency domain is defined as

$$X_A(\Omega) = \begin{cases} 2X(\Omega), & \Omega \geq 0; \\ 0, & \Omega < 0. \end{cases} \qquad (3.228)$$

For $x(t) = s(t)\cos(\Omega_1 t)$, we have that

$$X_A(\Omega) = \begin{cases} S(\Omega - \Omega_1), & \Omega \geq 0; \\ 0, & \Omega < 0, \end{cases} \qquad (3.229)$$

where $S(\Omega)$ denotes the Fourier transform of $s(t)$. The computation of the analytic signal is later described in Section 7.4.3. Employing the analytic signal, the resulting ambiguity function $A_{x_A}(\tau, \nu)$ no longer contains the two terms at $\pm 2\Omega_1$.

For the previous example in (3.223)–(3.225), the analytic signal is given by

$$x_A(t) = s(t)e^{j\Omega_1 t} \tag{3.230}$$

and

$$A_{x_A}(\tau, \nu) = A_s(\tau, \nu)e^{j\Omega_1 \tau}. \tag{3.231}$$

From now on, it will therefore be assumed that the analytic signal $x_A(t)$, rather than the original real signal $x(t)$, is analyzed although the denotation $x(t)$ is still used.

We now consider the case of a signal composed of two different components,

$$x(t) = x_1(t) + x_2(t), \tag{3.232}$$

where

$$x_1(t) = s_1(t)e^{j\Omega_1 t}, \tag{3.233}$$
$$x_2(t) = s_2(t)e^{j\Omega_2 t}. \tag{3.234}$$

Thus, the ambiguity function becomes

$$A_x(\tau, \nu) = \int_{-\infty}^{\infty} \left(x_1^*(t - \frac{\tau}{2}) + x_2^*(t - \frac{\tau}{2}) \right) \left(x_1(t + \frac{\tau}{2}) + x_2(t + \frac{\tau}{2}) \right) e^{j\nu t} dt$$
$$= A_{x_1}(\tau, \nu) + A_{x_2}(\tau, \nu) + \text{cross-terms}. \tag{3.235}$$

Apart from the two terms depending only on the individual signals (the so-called auto-terms), the ambiguity function in (3.235) also includes cross-terms which reflect the correlation between $x_1(t)$ and $x_2(t)$. The separation of these two types of terms in the τ-ν domain turns out to be a very important property to exploit when certain limitations of the WVD are to be addressed (Section 3.6.4).

The properties of the ambiguity domain are illustrated by a signal consisting of two components, $x_1(t)$ and $x_2(t)$, whose envelopes are of Gaussian shape. The two components are shifted in time by t_1 and t_2, respectively, and have center frequencies at Ω_1 and Ω_2, respectively,

$$x_1(t) = e^{-(t-t_1)^2/\sigma}e^{j\Omega_1 t}, \tag{3.236}$$
$$x_2(t) = e^{-(t-t_2)^2/\sigma}e^{j\Omega_2 t}, \tag{3.237}$$

see Figure 3.32(a). Hence, the two components are concentrated at the points (t_1, Ω_1) and (t_2, Ω_2), respectively, in the t-Ω domain—a fact which hopefully is reflected by any method designed for time–frequency analysis. In the τ-ν domain, on the other hand, $A_{x_1}(\tau, \nu)$ and $A_{x_2}(\tau, \nu)$ are both concentrated at the origin ($|A_{x_1}(\tau, \nu)|^2$ is displayed in Figure 3.32(b)), while the two cross-terms can be shown to have their concentration at the points $(-(t_2-t_1), -(\Omega_2-\Omega_1))$ and $(t_2-t_1, \Omega_2-\Omega_1)$, respectively, see Figure 3.32(c). The cross-terms are thus clearly separated from the auto-terms at the origin.

3.6.3 The Wigner–Ville Distribution

The continuous-time definition of the WVD is given by the two-dimensional Fourier transform of the recently introduced ambiguity function $A_x(\tau, \nu)$,

$$W_x(t, \Omega) = \frac{1}{2\pi} \int_{-\infty}^{\infty} \int_{-\infty}^{\infty} A_x(\tau, \nu) e^{-j\nu t} e^{-j\Omega\tau} d\nu d\tau. \tag{3.238}$$

The definition of the WVD can be given a more intuitively appealing form by inserting the definition of $A_x(\tau, \nu)$ in (3.218) into (3.238), yielding

$$W_x(t, \Omega) = \frac{1}{2\pi} \int_{-\infty}^{\infty} \int_{-\infty}^{\infty} \int_{-\infty}^{\infty} x^*(s - \frac{\tau}{2}) x(s + \frac{\tau}{2}) e^{-j\nu(t-s)} e^{-j\Omega\tau} ds d\nu d\tau$$

$$= \int_{-\infty}^{\infty} x^*(t - \frac{\tau}{2}) x(t + \frac{\tau}{2}) e^{-j\Omega\tau} d\tau. \tag{3.239}$$

This expression shows more explicitly how the WVD is related to $x(t)$, namely, the WVD is obtained by computing the Fourier transform of the instantaneous, time-dependent correlation $x^*\left(t - \frac{\tau}{2}\right) x\left(t + \frac{\tau}{2}\right)$ with respect to the time lag variable τ. The WVD is referred to as a quadratic time–frequency representation because $x(t)$ enters quadratically into the definition in (3.239). Moreover, it is obvious from the definition that the WVD is a nonlocal transform since it weighs times which are far apart equally to those which are close. The WVD can alternatively be expressed as a correlation in the frequency domain by invoking Parseval's relation,

$$W_x(t, \Omega) = \int_{-\infty}^{\infty} X^*(\Omega - \frac{\nu}{2}) X(\Omega + \frac{\nu}{2}) e^{-j\nu t} d\nu. \tag{3.240}$$

The WVD possesses several attractive properties of which we will mention the most important ones. The WVD is *real-valued* since

$$W_x(t, \Omega) = W_x^*(t, \Omega). \tag{3.241}$$

Time support is a desirable property of any time–frequency distribution and means that if $x(t) = 0$ for $|t| > t_0$, then $W_x(t, \Omega) = 0$ for $|t| > t_0$; expressed

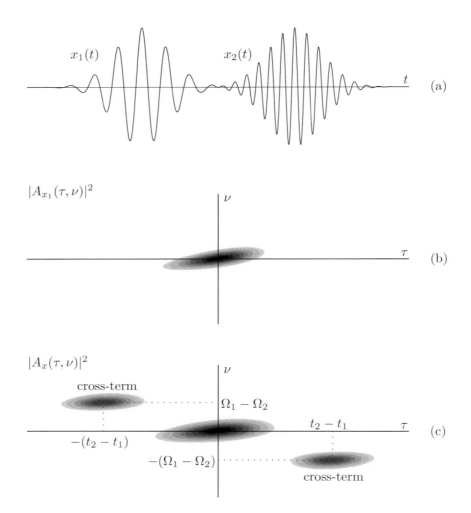

Figure 3.32: (a) The signal $x(t)$ is the sum of two modulated Gaussians $x_1(t)$ and $x_2(t)$. Using the analytic signal representation, the ambiguity domain is shown for (b) $x_1(t)$ and (c) $x(t) = x_1(t) + x_2(t)$, where the auto-term and the two cross-terms are well-separated from one another.

in words this means that if the signal is zero outside an interval, then the distribution is also zero outside an interval. In a similar way, the frequency support property means that if $X(\Omega) = 0$ for $|\Omega| > \Omega_0$, then $W_x(t, \Omega) = 0$ for $|\Omega| > \Omega_0$. While the WVD has support both in time and frequency, the spectrogram does not.

Two other properties of the WVD are related to what happens when the WVD is integrated with respect to time or frequency. It can be shown that by summing up the WVD for all values of t, we obtain, what one would hope for, the total spectrum $|X(\Omega)|^2$,

$$\int_{-\infty}^{\infty} W_x(t, \Omega)dt = |X(\Omega)|^2. \qquad (3.242)$$

This condition is usually referred to as the *frequency marginal condition*. The corresponding operation with respect to time yields the instantaneous signal energy $|x(t)|^2$,

$$\frac{1}{2\pi}\int_{-\infty}^{\infty} W_x(t, \Omega)d\Omega = |x(t)|^2, \qquad (3.243)$$

and is referred to as the *time marginal condition*.

The most important property of the WVD, and certainly the main reason why the WVD has received so much attention, is its excellent joint resolution in time and frequency [167]. This property can be established by introducing a different uncertainty principle, applicable to both the WVD and the spectrogram [184]. Since the derivation of this result is beyond the scope of this text, the improved resolution is instead illustrated by an example in which the WVD is compared to the spectrogram using the modulated, Gaussian signal in (3.236), i.e., a monocomponent signal. It can be observed from Figure 3.33 that the corresponding spectral component is much more concentrated in the WVD than in the spectrogram.

Unfortunately, the WVD exhibits some undesirable properties which are important to be aware of. In contrast to the spectrogram, the WVD is not always a positive-valued distribution even if it is supposed to reflect signal energy in the time–frequency domain (the only signal for which the WVD is always positive-valued is the Gaussian modulated chirp). In practice, this property is not a particularly serious problem since the positive-valued regions of the WVD have been found to correspond well with the time–frequency structure one would expect [167].

Another, more annoying property is the presence of cross-terms which arise because the WVD is a quadratic time–frequency distribution. For the two-component signal in (3.232), it can be shown that the WVD satisfies

$$W_x(t, \Omega) = W_{x_1}(t, \Omega) + W_{x_2}(t, \Omega) + 2\Re\{W_{x_1, x_2}(t, \Omega)\}, \qquad (3.244)$$

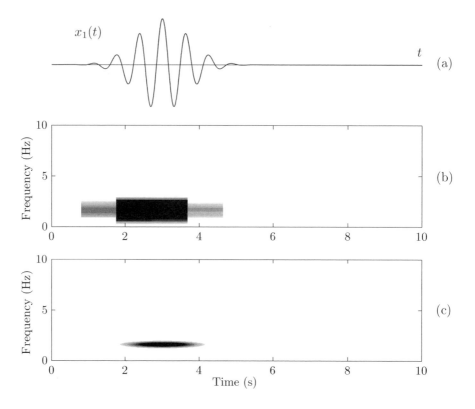

Figure 3.33: (a) The monocomponent signal $x_1(t)$ in (3.236) and the corresponding (b) spectrogram and (c) WVD.

where the first two terms are quadratic in each signal component, while the last term is a cross-term between $x_1(t)$ and $x_2(t)$, defined by

$$W_{x_1,x_2}(t,\Omega) = \int_{-\infty}^{\infty} x_1^*(t - \frac{\tau}{2})x_2(t + \frac{\tau}{2})e^{-j\Omega\tau}d\tau. \qquad (3.245)$$

The function $W_{x_1,x_2}(t,\Omega)$ is referred to as the *cross Wigner–Ville distribution*. While quadratic distributions provide good concentration of a signal in the time–frequency domain, the presence of cross-terms limits their practical use when multicomponent signals are encountered.

The presence of cross-terms is illustrated for the two-component signal consisting of two modulated Gaussians shown in Figure 3.34(a). The corresponding WVD is presented in Figure 3.34(c) where the cross-term in between the two components can be clearly discerned (striped pattern); for comparison, the spectrogram is shown in Figure 3.34(b). In this case, the cross-term occurs midway between the frequencies Ω_1 and Ω_2 at $(\Omega_1+\Omega_2)/2$ and has an amplitude oscillating with frequency $(\Omega_1 - \Omega_2)$. When $x(t)$ con-

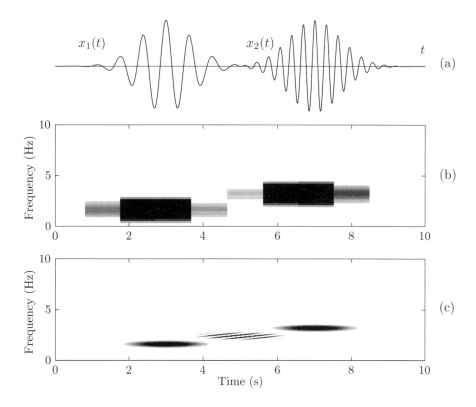

Figure 3.34: (a) The two-component signal of Figure 3.32 and the corresponding (b) spectrogram and (c) WVD.

sists of more than two components, one cross-term will arise for every pair of components, thereby causing substantial difficulties in interpreting the time–frequency distribution.

From the definition of the WVD in (3.239) it is clear that every point in time of a signal is assigned equal importance, and, therefore, the WVD produces a highly nonlocal description of a signal. In certain cases, however, it may be desirable to emphasize the signal properties around time t and thus to deemphasize times which are farther away. Such weighting can be done by multiplying the instantaneous, time-dependent correlation by a window $w(\tau)$ centered around the time $\tau = 0$,

$$\breve{W}_x(t, \Omega) = \int_{-\infty}^{\infty} x^*(t - \frac{\tau}{2}) x(t + \frac{\tau}{2}) w(\tau) e^{-j\Omega\tau} d\tau. \qquad (3.246)$$

This modification of the WVD is usually referred to as the *pseudo Wigner–Ville distribution* (PWVD), or simply the windowed WVD.

In situations where $x(t)$ is a monocomponent signal, it may be of interest to extract the predominant frequency, i.e., the frequency which defines the main ridge of the time–frequency distribution. This information can be extremely valuable for a neurophysiologist since it describes the temporal variation of the dominant frequency. Such variation patterns often convey more valuable clinical information than do the fine details of the power spectrum at every time instant. In order to study the dominant frequency, we make use of the complex signal representation,

$$x(t) = s(t)e^{j\varphi(t)}, \tag{3.247}$$

where $s(t)$ denotes the magnitude ("envelope") modulated by a complex exponential function, defined by the phase $\varphi(t)$. Both magnitude and phase are real-valued functions. The first derivative of the phase, $\varphi'(t)$, is defined as the *mean instantaneous frequency*. This definition is reasonable since in the case when $s(t)$ is modulated by a cosine with frequency Ω_0,

$$\Re\{x(t)\} = s(t)\cos(\Omega_0 t) = \Re\{s(t)e^{j\Omega_0 t}\}, \tag{3.248}$$

the mean instantaneous frequency is equal to $\varphi'(t) = \Omega_0$.

Another attractive property of the WVD is that the mean instantaneous frequency $\varphi'(t)$ of a signal can be exactly determined from the WVD by computing the mean frequency $\overline{\Omega}(t)$ for each value of t. Similar to the computation of the mean $\overline{\Omega}$ in (3.212), we obtain that

$$\overline{\Omega}(t) = \frac{\displaystyle\int_{-\infty}^{\infty} \Omega W_x(t, \Omega)d\Omega}{\displaystyle\int_{-\infty}^{\infty} W_x(t, \Omega)d\Omega} = \varphi'(t). \tag{3.249}$$

A proof of this result can be found in [167, 168], see also Problem 3.21. It can be shown that the result in (3.249) does not carry over to the cases when $W_x(t, \Omega)$ is replaced by, e.g., the spectrogram or the PWVD, although the resulting $\overline{\Omega}(t)$ may still have some practical utility.

The computation of $\overline{\Omega}(t)$, and the related interpretation as mean instantaneous frequency, is meaningful for signals classified as monocomponent signals, exemplified by the EEG signal shown in Figure 3.31. By determining $\overline{\Omega}(t)$ in that example, variations in the signal's dominant frequency can be quantified so that we can easily determine, e.g., the size and duration of the decline in frequency during the epileptic event. However, the presence of multicomponent signals renders the analysis difficult since $\overline{\Omega}(t)$ looses its interpretation as mean instantaneous frequency.

3.6.4   General Time–Frequency Distributions—Cohen's Class

A general class of time–frequency distributions has been introduced whose degrees of freedom can be exploited for mitigating the cross-term problem [185]. A two-dimensional kernel function $g(\tau, \nu)$ weights the ambiguity function in such a way that undesired cross-terms, being far away from the origin, are suppressed, whereas the auto-terms remain essentially unaffected; cf. the properties of the ambiguity function illustrated in Figure 3.32(c). The general time–frequency distribution is defined as

$$C_x(t, \Omega) = \frac{1}{2\pi} \int_{-\infty}^{\infty} \int_{-\infty}^{\infty} g(\tau, \nu) A_x(\tau, \nu) e^{-j\nu t} e^{-j\Omega\tau} d\nu d\tau \qquad (3.250)$$

and is known as *Cohen's class* [167].

The time–frequency distributions earlier described are all members of Cohen's class. The time–frequency distribution $C_x(t, \Omega)$ simplifies to the definition of the WVD in (3.238) when the kernel function is chosen as

$$g(\tau, \nu) = 1. \qquad (3.251)$$

The PWVD results from choosing the kernel function as the window function $w(t)$ in (3.246),

$$g(\tau, \nu) = w(\tau). \qquad (3.252)$$

It can be shown that the spectrogram, defined by (3.207) and (3.208), also belongs to Cohen's class with the kernel function [183]

$$g(\tau, \nu) = \int_{-\infty}^{\infty} w^*(t - \frac{\tau}{2}) w(t + \frac{\tau}{2}) e^{j\nu t} dt, \qquad (3.253)$$

where $w(t)$ denotes the STFT window function in (3.207). Hence, the kernel function associated with the spectrogram can, in itself, be understood as an ambiguity function $A_w(\tau, \nu)$ of the window $w(t)$.

The suppression of cross-terms comes, however, at a price since $g(\tau, \nu)$ introduces smoothing of the WVD which, by necessity, reduces the time–frequency resolution. The smoothing effect is due to the multiplication of $A_x(\tau, \nu)$ with $g(\tau, \nu)$ in (3.250), thus corresponding to a two-dimensional convolution of the WVD with the Fourier transform of the kernel function given by

$$C_x(t, \Omega) = \frac{1}{2\pi} \int_{-\infty}^{\infty} \int_{-\infty}^{\infty} W_x(t', \Omega') G(t - t', \Omega - \Omega') dt' d\Omega'. \qquad (3.254)$$

Here, the two-dimensional Fourier transform of the kernel function is obtained by

$$G(t, \Omega) = \frac{1}{2\pi} \int_{-\infty}^{\infty} \int_{-\infty}^{\infty} g(\tau, \nu) e^{-j\nu t} e^{-j\Omega \tau} d\nu d\tau. \qquad (3.255)$$

This function can be viewed as a lowpass filter since $g(\tau, \nu)$ is assumed to have its maximum at origin with $g(0,0) = 1$, which assures that the energy is unaffected, cf. (3.221).

A kernel function $g(\tau, \nu)$ should exhibit some other properties to make the resulting time–frequency distribution attractive. The frequency marginal condition in (3.243) remains valid for time–frequency distributions in Cohen's class,

$$\int_{-\infty}^{\infty} C_x(t, \Omega) dt = |X(\Omega)|^2, \qquad (3.256)$$

when the kernel function satisfies the condition

$$g(\tau, 0) = 1. \qquad (3.257)$$

Similarly, the time marginal condition in (3.242) is satisfied when

$$g(0, \nu) = 1. \qquad (3.258)$$

In order to assure that the distribution is real-valued, i.e., $C_x(t, \Omega) = C_x^*(t, \Omega)$, we must also require that

$$g(\tau, \nu) = g(-\tau, -\nu). \qquad (3.259)$$

A large number of kernel functions have been presented [186], and we will here confine ourselves to only mentioning the most popular member of Cohen's class—the *Choi–Williams distribution* (CWD). This time–frequency distribution, extensively applied in biomedical signal processing, is defined by the exponential kernel [187, 188]

$$g(\tau, \nu) = e^{-\nu^2 \tau^2 / (4\pi^2 \sigma)}, \quad \sigma > 0, \qquad (3.260)$$

where σ is a design parameter that determines the degree of cross-term interference reduction and related smoothing effect. Figure 3.35 presents the shape of the exponential kernel for two different values of σ: for a small value of σ, the kernel is concentrated around the origin in the ambiguity domain (except for the τ and ν axes), whereas the resulting distribution tends to the WVD for large values of σ. With a suitable choice of σ, the kernel can be used to reduce the influence of cross-terms, while essentially

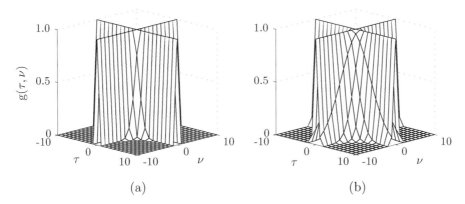

Figure 3.35: The exponential kernel function in (3.260) that defines the CWD plotted for two different values of σ, namely, (a) 0.1 and (b) 1.

preserving the auto-terms. It can be easily verified that the exponential kernel satisfies the marginal conditions in (3.257), (3.258), and (3.259).

The performance improvement associated with the CWD is illustrated in Figure 3.36 where the Gaussian two-component signal is analyzed again. It is evident that the cross-term of the WVD is now suppressed in the CWD. We also revisit the example in Figure 3.31 of an EEG signal, recorded at the onset of a seizure, by comparing the corresponding spectrogram, WVD, and CWD (Figure 3.37). It is noted that the abundance of cross-terms present in the WVD (Figure 3.37(c)) is largely suppressed in the CWD (Figure 3.37(d)). Of the three time–frequency representations, the CWD is thus the one which, for this particular example, gives the most clear-cut description of the spectral change.

The results presented in Figure 3.37(d) may also serve as an illustration of a problem associated with the exponential kernel that defines the CWD, as well as any kernel which fulfils the marginal conditions in (3.258) and (3.257). Since the exponential kernel does not offer any cross-term reduction along the τ-axis, or the ν-axis, the CWD contains as a result horizontal and vertical ripple. The horizontal ripple is due to auto-terms centered around the same frequency but occurring at different times, whereas the vertical ripple is due to auto-terms with the same time center but occurring at different frequencies. In Figure 3.37(d), some vertical ripple can be noted, although its presence is hardly decisive for the overall interpretation of the time–frequency analysis.

Another step taken to improve the performance of time–frequency analysis is to make use of a signal-dependent kernel rather than a fixed kernel such as the exponential one in (3.260). This step is motivated by the fact

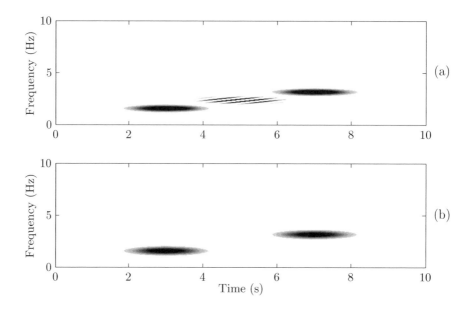

Figure 3.36: (a) The WVD and (b) the CWD for the two-component Gaussian signal in Figure 3.32. The parameter σ that determines the degree of cross-term interference reduction is equal to one.

that the locations of auto-terms and cross-terms in the ambiguity domain are dependent on the analyzed signal. With a signal-dependent kernel it is possible to reduce the presence of the above-mentioned horizontal/vertical ripple. The search for the "best" signal-dependent kernel has been an important research topic, and the reader is referred to [189–191] for details on how such kernels can be designed.

Time–frequency analysis has been extensively used in EEG signal processing for detection, characterization, and classification of epileptic signals, see, e.g., [176, 192–196]. Such analysis has also been considered for monitoring the depth of anesthesia; based on animal experiments, it was found that features derived from the CWD performed better than did features from traditional Fourier-based analysis [197]. Furthermore, time–frequency analysis has been used for investigating EEG signals recorded during sleep. The scope of such analysis may range from estimation of the instantaneous frequency for the purpose of sleep staging to the more advanced task of determining the anatomical location of cortical regions which generate sleep spindles [198–200].

Another application of time–frequency analysis in EEG signal processing is the evaluation of common energy between two signals, employing a cross time–frequency distribution. Such a distribution determines whether

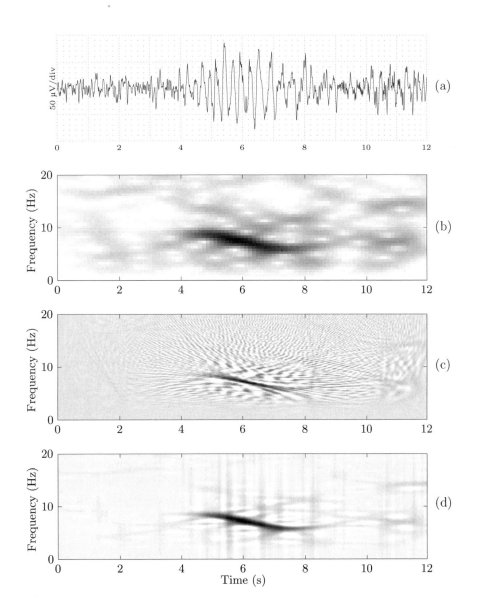

Figure 3.37: (a) The EEG at the onset of an epileptic seizure, previously shown in Figure 3.31(a), and the corresponding (b) spectrogram (imported from Figure 3.31(b)), (c) WVD, and (d) CWD.

the time–frequency components of one signal are related to the components of another signal. An example is the cross Wigner–Ville distribution $W_{x_1,x_2}(t, \Omega)$, defined in (3.245), which has been used for studying the relationship between pairs of EEG channels recorded during a seizure [193]. In such studies, the signals $x_1(t)$ and $x_2(t)$ are derived from two separate channels, rather than signifying two components of a single signal as was done in (3.232). In another study, an improved version of the "cross-spectrogram" was used for investigating signal transmission properties between two different regions of the brain [116, 201].

With several approaches to nonparametric time–frequency analysis available, the following question is inevitable: which type of time–frequency analysis is to be preferred? The answer to this question is unfortunately not easily delineated, although it is fair to say that the CWD, or a signal-dependent kernel, is often preferable thanks to its better time–frequency resolution. It is important to remember, however, that the spectrogram in its simplicity may be sufficiently adequate for analyzing certain types of biomedical signals. Recent research has dealt with the issue to define measures for evaluating the performance of different approaches to nonparametric, quadratic time–frequency analysis [202]. Although such measures are very helpful in comparing the performance when the observed signal is simulated, or available in a "stylized" form, the results cannot be directly translated to the performance on EEG signals but should be viewed as a guideline. The preferred time–frequency analysis is ultimately the one which yields the better performance in clinical terms.

3.6.5   Model-based Analysis of Slowly Varying Signals

Joint time–frequency analysis has, so far, been synonymous with nonparametric approaches. However, such analysis may as well be based on a parametric model of the signal. In this section, we will outline how the parameters of a time-varying AR model,

$$x(n) = -a_1(n)x(n-1) - \cdots - a_p(n)x(n-p) + v(n), \qquad (3.261)$$

can be estimated under the assumption that temporal variations are relatively slow. In addition to the parameters $a_1(n), \ldots, a_p(n)$, the input noise $v(n)$ is also assumed to be time-varying with variance $\sigma_v^2(n)$. Hence, the time-varying power spectrum is given by

$$S_x(e^{j\omega}, n) = \frac{\sigma_v^2(n)}{\left| 1 + \sum\limits_{k=1}^{p} a_k(n)e^{-j\omega k} \right|^2}, \qquad (3.262)$$

offering another alternative to time–frequency representation of the observed signal. It is often sufficient to compute the sequence of spectra in (3.262) for a subset of the samples, for example, for every fifth sample, which helps to reduce the amount of computations. However, methods which estimate the values of the AR parameters need to operate sequentially so that the estimator must nonetheless be updated every time a new sample becomes available.

We will present two adaptive methods below which minimize sequentially different definitions of the prediction error criterion. The first method minimizes the forward prediction error variance and leads to the LMS algorithm. The second method minimizes the forward and backward prediction error variances, assuming a lattice structure of the predictor, and leads to the gradient adaptive algorithm. In both cases, the gradient descent approach is employed to find the desired algorithm.

LMS Algorithm. The derivation of the LMS algorithm is done by first recalling the definition of the forward prediction error variance from (3.119),

$$E\left[e_p^2(n)\right] = E\left[\mathbf{a}_p^T \tilde{\mathbf{x}}_p(n)\tilde{\mathbf{x}}_p^T(n)\mathbf{a}_p\right].$$

This error is minimized sequentially by using the LMS algorithm, previously introduced in (3.48), which, with notation in terms of the linear predictor, becomes

$$\mathbf{a}_p'(n+1) = \mathbf{a}_p'(n) - \frac{1}{2}\mu\nabla_{\mathbf{a}_p'}E\left[e_p^2(n)\right], \tag{3.263}$$

where $\mathbf{a}_p'(n)$ denotes the coefficient vector with the fixed element $a_0 = 1$ excluded,

$$\mathbf{a}_p'(n) = \begin{bmatrix} a_1 \\ a_2 \\ \vdots \\ a_p \end{bmatrix}. \tag{3.264}$$

The error gradient is given by

$$\begin{aligned} \nabla_{\mathbf{a}_p}E\left[e_p^2(n)\right] &= 2E\left[\tilde{\mathbf{x}}_p(n)\tilde{\mathbf{x}}_p^T(n)\mathbf{a}_p\right] \\ &= 2E\left[e_p(n)\tilde{\mathbf{x}}_p(n)\right], \end{aligned} \tag{3.265}$$

where use has been made of (3.116) in order to obtain the second equality. The resulting LMS algorithm for estimating the predictor coefficients is then given by

$$\mathbf{a}_p'(n+1) = \mathbf{a}_p'(n) - \mu e_p(n)\tilde{\mathbf{x}}_p'(n), \tag{3.266}$$

where the expected value in (3.265) has been replaced by its instantaneous estimate, cf. (3.52); note that the vector $\tilde{\mathbf{x}}'_p(n)$ is defined similarly to $\mathbf{a}'_p(n)$ in (3.264). For the steady-state situation, the LMS algorithm converges if

$$0 < \mu < \frac{2}{\mathrm{tr}(\tilde{\mathbf{R}}_x)}. \tag{3.267}$$

Gradient Adaptive Lattice Algorithm. The other adaptive algorithm considered is related to Burg's method and is known as the *gradient adaptive lattice* (GAL) predictor. The derivation of the GAL algorithm is based on the error criterion in (3.140) and involves both forward and backward prediction error variances,

$$\sigma_e^2(n) = E\left[|e_j^+(n)|^2\right] + E\left[|e_j^-(n)|^2\right]. \tag{3.268}$$

The scalar update equation for each FIR lattice coefficient γ_j is again given by the gradient descent approach,

$$\gamma_j(n+1) = \gamma_j(n) - \frac{1}{2}\mu_j\frac{\partial\sigma_e^2(n)}{\partial\gamma_j}, \quad j = 1,\ldots,p. \tag{3.269}$$

The step size parameter μ_j is here allowed to be different for each lattice coefficient. From the lattice filter equations in (3.153) and (3.154) it follows that the gradient is

$$\frac{\partial\sigma_e^2(n)}{\partial\gamma_j} = -2\left(E\left[e_j^+(n)e_{j-1}^-(n-1)\right] + E\left[e_j^-(n)e_{j-1}^+(n)\right]\right). \tag{3.270}$$

Replacing the expected values with their corresponding instantaneous estimates, we obtain the GAL algorithm,

$$\gamma_j(n+1) = \gamma_j(n) + \mu_j\left(e_j^+(n)e_{j-1}^-(n-1) + e_j^-(n)e_{j-1}^+(n)\right). \tag{3.271}$$

It can be shown that $\gamma_j(n)$ converges to Burg's solution in (3.159) if the step size is chosen such that [103]

$$0 < \mu_j < \frac{2}{E\left[|e_{j-1}^+(n)|^2\right] + E\left[|e_{j-1}^-(n-1)|^2\right]}. \tag{3.272}$$

In order to compute the power spectrum from the resulting lattice coefficient estimates, we insert $\gamma_j(n)$ into the Levinson–Durbin recursion in (3.130) to obtain $a_1(n),\ldots,a_j(n)$, which then are used in (3.262).

Often, the step size μ_j is made time-varying and is chosen so that the gradient is normalized with respect to the total energy $\xi_j(n)$ of the forward

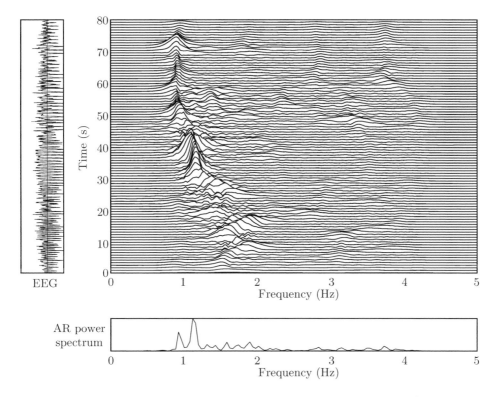

Figure 3.38: Time–frequency analysis of an EEG signal recorded during a seizure using a time-varying AR modeling approach (main panel). The left panel displays the EEG signal, and the bottom panel displays the power spectrum that results from the assumption of an AR model with time-invariant parameters. (Reprinted from Celka et al. [94] with permission.)

and backward prediction errors, i.e., $\mu_j(n) \sim 1/\xi_j(n)$. The energy $\xi_j(n)$ is usually defined by an exponentially weighted sum in which the most recent prediction errors are assigned the largest weights,

$$\xi_j(n) = \beta \sum_{i=1}^{n} (1-\beta)^{n-i} \left(|e_{j-1}^{+}(i)|^2 + |e_{j-1}^{-}(i-1)|^2 \right), \qquad (3.273)$$

where β denotes an exponential weighting factor ($0 < \beta < 1$). The weighted sum in (3.273) can be efficiently computed using the following recursion,

$$\xi_j(n) = \beta \xi_j(n-1) + (1-\beta) \left(|e_{j-1}^{+}(n)|^2 + |e_{j-1}^{-}(n-1)|^2 \right). \qquad (3.274)$$

Time–frequency analysis based on a time-varying AR model is illustrated in Figure 3.38 for an EEG signal recorded during a seizure. The time–frequency representation was obtained using a slightly modified version of the above LMS algorithm, assuming an AR model order of $p = 8$ [94]. Figure 3.38 shows that the dominant frequency of the seizure signal increases during the first 30 s, after which it declines rapidly in frequency during the next 10 s. The dominant frequency then reaches a steady state which lasts during the remaining 40 s of the recording. The graphical format of Figure 3.38 differs from those previously used for time–frequency representations in that both the signal and its power spectrum are included. This format illustrates clearly that the AR power spectrum provides insufficient information about the behavior of the signal. Finally, we note that several other studies have made use of time-varying AR models in the context of EEG signal processing [175, 203–205].

Bibliography

[1] T. Gasser, "General characteristics of the EEG as a signal," in *EEG Informatics. A Didactic Review of Methods and Applications of EEG Data Processing* (A. Rémond, ed.), pp. 37–55, Amsterdam: Elsevier Biomedical Press, 1977.

[2] F. H. Lopes da Silva, "Dynamics of EEGs as signals of neuronal populations: Models and theoretical considerations," in *Electroencephalography. Basic Principles, Clinical Applications and Related Fields* (E. Niedermayer and F. Lopes da Silva, eds.), pp. 76–92, Baltimore: Williams & Wilkins, 1999.

[3] J. S. Barlow, *The Electroencephalogram. Its Patterns and Origins.* Cambridge, Mass.: The MIT Press, 1993.

[4] B. H. Jansen, "Quantitative analysis of electroencephalograms: Is there chaos in the future," *Int. J. Biomed. Comput.*, vol. 27, pp. 95–123, 1991.

[5] W. S. Pritchard and D. W. Duke, "Measuring chaos in the brain: A tutorial review of nonlinear dynamical EEG analysis," *Int. J. Neurosci.*, vol. 67, pp. 31–80, 1992.

[6] A. Papoulis, *Probability, Random Variables, and Stochastic Processes.* New York: McGraw–Hill, 3rd ed., 1991.

[7] A. Siegel, "Stochastic aspects of the generation of the electroencephalogram," *J. Theor. Biol.*, vol. 92, pp. 317–339, 1981.

[8] M. G. Saunders, "Amplitude probability density studies on alpha and alpha-like patterns," *Electroencephal. Clin. Neurophysiol.*, vol. 15, pp. 761–767, 1963.

[9] J. Campbell, E. Bower, S. J. Dwyer, and G. V. Lado, "On the sufficiency of autocorrelation functions as EEG descriptors," *IEEE Trans. Biomed. Eng.*, vol. 14, pp. 49–52, 1967.

[10] R. Elul, "Gaussian behavior of the electroencephalogram," *Science*, vol. 164, pp. 328–331, 1969.

[11] G. Dumermuth, W. Walz, G. Scollo-Lavizzari, and B. Kleiner, "Spectral analysis of EEG activity during sleep stages in normal adults," *Eur. Neurol.*, vol. 7, pp. 265–296, 1971.

[12] J. A. McEwen and G. B. Anderson, "Modeling the stationarity and Gaussianity of spontaneous electroencephalographic activity," *IEEE Trans. Biomed. Eng.*, vol. 22, pp. 361–369, 1975.

[13] A. S. Gevins, "Correlation analysis," in *Handbook of Electroencephalography and Clinical Neurophysiology: Methods of Analysis of Brain Electrical and Magnetic Signals* (A. S. Gevins and A. Rémond, eds.), ch. 6, pp. 171–193, Amsterdam/New York: Elsevier, 1987.

[14] N. Kawabata, "A nonstationary analysis of the electroencephalogram," *IEEE Trans. Biomed. Eng.*, vol. 20, pp. 444–452, 1973.

[15] G. Bodenstein and H. M. Praetorius, "Feature extraction from electroencephalogram by adaptive segmentation," *Proc. IEEE*, vol. 65, no. 5, pp. 642–652, 1977.

[16] F. H. Lopes da Silva, K. van Hulten, J. G. Lommen, W. Storm van Leeuwen, C. W. M. van Veelen, and W. Vliegenhart, "Automatic detection and localization of epileptic foci," *Electroencephal. Clin. Neurophysiol.*, vol. 43, pp. 1–13, 1977.

[17] J. D. Frost, "Automatic recognition and characterization of epileptiform discharges in the human EEG," *J. Clin. Neurophysiol.*, vol. 2, pp. 231–250, 1985.

[18] K. Arakawa, D. H. Fender, H. Harashima, H. Miyakawa, and Y. Saitoh, "Separation of a nonstationary component from the EEG by a nonlinear digital filter," *IEEE Trans. Biomed. Eng.*, vol. 33, pp. 724–726, 1986.

[19] J. Qian and J. S. Barlow, "A simplified arithmetic detector for EEG sharp transients: Preliminary results," *IEEE Trans. Biomed. Eng.*, vol. 35, pp. 11–18, 1988.

[20] J. R. Glover Jr., N. Raghavan, P. Y. Ktonas, and J. D. Frost, "Context-based automated detection of epileptogene sharp transient in the EEG: Elimination of false positives," *IEEE Trans. Biomed. Eng.*, vol. 36, pp. 519–527, 1989.

[21] J. R. Glover Jr., P. Y. Ktonas, N. Raghavan, J. M. Urunuela, S. S. Velamuri, and E. L. Reilly, "A multichannel signal processor for the detection of epileptogenic sharp transients in the EEG," *IEEE Trans. Biomed. Eng.*, vol. 33, pp. 1121–1128, 1989.

[22] J. Gotman and L. Y. Wang, "State-dependent spike detection: Concepts and preliminary results," *Electroencephal. Clin. Neurophysiol.*, vol. 79, pp. 11–19, 1991.

[23] A. J. Gabor and M. Seyal, "Automated interictal EEG spike detection using artificial neural networks," *Electroencephal. Clin. Neurophysiol.*, vol. 83, pp. 271–280, 1992.

[24] A. A. Dingle, R. D. Jones, G. J. Carroll, and W. R. Fright, "A multistage system to detect epileptiform activity in the EEG," *IEEE Trans. Biomed. Eng.*, vol. 40, pp. 1260–1268, 1993.

[25] W. R. S. Webber, B. Litt, K. Wilson, and R. P. Lesser, "Practical detection of epileptiform discharges (ED's) in the EEG using an artificial neural network: A comparison of raw and parameterized EEG data," *Electroencephal. Clin. Neurophysiol.*, vol. 91, pp. 194–204, 1994.

[26] L. Senhadji, J. L. Dillenseger, F. Wendling, C. Rocha, and A. Kinie, "Wavelet analysis of EEG for three-dimensional mapping of epileptic events," *Ann. Biomed. Eng.*, vol. 23, pp. 543–551, 1995.

[27] C. Kurth, F. Gilliam, and B. Steinhoff, "EEG spike detection with a Kohonen feature map," *Ann. Biomed. Eng.*, vol. 28, pp. 1362–1369, 2000.

[28] M. Adjouadi, D. Sanchez, M. Cabrerizo, M. Ayala, P. Jayakar, I. Yaylali, and A. Barreto, "Interictal spike detection using the Walsh transform," *IEEE Trans. Biomed. Eng.*, vol. 51, pp. 868–872, 2004.

[29] T. Gasser and G. Dumermuth, "Non-Gaussianity and non-linearity in electroencephalographic time series," in *Stochastic Control Theory and Stochastic Differential Equations* (M. Kohlman and W. Vogel, eds.), pp. 373–386, Berlin: Springer, 1979.

[30] J. D. Bronzino, "Principles of electroencephalography," in *The Biomedical Engineering Handbook* (J. D. Bronzino, ed.), pp. 201–212, Boca Raton, FL: CRC Press, 1995.

[31] C. L. Nikias and A. P. Petropulu, *Higher-Order Spectral Analysis. A Nonlinear Signal Processing Framework.* New Jersey: Prentice-Hall, 1993.

[32] T. P. Barnett, L. C. Johnson, P. Naitoh, N. Hicks, and C. Nute, "Bispectrum analysis of electroencephalogram signals during waking and sleeping," *Science*, vol. 172, pp. 401–402, 1971.

[33] G. Dumermuth, P. J. Huber, B. Kleiner, and T. Gasser, "Analysis of interrelations between frequency bands of the EEG by means of the bispectrum—preliminary results," *Electroencephal. Clin. Neurophysiol.*, vol. 31, pp. 137–148, 1971.

[34] T. Ning and J. D. Bronzino, "Bispectral analysis of the rat EEG during various vigilance states," *IEEE Trans. Biomed. Eng.*, vol. 36, pp. 497–499, 1989.

[35] T. Ning and J. D. Bronzino, "Autoregressive bispectral analysis techniques: EEG applications," *IEEE Eng. Med. Biol. Mag.*, vol. 9, pp. 47–50, 1990.

[36] T. Akgül, M. Sun, R. J. Sclabassi, and A. E. Çetin, "Characterization of sleep spindles using higher order statistics and spectra," *IEEE Trans. Biomed. Eng.*, vol. 47, pp. 997–1009, 2000.

[37] A. Isaksson, A. Wennberg, and L. H. Zetterberg, "Computer analysis of EEG signals with parametric models," *Proc. IEEE*, vol. 69, pp. 451–641, April 1981.

[38] L. H. Zetterberg, "Means and methods for processing of physiological signals with emphasis on EEG analysis," in *Advances in Biology and Medical Physics* (J. H. Lawrence et al., ed.), pp. 41–91, New York: Academic Press, 1977.

[39] L. H. Zetterberg, "Estimation of parameters for a linear difference equation with application to EEG analysis," *Math. Biosci.*, vol. 5, pp. 227–275, 1969.

[40] A. Wennberg and L. H. Zetterberg, "Application of a computer-based model for EEG analysis," *Electroencephal. Clin. Neurophysiol.*, vol. 31, pp. 457–468, 1971.

[41] P. B. C. Fenwick, P. Mitchie, J. Dollimore, and G. W. Fenton, "Application of autoregressive model to E.E.G. analysis," *Agressologie*, vol. 10, pp. 553–564, 1969.

[42] P. B. C. Fenwick, P. Mitchie, J. Dollimore, and G. W. Fenton, "Mathematical simulation of the electroencephalogram using an autoregressive model," *Bio-Med. Comput.*, vol. 2, pp. 281–307, 1971.

[43] W. Gersch, "Spectral analysis of EEGs by autoregressive spectral decomposition of time series," *Math. Biosci.*, vol. 7, pp. 205–222, 1970.

[44] W. Gersch, "Non-stationary multichannel time series analysis," in *Handbook of Electroencephalography and Clinical Neurophysiology: Clinical Applications of Computer Analysis of EEG and Other Neurophysiological Signals* (F. Lopes da Silva, W. Storm van Leeuwen, and A. Rémond, eds.), ch. 10, pp. 261–296, Amsterdam/New York: Elsevier, 1986.

[45] N. Amir and I. Gath, "Segmentation of EEG during sleep using time varying autoregressive modeling," *Biol. Cybern.*, vol. 61, pp. 447–455, 1989.

[46] J. P. Kaipio and P. A. Karjalainen, "Estimation of event-related synchronization changes by a new TVAR method," *IEEE Trans. Biomed. Eng.*, vol. 44, pp. 649–656, 1997.

[47] W. Gersch and J. Yonemoto, "Parametric time series models for multivariate EEG analysis," *Comput. Biomed. Res.*, vol. 10, pp. 113–125, 1977.

[48] W. Gersch and J. Yonemoto, "Automatic classification of multivariate EEGs using an amount of information measure and the eigenvalues of parametric time series model features," *Comput. Biomed. Res.*, vol. 10, pp. 297–318, 1977.

[49] P. J. Franaszczuk, K. J. Blinowska, and M. Kowalczyk, "The application of parametric multichannel spectral estimates in the study of brain electrical activity," *Biol. Cybern.*, vol. 51, pp. 239–247, 1985.

[50] M. J. Kaminski and K. J. Blinowska, "A new method of the description of the information flow in the brain structures," *Biol. Cybern.*, vol. 65, pp. 203–210, 1991.

[51] J. C. Jimenez, R. Biscay, and O. Montoto, "Modeling the electroencephalogram by means of spatial spline smoothing and temporal autoregression," *Biol. Cybern.*, vol. 72, pp. 249–259, 1995.

[52] C. W. Anderson, E. A. Stoltz, and S. Shamsunder, "Multivariate autoregressive models for classification of spontaneous electroencephalographic signals during mental tasks," *IEEE Trans. Biomed. Eng.*, vol. 45, pp. 277–286, 1998.

[53] A. C. da Rosa, B. Kemp, T. Paiva, F. H. Lopes da Silva, and H. A. C. Kamphuisen, "A model-based detector of vertex waves and K complexes in sleep electroencephalogram," *Electroencephal. Clin. Neurophysiol.*, vol. 78, pp. 71–79, 1991.

[54] F. L. Lopes da Silva, A. Hoek, H. Smith, and L. H. Zetterberg, "Model of brain rhythmic activity," *Kybernetic*, vol. 15, pp. 27–37, 1974.

[55] L. H. Zetterberg, L. Kristiansson, and K. Mossberg, "Performance of a model for a local neuron population," *Biol. Cybern.*, vol. 31, pp. 15–26, 1978.

[56] B. H. Jansen, "Nonlinear dynamics and quantitative EEG analysis," *Electroencephal. Clin. Neurophysiol.*, vol. 45 (Suppl.), pp. 39–56, 1996.

[57] F. Wendling, J. J. Bellanger, F. Bartolomei, and P. Chauvel, "Relevance of nonlinear lumped-parameter models in the analysis of depth-EEG epileptic signals," *Biol. Cybern.*, vol. 83, pp. 367–378, 2000.

[58] F. Wendling, F. Bartolomei, J. J. Bellanger, and P. Chauvel, "Epileptic fast activity can be explained by a model of impaired GABAergic dendritic inhibition," *Eur. J. Neurosci.*, vol. 15, pp. 1499–1508, 2002.

[59] B. H. Jansen, G. Zouridakis, and M. E. Brandt, "A neurophysiologically-based mathematical model of flash evoked potentials," *Biol. Cybern.*, vol. 68, pp. 275–283, 1993.

[60] B. H. Jansen and V. G. Rit, "Electroencephalogram and visual potential generation in a mathematical model of coupled cortical columns," *Biol. Cybern.*, vol. 73, pp. 357–366, 1995.

[61] M. G. Saunders, "Artifacts: Activity of noncerebral origin in the EEG," in *Current Practice of Clinical Electroencephalography* (D. W. Klass and D. D. Daly, eds.), pp. 37–67, New York: Raven Press, 1979.

[62] F. S. Tyner, J. R. Knott, and W. B. Mayer, *Fundemantals of EEG Technology. Vol. 1. Basic Concepts and Methods.* New York: Raven Press, 1983.

[63] J. S. Barlow, "Artefact processing (rejection and minimization) in EEG data processing," in *Handbook of Electroencephalography and Clinical Electrophysiology: Clinical Applications of Computer Analysis of EEG and Other Neurophysiological Signals* (F. H. Lopes da Silva, W. Storm van Leeuwen, and A. Rémond, eds.), ch. 1, pp. 15–62, Elsevier, 1986.

[64] J. W. Clark, "The origin of biopotentials," in *Medical Instrumentation. Application and Design* (J. G. Webster, ed.), pp. 121–182, New York: John Wiley & Sons, 1998.

[65] J. V. Basmajian and C. J. De Luca, *Muscles Alive. Their Functions Revealed by Electromyography*. Baltimore: Williams & Wilkins, 1985.

[66] P. K. H. Wong, *Digital EEG in Clinical Practice*. Philadelphia: Lippincott - Raven, 1996.

[67] D. P. Burbank and J. G. Webster, "Reducing skin potential motion artifact by skin abrasion," *Med. Biol. Eng. & Comput.*, vol. 16, pp. 31–38, 1978.

[68] L. A. Geddes, *Electrodes and the Measurement of Bioelectric Events*. New York: John Wiley & Sons, 1972.

[69] A. S. Gevins, C. L. Yeager, G. M. Zeitlin, S. Ancoli, and M. Dedon, "On-line computer rejection of EEG artifact," *Electroencephal. Clin. Neurophysiol.*, vol. 42, pp. 267–274, 1977.

[70] A. S. Gevins, G. M. Zeitlin, S. Ancoli, and C. L. Yeager, "On-line computer rejection of EEG artifact. II. Contamination by drowsiness," *Electroencephal. Clin. Neurophysiol.*, vol. 43, pp. 31–42, 1977.

[71] E. R. John, H. Ahn, L. Prichep, M. Trepetin, D. Brown, and H. Kaye, "Developmental equations for the electroencephalogram," *Science*, vol. 210, pp. 1255–1258, 1980.

[72] A. V. Oppenheim, R. W. Schafer, and T. G. Stockham, "Nonlinear filtering of multiplied and convolved signals," *Proc. IEEE*, vol. 56, pp. 1264–1291, 1968.

[73] A. V. Oppenheim and R. W. Schafer, *Digital Signal Processing*. New Jersey: Prentice-Hall, 1975.

[74] R. C. Kemerait and D. G. Childers, "Signal detection and extraction using cepstrum techniques," *IEEE Trans. Inform. Theory*, vol. 18, pp. 745–759, 1972.

[75] B. Salzberg, "The potential role of cepstral analysis in EEG research in epilepsy," in *Quantitative Analytic Studies in Epilepsy* (P. Kellaway and I. Petersén, eds.), pp. 559–563, New York: Raven Press, 1976.

[76] J. Gotman, J. R. Ives, and P. Gloor, "Frequency content of EEG and EMG at seizure onset: Possibility of removal of EMG artifact by digital filtering," *Electroencephal. Clin. Neurophysiol.*, vol. 52, pp. 626–639, 1981.

[77] P. P. Lawrence, A. W. Juhn, and P. B. Michael, "Practical digital filters for reducing EMG artifact in EEG seizure recording," *Electroencephal. Clin. Neurophysiol.*, vol. 72, pp. 268–276, 1989.

[78] T. L. Johnson, S. C. Wright, and A. Segall, "Filtering of electrode and muscle artifact from the electroencephalogram," *IEEE Trans. Biomed. Eng.*, vol. 26, pp. 556–563, 1979.

[79] C. Zheng, J. C. Sackellares, W. J. Williams, A. Tornow, and R. Kushwaha, "Reducing EMG artifact in EEG recording with standard median filter and FIR-median hybrid filter," in *Proc. Conf. IEEE Eng. Med. Biol. Soc. (EMBS)*, pp. 847–848, IEEE, 1990.

[80] M. Akay and J. A. Daubenspeck, "Investigating the contamination of electroencephalograms by facial muscle electromyographic activity using matching pursuit," *Brain and Language*, vol. 66, pp. 184–200, 1999.

[81] M. van der Velde, G. van Erp, and P. J. M. Cluitmans, "Detection of muscle artefact in the normal human awake EEG," *Electroencephal. Clin. Neurophysiol.*, vol. 107, pp. 149–158, 1998.

[82] D. G. Girton and J. Kamiya, "A simple on-line technique for removing eye movement artifacts from the EEG," *Electroencephal. Clin. Neurophysiol.*, vol. 34, pp. 212–216, 1973.

[83] J. L. Whitton, F. Lue, and H. Moldofsky, "A spectral method for removing eye movement artifacts from the EEG," *Electroencephal. Clin. Neurophysiol.*, vol. 44, pp. 735–741, 1978.

[84] R. Verleger, T. Gasser, and J. Möcks, "Correction of EOG artifacts in event-related potentials of the EEG: Aspects of reliability and validity," *Psychophysiology*, vol. 19, pp. 472–480, 1982.

[85] C. Fortgens and M. P. DeBruin, "Removal of eye movement and EEG artifacts from the non-cephalic reference EEG," *Electroencephal. Clin. Neurophysiol.*, vol. 56, pp. 90–96, 1983.

[86] J. S. Barlow and A. Rémond, "Eye movement artefact nulling in EEGs by multichannel on-line EOG subtraction," *Electroencephal. Clin. Neurophysiol.*, vol. 52, pp. 418–423, 1981.

[87] B. W. Jervis, E. C. Ifeachor, and E. M. Allen, "The removal of ocular artefacts from the electroencephalogram: A review," *Med. Biol. Eng. & Comput.*, vol. 26, pp. 2–12, 1988.

[88] E. C. Ifeachor, B. W. Jervis, E. M. Allen, E. L. Morris, D. E. Wright, and N. R. Hudson, "Investigation and comparison of some models for removing ocular artifacts from EEG signals: Part 1. Review of models and data analysis," *Med. Biol. Eng. & Comput.*, vol. 26, pp. 584–590, 1988.

[89] E. C. Ifeachor, B. W. Jervis, E. M. Allen, E. L. Morris, D. E. Wright, and N. R. Hudson, "Investigation and comparison of some models for removing ocular artifacts from EEG signals: Part 2. Quantitative and pictorial comparison of models," *Med. Biol. Eng. & Comput.*, vol. 26, pp. 591–598, 1988.

[90] E. C. Ifeachor, B. W. Jervis, E. L. Morris, E. M. Allen, and N. R. Hudson, "A new microcomputer-based ocular artefact removal OAR system," *IEE Proc. Pt. A*, vol. 133, pp. 291–300, 1986.

[91] R. G. Bickford, J. K. Sims, T. W. Billinger, and M. H. Aung, "Problems in EEG estimation of brain death and use of computer techniques for their solution," *Trauma*, vol. 12, pp. 61–95, 1971.

[92] J. S. Barlow and J. Dubinsky, "EKG-artifact minimization in referential EEG recordings by computer," *Electroencephal. Clin. Neurophysiol.*, vol. 48, pp. 470–472, 1980.

[93] H. Witte, S. Glaser, and M. Rother, "New spectral detection and elimination test algorithms of ECG and EOG artefacts in neonatal EEG recordings," *Med. Biol. Eng. & Comput.*, vol. 25, pp. 127–130, 1987.

[94] P. Celka, B. Boashash, and P. Colditz, "Preprocessing and time–frequency analysis of newborn EEG seizures," *IEEE Eng. Med. Biol. Mag.*, vol. 20, pp. 30–39, 2001.

[95] E. C. Ifeachor, B. W. Jervis, E. L. Morris, E. M. Allen, and N. R. Hudson, "New on-line method for removing ocular artefacts from the EEG signals," *Med. Biol. Eng. & Comput.*, vol. 24, pp. 356–364, 1986.

[96] P. He, G. Wilson, and C. Russell, "Removal of ocular artifacts from electroencephalogram by adaptive filtering," *Med. Biol. Eng. & Comput.*, vol. 42, pp. 407–412, 2004.

[97] B. Widrow and S. D. Stearns, *Adaptive Signal Proccessing.* New Jersey: Prentice-Hall, 1985.

[98] S. Haykin, *Adaptive Filter Theory.* New Jersey: Prentice-Hall, 4th ed., 2002.

[99] J. C. Woestenburg, M. N. Verbaten, and J. L. Slangen, "The removal of the eye-movement artifact from the EEG by regression analysis in the frequency domain," *Biol. Psychol.*, vol. 16, pp. 127–147, 1983.

[100] T. Gasser, L. Sroka, and J. Möcks, "The transfer of EOG activity into the EEG for eyes open and closed," *Electroencephal. Clin. Neurophysiol.*, vol. 61, pp. 181–193, 1985.

[101] T. Gasser, L. Sroka, and J. Möcks, "The correction of EOG artifacts by frequency dependent and frequency independent methods," *Psychophysiology*, vol. 23, pp. 704–712, 1986.

[102] W. Du, H. M. Leong, and A. S. Gevins, "Ocular artifact minimization by adaptive filtering," in *Proc. IEEE Workshop on Statistical Signal and Array Proc.*, pp. 433–436, 1994.

[103] M. Hayes, *Statistical Digital Signal Proccessing and Modeling.* New York: John Wiley & Sons, 1996.

[104] G. Dumermuth and H. Flühler, "Some modern aspects in numerical spectrum analysis of multichannel electroencephalographic data," *Med. Biol. Eng.*, vol. 5, pp. 319–331, 1967.

[105] F. H. Lopes da Silva, "EEG analysis: Theory and practice," in *Electroencephalography. Basic Principles, Clinical Applications and Related Fields* (E. Niedermayer and F. Lopes da Silva, eds.), pp. 1135–1163, Baltimore: Williams & Wilkins, 1999.

[106] N. Kawabata, "Test of statistical stability of the electroencephalogram," *Biol. Cybern.*, vol. 22, pp. 235–238, 1976.

[107] A. Sances and B. A. Cohen, "Stationarity of the human electroencephalogram," *Med. Biol. Eng. & Comput.*, vol. 15, pp. 513–518, 1977.

[108] J. Möcks and T. Gasser, "How to select epochs of the EEG at rest for quantitative analysis," *Electroencephal. Clin. Neurophysiol.*, vol. 58, pp. 89–92, 1984.

[109] S. Blanco, H. García, R. Quian Quiroga, L. Romanelli, and O. A. Rosso, "Stationarity of the EEG series," *IEEE Eng. Med. Biol. Mag.*, vol. 14, pp. 395–399, 1995.

[110] G. Dietsch, "Fourier analyse von elektroenzephalogrammen des menschen," *Pflügers Arch. Ges. Physiol.*, vol. 230, pp. 106–112, 1932.

[111] A. M. Grass and F. A. Gibbs, "A Fourier transform of the electroencephalogram," *J. Neurophysiol.*, vol. 1, pp. 521–526, 1938.

[112] W. G. Walter, "Automatic low frequency analyzer," *Electron Eng.*, vol. 16, pp. 3–13, 1943.

[113] J. G. Proakis and D. G. Manolakis, *Digital Signal Processing. Principles, Algorithms, and Applications.* New Jersey: Prentice-Hall, 3rd ed., 1996.

[114] P. Stoica and R. Moses, *Introduction to Spectral Analysis.* New Jersey: Prentice-Hall, 1997.

[115] D. J. Thomson, "Spectrum estimation and harmonic analysis," *Proc. IEEE*, vol. 70, pp. 1055–1096, 1982.

[116] Y. Xu, S. Haykin, and R. J. Racine, "Multiple window time–frequency distribution and coherence of EEG using Slepian sequences and Hermite functions time–frequency analysis of EEG activity," *IEEE Trans. Biomed. Eng.*, vol. 46, pp. 861–866, 1999.

[117] M. Hansson and M. Lindgren, "Multiple window spectrogram of transient peaks in the electroencephalogram," *IEEE Trans. Biomed. Eng.*, vol. 48, pp. 284–293, 2001.

[118] M. Matousek and I. Petersén, "Automatic evaluation of EEG background activity by means of age-dependent EEG quotients," *Electroencephal. Clin. Neurophysiol.*, vol. 35, pp. 603–612, 1973.

[119] F. H. Lopes da Silva, W. Storm van Leeuwen, and A. Rémond, *Handbook of Electroencephalography and Clinical Ceurophysiology: Clinical Applications of Computer Analysis of EEG and Other Neurophysiological Signals*, vol. 2. Amsterdam/New York: Elsevier, 1986.

[120] J. Gotman, D. R. Skuce, C. J. Thompson, P. Gloor, J. R. Ives, and W. F. Ray, "Clinical applications of spectral analysis and extraction of features from electroencephalograms with slow waves in adult patients," *Electroencephal. Clin. Neurophysiol.*, vol. 35, pp. 225–235, 1973.

[121] P. Matthis, D. Scheffner, and C. Benninger, "Spectral analysis of the EEG: Comparison of various spectral parameters," *Electroencephal. Clin. Neurophysiol.*, vol. 52, pp. 218–221, 1981.

[122] F. H. Lopes da Silva, "Computer-assisted EEG diagnosis: Pattern recognition and brain mapping," in *Electroencephalography. Basic Principles, Clinical Applications and Related Fields* (E. Niedermayer and F. Lopes da Silva, eds.), pp. 1164–1189, Baltimore: Williams & Wilkins, 1999.

[123] G. Dumermuth and L. Molinari, "Spectral analysis of EEG background activity," in *Handbook of Electroencephalography and Clinical Neurophysiology: Methods of Analysis of Brain Electrical and Magnetic Signals* (A. S. Gevins and A. Rémond, eds.), ch. 4, pp. 85–130, Amsterdam/New York: Elsevier, 1987.

[124] B. Hjorth, "EEG analysis based on time domain properties," *Electroencephal. Clin. Neurophysiol.*, vol. 29, pp. 306–310, 1970.

[125] B. Hjorth, "The physical significance of time domain descriptors in EEG analysis," *Electroencephal. Clin. Neurophysiol.*, vol. 34, pp. 321–325, 1973.

[126] M. Walmsley, "On normalized slope descriptor method of quantifying electroencephalograms," *IEEE Trans. Biomed. Eng.*, vol. 31, pp. 720–723, 1984.

[127] B. Saltzberg, W. D. Burton Jr., J. S. Barlow, and N. R. Burch, "Moments of the power spectral density estimated from samples of the autocorrelation function (a robust procedure for monitoring changes in the statistical properties of non-stationary time series such as the EEG)," *Electroencephal. Clin. Neurophysiol.*, vol. 61, pp. 89–93, 1985.

[128] I. I. Goncharova and J. S. Barlow, "Changes in EEG mean frequency and spectral purity during spontaneous alpha blocking," *Electroencephal. Clin. Neurophysiol.*, vol. 76, pp. 197–204, 1990.

[129] C. W. Therrien, *Discrete Random Signals and Statistical Signal Processing*. New Jersey: Prentice-Hall, 1992.

[130] S. M. Kay, *Modern Spectral Estimation. Theory and Application*. New Jersey: Prentice-Hall, 1988.

[131] S. L. Marple Jr., *Digital Spectral Analysis with Applications*. New Jersey: Prentice-Hall, 1987.

[132] A. Wennberg and A. Isaksson, "Simulation of nonstationary EEG signals as a means of objective clinical interpretation of EEG," in *Quantitative Analytic Studies in Epilepsy* (P. Kellaway and I. Petersén, eds.), pp. 493–508, New York: Raven Press, 1976.

[133] S. V. Narasimhan and D. N. Dutt, "Software simulation of the EEG," *J. Biomed. Eng.*, vol. 7, pp. 275–281, 1985.

[134] B. H. Jansen, J. R. Bourne, and J. W. Ward, "Autoregressive estimation of short segment spectra for computerized EEG analysis," *IEEE Trans. Biomed. Eng.*, vol. 28, pp. 630–638, 1981.

[135] S. Cerutti, D. Liberati, and P. Mascellani, "Parameter extraction in EEG processing during riskful neurosurgical operations," *Signal Proc.*, vol. 9, pp. 25–35, 1985.

[136] C. W. Anderson, E. A. Stolz, and S. Shamsunder, "Discriminating mental tasks using EEG represented by AR models," in *Proc. Conf. IEEE Eng. Med. Biol. Soc. (EMBS)*, pp. 875–877, IEEE, 1995.

[137] V. Goel, A. M. Brambrink, A. Baykal, R. C. Koehler, D. F. Hanley, and N. V. Thakor, "Dominant frequency analysis of EEG reveals brain's response during injury and recovery," *IEEE Trans. Biomed. Eng.*, vol. 43, pp. 1083–1092, 1996.

[138] H. Akaike, "A new look at statistical model identification," *IEEE Trans. Autom. Control*, vol. 19, pp. 716–723, 1974.

[139] J. Rissanen, "Modeling of shortest data description," *Automatica*, vol. 14, pp. 465–471, 1978.

[140] G. E. Birch, P. D. Lawrence, J. C. Lind, and R. D. Hare, "Application of prewhitening to AR spectral estimation of EEG," *IEEE Trans. Biomed. Eng.*, vol. 35, pp. 640–645, 1988.

[141] F. L. Lopes da Silva, "Analysis of EEG nonstationarities," *Electroencephal. Clin. Neurophysiol.*, vol. 34, pp. 163–179, 1978.

[142] B. H. Jansen, A. Hasman, and R. Lenten, "Piece-wise analysis of EEGs using AR modeling and clustering," *Comput. Biomed. Res.*, vol. 14, pp. 168–178, 1981.

[143] B. H. Jansen, A. Hasman, and R. Lenten, "Piece-wise EEG analysis: An objective evaluation," *Int. J. Biomed. Comput.*, vol. 12, pp. 17–27, 1981.

[144] Z. A. Keirn and J. I. Aunon, "A new mode of communication between men and his surroundings," *IEEE Trans. Biomed. Eng.*, vol. 37, pp. 1209–1214, 1990.

[145] B. H. Jansen, "Analysis of biomedical signals by means of linear modeling," *CRC Crit. Rev. Biomed. Eng.*, vol. 12, no. 4, pp. 343–392, 1985.

[146] A. Isaksson, K. Lagergren, and A. Wennberg, "Visible and non-visible EEG changes demonstrated by spectral parameter analysis," *Electroencephal. Clin. Neurophysiol.*, vol. 41, pp. 225–236, 1976.

[147] W. D. Smith and D. L. Lager, "Evaluation of simple algorithms for spectral parameter analysis of the electroencephalogram," *IEEE Trans. Biomed. Eng.*, vol. 33, pp. 352–358, 1986.

[148] J. S. Barlow, "Methods of analysis of nonstationary EEGs, with emphasis on segmentation techniques: A comparative review," *J. Clin. Neurophysiol.*, vol. 2, pp. 267–304, 1985.

[149] U. Appel and A. van Brandt, "A comparative study of three sequential time series segmentation algorithms," *Signal Proc.*, vol. 6, pp. 45–60, 1984.

[150] M. Basseville and I. V. Nikiforov, *Detection of Abrupt Changes. Theory and Applications*. New Jersey: Prentice-Hall, 1993.

[151] F. Gustafsson, *Adaptive Filtering and Change Detection*. New York: John Wiley & Sons, 2000.

[152] G. Bodenstein, W. Schneider, and C. van der Malsburg, "Computerized EEG pattern classification by adaptive segmentation and probability-density-function classification: Description of the method," *Comput. Biol. Med.*, vol. 15, pp. 297–313, 1985.

[153] D. Michael and J. Houchin, "Automatic EEG analysis: A segmentation procedure based on the autocorrelation functions," *Electroencephal. Clin. Neurophysiol.*, vol. 46, pp. 232–235, 1979.

[154] J. S. Barlow, O. D. Creutzfeldt, D. Michael, J. Houchin, and H. Epelbaum, "Automatic adaptive segmentation of clinical EEGs," *Electroencephal. Clin. Neurophysiol.*, vol. 51, pp. 512–525, Dec. 1981.

[155] H. M. Praetorius, G. Bodenstein, and O. D. Creutzfeldt, "Adaptive segmentation of EEG records: A new approach to automated EEG analysis," *Electroencephal. Clin. Neurophysiol.*, vol. 42, pp. 84–94, 1977.

[156] I. Gath and E. Bar-On, "Computerized method for scoring of polysomnographic sleep recordings," *Comput. Prog. Biomed.*, vol. 11, pp. 217–223, 1980.

[157] J. Segen, A. C. Sanderson, and E. Richey, "Detecting change in a time series," *IEEE Trans. Inform. Theory*, vol. 26, pp. 249–255, 1980.

[158] A. C. Sanderson, J. Segen, and E. Richey, "Hierarchial modeling of EEG signals," *IEEE Trans. Pattern Anal. Mach. Intell.*, vol. 2, pp. 405–414, 1980.

[159] R. Aufrichtig, S. B. Pedersen, and P. Jennum, "Adaptive segmentation of EEG signals," in *Proc. Conf. IEEE Eng. Med. Biol. Soc. (EMBS)*, pp. 453–454, IEEE, 1991.

[160] O. D. Creutzfeldt, G. Bodenstein, and J. S. Barlow, "Computerized EEG pattern classification by adaptive segmentation and probability-density-function classification: Clinical evaluation," *Electroencephal. Clin. Neurophysiol.*, vol. 60, pp. 373–393, 1985.

[161] V. Krajča, S. Petranek, I. Patakova, and A. Varri, "Automatic identification of significant graphoelements in multichannel EEG recordings by adaptive segmentation and fuzzy clustering," *Int. J. Biomed. Comput.*, vol. 28, pp. 71–89, 1991.

[162] M. Arnold, A. Doering, H. Witte, M. Eiselt, and J. Dörschel, "Use of adaptive Hilbert transformation for EEG segmentation and calculation of instantaneous respiration rate in neonates," *J. Clin. Monitoring*, vol. 12, pp. 43–60, 1996.

[163] F. Wendling, G. Carrault, and J. M. Badier, "Segmentation of depth-EEG seizure signals: Method based on a physiological parameter and comparative study," *Ann. Biomed. Eng.*, vol. 25, pp. 1026–1039, 1997.

[164] S. D. Cranstoun, H. C. Ombao, R. von Sachs, W. Guo, and B. Litt, "Time–frequency spectral estimation of multichannel EEG using the auto-SLEX method," *IEEE Trans. Biomed. Eng.*, vol. 49, pp. 988–996, 2002.

[165] B. Boashash, *Time–Frequency Signal Analysis. Methods and Applications.* Melbourne: Longman-Cheshire, 1992.

[166] P. Flandrin, *Time–Frequency/Time–Scale Analysis.* San Diego: Academic Press, 1999. (Translated from French, *Temps-Frequence.* Paris: Hermes, 1993).

[167] L. Cohen, *Time–Frequency Analysis.* New Jersey: Prentice-Hall, 1995.

[168] S. Qian and D. Chen, *Joint Time–Frequency Analysis. Methods and Applications.* New Jersey: Prentice-Hall, 1996.

[169] M. Akay (ed.), *Time Frequency and Wavelets in Biomedical Signal Processing*. New York: IEEE Press, 1996.

[170] B. Boashash (ed.), *Time–Frequency Analysis*. Amsterdam/New York: Elsevier, 2003.

[171] W. Martin and P. Flandrin, "Wigner-Ville spectral analysis of non-stationary processes," *IEEE Trans. Acoust. Speech Sig. Proc.*, vol. 33, pp. 1461–1470, 1985.

[172] R. G. Bickford, J. Brimm, L. Berger, and M. Aung, "Application of compressed spectral array in clinical EEG," in *Automation of Clinical Electroencephalography* (P. Kellaway and I. Petersén, eds.), pp. 55–64, New York: Raven, 1973.

[173] M. L. Scheuer, "Continuous EEG monitoring in the intensive care unit," *Epilepsia*, vol. 43, pp. 114–127, 2002.

[174] N. Löfgren, K. Lindecrantz, M. Thordstein, A. Hedström, B. G. Wallin, S. Andreasson, A. Flisberg, and I. Kjellmer, "Remote sessions and frequency analysis for improved insight into cerebral function during paediatric and neonatal intensive care," *IEEE Trans. Info. Tech. Biomed.*, vol. 7, pp. 283–290, 2003.

[175] I. Gath, C. Feuerstein, D. T. Pham, and G. Rondouin, "On the tracking of rapid dynamic changes in seizure EEG," *IEEE Trans. Biomed. Eng.*, vol. 39, pp. 952–958, 1992.

[176] S. Blanco, S. Kochen, O. A. Rosso, and P. Salgado, "Applying time–frequency analysis to seizure EEG activity," *IEEE Eng. Med. Biol. Mag.*, vol. 16, pp. 64–71, 1997.

[177] P. J. Franaszczuk, G. K. Bergey, P. J. Durka, and H. M. Eisenberg, "Time–frequency analysis using the matching pursuit algorithm applied to seizures originating from the mesial temporal lobe," *Electroencephal. Clin. Neurophysiol.*, vol. 106, pp. 513–521, 1998.

[178] E. P. Wigner, "On the quantum correction for thermodynamic equilibrium," *Physical Rev.*, vol. 40, pp. 749–759, 1932.

[179] J. Ville, "Theorie et applications de la notion de signal analytique," *Cables et Transmissions*, vol. 2A, pp. 61–74, 1948.

[180] T. A. C. M. Claasen and W. F. G. Mecklenbräuker, "The Wigner distribution. A tool for time–frequency signal analysis—Part I: Continuous-time signals," *Philips J. Res.*, vol. 35, pp. 217–250, 1980.

[181] T. A. C. M. Claasen and W. F. G. Mecklenbräuker, "The Wigner distribution. A tool for time–frequency signal analysis—Part II: Discrete-time signals," *Philips J. Res.*, vol. 35, pp. 276–300, 1980.

[182] T. A. C. M. Claasen and W. F. G. Mecklenbräuker, "The Wigner distribution. A tool for time–frequency signal analysis—Part III: Relations with other time–frequency signal transformations," *Philips J. Res.*, vol. 35, pp. 372–389, 1980.

[183] A. Mertins, *Signal Analysis*. Chichester: John Wiley, 1999.

[184] K. Gröchenig, *Foundations of Time–Frequency Analysis*. Boston: Birkhäuser, 2000.

[185] L. Cohen, "Generalized phase-space distribution functions," *J. Math. Phys.*, vol. 7, pp. 781–786, 1966.

[186] F. Hlawatsch and G. F. Boudreaux-Bartels, "Linear and quadratic time–frequency signal representations," *IEEE Signal Proc. Mag.*, vol. 9, pp. 21–67, 1992.

[187] H. Choi and W. J. Williams, "Improved time–frequency representation of multicomponent signals using exponential kernels," *IEEE Trans. Acoust. Speech Sig. Proc.*, vol. 37, pp. 862–871, 1989.

[188] W. J. Williams, "Reduced interference distributions: Biological applications and interpretations," *Proc. IEEE*, vol. 84, pp. 1264–1280, 1996.

[189] D. L. Jones and T. W. Parks, "A high resolution data-adaptive time–frequency representation," *IEEE Trans. Acoust. Speech Sig. Proc.*, vol. 38, pp. 2127–2135, 1990.

[190] R. G. Baraniuk and D. L. Jones, "A signal-dependent time–frequency representation: Optimal kernel design," *IEEE Trans. Signal Proc.*, vol. 41, pp. 1589–1602, 1993.

[191] D. L. Jones and R. G. Baraniuk, "An adaptive optimal-kernel time–frequency representation," *IEEE Trans. Signal Proc.*, vol. 43, pp. 2361–2371, 1995.

[192] H. P. Zaveri, W. J. Williams, L. D. Iasemidis, and J. C. Sackellares, "Time–frequency representation of electrocorticograms in temporal lobe epilepsy," *IEEE Trans. Biomed. Eng.*, vol. 39, pp. 502–509, 1992.

[193] W. J. Williams, H. P. Zaveri, and J. C. Sackellares, "Time–frequency analysis of electrophysiology signals in epilepsy," *IEEE Eng. Med. Biol. Mag.*, vol. 14, pp. 133–143, 1995.

[194] R. Quian Quiroga, S. Blanco, O. Rosso, H. García, and A. Rabinowicz, "Searching for hidden information with Gabor transform in generalized tonic-clonic seizures," *Electroencephal. Clin. Neurophysiol.*, vol. 103, pp. 434–439, 1997.

[195] F. Wendling, M. B. Shamsollahi, J. M. Badier, and J. J. Bellanger, "Time–frequency matching of warped depth-EEG seizure observations," *IEEE Trans. Biomed. Eng.*, vol. 46, pp. 601–605, 1999.

[196] B. Boashash and M. Mesbah, "A time–frequency approach for newborn seizure detection," *IEEE Eng. Med. Biol. Mag.*, vol. 20, pp. 54–64, 2001.

[197] A. Nayak, R. J. Roy, and A. Sharma, "Time–frequency spectral representation of the EEG as an aid in the detection of anesthesia depth," *Ann. Biomed. Eng.*, vol. 22, pp. 501–513, 1994.

[198] M. S. Scher, M. Sun, G. M. Hatzilabrou, N. L. Greenberg, G. Cebulka, D. Krieger, R. D. Guthrie, and R. J. Sclabassi, "Computer analyses of EEG-sleep in the neonate: Methodological considerations," *J. Clin. Neurophysiol.*, vol. 7, pp. 417–441, 1989.

[199] R. J. Sclabassi, M. Sun, D. N. Krieger, P. Jasiukaitis, and M. S. Scher, "Time–frequency domain problems in neurosciences," in *Time–Frequency Signal Analysis. Methods and Applications* (B. Boashash, ed.), ch. 23, pp. 498–519, Melbourne: Longman Cheshire, 1992.

[200] M. Sun, S. Qian, X. Yan, S. B. Baumann, X. Xia, R. E. Dahl, N. D. Ryan, and R. J. Sclabassi, "Localizing functional activity in the brain through time–frequency analysis and synthesis of the EEG," *Proc. IEEE*, vol. 84, pp. 1302–1311, 1996.

[201] S. Haykin, R. J. Racine, Y. Xu, and C. A. Chapman, "Monitoring neural oscillations and signal transmission between cortical regions using time–frequency analysis of EEG activity," *Proc. IEEE*, vol. 84, pp. 1295–1301, 1996.

[202] B. Boashash and V. Sucic, "Resolution measure criteria for the objective assessment of the performance of quadratic time–frequency distributions," *IEEE Trans. Signal Proc.*, vol. 51, pp. 1253–1263, 2003.

[203] D. Popivanov and A. M. J. Duhanova, "Tracking EEG signal dynamics during mental tasks. A combined linear/nonlinear approach," *IEEE Eng. Med. Biol. Mag.*, vol. 17, pp. 89–94, 1998.

[204] A. Schlögl, D. Flotzinger, and G. Pfurtscheller, "Adaptive autoregressive modeling used for single-trial EEG classification," *Biomed. Tech.*, vol. 42, pp. 162–167, 1997.

[205] G. Pfurtscheller, C. Neuper, A. Schlögl, and K. Lugger, "Separability of EEG signals recorded during right and left motor imagery using adaptive autoregressive parameters," *IEEE Trans. Rehab. Eng.*, vol. 6, pp. 316–325, 1998.

Problems

3.1 Higher-order moments are sometimes used to characterize the EEG signal. For example, the univariate third- and fourth-order moments, i.e., the skewness γ_s and kurtosis γ_k, respectively, can be estimated by [30]

$$\gamma_s = \frac{1}{(\sigma_x^2)^{3/2}} \cdot \frac{1}{N} \sum_{n=0}^{N-1} (x(n) - \bar{x})^3,$$

and

$$\gamma_k = \frac{1}{(\sigma_x^2)^2} \cdot \frac{1}{N} \sum_{n=0}^{N-1} (x(n) - \bar{x})^4 - 3,$$

where $\bar{x} = \frac{1}{N} \sum_{n=0}^{N-1} x(n)$ and $\sigma_x^2 = \frac{1}{N} \sum_{n=0}^{N-1} (x(n) - \bar{x})^2$. In order to illustrate the significance of γ_s and γ_k, we assume that two different signals, $x_1(n)$ and $x_2(n)$, are characterized by PDFs which constrain the signals to only assume fixed values. For $x_1(n)$ the fixed values are $\pm A$, and the corresponding PDF is given by

$$p(x_1(n)) = \frac{1}{2}\delta(x_1(n) - A) + \frac{1}{2}\delta(x_1(n) + A),$$

whereas for $x_2(n)$ the fixed values are $\pm A/2$ and $\pm\sqrt{13}A/2$, and

$$p(x_2(n)) = \frac{3}{8}\delta\left(x_2(n) - \frac{A}{2}\right) + \frac{3}{8}\delta\left(x_2(n) + \frac{A}{2}\right)$$
$$+ \frac{1}{8}\delta\left(x_2(n) - \frac{\sqrt{13}A}{2}\right) + \frac{1}{8}\delta\left(x_2(n) + \frac{\sqrt{13}A}{2}\right).$$

Compute the mean, variance, skewness, and kurtosis for $x_1(n)$ and $x_2(n)$, and interpret the results.

3.2 We are interested in artifact cancellation using filtered reference signals when a priori information on the impulse response $\bar{\mathbf{h}}_i$ is available. Show that the filter solution $\mathbf{h}_i$, $i = 1, \ldots, M$, minimizing the MSE can be expressed as the solution of the matrix equation in (3.77),

$$\left(\begin{bmatrix} \mathbf{R}_{\mathbf{v}_1\mathbf{v}_1} & \cdots & \mathbf{R}_{\mathbf{v}_1\mathbf{v}_M} \\ \vdots & \ddots & \vdots \\ \mathbf{R}_{\mathbf{v}_M\mathbf{v}_1} & \cdots & \mathbf{R}_{\mathbf{v}_M\mathbf{v}_M} \end{bmatrix} - 2\nu\mathbf{I}\right) \begin{bmatrix} \mathbf{h}_1 \\ \vdots \\ \mathbf{h}_M \end{bmatrix} = \begin{bmatrix} \mathbf{r}_{xv_1} - 2\nu\bar{\mathbf{h}}_1 \\ \vdots \\ \mathbf{r}_{xv_M} - 2\nu\bar{\mathbf{h}}_M \end{bmatrix}.$$

3.3 Derive the mean of the periodogram given in (3.82).

3.4 Given the power spectrum $S_x(e^{j\omega})$, find the least-squares estimate of the spectral slope parameter b used in the approximate decomposition

$$\log S_x(e^{j\omega}) \approx \log S_x^r(e^{j\omega}) + b|\omega|.$$

3.5 Derive the time domain expressions of the spectral moments in (3.100)–(3.102). *Hint:* Use the Fourier transform pair

$$(\jmath\Omega)^n S_x(\Omega) \quad\overset{\mathcal{FT}}{\longleftrightarrow}\quad \frac{\partial^n r_x(\tau)}{\partial \tau^n}.$$

3.6 A continuous-time EEG signal is modeled by the following power spectrum with bandpass characteristic,

$$S_x(\Omega) = \begin{cases} 1, & 1/4 < |\Omega| < 1/2; \\ 0, & \text{otherwise.} \end{cases}$$

    a. Determine the value of the mobility descriptor $\mathcal{H}_1$, and decide if the value closely approximates the dominant frequency of the signal.

    b. Determine the value of the complexity descriptor $\mathcal{H}_2$, and decide if the value closely approximates half the spectral bandwidth.

    c. Determine the value of the spectral purity index Γ_{SPI}, and judge its ability to describe spectral concentration.

3.7 An amplitude-modulated sinusoid with frequency 9 Hz is subjected to spectral analysis which shows that two frequencies at 8 and 10 Hz are present. Determine the frequency of the modulating signal, and explain the results of the spectral analysis [3].

3.8 Show that an $\text{AR}(p)$ model analyzed with a model order of $p' > p$ yields the same parameter estimates as when analyzed with a model order of p. *Hint:* Consider the matrix equation in (3.126) for different model orders.

3.9 Derive the normal equations for backward prediction, and verify that $\mathbf{b}_p = \mathbf{a}_p$ for the real-valued case.

3.10 Sketch the AR autocorrelation function as a function of sampling rate by sampling a continuous-time signal at various rates. Comment on how the sampling rate influences the model order selection in the parameter estimation procedure.

3.11 Show that the solution of the normal equations for AR parameter estimation can be expressed as

$$
\begin{bmatrix} a_1 \\ a_2 \\ \vdots \\ a_p \end{bmatrix} = \begin{bmatrix} r_x(0) & \cdots & r_x(p-1) \\ r_x(1) & \cdots & r_x(p-2) \\ & \vdots & \\ r_x(p-1) & \cdots & r_x(0) \end{bmatrix}^{-1} \begin{bmatrix} -r_x(1) \\ -r_x(2) \\ \vdots \\ -r_x(p) \end{bmatrix},
$$

followed by a second step which produces an estimate of the variance σ_e^2,

$$
\sigma_e^2 = r_x(0) + \sum_{i=1}^{p} a_i r_x(i).
$$

3.12 Show that the pair of coefficients in the partial fraction expansion of (3.174) satisfies $c_{2i-1} = c_{2i}^*$ when these belong to a pair of complex-conjugate poles d_{2i} and d_{2i-1}.

3.13 Assuming that AR modeling is used for power spectral analysis, show that the power P_i of the ith spectral component can be calculated by the following expression,

$$
\begin{aligned}
P_i &= r_{x_i}(0) \\
&= \frac{8\sigma_v^2}{1 - |d_{2i}|^2} \left[\Re \left(\frac{\Re^2(c_{2i}) + \Re^2(c_{2i}d_{2i}^*) - \Re(c_{2i})\Re(c_{2i}d_{2i}^*)(d_{2i} + d_{2i}^{-1})}{1 + |d_{2i}|^2 - d_{2i}^2 - d_{2i}^* d_{2i}^{-1}} \right) \right],
\end{aligned}
$$

which results from evaluating the residues of the spectral component $S_{x_i}(z)$.

3.14 (cont'd) Assuming that AR modeling is considered for power spectral analysis, show that an estimate of the peak frequency of the ith spectral component can be obtained by

$$
\omega_i = \arccos \left(\frac{1 + r_i^2}{2r_i} \cos \phi_i \right).
$$

3.15 When the peaks of an AR power spectrum are not well-separated, i.e., the assumption in (3.178) no longer holds, it is not possible to determine the power from individual spectral components $S_{x_i}(z)$ associated with pairs of poles, cf. (3.184). Instead, the power must be determined from the total power spectrum by making use of the z-transform integral inversion formula

$$r_x(k) = \frac{1}{2\pi\jmath} \oint_C S_x(z) z^{k-1} dz$$

$$= \sum_{j=1}^{p} \text{Res}\left[S_x(z) z^{-1}, d_j \right] d_j^k,$$

which, when solving for $k = 0, 1, 2, \ldots$ and applying symmetry, results in

$$r_x(k) = \sum_{j=1}^{p} \gamma_j d_j^{|k|}.$$

a. For an AR power spectrum with p distinct poles,

$$S_x(z) = \frac{\sigma^2}{A(z)A(z^{-1})} = \frac{\sigma^2}{\displaystyle\prod_{j=1}^{p} \left(1 - d_j z^{-1}\right)\left(1 - d_j^* z\right)}.$$

Determine the value of each residual γ_j so that the signal power can be computed as the autocorrelation at zero lag, given by

$$r_x(0) = \sum_{j=1}^{p} \gamma_j.$$

b. With the definition of the complex power spectrum $S_x(z)$,

$$S_x(z) = \sum_{k=-\infty}^{\infty} r_x(k) z^{-k},$$

interpret the total power as the sum of p terms for each pole.

3.16 Modify the error measure $\Delta_2(n)$ in (3.199) such that an increase or a decrease in signal power is treated in the same way; compare with the modification done when $\Delta_1(n)$ was introduced in (3.193).

3.17 Find a time domain expression of $\Delta_2'(n)$ derived in Problem 3.16 by making use of the approach for deriving $\Delta_2(n)$ in (3.203).

3.18 Show that the uncertainty principle in (3.214) for the continuous-time case also holds for the discrete-time case such that

$$\Delta_n \Delta_\omega \geq \frac{1}{2},$$

where

$$\Delta_n^2 = \frac{\sum\limits_{n=-\infty}^{\infty} (n - \overline{n})^2 x^2(n)}{\sum\limits_{n=-\infty}^{\infty} x^2(n)}, \quad \Delta_\omega^2 = \frac{\frac{1}{2\pi}\int_0^\pi (\omega - \overline{\omega})^2 |X(e^{j\omega})|^2 d\omega}{\frac{1}{2\pi}\int_0^\pi |X(e^{j\omega})|^2 d\omega},$$

and

$$\overline{n} = \frac{\sum\limits_{n=-\infty}^{\infty} n x^2(n)}{\sum\limits_{n=-\infty}^{\infty} x^2(n)}, \quad \overline{\omega} = \frac{\frac{1}{2\pi}\int_0^\pi \omega |X(e^{j\omega})|^2 d\omega}{\frac{1}{2\pi}\int_0^\pi |X(e^{j\omega})|^2 d\omega}.$$

3.19 Show that the short-time Fourier transform, defined in (3.207), can be viewed as a linear filtering operation, and determine the impulse response of the linear filter.

3.20 The WVD is usually described in continuous-time where it can be more easily perceived. The WVD can, of course, also be described in discrete-time by, for example, introducing sampling into the following form of the WVD,

$$W_x(t, \Omega) = \int_{-\infty}^{\infty} x^*\left(t - \frac{\tau}{2}\right) x\left(t + \frac{\tau}{2}\right) e^{-j\Omega\tau} d\tau.$$

It is assumed that the sampling rate F_s is equal to the Nyquist frequency, i.e., $F_s = 2B$, where B is the bandwidth of $x(t)$. Show that folding arises in the WVD at *half* the Nyquist frequency of $x(t)$, i.e., at B rather than at $2B$ (the latter being the folding frequency of the power spectrum of a discrete-time signal). This result necessitates either that $x(t)$ is sampled at a rate higher than twice the Nyquist frequency or that use is made of the analytic signal in the WVD.

3.21 Show the result in (3.249), which states that $\overline{\Omega}(t) = \varphi'(t)$ for the WVD when the signal is given by $x_c(t) = s(t)e^{j\varphi(t)}$.

Chapter 4

Evoked Potentials

Evoked potentials (EPs) constitute an *event-related* activity which occurs as the electrical response from the brain or the brainstem to various types of sensory stimulation of nervous tissues; auditory and visual stimulation are commonly used. The recording of such electrical potentials represents a non-invasive objective test which provides information on, e.g., sensory pathways abnormalities, the localization of lesions affecting the sensory pathways, and disorders related to language and speech. Evoked potentials are recorded from the scalp using an electrode configuration similar to that of an EEG recording. The potentials typically manifest themselves as a transient waveform whose morphology depends on the type and strength of the stimulus and the electrode positions on the scalp. The mental state of the subject, exemplified by attention, wakefulness, and expectation, also influences the waveform morphology.

Individual EPs have a very low amplitude, ranging from 0.1 to 10 μV, and are, accordingly, hidden in the ongoing EEG background activity, having an amplitude on the order of 10 to 100 μV. In contrast to Chapter 3, the EEG in the present chapter is viewed as "noise" whose influence should be minimized so that the EP waveform can be subjected to reliable scrutiny. As a result, noise reduction is one of the most frequently addressed signal processing issues in the analysis of EPs. Fortunately, an EP usually occurs after a time interval related to the time of stimulus presentation, whereas the background EEG activity and non-neural noise occur in a more random fashion. The stimulus and response property means that repetitive stimulation can be used in combination with *ensemble averaging* techniques to help reduce the noise level (Section 4.3) [1, 2]. With a sufficiently low noise level, the time delay (*latency*) and amplitude of each constituent wave of the EP can be accurately estimated and interpreted in suitable clinical terms (Figure 4.1).

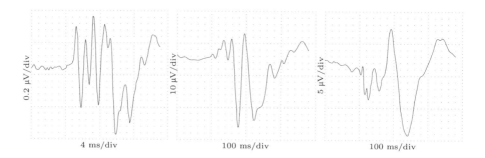

Figure 4.1: Various morphologies of evoked potentials. The duration, amplitude, and morphology differ considerably from potential to potential.

The use of ensemble averaging is, however, not without complications, since the evoked response, in certain situations, undergoes dynamic changes, thereby violating the averaging assumption of a response exhibiting fixed waveform morphology. One such situation occurs during neurosurgical procedures in which it is important to detect time-varying EP changes related to neurological injury. Considerable research has been directed toward finding techniques which can track dynamic changes, while at the same time providing sufficient noise reduction. One popular approach is to introduce certain prior information on the behavior of EP morphology, for example, by assuming that each response can be modeled as a linear combination of a subset of orthogonal basis functions. A noise-reduced response is obtained by "reconstructing" the response from a small number of basis functions; the weights of the linear combination result from fitting the basis functions to the observed response. The analysis of time-varying EP changes is commonly referred to as *single-trial analysis* and is described in Sections 4.5 and 4.6.

By convention, the peak/trough wave components of the EP are referred to by the letters P (positive amplitude) and N (negative amplitude). A number is appended to the letter reflecting the latency in milliseconds from the time at which the stimulus was elicited. Alternatively, the appended number may reflect the temporal order of the component, and is then less than ten. For example, P300 signifies that a positive peak occurred at 300 ms, whereas N3 implies that the third waveform component had a negative amplitude. It should be noted that EPs, by odd convention, are usually plotted with reversed polarity so that P300 actually corresponds to a trough, and vice versa (this convention is shared by some other bioelectrical signals).

Evoked potentials resulting from *auditory* (AEP), *visual* (VEP), and *somatosensory* (SEP) stimulation are the most commonly used modalities in clinical routine, with each modality being described in further detail in Section 4.1. For all modalities, measurements on latency and amplitude are extracted from the waves of the averaged EP and are compared to normative values in order to discriminate normal, healthy subjects from subjects with various kinds of neurological impairment [3]. Normative values are strongly dependent on age, and, therefore, different values have been determined for newborns and adults. Factors which suggest that an EP should be interpreted as abnormal include waves which

- have increased latency,

- have decreased amplitude, or

- are missing.

Evoked potentials are often analyzed in individual channels with respect to temporal and amplitude waveform properties without involving information recorded in other channels—an analysis perspective also adopted in this chapter. However, additional information can be derived on the *spatial distribution* of voltages on the scalp by simultaneously analyzing all the data in a multichannel recording. The analysis of such a recording is illustrated in Figure 4.2(a) where the EPs resulting from averaging of data, obtained at different electrode positions, are displayed. From these EPs, it is possible to construct a series of maps which reflects how the spatial distribution of voltages evolves in time, see Figure 4.2(b). Each map is created by connecting points on the scalp with equal voltages, i.e., an isopotential map; interpolation is usually performed to make the map continuous-looking for ease of interpretation. The resulting series of maps can be used to localize the underlying sources that would generate the maps, sources which usually are considered in terms of a dipole model [4]. The calculation of maps is, of course, not restricted to EPs, but can also be done with EEGs, recorded without any external stimulus, for the purpose of, for example, locating an epileptic focus [5].

4.1 Evoked Potential Modalities

A stimulus elicits electrical impulses in local sensory nerve cells which propagate along the nerve fibers to the brain. The sum of all resulting impulses, in combination with the ongoing electrical activity of the brain, constitutes the stimulus response. The elicited impulse is initially spike-shaped with a very short duration, but is prolonged by various factors when recorded by the surface electrode on the scalp. The change in waveform morphology is partly caused by impulse propagation in several parallel nerve fibers with

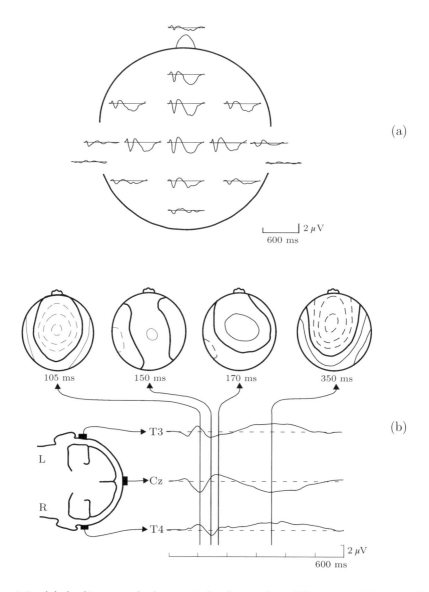

Figure 4.2: (a) Auditory evoked potentials obtained at different positions on the scalp. (b) Scalp distribution maps of voltages, calculated at four different latencies, for three of the AEPs shown in (a). The heavy lines in the maps represent the zero voltage level. The dashed lines represent contours for negative voltages, and the thin lines represent contours for positive voltages. The maps were computed for latencies at which the waveforms either had a peak or a trough. (Reprinted from Picton et al. [4] with permission.)

slightly different conduction velocities. Another determining factor is that the surface electrode "views" the neural activity over a rather large region, and, therefore, the morphology of the EP is smoothed. The resulting response reflects impulse propagation from the entrance of the brainstem into various parts of the cortex, see Figure 2.2. However, the amplitude of the brainstem response is smaller than that of the cortex since it originates from a more distant part of the brain with respect to the electrode. During identical stimulus and recording conditions, the brainstem response is about one tenth the size of the cortical response.

Another general property of an EP is that the waves exhibit a gradual slowdown as the response propagates toward the more complex structures of the cortex, i.e., the interpeak latencies are prolonged with time. Therefore, interpeak latencies of the brainstem response are on the order of a few milliseconds, while late cortical responses have an interpeak latency of more than 100 ms. Often, short latencies are associated with low amplitudes, while longer latencies are associated with larger amplitudes.

4.1.1   Auditory Evoked Potentials

Auditory EPs are generated in response to an auditory stimulation usually produced by a short sound wave. This type of evoked response reflects how neural information propagates from the acoustic nerve in the ear to the cortex. The response can be divided into three different intervals according to latency: the brainstem response, constituting the earliest part, followed by the middle and late cortical responses. Brainstem auditory evoked potentials (BAEPs) have primarily been used for the evaluation of different types of hearing loss ("audiometry"), diagnosis of certain brainstem disorders, and intraoperative monitoring in order to prevent neurological damage during surgery [6].

The waveform characteristics of the middle latency AEP are useful for monitoring the depth of anesthesia during surgery [7, 8]. Since a change in concentration of the anesthetic dose has been found to be closely related to latency, appropriate anesthetic depth can be maintained by continuously tracking changes in latency.

Recording setup.   Auditory EPs are elicited by a short duration click sound delivered to the subject through a conventional set of stereo headphones. One ear is stimulated at a time, while the other ear is masked with bandlimited noise ("pink noise"). The click sound is usually produced by a 0.1-ms square wave pulse, having a repetition rate of 8–10 clicks per second. The stimulus intensity is commonly defined in units of peak equivalent sound

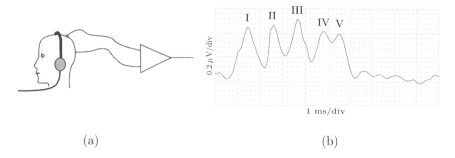

(a)   (b)

Figure 4.3: Auditory evoked potentials. (a) Recording setup and (b) typical morphology of a brainstem auditory evoked potential.

pressure level and can vary between 40 and 120 decibels (dB) [9]; the dB scale is logarithmic with 0 dB defined as a sound pressure of 20 μPa.

Auditory EPs are usually recorded by placing electrodes behind the left and right ear and at the vertex. The placement is identical to that used in EEG recordings, i.e., the standardized 10/20 electrode placement system described in Section 2.3.

Waveform characteristics. The three parts of the AEP exhibit considerable differences in signal properties. The BAEP has a very low amplitude, ranging from 0.1 to 0.5 μV, and occurs from 2 to 12 ms after stimulus. Due to its low amplitude, several thousands of stimuli are required to achieve an acceptable noise level by averaging. The short duration of the BAEP implies that most of its spectral content is contained in the interval from 500 Hz to about 1.5 kHz [10, 11]. In a normal subject, the BAEP consists of up to seven waves, generated by various neural structures in the auditory pathways. By convention, these waves are labeled with Roman numbers, see Figure 4.3. The loss or reduction of individual waves provides clinically important information, as do absolute and interpeak latencies.

The middle AEP occurs from 12 to 50 ms, and is followed by the late response [6]. The amplitudes of these later components are considerably larger (1–20 μV) than those of the BAEP and increase with latency. One hundred to 1000 stimuli are usually sufficient for adequate noise reduction. While the early brainstem response is quite reproducible from stimulus to stimulus, the middle and late responses can exhibit considerable variability in morphology.

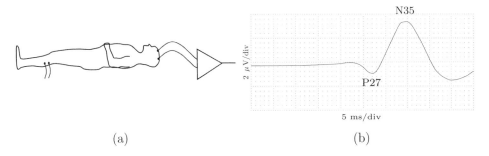

(a) (b)

Figure 4.4: Somatosensory evoked potentials. (a) Recording setup and (b) typical SEP morphology when the peroneal nerve of the leg is stimulated.

4.1.2 Somatosensory Evoked Potentials

Somatosensory evoked potentials (SEPs) are elicited by electrical stimulation from the body surface of a particular peripheral nerve, usually from an arm or a leg, see Figure 4.4.[1] This type of stimulation provides valuable information about nerve conduction functionality between the selected stimulation point via the spinal cord to the cerebral cortex.

Somatosensory EPs can be used to identify blocked or impaired conduction in the sensory pathways, produced by certain neurological disorders such as multiple sclerosis [12]. Another application of the SEP is intraoperative monitoring during spine surgery; an unchanged waveform morphology throughout surgery suggests that no deterioration in neurological function has taken place.

Recording setup. Stimulation is performed by delivering a brief electrical impulse via two stimulus electrodes positioned close to the sensory nerve fiber. Similar to the recording of AEPs and VEPs, SEPs are recorded by placing electrodes over the motorsensory cortex at predefined locations. However, a number of additional electrodes are needed, and these are positioned along the nerve pathway to the cortex, e.g., on the knee and the spinal cord.

In clinical routine, SEPs are usually recorded by stimulation of three different nerves: the median nerve in the arm and the tibial and peroneal nerves which are both in the leg.

Waveform characteristics. The SEP has most of its spectral content in an interval located well above 100 Hz. The total SEP duration is about

---

[1]The prefix *somato* implies that electrical stimulation is possible from almost any nerve on the human body.

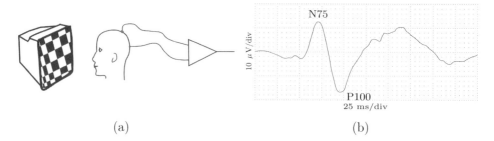

(a)   (b)

Figure 4.5: Visual evoked potentials. (a) Recording setup where pattern reversal is used as the stimulation method and (b) typical VEP morphology.

400 ms; however, only the first 40 ms are commonly recorded and analyzed because the long-latency response exhibits large variability. Similar to the AEP, the SEP amplitude has substantial intersubject variability and, therefore, is of limited clinical value. Important diagnostic information derived from the SEP waveform is conveyed by the absence of peaks, slow conduction velocities, and electrode-to-electrode variations in conduction velocity.

4.1.3   Visual Evoked Potentials

The electrical response elicited by visual stimuli can be recorded from the occipital region of the scalp for the evaluation of visual pathway functionality. Two different types of stimulus are used, pattern reversal and flashing, depending on the suspected disorder and the ability of the subject to cooperate during the recording procedure, see Figure 4.5. The clinically useful information from VEPs is extracted from the later parts of the response, starting at about 75 ms, and, accordingly, the VEP is referred to as a long-latency response [13].

Visual EPs are used for investigating ocular and retinal disorders and for detecting visual field defects and optic nerve pathology. It has also been suggested that the VEP be used for intraoperative monitoring where the aim is to detect early changes in waveform morphology in order to avoid visual loss and damage to the optic nerve.

Recording setup.   The recording of a VEP is often based on a pattern reversal stimulus, generated by a chessboard pattern being displayed on a video screen. During the investigation, the patient is required to focus on a point in the center of the screen while the black-and-white squares are reversed at a fixed repetition rate so that the white squares become black, and vice versa. Typically, a rate of two reversals per second is used. The size of the chessboard squares, the luminance and contrast of the squares, and the repetition rate exemplify factors which influence the waveform amplitude and latency. These factors must therefore be taken into account when interpreting the VEP; the factors can obviously be manipulated in order to infer additional information from the VEP.

The use of flash stimulus is considered when the patient is unable to either focus or maintain the level of fixation required for pattern reversal stimulation. For example, flash stimulation can be helpful when suspected vision disorders are investigated in neonates. Flash stimulus is delivered at a rate of five to seven flashes per second. Although the eyes are closed during this procedure, a sufficient amount of light will pass through the eyelids to activate the retina.

The recording electrodes are positioned at locations close to the visual cortex, and the reference electrode is placed at the vertex.

Waveform characteristics.   The VEP has an amplitude which is considerably larger than that of an AEP or SEP, ranging up to 20 μV. As a consequence, the VEP is the only type of EP that, at best, can be observed directly in the EEG without prior noise reduction. However, such reduction is performed in clinical routine in order to assure a sufficiently low noise level; typically 100 stimuli are needed to achieve that level.

The spectral components of a VEP range, in rough terms, from 1 to 300 Hz. The P100 wave can occasionally exhibit a split ("bifid") morphology, which may be indicative of abnormality. From a signals analyst's point of view, the presence of a bifid morphology implies that the high-frequency content of a VEP increases.

In a normal subject, the VEP waveform configuration is described by a small positive peak, a larger negative peak occurring about 75 ms after stimulus (N75), and a large positive peak about 100 ms after stimulus (P100). The duration of the response may extend beyond 300 ms. The absolute latency of P100 as well as differences in P100 latency between the left and right eye are important measurements which are useful for diagnosis.

4.1.4   Evoked Potentials and Cognition

The above three EP modalities represent different types of response to physical stimuli and are therefore referred to as "exogenous" responses. However, EPs can also be elicited by various cognitive factors ("endogenous" responses), resulting in a late response with latencies of 300 ms and longer [14, 15]. The most well-known, late-latency peak is P300 which is considered to reflect the cognitive capability of a subject, involving higher mental functions such as attention and memory processes. The P300 latency is related to the time required for memory updating associated with a given task; in general, latency decreases for increased cognitive performance [16].

The P300 is commonly elicited by means of an "oddball" task in which two different stimuli, for example, two tones with different pitch, are presented at random to the subject, although one of the stimuli occurs more infrequently. The task of the subject is to indicate, by pressing a button, when the infrequent stimulus occurs (i.e., the oddball) but not when the frequent one occurs. The responses related to the infrequent stimulus are then averaged, and the resulting waveform is analyzed [17].

Evoked potentials have also been considered in the study of language comprehension, where it is of interest to understand how the normal brain constructs meaning from words in real time. It has been shown that a negative peak at around 400 ms (N400) varies systematically when semantic information is being processed [18–20]. For example, it has been observed that a stimulus defined by a semantically anomalous word in a sentence context produces an electrical response. The amplitude of the N400 in response to such an "outlier" word is sensitive to the local context in which it occurs; words which are in context are easier to process for the brain and, therefore, elicit smaller N400 amplitudes than do words which are out of context (Figure 4.6). The amplitude of the N400 peak has also been found to be sensitive to the ease with which information is accessed from long-term memory.

4.2   Noise Characteristics

In the analysis of EPs, noise is essentially synonymous with the spontaneous background EEG activity whose signal properties have been described in Section 2.2. Thus, the EEG is the target activity when methods for noise reduction are discussed below. Noncerebral noise sources, hampering the success of EEG signal processing, must also be taken into account when EPs are processed. We reiterate the insight expressed in Section 3.2, that the main noncerebral sources are eye blinks, eye and eyelid movements, muscle activity, 50/60 Hz powerline interference, instrumentation noise, and poor electrode attachment.

Linear, time-invariant, bandpass filtering is sometimes used to remove noise whose spectral content is outside that of the EP. It is imperative to use filters with a linear phase response in order to avoid distorting the interpeak latencies [22–24]. For example, the analysis of BAEPs usually includes bandpass filtering of the averaged signal; the lower and upper cut-off frequencies of the filter are located approximately at 25 and 2000 Hz, respectively. As we will see later, the bandpass characteristic arises naturally when the aim is to design filters which maximize the signal-to-noise ratio (SNR) of EPs.

'They wanted to make the hotel look like a tropical resort.
So along the driveway they planted rows of ....'

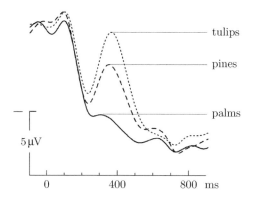

Figure 4.6: The amplitude of the N400 response is sensitive to semantic relationships. The sentence above is presented with three different endings, of which two are unexpected and associated with increased amplitude in the interval between 250 and 500 ms after the final word. The ending "pines", which is within the same category (i.e., trees) as the expected ending, produced an N400 peak with lower amplitude than the ending "tulips", which violates the tree category. (Reprinted from Federmaier and Kutas [21] with permission.)

Certain types of noise and artifacts are related in time to the stimulus and, therefore, will degrade the performance of ensemble averaging methods. Eye movement represents the most important source of such artifacts and distorts the signal to various degrees depending on the direction of the eye movement and the electrode position. The influence of such ocular activity can be substantially reduced by computing an ensemble average of the ocular activity using the EOG which reflects the electrical activity associated with eye movement. The averaging technique requires that eye movement first be reliably detected in an EOG. A signal, produced by suitably combining selected weights of the average ocular activity, is then subtracted from the EEG channel, and the "corrected" EEG is used for analysis of EPs [25–28]. The weights used for subtraction can be determined using the linear cancellation method described in Section 3.2.4.

The electrical activity of the heart is another noise source whose contribution is difficult to cancel with ensemble averaging techniques, especially when the heart rate happens to coincide with the stimulus rate. Information on heartbeat timing can, however, be acquired so that a stimulus rate can be selected which precludes this type of time-locked artifact [29]. Another technique for avoiding such artifacts is to use an aperiodic presentation of the stimulus [30], [31, p. 192ff].

4.3   Noise Reduction by Ensemble Averaging

The starting point for noise reduction in event-related signals is the observation that a stimulus causes a brain response time-synchronized to the stimulus. The stimulus is assumed to be elicited at equidistant points in time, and the resulting EPs are contained in a signal recorded from a suitably positioned scalp electrode. The observed EEG signal can be transformed into an ensemble of M different potentials, with each potential $x_i(n)$ described by N samples,

$$x_i(n), \quad i = 1, \ldots, M; \; n = 0, \ldots, N - 1. \tag{4.1}$$

The ensemble is the natural starting point for the discussion in this section and is conveniently represented by the $N \times M$ matrix $\mathbf{X}$,

$$\mathbf{X} = \begin{bmatrix} \mathbf{x}_1 & \mathbf{x}_2 & \cdots & \mathbf{x}_M \end{bmatrix}, \tag{4.2}$$

where the ith potential is contained in the column vector

$$\mathbf{x}_i = \begin{bmatrix} x_i(0) \\ x_i(1) \\ \vdots \\ x_i(N - 1) \end{bmatrix}. \tag{4.3}$$

Figure 4.7 illustrates the formation of an ensemble with EPs.

The assumption of perfect time synchronization between stimulus and response, being implicit in (4.1), is not always valid; variations in latency can be attributed, to various degrees, to the inherent phenomenon of biological variability. It may, therefore, be necessary to introduce techniques that can estimate and compensate for such variations prior to ensemble averaging (see Section 4.3.7). Other complications are related to the fact that EPs can differ in amplitude and morphology and are sometimes even completely absent; some of these aspects are discussed in Section 4.3.4.

Although noise reduction by ensemble averaging will be discussed within the context of EP analysis, identical approaches have been considered in many applications in the area of biomedical signal processing. For example, reliable characterization of cardiac late potentials in the ECG requires that averaging of several cardiac cycles first be performed (Section 6.6.5). In that particular application, no prior time reference is available for the heartbeats, but it must be inferred from the signal itself employing a suitable detection/estimation procedure.

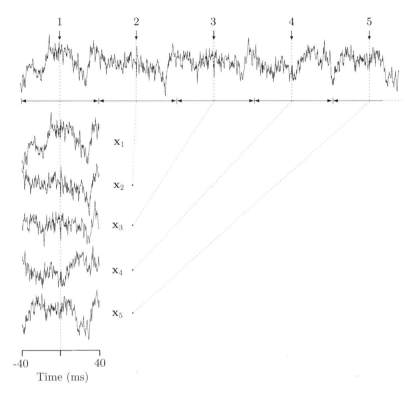

Figure 4.7: Formation of an ensemble with EPs of somatosensory origin. The stimulus is elicited periodically at the instants indicated by the arrows. The signal $\mathbf{x}_1$ contains 40 ms of data preceding the first stimulus (prestimulus data), 40 ms of data following the first stimulus, and so on. Stimulation can, if required, be done aperiodically.

4.3.1 Averaging of Homogeneous Ensembles

Ensemble averaging is based on a simple signal model in which the potential $\mathbf{x}_i$ of the ith stimulus is assumed to be additively composed of a deterministic, evoked signal component $\mathbf{s}$ and random noise $\mathbf{v}_i$ which is asynchronous to the stimulus,

$$\mathbf{x}_i = \mathbf{s} + \mathbf{v}_i, \tag{4.4}$$

where

$$\mathbf{s} = \begin{bmatrix} s(0) \\ s(1) \\ \vdots \\ s(N-1) \end{bmatrix}. \tag{4.5}$$

The noise $\mathbf{v}_i$ of the ith EP,

$$\mathbf{v}_i = \begin{bmatrix} v_i(0) \\ v_i(1) \\ \vdots \\ v_i(N-1) \end{bmatrix}, \tag{4.6}$$

is assumed to derive from the ongoing "noise" process $v(n)$ which, in this model, is a stationary, zero-mean process,

$$E[v(n)] = 0. \tag{4.7}$$

The noise is characterized by its correlation function $r_v(k)$,

$$r_v(k) = E[v(n)v(n-k)]. \tag{4.8}$$

Consequently, the noise variance is fixed and identical in all potentials,

$$r_v(0) = E[v_i^2(n)] = \sigma_v^2, \quad i = 1, \ldots, M. \tag{4.9}$$

Ensemble averaging does not exploit the detailed properties of $r_v(k)$, except that $r_v(k)$ is assumed to decay to zero so fast that the background noise can be considered to be uncorrelated from potential to potential,

$$E[v_i(n)v_j(n-k)] = r_v(k)\delta(i-j), \tag{4.10}$$

where

$$\delta(i) = \begin{cases} 1, & i = 0; \\ 0, & i \neq 0. \end{cases} \tag{4.11}$$

However, the detailed properties of $r_v(k)$ will be investigated in Section 4.4 for the purpose of noise reduction using linear filtering techniques.

Ensemble averaging is a straightforward approach to estimate the deterministic signal $\mathbf{s}$ and produces the estimate $\hat{\mathbf{s}}_a$,

$$\hat{\mathbf{s}}_a = \frac{1}{M}(\mathbf{x}_1 + \mathbf{x}_2 + \cdots + \mathbf{x}_M) = \frac{1}{M}\mathbf{X1} \tag{4.12}$$

$$= \mathbf{s} + \frac{1}{M}\mathbf{V1},$$

where the noise components, represented by the column matrix

$$\mathbf{V} = \begin{bmatrix} \mathbf{v}_1 & \mathbf{v}_2 & \cdots & \mathbf{v}_M \end{bmatrix}, \tag{4.13}$$

are attenuated by the factor $1/M$. The column vector $\mathbf{1}$ has the value one in all entries. The exact notation is $\hat{\mathbf{s}}_{a,M}$, where M indicates the size of the

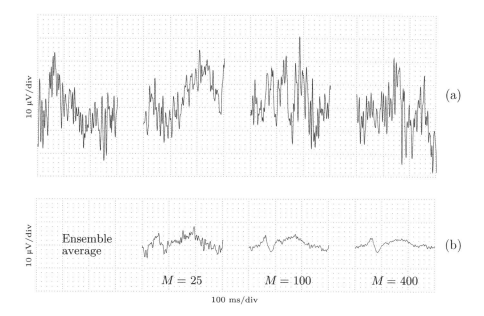

Figure 4.8: Noise reduction by ensemble averaging. (a) Four of the M different VEPs contained in the ensemble and (b) the resulting ensemble average based on different ensemble sizes.

ensemble; however, M is omitted unless it is explicitly required. The more familiar expression for ensemble averaging,

$$\hat{s}_a(n) = \frac{1}{M} \sum_{i=1}^{M} x_i(n),\tag{4.14}$$

is obtained by separating out the nth element of $\hat{\mathbf{s}}_a$ in (4.12). The performance of ensemble averaging is illustrated in Figure 4.8, where $\hat{\mathbf{s}}_a$ is computed from ensembles of different sizes, with each individual potential having a very low SNR.

It is of great interest to determine the statistical properties of $\hat{s}_a(n)$ as characterized, for example, in terms of its mean and variance. Since the noise is zero-mean, it is easily shown that the ensemble average $\hat{s}_a(n)$ is an unbiased estimator,

$$E[\hat{s}_a(n)] = s(n).\tag{4.15}$$

Combining this result with (4.10), which states that the noise is uncorrelated from potential to potential, it is easily shown that $\hat{s}_a(n)$ has a variance

$V[\hat{s}_a(n)]$ being inversely proportional to the number of averaged potentials,

$$V[\hat{s}_a(n)] = E\big[(\hat{s}_a(n) - E[\hat{s}_a(n)])^2\big]$$

$$= \frac{1}{M^2} \sum_{i=1}^{M} \sum_{j=1}^{M} E[v_i(n)v_j(n)]$$

$$= \frac{\sigma_v^2}{M}. \tag{4.16}$$

Since the variance approaches zero with an increasing value of M, the estimator is referred to as consistent. From (4.16) it is obvious that the expected magnitude of the noise is reduced by a factor $\sqrt{M}$. Accordingly, a fourfold number of potentials is required to reduce the noise level, defined by σ_v, by a factor of two.

The reduction in noise variance rests upon a number of model assumptions whose validity must be examined. The discussion also provides important background to the development of other methods described later in this chapter.

1. Starting with the noise, the assumption of zero-mean in (4.7) is the least critical of the model; a nonzero mean can easily be estimated from the observed signal and subtracted.

2. Large, slowly changing EEG components may invalidate the assumption in (4.10), which states that the noise is uncorrelated from potential to potential. However, suitable use of linear phase, highpass filtering can often remedy this problem without distorting the amplitudes and latencies of the EP waveform. Any remaining interpotential correlation will reduce the effectiveness of ensemble averaging. Decreasing the stimulus rate reduces this problem although at the expense of prolonged acquisition time.

   Although the ongoing EEG activity may be viewed as a stationary process during a single potential, its statistical properties change considerably during the course of an EP investigation. One way to account for such changes is to assume that the noise variance is response-dependent; weighted averaging is a methodology that exploits such an assumption, see Section 4.3.4.

3. The assumption of a signal component $s(n)$ being fixed from potential to potential is inadequate in certain applications, for example, in intra-operative monitoring where sudden changes in waveform morphology may occur. Even under less strenuous conditions, various factors, due to subject expectation and habituation or environmental activity, may

introduce potential-to-potential variations. Techniques for handling such variations are presented in Section 4.3.7.

Although ensemble averaging can be based on shorter, consecutive sub-intervals ("subaveraging") for improved tracking of waveform changes, such an improvement comes at the expense of less efficient noise reduction. Therefore, various methods involving model-based signal processing and adaptive estimation techniques have been developed for single-trial analysis (Section 4.5). Even the analysis of unaveraged EP data has recently been advocated as a better way of viewing and describing event-related brain dynamics [32].

4. In addition to inter-response changes in waveform morphology, the signal $s(n)$ may exhibit correlation with the noise $v_i(n)$. This situation may arise when the patient is aware of the experiment and, therefore, expects a stimulus. The additive model in (4.4) was actually seriously questioned by Sayers et al. [33, 34] who hypothesized that the EP is not an additive component, but rather a phase reorganization of the ongoing EEG. For auditive stimuli, they showed, using Fourier-based spectral analysis, that the phase spectra changed from the prestimulus EEG to the EP, while the amplitude spectra and the total energy content of the signal largely remained unchanged. A more recent study was, however, not able to confirm these findings, but concluded that additive energy was introduced during the presence of an EP [35]. In another experiment, a certain interaction between the EP and the ongoing EEG activity was observed for visual stimuli [36]. In particular, it was found that the EEG amplitude decreased by 5–15% during massive visual stimulation.

   While correlation between $s(n)$ and $v_i(n)$ implies less efficient noise reduction, the ensemble averaging procedure is still applicable and will enhance a recurrent signal component.

5. Ensemble averaging, as presented above, does not involve any assumption on the statistical distribution of the noise. However, the computation of the mean and the variance in (4.15) and (4.16), respectively, possibly suggests that the noise is Gaussian since these two quantities provide complete characterization of $\hat{s}_a(n)$ when the noise is white and Gaussian. Not really surprisingly, it will later be shown that the ensemble average constitutes an optimal estimator of $s(n)$ when the noise is Gaussian. The appropriateness of modeling the spontaneous EEG as a Gaussian process has been the subject of investigation, see Section 3.1.2 and [37, 38]. Although published results are conflicting, it can be concluded that the EEG is close to a Gaussian distribution

when considered as an ensemble of waveforms [39]. Intermittently occurring noise and artifacts, for example, due to blinking, may suggest
that a noise distribution with longer tails is more appropriate. Since
the ensemble average in (4.12) cannot handle such noise distributions
in a graceful way, it is desirable to study more robust methods for the
estimation of $s(n)$; the topic of robust averaging is briefly described in
Section 4.3.5.

Once the assumptions associated with ensemble averaging are judged to
be reasonably valid, it may be of interest to estimate the variance σ_v^2 in order
to assess the reliability of the ensemble average. A simple approach to variance estimation is to use the "silent" prestimulus interval which essentially
contains only noise. However, this approach becomes unreliable when the
prestimulus interval is very short and fails completely when the timing of
stimuli is such that late EP components overlap the early components of the
subsequent response.

Another technique for estimating σ_v^2 is to compute the ensemble variance
for each sample n using the following expression,[2]

$$\hat{\sigma}_v^2(n) = \frac{1}{M} \sum_{i=1}^{M} (x_i(n) - \hat{s}_{a,M}(n))^2 . \tag{4.17}$$

Although the ensemble variance is a function of time, it is reasonable to estimate σ_v^2 by averaging the samples of $\hat{\sigma}_v^2(n)$ within the observation interval
$[0, N-1]$ if $s(n)$ is fixed and the noise is stationary. Figure 4.9 shows the
reduction in noise level, as characterized by the ensemble standard deviation $\hat{\sigma}_v/\sqrt{M}$, for an increasing number of potentials M.

It should be pointed out that the ensemble variance is not only used
for estimating σ_v^2, but can also be considered for measuring the degree of
morphologic waveform variability, as produced by various underlying physiological mechanisms. For example, latency variation of a certain peak is
manifested by a local peak in $\hat{\sigma}_v^2(n)$ which stands out from the background
variability due to noise.

Yet another approach to estimate σ_v^2 is *split trial assessment*, being a
computationally much less costly technique than the ensemble variance defined in (4.17) [40, 41]. The ensemble is split into two parts of equal size
obtained by grouping together odd- and even-numbered potentials and com-

---

[2]Although this is an asymptotically unbiased estimator, it is used for reasons of similarity with other estimators presented in this chapter. In the unbiased variance estimator
the factor $1/M$ is replaced by $1/(M-1)$.

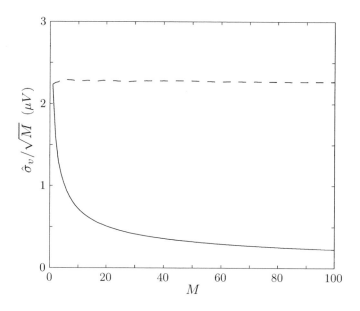

Figure 4.9: The reduction in noise level of the ensemble average as a function of the number of potentials M; the data are identical to those used in Figure 4.8. The noise level is defined by the ensemble standard deviation $\hat{\sigma}_v$, obtained by averaging $\hat{\sigma}_v(n)$ over the interval $[0, N-1]$, cf. (4.17). The dashed line displays the noise estimate $\hat{\sigma}_v$ before division by the factor $1/\sqrt{M}$.

puting the corresponding subaverages,

$$\hat{s}_{a_l}(n) = \frac{2}{M} \sum_{i=1}^{M/2} x_{2i-l}(n), \quad l = 0, 1, \tag{4.18}$$

where M is assumed to be an even integer. The ensemble can, of course, be divided in many other ways, e.g., using the first and second half of the ensemble, respectively, to compute the subaverage,

$$\hat{s}_{a_l}(n) = \frac{2}{M} \sum_{i=1}^{M/2} x_{i+lM/2}(n), \quad l = 0, 1. \tag{4.19}$$

The estimation of σ_v^2 is based on the difference signal $\Delta \hat{s}_a(n)$ between the two subaverages,

$$\Delta \hat{s}_a(n) = \hat{s}_{a_0}(n) - \hat{s}_{a_1}(n), \tag{4.20}$$

which only contains noise since the signal contribution is cancelled out. Using either of the subaverage definitions in (4.18) and (4.19), the variance of

$\Delta\hat{s}_a(n)$ is given by

$$V\left[\Delta\hat{s}_a(n)\right] = \frac{4\sigma_v^2}{M}. \tag{4.21}$$

Thus, an estimate of σ_v^2 can be determined by first computing the variance of $\Delta\hat{s}_a(n)$ followed by multiplication by $M/4$,

$$\hat{\sigma}_v^2 = \frac{M}{4}\,\hat{\sigma}_{\Delta\hat{s}_a}^2$$

$$= \frac{M}{4}\frac{1}{N}\sum_{n=0}^{N-1}(\Delta\hat{s}_a(n))^2. \tag{4.22}$$

It should be emphasized that the variance estimation in (4.22) is computed in time for consecutive samples, whereas the variance estimation in (4.17) is computed across the ensemble for a fixed instant in time.

Subaveraging has also been employed for defining performance measures that reflect the SNR of the ensemble average. One such measure is the cross-correlation coefficient ρ between the two subaverages $\hat{s}_{a_0}$ and $\hat{s}_{a_1}$ [42],

$$\rho = \frac{\hat{s}_{a_0}^T\hat{s}_{a_1}}{\sqrt{\hat{s}_{a_0}^T\hat{s}_{a_0}}\sqrt{\hat{s}_{a_1}^T\hat{s}_{a_1}}}. \tag{4.23}$$

By use of the Cauchy–Schwarz inequality, it can be shown that $|\rho| \leq 1$. The cross-correlation coefficient ρ approaches one when $\hat{s}_{a_0}$ and $\hat{s}_{a_1}$ become increasingly similar in morphology, and, therefore, the cross-correlation coefficient can be interpreted as a normalized SNR measure. In order to make the computation of ρ meaningful, the DC levels of $\hat{s}_{a_0}$ and $\hat{s}_{a_1}$ must first be removed. The SNR measure in (4.23) can be extended from being a cross-correlation between two subaverages to a cross-correlation between all M potentials—a measure which is referred to as "ensemble correlation" (Section 4.3.8). Other SNR definitions have been suggested in the literature which are based on a pair of subaverages [43–45].

In addition to the definition of an SNR, subaveraging represents a standard technique for visually assessing the reproducibility of an EP investigation. The two subaverages are graphically superimposed on each other to help reveal morphologic deviations that may indicate variability in signal morphology or noise with nonstationary properties.

4.3.2   Ensemble Averaging Interpreted as Linear Filtering

Ensemble averaging can be interpreted in terms of linear, time-invariant filtering and can therefore be characterized by a filter impulse response $h(n)$.

The derivation of $h(n)$ and its Fourier transform provide additional insight into the mechanisms of ensemble averaging. The ensemble of EPs can be viewed as a composite signal $x(n)$ that results from periodic stimulation with a period length N,

$$\hat{s}_a(n) = \frac{1}{M} \sum_{i=0}^{M-1} x(n - iN). \tag{4.24}$$

The signal $x(n)$ is obtained from concatenation of successive potentials $x_1(n), \ldots, x_M(n)$,

$$x(n) = x_{\lfloor \frac{n}{N} \rfloor + 1}\left(n - \left\lfloor \frac{n}{N} \right\rfloor N\right), \quad n = 0, \ldots, NM - 1, \tag{4.25}$$

where $\lfloor \cdot \rfloor$ denotes the integer part of the argument. Using the fundamental property which allows a discrete-time signal $x(n)$ to be expressed as a series expansion of unit impulse functions $\delta(n)$,

$$x(n) = \sum_{l=-\infty}^{\infty} x(l)\delta(n - l), \tag{4.26}$$

the ensemble average in (4.24) can be rewritten as a convolution between $x(n)$ and $h(n)$,

$$\hat{s}_a(n) = \frac{1}{M} \sum_{i=0}^{M-1} \sum_{l=-\infty}^{\infty} x(l)\delta(n - l - iN)$$

$$= \sum_{l=-\infty}^{\infty} x(l)h(n - l), \tag{4.27}$$

where the impulse response is defined by

$$h(n) = \frac{1}{M} \sum_{i=0}^{M-1} \delta(n - iN). \tag{4.28}$$

Ensemble averaging is thus synonymous with causal filtering of the M potentials, using an FIR filter whose coefficients are all identical to $1/M$. The frequency response $H(e^{j\omega})$ of the filter is given by the Fourier transform of $h(n)$,

$$H(e^{j\omega}) = \frac{1}{M} \sum_{n=0}^{\infty} \sum_{i=0}^{M-1} \delta(n - iN)e^{-j\omega n}. \tag{4.29}$$

The calculation of $H(e^{\jmath\omega})$ can be performed by first calculating the Fourier transform of

$$h'(n) = \frac{1}{M} \sum_{i=0}^{M-1} \delta(n-i) \tag{4.30}$$

and then taking care of the factor N. Since $h'(n)$ represents a rectangular signal of duration M, its Fourier transform is given by

$$\begin{aligned}
H'(e^{\jmath\omega}) &= \frac{1}{M} \sum_{n=0}^{\infty} \sum_{i=0}^{M-1} \delta(n-i)e^{-\jmath\omega n} \\
&= \frac{\sin(\omega M/2)}{M\sin(\omega/2)} \exp\left[-\frac{\jmath\omega(M-1)}{2}\right].
\end{aligned} \tag{4.31}$$

Next, we note that the factor N in (4.28) corresponds to a zero-filling operation of the impulse response $h'(n)$ in which $N-1$ zeros are inserted between the successive unit impulse functions, represented by the term $\delta(n-iN)$. It is well-known from multirate signal processing that this operation is synonymous with upsampling which results in an N-fold periodic repetition of $H'(e^{\jmath\omega})$ [46]. Hence, the desired frequency function $H(e^{\jmath\omega})$ is given by

$$\begin{aligned}
H(e^{\jmath\omega}) &= H'(e^{\jmath\omega N}) \\
&= \frac{\sin(\omega NM/2)}{M\sin(\omega N/2)} \exp\left[-\frac{\jmath\omega N(M-1)}{2}\right].
\end{aligned} \tag{4.32}$$

The magnitude response of $H(e^{\jmath\omega})$ is presented for various values of M and N in Figure 4.10. Since $|H(e^{\jmath\omega})|$ is a periodic function of ω with period $2\pi/N$, exhibiting a comb-like appearance, this type of filter is commonly referred to as a *comb filter*. It is obvious from Figures 4.10(a) and (b) that the amplitude and width of the sidelobes are inversely proportional to the number of potentials M. Hence, improved noise suppression is achieved when the sidelobes become increasingly smaller for increasing values of M. Problems arise, however, when the background EEG rhythm, or various types of artifact, has a spectral content close to the stimulus frequency and/or its harmonics. For example, the stimulus rate should not be selected as a harmonic of the powerline interference, i.e., the 50/60 Hz powerline frequency.

4.3.3   Exponential Averaging

A disadvantage of ensemble averaging is that the resulting estimate cannot track dynamic changes occurring in the observed signal. In order to derive such an estimator, we will start by describing how the ensemble average $\hat{s}_{a,M}$

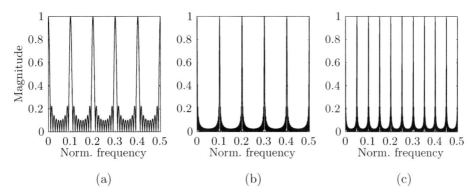

Figure 4.10: The magnitude response $|H(e^{j\omega})|$ of the ensemble averager $h(n)$ in (4.28) is shown for three different combinations of the number of potentials M and the length N of each potential: (a) $(M, N) = (10, 10)$, (b) $(M, N) = (40, 10)$, and (c) $(M, N) = (40, 20)$.

can be computed recursively. This is done by updating the existing estimate as soon as a new potential $\mathbf{x}_M$ becomes available. The update equation is then given by

$$\hat{\mathbf{s}}_{a,M} = \frac{1}{M}\mathbf{X}_M\mathbf{1}_M$$
$$= \frac{1}{M}(\mathbf{X}_{M-1}\mathbf{1}_{M-1} + \mathbf{x}_M)$$
$$= \hat{\mathbf{s}}_{a,M-1} + \frac{1}{M}(\mathbf{x}_M - \hat{\mathbf{s}}_{a,M-1}), \quad M \geq 1. \tag{4.33}$$

The recursion is initialized by setting

$$\hat{\mathbf{s}}_{a,0} = \mathbf{0}, \tag{4.34}$$

where $\mathbf{0}$ denotes a vector whose entries are all zero.

The recursive approach to computing the ensemble average suggests a technique for tracking slow changes in waveform morphology, namely, the one obtained by simply replacing the factor $1/M$ in (4.33) with a fixed weight factor α,

$$\hat{\mathbf{s}}_{e,M} = \hat{\mathbf{s}}_{e,M-1} + \alpha(\mathbf{x}_M - \hat{\mathbf{s}}_{e,M-1}). \tag{4.35}$$

This type of estimator is referred to as the *exponential averager* and is commonly used. The weight factor α should be chosen such that $0 < \alpha < 1$ in order to assure stability and an asymptotically unbiased estimator $\hat{\mathbf{s}}_{e,M}$ (see below). Evidently, small values of α imply that less new information is introduced into $\hat{\mathbf{s}}_{e,M}$ and thus slower tracking of morphologic changes results.

On the other hand, an α value close to one results in a noisy estimate since only the most recent potentials are considered.

In order to study the performance of the exponential averager, we will investigate the bias and the variance of $\hat{s}_{e,M}$ as the number of potentials approaches infinity. The bias is given by

$$E[\hat{s}_{e,M}(n)] = E\left[\sum_{i=0}^{M-1} \alpha(1-\alpha)^i x_{M-i}(n)\right]$$

$$= \sum_{i=0}^{M-1} \alpha(1-\alpha)^i (s(n) + E[v_{M-i}(n)])$$

$$= \left(1 - (1-\alpha)^M\right) s(n), \tag{4.36}$$

where initialization is assumed to be given by $\hat{s}_{e,0}(n) = 0$. For $0 < \alpha < 1$, $\hat{s}_{e,M}(n)$ is asymptotically unbiased since

$$\lim_{M\to\infty} E[\hat{s}_{e,M}(n)] = s(n). \tag{4.37}$$

The variance of $\hat{s}_{e,M}(n)$ is given by

$$V[\hat{s}_{e,M}(n)] = E\left[(\hat{s}_{e,M}(n) - E[\hat{s}_{e,M}(n)])^2\right]$$

$$= \alpha^2 \frac{1 - (1-\alpha)^{2M}}{1 - (1-\alpha)^2} \sigma_v^2. \tag{4.38}$$

The derivation of this expression is treated in Problem 4.7. The asymptotic variance is easily found to be

$$\lim_{M\to\infty} V[\hat{s}_{e,M}(n)] = \frac{\alpha}{2-\alpha}\sigma_v^2. \tag{4.39}$$

It may be of interest to compare the asymptotic variance in (4.39) to that of the ensemble averager. In doing so, a Taylor series approximation can be employed for the common case when α is close to zero,

$$\lim_{M\to\infty} V[\hat{s}_{e,M}(n)] = \frac{\alpha}{2}\left(1 + \frac{\alpha}{2} + \frac{\alpha^2}{4} + \cdots\right)\sigma_v^2$$

$$\approx \frac{\alpha}{2}\sigma_v^2. \tag{4.40}$$

Consequently, the asymptotic variance in (4.40) is approximately equal to that of the ensemble average $\hat{s}_{a,M}(n)$, given in (4.16), when the weight factor α is chosen such that

$$\alpha = \frac{2}{M}. \tag{4.41}$$

Similar to ensemble averaging, we can interpret exponential averaging in terms of linear, time-invariant filtering. Assuming that subsequent exponential averages form a repetitive signal $y(n)$,

$$y(n) = \hat{s}_{e,(\lfloor \frac{n}{N} \rfloor + 1)}\left(n - \left\lfloor \frac{n}{N} \right\rfloor N\right), \quad n = 0, \ldots, NM - 1, \qquad (4.42)$$

we can write

$$y(n) = y(n - N) + \alpha\left(x(n) - y(n - N)\right). \qquad (4.43)$$

The transfer function of the exponential averager is given by the Fourier transform of (4.43), which is

$$H(e^{j\omega}) = \frac{\alpha}{1 + (\alpha - 1)e^{-j\omega N}}. \qquad (4.44)$$

The magnitude function of $H(e^{j\omega})$ is shown in Figure 4.11 for two different values of the weight factor α.

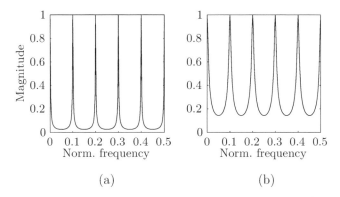

(a)   (b)

Figure 4.11: The magnitude response $|H(e^{j\omega})|$ of the exponential averager with the weight factor (a) $\alpha = 0.05$ and (b) $\alpha = 0.25$, for $N = 10$.

The performance of the exponential averager is illustrated in Figure 4.12 by a case where the EP amplitude suddenly (and by artificial means) increases by a factor of two. The corresponding exponential averages demonstrate the trade-off that must be made between achieving a low noise level, obtained for small values of α, and sufficiently rapid tracking of amplitude changes, obtained for large values of α.

The exponential averager will reappear later on in Section 4.6, where noise reduction is accomplished by combining adaptive filtering techniques with the modeling of EPs using a series expansion of orthogonal basis functions.

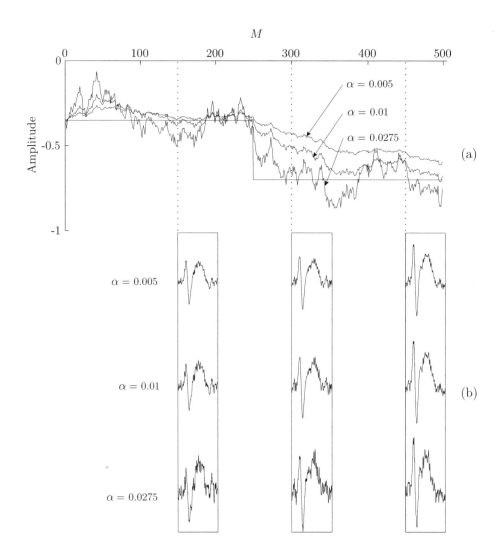

Figure 4.12: The performance of exponential averaging for a sudden increase in amplitude occurring at $M = 250$. (a) The amplitude of the trough of $\hat{\mathbf{s}}_{e,M}$ for α equal to 0.005, 0.01, and 0.0275, respectively, has been computed from (b) the signal estimate $\hat{\mathbf{s}}_{e,M}$.

4.3.4 Averaging of Inhomogeneous Ensembles

The assumption of a fixed noise variance in all EPs of the ensemble, as defined in (4.9), can be questioned since the ongoing EEG activity generally must be viewed as a nonstationary process. It may therefore be more appropriate to consider the ensemble of potentials as "inhomogeneous", signifying the presence of various types of noise and artifact such as:

- The presence of short-duration artifacts;

- EP-to-EP variations in noise level; and

- a non-Gaussian noise distribution.

Since such noise and artifacts cause performance degradation of ensemble averaging, it is important to develop methods which better handle these problems. One such method is weighted averaging, described below. Before going into the details of weighted averaging, it should be pointed out that simple measures such as the rejection of occasional EPs with unreasonably large amplitudes, for example, exceeding the limits of the A/D converter, can improve the accuracy of averaging [45]. Care must, however, be exercised not to reject too many EPs, since this leads to inadequate noise reduction.

Large EP-to-EP variations in noise level may persist despite the fact that artifact rejection has been performed. In order to account for such variations, the model in (4.4) is extended so that the noise variance is allowed to change from potential to potential; the noise within each potential is still considered to have fixed variance. This model extension implies that the fixed weights $1/M$ of ensemble averaging are replaced by weights adapted to the noise level of each potential $\mathbf{x}_i$: an EP with a high noise level is assigned a smaller weight than is an EP with a low noise level [47–50]. We will also describe the "dual" situation in which the amplitude of $\mathbf{s}$ is assumed to vary from EP to EP, while the noise variance is fixed.

The *weighted average* $\hat{\mathbf{s}}_w$ is computed by weighting the EPs in the ensemble $\mathbf{X}$ with the vector $\mathbf{w}$,

$$\hat{\mathbf{s}}_w = \mathbf{X}\mathbf{w}, \qquad (4.45)$$

where

$$\mathbf{w} = \begin{bmatrix} w_1 \\ w_2 \\ \vdots \\ w_M \end{bmatrix}. \qquad (4.46)$$

Each weight w_i must be positive-valued and chosen such that $\hat{\mathbf{s}}_w$ is unbiased,

$$E[\hat{\mathbf{s}}_w] = \mathbf{s}. \tag{4.47}$$

In order to determine the exact structure of the weight vector $\mathbf{w}$, we define the following two entities:

1. A model for the ensemble data $\mathbf{X}$; and

2. a performance criterion which yields the desired weight vector.

The model of all signals in the ensemble is described by

$$\mathbf{X} = \mathbf{s}\mathbf{a}^T + \mathbf{V}, \tag{4.48}$$

where $\mathbf{s}$ is a deterministic waveform whose amplitude can, if desired, be made to differ from EP to EP by the vector

$$\mathbf{a} = \begin{bmatrix} a_1 \\ a_2 \\ \vdots \\ a_M \end{bmatrix}. \tag{4.49}$$

Each amplitude a_i must be positive-valued to make the averaging operation meaningful. The additive noise $\mathbf{V}$ is completely characterized by its $M \times M$ correlation matrix

$$\mathbf{R}_V = E[\mathbf{V}^T\mathbf{V}], \tag{4.50}$$

which defines the noise correlation between different potentials. It should be emphasized that the matrix definition in (4.50) differs from the matrix $\mathbf{R}_v$ which describes the correlation between the samples $v(n)$ at different time lags. The diagonal elements of $\mathbf{R}_V$ describe the noise variance in each individual EP, while the remaining elements describe the degree to which the noise variances in two EPs are correlated. It is obvious from (4.48) and (4.50) that this model allows us to account for both the cases with varying noise variance and varying amplitude of $\mathbf{s}$.

Several criteria have been considered for determination of the weight vector $\mathbf{w}$, i.e., *maximum likelihood* (ML) estimation, *minimum mean-square error* (MMSE), and *SNR maximization*. Interestingly, all three criteria will, under certain conditions, produce the same weight vector as the optimal solution. Here, we will focus on SNR maximization since it does not require any assumptions on the noise distribution; the relationship between SNR maximization and ML estimation is then briefly reviewed. The derivation based on the MMSE is considered in Problem 4.16.

We begin our derivation by noting that the signal part of the weighted average is given by $\mathbf{sa}^T\mathbf{w}$ and the noise part is given by $\mathbf{Vw}$. The SNR of the weighted average $\hat{\mathbf{s}}_w$ is defined as the ratio of the signal energy and the noise energy,

$$\text{SNR} = \frac{\mathbf{w}^T\mathbf{as}^T\mathbf{sa}^T\mathbf{w}}{E[\mathbf{w}^T\mathbf{V}^T\mathbf{Vw}]} = \frac{\mathbf{w}^T\mathbf{as}^T\mathbf{sa}^T\mathbf{w}}{\mathbf{w}^T\mathbf{R}_V\mathbf{w}}, \tag{4.51}$$

where the expectation operator in the denominator is required since the noise is random. We can, without sacrificing generality, assume that the energy of $\mathbf{s}$ is normalized to unity,

$$\mathbf{s}^T\mathbf{s} = 1, \tag{4.52}$$

since the amplitude $\mathbf{a}$ and the noise variance, characterized by $\mathbf{R}_V$, are both allowed to vary.

The goal is now to find that weight vector $\mathbf{w}$ which maximizes the SNR. In order to perform the maximization, the numerator of (4.51) is maximized while the denominator is kept fixed, constraining it to

$$\mathbf{w}^T\mathbf{R}_V\mathbf{w} = 1. \tag{4.53}$$

This constraint is necessary since, otherwise, any factor of $\mathbf{w}$ will also maximize the SNR. Optimization problems involving one or several linear constraints are usually solved by the use of Lagrange multipliers (see Appendix A). The function to be maximized is augmented with the constraint in (4.53), multiplied by the Lagrange multiplier λ,

$$\mathcal{L} = \mathbf{w}^T\mathbf{Aw} + \lambda(1 - \mathbf{w}^T\mathbf{R}_V\mathbf{w}), \tag{4.54}$$

where $\mathbf{A}$ denotes a rank-one, amplitude "correlation matrix",

$$\mathbf{A} = \mathbf{aa}^T. \tag{4.55}$$

When the constraint is satisfied, the extra term in (4.54) is zero, so that maximizing $\mathcal{L}$ is equivalent to maximizing the SNR. Differentiation of $\mathcal{L}$ with respect to $\mathbf{w}$ yields

$$\nabla_{\mathbf{w}}\mathcal{L} = 2\mathbf{Aw} - 2\lambda\mathbf{R}_V\mathbf{w} = \mathbf{0}, \tag{4.56}$$

which is equal to

$$\mathbf{Aw} = \lambda\mathbf{R}_V\mathbf{w}. \tag{4.57}$$

This equation is recognized as the *generalized eigenvalue problem* [51]. In order to maximize the SNR, we should choose the eigenvector $\mathbf{w}$ associated with the largest eigenvalue $\lambda_{\max}$. This solution can be obtained by multiplying (4.57) by $\mathbf{w}^T$,

$$\mathbf{w}^T \mathbf{A} \mathbf{w} = \lambda \mathbf{w}^T \mathbf{R}_V \mathbf{w} = \lambda, \tag{4.58}$$

and noting that the weight vector that maximizes the SNR corresponds to the generalized eigenvector with the largest eigenvalue. The last equality in (4.58) follows from the constraint in (4.53).

Naturally, the solution of the generalized eigenvalue problem depends on the particular structure assigned to the quantities $\mathbf{a}$ and $\mathbf{R}_V$. In the following, we will consider the two cases "varying noise variance" and "varying signal amplitude", both having attracted attention in the EP literature.[3]

Case 1: Varying noise variance.   In this case, we assume that the amplitude is fixed for all stimulus responses and equal to a_0 [47, 52],

$$\mathbf{a} = a_0 \mathbf{1}. \tag{4.59}$$

The noise variance is modeled by the diagonal matrix $\mathbf{R}_V$,

$$\mathbf{R}_V = N \begin{bmatrix} \sigma_{v_1}^2 & 0 & \cdots & 0 \\ 0 & \sigma_{v_2}^2 & \cdots & 0 \\ \vdots & \vdots & & \vdots \\ 0 & 0 & \cdots & \sigma_{v_M}^2 \end{bmatrix}, \tag{4.60}$$

where $\sigma_{v_i}^2$ denotes the noise variance of the ith potential. Insertion of the amplitude and the noise variance into (4.57) yields the following ordinary eigenvalue problem,

$$a_0^2 \mathbf{R}_V^{-1} \mathbf{1} \mathbf{1}^T \mathbf{w} = \lambda \mathbf{w}. \tag{4.61}$$

While the solution to this equation, in general, consists of M different eigenvectors, the matrix $\mathbf{R}_V^{-1} \mathbf{1} \mathbf{1}^T$ has rank one, and, therefore, only one eigenvalue, λ_1, is nonzero, which thus must be the largest value, i.e., $\lambda_{\max}$; the remaining eigenvalues are $\lambda_2 = \ldots = \lambda_M = 0$.[4] In order to find the nonzero eigenvalue, and the corresponding eigenvector, we will make use of the relation that (cf. (A.39))

$$a_0^2 \, \text{tr}(\mathbf{R}_V^{-1} \mathbf{1} \mathbf{1}^T) = \sum_{i=1}^{M} \lambda_i = \lambda_1 = \lambda_{\max}. \tag{4.62}$$

---

[3]While these two cases were originally analyzed using different performance criteria, we will treat both cases within the context of SNR maximization and, accordingly, as solutions to the generalized eigenvalue problem.

[4]The reader may want to consult Appendix A, page 638 and onwards, for a brief review of matrix eigendecomposition.

For this particular case, the trace of the matrix can be expressed as a quadratic form using (A.30),

$$\lambda_1 = a_0^2 \, \text{tr}(\mathbf{R}_V^{-1}\mathbf{1}\mathbf{1}^T) = a_0^2 \mathbf{1}^T \mathbf{R}_V^{-1} \mathbf{1}. \tag{4.63}$$

By insertion of λ_1 into (4.61), it can be shown that the corresponding eigenvector is given by

$$\mathbf{w} = c_w \mathbf{R}_V^{-1} \mathbf{1}, \tag{4.64}$$

where c_w denotes a constant chosen such that the requirement of unbiased estimation in (4.47) is fulfilled. We have

$$
\begin{aligned}
E[\hat{\mathbf{s}}_w] = E[\mathbf{X}\mathbf{w}] &= E[\mathbf{X}c_w \mathbf{R}_V^{-1}\mathbf{1}] \\
&= E[(\mathbf{s}a_0\mathbf{1}^T + \mathbf{V})c_w \mathbf{R}_V^{-1}\mathbf{1}] \\
&= E[a_0\mathbf{s}\mathbf{1}^T c_w \mathbf{R}_V^{-1}\mathbf{1}] \\
&\overset{?}{=} a_0 \mathbf{s},
\end{aligned}
\tag{4.65}
$$

which implies that

$$c_w = \frac{1}{\mathbf{1}^T \mathbf{R}_V^{-1}\mathbf{1}}. \tag{4.66}$$

Note that this choice of c_w also fulfils the constraint in (4.53). The optimal weight vector is then found by combining the expression of c_w with $\mathbf{w}$ in (4.64), yielding

$$\mathbf{w} = \frac{\mathbf{R}_V^{-1}\mathbf{1}}{\mathbf{1}^T \mathbf{R}_V^{-1}\mathbf{1}} = \frac{1}{\displaystyle\sum_{i=1}^{M} \frac{1}{\sigma_{v_i}^2}} \begin{bmatrix} \frac{1}{\sigma_{v_1}^2} \\ \frac{1}{\sigma_{v_2}^2} \\ \vdots \\ \frac{1}{\sigma_{v_M}^2} \end{bmatrix}. \tag{4.67}$$

It is obvious that the weights are inversely proportional to the noise variance of $\mathbf{x}_i$. This result is, of course, satisfactory from an intuitive point of view since noisy EPs are assigned smaller weights than reliable EPs.

In order to gain some insight into the performance of weighted averaging based on the varying noise variance assumption, we will calculate the variance of $\hat{s}_w(n)$. The variance is given by

$$
V\left[\hat{s}_w(n)\right] = V \left[\frac{1}{\displaystyle\sum_{i=1}^{M} \frac{1}{\sigma_{v_i}^2}} \sum_{i=1}^{M} \frac{x_i(n)}{\sigma_{v_i}^2} \right]
$$

$$
= \frac{1}{\displaystyle\sum_{i=1}^{M} \frac{1}{\sigma_{v_i}^2}}. \tag{4.68}
$$

The variance reduction associated with weighted averaging is now compared to that of ensemble averaging. To do this, we consider an inhomogeneous ensemble where ϵM EPs of subset A have a variance of $\sigma_{v_A}^2$ and the other $(1 - \epsilon)M$ EPs of subset B have the variance $\sigma_{v_B}^2$ where $0 \le \epsilon \le 1$ [52]; for simplicity, we assume that ϵ is chosen such that the values of both ϵM and $(1 - \epsilon)M$ are integers. The resulting variance of the classical ensemble average $\hat{s}_a(n)$ is given by

$$
V[\hat{s}_a(n)] = V\left[\epsilon \hat{s}_A(n) + (1 - \epsilon)\hat{s}_B(n)\right]
$$

$$
= \frac{1}{M}\left(\epsilon \sigma_{v_A}^2 + (1 - \epsilon)\sigma_{v_B}^2\right). \tag{4.69}
$$

The variance of the weighted average $\hat{s}_w(n)$ is obtained using the optimal weights in (4.64),

$$
V[\hat{s}_w(n)] = \frac{1}{\dfrac{\epsilon M}{\sigma_{v_A}^2} + \dfrac{(1 - \epsilon)M}{\sigma_{v_B}^2}}. \tag{4.70}
$$

The variance ratio is then given by

$$
\frac{V[\hat{s}_a(n)]}{V[\hat{s}_w(n)]} = 1 + \epsilon(1 - \epsilon)\frac{(\sigma_{v_A}^2 - \sigma_{v_B}^2)^2}{\sigma_{v_A}^2 \sigma_{v_B}^2}. \tag{4.71}
$$

Since the second term on the right-hand side in (4.71) is always positive, it can be concluded that weighted averaging is associated with a lower variance than ensemble averaging. The variance ratio is equal to one only when averaging of a homogeneous ensemble is performed, i.e., for $\epsilon = 0, 1$ or $\sigma_{v_A}^2 = \sigma_{v_B}^2$.

The variance ratio $V[\hat{s}_a(n)]/V[\hat{s}_w(n)]$ gives an indication of the kind of ensemble heterogeneity that results in a variance reduction for weighted averaging. The variance ratio is presented in Figure 4.13 as a function of ϵ

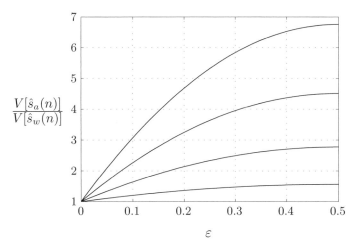

Figure 4.13: Variance ratio of the ensemble average and the weighted average, displayed as a function of ε, which describes the degree of ensemble inhomogeneity. The variance ratio is computed for $\sigma_{v_A}^2/\sigma_{v_B}^2$ equal to 2, 3, 4, and 5 (bottom to top). The variance ratio is only presented for $0 \le \epsilon \le 0.5$ since the ratio is symmetric with respect to $\epsilon = 0.5$.

for different noise variance ratios $\sigma_{v_A}^2/\sigma_{v_B}^2$. It is clear from this diagram that the weighted average is much more efficient when large differences in noise variance exist between the subsets A and B; this property is particularly pronounced for subsets of similar sizes, i.e., when ϵ is close to 0.5.

Figure 4.14 illustrates weighted averaging of an inhomogeneous EP ensemble, characterized by $\epsilon = 0.8$. It is evident that the noise level of the weighted average is considerably lower than that of the ensemble average. Moreover, the peak amplitude of $\hat{s}_{a,100}$ is overestimated due to noise components; however, the overestimation becomes less and less pronounced as M increases.

A fundamental difference between weighted averaging and ensemble averaging is that the former technique requires knowledge of the noise variance of individual EPs. Therefore, the noise variances $\sigma_{v_i}^2$ *must* be estimated from the ensemble $\mathbf{X}$ before weighted averaging becomes practical.

The prestimulus interval may be used to find an estimate of $\sigma_{v_i}^2$, provided that the stimulus repetition rate is slow enough. Then, the following model is assumed,

$$x_i(n) = \begin{cases} v_i(n), & -D \le n \le -1; \\ s(n) + v_i(n), & 0 \le n \le N-1, \end{cases} \tag{4.72}$$

where the interval $[-D, -1]$, immediately preceding the elicited stimulus at $n = 0$, is used.

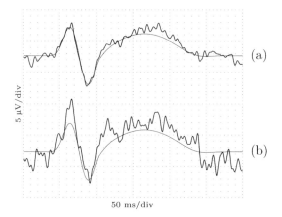

Figure 4.14: (a) The weighted average $\hat{\mathbf{s}}_{w,100}$ and (b) the ensemble average $\hat{\mathbf{s}}_{a,100}$ using the data of Figure 4.8, but with the amplitude of the added noise (i.e., the background EEG) scaled such that $\sigma_{v_A}^2 = 1$ for 80 EPs and $\sigma_{v_B}^2 = 20$ for the remaining ones (i.e., $\epsilon = 0.8$). The noise-free EP is indicated by the thin line.

The response interval $[0, N-1]$ itself can also be used for variance estimation provided that the SNR is very low. The noise variance is then simply obtained by

$$\hat{\sigma}_{v_i}^2 = \frac{1}{N}\mathbf{x}_i^T\mathbf{x}_i, \tag{4.73}$$

where $\mathbf{x}_i$ is assumed to be zero-mean. The variance estimate in (4.73) has been found adequate in weighted averaging of BAEPs and SEPs, whereas it should be avoided in VEP analysis where the SNR may be too high.

Yet another approach investigated is to adaptively estimate the weights $\mathbf{w}$ in (4.67) using the LMS algorithm, assuming that one of the EPs constitutes the reference signal [53]. The details of this approach are developed in Problem 4.19.

Case 2: Varying signal amplitude. Another approach to weighted averaging is to assume that the signal amplitude $\mathbf{a}$ differs from EP to EP, while the noise variance remains constant in all EPs [54],

$$\mathbf{a} = \begin{bmatrix} a_1 \\ a_2 \\ \vdots \\ a_M \end{bmatrix}, \tag{4.74}$$

$$\mathbf{R}_V = N\sigma_v^2\mathbf{I}. \tag{4.75}$$

For these model assumptions, the generalized eigenvalue problem in (4.57), yielding the optimal weights, is again reduced to the ordinary eigenvalue problem,

$$\mathbf{A}\mathbf{w} = \lambda N \sigma_v^2 \mathbf{w}, \tag{4.76}$$

where $\mathbf{A} = \mathbf{a}\mathbf{a}^T$. Similar to the previous case with varying noise variance, a closed-form solution can be obtained since $\mathbf{A}$ is a rank-one matrix. All eigenvalues are equal to zero, except λ_1 which equals

$$\lambda_1 = \frac{\mathbf{a}^T\mathbf{a}}{N\sigma_v^2}. \tag{4.77}$$

The optimal weight vector $\mathbf{w}$ is proportional to the corresponding eigenvector given by $\mathbf{a}$,

$$\mathbf{w} = c_w\mathbf{a}.$$

The weight vector must be normalized such that

$$\mathbf{w} = \frac{1}{\mathbf{a}^T\mathbf{a}}\mathbf{a} \tag{4.78}$$

in order to assure that the weighted average for the case of varying signal amplitude is unbiased.

Again, it is necessary to first estimate the amplitude $\mathbf{a}$ from $\mathbf{X}$ before (4.78) can be used in practice. One approach which produces such an estimate is the cross-correlation of the EPs in $\mathbf{X}$ to the ensemble average $\hat{\mathbf{s}}_a$,

$$\hat{\mathbf{a}} = \mathbf{X}^T\hat{\mathbf{s}}_a. \tag{4.79}$$

The rationale behind the cross-correlation approach can be understood by studying the cross-correlation for one individual weight,

$$\begin{aligned}
\hat{\mathbf{s}}_a^T\mathbf{x}_i &= \hat{\mathbf{s}}_a^T(a_i\mathbf{s} + \mathbf{v}_i) \\
&= a_i\hat{\mathbf{s}}_a^T\mathbf{s} + \hat{\mathbf{s}}_a^T\mathbf{v}_i.
\end{aligned} \tag{4.80}$$

Provided that sufficiently many potentials have been included in $\hat{\mathbf{s}}_a$, it is reasonable to assume that $\hat{\mathbf{s}}_a$ and $\mathbf{v}_i$ are approximately uncorrelated and that $\hat{\mathbf{s}}_a$ and $\mathbf{s}$ have approximately the same morphology, implying that $\hat{\mathbf{s}}_a^T\mathbf{s} \approx 1$ because of (4.52). Thus, the expected value of $\hat{\mathbf{s}}_a^T\mathbf{x}_i$ is approximately equal to the optimal weight in (4.78),

$$E\left[\hat{\mathbf{s}}_a^T\mathbf{x}_i\right] = E\left[a_i\hat{\mathbf{s}}_a^T\mathbf{s} + \hat{\mathbf{s}}_a^T\mathbf{v}_i\right] \approx a_i. \tag{4.81}$$

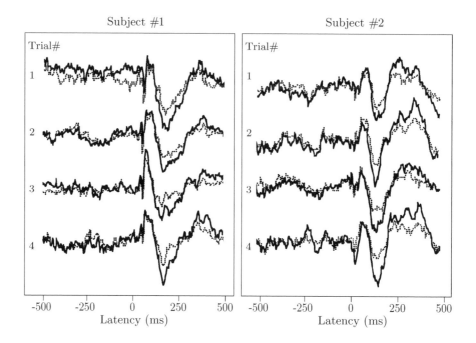

Figure 4.15: Weighted averaging of AEPs recorded from two different subjects. Each diagram shows the results of four trials of each subject. The ensemble average and the weighted average are plotted with dotted and solid lines, respectively. (Reprinted from Davila and Mobin [54] with permission.)

While the optimal weight vector in (4.78) produces an SNR which is better than that of ensemble averaging, it is not evident that the weight vector estimate in (4.79) has the same property. However, it can be shown that this weight vector estimate actually exhibits a similar SNR-enhancing property; the details of the proof can be found in [54].

The performance of weighted averaging, assuming a varying signal amplitude, is illustrated in Figure 4.15 for AEPs recorded from two different subjects. The most notable difference observed in these AEPs is that the amplitude of the weighted average is considerably larger than that of the ensemble average.

Gaussian noise with varying variance—The ML approach.   We will now show that the weights resulting from maximization of the SNR criterion in (4.51), assuming fixed amplitude and varying noise variance, can also be obtained by employing ML estimation. This type of estimation assumes that the probability density function $p(\mathbf{x}; \theta)$ for the observations $\mathbf{x}$ has a known form, and depends on a fixed but unknown parameter θ to be estimated.

The ML estimate is defined as that value of θ which maximizes the PDF, provided that the samples of the observed signal $\mathbf{x}$ are *fixed*,

$$\hat{\theta} = \arg \max_{\theta} p(\mathbf{x}; \theta). \qquad (4.82)$$

In this expression, the PDF $p(\mathbf{x}; \theta)$ should not be interpreted, as is normally done, as a function of the random quantity $\mathbf{x}$, but rather as a function of θ. The function $p(\mathbf{x}; \theta)$ is therefore referred to as a *likelihood function* [55, 56].

For weighted averaging with varying noise variance, the observation model is given by

$$x_i(n) = s(n) + v_i(n), \quad n = 0, 1, \ldots, N - 1, \qquad (4.83)$$

where it is assumed that the noise $v_i(n)$ is not only zero-mean and with variance $\sigma_{v_i}^2$, but now also Gaussian. The noise is considered to be uncorrelated from EP to EP, cf. (4.10). For time n, the joint PDF of the noise for the ensemble of M different EPs is given by

$$p_v(v_1(n), \ldots, v_M(n)) = \prod_{i=1}^{M} p_v(v_i(n))$$

$$= \prod_{i=1}^{M} \frac{1}{\sqrt{2\pi\sigma_{v_i}^2}} \exp\left[-\frac{v_i^2(n)}{2\sigma_{v_i}^2} \right]. \qquad (4.84)$$

The joint PDF of $x_i(n)$ at time n is identical to that of $v_i(n)$ but with the mean value equal to $s(n)$ (recall from the additive noise model that $v_i(n) = x_i(n) - s(n)$),

$$p_v(x_1(n), \ldots, x_M(n); s(n)) = \prod_{i=1}^{M} \frac{1}{\sqrt{2\pi\sigma_{v_i}^2}} \exp\left[-\frac{(x_i(n) - s(n))^2}{2\sigma_{v_i}^2} \right]. \qquad (4.85)$$

In order to find the ML estimator of $s(n)$, which thus represents our desired parameter θ, we maximize the logarithm of the likelihood function,[5]

$$\ln p_v(x_1(n), \ldots, x_M(n); s(n)) = -\frac{1}{2} \sum_{i=1}^{M} \ln\left(2\pi\sigma_{v_i}^2\right) - \sum_{i=1}^{M} \frac{(x_i(n) - s(n))^2}{2\sigma_{v_i}^2}.$$

$$(4.86)$$

---

[5]Maximization of (4.85) is equivalent to maximization of the log-likelihood function because the logarithm is a monotonic function. The log-likelihood function is often considered since maximization can be more easily performed.

By taking the derivative with respect to $s(n)$ and setting the result equal to zero,

$$\frac{\partial \ln p_v\left(x_1(n),\ldots,x_M(n);s(n)\right)}{\partial s(n)} = \sum_{i=1}^{M} \frac{(x_i(n) - s(n))}{\sigma_{v_i}^2} = 0, \qquad (4.87)$$

the ML estimator of $s(n)$ is obtained and is found to equal the weighted average of $x_i(n)$,

$$\hat{s}_w(n) = \frac{1}{\displaystyle\sum_{i=1}^{M} \frac{1}{\sigma_{v_i}^2}} \sum_{i=1}^{M} \frac{x_i(n)}{\sigma_{v_i}^2}. \qquad (4.88)$$

Hence, each EP is weighted by

$$w_i = \frac{\dfrac{1}{\sigma_{v_i}^2}}{\displaystyle\sum_{j=1}^{M} \frac{1}{\sigma_{v_j}^2}}, \qquad (4.89)$$

which is identical to the weight in (4.67) which resulted from maximization of the SNR criterion.

Moreover, it should be noted that the expression in (4.88) simplifies to the ensemble average when the noise is considered to be fixed and identical in all EPs, i.e., $\sigma_{v_i}^2 \equiv \sigma_v^2$,

$$\hat{s}_a(n) = \frac{1}{M} \sum_{i=1}^{M} x_i(n). \qquad (4.90)$$

The ensemble average is thus the optimal estimator of a fixed waveform $s(n)$ disturbed by white, Gaussian noise.

Finally, we should mention that the weighted average $\hat{s}_{w,M}(n)$ can, as in the case of the ensemble average $\hat{s}_{a,M}(n)$, be computed recursively from

$$\hat{s}_{w,M}(n) = \hat{s}_{w,M-1}(n) + \alpha_M(x_M(n) - \hat{s}_{w,M-1}(n)), \qquad (4.91)$$

where the gain α_M is identical to the weight defined in (4.67),

$$\alpha_M = w_M = \frac{\dfrac{1}{\sigma_{v_M}^2}}{\displaystyle\sum_{j=1}^{M} \frac{1}{\sigma_{v_j}^2}}. \qquad (4.92)$$

The derivation of the recursion in (4.91), as well as that for the associated variance $V[\hat{s}_{w,M}(n)]$, is considered in Problems 4.12 and 4.14.

4.3.5 Spike Artifacts and Robust Averaging

The ensemble averager, the exponential averager, and the weighted averager represent linear techniques and, as such, perform well when the noise is Gaussian. Their performance will, however, become markedly poorer when occasional spike artifacts ("outlier samples") occur with atypically large amplitudes, as linear techniques offer no means of limiting the influence of such disturbances. Although weighted averaging mitigates this type of problem to a certain degree, by weighting each EP inversely proportionally to its noise variance, the model in (4.83) involves the assumption of Gaussian noise and thus cannot handle such spike artifacts sufficiently well.

One approach used to derive more robust averaging methods, capable of handling impulse disturbances, is to assume that the noise in (4.83) is modeled by a PDF whose tails account for the presence of outlier samples. Below, some well-known, robust methods possessing a nonlinear structure are described, suitable for either batch or recursive EP processing.

Ensemble averaging with outlier rejection.   Improved statistical modeling of spike artifacts may be obtained by considering the *generalized Gaussian* PDF—this is actually a family of PDFs whose tail decay rates are determined by the shape parameter ν. The zero-mean, generalized Gaussian PDF is completely characterized by ν and the standard deviation σ_v,

$$p_\nu(v_i(n)) = \frac{\nu}{2\sigma_v \Gamma(1/\nu)b(\nu)} \exp\left[-\left(\frac{|v_i(n)|}{\sigma_v b(\nu)}\right)^\nu\right], \qquad (4.93)$$

where

$$b(\nu) = \sqrt{\frac{\Gamma(1/\nu)}{\Gamma(3/\nu)}}$$

and the Gamma function $\Gamma(\nu)$ is defined by

$$\Gamma(\nu) = \int_0^\infty t^{\nu-1} e^{-t} dt.$$

Figure 4.16 presents the shape of the PDF for $\nu = 1$, 1.5, and 2 using a logarithmic format; the tails of the PDF become increasingly larger as ν decreases. It should be noted that the Gaussian PDF ($\nu = 2$) and the uniform PDF ($\nu \to -1$) are both special cases of the generalized Gaussian PDF.

Another important special case is the Laplacian PDF for $\nu = 1$ which, thanks to its heavy tails, represents a reasonable, and analytically tractable, model of spike artifacts. The Laplacian PDF of a noise sample $v_i(n)$ is given

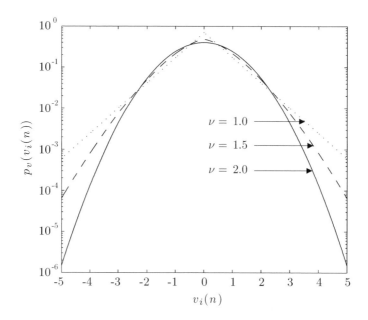

Figure 4.16: The generalized Gaussian PDF, with zero-mean and unit variance, displayed for three different values of ν. The Laplacian and Gaussian PDFs are obtained for $\nu = 1$ and 2, respectively. Note that the scale of the vertical axis is logarithmic.

by

$$p_v(v_i(n)) = \frac{1}{\sqrt{2\sigma_{v_i}^2}} \exp\left[-\sqrt{\frac{2}{\sigma_{v_i}^2}}|v_i(n)|\right]. \qquad (4.94)$$

Our primary goal here is to determine the optimal ML estimator of $s(n)$ based on the available noisy observations $x_1(n), \ldots, x_M(N)$, each characterized by a Laplacian PDF. As before, we assume that the noise is uncorrelated from EP to EP and that the noise variance is fixed and identical in all EPs (i.e., $\sigma_{v_i}^2 \equiv \sigma_v^2$). The ML estimator is obtained by maximization of the following likelihood function,

$$p_v(x_1(n), \ldots, x_M(n); s(n)) = \prod_{i=1}^{M} \frac{1}{\sqrt{2\sigma_v^2}} \exp\left[-\sqrt{\frac{2}{\sigma_v^2}}|x_i(n) - s(n)|\right], \quad (4.95)$$

with respect to $s(n)$; for now, M is assumed to be odd. Taking the logarithm of the likelihood function, the estimate that maximizes (4.95) is found by differentiating the expression and setting the result equal to zero,

$$\frac{\partial \ln p_v\left(x_1(n), \ldots, x_M(n); s(n)\right)}{\partial s(n)} = -\frac{\partial}{\partial s(n)}\left(\sqrt{\frac{2}{\sigma_v^2}}\sum_{i=1}^{M}|x_i(n) - s(n)|\right) = 0.$$

$$(4.96)$$

The function to be maximized is denoted $J(n)$,

$$J(s(n)) = \sum_{i=1}^{M}|x_i(n) - s(n)|$$

$$= \sum_{i=1}^{M}\sqrt{(x_i(n) - s(n))^2}, \qquad (4.97)$$

which, when differentiated, yields

$$\frac{\partial J(s(n))}{\partial s(n)} = \sum_{i=1}^{M}\frac{x_i(n) - s(n)}{\sqrt{(x_i(n) - s(n))^2}}$$

$$= \sum_{i=1}^{M}\frac{x_i(n) - s(n)}{|x_i(n) - s(n)|} = 0. \qquad (4.98)$$

By introducing the sgn(x) function, defined by[6]

$$\mathrm{sgn}(x) = \frac{x}{|x|} = \begin{cases} 1, & x > 0; \\ 0, & x = 0; \\ -1, & x < 0, \end{cases} \qquad (4.99)$$

the equation which describes the ML estimator of $s(n)$ can be written as

$$\sum_{i=1}^{M}\mathrm{sgn}(x_i(n) - s(n)) = 0. \qquad (4.100)$$

To make sure that the sum in (4.100) is equal to zero, we must choose $s(n)$ so that exactly half of the sample values are greater than $s(n)$ and the remaining values are smaller. This procedure is identical to finding the *ensemble median* of the data. Hence, the median constitutes the ML estimator of $s(n)$ when the noise is Laplacian.

---

[6]Although the function $|x|$ does not have a derivative at $x = 0$, it is nevertheless reasonable to use the value zero since it agrees with the customary definition of the sgn function.

The procedure for computing the median consists of sorting the sequence of observed samples in order of magnitude,

$$\{x_1(n), x_2(n), \ldots, x_M(n)\} \xrightarrow{\text{sort}} \{x_{(1)}(n), x_{(2)}(n), \ldots, x_{(M)}(n)\}, \quad (4.101)$$

where $x_{(1)}(n) < x_{(2)}(n) < \cdots < x_{(M)}(n)$, followed by selection of the midpoint at $(M+1)/2$ when M is odd or the average of the two midpoints at $M/2$ and $M/2 + 1$ when M is even,

$$\hat{s}_{\text{med}}(n) = \begin{cases} x_{\left(\frac{M+1}{2}\right)}(n), & M \text{ odd}; \\ \frac{1}{2}(x_{\left(\frac{M}{2}\right)}(n) + x_{\left(\frac{M}{2}+1\right)}(n)), & M \text{ even}. \end{cases} \quad (4.102)$$

The benefits of computing the ensemble median instead of the ensemble average were realized at an early stage of computerized EP analysis [57]. The performance of these two estimators is illustrated in Figure 4.17, where either Gaussian or Laplacian noise is added to an ensemble of simulated EPs. The noise is white and has a variance of $\sigma_v^2 = 0.25$ for both PDFs. Since the noise-free signal is available, the performance can be quantified in terms of noise variance of the resulting estimate. For Laplacian noise, the ensemble median is better than the ensemble average since $\hat{\sigma}_v^2 = 0.0014$ and 0.0025, respectively, whereas the ensemble average is better than the median for Gaussian noise, $\hat{\sigma}_v^2 = 0.0025$ (which equals the theoretically predicted value, $\sigma_v^2/100 = 0.0025$) and 0.0038, respectively. A theoretical comparison of these two estimators shows that the noise variance increases by approximately 50% when the median is used in Gaussian noise [58]; this increase can also be observed from the simulated EP data in Figure 4.17.

The ensemble average and the ensemble median can both be viewed as special cases of a more general family of estimators, commonly referred to as *trimmed means* [59]. In contrast to the median, which trims away all samples except the midpoint of the ordered samples, this family of estimators makes use of all the ordered samples except a certain fraction of the smallest and largest sample values which are trimmed away. The number of samples to be excluded is related to the trimming factor γ, which is constrained to the interval $0 \le \gamma < 0.5$. The trimmed mean is computed by averaging the remaining samples,

$$\hat{s}_{\text{tri}}(n) = \frac{1}{M - 2K} \sum_{i=K+1}^{M-K} x_{(i)}(n), \quad (4.103)$$

where K denotes the number of samples to be trimmed away, given by the largest integer less than or equal to γM. The trimming factor can either be fixed or adapted to a measure reflecting the tail behavior of the PDF [60].

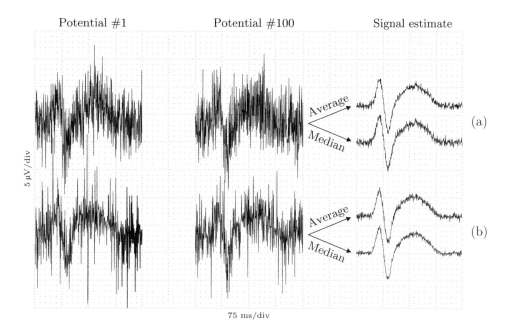

Figure 4.17: Noise reduction achieved by the ensemble average and the ensemble median, using an ensemble of 100 simulated EPs. Each EP is additively corrupted by white noise characterized by (a) a Gaussian PDF or (b) a Laplacian PDF.

It should be noted that the ensemble average and the ensemble median correspond to $\gamma = 0$ and 0.5, respectively.

The use of robust averaging techniques in the analysis of VEPs has not been found to offer significant improvements in noise reduction [39]; however, the improvements will be more pronounced in recording situations where it is probable that outliers will often occur.

Recursive, robust averaging with outlier rejection. We will just briefly mention the possibility of performing robust estimation recursively. The general structure of such a recursive estimator is closely related to that of the exponential averager, but with the update part modified by the *influence function* $\eta(x)$, so that

$$\hat{\mathbf{s}}_{r,M} = \hat{\mathbf{s}}_{r,M-1} + \alpha_M \cdot \eta(\mathbf{x}_{r,M} - \hat{\mathbf{s}}_{r,M-1}), \qquad (4.104)$$

where the subscript "r" denotes robust. The function $\eta(x)$ is designed to reduce the influence of outlier values on the estimate $\hat{\mathbf{s}}_{r,M}$; whether $\mathbf{x}_M$ should be considered as an outlier value or not is judged by its relation to the most recent estimate $\hat{\mathbf{s}}_{r,M-1}$.

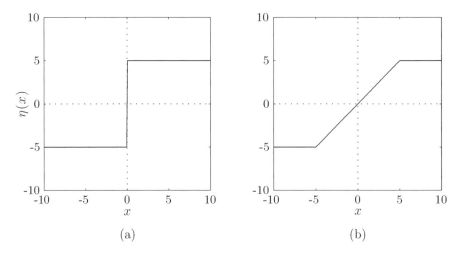

(a)   (b)

Figure 4.18: Two examples of influence functions for use in robust, recursive averaging: (a) the sign limiter and (b) the hard limiter. The increment/threshold parameter η_0 was set to 5.

An example of a simple influence function is the one which adds or subtracts a fixed increment η_0 from the current estimate depending on the sign of the update (see Figure 4.18(a)),

$$\eta(x) = \eta_0 \cdot \text{sgn}(x) = \begin{cases} \eta_0, & x > 0; \\ 0, & x = 0; \\ -\eta_0, & x < 0. \end{cases} \tag{4.105}$$

It can be shown that the recursion in (4.104), when combined with this influence function, tends to converge to the ensemble median [61], see Problem 4.22. This function is apparently insensitive to intermittently occurring spikes but provides, on the other hand, very conservative tracking of changes in EP morphology.

Another example of an influence function is the hard limiter in which deviations with a magnitude below a certain threshold η_0 are treated in the same was as in the exponential averager, but otherwise limited to the threshold values $\pm\eta_0$ (see Figure 4.18(b)),

$$\eta(x) = \begin{cases} \eta_0, & x > \eta_0; \\ x, & -\eta_0 \le x \le \eta_0; \\ -\eta_0, & x < -\eta_0. \end{cases} \tag{4.106}$$

The exponential averager is defined by the function $\eta(x) = x$ which, evidently, does not offer any protection against outlier values.

4.3.6   The Effect of Latency Shifts

The latencies of an EP have, so far, been considered as a quantity which is fixed from EP to EP. However, variations in latency may occur which introduce distortion into the resulting ensemble average. The nature of such distortion can be perceived from the simulation example in Figure 4.19, where ten noise-free EPs are shown with identical morphology but with slightly different latencies. The ensemble average of the ten EPs is shown at the bottom of Figure 4.19, together with the ensemble average for the case when no variation in latency is present. The effect of latency shifts is primarily manifested as a significant reduction in amplitude of the ensemble average.

We will now take a closer look at two types of time shift, namely, those modeled by continuous- and discrete-valued random variables. The first case reflects the fact that latency shifts caused by various biological mechanisms are not constrained to the time grid imposed by sampling. The second, discrete-time case is useful when studying the effect of variations that take place in the sampled signal.

Shifts in continuous-time signals.   The influence of latency shifts on the ensemble average can be studied in terms of the earlier adopted "signal-plus-noise" model in (4.4), but modified to account for an unknown latency shift τ. Since τ is continuous-valued, we will consider the continuous-time counterpart to the model in (4.4),

$$x_i(t) = s(t - \tau_i) + v_i(t), \tag{4.107}$$

where t denotes time and τ_i, $i = 1, \ldots, M$ are samples of the random variable τ which is completely characterized by the PDF $p_\tau(\tau)$. Based on the observation model in (4.107), the expected value of the ensemble average $\hat{s}_a(t)$ is given by

$$E[\hat{s}_a(t)] = \frac{1}{M} \sum_{i=1}^{M} E[s(t - \tau_i)]$$

$$= \int_{-\infty}^{\infty} s(t - \tau)\, p_\tau(\tau)\, d\tau. \tag{4.108}$$

Introducing the *characteristic function* of $p_\tau(\tau)$ [62, p. 115],

$$P_\tau(\Omega) = \int_{-\infty}^{\infty} p_\tau(\tau) e^{j\Omega\tau} d\tau, \tag{4.109}$$

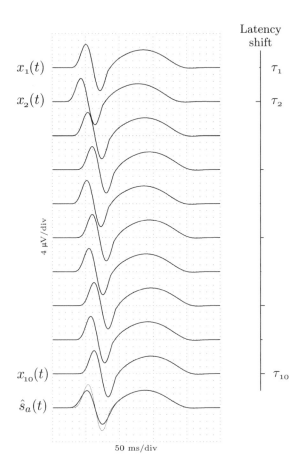

Figure 4.19: The effect of latency shifts on the ensemble average. The shifts are introduced in a simulated, noise-free signal, and the size of the shifts are shown by the horizontal bars to the right. At the bottom, the resulting ensemble average $\hat{s}_a(t)$ is shown together with the true signal $s(t)$ superimposed (thin line).

the convolution integral in (4.108) can be expressed as a product in the frequency domain,

$$E[\hat{S}_a(\Omega)] = S(\Omega)P_\tau^*(\Omega), \qquad (4.110)$$

where $\hat{S}_a(\Omega)$ and $S(\Omega)$ are the continuous-time Fourier transforms of $\hat{s}_a(t)$ and $s(t)$, respectively; $\Omega = 2\pi F$ where F denotes analog frequency and the asterisk $(*)$ denotes the complex-conjugate.[7]

In most cases of practical interest, the PDF $p_\tau(\tau)$ can be assumed to be symmetric around $\tau = 0$, with tails that decrease monotonically to zero. As a result, the effect of $P_\tau(\Omega)$ on the original signal $S(\Omega)$ in (4.110) is equivalent to filtering of $s(t)$ with a linear, time-invariant, lowpass filter whose impulse response is given by $p_\tau(\tau)$. The ensemble average computed in the presence of latency shifts is thus biased and will not approach $s(t)$ as the number of EPs increases.

An example of $p_\tau(\tau)$ is the zero-mean, Gaussian PDF with a variance σ_τ^2 whose characteristic function is

$$P_\tau(\Omega) = e^{-\frac{1}{2}\Omega^2\sigma_\tau^2}. \qquad (4.111)$$

This type of latency shift acts as a lowpass filter on $s(t)$ to a degree determined by σ_τ. It is of particular interest to compute the –3 dB cut-off frequency, denoted F_c, as a function of σ_τ:

$$e^{-2(\pi F_c\sigma_\tau)^2} = \frac{1}{\sqrt{2}},$$

or

$$F_c = \frac{\sqrt{\ln 2}}{2\pi\sigma_\tau}. \qquad (4.112)$$

Figure 4.20(a) displays the relationship between F_c and σ_τ. For example, it can be seen that a dispersion of $\sigma_\tau = 1$ ms corresponds to lowpass filtering with a cut-off frequency of 133 Hz.

Another type of latency shift is that caused by sampling ("sampling jitter"), typically assumed to have a uniform PDF over the sampling interval T,

$$p_\tau(\tau) = \begin{cases} \dfrac{1}{T}, & -T/2 \leq \tau \leq T/2; \\ 0, & \text{otherwise.} \end{cases} \qquad (4.113)$$

---

[7]The definition of the characteristic function is identical to the Fourier transform of $p_\tau(\tau)$, except that $e^{j\Omega\tau}$ is used instead of $e^{-j\Omega\tau}$ in the integrand.

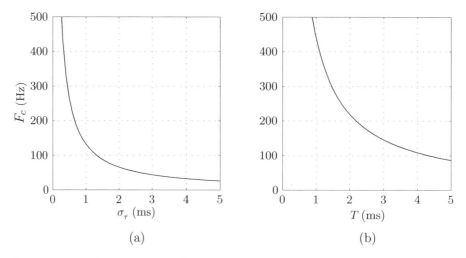

Figure 4.20: The –3 dB cut-off frequency F_c associated with the lowpass filtering effect due to latency shifts; F_c is plotted as a function of (a) the standard deviation σ_τ of a Gaussian PDF and (b) the sampling interval T of a uniform PDF.

The lowpass filtering effect due to sampling is described by the corresponding sinc characteristic function

$$P_\tau(\Omega) = \frac{\sin \frac{1}{2}\Omega T}{\frac{1}{2}\Omega T}. \tag{4.114}$$

In contrast to the Gaussian case in (4.112), it is difficult to derive a closed-form expression from (4.114) relating F_c to the dispersion parameter T. The desired relationship is, however, easily calculated by numerical techniques and is presented in Figure 4.20(b).

When $p_\tau(\tau)$ is known, the influence of latency shifts on the ensemble average can be determined using (4.110). However, it is actually possible to obtain certain information on the statistics of τ without knowledge of the PDF. For example, the variance of τ can be estimated from the ensemble variance [63].

Shifts in discrete-time signals. The discrete-time counterpart of the observation model in (4.107) is defined by

$$x_i(n) = s(n - \theta_i) + v_i(n), \tag{4.115}$$

where θ_i is an integer-valued random variable characterized by the PDF $p_\theta(\theta)$. The discrete-time case may be of special interest when studying the performance of different methods developed for compensation of random latency shifts. Proceeding in a way similar to the continuous-time case, the expected value of the discrete-time ensemble average $\hat{s}_a(n)$ is

$$E\left[\hat{s}_a(n)\right] = \sum_{\theta=-\infty}^{\infty} s(n-\theta)\,p_\theta(\theta). \qquad (4.116)$$

Introducing the discrete-time characteristic function of $p_\theta(\theta)$,

$$P_\theta(e^{\jmath\omega}) = \sum_{\theta=-\infty}^{\infty} p_\theta(\theta)e^{\jmath\omega\theta}, \qquad (4.117)$$

the convolution sum in (4.116) can be expressed as a product in the frequency domain,

$$E[\hat{S}_a(e^{\jmath\omega})] = S(e^{\jmath\omega})P_\theta^*(e^{\jmath\omega}), \qquad (4.118)$$

where $\hat{S}_a(e^{\jmath\omega})$ and $S(e^{\jmath\omega})$ are the discrete-time Fourier transforms of $\hat{s}_a(n)$ and $s(n)$, respectively.

4.3.7 Estimation of Latency Shifts

It is desirable to develop methods which can compensate for variations in latency so that the ensemble average becomes more accurate. A tempting approach is to simply perform deconvolution of the ensemble average $\hat{S}_a(e^{\jmath\omega})$ by inverse filtering with $P_\theta^*(e^{\jmath\omega})$ [64],

$$\hat{S}(e^{\jmath\omega}) = \frac{\hat{S}_a(e^{\jmath\omega})}{P_\theta^*(e^{\jmath\omega})}. \qquad (4.119)$$

This technique, which operates directly on the ensemble average, is, however, associated with serious problems, making it less suitable for correction of latency shifts. The characteristic function $P_\theta(\theta)$ is not known a priori and cannot be easily estimated from the ensemble of EPs. Even if $P_\theta(e^{\jmath\omega})$ is known, inverse filtering implies that the high-frequency content of $\hat{S}(e^{\jmath\omega})$ may be overemphasized since $P_\theta(e^{\jmath\omega})$ has values close to zero for higher frequencies.

Another, much more useful approach is to find an estimate of the latency shift θ_i in each individual EP followed by computation of the ensemble average from the latency-corrected EPs or by any other technique for noise reduction. Of the many methods developed for latency correction [65–71], the most well-known method is the *Woody method* [72, 73] which has its starting point in the observation model defined in (4.115). The main ingredients of this method are:

- The estimation of θ_i using a matched filter;

- the related problem of estimating the impulse response of the matched filter; and

- an iterative procedure for improving the latency estimates.

The final product of the Woody method is the *latency-corrected ensemble average*.

Initially, we assume that $s(n)$ is a known waveform, additively disturbed by zero-mean, white, Gaussian noise $v_i(n)$ with variance σ_v^2. Since both these assumptions on signal and noise properties can be questioned, we will relax the assumptions so that $s(n)$ is estimated from the ensemble of data $\mathbf{X}$, and the color of the noise is characterized by the correlation matrix $\mathbf{R}_v$. Each observed signal $x_i(n)$ is associated with an unknown latency shift θ_i so that

$$
x_i(n) = \begin{cases} v_i(n), & n = 0, \ldots, \theta_i - 1; \\ s(n - \theta_i) + v_i(n), & n = \theta_i, \ldots, \theta_i + D - 1; \\ v_i(n), & n = \theta_i + D, \ldots, N - 1, \end{cases} \tag{4.120}
$$

where D denotes the duration of $s(n)$.[8] The latency shift θ_i is assumed to be constrained to the interval $[0, N - D]$ so that $s(n)$ is always completely contained in the observation interval and the energy of $s(n)$ is fixed for all values of θ_i,

$$
E_s = \sum_{n=\theta_i}^{\theta_i+D-1} s^2(n - \theta_i) = \sum_{n=0}^{N-1} s^2(n). \tag{4.121}
$$

Maximum likelihood estimation is now considered to find the estimator of the latency θ_i. Since the noise is assumed to be white and Gaussian, the PDF of the observed signal is given by

$$
\begin{aligned}
p_v(\mathbf{x}_i; \theta_i) = & \prod_{n=0}^{\theta_i-1} \frac{1}{\sqrt{2\pi\sigma_v^2}} \exp\left[-\frac{x_i^2(n)}{2\sigma_v^2} \right] \\
& \cdot \prod_{n=\theta_i}^{\theta_i+D-1} \frac{1}{\sqrt{2\pi\sigma_v^2}} \exp\left[-\frac{(x_i(n) - s(n - \theta_i))^2}{2\sigma_v^2} \right] \\
& \cdot \prod_{n=\theta_i+D}^{N-1} \frac{1}{\sqrt{2\pi\sigma_v^2}} \exp\left[-\frac{x_i^2(n)}{2\sigma_v^2} \right],
\end{aligned} \tag{4.122}
$$

---

[8]The model in (4.120) can be extended from including only one EP to account for all the M different EPs of the ensemble so that the latency shifts $\theta_1, \ldots, \theta_M$ are jointly estimated [74]. Using an ML approach, the optimal estimator can be derived, but is found to require a massive amount of computation. It is therefore of interest to develop suboptimal approaches of which the Woody method represents one.

where $\mathbf{x}_i = \begin{bmatrix} x_i(0) & x_i(1) & \cdots & x_i(N-1) \end{bmatrix}^T$. Evaluation of the exponent $(x_i(n) - s(n - \theta_i))^2$ yields the following expression for the PDF,

$$
\begin{aligned}
p_v(\mathbf{x}_i; \theta_i) = \prod_{n=0}^{N-1} \frac{1}{\sqrt{2\pi\sigma_v^2}} \exp\left[-\frac{x_i^2(n)}{2\sigma_v^2}\right] \\
\cdot \prod_{n=\theta_i}^{\theta_i+D-1} \frac{1}{\sqrt{2\pi\sigma_v^2}} \exp\left[\frac{x_i(n)s(n-\theta_i)}{\sigma_v^2}\right] \\
\cdot \prod_{n=\theta_i}^{\theta_i+D-1} \frac{1}{\sqrt{2\pi\sigma_v^2}} \exp\left[-\frac{s^2(n-\theta_i)}{2\sigma_v^2}\right],
\end{aligned}
\tag{4.123}
$$

from which it is obvious that the first product factor is not a function of θ_i and that the third product factor contains the energy E_s which is also independent of θ_i due to (4.121). The logarithm of the likelihood function is given by

$$
\ln p_v(\mathbf{x}_i; \theta_i) = \text{constant} + \frac{1}{\sigma_v^2} \sum_{n=\theta_i}^{\theta_i+D-1} x_i(n)s(n-\theta_i),
\tag{4.124}
$$

where the constant collects all the terms that are independent of θ_i. The ML estimate is given by that value of θ_i which maximizes the sum on the right-hand side of (4.124),

$$
\hat{\theta}_i = \arg\max_{\theta_i} \left(\sum_{n=\theta_i}^{\theta_i+D-1} x_i(n)s(n-\theta_i) \right).
\tag{4.125}
$$

Hence, the estimate of θ_i is given by the time at which the best cross-correlation between $s(n)$ and $x_i(n)$ is achieved. Alternatively, the maximization in (4.125) can be interpreted as a filtering operation,

$$
y_i(\theta_i) = \sum_{n=\theta_i}^{\theta_i+D-1} x_i(n)h(\theta_i - n),
\tag{4.126}
$$

where the impulse response of the filter $h(k)$ is equal to the time-reversed version of $s(n)$,

$$
h(k) = \begin{cases} s(D-1-k), & k = 0, \ldots, D-1; \\ 0, & \text{otherwise.} \end{cases}
\tag{4.127}
$$

In terms of the earlier introduced vector notation (see page 88), the impulse response of the filter is given by

$$\mathbf{h} = \begin{bmatrix} s(D-1) \\ s(D-2) \\ \vdots \\ s(0) \end{bmatrix} = \tilde{\mathbf{s}}. \tag{4.128}$$

Thus, the ML estimator of θ_i is that particular time at which the maximum peak occurs in the filter output $y_i(\theta_i)$. The filter in (4.127) is the well-known *matched filter* and constitutes an important building block in many detection schemes [75].

In practice, the waveform $s(n)$ that defines the matched filter has to be estimated from $\mathbf{X}$. Under the assumption that the latency variations are relatively small, the ensemble average $\hat{s}_a(n)$ can be used as an initial estimate of $s(n)$ for the estimation procedure in (4.125),

$$\hat{s}_a^{(0)}(n) = \hat{s}_a(n). \tag{4.129}$$

In the case of large latency variations, it may be more appropriate to use a predefined pulse-shaped waveform, for example, a triangular waveform resembling the overall shape of the response.[9]

A new, latency-corrected ensemble average, denoted $\hat{s}_a^{(1)}(n)$, is obtained after all the EPs have been corrected by $\hat{\theta}_i^{(1)}$,

$$\hat{s}_a^{(1)}(n) = \frac{1}{M} \sum_{i=1}^{M} x_i(n + \hat{\theta}_i^{(1)}). \tag{4.130}$$

Evidently, the estimation of $\theta_i^{(1)}$ can be repeated using $\hat{s}_a^{(1)}(n)$ instead of $\hat{s}_a(n)$ as the matched filter; such a step will, almost certainly, improve the earlier latency-corrected ensemble average $\hat{s}_a^{(1)}(n)$. In fact, the Woody method was originally designed to iteratively update the ensemble average

$$\hat{s}_a^{(j)}(n) = \frac{1}{M} \sum_{i=1}^{M} x_i(n + \hat{\theta}_i^{(j)}) \tag{4.131}$$

until a suitable termination criterion was fulfilled. In (4.131), $\hat{s}_a^{(j)}(n)$ denotes the ensemble average that results from the jth iteration making use of the latency estimate $\theta_i^{(j)}$.

---

[9]A related approach is the one based on the Karhunen–Loève expansion in which the most significant eigenvector is used as an estimate of $s(n)$ [76]; this type of series expansion is described later on in this chapter.

The iterative procedure can be terminated in various ways. One approach is to terminate when the maximum of successive differences in the latency estimates drops below a certain threshold η_θ,

$$\max_i \left| \hat{\theta}_i^{(j)} - \hat{\theta}_i^{(j-1)} \right| \leq \eta_\theta, \quad j \geq 1, \tag{4.132}$$

where η_θ is a positive-valued integer. Another approach is based on similarity in morphology, quantified by the cross-correlation coefficient defined in (4.23). Since the morphology of several EPs should be taken into account, the average of the cross-correlation coefficient $\rho_i^{(j)}$ between $\hat{\mathbf{s}}_a^{(j)}$ and $\mathbf{x}_i$ is computed,

$$\overline{\rho}^{(j)} = \frac{1}{M} \sum_{i=1}^M \rho_i^{(j)}$$

$$= \frac{1}{M} \sum_{i=1}^M \frac{\mathbf{x}_i^T \hat{\mathbf{s}}_a^{(j)}}{\sqrt{\mathbf{x}_i^T \mathbf{x}_i} \sqrt{\left(\hat{\mathbf{s}}_a^{(j)} \right)^T \hat{\mathbf{s}}_a^{(j)}}}. \tag{4.133}$$

The iterative procedure is terminated when

$$\left| \overline{\rho}^{(j)} - \overline{\rho}^{(j-1)} \right| \leq \eta_\rho, \tag{4.134}$$

where the threshold η_ρ should be chosen such that $0 < \eta_\rho < 1$. A value of ρ_j equal to one corresponds to the case when the SNR is infinite and all waveforms have identical morphologies. Although no general proof of convergence has been presented for the Woody method, experimental results have shown that convergence is usually achieved within a few iterations, provided that the waveforms are initially reasonably well-aligned and that the SNR is reasonably good.

A block diagram summarizing the Woody method is presented in Figure 4.21. The performance of the method is illustrated in Figure 4.22, which shows the results of processing signals with different SNRs. Ensemble averages are computed before and after latency correction. For moderate to high SNRs, the peak amplitudes of the latency-corrected averages are considerably larger than those of the uncorrected averages, see Figures 4.22(a) and (b). When the SNR deteriorates, however, the main waves of the latency-corrected EP become smeared, and reliable amplitude measurements can no longer be made. This type of behavior is caused by various undesired EEG components, such as alpha activity, which are aligned and included in the ensemble average. Several studies have reported on the performance of the Woody method, including the observation of limiting behavior at low

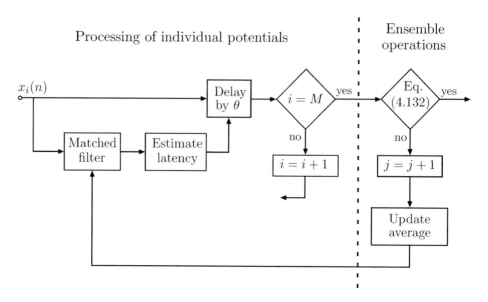

Figure 4.21: Block diagram of the Woody method for latency correction of EPs. The correction procedure is iterative and is terminated when either the criteria defined in (4.132) or (4.134) is fulfilled or, possibly, when both criteria are fulfilled.

SNRs [77–82]. Ultimately, the usefulness of the Woody method depends on the application of interest: analysis of VEPs stands out as the most appropriate one due to its relatively high SNR.

Although the matched filter in (4.127), derived from the white noise assumption, can be used for latency estimation in a colored noise situation, its performance will be inferior to that achieved by a filter especially designed for the colored noise situation. Such noise is characterized by the correlation matrix $\mathbf{R}_v$ which, for example, can be estimated from the prestimulus interval containing only the background EEG. For the analysis of AEPs, it has been found that the variance of the latency estimates $\hat{\theta}_i$ was considerably lower when a filter matched to colored noise was applied [80].

The ML estimator of θ_i for the case of colored noise can be compactly expressed using matrix notation. The necessary notation may be introduced by first recasting the "white noise" ML estimator in (4.125) into

$$\hat{\theta}_i = \arg\max_{\theta_i}(\mathbf{x}_i^T \mathbf{s}_{\theta_i}),$$

where the signal vector $\mathbf{s}_{\theta_i}$ is defined by

$$\mathbf{s}_{\theta_i} = \begin{bmatrix} \mathbf{0}_{\theta_i} \\ \mathbf{s} \\ \mathbf{0}_{N-D-\theta_i} \end{bmatrix}, \qquad (4.135)$$

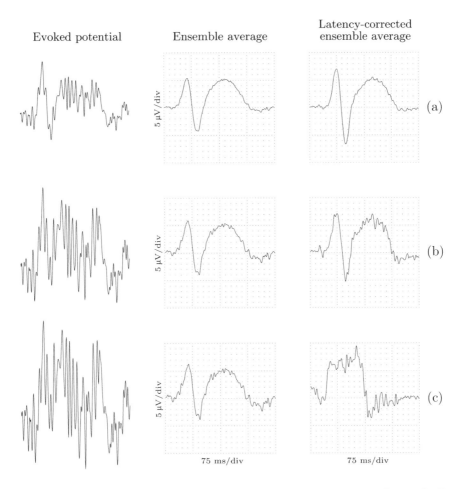

Figure 4.22: The performance of the Woody method illustrated at different SNRs: (a) a high SNR, (b) an intermediate SNR, and (c) a low SNR. For each of the three SNRs, one of the EPs in the ensemble is displayed together with the corresponding ensemble average, obtained either before or after latency correction. The averages are based on an ensemble of 100 EPs.

and $\mathbf{0}_k$ denotes a column vector with k zeros. Assuming that the number of observed samples N is considerably larger than the maximum nonzero correlation lag of the colored noise, it can be shown that the ML estimator is given by (see Problem 4.26),

$$\hat{\theta}_i = \arg \max_{\theta_i}(\mathbf{x}_i^T \mathbf{R}_v^{-1}\mathbf{s}_{\theta_i}). \tag{4.136}$$

Thus, the signal vector $\mathbf{s}_{\theta_i}$ is modified by the inverse of the noise correlation matrix $\mathbf{R}_v^{-1}$ before it is correlated with the signal $\mathbf{x}_i$; the resulting product is then used for determining the optimal value of θ_i.

Reality is actually more complicated than what was suggested by the model in (4.120) because the EP peak components may have latencies which vary independently of one another. This property cannot be handled by the Woody method since it is based on the assumption that the latencies of all peaks are identical. Consequently, attempts have been made to develop methods for latency correction in which the EP peak components are subjected to segmentation, followed by alignment and averaging of each individual peak within the ensemble [83–85]. These approaches suffer, however, from the drawback that the latency-corrected average is a signal defined by a series of disjointed segments which, due to the gaps, may be difficult to interpret. The problem of disjointed segments has been addressed by introducing a nonlinear procedure in which the time axis of each EP is subjected to expansion or contraction before alignment is performed [86, 87].

We conclude this section on estimation of latency shifts by commenting on the jitter that results from time discretization of the signal. In certain situations, such sampling jitter must be taken into account when designing a method for latency correction. If this is not done, jitter may seriously deteriorate the accuracy of certain ensemble-related clinical measures. It has been shown that an appreciably higher sampling rate (about three times) must be used when the ensemble variance, rather than the ensemble average, represents the measure conveying clinical information [88]. Choosing the Nyquist rate as the sampling rate is insufficient when the goal is to obtain accurate measurements on ensemble variance. The simplest way to account for this result is to precede the estimation of latency shifts by a block which increases the sampling rate by interpolating between the existing samples of the signal. Unfortunately, an increased sampling rate not only implies better performance, but also an increased amount of computation.

4.3.8 Weighting of Averaged Evoked Potentials Using Ensemble Correlation

Although ensemble averaging reduces the noise level, it is obvious that certain segments of the ensemble average will contain only noise simply because no activity is present. One approach to further reduce the noise level in such noisy segments is to analyze the correlation properties across the ensemble for each sample n and then to apply a weight function to the ensemble average which reflects the ensemble correlation information. A high correlation value indicates that activity is present, and, therefore, the sample in the ensemble average should be weighted by a value close to one. A low correlation value indicates, on the other hand, that activity is essentially absent, and the corresponding weight should therefore be close to zero.

Sample-by-sample weighting of the ensemble average by a weight function $w(n)$ can be useful when the aim is to find a transition from a noise-only segment to a segment that contains activity, or vice versa. Although this technique has found no particular interest in EP analysis, the idea behind ensemble correlation presented below nevertheless helps to bridge the gap between previous concepts on ensemble averaging and those presented in the next section on noise reduction by linear filtering. The idea of applying a weight function based on ensemble correlation originally stems from the area of high-resolution ECG analysis where an important task is to detect the end point in time of late potentials after which only noise is considered to be present [89].

We will once again study the signal-plus-noise model

$$x_i(n) = s(n) + v_i(n) \tag{4.137}$$

in order to derive an optimal weight function. However, the present model differs, in certain respects, from the one considered earlier for averaging of homogeneous ensembles. While the signal part $s(n)$ is still assumed to be identical in all potentials, it is now *random* in nature and characterized by its first- and second-order moments,

$$E[s(n)] = 0, \tag{4.138}$$
$$E[s^2(n)] = \sigma_s^2(n), \tag{4.139}$$

for $n = 0, \ldots, N - 1$. The variance $\sigma_s^2(n)$ is now a function of time in order to account for the fact that $s(n)$ has varying strength. The two assumptions in (4.138) and (4.139) provide, of course, a minimalistic statistical characterization of $s(n)$ but will suffice for the purposes of this section. Similar to averaging of homogeneous ensembles, the noise $v_i(n)$ is considered to be zero-mean, with a fixed variance σ_v^2 for all EPs, and is uncorrelated from EP to EP. We conclude the model description by pointing out that the signal and noise are mutually uncorrelated,

$$E[s(n)v_i(n)] = 0, \quad i = 1, \ldots, M. \tag{4.140}$$

The weight $w(n)$ should be chosen such that the error between the signal $s(n)$ and the weighted ensemble average $w(n)\hat{s}_a(n)$ is minimized. Employing the MSE criterion,

$$E\left[(s(n) - w(n)\hat{s}_a(n))^2\right],\qquad(4.141)$$

it is easily shown that the weight minimizing this criterion is given by

$$w(n) = \frac{\sigma_s^2(n)}{\sigma_s^2(n) + \dfrac{\sigma_v^2}{M}}.\qquad(4.142)$$

The optimal weight $w(n)$ can be interpreted as a measure of the SNR whose values are normalized to the interval $[0, 1]$. Consequently, the amplitude of the weighted ensemble average $w(n)\hat{s}_a(n)$ remains unaltered only if the observed data are noise-free, while it is otherwise reduced depending on the local SNR at time n.

Gaussian assumption. The weight suggested in (4.142) is, unfortunately, of limited value since the quantities $\sigma_s^2(n)$ and σ_v^2 are not known a priori. Somewhat surprisingly, it is possible to estimate the right-hand side of (4.142) from the ensemble of data when both $s(n)$ and $v_i(n)$ are assumed to be Gaussian. As a starting point for our derivation, we observe that the cross-correlation coefficient of $x_i(n)$ and $x_j(n)$,

$$\rho_{ij}(n) = \frac{E\left[x_i(n)x_j(n)\right]}{\sqrt{E[x_i^2(n)]}\sqrt{E[x_j^2(n)]}},\qquad(4.143)$$

for the above model is almost identical to (4.142). We obtain

$$\rho_{ij}(n) = \begin{cases} 1, & i = j; \\[2mm] \dfrac{\sigma_s^2(n)}{\sigma_s^2(n) + \sigma_v^2}, & i \neq j, \end{cases}\qquad(4.144)$$

and thus $\rho_{ij}(n) = \rho(n)$ for $i \neq j$. The terms due to noise obviously differ in (4.142) and (4.144) because ensemble averaging reduces the noise variance σ_v^2 by a factor of M. It is, however, possible to express $w(n)$ in terms of $\rho(n)$ using the following simple transformation,

$$w(n) = \frac{\rho(n)}{\rho(n)\left(1 - \dfrac{1}{M}\right) + \dfrac{1}{M}} \left(= \frac{\sigma_s^2(n)}{\sigma_s^2(n) + \dfrac{\sigma_v^2}{M}}\right).\qquad(4.145)$$

The vector $\mathbf{x}(n)$ contains all the M samples of the ensemble at a certain time n,

$$\mathbf{x}(n) = \begin{bmatrix} x_1(n) \\ x_2(n) \\ \vdots \\ x_M(n) \end{bmatrix}, \tag{4.146}$$

and follows a zero-mean, Gaussian PDF defined by

$$p(\mathbf{x}(n)) = \frac{1}{\sqrt{(2\pi)^M |\mathbf{R}_x|}} \exp\left[-\frac{1}{2}\mathbf{x}^T(n)\mathbf{R}_x^{-1}\mathbf{x}(n)\right]. \tag{4.147}$$

This density function is completely characterized by its correlation matrix $\mathbf{R}_x$,

$$\mathbf{R}_x = \begin{bmatrix} E[x_1^2(n)] & E[x_1(n)x_2(n)] & \cdots & E[x_1(n)x_M(n)] \\ E[x_2(n)x_1(n)] & E[x_2^2(n)] & \cdots & E[x_2(n)x_M(n)] \\ \vdots & \vdots & & \vdots \\ E[x_M(n)x_1(n)] & E[x_1(n)x_2(n)] & \cdots & E[x_M^2(n)] \end{bmatrix}, \tag{4.148}$$

which, for the model in (4.137), is equal to

$$\mathbf{R}_x = \left(\sigma_s^2(n) + \sigma_v^2\right) \begin{bmatrix} 1 & \rho(n) & \cdots & \rho(n) \\ \rho(n) & 1 & \cdots & \rho(n) \\ \vdots & \vdots & \ddots & \vdots \\ \rho(n) & \rho(n) & \cdots & 1 \end{bmatrix}. \tag{4.149}$$

Maximum likelihood estimation. Next, ML estimation is employed to obtain an estimate of $\rho(n)$, and the resulting estimate is then inserted into (4.145) in order to produce the desired weight $w(n)$. The ML estimator is derived by finding $\rho(n)$ that maximizes the log-likelihood function $\ln\left(p(\mathbf{x}(n); \rho(n))\right)$,

$$\hat{\rho}(n) = \arg\max_{\rho(n)} \ln\left(p(\mathbf{x}(n); \rho(n))\right). \tag{4.150}$$

The different steps involved in the derivation of $\hat{\rho}(n)$, such as finding the determinant $|\mathbf{R}_x|$ and the inverse $\mathbf{R}_x^{-1}$ in (4.147) and the differentiation of the log-likelihood function, can be found in [89], see also Problem 4.29. The resulting ML estimate is given by

$$\hat{\rho}(n) = \frac{\displaystyle\sum_{i=1}^{M}\sum_{\substack{j=1 \\ i\neq j}}^{M} x_i(n)x_j(n)}{(M-1)\displaystyle\sum_{i=1}^{M} x_i^2(n)}. \tag{4.151}$$

Here, the numerator provides us with an estimate of the signal variance $\sigma_s^2(n)$, since averaging is performed over all possible cross-term combinations of $x_i(n)$ and $x_j(n)$. The sum in the denominator of (4.151) provides an estimate of the total energy of the observed signals $x_i(n)$, i.e., $\sigma_s^2(n) + \sigma_v^2$.

Finally, the sample-by-sample weighted ensemble average $\check{s}_a(n)$ is computed from

$$\check{s}_a(n) = w(n)\hat{s}_a(n), \quad n = 0, \dots, N-1. \tag{4.152}$$

Again, we note that the weight $w(n)$ is related to the SNR of the ensemble average $\hat{s}_a(n)$ at time n.

Estimator properties. The ensemble correlation estimator in (4.151) possesses certain undesirable properties, namely, negative values and large variance. While the optimal weight in (4.142) is always positive-valued, this property is not carried over to $\hat{\rho}(n)$. It can be shown that $\hat{\rho}(n)$ is bounded by $-1/M \leq \hat{\rho}(n) \leq 1$ for even values of M and by $-1/(M-1) \leq \hat{\rho}(n) \leq 1$ for odd values of M. A simple remedy to this problem is to set negative values of $\hat{\rho}(n)$ equal to zero.

More seriously, the estimate $\hat{\rho}(n)$ is associated with a large variance unless M is large. A straightforward approach to reduce the variance is to replace the numerator and denominator expressions in (4.151) with time averages computed locally around the time n,

$$\hat{\rho}(n) = \frac{\displaystyle\sum_{k=n-W}^{n+W}\left[\sum_{i=1}^{M}\sum_{\substack{j=1 \\ i\neq j}}^{M} x_i(n)x_j(n)\right]}{(M-1)\displaystyle\sum_{k=n-W}^{n+W}\sum_{i=1}^{M} x_i^2(k)}, \tag{4.153}$$

where $2W+1$ is the length of the averaging window. Although temporal detail of $\hat{\rho}(n)$ has to be traded for variance reduction, "local" averaging can be acceptable since $\sigma_s^2(n)$ often changes slowly. It should be noted that

(4.153) has to be modified in some way to handle averaging at the interval end points, i.e., for $n < W$ and $n \geq N - W$.

Weighting by ensemble correlation is illustrated in Figure 4.23 for a simulated signal which exhibits a gradually decreasing amplitude. It is obvious from Figure 4.23(d) that weighting of $\hat{s}_a(n)$ with $w(n)$, thus resulting in $\check{s}_a(n)$, reduces the noise level in the intervals adjacent to the signal interval in comparison with the (unweighted) ensemble average $\hat{s}_a(n)$ displayed in Figure 4.23(b).

The estimation of ensemble correlation is here based on a homogeneous ensemble of data, i.e., with fixed noise variance in all EPs. This type of analysis can, however, be extended to process inhomogeneous ensembles of data with varying noise variances, a case previously considered in Section 4.3.4 [90].

4.4 Noise Reduction by Linear Filtering

Ensemble averaging relies on the assumption that the noise is uncorrelated from EP to EP. No information is included that reflects the degree to which successive samples in an EP are correlated. Hence, the noise level is reduced by averaging, while the stimulus-related EP (ideally) remains unchanged. It seems, however, plausible to assume that the noise in the ensemble average can be even further reduced by exploiting the property that the signal and the noise are, to various degrees, correlated in time; a white noise assumption applying to both these quantities is evidently rather far-fetched. Once such correlation information becomes available, the extensive knowledge available in the area of optimal filtering may help us to develop a method for extracting the signal part from the noisy observations. The main ingredients of optimal filtering are the statistical signal model, incorporating the correlation information, and the optimization of an error criterion, ensuring that the filtered signal resembles the desired signal as much as possible, for example, in the MSE sense. Comprehensive presentations of optimal filtering techniques can be found in [55, 56, 91].

Below, we will describe EP modeling in terms of stationary processes and the design of an optimal, linear, time-invariant filter to be applied to the ensemble average $\hat{s}_a(n)$. Since the correlation information is not a priori available, it must be estimated a posteriori from the ensemble of EPs once the recording procedure is finished; hence, the resulting filter is commonly referred to as an a posteriori filter.

It is essential to underline that the word "optimal"—as in "optimal filtering"—must be used with great caution since optimality is meaning-

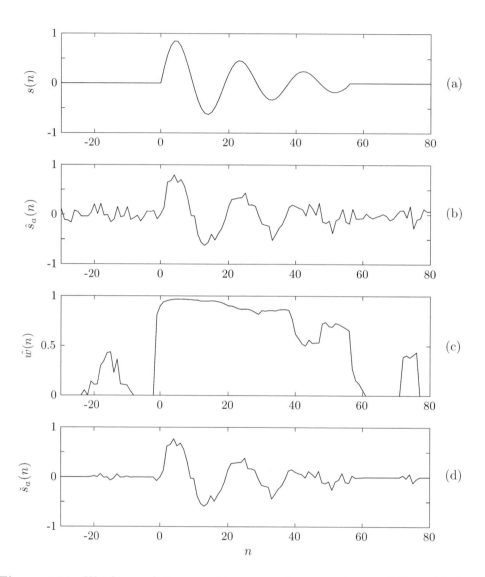

Figure 4.23: Weighting of the ensemble average by ensemble correlation. (a) A noise-free, simulated signal $s(n)$, (b) the ensemble average $\hat{s}_a(n)$ resulting from 100 EPs with added white, Gaussian noise with $\sigma_v^2 = 1$, (c) the estimated weight function $\hat{w}(n)$ describing the ensemble correlation, and (d) the weighted ensemble average $\check{s}_a(n)$.

ful only as long as the model assumptions are fulfilled (which, unfortunately, is rarely the case in biomedical signal processing).

4.4.1 Time-Invariant, A Posteriori Filtering

As several times before in this chapter, we adopt the signal-plus-noise model in (4.4) as the starting point for our presentation,

$$x_i(n) = s(n) + v_i(n), \quad i = 1, \dots, M, \tag{4.154}$$

where $s(n)$ and $v_i(n)$ are both considered to be zero-mean, stationary processes and thus completely characterized by their respective correlation functions $r_s(k)$ and $r_v(k)$. Alternatively, they are completely characterized by their respective power spectra $S_s(e^{j\omega})$ and $S_v(e^{j\omega})$. The noise $v_i(n)$, present in each of the M different EPs, is described by the same correlation function $r_v(k)$ and, hence, the variance $r_v(0) = \sigma_v^2$. Moreover, it is assumed that both $s(n)$ and $v_i(n)$ are uncorrelated,

$$E[s(n)v_i(n)] = 0, \quad i = 1, \dots, M. \tag{4.155}$$

The reader is reminded of the relationships between the correlation function and the power spectrum:

$$S_x(e^{j\omega}) = \sum_{k=-\infty}^{\infty} r_x(k)e^{-j\omega k}$$

and

$$r_x(k) = \frac{1}{2\pi} \int_{-\pi}^{\pi} S_x(e^{j\omega})e^{j\omega k}d\omega.$$

Either of these representations is used below, depending on which is considered to be the most suitable.

Correlation functions. In practice, the correlation functions $r_s(k)$ and $r_v(k)$ are unknown and must be inferred, in some way, from the ensemble of EPs. One straightforward approach is to consider the correlation function $r_{\hat{s}_a}(k)$ of the ensemble average $\hat{s}_a(n)$, which can be expressed in terms of $r_s(k)$ and $r_v(k)$,

$$r_{\hat{s}_a}(k) = r_s(k) + \frac{1}{M}r_v(k). \tag{4.156}$$

Similarly, the average of the correlation functions $r_{x_i}(k)$, obtained from each of the EPs $x_i(n)$, is given by

$$\bar{r}_x(k) = \frac{1}{M} \sum_{i=1}^{M} r_{x_i}(k) = r_s(k) + r_v(k). \tag{4.157}$$

By combining (4.156) and (4.157), the desired correlation functions can be expressed in terms of $r_{\hat{s}_a}(k)$ and $\bar{r}_x(k)$ such that

$$r_s(k) = \frac{M}{M-1} \left(r_{\hat{s}_a}(k) - \frac{1}{M} \bar{r}_x(k) \right) \tag{4.158}$$

and

$$r_v(k) = \bar{r}_x(k) - r_s(k). \tag{4.159}$$

It is straightforward to compute the correlation functions $r_{x_1}(k), \ldots, r_{x_M}(k)$ and $r_{\hat{s}_a}(k)$ from the ensemble of data using the estimator previously given in (3.78). This a posteriori approach was originally suggested in [92, 93], see also [94–97].

Another a posteriori approach to estimate the signal and noise correlation functions is to employ alternate ensemble averaging in which every second EP is included with opposite sign [98] (see also page 198). The alternate ensemble average is defined by

$$\bar{v}(n) = \frac{1}{M} \sum_{i=1}^{M} (-1)^i \, x_i(n), \tag{4.160}$$

where M is assumed to be an even integer. This technique cancels the signal part $s(n)$, assumed to be invariant from EP to EP, whereas a residual noise component persists in $\bar{v}(n)$. Since the correlation functions of $v(n)$ and $\bar{v}(n)$ are related by

$$r_v(k) = M r_{\bar{v}}(k), \tag{4.161}$$

we have from (4.156) that

$$r_s(k) = r_{\hat{s}_a}(k) - r_{\bar{v}}(k). \tag{4.162}$$

An obvious advantage of the procedure defined by (4.161) and (4.162) is that these expressions require a far smaller amount of computation than do the corresponding ones in (4.158) and (4.159); the correlation function needs to be computed only twice instead of $M+1$ times. Unfortunately, the saving in computation time is accompanied by an increased variance in the correlation function estimates based on alternate ensemble averaging [98].

Yet another approach is to make use of the prestimulus interval to estimate the noise correlation $r_v(k)$ [99], cf. the definition in (4.72). The validity of this approach relies on the assumption that the correlation properties of the EEG remain the same throughout the EP; an assumption which is not entirely valid, as was already pointed out on page 197. In contrast to the above two a posteriori techniques, which rely on the ensemble properties to find estimates of $r_s(k)$ and $r_v(k)$, the prestimulus technique makes it possible to use estimates of the noise correlation function $r_{v_i}(k)$ that differ from EP to EP.

Noncausal infinite impulse response filtering. We will now consider the design of a linear filter $h(k)$ that processes the ensemble average $\hat{s}_a(n)$ in order to produce an estimate of the desired signal $s(n)$. Initially, we will assume that the observation interval is infinite and that the filter is allowed to be noncausal so that both past and future samples are used for filtering. The transfer function of the filter is described by its z-transform

$$H(z) = \sum_{k=-\infty}^{\infty} h(k)z^{-k}.$$

The output of the filter is obtained from the following convolution sum,

$$\hat{s}(n) = \sum_{k=-\infty}^{\infty} h(k)\hat{s}_a(n-k). \tag{4.163}$$

The filter is designed with reference to the MSE criterion, which involves the desired signal $s(n)$ and the filtered signal $\hat{s}(n)$,

$$\mathcal{E} = E\left[(s(n) - \hat{s}(n))^2\right]. \tag{4.164}$$

Minimization of this error criterion is achieved by differentiating $\mathcal{E}$ with respect to each of the filter coefficients $h(k)$ and setting the resulting derivatives equal to zero,

$$\frac{\partial}{\partial h(k)} E\left[\left(s(n) - \sum_{k=-\infty}^{\infty} h(k)\hat{s}_a(n-k)\right)^2\right] = 0, \quad -\infty < k < \infty. \tag{4.165}$$

Performing the differentiation, we obtain an infinite set of equations with an infinite number of unknowns,

$$\sum_{l=-\infty}^{\infty} h(l)r_{\hat{s}_a}(k-l) = r_{s\hat{s}_a}(k), \quad -\infty < k < \infty, \tag{4.166}$$

where the cross-correlation between $s(n)$ and $\hat{s}_a(n)$ is

$$r_{s\hat{s}_a}(k) = E[s(n)\hat{s}_a(n-k)]. \tag{4.167}$$

The equations in (4.166) are well-known as the *Wiener–Hopf equations* of the noncausal, infinite impulse response (IIR) Wiener filter [100]. Their solution is straightforward, since the left-hand side of (4.166) is a convolution of $h(k)$ with $r_{\hat{s}_a}(k)$ in the time domain, and, therefore, can be expressed as a multiplication in the frequency domain. Hence, the Wiener–Hopf equations can be expressed as

$$H(e^{\jmath\omega})S_{\hat{s}_a}(e^{\jmath\omega}) = S_{s\hat{s}_a}(e^{\jmath\omega}),$$

where $S_{s\hat{s}_a}(e^{\jmath\omega})$ is the cross-power spectrum of $s(n)$ and $\hat{s}_a(n-k)$. The frequency response of the optimal, noncausal Wiener filter is given by

$$H(e^{\jmath\omega}) = \frac{S_{s\hat{s}_a}(e^{\jmath\omega})}{S_{\hat{s}_a}(e^{\jmath\omega})}. \tag{4.168}$$

Since the noise $v_i(n)$ is assumed to be zero-mean and uncorrelated with $s(n)$, the cross-correlation between $s(n)$ and $\hat{s}_a(n)$ becomes

$$E[s(n)\hat{s}_a(n-k)] = E[s(n)s(n-k)] + \frac{1}{M}\sum_{i=1}^{M}E[s(n)v_i(n-k)]$$

$$= E[s(n)s(n-k)] + 0 = r_s(k). \tag{4.169}$$

Hence, the frequency response in (4.168) can be expressed in terms of the signal and noise spectra,

$$H(e^{\jmath\omega}) = \frac{S_s(e^{\jmath\omega})}{S_s(e^{\jmath\omega}) + \dfrac{1}{M}S_v(e^{\jmath\omega})}. \tag{4.170}$$

Since the power spectrum of a stationary process is always non-negative, i.e., $S(e^{\jmath\omega}) \geq 0$, the frequency response $H(e^{\jmath\omega})$ is restricted such that

$$0 < H(e^{\jmath\omega}) < 1 \tag{4.171}$$

for all values of ω. Filtering with $H(e^{\jmath\omega})$ does not introduce any phase distortion since the power spectrum is a real function, and, therefore, the phase function is equal to zero. It is noted that the signal passes through the filter almost unattenuated for values of ω at which the SNR is high, i.e., for $S_s(e^{\jmath\omega}) \gg S_v(e^{\jmath\omega})$, since $|H(e^{\jmath\omega})| \approx 1$. On the other hand, the filter suppresses the noise at low SNRs, $S_s(e^{\jmath\omega}) \ll S_v(e^{\jmath\omega})$, since $|H(e^{\jmath\omega})|$ has

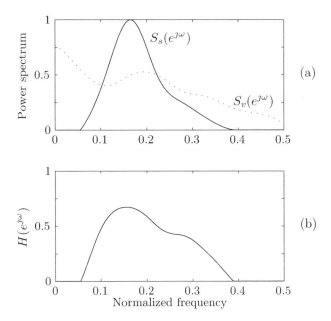

Figure 4.24: Design of a Wiener filter. (a) The signal and noise power spectra $S_s(e^{j\omega})$ and $S_v(e^{j\omega})$ and (b) the frequency response of the corresponding Wiener filter $H(e^{j\omega})$, defined in (4.170). The noise is completely attenuated by the filter in frequency intervals without signal content.

almost zero gain. These filter characteristics resemble those of the sample-by-sample weight function, presented previously in (4.142), in the sense that the weight is close to one for a high SNR and close to zero for a low SNR; however, the weight function in (4.142) does not depend on the frequency ω. Figure 4.24 presents the frequency response of a Wiener filter resulting from known signal and noise spectra.

An undesirable implication of the filter gain property in (4.171) is that amplitude measurements from the filtered ensemble average are systematically underestimated. Hence, the estimate $\hat{s}(n)$ that results from the Wiener filter in (4.170) is, in general, a biased estimate of $s(n)$, although with reduced variance. As a consequence, amplitude measurements obtained from the filtered ensemble average cannot be directly compared to normal values of EP amplitude obtained from ensemble average measurements. Instead, such values have to be redeveloped using filtered ensemble averages obtained from a population of normal subjects. Normal values of EP latencies are, on the other hand, less influenced by filtering thanks to the zero-phase property.

Having derived the Wiener filter under the assumption of known power spectra, we now consider how the filter design is modified when the a pos-

teriori estimation technique is incorporated. From (4.158) and (4.159) we
know that estimates of the signal and noise power spectra are given by

$$\hat{S}_s(e^{\jmath\omega}) = \frac{M}{M-1}\left(S_{\hat{s}_a}(e^{\jmath\omega}) - \frac{1}{M}\overline{S}_x(e^{\jmath\omega})\right), \tag{4.172}$$

$$\hat{S}_v(e^{\jmath\omega}) = \overline{S}_x(e^{\jmath\omega}) - \hat{S}_s(e^{\jmath\omega}), \tag{4.173}$$

respectively, where $\overline{S}_x(e^{\jmath\omega})$ is related to $\overline{r}_x(k)$. Insertion of these estimates
into $H(e^{\jmath\omega})$, as given in (4.170), yields the frequency response

$$\hat{H}_1(e^{\jmath\omega}) = \frac{M}{M-1}\left(1 - \frac{1}{M}\frac{\overline{S}_x(e^{\jmath\omega})}{S_{\hat{s}_a}(e^{\jmath\omega})}\right). \tag{4.174}$$

It is important to realize that this frequency response represents an *estimate*
of the Wiener filter and can no longer be considered optimal for the model
of interest. Considering the power spectra based on the alternate ensemble
average, we obtain the following frequency response:

$$\hat{H}_2(e^{\jmath\omega}) = 1 - \frac{S_{\overline{v}}(e^{\jmath\omega})}{S_{\hat{s}_a}(e^{\jmath\omega})}. \tag{4.175}$$

Both of the above two frequency response estimates suffer from certain
problems which must be dealt with in order to avoid serious degradation of
performance. Perhaps, the most striking problem is that negative values may
occur in the frequency response because the terms $\overline{S}_x(e^{\jmath\omega})/(MS_{\hat{s}_a}(e^{\jmath\omega}))$ and
$S_{\overline{v}}(e^{\jmath\omega})/S_{\hat{s}_a}(e^{\jmath\omega})$, being subtracted in (4.174) and (4.175), respectively, may
be greater than one; this property is in contrast to the frequency response
of the Wiener filter, which is always positive-valued. A simple remedy is to
clip any value that is less than zero [101, 102],

$$\hat{H}_1^c(e^{\jmath\omega}) = \begin{cases} \hat{H}_1(e^{\jmath\omega}), & \hat{H}_1(e^{\jmath\omega}) \geq 0; \\ 0, & \hat{H}_1(e^{\jmath\omega}) < 0. \end{cases} \tag{4.176}$$

A similar procedure has also been suggested for the sample-by-sample weight
function in (4.152) which exhibits the same type of problem.

Another problem is that the estimates of $\overline{S}_x(e^{\jmath\omega})$ and $S_{\hat{s}_a}(e^{\jmath\omega})$ are periodograms and, therefore, exhibit substantial variance, as indicated by the
expression in (3.86). The variance can be reduced, to a certain degree, by
smoothing of each spectrum (i.e., lowpass filtering of the spectrum) or by
splitting the ensemble into several subensembles accompanied by averaging
of the resulting subensemble power spectra [98]. Figure 4.25 presents a typical example of an estimated frequency response and the effect of clipping and
spectral smoothing. It can be seen that the combined use of smoothing and
clipping yields an estimate of the frequency response which is considerably

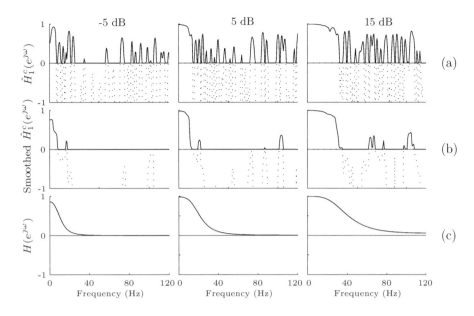

Figure 4.25: The frequency response of the time-invariant, a posteriori filter in (4.174) and its improvements. The filters are estimated from an ensemble for which the signal-to-noise ratio of the average is presented. (a) The frequency response before (dotted line) and after clipping, (b) the smoothed frequency response before and after clipping, and (c) the theoretical frequency response determined from prior knowledge of the signal and noise power spectra.

closer to the a priori frequency response than is the unprocessed frequency response. The frequency responses in Figure 4.25 are estimated from the simulated data used in Figure 4.26.

As the number M of EPs increases, the a posteriori filter becomes an increasingly better approximation of the a priori Wiener filter because the accuracy of the power spectrum estimates improves. However, the SNR of the ensemble average $\hat{s}_a(n)$ to be filtered also increases, and, consequently, the need for a posteriori filtering diminishes. On the other hand, the a posteriori filter becomes less and less reliable when it is really needed at lower SNRs [103]. This property limits the overall usefulness of the a posteriori filter.

The performance of a posteriori filtering is illustrated by the simulation example in Figure 4.26. The filters are estimated using (4.174) in combination with an ensemble of 100 identical EPs at three different SNRs. The ensemble averages plotted in Figure 4.26(a) were processed with estimated filters whose frequency responses have either been clipped (Figure 4.26(b)), or smoothed and clipped (Figure 4.26(c)). Since both the signal and noise

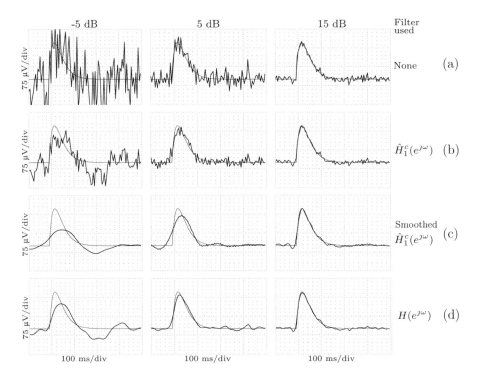

Figure 4.26: Performance of time-invariant, a posteriori "Wiener" filtering for different signal-to-noise ratios. (a) The ensemble average and the desired signal (thin line). The filtered ensemble average results from a filter whose frequency response is either (b) clipped or (c) smoothed and clipped. (d) The filtered ensemble average using the optimal filter, defined in (4.170).

power spectra are known a priori in this example, filtered ensemble averages can also be computed using the optimal filter in (4.170), see Figure 4.26(d). A considerable reduction in variance can be observed in the a posteriori filtered ensemble averages when compared to the corresponding ensemble averages; this is especially noticeable for lower SNRs. However, this improvement comes at the expense of underestimated peak amplitudes of the waveform whether a priori or a posteriori filtering is employed.

Causal FIR filtering. The derivation of the noncausal IIR Wiener filter displays in a clear way several important issues related to the usefulness of a posteriori filtering. However, a causal filter with FIR structure is often preferable [83]. Although an IIR filter can be subjected to truncation, the resulting FIR filter no longer has an obvious relation to the MMSE. In deriving the FIR Wiener filter, the following error criterion should instead be minimized,

$$\mathcal{E} = E\left[\left(s(n) - \sum_{k=0}^{L-1} h(k)\hat{s}_a(n-k)\right)^2\right], \qquad (4.177)$$

where the upper limit of the convolution sum is determined by the FIR filter length L. By introducing the following vector notation (cf. (3.62)),

$$\hat{\mathbf{s}}_a(n) = \begin{bmatrix} \hat{s}_a(n-L+1) \\ \vdots \\ \hat{s}_a(n-1) \\ \hat{s}_a(n) \end{bmatrix}, \quad \mathbf{h} = \begin{bmatrix} h(0) \\ h(1) \\ \vdots \\ h(L-1) \end{bmatrix}, \qquad (4.178)$$

the gradient of (4.177) can be calculated with respect to $\mathbf{h}$ and set equal to zero,

$$\nabla_{\mathbf{h}} E\left[\left(s(n) - \mathbf{h}^T \tilde{\mathbf{s}}_a(n)\right)^2\right] = 0 \qquad (4.179)$$

or

$$2E\left[\tilde{\mathbf{s}}_a(n)\left(s(n) - \tilde{\mathbf{s}}_a^T(n)\mathbf{h}\right)\right] = 0. \qquad (4.180)$$

The resulting equation can be written as

$$\tilde{\mathbf{R}}_{\hat{s}_a}\mathbf{h} = \mathbf{r}_{s\hat{s}_a}, \qquad (4.181)$$

where

$$\mathbf{R}_{\hat{s}_a} = E\left[\hat{\mathbf{s}}_a(n)\hat{\mathbf{s}}_a^T(n)\right] \qquad (4.182)$$

and

$$\mathbf{r}_{s\hat{s}_a} = E\left[s(n)\tilde{\mathbf{s}}_a(n)\right]. \qquad (4.183)$$

Equation (4.181) is the matrix form of the Wiener–Hopf equations and constitutes the causal FIR filter counterpart of the noncausal IIR filter in (4.168). Since $\hat{s}_a(n)$ was assumed to be a stationary process, the matrix $\mathbf{R}_{\hat{s}_a}$ is symmetric and Toeplitz, and, accordingly, the reversed correlation matrix is $\tilde{\mathbf{R}}_{\hat{s}_a} = \mathbf{R}_{\hat{s}_a}$.

Similar to the noncausal filtering case, $\mathbf{R}_{\hat{s}_a}$ and $\mathbf{r}_{s\hat{s}_a}$ can be expressed in terms of signal and noise correlation using the relationships given in (4.156) and (4.169), respectively, and thus,

$$\mathbf{h} = \left(\mathbf{R}_s + \frac{1}{M}\mathbf{R}_v\right)^{-1}\mathbf{r}_s, \qquad (4.184)$$

where

$$\mathbf{r}_{s\hat{s}_a} = \mathbf{r}_s = \begin{bmatrix} r_s(0) \\ r_s(1) \\ \vdots \\ r_s(L-1) \end{bmatrix}. \tag{4.185}$$

Finally, an estimate of $\mathbf{h}$ is obtained by replacing $\mathbf{R}_s, \mathbf{R}_v$, and $\mathbf{r}_s$ in (4.184) with their a posteriori estimated counterparts in (4.158) and (4.159).

4.4.2   Limitations with Time-Invariant, A Posteriori Filtering

The value of time-invariant, a posteriori filtering for improving the SNR of averaged EPs has not been unanimously agreed upon in the literature; instead, the performance results range from "inefficient" [101, 104–106] to "significant improvements" [107–109]. One plausible explanation of why such considerable differences in results have been reported is due to the fact that different EP modalities have been investigated [103]. For example, an AEP or VEP consists of several, consecutive waves which are relatively well modeled by a stationary process. On the other hand, an SEP has a more transient character and is therefore poorly modeled by a stationary process. For any transient waveform, however, the assumption of stationarity is not very well-founded, and, therefore, the motivation to develop methods that replace the above power spectral characterization has been strong. In fact, several approaches have been presented which account for the nonstationary behavior of EPs, most notably by generalizing a posteriori filtering to a time-varying spectral description. In such cases, the ensemble average $\hat{s}_a(n)$ is processed by a linear filter whose impulse response $h(l, n)$ is time-variant. The output $\hat{s}(n)$ is given by the convolution sum

$$\hat{s}(n) = \sum_l h(n - l, n)\hat{s}_a(l). \tag{4.186}$$

The filter $h(l, n)$ can be designed to be either causal or noncausal. The design of time-varying filters, as well as their performance, has been investigated at length [110–115]. We note that the sample-by-sample weight function related to ensemble correlation in (4.142) is one particular simple case of a time-varying filter whose impulse response equals

$$h(l, n) = w(n)\delta(l).$$

Another weak point of time-invariant, a posteriori filtering is that the signal $s(n)$ is assumed to be deterministic when considering the reduction in noise level of the ensemble average (e.g., used in the derivation of (4.156)), whereas it is assumed to be stochastic and stationary in the development of the Wiener filter [103].

The reader may question why several pages have been spent on describing a technique which has some serious limitations. It is, however, important to be aware of possible pitfalls in any area of application where biomedical signal processing is of interest. In this case it has been demonstrated that great care must be exercised before the assumption of stationarity is incorporated into a signal model.

4.5   Single-Trial Analysis Using Basis Functions

The tracking of transient changes in EP amplitude and latency is important in monitoring during neurosurgical procedures. Such tracking can, at an early stage, detect changes indicative of injury to the central nervous system. Previously described techniques for noise reduction usually require a substantial number of EPs for satisfactory performance and should therefore be replaced by signal processing techniques operating on a single-trial or, possibly, a few-trial basis.

Another application of single-trial noise reduction is the processing of EPs elicited when a subject performs a mental task. For example, brain function can be studied by asking a subject to verify the content of different sentences; following each verification, the single EP is analyzed and classified [116]. The accuracy in classifying each of the EPs is likely to improve once the data has been subjected to noise reduction.

Yet another reason for studying single-trial analysis is the recent observation that EPs are composed of a reproducible stimulus response and a dynamically changing background EEG activity, likely to reflect varying brain states, see, e.g., [117]. Thus, the background activity is likely to influence EP morphology to a larger extent than was initially believed when the "fixed signal-plus-noise" hypothesis was developed. The assumption that the background EEG can be modeled as additive noise may therefore have to be revised.

Noise reduction in a single trial can be achieved by introducing certain prior information on EP morphology which constrains the morphologic degrees of freedom, for example, by requiring the EP to be pulse shaped. In this section, we will describe single-trial techniques for estimating EP morphology that rely on a set of orthonormal basis functions for signal representation. The orthonormality property is very attractive since the components of the EP associated with a certain basis function do not interfere with the other basis functions. A straightforward approach is to apply sines/cosines as basis functions—the well-known Fourier series representation of a signal—followed by truncation of the series expansion so that only lower frequencies are allowed to model the EPs. In this case, noise is considered to be concentrated to the basis functions representing higher frequencies. Another approach is to design the basis functions so that the truncated expansion provides the most efficient representation, in the MSE sense, to an ensemble of different EP morphologies; this approach is known as the *discrete Karhunen–Loève expansion*.

In certain situations, it is acceptable to make use of the most recently acquired trials to achieve better noise reduction—a property which has already been mentioned in relation to exponential averaging. Adaptive filtering techniques, in combination with a basis function representation, have been found useful when tracking time-varying changes in EP morphology in noisy signals (Section 4.6).

Single-trial analysis is, of course, not limited to techniques relying on basis functions, although this approach represents the main focus in this section, but other techniques have been investigated incorporating prior information through parametric modeling of the EP and the background activity. For example, it has been suggested that each EP can be modeled as a filtered version of the ensemble average, while the background activity is modeled as an AR process [118, 119], cf. EEG modeling on page 65. The filter parameters that model the EP, as well as the AR model parameters, are estimated from each individual trial and are used to produce a noise-reduced estimate of the EP. Several other single-trial approaches can also be found in the literature [120–123].

4.5.1 Orthogonal Expansions

An EP $\mathbf{x}_i$, composed of both signal and noise, is modeled as a stochastic process which can be represented by a linear combination (series expansion) of basis functions $\boldsymbol{\varphi}_k$,

$$\mathbf{x}_i = \sum_{k=1}^{N} w_{i,k} \boldsymbol{\varphi}_k, \tag{4.187}$$

where each basis function is represented by a vector with N elements

$$\boldsymbol{\varphi}_k = \begin{bmatrix} \varphi_k(0) \\ \varphi_k(1) \\ \vdots \\ \varphi_k(N-1) \end{bmatrix}, \quad k = 1, \ldots, N. \tag{4.188}$$

The signal $\mathbf{x}_i$ is said to be an element of the space $\mathcal{X}$ spanned by the N basis functions φ_k. The space $\mathcal{X}$ is defined by the set of all vectors which can be represented by linear combinations of the basis $\{\varphi_1, \varphi_2, \ldots, \varphi_N\}$, denoted

$$\mathcal{X} = \text{span}\{\varphi_1, \varphi_2, \ldots, \varphi_N\}. \tag{4.189}$$

The coefficient (weight) vector $\mathbf{w}_i$ is the representation of $\mathbf{x}_i$ in terms of the basis $\{\varphi_1, \varphi_2, \ldots, \varphi_N\}$,

$$\mathbf{w}_i = \begin{bmatrix} w_{i,1} \\ w_{i,2} \\ \vdots \\ w_{i,N} \end{bmatrix}. \tag{4.190}$$

Introducing the matrix notation $\boldsymbol{\Phi}$ to represent the set of basis functions,

$$\boldsymbol{\Phi} = \begin{bmatrix} \varphi_1 & \varphi_2 & \cdots & \varphi_N \end{bmatrix}, \tag{4.191}$$

we can write the series expansion in (4.187) more compactly as

$$\mathbf{x}_i = \boldsymbol{\Phi} \mathbf{w}_i. \tag{4.192}$$

The orthonormality property implies that the basis functions are mutually orthogonal, and with their energy normalized to one,

$$\varphi_k^T \varphi_l = \begin{cases} 1, & k = l; \\ 0, & k \neq l. \end{cases} \tag{4.193}$$

Since the columns of $\boldsymbol{\Phi}$ are orthogonal, we have $\boldsymbol{\Phi}\boldsymbol{\Phi}^T = \boldsymbol{\Phi}^T\boldsymbol{\Phi} = \mathbf{I}$; such a matrix is said to be orthogonal (see Appendix A). By premultiplying both sides of (4.192) by $\boldsymbol{\Phi}^T$, the coefficient vector $\mathbf{w}_i$ can be calculated from $\mathbf{x}_i$ using the relation

$$\mathbf{w}_i = \boldsymbol{\Phi}^T \mathbf{x}_i. \tag{4.194}$$

Thus, each weight $w_{i,k}$ results from a correlation operation between $\mathbf{x}_i$ and the basis function φ_k—the *inner product*:

$$w_{i,k} = \varphi_k^T \mathbf{x}_i = \sum_{n=0}^{N-1} \varphi_k(n) x_i(n). \tag{4.195}$$

The calculation of $\mathbf{w}_i$ can be treated from an estimation point of view in which $\mathbf{w}_i$ is chosen such that the MSE is minimized. Each EP $\mathbf{x}_i$ is modeled by

$$\mathbf{x}_i = \mathbf{s}_i + \mathbf{v}_i, \tag{4.196}$$

where $E[\mathbf{x}_i] = \mathbf{s}_i$ since the noise $\mathbf{v}_i$ is assumed to be zero-mean. The correlation matrix for the ith EP is

$$\mathbf{R}_{x_i} = E\left[\mathbf{x}_i\mathbf{x}_i^T\right]. \tag{4.197}$$

The modeling assumptions are now more general than those presented in (4.4) because $\mathbf{s}_i$ is allowed to change from EP to EP.

A suitable criterion to minimize would be the following MSE,

$$E\left[\|\mathbf{s}_i - \boldsymbol{\Phi}\mathbf{w}_i\|^2\right] = E\left[(\mathbf{s}_i - \boldsymbol{\Phi}\mathbf{w}_i)^T(\mathbf{s}_i - \boldsymbol{\Phi}\mathbf{w}_i)\right], \tag{4.198}$$

but since $\mathbf{s}_i$ is unknown it is not useful. However, it is easily shown that the minimization of the MSE $\mathcal{E}_{\mathbf{w}_i}$ between $\mathbf{x}_i$ and the series expansion representation $\boldsymbol{\Phi}\mathbf{w}_i$,

$$\mathcal{E}_{\mathbf{w}_i} = E\left[\|\mathbf{x}_i - \boldsymbol{\Phi}\mathbf{w}_i\|^2\right], \tag{4.199}$$

is equivalent to the minimization of (4.198) since

$$\mathcal{E}_{\mathbf{w}_i} = E\left[\|\mathbf{s}_i - \boldsymbol{\Phi}\mathbf{w}_i\|^2\right] + E\left[\|\mathbf{v}_i\|^2\right], \tag{4.200}$$

where the noise term can be neglected since it is independent of $\mathbf{w}_i$. In order to minimize $\mathcal{E}_{\mathbf{w}_i}$, it is expanded,

$$\mathcal{E}_{\mathbf{w}_i} = E\left[\mathbf{x}_i^T\mathbf{x}_i\right] - 2E\left[\mathbf{x}_i^T\right]\boldsymbol{\Phi}\mathbf{w}_i + \mathbf{w}_i^T\boldsymbol{\Phi}^T\boldsymbol{\Phi}\mathbf{w}_i,$$

and differentiated with respect to $\mathbf{w}_i$ (see Appendix A for vector differentiation rules), resulting in

$$\nabla_{\mathbf{w}_i}\mathcal{E}_{\mathbf{w}_i} = -2\boldsymbol{\Phi}^T E[\mathbf{x}_i] + 2\boldsymbol{\Phi}^T\boldsymbol{\Phi}\mathbf{w}_i, \tag{4.201}$$

which, when set to zero, yields the MMSE estimator

$$\mathbf{w}_i = (\boldsymbol{\Phi}^T\boldsymbol{\Phi})^{-1}\boldsymbol{\Phi}^T E[\mathbf{x}_i] = \boldsymbol{\Phi}^T\mathbf{s}_i. \tag{4.202}$$

This estimator is, however, not very practical since $\mathbf{s}_i$ is exactly the information we are looking for in single-trial analysis. Our best guess of $\mathbf{s}_i$ is simply to replace it with the observed signal $\mathbf{x}_i$, yielding the approximate MSE estimator

$$\hat{\mathbf{w}}_i = \boldsymbol{\Phi}^T\mathbf{x}_i, \tag{4.203}$$

an expression which turns out to be identical to the one in (4.194). This particular type of estimation is sometimes referred to as *inner product estimation* since each coefficient is obtained as the inner product of the basis functions $\boldsymbol{\varphi}_k$ and $\mathbf{x}_i$.

The approximate estimator in (4.203) suggests that the estimate of the coefficient vector $\mathbf{w}_i$ would exhibit considerable variance since only one trial $\mathbf{x}_i$ is involved. By also making use of the most recent EPs $\mathbf{x}_{i-1}, \mathbf{x}_{i-2}, \ldots$ to estimate $\mathbf{w}_i$, the variance can be effectively reduced. This extension of the method is considered in the next section within the context of adaptive filtering, where the signal morphology is assumed to change relatively slowly.

Truncation. So far, the orthogonal series expansion in (4.192) has been considered to represent the observed signal $\mathbf{x}_i$, thus including both signal and noise. However, the underlying idea of signal estimation through a truncated series expansion is that a subset of basis functions can provide an adequate representation of the signal part $\mathbf{s}_i$. Therefore, the matrix $\boldsymbol{\Phi}$ is decomposed into two matrices, $\boldsymbol{\Phi}_s$ and $\boldsymbol{\Phi}_v$, whose columns represent the signal and the noise parts, respectively,

$$\boldsymbol{\Phi} = \begin{bmatrix} \boldsymbol{\Phi}_s & \boldsymbol{\Phi}_v \end{bmatrix}, \qquad (4.204)$$

where $\boldsymbol{\Phi}_s$ is $N \times K$ and $\boldsymbol{\Phi}_v$ is $N \times (N - K)$; K denotes the number of basis functions that approximates the signal $(K < N)$. In terms of the series expansion in (4.187), we are thus interested in using the first sum on right-hand side of

$$\mathbf{x}_i = \sum_{k=1}^{K} w_{i,k} \boldsymbol{\varphi}_k + \sum_{k=K+1}^{N} w_{i,k} \boldsymbol{\varphi}_k \qquad (4.205)$$

to estimate $\mathbf{s}_i$.

It should be emphasized that the decomposition in (4.204), by necessity, is somewhat arbitrary since, in practice, we rarely have knowledge on which basis functions represent the signal and noise parts. In fact, these parts often overlap to a considerable degree. The estimate of the signal is obtained from

$$\hat{\mathbf{s}}_i = \boldsymbol{\Phi}_s \hat{\mathbf{w}}_i = \boldsymbol{\Phi}_s \boldsymbol{\Phi}_s^T \mathbf{x}_i. \qquad (4.206)$$

It should be pointed out that for truncated expansions $\boldsymbol{\Phi}_s \boldsymbol{\Phi}_s^T \neq \mathbf{I}$, whereas $\boldsymbol{\Phi}_s^T \boldsymbol{\Phi}_s = \mathbf{I}$. Note also that the coefficient vector $\hat{\mathbf{w}}_i$ in (4.206) now contains only K elements due to truncation. The estimation procedure is illustrated by the block diagram in Figure 4.27.

In terms of spaces, truncation may be related to what is called *direct decomposition* of $\mathcal{X}$ into the subspaces $\mathcal{X}_s$ and $\mathcal{X}_v$,

$$\mathcal{X} = \mathcal{X}_s \oplus \mathcal{X}_v, \qquad (4.207)$$

where $\oplus$ denotes the *direct sum* of the two subspaces

$$\mathcal{X}_s = \text{span}\{\boldsymbol{\varphi}_1, \boldsymbol{\varphi}_2, \ldots, \boldsymbol{\varphi}_K\}, \qquad (4.208)$$
$$\mathcal{X}_v = \text{span}\{\boldsymbol{\varphi}_{K+1}, \boldsymbol{\varphi}_2, \ldots, \boldsymbol{\varphi}_N\}. \qquad (4.209)$$

The decomposition into subspaces is used below when wavelets are considered for multiresolution signal analysis.

A truncated series expansion of two basis functions $(K = 2)$ is illustrated in Figure 4.28. In this case, the series expansion is completely characterized by the coefficients $w_{i,1}$ and $w_{i,2}$ or, equivalently, by polar coordinates which sometimes may be preferable.

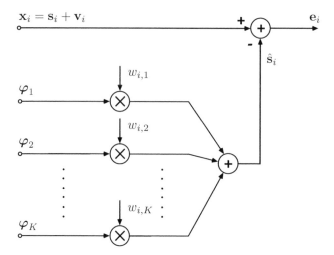

Figure 4.27: Estimation of the coefficient vector $\mathbf{w}$, which defines the linear combination of basis functions used to estimate the signal part $\mathbf{s}_i$. The output error $\mathbf{e}_i = \mathbf{x}_i - \mathbf{\Phi}\mathbf{w}_i$ is minimized in the mean-square error sense.

Signal-to-noise ratio. A definition of the SNR is helpful when evaluating the performance of the signal estimation in (4.206). The definition of the SNR for the input signal is related to the signal and noise energy,

$$E\left[\mathbf{x}_i^T \mathbf{x}_i\right] = \mathbf{s}_i^T \mathbf{s}_i + E\left[\mathbf{v}_i^T \mathbf{v}_i\right] + 2\mathbf{s}_i^T E\left[\mathbf{v}_i\right]$$
$$= \mathbf{s}_i^T \mathbf{s}_i + E\left[\mathbf{v}_i^T \mathbf{v}_i\right] \tag{4.210}$$

and

$$\mathrm{SNR}_{x_i} = \frac{\mathbf{s}_i^T \mathbf{s}_i}{E\left[\mathbf{v}_i^T \mathbf{v}_i\right]}. \tag{4.211}$$

The SNR of the resulting estimate $\hat{\mathbf{s}}_i$ depends on the accuracy of the coefficients $\hat{\mathbf{w}}_i$, estimated from the noisy observations $\mathbf{x}_i$. The coefficient estimate can be decomposed into two components,

$$\hat{\mathbf{w}}_i = \mathbf{w}_i + \Delta\mathbf{w}_i, \tag{4.212}$$

where the coefficient error vector $\Delta\mathbf{w}_i$ can be viewed as the bias in estimating $\mathbf{w}_i$ (i.e., $\Delta\mathbf{w}_i = \hat{\mathbf{w}}_i - \mathbf{w}_i$). The energy of the estimate $\hat{\mathbf{s}}_i$ is given by

$$E\left[\hat{\mathbf{s}}_i^T \hat{\mathbf{s}}_i\right] = \mathbf{s}_i^T \mathbf{\Phi}_s \mathbf{\Phi}_s^T \mathbf{s}_i + E\left[\Delta\mathbf{w}_i^T \Delta\mathbf{w}_i\right] + 2E\left[\Delta\mathbf{w}_i^T\right] \mathbf{\Phi}_s^T \mathbf{s}_i, \tag{4.213}$$

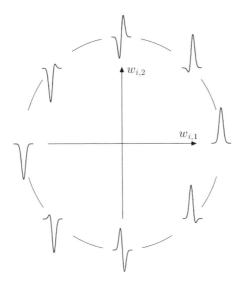

Figure 4.28: Signal morphologies resulting from a linear combination of two orthogonal basis functions φ_1 and φ_2 with mono- and biphasic morphology, respectively. For signals on the circle, the total energy is fixed, i.e., the radius $w_{i,1}^2 + w_{i,2}^2$ is equal to a constant.

and, consequently, a general SNR definition of $\hat{s}_i$ can be expressed as

$$\text{SNR}_{\hat{s}_i} = \frac{\mathbf{s}_i^T \mathbf{\Phi}_s \mathbf{\Phi}_s^T \mathbf{s}_i}{E\left[\Delta \mathbf{w}_i^T \Delta \mathbf{w}_i\right]}, \qquad (4.214)$$

since the last term in (4.213) is equal to zero. In order to derive a more specific expression of the coefficient error vector $\Delta \mathbf{w}_i$, we note that

$$\begin{aligned}
\hat{\mathbf{w}}_i &= \mathbf{\Phi}_s^T \mathbf{x}_i \\
&= \mathbf{\Phi}_s^T (\mathbf{s}_i + \mathbf{v}_i) \\
&= \mathbf{w}_i + \mathbf{\Phi}_s^T \mathbf{v}_i,
\end{aligned} \qquad (4.215)$$

and thus

$$\Delta \mathbf{w}_i = \mathbf{\Phi}_s^T \mathbf{v}_i. \qquad (4.216)$$

Inserting this expression into (4.214), the SNR definition becomes

$$\text{SNR}_{\hat{s}_i} = \frac{\mathbf{s}_i^T \mathbf{\Phi}_s \mathbf{\Phi}_s^T \mathbf{s}_i}{E\left[\mathbf{v}_i^T \mathbf{\Phi}_s \mathbf{\Phi}_s^T \mathbf{v}_i\right]}. \qquad (4.217)$$

For the special case when the noise $\mathbf{v}_i$ is white, i.e., for $\mathbf{R}_{v_i} = \sigma_{v_i}^2 \mathbf{I}$, we obtain

$$
\begin{aligned}
\mathrm{SNR}_{\hat{s}_i} &= \frac{\mathbf{s}_i^T \mathbf{\Phi}_s \mathbf{\Phi}_s^T \mathbf{s}_i}{\mathrm{tr}\left(\mathbf{\Phi}_s^T E\left[\mathbf{v}_i \mathbf{v}_i^T\right] \mathbf{\Phi}_s\right)} \\
&= \frac{\mathbf{s}_i^T \mathbf{\Phi}_s \mathbf{\Phi}_s^T \mathbf{s}_i}{\mathrm{tr}\left(\mathbf{\Phi}_s^T \sigma_{v_i}^2 \mathbf{I} \mathbf{\Phi}_s\right)} \\
&= \frac{\mathbf{s}_i^T \mathbf{\Phi}_s \mathbf{\Phi}_s^T \mathbf{s}_i}{K \sigma_{v_i}^2}.
\end{aligned} \tag{4.218}
$$

It is important to realize that not only does the noise part of $\mathrm{SNR}_{\hat{s}_i}$ depend on K, but so does the signal part in the numerator. Hence, the truncation value K that produces the highest SNR is determined by choosing that value of K which maximizes $\mathrm{SNR}_{\hat{s}_i}$.

4.5.2 Sets of Basis Functions

The complex exponential functions represent the most well-known set of complete orthonormal basis functions since they constitute the cornerstone of Fourier analysis. The basis functions are defined by

$$
\varphi_k = \frac{1}{\sqrt{N}} \begin{bmatrix} 1 \\ e^{j\omega_k} \\ \vdots \\ e^{j\omega_k(N-1)} \end{bmatrix}, \quad k = 1, \ldots, N.
$$

The frequencies $(\omega_k = 2\pi f_k)$ must be harmonically related to the fundamental $f_1 = 1/N$ by $f_k = k/N$, otherwise orthogonality does not apply to this set of basis functions. Since we have, in general, avoided describing the complex-valued version of a method, we will consider the closely related, but real-valued, representation in which sines and cosines define the set of basis functions, i.e.,

$$
\varphi_{2k+1} = \sqrt{\frac{2}{N}} \begin{bmatrix} 1 \\ \cos\left(\frac{2\pi k}{N}\right) \\ \vdots \\ \cos\left(\frac{2\pi k(N-1)}{N}\right) \end{bmatrix}, \quad k = 0, \ldots, N/2 - 1, \tag{4.219}
$$

$$
\varphi_{2k} = \sqrt{\frac{2}{N}} \begin{bmatrix} 0 \\ \sin\left(\frac{2\pi k}{N}\right) \\ \vdots \\ \sin\left(\frac{2\pi k(N-1)}{N}\right) \end{bmatrix}, \quad k = 1, \ldots, N/2, \tag{4.220}
$$

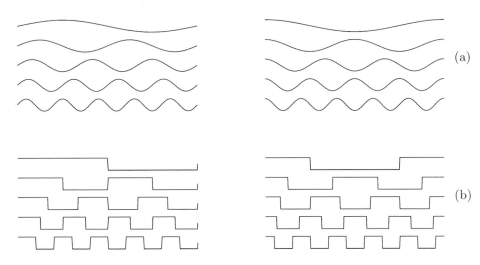

Figure 4.29: (a) Sine and cosine basis functions and (b) the Walsh sal and cal basis functions. Although these sequences are discrete-time, they are, for ease of interpretation, presented as continuous-time curves.

where N is an integer assumed to be even. It can be verified that the basis functions in (4.219) and (4.220) constitute an orthonormal set since

$$\boldsymbol{\varphi}_k^T \boldsymbol{\varphi}_l = \begin{cases} 1, & k = l; \\ 0, & k \neq l. \end{cases}$$

The sinusoidal basis functions with the lowest frequencies are shown in Figure 4.29(a). The coefficients of the series expansion are calculated by the following two inner products,

$$w_{2k+1} = \mathbf{x}^T \boldsymbol{\varphi}_{2k+1}$$
$$= \sqrt{\frac{2}{N}} \sum_{n=0}^{N-1} x(n) \cos\left(\frac{2\pi kn}{N}\right), \quad k = 0, \dots, N/2 - 1, \qquad (4.221)$$

and

$$w_{2k} = \mathbf{x}^T \boldsymbol{\varphi}_{2k}$$
$$= \sqrt{\frac{2}{N}} \sum_{n=0}^{N-1} x(n) \sin\left(\frac{2\pi kn}{N}\right), \quad k = 1, \dots, N/2. \qquad (4.222)$$

These two expressions are well-known since they yield the coefficients of the Fourier series representation of $x(n)$.

Estimation of EP morphology can be considered in terms of *lowpass modeling* using the truncated Fourier series [124–126]. Higher harmonics are

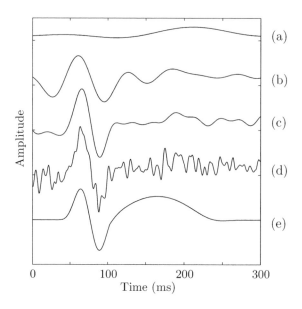

Figure 4.30: Truncated Fourier series modeling of EPs using different numbers of basis functions: (a) $K = 3$, (b) $K = 7$, (c) $K = 12$ (found to produce the smallest error, see Figure 4.31(b)), (d) $K = 500$, i.e., the complete set of basis functions and thus identical to the observed signal with noise, and (e) the simulated EP waveform before the recorded background EEG activity was added.

discarded since these are considered to contain negligible information for representation of latency and amplitude features with diagnostic significance. Figure 4.30 presents an example in which a simulated EP, disturbed by recorded background EEG activity, is estimated by the expression in (4.206) using the sine/cosine basis functions for different truncation values K. For this particular example, it is evident that K should be chosen to be at least 10 in order to give an acceptable representation of the EP morphology while, at the same time, providing suppression of high-frequency noise. Although the main deflection at 100 ms (P100) is relatively well-represented by $K = 12$, the subsequent, much slower wave components are poorly modeled. Therefore, this example illustrates the fact that periodic basis functions are rather ill-suited for modeling of transient waveforms. After all, the sine/cosine basis functions are highly localized in the frequency domain, but not localized at all in the time domain.

The P100 amplitude of the EP estimate is presented in Figure 4.31(a) for different values of K. The amplitude is accurate for values of K ranging from 10 to around 30, whereas larger values of K make the P100 amplitude biased; this behavior is caused by noise included in the EP estimate for large values

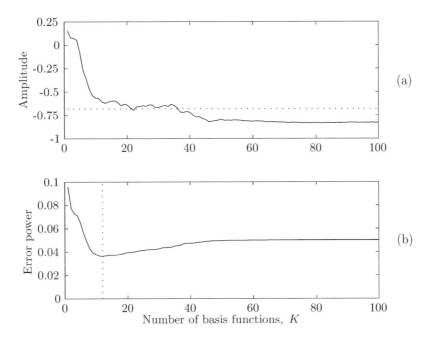

Figure 4.31: (a) The P100 amplitude shown as a function of the number of basis functions for the EP in Figure 4.30. The true P100 amplitude, taken from the EP without added noise, is indicated by the dotted line. (b) The error power reaches its minimum for $K = 12$ basis functions, cf. the corresponding signal estimate in Figure 4.30(c).

of K. Since the EP signal is simulated, it is possible to compute the power of the error $\mathbf{e}$ between the noise-free signal $\mathbf{s}$, shown in Figure 4.30(e), and its estimate $\hat{\mathbf{s}}$. For this example, the error power is found to have its minimum for $K = 12$, see Figure 4.31. However, the minimum is relatively flat, and additional basis functions do not significantly increase the error power. The optimal value of K will, in general, vary from morphology to morphology, and, therefore, the choice of K usually involves compromise. The necessity of a compromise choice is even further emphasized by recalling the basic fact that a noise-free EP is unavailable in practice.

The discrete *Walsh functions* constitute another set of basis functions of rectangular shape, assuming only the values one and zero and with different periodicities [127, Ch. 5]. The Walsh functions can be divided into odd functions, denoted *sal*, and even functions, denoted *cal*, in analogy with the sine and cosine functions. The first four of the sal and cal functions are displayed in Figure 4.29(b), excluding the function representing the DC level. The Walsh functions are primarily of interest when the computational aspects are critical, since these functions can be implemented very efficiently. On the downside, even more basis functions are required to adequately model EPs than when sine/cosine functions are used [128].

4.5.3 Karhunen–Loève Expansion—Optimal Basis Functions

A powerful alternative to the sets of basis functions just described is the Karhunen–Loève expansion which offers the property of optimal representation of a random process by a truncated, orthonormal series expansion: no other expansion yields a lower approximation error in the MSE sense. This property assumes that the correlation matrix $\mathbf{R}_x$, characterizing the ensemble of signals, is either known or can be estimated from available data (not necessarily stationary). The Karhunen–Loève expansion does not produce basis functions which are already labeled as belonging to the signal or the noise part. However, certain basis functions are, as we will see later, well-suited for representing the signal part when this part is consistently present in the ensemble of data employed for determining the basis functions. In contrast to the previous section, the basis functions that constitute the Karhunen–Loève expansion do not have an analytical description.

How should the basis functions φ_k be designed so that the signal part is efficiently represented with a small number of functions? Again, we start our derivation by decomposing the series expansion of $\mathbf{x}$ into two sums (for clarity of presentation, we will temporarily omit the index i),

$$\mathbf{x} = \sum_{k=1}^{K} w_k \varphi_k + \sum_{k=K+1}^{N} w_k \varphi_k = \hat{\mathbf{s}} + \hat{\mathbf{v}}, \qquad (4.223)$$

where the K first basis functions produce an estimate of $\mathbf{s}$, and the remaining $(N - K)$ terms produce the noise estimate $\hat{\mathbf{v}}$. Our aim is now to find the set of φ_k's that makes $\hat{\mathbf{s}}$ resemble $\mathbf{s}$ as closely as possible. This objective can be achieved by minimizing the noise power estimate in the MSE sense,

$$\mathcal{E} = E\left[\hat{\mathbf{v}}^T \hat{\mathbf{v}}\right] = E\left[(\mathbf{x} - \hat{\mathbf{s}})^T (\mathbf{x} - \hat{\mathbf{s}})\right], \qquad (4.224)$$

which, with the model that the observed signal is composed of

$$\mathbf{x} = \mathbf{s} + \mathbf{v},$$

becomes

$$\mathcal{E} = E\left[(\mathbf{s} - \hat{\mathbf{s}})^T (\mathbf{s} - \hat{\mathbf{s}})\right] + 2E\left[(\mathbf{s} - \hat{\mathbf{s}})^T \mathbf{v}\right] + E\left[\mathbf{v}^T \mathbf{v}\right]. \qquad (4.225)$$

In order to proceed, we will make use of the assumptions that the signal and noise are uncorrelated and that the noise is white, i.e., characterized by the correlation matrix

$$\mathbf{R}_v = E[\mathbf{v}\mathbf{v}^T] = \sigma_v^2 \mathbf{I}. \tag{4.226}$$

The cross-terms in (4.225) can then be expressed as

$$
\begin{aligned}
E\left[(\mathbf{s} - \hat{\mathbf{s}})^T \mathbf{v}\right] &= 0 - E\left[\left(\sum_{k=1}^{K} w_k \boldsymbol{\varphi}_k\right)^T \mathbf{v}\right] \\
&= -E\left[\left(\sum_{k=1}^{K} (\boldsymbol{\varphi}_k^T \mathbf{x}) \boldsymbol{\varphi}_k\right)^T \mathbf{v}\right] \\
&= -E\left[\left(\sum_{k=1}^{K} \boldsymbol{\varphi}_k^T (\mathbf{s} + \mathbf{v}) \boldsymbol{\varphi}_k\right)^T \mathbf{v}\right],
\end{aligned}
\tag{4.227}
$$

which, by use of (4.226), becomes

$$
\begin{aligned}
E\left[(\mathbf{s} - \hat{\mathbf{s}})^T \mathbf{v}\right] &= -\sum_{k=1}^{K} \boldsymbol{\varphi}_k^T E\left[\mathbf{v}\mathbf{v}^T\right] \boldsymbol{\varphi}_k \\
&= -K\sigma_v^2.
\end{aligned}
\tag{4.228}
$$

As a result, $\mathcal{E}$ can be expressed as

$$\mathcal{E} = E\left[\hat{\mathbf{v}}^T \hat{\mathbf{v}}\right] = E\left[(\mathbf{s} - \hat{\mathbf{s}})^T (\mathbf{s} - \hat{\mathbf{s}})\right] + (N - 2K)\sigma_v^2, \tag{4.229}$$

from which we conclude that the noise power term can be neglected since it does not depend on $\boldsymbol{\varphi}_k$ (this is only possible for the white noise situation). Consequently, minimization of $\mathcal{E}$ in (4.224) is *equivalent* to making $\hat{\mathbf{s}}$ resemble $\mathbf{s}$. Thus, we have

$$\mathcal{E} = E\left[\hat{\mathbf{v}}^T \hat{\mathbf{v}}\right] = E\left[\left(\sum_{k=K+1}^{N} w_k \boldsymbol{\varphi}_k\right)^T \left(\sum_{l=K+1}^{N} w_l \boldsymbol{\varphi}_l\right)\right], \tag{4.230}$$

which, due to orthonormality of the basis functions $\boldsymbol{\varphi}_k$, can be concisely expressed as

$$\mathcal{E} = \sum_{k=K+1}^{N} E\left[w_k^2\right]. \tag{4.231}$$

Furthermore, since each coefficient w_k is obtained as the inner product $\boldsymbol{\varphi}_k^T \mathbf{x}$ in (4.194), we have

$$E\left[w_k^2\right] = E\left[\boldsymbol{\varphi}_k^T \mathbf{x}\mathbf{x}^T \boldsymbol{\varphi}_k\right] = \boldsymbol{\varphi}_k^T \mathbf{R}_x \boldsymbol{\varphi}_k, \tag{4.232}$$

and the error can be written as

$$\mathcal{E} = \sum_{k=K+1}^{N} \boldsymbol{\varphi}_k^T \mathbf{R}_x \boldsymbol{\varphi}_k. \tag{4.233}$$

Minimization of the error $\mathcal{E}$ must be performed subject to the constraints that $\boldsymbol{\varphi}_{K+1}, \ldots, \boldsymbol{\varphi}_N$ are orthonormal. Similar to the SNR maximization problem dealt with in connection with weighted averaging, see page 209, we can solve the minimization problem by using the Lagrange multiplier technique. The function to be minimized is defined by

$$\mathcal{L} = \sum_{k=K+1}^{N} \boldsymbol{\varphi}_k^T \mathbf{R}_x \boldsymbol{\varphi}_k + \sum_{k=K+1}^{N} \lambda_k (1 - \boldsymbol{\varphi}_k^T \boldsymbol{\varphi}_k), \tag{4.234}$$

where the λ_k's are Lagrange multipliers related to each of the constraints. Taking the gradient of $\mathcal{L}$ with respect to $\boldsymbol{\varphi}_k$ and setting the result to zero yields

$$\nabla_{\boldsymbol{\varphi}_k} \mathcal{L} = \mathbf{R}_x \boldsymbol{\varphi}_k - \lambda_k \boldsymbol{\varphi}_k = \mathbf{0},$$

or

$$\mathbf{R}_x \boldsymbol{\varphi}_k = \lambda_k \boldsymbol{\varphi}_k, \quad k = K+1, \ldots, N, \tag{4.235}$$

recognized as the ordinary eigenvalue problem which is solved when a square matrix is diagonalized (its solution is briefly described in Appendix A). Equation (4.235) establishes the very important finding that the basis functions $\boldsymbol{\varphi}_k$ should be chosen as the eigenvectors of $\mathbf{R}_x$. Since $\mathbf{R}_x$ is a correlation matrix, we recall that all of its eigenvalues λ_k are positive-valued, or possibly equal to zero. The eigenvalues are arranged in decreasing order

$$\lambda_1 > \lambda_2 > \cdots > \lambda_M. \tag{4.236}$$

Inserting (4.235) into (4.233) allows the MSE to be expressed as

$$\mathcal{E} = \sum_{k=K+1}^{N} \boldsymbol{\varphi}_k^T \mathbf{R}_x \boldsymbol{\varphi}_k$$

$$= \sum_{k=K+1}^{N} \boldsymbol{\varphi}_k^T (\lambda_k \boldsymbol{\varphi}_k)$$

$$= \sum_{k=K+1}^{N} \lambda_k, \tag{4.237}$$

and thus $\mathcal{E}$ is minimized when the $N - K$ *smallest* eigenvalues are chosen since the sum of the eigenvalues then reaches its minimum value. More importantly, this choice implies that the eigenvectors corresponding to *the K largest eigenvalues* should be used as the basis functions in the truncated series expansion in order to achieve the optimal representation property in the MMSE sense [55].

The average energy of the coefficients w_k is related to the eigenvalues λ_k of $\mathbf{R}_x$, a relationship established by first expressing the system of equations in (4.235) in a compact matrix form,

$$\mathbf{R}_x \mathbf{\Phi} = \mathbf{\Phi} \mathbf{\Lambda}, \tag{4.238}$$

where $\mathbf{\Phi}$ is defined in (4.191), and $\mathbf{\Lambda}$ is a diagonal matrix whose diagonal elements equal the eigenvalues $\lambda_1, \ldots, \lambda_N$; all N equations in (4.235) are included. Since $\mathbf{\Phi}$ is orthogonal, we can express $\mathbf{R}_x$ as

$$\mathbf{R}_x = \mathbf{\Phi} \mathbf{\Lambda} \mathbf{\Phi}^T. \tag{4.239}$$

Next, we express the correlation matrix of $\mathbf{w}$ in terms of $\mathbf{R}_x$, making use of (4.192),

$$E\left[\mathbf{w}\mathbf{w}^T\right] = \mathbf{\Phi}^T E\left[\mathbf{x}\mathbf{x}^T\right]\mathbf{\Phi} = \mathbf{\Phi}^T \mathbf{R}_x \mathbf{\Phi}, \tag{4.240}$$

which becomes

$$E\left[\mathbf{w}\mathbf{w}^T\right] = \mathbf{\Phi}^T \mathbf{\Phi} \mathbf{\Lambda} \mathbf{\Phi}^T \mathbf{\Phi} = \mathbf{\Lambda} \tag{4.241}$$

after use of (4.238), i.e., the equation which is crucial for the optimal representation property. The average energy associated with each coefficient w_k thus equals λ_k,

$$E[w_k w_l] = \begin{cases} \lambda_k, & k = l; \\ 0, & k \neq l, \end{cases} \tag{4.242}$$

whereas the coefficients w_k and w_l for $k \neq l$ are uncorrelated. The important implication of the result in (4.242) is that the set of basis functions associated with Karhunen–Loève expansion produces mutually uncorrelated coefficients.

Given an ensemble of signals characterized by $\mathbf{R}_x$, a performance index $\mathcal{R}_K$ can be defined which reflects how well the truncated series expansion approximates the ensemble in energy terms,

$$\mathcal{R}_K = \frac{\sum_{k=1}^{K} \lambda_k}{\sum_{k=1}^{N} \lambda_k}. \tag{4.243}$$

The performance index $\mathcal{R}_K$ is normalized to the interval $[0, 1]$, because all eigenvalues are positive-valued, and therefore approaches one as the number of basis functions increases.

In practice, the correlation matrix $\mathbf{R}_x$ cannot be estimated from a single potential but must be estimated from the ensemble $\mathbf{x}_1, \mathbf{x}_2, \ldots, \mathbf{x}_M$. The estimation of $\mathbf{R}_x$ is commonly achieved by simply replacing the expected value in the definition of $\mathbf{R}_x$ (cf. (3.6)) by averaging the M rank-one correlation matrices $\mathbf{x}_i \mathbf{x}_i^T$ for each of the EPs,

$$\hat{\mathbf{R}}_x = \frac{1}{M} \sum_{i=1}^{M} \mathbf{x}_i \mathbf{x}_i^T. \tag{4.244}$$

The estimation of $\mathbf{R}_x$ from the entire ensemble implies that the resulting basis functions will not be matched to the statistics of individual EPs $\mathbf{s}_i$, but to a random process that characterizes the variations found within the ensemble. The basis functions can be determined either from the data for each individual subject or from a database consisting of several subjects; in the latter case, the EPs of all subjects are merged into the computation of $\hat{\mathbf{R}}_x$ in (4.244).

The signal concentration property of the Karhunen–Loève expansion, as well as the related noise-reducing property, can be illustrated by studying the properties of the well-known model

$$\mathbf{x}_i = \mathbf{s} + \mathbf{v}_i \tag{4.245}$$

and by finding the related basis functions. In this model, the signal $\mathbf{s}$ is deterministic with energy $\mathbf{s}^T \mathbf{s} = E_s$, whereas the noise $\mathbf{v}_i$ is a white, stationary process that is fully characterized by

$$\mathbf{R}_{v_i} = \mathbf{R}_v = \sigma_v^2 \mathbf{I}, \quad i = 1, \ldots, M.$$

The correlation matrix of the observed signals $\mathbf{x}_i$ is given by

$$\mathbf{R}_x = \mathbf{s}\mathbf{s}^T + \sigma_v^2 \mathbf{I}. \tag{4.246}$$

The eigenvalues and eigenvectors of $\mathbf{R}_x$ are found by solving the equations in (4.235),

$$(\mathbf{s}\mathbf{s}^T + \sigma_v^2 \mathbf{I})\boldsymbol{\varphi}_k = \lambda_k \boldsymbol{\varphi}_k. \tag{4.247}$$

The signal $\mathbf{s}$ is proportional to one of the eigenvectors because

$$\mathbf{R}_x \mathbf{s} = \mathbf{s}\mathbf{s}^T \mathbf{s} + \sigma_v^2 \mathbf{s} = (E_s + \sigma_v^2)\mathbf{s}, \tag{4.248}$$

and the corresponding eigenvalue is equal to $E_s + \sigma_v^2$. The remaining eigenvectors $\boldsymbol{\varphi}_k$, which must be orthonormal to $\mathbf{s}$ as well as mutually orthonormal, are determined by

$$\mathbf{R}_x \boldsymbol{\varphi}_k = \mathbf{s}\mathbf{s}^T \boldsymbol{\varphi}_k + \sigma_v^2 \boldsymbol{\varphi}_k = \sigma_v^2 \boldsymbol{\varphi}_k, \qquad (4.249)$$

where $\mathbf{s}^T \boldsymbol{\varphi}_k = 0$ is used to arrive at the last step. The corresponding eigenvalues are all equal to σ_v^2 and are thus smaller than $E_s + \sigma_v^2$. The remaining eigenvectors can be chosen arbitrarily. However, there is no need to compute these eigenvectors because they are only needed for representation of the noise.

For the model in (4.245), it can thus be concluded that the signal information is concentrated to one single eigenvector corresponding to the largest eigenvalue,

$$\boldsymbol{\varphi}_1 = \frac{1}{\sqrt{E_s}} \mathbf{s} \qquad (4.250)$$

and

$$\lambda_1 = \lambda_{\max} = E_s + \sigma_v^2. \qquad (4.251)$$

For the more general case of a stochastic signal $\mathbf{s}$ disturbed by colored noise, several basis functions are typically required to achieve adequate signal representation.

Noise reduction by a truncated series expansion using the Karhunen–Loève basis functions is illustrated in Figure 4.32. In this example, the basis functions were determined from an ensemble of EPs acquired during an experiment on one subject. The basis functions corresponding to the five largest eigenvalues are presented in Figure 4.32(a); however, only the first two eigenvectors are essential to obtain a good representation of the EPs because λ_3, λ_4, and λ_5 are very small, see Figure 4.32(c). Three EPs are plotted together with the corresponding signal estimates, based on two basis functions, in Figure 4.32(b). These basis functions yield a considerable reduction in the noise level, while still not obscuring the delay in latency that occurs from $\mathbf{x}_1$ to $\mathbf{x}_3$.

Viewing the signal model in (4.245) in light of the results shown in Figure 4.32(c), the distribution of the eigenvalues, as reflected by $\mathcal{R}_K$, indicates that the assumption of a fixed signal $\mathbf{s}$ is inappropriate for this particular data set. Such an assumption would, apart from the eigenvalue λ_1 of the signal, have implied a straight line in Figure 4.32(c). Since the energy is unevenly spread over several eigenvalues, this indicates either that there is considerable morphologic variability within the data ensemble or, possibly, that the noise is colored.

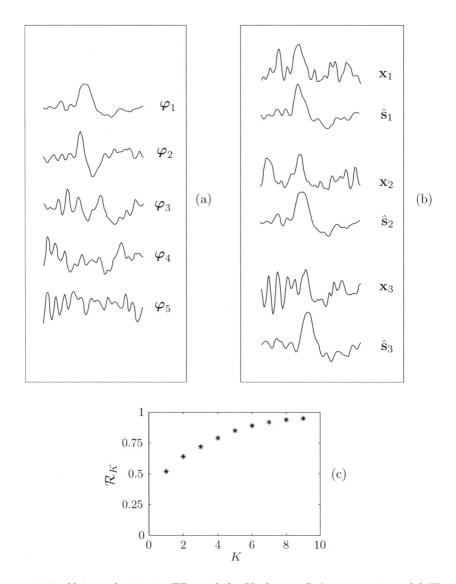

Figure 4.32: Noise reduction in EPs and the Karhunen–Loève expansion. (a) The five most significant basis functions, (b) three different EPs $\mathbf{x}_i$ and their corresponding estimates $\hat{\mathbf{s}}_i$ reconstructed from the basis functions φ_1 and φ_2 displayed in (a), and (c) the performance index $\mathcal{R}_K$ as a function of the number of eigenvalues K. (Reprinted from Lange and Inbar [129] with permission.)

4.5.4 Interpretation as Linear, Time-Variant Filtering

Signal modeling based on a truncated orthonormal series expansion can be interpreted in terms of linear, time-variant filtering [130]. This is an attractive property which allows us to compute the frequency response at different instants, thereby providing valuable information on how the different waves of $\mathbf{x}_i$ are spectrally shaped when $\mathbf{s}_i$ is estimated by basis functions. The derivation of the time-variant impulse response $h(l, n)$ begins by rewriting the relationship in (4.206),

$$\hat{\mathbf{s}}_i = \mathbf{\Phi}_s \mathbf{\Phi}_s^T \mathbf{x}_i,$$

and realizing that each row of the matrix $\mathbf{\Phi}_s \mathbf{\Phi}_s^T$ can be interpreted as a noncausal impulse response that differs from row to row. Consequently, we can express this relationship as a function of time n,

$$
\begin{aligned}
\hat{s}_i(n) &= \sum_{k=1}^{K} \sum_{l=0}^{N-1} \varphi_k(n) \varphi_k(l) x_i(l) \\
&= \sum_{l=0}^{N-1} g(l, n) x_i(l), \quad n = 0, \ldots, N-1,
\end{aligned}
\tag{4.252}
$$

where

$$
g(l, n) = \sum_{k=1}^{K} \varphi_k(l) \varphi_k(n), \quad l, n = 0, \ldots, N-1.
\tag{4.253}
$$

In order to express (4.252) as a convolution sum, we have to extend the definition of the basis functions so that these become periodic,

$$
\varphi_k(n + N) = \varphi_k(n),
\tag{4.254}
$$

and thus

$$
g(l + N, n) = g(l, n + N) = g(l, n).
\tag{4.255}
$$

This extension is straightforward since the basis functions are deterministic. The following definition of a finite duration impulse response $h(l, n)$ is then introduced,

$$
h(l, n) = \begin{cases} g(n - l, n), & l = 0, \ldots, N-1; \\ 0, & \text{otherwise,} \end{cases}
\tag{4.256}
$$

where $n = 0, \ldots, N-1$; the index l denotes local time within the impulse response, while the index n denotes the time at which the impulse response is valid. With $h(l, n)$, we are now in a position to express the input/output relationship as a convolution,

$$\hat{s}_i(n) = \sum_{l=0}^{N-1} h(n-l, n)x_i(l), \quad n = 0, \ldots, N-1. \tag{4.257}$$

The instantaneous frequency response $H(e^{\jmath\omega}, n)$ is obtained as the Fourier transform of $h(l, n)$ at time n,

$$
\begin{aligned}
H(e^{\jmath\omega}, n) &= \sum_{l=-\infty}^{\infty} h(l, n)e^{-\jmath\omega l} \\
&= \sum_{l=0}^{N-1} g(n-l, n)e^{-\jmath\omega l} \\
&= e^{-\jmath\omega n}G^*(e^{\jmath\omega}, n).
\end{aligned} \tag{4.258}
$$

Hence, the frequency response depends on the basis functions in a simple way through the Fourier transform of $g(l, n)$.

The interpretation of signal modeling by a truncated series expansion in terms of a linear, time-variant filter is demonstrated by the example presented in Figure 4.33. The basis functions were determined from an ensemble of pulse-shaped waveforms using the Karhunen–Loève expansion; only the basis functions corresponding to the ten largest eigenvalues are displayed. The instantaneous frequency response $H(e^{\jmath\omega}, n)$ is presented for three different times, $n = 25$, 100, and 150, and is based on either $K = 4$ or 10 basis functions. The characteristics of the frequency responses show that the low-pass filtering effect is more pronounced during the flatter parts of the signal (i.e., around $n = 25$) than during the transient ones. The bandwidth of the time-varying filter increases as the number of basis functions increases. In the limit, the signal will, of course, be left unaffected when the complete set of basis functions is applied, i.e., for $K = N$.

4.5.5 Modeling with Damped Sinusoids

By expanding the sinusoidal basis function model described on page 260 to include exponential damping, a wider variety of transient waveform morphologies can be modeled more efficiently. Such an approach takes its starting point in a deterministic signal model defined by the following expansion of damped complex exponentials,

$$x(n) = \sum_{k=1}^{K} w_k e^{\rho_k n} e^{\jmath(\omega_k n + \phi_k)}, \tag{4.259}$$

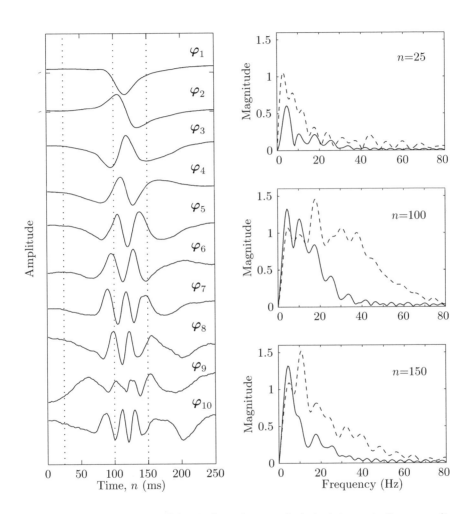

Figure 4.33: Orthonormal basis functions and their interpretation as a linear, time-varying filter with an instantaneous frequency response. The frequency response is presented for three different times, indicated by the dotted lines in the left panel. For each time, the response is computed for $K = 4$ (solid line) or 10 basis functions (dashed line). The sampling rate is 1000 Hz.

for $n = 0, \ldots, N - 1$. Each term of the expansion is characterized by its amplitude w_k, frequency ω_k, phase ϕ_k, and a damping factor ρ_k ($\rho_k < 0$) which determines the decay in amplitude. In contrast to the other signal models considered in this textbook, the observed signal $x(n)$ in (4.259) is assumed to be noise-free. This apparent oversimplification facilitates the development of the quintessential ideas of the method for estimating the model parameters $w_k, \omega_k, \phi_k,$ and ρ_k. Similar methods modified to deal with the presence of additive noise have been suggested but will not be described here due to their much more involved structure, see instead [131].

Since $x(n)$ is a real-valued signal in our context, it is necessary that the complex exponentials in (4.259) occur in complex-conjugate pairs of equal amplitude so that the model can be written

$$x(n) = \sum_{k=1}^{K/2} 2w_k e^{\rho_k n} \cos(\omega_k n + \phi_k). \qquad (4.260)$$

We note that $x(n)$ contains $K/2$ damped cosines when K is even, while it contains $(K-1)/2$ damped cosines and one purely damped exponential when K is odd.

The power of modeling EPs with damped sinusoids is illustrated in Figure 4.34, where a variant of the method presented below is used to decompose the signal [132]. From this example, it is evident that damping, as an additional degree of freedom, makes it possible to model EPs with much fewer basis functions than required with sine/cosine basis functions, see Figure 4.30. Other studies which make use of the damped-sinusoid model for the analysis of single-trial EPs can be found in [133, 134].

Although the model in (4.259) can be expressed in terms of basis functions, the resulting basis functions do not constitute an orthonormal set. Accordingly, they do not lend themselves to the simple inner product calculation of the coefficient w_k in (4.194). Another approach would be to fit the model to $x(n)$ by jointly minimizing the related least-squares error with respect to the parameters $w_k, \phi_k, \rho_k,$ and ω_k. Unfortunately, the resulting set of equations is nonlinear in nature, and their solution is difficult. Therefore, suboptimal approaches are presented below.

The original Prony method. The original solution to this problem was provided by Prony who recognized that the model in (4.259) can be viewed as the homogeneous solution to a linear difference equation with fixed parameters [131]; the method is therefore referred to as the *original Prony method*. This solution is more easily perceived by rewriting (4.259) so that each term in the expansion is composed of two parameters, either dependent

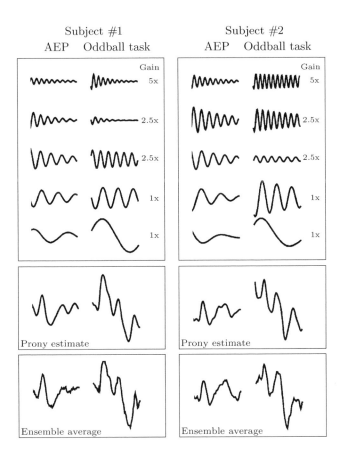

Figure 4.34: Decomposition of EPs into damped sinusoids using Prony's method. The EPs were recorded from two different subjects during auditory stimulation and an oddball task. The EP estimate ("Prony estimate") was obtained by summing the five damped sinusoids shown in the upper panel. The duration of the signals is 500 ms, and the damped sinusoids are displayed with different amplitude gains. (Reprinted from Demiralp et al. [132] with permission.)

on time n or independent,

$$x(n) = \sum_{k=1}^{K} h_k z_k^n,$$ (4.261)

where

$$h_k = w_k e^{\jmath \phi_k},$$ (4.262)

$$z_k = e^{\rho_k + \jmath \omega_k}.$$ (4.263)

Based on this model formulation, Prony's method comprises a three-step procedure in which the complex parameters $z_1, \ldots, z_K$ are first estimated, then the complex parameters $h_1, \ldots, h_K$ are estimated, and finally the $4K$ real parameters w_k, ϕ_k, ρ_k, and ω_k are determined from the $2K$ complex parameter estimates. Thus, the estimates do not result from the optimization of a performance criterion.

The expression in (4.261) can be viewed as the solution to the homogeneous linear difference equation [135]

$$x(n) + a_1 x(n-1) + \ldots + a_K x(n-K) = 0.$$ (4.264)

In Prony's original method, it was assumed that the number of samples available is equal to the number of unknown parameters, i.e., $N = 2K$, and, therefore, the difference equation is valid for $n = K, \ldots, 2K-1$. The related characteristic polynomial $A(z)$ is given by

$$A(z) = \sum_{l=0}^{K} a_l z^{K-l},$$ (4.265)

where $a_0 = 1$. In order to find the roots $z_1, \ldots, z_K$ of the polynomial $A(z)$, defined by

$$A(z) = \prod_{k=1}^{K} (z - z_k),$$ (4.266)

we first need to determine the coefficients $a_1, \ldots, a_K$ from the $2K$ samples $x(n)$. This can be done by expressing the difference equation at different times as a $K \times K$ matrix equation:

$$\begin{bmatrix} x(K-1) & x(K-2) & \cdots & x(0) \\ x(K) & x(K-1) & \cdots & x(1) \\ \vdots & \vdots & \ddots & \vdots \\ x(2K-2) & x(2K-3) & \cdots & x(K-1) \end{bmatrix} \begin{bmatrix} a_1 \\ a_2 \\ \vdots \\ a_K \end{bmatrix} = - \begin{bmatrix} x(K) \\ x(K+1) \\ \vdots \\ x(2K-1) \end{bmatrix}.$$

(4.267)

Once the coefficients $a_1, \ldots, a_K$ are available, the roots of the characteristic polynomial in (4.265) can be computed. The unknown parameters $h_1, \ldots, h_K$ are determined from the model equation in (4.261) for the times $n = 0, \ldots, K-1$,

$$
\begin{bmatrix}
z_1^0 & z_2^0 & \cdots & z_K^0 \\
z_1^1 & z_2^1 & \cdots & z_K^1 \\
\vdots & \vdots & & \vdots \\
z_1^{K-1} & z_2^{K-1} & \cdots & z_K^{K-1}
\end{bmatrix}
\begin{bmatrix}
h_1 \\
h_2 \\
\vdots \\
h_K
\end{bmatrix}
=
\begin{bmatrix}
x(0) \\
x(1) \\
\vdots \\
x(K-1)
\end{bmatrix}.
\tag{4.268}
$$

Finally, the real parameters ρ_k and ω_k are computed from z_k by

$$
\rho_k = \ln |z_k|,
\tag{4.269}
$$

$$
\omega_k = \arctan \left(\frac{\Im(z_k)}{\Re(z_k)} \right).
\tag{4.270}
$$

The two remaining real parameters w_k and ϕ_k are computed from h_k by

$$
w_k = |h_k|,
\tag{4.271}
$$

$$
\phi_k = \arctan \left(\frac{\Im(h_k)}{\Re(h_k)} \right).
\tag{4.272}
$$

While the parameters w_k, ϕ_k, ρ_k, and ω_k are determined for $k = 1, \ldots, K$, only $K/2$ values are required for real signals since h_k and z_k occur in complex-conjugate pairs.

The least-squares Prony method. For practical use, it is necessary to modify the above procedure to deal with the situation when $N > 2K$. Since finding $z_1, \ldots, z_K$ now becomes part of an overdetermined problem (i.e., more equations than unknowns), it is necessary to relax the requirement that the difference equation in (4.264) be exactly zero and to introduce an error $e(n)$ such that

$$
x(n) + a_1 x(n-1) + \ldots + a_K x(n-K) = e(n),
\tag{4.273}
$$

now valid for $n = K, \ldots, N-1$. Then, the problem is to minimize a quadratic function of the error $e(n)$ with respect to the parameters $a_1, \ldots, a_K$: a problem which we have actually already encountered in the context of AR-based spectral analysis when solving the forward linear prediction problem. Thus, we can make use of the solution to the normal equation in (3.126) in combination with the covariance method for finding $a_1, \ldots, a_K$.

In cases of $N > 2K$, (4.268), yielding the parameters $h_1, \ldots, h_K$, has an overdetermined solution since

$$
\begin{bmatrix}
z_1^0 & z_2^0 & \cdots & z_K^0 \\
z_1^1 & z_2^1 & \cdots & z_K^1 \\
\vdots & \vdots & & \vdots \\
z_1^{N-2} & z_2^{N-2} & \cdots & z_K^{N-2} \\
z_1^{N-1} & z_2^{N-1} & \cdots & z_K^{N-1}
\end{bmatrix}
\begin{bmatrix}
h_1 \\
h_2 \\
\vdots \\
h_K
\end{bmatrix}
=
\begin{bmatrix}
x(0) \\
x(1) \\
\vdots \\
x(N-2) \\
x(N-1)
\end{bmatrix}.
\tag{4.274}
$$

The commonly used approach to handle this situation is to find the least-squares solution, i.e., those values of h_k which minimize the Euclidean norm of the error

$$
\| \mathbf{x}_N - \mathbf{Z}_N \mathbf{h}_K \|_2^2 = (\mathbf{x}_N - \mathbf{Z}_N \mathbf{h}_K)^H (\mathbf{x}_N - \mathbf{Z}_N \mathbf{h}_K),
\tag{4.275}
$$

where $\mathbf{Z}_N, \mathbf{h}_K$, and $\mathbf{x}_N$ denote the matrix and the two vectors in (4.274), respectively. The definition of the Euclidean norm is given in (A.32). The least-squares solution is (see (A.52))

$$
\mathbf{h}_K = (\mathbf{Z}_N^H \mathbf{Z}_N)^{-1} \mathbf{Z}_N^H \mathbf{x}_N,
\tag{4.276}
$$

where $\mathbf{Z}_N^H$ denotes the Hermitian transpose of the complex matrix $\mathbf{Z}_N$, i.e., the complex-conjugate of the transpose of $\mathbf{Z}_N$. Once estimates of the complex parameters are available, we can, as before, compute the desired parameters w_k, ϕ_k, ρ_k, and ω_k with (4.269)–(4.272).

Certain additional issues need to be addressed before the least-squares Prony method becomes really useful. Most importantly, it has been found that the above procedure is sensitive to noise, and, therefore, the model in (4.260) needs to be modified to incorporate additive noise. More robust methods have been developed for estimating the damped sinusoids in the presence of white [131] or colored noise [133]. Another issue is that the model order K is not known a priori, but must be estimated from the observed signal, for example, using techniques similar to AR model order estimation, see page 118.

4.6 Adaptive Analysis Using Basis Functions

We will present two different approaches to adaptive analysis of EPs, both relying on the use of basis functions.

- The *instantaneous LMS algorithm*, in which the weights of the series expansion are adapted at every time instant, thereby producing a weight vector $\mathbf{w}(n)$ [124, 136–140].

- The *block LMS algorithm*, in which the weights are adapted only once for each EP ("block"), thereby producing a weight vector $\mathbf{w}_i$ that corresponds to the ith potential [141].

The instantaneous LMS algorithm is the *deterministic reference input* counterpart of the adaptive algorithm described in Section 3.2.5 for EOG artifact cancellation, since now the reference input is given by a set of basis functions $\mathbf{\Phi}_s$ rather than by input signals modeled as stochastic processes.[10] The block LMS algorithm can be viewed as a marriage of single-trial analysis, relying on the inner product computation in (4.203), and the LMS algorithm. In the following two subsections, we will describe these two algorithms and provide some insight into their relative performance.

4.6.1   The Instantaneous LMS Algorithm

Since the instantaneous LMS algorithm operates on a sample-by-sample basis, we assume that the observed signal $x(n)$ results from concatenation of successive EPs such that

$$x(n) = x_{\lfloor \frac{n}{N} \rfloor + 1}\left(n - \left\lfloor \frac{n}{N} \right\rfloor N\right).$$

In the same way, the signal $s(n)$ and the noise $v(n)$ are obtained by concatenation. The LMS algorithm is derived along the same lines as in Section 3.2.5, i.e., by minimization of the time-dependent MSE criterion

$$\mathcal{E}_{\mathbf{w}}(n) = E\left[(x(n) - \boldsymbol{\varphi}_s^T(n)\mathbf{w}(n))^2\right]. \tag{4.277}$$

As before, the desired signal estimate $\hat{s}(n)$ is calculated as the linear combination of the truncated series expansion $\boldsymbol{\varphi}_s(n)$ and the time-varying weight vector $\mathbf{w}(n)$,

$$\hat{s}(n) = \boldsymbol{\varphi}_s^T(n)\mathbf{w}(n), \tag{4.278}$$

where $\boldsymbol{\varphi}_s(n)$ contains the values of the K basis functions at time n,

$$\boldsymbol{\varphi}_s(n) = \begin{bmatrix} \varphi_1(n) \\ \varphi_2(n) \\ \vdots \\ \varphi_K(n) \end{bmatrix} \tag{4.279}$$

---

[10]The stochastic input case has also been considered for noise reduction of EPs, see, e.g., [142–146].

and, consequently,

$$\mathbf{\Phi}_s = \begin{bmatrix} \boldsymbol{\varphi}_s^T(0) \\ \vdots \\ \boldsymbol{\varphi}_s^T(n) \\ \vdots \\ \boldsymbol{\varphi}_s^T(N-1) \end{bmatrix}.$$

Since the adaptive algorithm tries to estimate $s(n)$ at every instant, not only in the interval $[0, N-1]$ over which the basis functions $\varphi_k(n)$ were originally defined (see (4.188)), it is necessary to make use of the periodic extension of $\varphi_k(n)$, defined in (4.254).

The method of steepest descent requires the gradient of the error $\mathcal{E}_{\mathbf{w}}(n)$, easily found to be

$$\nabla_{\mathbf{w}}\mathcal{E}_{\mathbf{w}}(n) = -2E[e(n)\boldsymbol{\varphi}_s(n)], \tag{4.280}$$

where

$$e(n) = x(n) - \hat{s}(n).$$

Replacing the expected value in (4.280) with its instantaneous estimate (cf. page 85), we obtain the update equation for the LMS algorithm,

$$\mathbf{w}(n+1) = \mathbf{w}(n) + \mu e(n)\boldsymbol{\varphi}_s(n). \tag{4.281}$$

The block diagram in Figure 4.35 illustrates how the time-varying weights of the truncated series expansion are updated adaptively by the instantaneous LMS algorithm.

The stability of the LMS algorithm can be investigated by rewriting the recursion in (4.281) such that $\mathbf{w}(n)$ is expressed in terms of the initial weight vector $\mathbf{w}(0)$ and the input signal $x(n)$. To do this, we rewrite the recursion so that the influence of $x(n)$ becomes evident,

$$\mathbf{w}(n+1) = \mathbf{w}(n) + \mu(x(n) - \boldsymbol{\varphi}_s^T(n)\mathbf{w}(n))\boldsymbol{\varphi}_s(n)$$
$$= (\mathbf{I} - \mu\boldsymbol{\varphi}_s(n)\boldsymbol{\varphi}_s^T(n))\mathbf{w}(n) + \mu x(n)\boldsymbol{\varphi}_s(n).$$

For $n = 0$, we have

$$\mathbf{w}(1) = (\mathbf{I} - \mu\boldsymbol{\varphi}_s(0)\boldsymbol{\varphi}_s^T(0))\mathbf{w}(0) + \mu x(0)\boldsymbol{\varphi}_s(0),$$

which is used to express $\mathbf{w}(2)$ in terms of $\mathbf{w}(0)$, $x(0)$, and $x(1)$,

$$\mathbf{w}(2) = (\mathbf{I} - \mu\boldsymbol{\varphi}_s(1)\boldsymbol{\varphi}_s^T(1))\mathbf{w}(1) + \mu x(1)\boldsymbol{\varphi}_s(1)$$
$$= (\mathbf{I} - \mu\boldsymbol{\varphi}_s(1)\boldsymbol{\varphi}_s^T(1))(\mathbf{I} - \mu\boldsymbol{\varphi}_s(0)\boldsymbol{\varphi}_s^T(0))\mathbf{w}(0)$$
$$\quad + (\mathbf{I} - \mu\boldsymbol{\varphi}_s(1)\boldsymbol{\varphi}_s^T(1))\mu x(0)\boldsymbol{\varphi}_s(0) + \mu x(1)\boldsymbol{\varphi}_s(1),$$

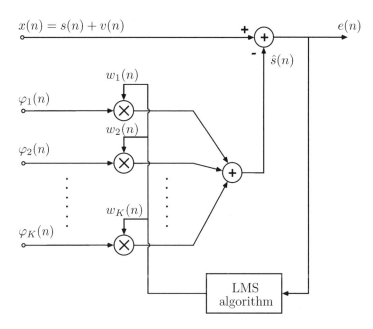

Figure 4.35: Adaptive linear combination of basis functions for the estimation of the weight vector $\mathbf{w}(n)$, using the observed signal $x(n)$.

and so on. The general expression for $\mathbf{w}(n)$ is given by [147]

$$\mathbf{w}(n) = \mathbf{F}_0(n-1)\mathbf{w}(0) + \mu \sum_{m=0}^{n-1} x(m)\mathbf{F}_{m+1}(n-1)\boldsymbol{\varphi}_s(m), \qquad (4.282)$$

where

$$\mathbf{F}_m(n) = \begin{cases} \displaystyle\prod_{j=m}^{n} \left(\mathbf{I} - \mu\boldsymbol{\varphi}_s(j)\boldsymbol{\varphi}_s^T(j)\right), & n \geq m; \\ \mathbf{I}, & n < m. \end{cases} \qquad (4.283)$$

In order to assure stability, the step size μ must be chosen such that the matrix $\mathbf{F}_0(n-1)$, being related to the initial condition $\mathbf{w}(0)$, approaches zero as n increases. In order to establish bounds on μ, we consider a matrix norm which assigns a number to the matrix characterizing its magnitude. The *Frobenius norm* of a matrix $\mathbf{A}$ is defined by

$$\|\mathbf{A}\|_F = \sqrt{\operatorname{tr}(\mathbf{A}^T\mathbf{A})}.$$

It can be shown that each of the matrices in the product $\mathbf{F}_m(n)$ has a norm that is less than that of the unit matrix,

$$\|\mathbf{I} - \mu\boldsymbol{\varphi}_s(j)\boldsymbol{\varphi}_s^T(j)\|_F < \|\mathbf{I}\|_F, \qquad (4.284)$$

if

$$0 < \mu < 2, \tag{4.285}$$

and thus stability is assured.

It is of interest to study the weight behavior in the mean for the steady-state situation in which the signal morphology is fixed, i.e., when $\mathbf{s}_i = \mathbf{s}$ for all i. The result, obtained by taking the expected value of (4.282) and performing the limit as n goes to infinity, has been shown to converge to a biased solution [147],

$$\lim_{n \to \infty} E[\mathbf{w}(n)] = \mathbf{w}^\mathrm{o} + \Delta \mathbf{w}^\mathrm{b}, \tag{4.286}$$

where $\mathbf{w}^\mathrm{o}$ denotes the optimal MMSE solution in (4.202). The exact expression of the bias $\Delta \mathbf{w}^\mathrm{b}$ is omitted since it is relatively complicated; instead we will confine ourselves to merely mentioning that the bias is increasingly insignificant for small values of μ, i.e., when the LMS algorithm is less capable of tracking dynamic signal changes.

Another important performance measure is the excess MSE $\mathcal{E}_\mathrm{ex}(n)$ which quantifies the increase in $\mathcal{E}_\mathrm{min}$ due to the fluctuations $\Delta \mathbf{w}(n)$ resulting from the adaptation process, cf. page 87. Assuming that a steady-state situation has been reached and that $\mathbf{s}$ is corrupted by white noise with a variance σ_v^2, the excess MSE can be approximated by the expression [148]

$$\mathcal{E}_\mathrm{ex}(\infty) \approx \frac{\mu K}{(2 - \mu)N} \left(\sigma_v^2 + \frac{\mathbf{s}^T \boldsymbol{\Phi}_v \boldsymbol{\Phi}_v^T \mathbf{s}}{N} \right). \tag{4.287}$$

This expression demonstrates that the selection of a suitable set of basis functions—usually associated with a small value of K—produces a smaller excess MSE and, accordingly, a better estimate of the signal. It should also be noted from (4.287) that the truncation error influences the excess MSE through the quadratic term $\mathbf{s}^T \boldsymbol{\Phi}_v \boldsymbol{\Phi}_v^T \mathbf{s}/N$ so that a large value of K makes this term smaller.

The performance of the instantaneous LMS algorithm is illustrated by an example in which a signal with fixed morphology is subjected to estimation using either sine/cosine or Karhunen–Loève basis functions, see Figures 4.36 and 4.37, respectively. In this example, the signal's maximum negative amplitude is the quantity of interest, and its estimate is presented as a function of the EP number. Comparing Figures 4.36 and 4.37, it is clear that the estimates based on the Karhunen–Loève basis functions are more accurate than those based on sines/cosines in spite of the fact that only 4 functions are used instead of 16. The superior performance of the Karhunen–Loève approach is, of course, explained by the fact that the basis functions are determined from the data ensemble itself so that the approximation error is minimized in the MSE sense, see page 264.

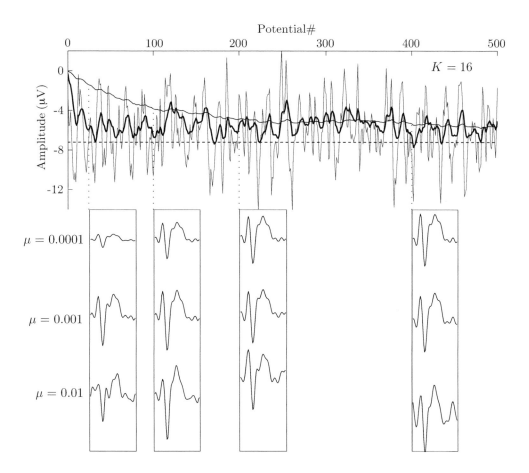

Figure 4.36: Signal estimation using the instantaneous LMS algorithm in combination with sine/cosine basis functions—the adaptive Fourier linear combiner. The top panel displays the maximum negative amplitude measured from successive EP estimates for three different values of μ; the dashed line indicates the true amplitude. The estimated EPs are shown in the bottom panels for some instants (the noise-free signal is shown in Figure 4.30(e)). The number of basis functions K is equal to 16.

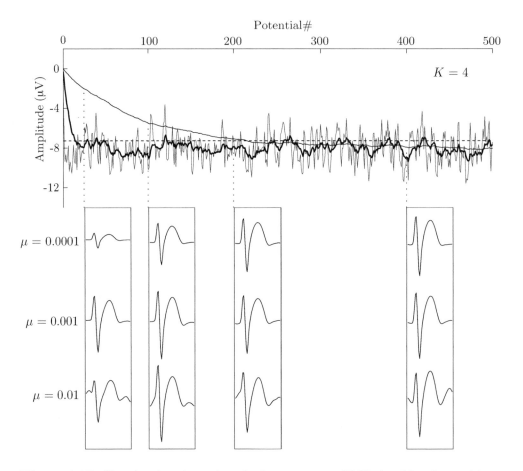

Figure 4.37: Signal estimation using the instantaneous LMS algorithm in combination with Karhunen–Loève basis functions, estimated from the ensemble of data. The top panel displays the maximum negative amplitude measured from successive EP estimates for three different values of μ; the dashed line indicates the true amplitude. The estimated EPs are shown in the bottom panels for some instants (the noise-free signal is shown in Figure 4.30(e)). The number of basis functions K is equal to 4.

4.6.2 The Block LMS Algorithm

Historically, the instantaneous LMS algorithm is the adaptive technique for noise reduction that has almost exclusively been considered in EP analysis. However, the block LMS algorithm represents a natural extension of single-trial analysis based on a truncated series expansion and is therefore discussed below [141]. In addition, the block LMS algorithm offers certain theoretical advantages over the instantaneous LMS algorithm, notably with respect to bias and excess MSE, making its use particularly attractive.

The MSE criterion in (4.277), constituting the starting point for the derivation of the instantaneous LMS algorithm, is replaced by an MSE criterion defined in vector terms:

$$\mathcal{E}_{\mathbf{w}_{i-1}} = E\left[(\mathbf{x}_i - \mathbf{\Phi}_s\mathbf{w}_{i-1})^T(\mathbf{x}_i - \mathbf{\Phi}_s\mathbf{w}_{i-1})\right]. \tag{4.288}$$

This criterion resembles the MSE criterion in (4.199) introduced for single-trial analysis, except that the weight vector $\mathbf{w}_{i-1}$, rather than $\mathbf{w}_i$, is used to define the error for the EP with index i. With this slight change of definition, it is straightforward to relate the recursive algorithm derived below to the inner product estimator in (4.203).

The block LMS algorithm iteratively finds the weight vector by making use of the steepest decent algorithm,

$$\mathbf{w}_i = \mathbf{w}_{i-1} - \frac{1}{2}\mu\nabla_{\mathbf{w}_{i-1}}\mathcal{E}_{\mathbf{w}_{i-1}}. \tag{4.289}$$

Recalling from (4.201) that the gradient is given by

$$\nabla_{\mathbf{w}_{i-1}}\mathcal{E}_{\mathbf{w}_{i-1}} = -2\mathbf{\Phi}_s^T E\left[\mathbf{x}_i\right] + 2\mathbf{\Phi}_s^T\mathbf{\Phi}_s\mathbf{w}_{i-1},$$

and then replacing the expected value with its instantaneous estimate, we obtain the block LMS algorithm

$$\mathbf{w}_i = (1 - \mu)\mathbf{w}_{i-1} + \mu\mathbf{\Phi}_s^T\mathbf{x}_i, \tag{4.290}$$

initialized by

$$\mathbf{w}_0 = \mathbf{0}. \tag{4.291}$$

This initialization seems natural since, apart from the step size μ, it leads to the inner product estimator of $\mathbf{w}_1$, i.e., $\mathbf{w}_1 = \mu\mathbf{\Phi}_s^T\mathbf{x}_1$. With the error notation

$$\mathbf{e}_i = \mathbf{x}_i - \mathbf{\Phi}_s\mathbf{w}_{i-1}, \tag{4.292}$$

the block LMS algorithm can alternatively be expressed as

$$\mathbf{w}_i = \mathbf{w}_{i-1} + \mu \boldsymbol{\Phi}_s^T \mathbf{e}_i. \tag{4.293}$$

Similar to the instantaneous LMS algorithm, the block LMS algorithm remains stable for $0 < \mu < 2$.

It may be worthwhile to point out certain relationships between the block LMS algorithm and the previously described algorithms for EP analysis. The block LMS algorithm reduces to single-trial analysis when $\mu = 1$, since (4.290) then becomes identical to (4.203). Moreover, when a complete series expansion is considered, i.e., $\boldsymbol{\Phi}\boldsymbol{\Phi}^T = \mathbf{I}$, the block LMS algorithm becomes identical to exponential averaging as defined in (4.33); this result is realized from

$$\hat{\mathbf{s}}_i = \boldsymbol{\Phi}\mathbf{w}_i$$
$$= \boldsymbol{\Phi}\left((1-\mu)\mathbf{w}_{i-1} + \mu\boldsymbol{\Phi}^T\mathbf{x}_i\right)$$
$$= \hat{\mathbf{s}}_{i-1} + \mu\left(\mathbf{x}_i - \hat{\mathbf{s}}_{i-1}\right).$$

For the case of most practical interest, i.e., when a truncated expansion is considered, the block LMS algorithm performs exponential averaging of the weight vector: an operation which produces an estimate of the weight vector which is less noisy but also less capable of tracking dynamic signal changes.

For the steady-state condition when $\mathbf{s}_i = \mathbf{s}$, the block LMS algorithm can, in contrast to the instantaneous LMS algorithm, be shown to produce a steady-state weight vector $\mathbf{w}_\infty$ which is an unbiased estimate of the optimal solution in (4.202) [141],

$$\lim_{i\to\infty} E[\mathbf{w}_i] = \mathbf{w}^\circ. \tag{4.294}$$

Another attractive property of the block LMS algorithm is that its excess MSE is given by [141],

$$\mathcal{E}_{\mathrm{ex}}(\infty) = \frac{\mu K}{(2-\mu)N}\sigma_v^2, \tag{4.295}$$

which does not involve any term due to the truncation error as did the excess MSE in (4.287) for the instantaneous LMS algorithm, i.e., the term $\mathbf{s}^T\boldsymbol{\Phi}_v\boldsymbol{\Phi}_v^T\mathbf{s}/N$. Hence, the block LMS algorithm is always associated with a lower excess MSE—a result which becomes particularly advantageous when the signal energy is concentrated to a few basis functions.

4.7   Wavelets

Due to the highly transient nature of EPs, their frequency content varies considerably over time. As a result, sines and cosines as basis functions are not well-suited for modeling EPs because these functions cannot account for information localized in time; sines and cosines are appropriate for signals which are periodic. A relatively large number of series expansion coefficients are required to achieve acceptable modeling, thereby also increasing the risk of modeling noise components.

Since the basis functions of the Karhunen–Loève expansion are derived from an ensemble of transient data, they are, by necessity, well-matched to signals with transient properties. As was shown in Section 4.5.4, noise reduction through truncation of a series expansion can be interpreted as linear, time-variant filtering, with frequency characteristics being more adjustable for Karhunen–Loève basis functions than for sines/cosines. However, the Karhunen–Loève functions are signal-dependent and, therefore, their use is associated with a considerable amount of computation. No fast algorithm is available which corresponds to the FFT for computing the discrete Fourier transform.

Since tracking of latency changes is a crucial aspect of EP analysis, it is important that the selected basis functions appropriately accommodate such temporal information. Although both sines/cosines and Karhunen–Loève functions, to various degrees, allow this, they lack flexibility for efficient tracking. We will in this section describe a very general and powerful class of basis functions, known as *wavelets*, which involve two parameters: one for *translation* in time and another for *scaling* in time. A wavelet is an oscillating function whose energy is concentrated in time in order to better represent transient, nonstationary signals. For a function to qualify as a wavelet, it must exhibit certain mathematical properties, one of which is to have bandpass filter characteristics.

An important aspect of wavelet analysis is the desire to achieve good localization in both time and frequency, similar to the motivation for studying time–frequency representations in Section 3.6. With two new degrees of freedom of the basis functions, scaling and translation, it is possible to analyze and resolve the joint presence of global waveforms ("large scale") as well as fine structures ("small scale") in signals using wavelet analysis. The fundamental idea of analyzing signals at different scales, with an increasing level of detail resolution, is referred to as a *multiresolution analysis* and will be further considered in Section 4.7.2. Such a representation serves the analysis of EPs particularly well because interpeak latencies usually increase with time (see page 185). Hence, waves occurring early with short durations require better time resolution than do waves with large durations occurring at later times.

The main reason for considering wavelet analysis is for the purpose of EP noise reduction, an operation which in the wavelet literature is commonly referred to as *denoising* (Section 4.7.6). However, it is important to be aware that wavelet analysis is equally useful for analyzing the background EEG for the purpose of detecting transient waveforms such as epileptic spikes and sleep spindles, extracting features to be used for classifying different signal components, and displaying information in a more clear-cut fashion than is possible in the original signal [149]. Yet another important application of basis functions, and wavelets in particular, is the compression of large amounts of data for storage or transmission; the data compression issue will be touched upon later within the context of ECG signal processing in Section 7.6. While wavelet analysis is certainly not the panacea to all types of problems, it nonetheless constitutes an extremely powerful tool for analyzing and processing a wide range of transient, nonstationary biomedical signals.

Comprehensive descriptions of wavelets employing mathematical rigor can be found in a number of well-written textbooks [150–152]; these books also cover the closely related topic of filter banks. Good overviews of wavelets and their application in biomedical signal processing can be found in [153–155].

4.7.1   The Wavelet Transform

We start our exposition by recalling that the fundamental operation in orthonormal basis function analysis is the correlation (inner product) between the observed signal $x(n)$ and the basis functions $\varphi_k(n)$ (cf. page 255),

$$w_k = \sum_{n=0}^{N-1} x(n)\varphi_k(n), \qquad (4.296)$$

where the index referring to the EP number has been omitted for convenience. The resulting coefficients w_k define the series expansion of basis functions that describe $x(n)$. In wavelet analysis, the two operations of scaling and translation in time are most simply introduced when the continuous-time description is adopted. Therefore, we mention the continuous-time version of the correlation in (4.296),

$$w_k = \int_{-\infty}^{\infty} x(t)\varphi_k(t)dt. \qquad (4.297)$$

A family of wavelets $\psi_{s,\tau}(t)$ is defined by scaling and translating the *mother wavelet* $\psi(t)$ with the continuous-valued parameters s (> 0) and τ,

$$\psi_{s,\tau}(t) = \frac{1}{\sqrt{s}}\psi\left(\frac{t-\tau}{s}\right), \qquad (4.298)$$

where the factor $1/\sqrt{s}$ assures that all scaled functions have equal energy. Thus, the wavelet is contracted for $s < 1$, whereas it is expanded for $s > 1$.

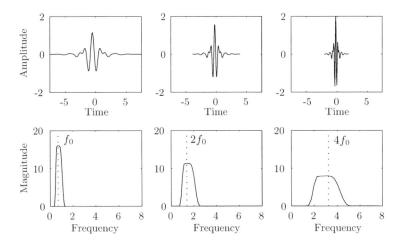

Figure 4.38: A wavelet shown at three different scales and the corresponding bandpass frequency responses (the Meyer wavelet). Note that the center frequency and bandwidth both increase as the wavelet is contracted in time.

The contraction of a wavelet to a smaller scale makes it more localized in time, while the corresponding frequency response is shifted to higher frequencies and the bandwidth is increased to become less localized in frequency; the reverse behavior is obtained when the wavelet is expanded in time, see Figure 4.38. The notation $\psi(t)$ is used here instead of the previously established $\varphi(t)$ for reasons which will shortly become evident.

The *continuous wavelet transform* (CWT) $w(s, \tau)$ of a continuous-time signal $x(t)$ is defined by the correlation between $x(t)$ and a scaled and translated version of $\psi(t)$,

$$w(s, \tau) = \int_{-\infty}^{\infty} x(t) \frac{1}{\sqrt{s}} \psi\left(\frac{t - \tau}{s}\right) dt, \qquad (4.299)$$

thus constituting two-dimensional mapping onto the time–scale domain. The CWT can be interpreted as a linear filtering operation since (4.299) defines the convolution between the signal $x(t)$ and a filter whose impulse response is $\psi(-t/s)/\sqrt{s}$.

The function $x(t)$ can be exactly recovered from $w(s, \tau)$ using the reconstruction equation [152]

$$x(t) = \frac{1}{C_\psi} \int_{-\infty}^{\infty} \int_{0}^{\infty} w(s, \tau) \frac{1}{\sqrt{s}} \psi\left(\frac{t - \tau}{s}\right) \frac{d\tau ds}{s^2}, \qquad (4.300)$$

where

$$C_\psi = \int_{0}^{\infty} \frac{|\Psi(\Omega)|^2}{|\Omega|} d\Omega < \infty, \qquad (4.301)$$

and $\Psi(\Omega)$ denotes the Fourier transform of $\psi(t)$. For the integral in (4.301) to exist, $\Psi(0)$ must equal zero, i.e., the DC gain must be zero,

$$\Psi(0) = \int_{-\infty}^{\infty} \psi(t)dt = 0. \tag{4.302}$$

Another requirement is that $|\Psi(\Omega)|$ must decrease to zero for $|\Omega| \to \infty$. These two requirements imply that the wavelet function $\psi(t)$ must have bandpass characteristics. Consequently, the CWT can be viewed as a type of bandpass analysis where the scaling parameter s modifies the center frequency and the bandwidth in the way illustrated in Figure 4.38.

Since the above requirements on the mother wavelet are relatively modest, it turns out to be a highly adjustable function which can be designed to suit various signal problems (this stands in sharp contrast to the Fourier transform where the basis functions are fixed once and for all). The simplest wavelet example is the *Haar wavelet* which is defined as

$$\psi(t) = \begin{cases} 1, & 0 \leq t < \frac{1}{2}; \\ -1, & \frac{1}{2} \leq t < 1; \\ 0, & \text{otherwise}, \end{cases} \tag{4.303}$$

and is related to the Walsh basis functions considered earlier on page 263. Another popular wavelet is the *Mexican hat*,

$$\psi(t) = (1 - t^2)e^{-t^2/2}, \quad -\infty < t < \infty, \tag{4.304}$$

being identical to the second derivative of a Gaussian function, apart from a normalization factor. The CWT is presented in Figure 4.39 for a composite signal using the Mexican hat wavelet. An essential difference between the above two wavelets is that the Haar wavelet has a duration which is limited in time (compact support), while the Mexican hat wavelet does not; in most situations, the former type of wavelet is preferred because of its superior time localization. Another significant difference is that the Haar wavelet has discontinuities while the Mexican hat is a smooth function. A wide variety of wavelets have been suggested, each one possessing some feature particularly suitable for certain applications. For example, the Haar wavelet does not resemble the shape of EPs, and, therefore, other, smoother types of wavelets are desirable; the problem of designing wavelets is briefly considered in Section 4.7.5.

The CWT is a two-dimensional function $w(s, \tau)$ which is highly redundant. It is, therefore, necessary to discretize the scaling and translation parameters s and τ according to a suitably chosen sampling grid. The most popular approach is to use *dyadic sampling* of the two parameters,

$$s = 2^{-j}, \quad \tau = k2^{-j}, \tag{4.305}$$

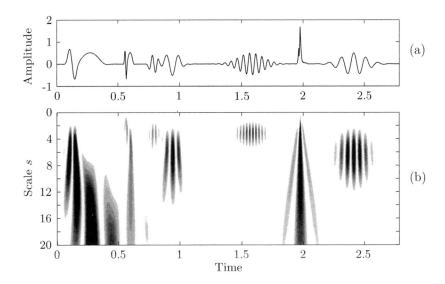

Figure 4.39: An example of (a) a composite signal and (b) the corresponding squared magnitude of the continuous-time wavelet transform $|w(s, \tau)|^2$—the *scalogram*; the mother wavelet is the Mexican hat. Note that the low-frequency waveforms are mostly reflected at coarser scales (large values of s) while transient waveforms are at finer scales. In comparison with the spectrogram, the scalogram offers better frequency resolution at low frequencies and better time resolution at high frequencies, but also poorer time resolution at low frequencies and poorer frequency resolution at high frequencies.

where j and k are both integers. Accordingly, the discretized wavelet function is defined by

$$\psi_{j,k}(t) = 2^{j/2}\psi(2^j t - k). \tag{4.306}$$

Inserting (4.306) into the CWT in (4.299), we obtain the *discrete wavelet transform* (DWT)

$$w_{j,k} = \int_{-\infty}^{\infty} x(t)\psi_{j,k}(t)dt. \tag{4.307}$$

It can be shown that with dyadic sampling it is still possible to exactly reconstruct $x(t)$ from the coefficients $w_{j,k}$ resulting from discretization of the CWT; a coarser sampling grid cannot reconstruct $x(t)$ [152]. The original signal is retrieved by the *inverse* DWT, or the wavelet series expansion

$$x(t) = \sum_{j=-\infty}^{\infty} \sum_{k=-\infty}^{\infty} w_{j,k}\psi_{j,k}(t), \tag{4.308}$$

where $\psi_{j,k}(t)$ is a set of orthonormal basis functions. In contrast to the series expansion of basis functions in (4.187), defined as the sum over one index, the wavelet series expansion is more flexible since it is the sum over two indices which are related to scaling and translation of the basis functions $\psi_{j,k}(t)$.

4.7.2   Multiresolution Signal Analysis

A signal can be viewed as the sum of a smooth ("coarse") part and a detailed ("fine") part: the smooth part reflects the main features of the signal, therefore called the approximation signal, whereas the faster fluctuations represent the details of the signal. The separation of a signal into two parts is determined by the resolution with which the signal is analyzed, i.e., by the scale below which no details can be discerned. A progressively better approximation of the signal is obtained by increasing the resolution so that finer and finer details are included in the smooth part.

The approximation of a signal $x(t)$ at scale (resolution level) j is denoted $x_j(t)$. At the next scale $j+1$, the approximation signal $x_{j+1}(t)$ is composed of $x_j(t)$ and the details $y_j(t)$ at that level such that

$$x_{j+1}(t) = x_j(t) + y_j(t). \tag{4.309}$$

By adding more and more detail to $x_j(t)$ we arrive, as the resolution approaches infinity, at a dyadic multiresolution representation of the original signal $x(t)$ which involves a smooth part and the sum of different details,

$$x(t) = x_j(t) + \sum_{l=j}^{\infty} y_l(t). \tag{4.310}$$

In the following, we will describe the means that will allow us to perform the multiresolution signal analysis suggested by (4.310) and which implements the DWT [152].

The scaling function.   The scaling function $\varphi(t)$ is introduced for the purpose of efficiently representing the approximation signal $x_j(t)$ at different resolution. This function, being related to a unique wavelet function $\psi(t)$, can be used to generate a set of scaling functions defined by different translations,

$$\varphi_{0,k}(t) = \varphi(t-k), \tag{4.311}$$

where the index "0" indicates that no time scaling is performed. The design of a scaling function $\varphi(t)$, being a complicated issue only briefly touched upon below, must be such that translations of $\varphi(t)$ constitute an orthonormal set of functions,

$$\int_{-\infty}^{\infty} \varphi_{0,k}(t)\varphi_{0,n}(t)dt = \int_{-\infty}^{\infty} \varphi(t-k)\varphi(t-n)dt = \begin{cases} 1, & k = n; \\ 0, & k \neq n. \end{cases} \quad (4.312)$$

Therefore, the scaling functions $\varphi_{0,k}(t)$ are said to span a subspace $\mathcal{V}_0$ of the whole space of square integrable functions denoted $L^2(\mathbf{R})$,

$$\mathcal{V}_0 = \operatorname*{span}_{k}\{\varphi_{0,k}(t)\}. \quad (4.313)$$

This subspace allows us to approximate $x(t)$ to a signal $x_0(t)$ described as a linear combination of $\varphi(t)$ at different translations $\varphi(t-n)$,[11]

$$x_0(t) = \sum_{n=-\infty}^{\infty} c_0(n)\varphi(t-n). \quad (4.314)$$

As before, the coefficients of the series expansion result from computing the inner product

$$c_0(k) = \int_{-\infty}^{\infty} x(t)\varphi_{0,k}(t)dt. \quad (4.315)$$

Analogously to dyadic sampling of the wavelet function $\psi(t)$, the scaling function in (4.311) can be generalized through dyadic sampling to generate a set of orthonormal scaling functions for approximations at different resolution,

$$\varphi_{j,k}(t) = 2^{j/2}\varphi(2^j t - k), \quad (4.316)$$

where the factor $2^{j/2}$ assures that the norm of $\varphi_{j,k}(t)$ is one for all indices j and k,

$$\sqrt{\int_{-\infty}^{\infty} |\varphi_{j,k}(t)|^2 dt} = 1. \quad (4.317)$$

Orthonormality applies only to different translation indices k for a fixed scale j, and the scaling functions are thus not required to be orthonormal between

---

[11]The scaling coefficients are indexed $c_0(k)$ rather than $c_{0,k}$ in order to indicate that k refers to time. The variable name c refers to coefficients that describe the *coarse* part of a signal (an approximation), whereas the variable name d, used later on, refers to coefficients that describe *detail*.

different scales. With these basis functions, the approximation signal $x_j(t)$ is given by

$$x_j(t) = \sum_{n=-\infty}^{\infty} c_j(n)\varphi_{j,n}(t)$$

$$= 2^{j/2} \sum_{n=-\infty}^{\infty} c_j(n)\varphi(2^j t - n), \qquad (4.318)$$

where

$$c_j(k) = \int_{-\infty}^{\infty} x(t)\varphi_{j,k}(t)dt. \qquad (4.319)$$

It is important to realize that, for $j > 0$, the span increases since $\varphi_{j,k}(t)$ contracts in time, thereby allowing details of $x(t)$ to be better represented by the approximation signal $x_j(t)$. On the other hand, only the coarser information can be represented for $j < 0$ since $\varphi_{j,k}(t)$ then expands.

The subspace $\mathcal{V}_j$ is spanned by $\varphi_{j,k}(t)$,

$$\mathcal{V}_j = \operatorname*{span}_k \{\varphi_{j,k}(t)\}, \qquad (4.320)$$

which has a time resolution only half as good as that of $\mathcal{V}_{j+1}$ since the scaling function in $\mathcal{V}_{j+1}$ is contracted by a factor of two, i.e., $\varphi(2^{j+1}t)$ in relation to $\varphi(2^j t)$. As a result, the orthonormal basis functions that span $\mathcal{V}_j$ are also part of $\mathcal{V}_{j+1}$, and the multiresolution property is consequently described by a set of nested signal subspaces,

$$\ldots \mathcal{V}_{-2} \subset \mathcal{V}_{-1} \subset \mathcal{V}_0 \subset \mathcal{V}_1 \subset \mathcal{V}_2 \subset \ldots. \qquad (4.321)$$

$$\leftarrow \text{coarser} \qquad \text{finer} \rightarrow$$

Each subspace is spanned by a different set of basis functions $\varphi_{j,k}(t)$, offering progressively better approximations such that $x_j(t)$ approaches $x(t)$ in the limit as $j \to \infty$,

$$\lim_{j \to \infty} x_j(t) = x(t), \qquad (4.322)$$

where $x(t)$ belongs to the space $L^2(\mathbf{R})$.

An important relation is the *refinement equation* which relates $\varphi(t)$, spanning $\mathcal{V}_0$, to $\varphi(2t)$, spanning $\mathcal{V}_1$. Since these two signal subspaces are such that $\mathcal{V}_0 \subset \mathcal{V}_1$, it is possible to express $\varphi(t)$ as a linear combination of the shifted versions of $\varphi(2t)$,

$$\varphi(t) = \sum_{n=-\infty}^{\infty} h_\varphi(n)\varphi_{1,n}(t)$$

$$= \sqrt{2} \sum_{n=-\infty}^{\infty} h_\varphi(n)\varphi(2t - n), \qquad (4.323)$$

where $h_\varphi(n)$ is a sequence of *scaling coefficients*. As we will see later, the design of a wavelet function is synonymous with the selection of the coefficients $h_\varphi(n)$. The relation between scaling functions at different scales, as expressed by the refinement equation, will be used to develop a technique with which the series expansion coefficients can be calculated.

The wavelet function. It is desirable to introduce the function $\psi(t)$ which complements the scaling function by accounting for the details of a signal, rather than its approximations. For this purpose, a set of orthonormal basis functions at scale j is given by

$$\psi_{j,k}(t) = 2^{j/2}\psi(2^j t - k), \qquad (4.324)$$

which spans the difference between the two subspaces $\mathcal{V}_j$ and $\mathcal{V}_{j+1}$. The functions $\psi_{j,k}(t)$ are related to the mother wavelet, introduced in (4.298), and subjected to dyadic sampling. At scale $j + 1$, the subspace describing signal detail is given by

$$\mathcal{W}_j = \operatorname*{span}_{k}\{\psi_{j,k}(t)\}, \qquad (4.325)$$

where the wavelet functions that span $\mathcal{W}_j$ are required to be orthonormal to the scaling functions of $\mathcal{V}_j$,

$$\int_{-\infty}^{\infty} \varphi_{j,k}(t)\psi_{j,l}(t)dt = 0, \qquad (4.326)$$

for all indices j and k. As before, orthonormality is advantageous since it simplifies the calculation of the series expansion coefficients.

In the subspace $\mathcal{V}_{j+1}$, $\mathcal{W}_j$ is said to constitute an orthogonal complement to $\mathcal{V}_j$ which is denoted

$$\mathcal{V}_{j+1} = \mathcal{V}_j \oplus \mathcal{W}_j, \qquad (4.327)$$

where $\oplus$ denotes the direct sum between two subspaces. Since (4.327) is valid for an arbitrary value of j, we also have that

$$\mathcal{V}_j = \mathcal{V}_{j-1} \oplus \mathcal{W}_{j-1}, \qquad (4.328)$$

which, when continued until a certain value j_0 $(\leq j)$ is reached, yields the decomposition

$$\mathcal{V}_{j+1} = \mathcal{V}_{j_0} \oplus \mathcal{W}_{j_0} \oplus \mathcal{W}_{j_0+1} \oplus \ldots \oplus \mathcal{W}_j. \tag{4.329}$$

As j approaches infinity, the subspace decomposition can be expressed as

$$x(t) = x_{j_0}(t) + \sum_{j=j_0}^{\infty} y_j(t), \tag{4.330}$$

where the detail signal $y_j(t)$ is determined by the detail coefficients $d_j(k)$, calculated as the inner product of $x(t)$ and $\psi_{j,k}(t)$,

$$
\begin{aligned}
y_j(t) &= \sum_{n=-\infty}^{\infty} d_j(n)\psi_{j,n}(t) \\
&= 2^{j/2} \sum_{n=-\infty}^{\infty} d_j(n)\psi(2^j t - n), \tag{4.331}
\end{aligned}
$$

where

$$d_j(k) = \int_{-\infty}^{\infty} x(t)\psi_{j,k}(t)dt. \tag{4.332}$$

It should be noted that the coefficients $d_j(k)$ are the same as the $w_{j,k}$ of the DWT in (4.307); however, $d_j(k)$ is the preferred notation for expressing detail at different scales. At the scale j_0, the signal $x(t)$ can be expressed as a *wavelet series expansion* in terms of the scaling coefficients $c_{j_0}(k)$ and the wavelet coefficients $d_j(k)$,

$$x(t) = \sum_{n=-\infty}^{\infty} c_{j_0}(n)\varphi_{j_0,n}(t) + \sum_{j=j_0}^{\infty}\sum_{n=-\infty}^{\infty} d_j(n)\psi_{j,n}(t). \tag{4.333}$$

Hence, $x(t)$ can be decomposed into a signal $x_{j_0}(t)$, being a lowpass approximation of $x(t)$, and a set of signals $y_j(t)$ which gives varying degrees of high-resolution details of $x(t)$. Furthermore, since the series expansion in (4.333) is expressed in terms of basis functions being mutually orthonormal, the coefficients $c_{j_0}(k)$ and $d_j(k)$ are easily calculated by their corresponding inner products in (4.319) and (4.332), respectively.

Just as the scaling function $\varphi(t)$ can be expressed as a linear combination of the shifted scaling functions with half the width, i.e., using the refinement equation in (4.323), the wavelet function $\psi(t)$ can be similarly expressed by the *wavelet equation*

$$\psi(t) = \sum_{n=-\infty}^{\infty} h_\psi(n)\sqrt{2}\varphi(2t - n). \tag{4.334}$$

The wavelet equation results from the property that $\mathcal{W}_j \subset \mathcal{V}_{j+1}$, which allows us to express $\psi(t)$ in terms of shifted versions of $\varphi(2t)$ similar to the procedure applied to $\varphi(t)$ in (4.323). The coefficients $h_\psi(n)$ constitute a sequence of *wavelet coefficients* that differ from the scaling coefficients $h_\varphi(n)$. However, it can be shown that $h_\psi(n)$ can be determined from $h_\varphi(n)$ such that when the number of coefficients N_φ is finite and even [151],

$$h_\psi(n) = (-1)^n h_\varphi(N_\varphi - 1 - n) \quad n = 0, \ldots, N_\varphi - 1. \tag{4.335}$$

The two types of coefficients are thus the same except that every other coefficient has the opposite sign.

We conclude this subsection by stating that the DWT is defined by the coefficients of the wavelet series expansion in (4.333). These coefficients can be viewed as the counterpart of the Fourier series coefficients, although their interpretation is no longer equally simple (i.e., no frequency interpretation) and the basis functions remain to be specified before the DWT can be calculated.

4.7.3 Multiresolution Signal Analysis—A Classical Example

Multiresolution signal analysis is illustrated by considering the classical *Haar functions* which constitute a set of shifted and scaled square wave functions, suitable for defining scaling and wavelet functions [156]. The Haar scaling function is defined as

$$\varphi(t) = \begin{cases} 1, & 0 \leq t < 1; \\ 0, & \text{otherwise.} \end{cases} \tag{4.336}$$

It is easily verified that this function satisfies the orthonormality condition in (4.312). Furthermore, the Haar scaling function is a solution of the refinement equation with two nonzero coefficients,

$$\begin{bmatrix} h_\varphi(0) & h_\varphi(1) \end{bmatrix} = \begin{bmatrix} \frac{1}{\sqrt{2}} & \frac{1}{\sqrt{2}} \end{bmatrix}. \tag{4.337}$$

Due to (4.334) and (4.335), the related Haar wavelet function is required to be

$$\psi(t) = \begin{cases} 1, & 0 \leq t < \frac{1}{2}; \\ -1, & \frac{1}{2} \leq t < 1; \\ 0, & \text{otherwise,} \end{cases} \tag{4.338}$$

with two nonzero coefficients

$$\begin{bmatrix} h_\psi(0) & h_\psi(1) \end{bmatrix} = \begin{bmatrix} \frac{1}{\sqrt{2}} & -\frac{1}{\sqrt{2}} \end{bmatrix}. \tag{4.339}$$

$$\varphi(2^j t - k)$$

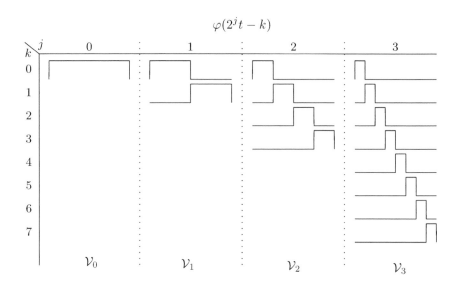

Figure 4.40: The Haar scaling functions that span the subspaces $\mathcal{V}_0, \mathcal{V}_1, \mathcal{V}_2$, and $\mathcal{V}_3$.

The scaling and wavelet functions are orthonormal.

Haar scaling functions that span different subspaces are shown in Figure 4.40. Using these functions, the EP shown at the top of Figure 4.41(a), which has appeared previously in other examples, is subjected to multiresolution analysis and approximated at different time resolutions. At the coarsest scale, defined to be $j = 0$, no signal detail is included since $x_0(t)$ is simply a constant equal to the mean of the original signal $x(t)$. However, the piecewise constant approximation signal $x_j(t)$ becomes increasingly better as more and more scales are included. In the limit as $j \to \infty$, the width of each "staircase" in $x_j(t)$ becomes so narrow so that $x_j(t)$ approaches $x(t)$.

The decomposition of the subspace $\mathcal{V}_1$ into its coarser approximation space $\mathcal{V}_0$ and detail space $\mathcal{W}_0$, i.e., $\mathcal{V}_1 = \mathcal{V}_0 \oplus \mathcal{W}_0$, is graphically presented in Figure 4.42 for the Haar scaling and wavelet functions. With this decomposition, we can calculate the detail signals $y_j(t)$ at increasing time resolution as illustrated in Figure 4.41(b). Moreover, the original signal $x(t)$ is returned by adding the detail signals $y_0(t), \ldots, y_8(t)$ to the approximation signal $x_0(t)$ as described by the decomposition in (4.330) (but now with the lower and upper summation limits equal to $j_0 = 0$ and 8, respectively). It is clear from Figure 4.41(b) that the noise is primarily found at the finest scales, while the EP waveform is at the coarser scales.

It should be noted that, for this example, we have focused on the two different signal representations—approximation and detail—while the prop-

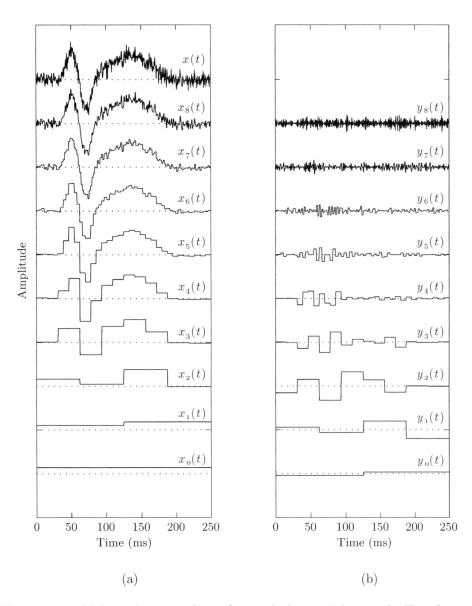

(a)   (b)

Figure 4.41: Multiresolution analysis of an evoked potential using the Haar functions. (a) The approximation signals $x_i(t)$ and (b) the detail signals $y_i(t)$ at different scales; the original signal $x(t)$ is shown at the top left of the figure.

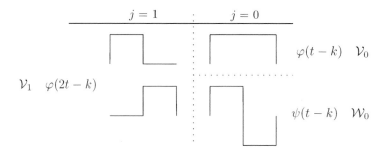

Figure 4.42: Decomposition of the subspace $\mathcal{V}_1 = \mathcal{V}_0 \oplus \mathcal{W}_0$ by the Haar scaling and wavelet functions.

erties of the associated coefficients $c_j(k)$ and $d_j(k)$ have not yet been looked at; this will be done in the following subsection.

4.7.4   Implementation of the Discrete Wavelet Transform Using Filter Banks

Analysis filter bank.   An important reason for the popularity of multi-resolution analysis is the efficient calculation of the scaling and wavelet coefficients. This can be done with a set of recursive equations whose implementation involves well-known, basic signal processing operations (i.e., filtering and down- or upsampling). Starting with the refinement equation in (4.323),

$$\varphi(t) = \sqrt{2} \sum_{n=-\infty}^{\infty} h_\varphi(n)\varphi(2t - n),$$

we have for an arbitrary scale j,

$$\varphi(2^j t - k) = \sqrt{2} \sum_{n=-\infty}^{\infty} h_\varphi(n)\varphi(2(2^j t - k) - n)$$

$$= \sqrt{2} \sum_{n=-\infty}^{\infty} h_\varphi(n)\varphi(2^{j+1} t - 2k - n). \qquad (4.340)$$

Making the substitution $l = 2k + n$, we obtain a relation between two time resolutions, taking us from finer to coarser resolution,

$$\varphi(2^j t - k) = \sqrt{2} \sum_{n=-\infty}^{\infty} h_\varphi(n - 2k)\varphi(2^{j+1} t - n), \qquad (4.341)$$

and

$$\varphi_{j,k}(t) = \sum_{n=-\infty}^{\infty} h_{\varphi}(n-2k)\varphi_{j+1,n}(t). \tag{4.342}$$

A recursive relation can be derived for the scaling coefficients $c_j(k)$ by multiplying both sides of (4.342) by $x(t)$ and integrating to obtain the inner products,

$$\int_{-\infty}^{\infty} x(t)\varphi_{j,k}(t)dt = \int_{-\infty}^{\infty} x(t)\sum_{n=-\infty}^{\infty} h_{\varphi}(n-2k)\varphi_{j+1,n}(t)dt, \tag{4.343}$$

which yield the convolution,

$$c_j(k) = \sum_{n=-\infty}^{\infty} h_{\varphi}(n-2k)c_{j+1}(n)$$

$$= h_{\varphi}(-n) * c_{j+1}(n)|_{n=2k}. \tag{4.344}$$

In an analogous manner, the wavelet coefficients $d_j(k)$ can be calculated by convolving the time-reversed coefficients $h_{\psi}(-n)$ with $c_{j+1}(n)$ and subsequent downsampling of the filtered output by a factor of two:

$$d_j(k) = \sum_{n=-\infty}^{\infty} h_{\psi}(n-2k)c_{j+1}(n)$$

$$= h_{\psi}(-n) * c_{j+1}(n)|_{n=2k}. \tag{4.345}$$

The calculation of the coefficients $c_j(k)$ and $d_j(k)$ can be implemented using the *two-channel analysis filter bank* shown in Figure 4.43(a), with which the coefficients at scale j are calculated from those at scale $j+1$. By repeatedly combining two-channel analysis filter banks to the output of $h_{\varphi}(-n)$, we obtain a dyadic tree structure which efficiently implements the DWT, see Figure 4.43(b). It is important to realize that the scaling and wavelet functions do not explicitly appear in the calculation of the DWT, but only the scaling and wavelet coefficients are required. As a result, the output of the filter bank is a set of coefficients used to calculate the approximation and detail signals with (4.318) and (4.331), respectively.

A frequency domain interpretation comes naturally for the filter parts of (4.344) and (4.345), which are defined by the scaling and wavelet coefficients, respectively. For the case of Haar functions, it is easily shown that the filter $h_{\varphi}(n)$ in (4.337) is lowpass because the average of two adjacent samples is calculated. The filter $h_{\psi}(n)$ in (4.339) is highpass because the difference between two samples is calculated. Both these filters have FIR

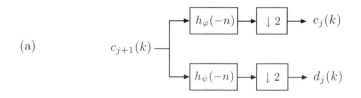

(a)

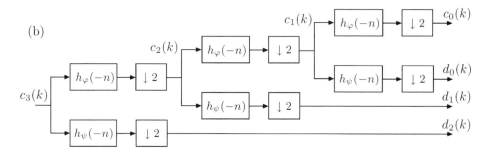

(b)

Figure 4.43: (a) A two-channel analysis filter bank for calculating the coefficients of the wavelet series expansion in (4.333). (b) The discrete wavelet transform based on the filter bank in (a), which, in this case, produces the coefficients that decompose the space $\mathcal{V}_3$ into $\mathcal{V}_0, \mathcal{W}_0, \mathcal{W}_1$, and $\mathcal{W}_2$.

structures. Having established these two filter characteristics, we realize that the analysis filter bank with its dyadic tree structure produces output signals which range from being highpass (the output of the bottom branch in Figure 4.43(b)) to lowpass (the output of the top branch), with various degrees of bandpass in between (the remaining branches). The detail coefficients that result from bandpass filtering involve filters, whose center frequency gradually decreases due to the increasing number of lowpass filters $h_\varphi(n)$ being cascaded to the highpass filter to form the overall filter. The coefficients $c_0(k)$ which describe the approximation signal in subspace $\mathcal{V}_0$ are obtained by cascaded lowpass filters only.

It can be shown that the characteristics of the filters that implement the Haar coefficients are equally valid for any filter $h_\varphi(n)$ and $h_\psi(n)$; the filter defined by the scaling coefficients $h_\varphi(n)$ must be lowpass, and $h_\psi(n)$ must be highpass [151]. With the requirement of $h_\varphi(n)$ being lowpass, the relation between the scaling and wavelet coefficients in (4.335) leads to the frequency function $H_\varphi(e^{j\omega})$ of $h_\varphi(n)$ being translated by π in order to yield $H_\psi(e^{j\omega})$,

$$|H_\psi(e^{j\omega})| = |H_\varphi(e^{j(\omega+\pi)})|, \qquad (4.346)$$

and is thus a highpass filter.

Before the set of recursive equations can be used to produce $c_j(k)$ and $d_j(k)$, we must devise a technique for their *initialization*. It is, of course, necessary that $x(t)$ enters the calculations; this applies in particular to its sampled counterpart $x(n)$, invariably constituting the signal to be analyzed. For a fine enough scale j, one may argue that the scaling function has become so very narrow that the coefficients $c_j(k)$, which initialize the recursion, result from an inner product in which $x(t)$ is multiplied by a delta function,

$$c_j(k) \approx \int_{-\infty}^{\infty} x(t)\delta(t-k)dt = x(k). \qquad (4.347)$$

Consequently, the signal samples $x(n)$ themselves would serve as good approximations of the coefficients $c_j(k)$, provided that the signal $x(t)$ has been sampled well above the Nyquist rate. Although this initialization procedure is the one which is normally used, other procedures exist which offer certain advantages [150, 151].

Hence, the recursion is initialized with the sampled signal $x(n)$, whose length is finite and given by N. Due to the very dyadic nature of the algorithm, it is natural to assume that the length is a power of two, i.e., $N = 2^J$. Accordingly, $J+1$ different scales can be analyzed, of which the finest scale is $j = J$ and described by N coefficients (i.e., the signal itself), while the coarsest scale is $j = 0$ with only one coefficient.[12]

The calculation of the DWT through successive decomposition of the approximation coefficients is illustrated in Figure 4.44, where the finest resolution is given by the scale $j = 3$. The procedure is initialized by setting the approximation coefficients $c_3(k)$ equal to the signal samples $x(n)$. In this example where $x(n)$ has a length of $N = 8$, the DWT is given by the coefficients $c_0(0), d_0(0), d_1(0), d_1(1), d_2(0), d_2(1), d_2(2)$, and $d_2(3)$. Thus, the resulting number of coefficients is identical to the length of the signal.

Synthesis filter bank. While the analysis filter bank decomposes the signal into a set of coefficients at different resolution, the purpose here is to perform the reverse operation of merging the coefficient sequences so as to implement the inverse DWT. The inverse transform can also be implemented with a filter bank, but with a structure that differs slightly from the one which implements the DWT.

In order to derive a set of equations which recursively determine $c_{j+1}(k)$ from $c_j(k)$ and $d_j(k)$, we start by expressing the approximation signal $x_{j+1}(t)$ as a linear expansion of the scaling function at scale $j+1$,

$$x_{j+1}(t) = 2^{(j+1)/2} \sum_{n=-\infty}^{\infty} c_{j+1}(n)\varphi(2^{j+1}t - n). \qquad (4.348)$$

---

[12]The reader should be aware that other conventions exist for enumerating scales where an increase in j implies a coarser scale, rather than a finer one as is assumed in this presentation.

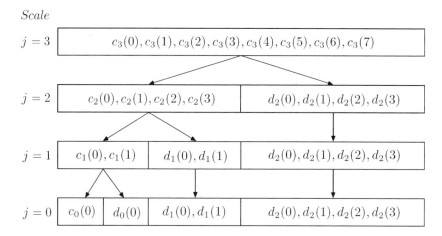

Figure 4.44: Calculation of the DWT for a signal of length $N = 8$. The final result is given by the coefficients at the bottom for $j = 0$. The vertical arrows indicate that the coefficients are simply copied down from the previous scale. The calculation is initialized by setting the coefficients $c_3(k)$ equal to the signal samples $x(k)$.

Relying on the decomposition in (4.327), stating that $\mathcal{V}_{j+1} = \mathcal{V}_j \oplus \mathcal{W}_j$, we can alternatively express $x_{j+1}(t)$ at scale j,

$$x_{j+1}(t) = 2^{j/2} \sum_{n=-\infty}^{\infty} c_j(n)\varphi(2^j t - n) + 2^{j/2} \sum_{n=-\infty}^{\infty} d_j(n)\psi(2^j t - n). \quad (4.349)$$

Now, making use of the scaling and wavelet equations in (4.323) and (4.334), respectively, we obtain

$$x_{j+1}(t) = \sum_{n=-\infty}^{\infty} c_j(n) \sum_{l=-\infty}^{\infty} h_\varphi(l) 2^{(j+1)/2} \varphi(2^{j+1} t - 2n - l)$$

$$+ \sum_{n=-\infty}^{\infty} d_j(n) \sum_{l=-\infty}^{\infty} h_\psi(l) 2^{(j+1)/2} \varphi(2^{j+1} t - 2n - l). \quad (4.350)$$

By multiplying both sides of (4.350) by $\varphi_{j+1,k}(t)$ and integrating to obtain the inner products, the following recursion is obtained for $c_{j+1}(k)$,

$$c_{j+1}(k) = \sum_{n=-\infty}^{\infty} c_j(n) h_\varphi(k - 2n) + \sum_{n=-\infty}^{\infty} d_j(n) h_\psi(k - 2n). \quad (4.351)$$

Alternatively, this equation can be expressed as

$$c_{j+1}(k) = c_j^u(k) * h_\varphi(k) + d_j^u(k) * h_\psi(k), \quad (4.352)$$

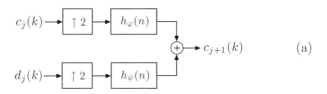

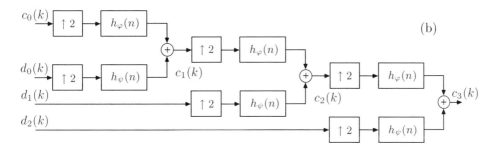

Figure 4.45: (a) A two-channel synthesis filter bank. (b) The inverse discrete wavelet transform based on the filter bank in (a) which, in this case, produces the coefficients of the space $\mathcal{V}_3$ based on $\mathcal{V}_0, \mathcal{W}_0, \mathcal{W}_1$, and $\mathcal{W}_2$.

where

$$c_j^u(k) = \begin{cases} c_j(k), & k \text{ even}; \\ 0, & k \text{ odd}, \end{cases} \tag{4.353}$$

and $d_j^u(k)$ is defined analogously to $c_j^u(k)$. Hence, the two sums in (4.351) can be interpreted in terms of upsampling by a factor of two, i.e., by inserting zeros as every other input sample, and filtering so that the calculation of the coefficients $c_{j+1}(k)$ is implemented by the *two-channel synthesis filter bank* shown in Figure 4.45(a). By repeatedly combining two-channel synthesis filter banks to merge signals at different resolutions, we obtain a dyadic tree structure which implements the inverse DWT, see Figure 4.45(b). The filters involved in the synthesis filter bank are the same as those used in the analysis filter bank, but with their impulse response reversed in time.

In practice, there is always a maximum scale J with a resolution so fine that the wavelet (detail) coefficients can be neglected. Therefore, the wavelet series expansion in (4.333) may be replaced by

$$x(t) = \sum_{n=-\infty}^{\infty} c_{j_0}(n)\varphi_{j_0,n}(t) + \sum_{j=j_0}^{J}\sum_{n=-\infty}^{\infty} d_j(n)\psi_{j,n}(t), \tag{4.354}$$

thus indicating the coefficients that must be calculated with (4.351) to obtain $x(t)$ (recall from (4.347) that $x(k) \approx c_j(k)$ for a sufficiently fine scale).

In certain situations, signal denoising can be achieved by selecting a lower maximum scale so that noise concentrated to the finest scales is removed; the same idea was actually introduced already on page 257, where the series expansion of basis functions was truncated for the purpose of improving the SNR.

Finally, we note that $x(t)$ can be expressed as a series expansion solely in terms of the wavelet functions when $c_{j_0}(n) \to -\infty$ for $j_0 \to -\infty$,

$$x(t) = \sum_{j=-\infty}^{J} \sum_{n=-\infty}^{\infty} d_j(n)\psi_{j,n}(t), \qquad (4.355)$$

which becomes the definition of the inverse DWT in (4.308), but with truncation of scales with negligible coefficients.

4.7.5   Wavelet Examples

The Haar wavelets offer the advantage of being very well-localized in time (compact support). However, these functions are discontinuous and, accordingly, introduce undesired high-frequency content. The *sinc scaling* function, defined by

$$\varphi(t) = \frac{\sin \pi t}{\pi t}, \qquad (4.356)$$

is, in a sense, dual to the Haar scaling function since its Fourier transform $\Phi(\Omega)$ is a box function, i.e., an ideal lowpass filter with a cut-off radian frequency at about π. Hence, the sinc scaling function is continuous (smooth) with excellent frequency resolution, but comes with poor time resolution due to its slow decay to zero. The orthonormality of the sinc scaling function to translations of itself is easily established by making use of Parseval's relation,

$$
\begin{aligned}
\int_{-\infty}^{\infty} \varphi(t-k)\varphi^*(t-n)dt &= \frac{1}{2\pi} \int_{-\infty}^{\infty} \Phi(\Omega)e^{-j\Omega k}\Phi^*(\Omega)e^{j\Omega n}d\Omega \\
&= \frac{1}{2\pi} \int_{-\pi}^{\pi} e^{-j\Omega(k-n)}d\Omega \\
&= \begin{cases} 1, & k = n; \\ 0, & k \neq n. \end{cases}
\end{aligned}
\qquad (4.357)
$$

The wavelet that belongs to the sinc scaling function is given by

$$\psi(t) = \frac{\sin \frac{\pi}{2}t}{\frac{\pi}{2}t} \cos \frac{3\pi}{2}t, \qquad (4.358)$$

whose Fourier transform is the ideal bandpass filter with lower and upper cut-off radian frequencies at about π and 2π, respectively. Again, Parseval's relation can be used to show that the sinc scaling and wavelet functions are orthogonal to each other.

While both the Haar and sinc wavelets serve as comprehensible examples, these wavelets are rarely used in practice due to the disadvantages of either poor frequency or time localization. Fortunately, methods are available for *designing* wavelets with compact support, while also offering good localization properties in frequency. Since a detailed presentation of such design methods is well beyond the scope of this text, we will only provide a sketch of how $h_\varphi(n)$ can be determined; the wavelet coefficients $h_\psi(n)$ are then obtained from (4.335).

For the refinement equation to be valid, it can be shown, by integrating both sides of (4.323), that $h_\varphi(n)$ must fulfil the requirements

$$\sum_{n=0}^{N_\varphi-1} h_\varphi(n) = \sqrt{2}, \tag{4.359}$$

and, due to the orthonormality of $\varphi(t)$ and its translates, we have

$$\sum_{n=0}^{N_\varphi-1} h_\varphi(n)h_\varphi(n-2k) = \delta(k), \tag{4.360}$$

where $\delta(k)$ denotes the delta function, and N_φ is the even-valued length of the filter $h_\varphi(n)$ [151]. For the case $N_\varphi = 2$, we obtain the following two equations,

$$h_\varphi(0) + h_\varphi(1) = \sqrt{2},$$
$$h_\varphi^2(0) + h_\varphi^2(1) = 1,$$

whose solution is

$$\begin{bmatrix} h_\varphi(0) & h_\varphi(1) \end{bmatrix} = \begin{bmatrix} \frac{1}{\sqrt{2}} & \frac{1}{\sqrt{2}} \end{bmatrix}. \tag{4.361}$$

These coefficients were encountered already in connection with the Haar functions. By increasing N_φ to 4, the following three equations determine the scaling coefficients

$$h_\varphi(0) + h_\varphi(1) + h_\varphi(2) + h_\varphi(3) = \sqrt{2},$$
$$h_\varphi^2(0) + h_\varphi^2(1) + h_\varphi^2(2) + h_\varphi^2(3) = 1,$$
$$h_\varphi(0)h_\varphi(2) + h_\varphi(1)h_\varphi(3) = 0.$$

These equations have several solutions of which one is given by

$$\begin{bmatrix} h_\varphi(0) & h_\varphi(1) & h_\varphi(2) & h_\varphi(3) \end{bmatrix} = \begin{bmatrix} \frac{1+\sqrt{3}}{4\sqrt{2}} & \frac{3+\sqrt{3}}{4\sqrt{2}} & \frac{3-\sqrt{3}}{4\sqrt{2}} & \frac{1-\sqrt{3}}{4\sqrt{2}} \end{bmatrix}. \tag{4.362}$$

For larger values of N_φ, numerical techniques are almost always needed to find the solution.

Although the scaling function $\varphi(t)$ is not explicitly required for calculation of the DWT, it is nevertheless important to assess whether its properties are suitable or not. One approach to calculating $\varphi(t)$ from $h_\varphi(n)$ is to insert the scaling coefficients into the refinement equation, but now modified into an iterative algorithm,

$$\varphi^{(i+1)}(t) = \sqrt{2} \sum_{n=0}^{N_\varphi-1} h_\varphi(n)\varphi^{(i)}(2t - n), \qquad (4.363)$$

where i denotes the iteration index. This algorithm, known as the *cascade algorithm*, produces successive approximations of $\varphi(t)$ so that $\varphi^{(i)}(t)$ approaches $\varphi(t)$ as the iteration index i increases. If the algorithm converges, the Fourier transform $\Phi(\Omega)$ of $\varphi(t)$ can be related to the scaling coefficients $h_\varphi(n)$ by iteratively applying the Fourier transform to (4.363),

$$\Phi(\Omega) = \Phi(0) \prod_{l=1}^{\infty} \frac{1}{\sqrt{2}} H_\varphi\left(e^{j\Omega/2^l}\right), \qquad (4.364)$$

where $H_\varphi(e^{j\Omega})$ denotes the discrete-time Fourier transform of $h_\varphi(n)$ and is a periodic function. Using the wavelet equation in (4.334), the Fourier transform $\Psi(\Omega)$ of the wavelet $\psi(t)$ can be expressed as

$$\Psi(\Omega) = \Phi(0) \frac{1}{\sqrt{2}} H_\psi(e^{j\Omega/2}) \prod_{l=2}^{\infty} \frac{1}{\sqrt{2}} H_\varphi(e^{j\Omega/2^l}). \qquad (4.365)$$

Since $\varphi(t)$ is assumed to have lowpass characteristics, the factor $\Phi(0)$ can be normalized such that

$$\Phi(0) = \int_{-\infty}^{\infty} \varphi(t)dt = 1. \qquad (4.366)$$

Hence, the outcome of the cascade algorithm in (4.363) depends only on the properties of the scaling coefficients and not on the shape of the initial $\varphi^{(0)}(t)$, except the factor $\Phi(0)$ which is invariant over the iterations.

To exemplify the calculation of $\varphi(t)$ when the cascade algorithm is employed, the four scaling coefficients in (4.362) are chosen. The algorithm converges to the scaling function, and related wavelet, shown in Figure 4.46(a). Both functions fulfil the requirement of orthonormality and have compact support. However, their shapes are relatively unsmooth and may therefore be less suitable for modeling and analysis of physiological signals. It is therefore desirable to introduce additional requirements which, as the degrees of freedom increase with N_φ, assure that the functions are smooth.

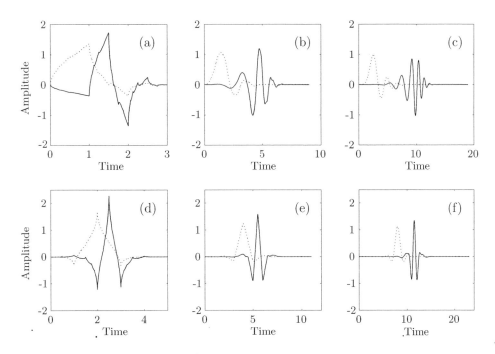

Figure 4.46: The scaling function (dotted line) and wavelet function (solid line) for (a) Daubechies–2, (b) Daubechies–5, (c) Daubechies–10, (d) Coiflet–1, (e) Coiflet–2, and (f) Coiflet–4. Note that the timescale differs between the diagrams.

A useful approach to designing smooth wavelets is to require that their moments $\overline{m}_k$, defined by

$$\overline{m}_k = \int_{-\infty}^{\infty} t^k \psi(t)dt, \qquad (4.367)$$

vanish up to a certain value $k = K_\psi$. Alternatively, the moment definition in (4.367) can be expressed in terms of its Fourier transform

$$\overline{m}_k = (-\jmath)^{-k} \left. \frac{\partial^k \Psi(\Omega)}{\partial \Omega^k} \right|_{\Omega=0}, \qquad (4.368)$$

which establishes that vanishing moments are synonymous with K_ψ derivatives of $\Psi(\Omega)$ at DC, i.e., $\Omega = 0$, being equal to zero. This requirement implies that $\psi(t)$ is smooth and may, if desired, be extended to embrace $\varphi(t)$ as well. Another consequence of vanishing wavelet moments is that the inner product between a polynomial signal $x(t) = \sum_k a_k t^k$ and $\psi(t)$ is zero, and thus the detail coefficients are zero. As a result, polynomial signals are well-represented by the approximation coefficients, and the detail coefficients can be discarded.

The expressions given for $\Phi(\Omega)$ and $\Psi(\Omega)$ in (4.364) and (4.365), respectively, suggest that smooth wavelets are directly connected to the behavior of the filters $h_\varphi(n)$ and $h_\psi(n)$: if $h_\varphi(n)$ is lowpass, then $\psi(t)$ will be smooth. Therefore, another approach to designing smooth wavelets is the one which requires the moments $\overline{\mu}_k$ of the wavelet coefficients,

$$\overline{\mu}_k = \sum_{n=0}^{N_\psi - 1} n^k h_\psi(n), \qquad (4.369)$$

to vanish.

The *Daubechies wavelets* are a family of wavelets designed so that the maximum number of moments $\overline{\mu}_k$ is equal to zero, which is $K_\psi = N_\psi/2$ moments [157]. The wavelet corresponding to $K_\psi = 2$ is actually one we have already studied in Figure 4.46(a), whose scaling coefficients were given in (4.362). As K_ψ increases, both the wavelet function and the scaling function become increasingly smooth, as illustrated in Figures 4.46(b) and (c), where the cases $K_\psi = 5$ and 10 are presented. A disadvantage of the members of this family is their highly asymmetric shape.

The *Coiflets* constitute another wavelet family with compact support, but designed such that $N_\psi/3 - 1$ moments of the scaling function and $K_\psi = N_\psi/3$ of the wavelet vanish. Figures 4.46(d)–(f) show Coiflets with $K_\psi = 1$, 2, and 4 vanishing moments, respectively. The Coiflets are more symmetric than are the Daubechies wavelets, a property that comes at the price of an increased filter length. Multiresolution analysis using Coiflets is exemplified in Figure 4.47; it is obvious that Coiflets are superior in producing smooth approximations when compared to the results of the Haar functions in Figure 4.41.

There are many other types of wavelets available in addition to those mentioned here, with each exhibiting its particular advantages. In biomedical signal processing, it is often desirable to have symmetric wavelets. However, scaling functions and wavelets cannot, in general, accommodate this property since they are required to be orthogonal (exceptions are the Haar and the sinc wavelets). By softening the orthogonality requirement between analysis and synthesis filters to, what is called, bi-orthogonality [150], it is possible to design symmetric wavelets which still implement the DWT and its inverse.

In EP analysis, work has been undertaken to design wavelets matched specifically to the shape of the expected waveforms [149]. While such a design approach may be highly appropriate in certain situations, it does not always guarantee a successful outcome of the multiresolution signal analysis. In fact, promising results have been achieved in EP analysis using a wide variety of wavelets, ranging from smooth to discontinuous, see, e.g., [158–162].

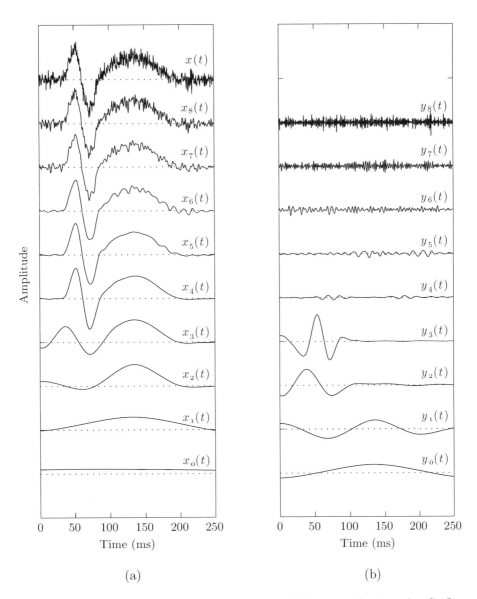

Amplitude

Time (ms)

(a)

(b)

Figure 4.47: Multiresolution analysis of an evoked potential using the Coiflet-4. (a) The approximation signals $x_i(t)$ and (b) the detail signals $y_i(t)$ at different scales; the original signal $x(t)$ is shown at the top left of the figure.

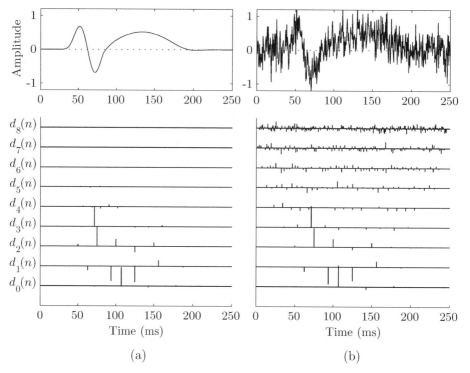

Figure 4.48: (a) An evoked potential and the related DWT using the Daubechies–4 wavelet. (b) The same signal as in (a), but with noise added. Note that the noise is concentrated in the detail coefficients of the finest scales.

4.7.6   Denoising

The detail coefficients of a noisy signal are often such that the coefficients of the signal are confined to coarser scales, while those of the noise are observed in finer scales. The separation of coefficients is illustrated by the example in Figure 4.48, where the detail coefficients $d_j(k)$ are shown for both a clean and a noisy signal. *Denoising* can be viewed as a nonlinear filtering operation in which the pattern of detail coefficients is exploited in order to produce a smoother signal. This operation involves three main steps, namely, calculating the DWT for the noisy signal, zeroing or modifying certain coefficients by a suitable rule, and reconstructing the signal from the modified coefficients.

A straightforward way to implement denoising is to simply set *all* coefficients equal to zero for scales with index larger (finer) than J_T,

$$\check{d}_j(k) = \begin{cases} d_j(k), & j \le J_T; \\ 0, & j > J_T, \end{cases} \tag{4.370}$$

for all appropriate values of k. The denoised signal results from calculating the inverse DWT using the modified detail coefficients $\check{d}_j(k)$ and the coarse coefficients; the denoising operation is equivalent to using the wavelet series expansion in (4.354) with $J = J_T$.

The truncation defined by (4.370) is, of course, a kindred spirit to the truncation earlier introduced in the context of basis function analysis, defined by (4.206)—both types of truncation are bound to produce smooth signal estimates. Truncation of the Fourier series expansion naturally excludes higher frequencies, while truncation of the Karhunen–Loève expansion in (4.223) usually leads to the signal energy being concentrated to a few (smooth) basis functions. While the Karhunen–Loève approach is optimal in the MSE sense for a specific ensemble of signals, it lacks some of the DWT's flexibility: in its capacity as a two-indices series expansion, the DWT offers good resolution in both time and frequency.

It was shown in Section 4.5.4 that a truncated series expansion of basis functions can be interpreted in terms of linear, time-variant filtering. A limitation on linear techniques, however, is that for noisy signals with fast changes (jumps), noise reduction can only be achieved at the price of considerable smoothing of the fast changes. On the other hand, the detail coefficients of the DWT can be subjected to nonlinear processing so that denoising is achieved without having to sacrifice too much of the fast changes in the signal.

Two common, nonlinear techniques remove coefficients of the DWT below a certain threshold. The inverse DWT of the thresholded coefficients is then performed to produce a denoised signal. To proceed, we introduce the wavelet transform vector $\mathbf{w}_N$ containing both the approximation and detail coefficients (i.e., the vector given by the bottom row in Figure 4.44):

$$
\begin{aligned}
\mathbf{w}_N &= \begin{bmatrix} w_1 & w_2 & \cdots & w_N \end{bmatrix}^T \\
&= \begin{bmatrix} c_0(0) & d_0(0) & d_1(0) & d_1(1) & d_2(0) & \cdots & d_{(\log_2 N - 1)}(\log_2 N) \end{bmatrix}^T,
\end{aligned}
\tag{4.371}
$$

where the finest details are described by the elements with the highest indices of $\mathbf{w}_N$, and so on. Denoising by *hard thresholding* is defined by

$$
\check{w}_i = \begin{cases} w_i, & |w_i| \geq \eta_T; \\ 0, & |w_i| < \eta_T, \end{cases}
\tag{4.372}
$$

where η_T is a threshold.

Denoising by *soft thresholding* is performed by thresholding the coefficients and shrinking them by the same amount as the threshold η_T [163],

$$
\check{w}_i = \begin{cases} \operatorname{sign}(w_i)(|w_i| - \eta_T), & |w_i| \geq \eta_T; \\ 0, & |w_i| < \eta_T. \end{cases}
\tag{4.373}
$$

The threshold η_T may be chosen as fixed, with a value based on some prior information that may exist on the signal. When a model of the signal and noise is available, one can also determine the particular η_T that produces the "best agreement" between the original and denoised signal. Best agreement is usually identical to the MSE criterion, computed for different values of η_T until the lowest MSE is obtained. The MSE as a performance measure should, however, be used with caution in biomedical signal processing since important physiological information may be lost or distorted, although a very low MSE has been achieved. Different aspects of signal distortion and performance measures are further considered in Section 7.6.7 in the context of ECG data compression.

Another approach is to relate η_T to the dispersion of the coefficients of the DWT vector $\mathbf{w}_N$. One such threshold, derived under the assumption that the noise is white with variance σ_v^2, is given by [152]

$$\eta_T = \sigma_v \sqrt{2 \ln N}, \tag{4.374}$$

where the factor $\sqrt{2 \ln N}$ is the expected maximum value of a white noise sequence of length N and unit standard deviation. Since σ_v is unknown in practice, it is often estimated using the median of the absolute deviation,

$$\hat{\sigma}_v = 1.483 \cdot \text{median}(|d_{J-1}(0)|, \dots, |d_{J-1}(N/2)|), \tag{4.375}$$

which avoids the influence of outlier values. The factor 1.483 is introduced to calibrate the median estimator with the standard deviation of a Gaussian PDF. Although η_T is applied to all scales, the estimation of σ_v in (4.375) typically involves only the coefficients of the finest scale $J - 1$, since this scale is the least influenced by the signal.

Denoising techniques can be made more sophisticated by introducing thresholds which are scale-dependent [164]. Moreover, the thresholds can be made dependent on time within each scale so that coefficients are thresholded in certain time intervals, while consistently set to zero in others; this technique is sometimes referred to as *time windowing*. Such windowing allows us to introduce information in the denoising process where certain signal components are more likely to occur in time. Figure 4.49 presents a number of single-trial EPs, acquired during visual stimulation with a checkerboard light pattern, and their denoised counterparts. Denoising is implemented using scale-dependent time intervals, whose locations are determined from the properties of the wavelet coefficients of the ensemble average [165, 166]. This procedure is motivated by the fact that EPs are time-locked to stimuli so that certain waveform components are expected to occur in certain time intervals. Rather than relating the time intervals to the properties of the ensemble average, scale-dependent time windowing can instead be related

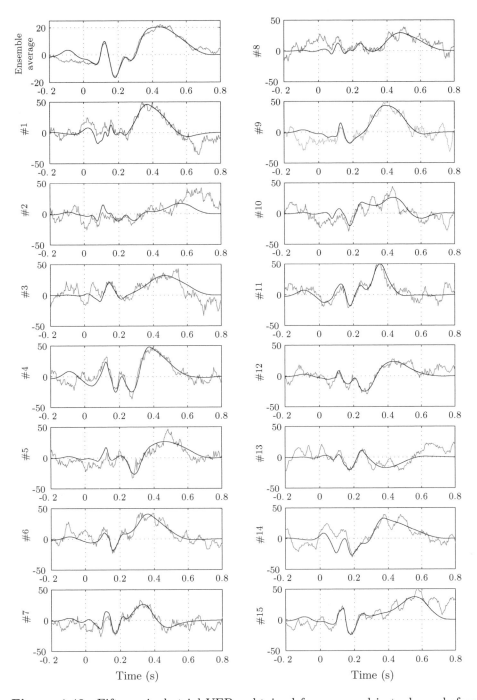

Figure 4.49: Fifteen single-trial VEPs, obtained from one subject, shown before (light line) and after (dark line) denoising. The ensemble averages, obtained without and with denoising, are shown in the top left diagram. (Reprinted from Quian Quiroga [165] with permission.)

to the ensemble correlation function, i.e., the correlation existing within the ensemble of signals, see Section 4.3.8 [167].

Signal denoising is usually employed for the purpose of producing accurate measurements of latency and amplitude and is of particular value when analyzing single-trial EPs. Denoising may also be helpful in uncovering signal patterns that are difficult to perceive directly from single-trial EPs, but which nevertheless are consistent within a certain group of subjects. This latter aspect of denoising has been explored in a study where VEPs were acquired from normal subjects during pattern reversal and from elderly patients suffering from dementia, see Figure 4.50(a) [168]. For the normal subjects, one would expect the VEP to exhibit peaks at latencies of 70, 100, and 130 ms (denoted N70–P100–N130), whereas these latencies are not expected for the pathological group. To check whether a normal latency pattern was present or not, the signals were subjected to multiresolution analysis using a so-called quadratic B-spline wavelet, having compact support, see Figure 4.51.

Each VEP was acquired at a sampling rate of 1 kHz during the 512 ms the pattern was presented. The VEP obtained from ensemble averaging of 60 responses was then decomposed into detail subspaces $\mathcal{W}_j$ which covered the frequency bands 250–500 Hz ($\mathcal{W}_8$), 125–250 Hz ($\mathcal{W}_7$), 62.5–125 Hz ($\mathcal{W}_6$), 31.3–62.5 Hz ($\mathcal{W}_5$), 15.6–31.3 Hz ($\mathcal{W}_4$), and 7.8–15.6 Hz ($\mathcal{W}_3$) and the approximation subspace which covered 0–7.8 Hz ($\mathcal{V}_3$). With this decomposition, the frequency bands related to $\mathcal{W}_3$ and $\mathcal{V}_3$ roughly correspond to the alpha and delta–theta activities of the EEG, respectively.

Following truncation of all the detail scales $\mathcal{W}_3$–$\mathcal{W}_8$, only the approximation scale $\mathcal{V}_3$ was used for reconstructing the signal, and, therefore, a very smooth signal resulted from denoising. Figure 4.50(b) presents the results of all normal subjects whose VEPs are superimposed in one diagram, while those of the patients are superimposed in another. It is striking that the normal VEPs have a strong phase coherence in the region of N70–P100–N130, whereas this phase coherence is absent for the pathological VEPs.

Another interesting prospect of signal denoising is its use in combination with the Woody method for latency correction, a technique which was described in Section 4.3.7. By denoising each EP of the ensemble prior to the estimation of latency shifts, improved alignment performance of the Woody method has been observed in the sense that the latency-corrected waveform is more reliable thanks to the improved SNR [166], see also [169]. It is also possible to modify the alignment method as such by replacing the filtering operation in (4.126) by a set of filters which operates at different scales [170]. In this case, the latency estimate is obtained from the combined output of the filters.

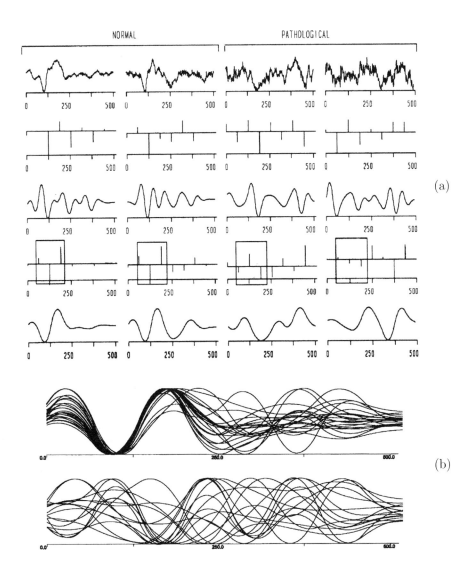

Figure 4.50: (a) Visual evoked potentials from two normal subjects and two patients with dementia, the coefficients of $\mathcal{W}_3$ and the reconstructed waveform, and the coefficients of $\mathcal{V}_3$ and the reconstructed waveform are shown from top to bottom. (b) Waveforms reconstructed from $\mathcal{V}_3$ and superimposed for 24 normal subjects (upper panel) and for 16 patients with dementia (lower panel). (Reprinted from Ademoglu et al. [168] with permission.)

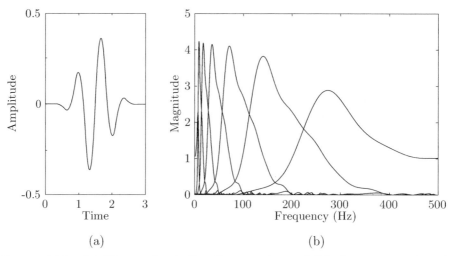

Figure 4.51: (a) The quadratic B-spline wavelet and (b) the corresponding frequency bands which describe the detail subspaces $\mathcal{W}_8, \ldots, \mathcal{W}_3$, assuming that the signal is sampled at a rate of 1 kHz [168].

We conclude this section by reiterating the fact that wavelet analysis has been found very useful, not only for signal denoising but for analysis and characterization of EPs [171–180]. This also applies to many other types of biomedical signals.

Bibliography

[1] G. D. Dawson, "A summation technique for detecting small signals in a large irregular background," *J. Physiol. (London)*, vol. 115, p. 2, 1951.

[2] G. D. Dawson, "A summation technique for the detection of small evoked potentials," *Electroencephal. Clin. Neurophysiol.*, vol. 6, pp. 65–84, 1954.

[3] R. J. Chabot and E. R. John, "Normative evoked potential data," in *Handbook of Electroencephalography and Clinical Electrophysiology: Clinical Applications of Computer Analysis of EEG and Other Neurophysiological Signals* (F. H. Lopes da Silva, W. Storm van Leeuwen, and A. Rémond, eds.), ch. 1, pp. 263–309, Amsterdam/New York: Elsevier, 1986.

[4] T. W. Picton, O. G. Lins, and M. Scherg, "The recording and analysis of event-related potentials," in *Handbook of Neurophysiology, Vol. 10* (F. Boller and J. Grafman, eds.), pp. 3–73, Baltimore: Elsevier Science, 1995.

[5] T. Lagerlund, "EEG source localization. Model-dependent and model-independent methods," in *Electroencephalography. Basic Principles, Clinical Applications and Related Fields* (E. Niedermayer and F. Lopes da Silva, eds.), pp. 809–822, Baltimore: Williams & Wilkins, 1999.

[6] G. G. Celesia and M. G. Brigell, "Auditory evoked potentials," in *Electroencephalography. Basic Principles, Clinical Applications and Related Fields* (E. Niedermayer and F. Lopes da Silva, eds.), pp. 994–1013, Baltimore: Williams & Wilkins, 1999.

[7] C. Thornton, "Evoked potentials in anaesthesia," *Eur. J. Anaesth.*, vol. 8, pp. 89–107, 1991.

[8] C. Thornton, M. Barrowcliffe, K. Konieczko, P. Ventham, C. Doré, D. Newton, and J. Jones, "The auditory evoked response as an indicator of awareness," *British J. Anaesth.*, vol. 63, pp. 113–115, 1989.

[9] American Electroencephalographic Society, "Guidelines in EEG and evoked potentials," *J. Clin. Neurophys.*, vol. 3 (suppl. 1), pp. 12–17, 1986.

[10] J. R. Boston, "Spectra of auditory brainstem responses and spontaneous EEG," *IEEE Trans. Biomed. Eng.*, vol. 28, pp. 334–341, 1981.

[11] R. A. Levine, "Auditory evoked potentials: Engineering considerations," in *Proc. Conf. IEEE Eng. Med. Biol. Soc. (EMBS)*, pp. 85–89, IEEE, 1982.

[12] R. J. Sclabassi, H. A. Risch, C. L. Hinman, J. S. Kroin, N. F. Enns, and N. S. Namerow, "Complex pattern evoked somatosensory responses in the study of multiple sclerosis," *Proc. IEEE*, vol. 65, pp. 626–633, 1977.

[13] G. G. Celesia and N. S. Peachey, "Visual evoked potentials and electroretinograms," in *Electroencephalography. Basic Principles, Clinical Applications and Related Fields* (E. Niedermayer and F. Lopes da Silva, eds.), pp. 968–993, Baltimore: Williams & Wilkins, 1999.

[14] A. S. Gevins and B. A. Cutillo, "Signals of cognition," in *Handbook of Electroencephalography and Clinical Electrophysiology: Clinical Applications of Computer Analysis of EEG and Other Neurophysiological Signals* (F. H. Lopes da Silva, W. Storm van Leeuwen, and A. Rémond, eds.), ch. 11, pp. 335–381, Amsterdam/New York: Elsevier, 1986.

[15] M. Kutas and A. Dale, "Electric and magnetic readings of mental functions," in *Cognitive Neuroscience* (M. D. Rugg, ed.), pp. 197–242, Cambridge, Mass.: Psychology Press, 1997.

[16] J. Polich, "P300 in clinical applications," in *Electroencephalography. Basic Principles, Clinical Applications and Related Fields* (E. Niedermayer and F. Lopes da Silva, eds.), pp. 1073–1091, Baltimore: Williams & Wilkins, 1999.

[17] T. W. Picton, S. Bentin, P. Berg, E. Donchin, S. A. Hillyard, R. Johnson Jr., G. A. Miller, W. Ritter, D. S. Ruchkin, M. D. Rugg, and M. J. Taylor, "Guidelines for using human event-related potentials to study cognition: Recording standards and publication criteria," *Psychophysiology*, vol. 37, pp. 127–152, 2000.

[18] M. Kutas and S. A. Hillyard, "Reading senseless sentences: Brain potentials reflect semantic incongruity," *Science*, vol. 207, pp. 203–205, 1980.

[19] I. Fischler, P. A. Bloom, D. G. Childers, S. E. Roucos, and N. W. Perry Jr., "Brain potentials related to stages of sentence verification," *Psychophysiology*, vol. 20, pp. 400–409, 1983.

[20] M. Kutas and K. D. Federmaier, "Electrophysiology reveals semantic memory use in language comprehension," *Trends in Cognitive Sciences*, vol. 4, pp. 463–470, 2000.

[21] K. D. Federmaier and M. Kutas, "A rose by any other name: Long-term memory structure and sentence processing," *J. Mem. Lang.*, vol. 41, pp. 469–495, 1999.

[22] D. G. Wastell, "The application of low-pass linear filters to evoked potential data: Filtering without phase distortion," *Electroencephal. Clin. Neurophysiol.*, vol. 46, pp. 355–356, 1979.

[23] J. R. Boston and P. J. Ainslie, "Effects of analog and digital filtering on brainstem auditory evoked potentials," *Electroencephal. Clin. Neurophysiol.*, vol. 48, pp. 361–364, 1980.

[24] O. Svensson, B. Almqvist, and K. E. Jönsson, "Effects of low-frequency components and analog filtering on auditory brainstem responses," *Scand. Audiol.*, vol. 16, pp. 43–47, 1987.

[25] R. Verleger, T. Gasser, and J. Möcks, "Correction of EOG artifacts in event-related potentials of the EEG: Aspects of reliability and validity," *Psychophysiology*, vol. 19, pp. 472–480, 1982.

[26] G. Gratton, M. G. H. Coles, and E. Donchin, "A new method for off-line removal of ocular artifact," *Electroencephal. Clin. Neurophysiol.*, vol. 55, pp. 468–484, 1983.

[27] H. V. Semlitsch, P. Anderer, P. Schuster, and O. Presslich, "A solution for reliable and valid reduction of ocular artifacts, applied to the P300 ERP," *Psychophysiology*, vol. 23, pp. 695–703, 1986.

[28] G. Gratton, "Dealing with artifacts: The EOG contamination of the event-related brain potential," *Behav. Res. Meth. Instrum. Comput.*, vol. 30, pp. 44–53, 1998.

[29] J. R. Wolpaw and C. C. Wood, "Scalp distribution of human auditory evoked potentials. I. Evaluation of reference electrode site," *Electroencephal. Clin. Neurophysiol.*, vol. 54, pp. 15–24, 1982.

[30] D. S. Ruchkin, "An analysis of average response computations based upon aperiodic stimuli," *IEEE Trans. Biomed. Eng.*, vol. 12, pp. 87–94, 1965.

[31] E. M. Glaser and D. S. Ruchkin, *Principles of Neurobiological Signal Analysis*. San Diego: Academic Press, 1976.

[32] S. Makeig, S. Debener, J. Onton, and A. Delorme, "Mining event-related brain dynamics," *Trends in Cognitive Sciences*, vol. 8, pp. 204–210, 2004.

[33] B. McA. Sayers, H. A. Beagley, and W. R. Henshall, "The mechanism of auditory evoked EEG responses," *Nature*, vol. 247, pp. 481–483, 1974.

[34] B. McA. Sayers and H. A. Beagley, "Objective evaluation of auditory evoked EEG responses," *Nature*, vol. 251, pp. 608–609, 1974.

[35] B. W. Jervis, M. J. Nichols, T. E. Johnson, E. Allen, and N. R. Hudson, "A fundamental investigation of the composition of auditory evoked potentials," *IEEE Trans. Biomed. Eng.*, vol. 30, pp. 43–50, 1983.

[36] P. C. M. Vijn, B. W. Van Dijk, and H. Spekreije, "Visual stimulation reduces EEG activity in man," *Brain Res.*, vol. 550, pp. 49–53, 1991.

[37] D. G. Childers, T. C. Doyle, A. G. Brinck, and N. W. Perry Jr., "Ensemble characteristics of human visual evoked response: Periodic and random stimulation," *IEEE Trans. Biomed. Eng.*, vol. 19, pp. 408–415, 1972.

[38] E. R. John, D. S. Ruchkin, and J. V. Vidal, "Measurement of event-related potentials," in *Event-Related Brain Potentials in Man* (E. Calloway, P. Tueting, and H. Koslow, eds.), pp. 93–138, San Diego: Academic Press, 1978.

[39] T. Gasser, J. Möcks, and W. Köhler, "Amplitude probability distribution of noise for flash-evoked potentials and robust response estimates," *IEEE Trans. Biomed. Eng.*, vol. 33, pp. 579–584, 1986.

[40] K. Lowy and B. Weiss, "Assessing the significance of averaged evoked potentials with an on-line computer: The single-sweep method," *Electroencephal. Clin. Neurophysiol.*, vol. 25, pp. 177–180, 1968.

[41] H. Schimmel, "The plus/minus reference: Accuracy of estimated mean components in average evoked response studies," *Science*, vol. 157, pp. 92–94, 1967.

[42] R. Coppola, R. Tabor, and M. Buschsbaum, "Signal to noise ratio and response variability measurements in single trial evoked potentials," *Electroencephal. Clin. Neurophysiol.*, vol. 44, pp. 212–222, 1978.

[43] N. J. Bershad and A. J. Rockmore, "On estimating the signal-to-noise ratio using the sample correlation coefficient," *IEEE Trans. Inform. Theory*, vol. 20, pp. 112–113, 1974.

[44] P. K. H. Wong and R. G. Bickford, "Brainstem auditory evoked potentials: The use of noise estimate," *Electroencephal. Clin. Neurophysiol.*, vol. 50, pp. 25–34, 1980.

[45] T. W. Picton, R. F. Hink, M. Perez-Abalo, R. D. Linden, and A. S. Wiens, "Evoked potentials: How now?," *J. Electrophysiol. Techn.*, vol. 10, pp. 177–221, 1984.

[46] S. K. Mitra, *Digital Signal Processing: A Computer-based Approach.* New York: McGraw–Hill, 1998.

[47] M. Hoke, B. Ross, R. Wickesberg, and B. Lütkenhöner, "Weighted averaging: Theory and application to electric response audiometry," *Electroencephal. Clin. Neurophysiol.*, vol. 57, pp. 579–584, 1984.

[48] M. Don, C. Elberling, and M. Waring, "Objective detection of averaged auditory brainstem response," *Scand. Audiol.*, vol. 13, pp. 219–228, 1984.

[49] C. Elberling and M. Don, "Quality estimation of averaged auditory brainstem response," *Scand. Audiol.*, vol. 13, pp. 187–197, 1984.

[50] C. Elberling and O. Wahlgreen, "Estimation of auditory brainstem response, ABR, by means of Bayesian inference," *Scand. Audiol.*, vol. 14, pp. 89–96, 1985.

[51] G. H. Golub and C. F. van Loan, *Matrix Computations.* Baltimore: The Johns Hopkins University Press, 2nd ed., 1989.

[52] B. Lütkenhöner and C. Pantev, "Possibilities and limitations of weighted averaging," *Biol. Cybern.*, vol. 52, pp. 409–416, 1985.

[53] E. Bataillou, E. Thierry, H. Rix, and O. Meste, "Weighted averaging using adaptive estimation of the weights," *Signal Proc.*, vol. 44, pp. 51–66, 1995.

[54] C. Davila and M. Mobin, "Weighted averaging of evoked potentials," *IEEE Trans. Biomed. Eng.*, vol. 39, pp. 338–345, 1992.

[55] C. W. Therrien, *Discrete Random Signals and Statistical Signal Processing.* New Jersey: Prentice-Hall, 1992.

[56] S. M. Kay, *Fundamentals of Statistical Signal Processing. Estimation Theory.* New Jersey: Prentice-Hall, 1993.

[57] R. P. Borda and J. D. Frost, "Error reduction in small sample averaging through the use of the median rather than the mean," *Electroencephal. Clin. Neurophysiol.*, vol. 25, pp. 391–392, 1968.

[58] D. S. Ruchkin, "Comparison of statistical errors of the median and average evoked responses," *IEEE Trans. Biomed. Eng.*, vol. 21, pp. 54–56, 1974.

[59] J. L. Rosenberger and M. Gasko, "Comparing location estimators: trimmed means, medians and trimeans," in *Understanding Robust and Exploratory Data Analysis*, ch. 10, pp. 297–338, New York: Wiley–Interscience, 1983.

[60] R. V. Hogg, "Adaptive robust procedures," *J. Am. Stat. Assoc.*, vol. 69, pp. 909–927, 1974.

[61] D. O. Walter, "Two approximations to the median evoked response," *Electroencephal. Clin. Neurophysiol.*, vol. 30, pp. 246–247, 1971.

[62] A. Papoulis, *Probability, Random Variables, and Stochastic Processes.* New York: McGraw–Hill, 3rd ed., 1991.

[63] O. Meste and H. Rix, "Jitter statistics estimation in alignment processes," *Signal Proc.*, vol. 51, pp. 41–53, 1996.

[64] C. D. McGillem, J. I. Auñon, and C. A. Pomalaza, "Improved waveform estimation procedures for event-related potentials," *IEEE Trans. Biomed. Eng.*, vol. 32, pp. 371–379, 1985.

[65] S. Senmoto and D. G. Childers, "Adaptive decomposition of a composite signal of identical wavelets in noise," *IEEE Trans. Systems Man Cybernetics*, vol. 2, pp. 59–66, 1972.

[66] K. C. McGill and L. J. Dorfman, "High-resolution alignment of sampled waveforms," *IEEE Trans. Biomed. Eng.*, vol. 31, pp. 462–468, 1984.

[67] J. Möcks, W. Köhler, T. Gasser, and D. T. Pham, "Novel approaches to the problem of latency jitter," *Psychophysiology*, vol. 25, pp. 217–226, 1988.

[68] X. Kong and N. V. Thakor, "Adaptive estimation of latency changes in evoked potentials," *IEEE Trans. Biomed. Eng.*, vol. 43, pp. 189–197, 1996.

[69] X. Kong and T. Qiu, "Adaptive estimation of latency change in evoked potentials by direct least mean p-norm time delay estimation," *IEEE Trans. Biomed. Eng.*, vol. 46, pp. 994–1003, 1999.

[70] X. Kong and T. Qiu, "Latency change estimation for evoked potentials via frequency selective adaptive phase spectrum analyzer," *IEEE Trans. Biomed. Eng.*, vol. 46, pp. 1004–1012, 1999.

[71] P. Wahlberg and G. Salomonsson, "Methods for alignment of multi-class signal sets," *Signal Proc.*, vol. 83, pp. 983–1000, 2003.

[72] C. D. Woody, "Characterization of an adaptive filter for the analysis of variable latency neuroelectric signals," *Med. Biol. Eng. & Comput.*, vol. 5, pp. 539–553, 1967.

[73] E. K. Harris and C. D. Woody, "Use of an adaptive filter to characterize signal-noise relationships," *Comput. Biomed. Res.*, vol. 2, pp. 242–273, 1969.

[74] W. Muhammad, O. Meste, and H. Rix, "Comparison of single and multiple time delay estimators: Application to muscle fiber conduction velocity estimation," *Signal Proc.*, vol. 82, pp. 925–940, 2002.

[75] H. L. van Trees, *Detection, Estimation and Modulation Theory. Part I.* New York: J. Wiley & Sons, 1968.

[76] R. A. Christensen and A. D. Hirschman, "Automatic phase alignment for the Karhunen-Loève expansion," *IEEE Trans. Biomed. Eng.*, vol. 26, pp. 94–99, 1979.

[77] C. D. Woody and M. J. Nahvi, "Application of optimum linear filter theory to the detection of cortical signals preceding facial movements in cat," *Exp. Brain Res.*, vol. 16, pp. 455–465, 1973.

[78] D. G. Wastell, "Statistical detection of individual evoked responses: An evaluation of Woody's adaptive filter," *Electroencephal. Clin. Neurophysiol.*, vol. 42, pp. 835–839, 1977.

[79] J. Auñon and R. W. Sencaj, "Comparison of different techniques for processing evoked potentials," *Med. Biol. Eng. & Comput.*, vol. 16, pp. 642–650, 1978.

[80] G. H. Steeger, O. Hermann, and M. Spreng, "Some improvements in the measurement of variable latency acoustically evoked potentials in human EEG," *IEEE Trans. Biomed. Eng.*, vol. 30, pp. 295–303, 1983.

[81] D. Ruchkin and C. Wood, "The measurement of event-related potentials," in *Human Event-Related Potentials* (T. W. Picton, ed.), Amsterdam: Elsevier, 1985.

[82] A. Puce, S. F. Berkovic, P. J. Cadusch, and P. F. Bladin, "P3 latency jitter assessed using two techniques. I. Simulated data and surface recordings in normal subjects," *Electroencephal. Clin. Neurophysiol.*, vol. 92, pp. 352–364, 1994.

[83] C. D. McGillem and J. I. Auñon, "Measurements of signal components in single visually evoked brain potentials," *IEEE Trans. Biomed. Eng.*, vol. 24, pp. 232–241, 1977.

[84] C. D. McGillem, J. I. Auñon, and K. B. Yu, "Signals and noise in evoked brain potentials," *IEEE Trans. Biomed. Eng.*, vol. 32, pp. 1012–1016, 1985.

[85] X. Yu, Y. Zhang, and Z. He, "Peak component latency-corrected average method for evoked potential waveform estimation," *IEEE Trans. Biomed. Eng.*, vol. 41, pp. 1072–1082, 1994.

[86] L. Gupta, D. L. Molfese, R. Tammana, and P. G. Simos, "Nonlinear alignment and averaging for estimating the evoked potential," *IEEE Trans. Biomed. Eng.*, vol. 43, pp. 348–356, 1996.

[87] S. Casarotto, A. M. Bianchi, S. Cerutti, and G. A. Chiarenza, "Dynamic time warping in the analysis of event-related potentials," *IEEE Eng. Med. Biol. Mag.*, vol. 24, pp. 68–77, 2005.

[88] P. Laguna and L. Sörnmo, "Sampling rate and the estimation of ensemble variability for repetitive signals," *Med. Biol. Eng. & Comput.*, vol. 38, pp. 540–546, 2000.

[89] R. Atarius and L. Sörnmo, "Cardiac late potentials and signal-to-noise ratio enhancement by ensemble correlation," *IEEE Trans. Biomed. Eng.*, vol. 42, pp. 1132–1137, 1995.

[90] R. Atarius and L. Sörnmo, "Detection of cardiac late potentials in nonstationary noise," *Med. Eng. & Physics*, vol. 19, pp. 291–298, 1997.

[91] M. Hayes, *Statistical Digital Signal Proccessing and Modeling*. New York: John Wiley & Sons, 1996.

[92] D. O. Walter, "A posteriori 'Wiener filtering' of averaged evoked responses," *IEEE Trans. Biomed. Eng.*, vol. 42, pp. 1132–1137, 1975.

[93] D. J. Doyle, "Some comments on the use of Wiener filtering for the estimation of evoked potentials," *Electroencephal. Clin. Neurophysiol.*, vol. 38, pp. 533–534, 1975.

[94] D. O. Walter, "A posteriori Wiener filtering of average evoked responses," *Electroencephal. Clin. Neurophysiol.*, vol. 27, pp. 61–70, 1969.

[95] T. Nogawa, K. Katayama, Y. Tabata, T. Kawahara, and T. Oshio, "Visual evoked potentials estimated by Wiener filtering," *IEEE Trans. Biomed. Eng.*, vol. 20, pp. 375–758, 1973.

[96] E. Başar, A. Gönder, Ç. Özesmi, and P. Ungan, "Dynamics of brain rhythmic and evoked potentials: Some computational methods for the analysis of electrical signals from the brain," *Biol. Cybern.*, vol. 20, pp. 137–143, 1975.

[97] V. Albrecht and T. Radil-Weiss, "Some comments on the derivation of the Wiener filter for average evoked potentials," *Biol. Cybern.*, vol. 24, pp. 43–46, 1976.

[98] J. P. C. de Weerd, G. J. H. Uijen, P. I. M. Johannesma, and W. L. J. Martens, "Estimation of signal and noise spectra by special averaging techniques with application to a posteriori Wiener filtering," *Biol. Cybern.*, vol. 27, pp. 153–164, 1979.

[99] S. Cerutti, V. Bersani, A. Carrera, and D. Liberati, "Analysis of visual evoked potentials through Wiener filtering applied to a small number of sweeps," *J. Biomed. Eng.*, vol. 9, pp. 3–12, 1987.

[100] N. Wiener, *Extrapolation, Interpolation, Smoothing of Stationary Time Series*. Cambridge, MA: MIT, 1964.

[101] J. Strackee and S. Cerri, "Some statistical aspects of digital Wiener filtering and detection of prescribed frequency components in time averaging of biological signals," *Biol Cybern*, vol. 28, pp. 55–61, 1977.

[102] J. P. C. de Weerd and W. L. J. Martens, "Theory and practice of a posteriori Wiener filtering of average evoked potentials," *Biol. Cybern.*, vol. 30, pp. 81–94, 1978.

[103] J. P. C. de Weerd, "Facts and fancies about a posteriori Wiener filtering," *IEEE Trans. Biomed. Eng.*, vol. 28, pp. 252–257, 1981.

[104] E. H. Carlton and S. Katz, "Is Wiener filtering an effective method of improving evoked potential estimation," *IEEE Trans. Biomed. Eng.*, vol. 27, pp. 187–192, 1980.

[105] P. Ungan and E. Basar, "Comparison of Wiener filtering and selective averaging of evoked potentials," *Electroencephal. Clin. Neurophysiol.*, vol. 40, pp. 516–520, 1976.

[106] D. G. Wastell, "When Wiener filtering is less than optimal: An illustrative application to the brainstem evoked potential," *Electroencephal. Clin. Neurophysiol.*, vol. 451, pp. 678–682, 1981.

[107] J. W. Hartwell and C. W. Erwin, "Evoked potential analysis: Online signal optimization using a mini-computer," *Electroencephal. Clin. Neurophysiol.*, vol. 41, pp. 416–421, 1976.

[108] V. Albrecht, P. Lánský, M. Indra, and T. Radil-Weiss, "Wiener filtration versus averaging of evoked responses," *Biol. Cybern.*, vol. 27, pp. 147–154, 1977.

[109] R. E. Kearney, "Evaluation of the Wiener filter applied to evoked EMG potentials," *Electroencephal. Clin. Neurophysiol.*, vol. 46, pp. 475–478, 1979.

[110] J. P. C. de Weerd, "A posteriori time-varying filtering of averaged evoked potentials: I. Introduction and conceptual basis," *Biol. Cybern.*, vol. 41, pp. 211–222, 1981.

[111] J. P. C. de Weerd, "A posteriori time-varying filtering of averaged evoked potentials: II. Mathematical and computational aspects," *Biol. Cybern.*, vol. 41, pp. 223–234, 1981.

[112] J. P. C. de Weerd and J. I. Kap, "Spectro-temporal representations and time-varying spectra of evoked potential," *Biol. Cybern.*, vol. 41, pp. 101–117, 1981.

[113] J. B. MacNeil, R. E. Kearney, and I. W. Hunter, "Identification of time-varying biological systems from ensemble data," *IEEE Trans. Biomed. Eng.*, vol. 39, pp. 1213–1225, 1992.

[114] M. Furst and A. Blau, "Optimal a posteriori time domain filter for average evoked potentials," *IEEE Trans. Biomed. Eng.*, vol. 38, no. 9, pp. 827–833, 1991.

[115] K. B. Yu and C. D. McGillem, "Optimum filters for estimating evoked potential waveforms," *IEEE Trans. Biomed. Eng.*, vol. 30, pp. 730–737, 1983.

[116] D. G. Childers, P. A. Bloom, A. A. Arroyo, S. E. Roucos, I. S. Fischler, T. Acharitya-paopan, and N. W. Perry, "Classification of cortical responses using features from single EEG records," *IEEE Trans. Biomed. Eng.*, vol. 29, pp. 423–438, 1982.

[117] A. Arieli, A. Sterkin, A. Grinvald, and A. Aertsen, "Dynamics of ongoing activity: Explanation of the large variability in evoked cortical responses," *Science*, vol. 273, pp. 1868–1871, 1996.

[118] S. Cerutti, G. Baselli, D. Liberati, and G. Pavesi, "Single sweep analysis of visual evoked potentials through a model of parametric identification," *Biol. Cybern.*, vol. 56, pp. 111–120, 1987.

[119] S. Cerutti, G. A. Chiarenza, D. Liberati, P. Mascellani, and G. Pavesi, "A parametric method of identification of single-trial event-related potentials in the brain," *IEEE Trans. Biomed. Eng.*, vol. 35, pp. 701–711, 1988.

[120] D. Krieger and W. Larimore, "Automatic enhancement of single evoked potentials," *Electroencephal. Clin. Neurophysiol.*, vol. 64, pp. 568–572, 1986.

[121] G. E. Birch, P. D. Lawrence, and R. D. Hare, "Single trial processing of event related potentials using outlier information," *IEEE Trans. Biomed. Eng.*, vol. 40, pp. 59–72, 1993.

[122] D. H. Lange and G. F. Inbar, "A robust parametric estimator for single-trial movement related brain potentials," *IEEE Trans. Biomed. Eng.*, vol. 43, pp. 341–347, 1996.

[123] D. H. Lange, H. Pratt, and G. F. Inbar, "Modeling and estimation of single evoked brain potential components," *IEEE Trans. Biomed. Eng.*, vol. 44, pp. 791–799, 1997.

[124] C. A. Vaz and N. V. Thakor, "Adaptive Fourier estimation of time-varying evoked potentials," *IEEE Trans. Biomed. Eng.*, vol. 36, pp. 448–455, 1989.

[125] N. V. Thakor, C. A. Vaz, R. W. McPherson, and D. F. Hanley, "Adaptive Fourier series modeling of time-varying evoked potentials," *Electroencephal. Clin. Neurophysiol.*, vol. 80, pp. 108–118, 1991.

[126] C. A. Vaz, X. Kong, and N. V. Thakor, "An adaptive estimation of periodic signals using a Fourier linear combiner," *IEEE Trans. Signal Proc.*, vol. 42, pp. 1–10, 1994.

[127] N. Ahmed and K. R. Rao, *Orthogonal Transforms for Digital Signal Processing*. Berlin: Springer-Verlag, 1975.

[128] N. V. Thakor, G. Xin-Rong, C. Vaz, P. Laguna, R. Jané, P. Caminal, H. Rix, and D. Hanley, "Orthonormal (Fourier and Walsh) models of time-varying evoked potentials in neurological injury," *IEEE Trans. Biomed. Eng.*, vol. 40, no. 3, pp. 213–221, 1993.

[129] D. H. Lange and G. F. Inbar, "Variable single-trial evoked potential estimation via principal component identification," in *Proc. Conf. IEEE Eng. Med. Biol. Soc. (EMBS)*, pp. 954–955, IEEE, 1996.

[130] S. Olmos, J. García, R. Jané, and P. Laguna, "Truncated orthogonal expansions of recurrent signals: Equivalence to a linear time-variant periodic filter," *IEEE Trans. Signal Proc.*, vol. 47, pp. 3164–3172, 1999.

[131] S. L. Marple Jr., *Digital Spectral Analysis with Applications*. New Jersey: Prentice-Hall, 1987.

[132] T. Demiralp, A. Ademoglu, Y. Istefanopulos, and H. Ö. Gülşür, "Analysis of event-related potentials (ERP) by damped sinusoids," *Biol. Cybern.*, vol. 78, pp. 487–493, 1998.

[133] M. Hansson, T. Gänsler, and G. Salomonsson, "Estimation of single event related potentials utilizing the Prony method," *IEEE Trans. Biomed. Eng.*, vol. 43, pp. 973–981, 1996.

[134] V. Garoosi and B. H. Jansen, "Development and evaluation of the piecewise Prony method for evoked potential analysis," *IEEE Trans. Biomed. Eng.*, vol. 47, pp. 1549–1554, 2000.

[135] J. G. Proakis and D. G. Manolakis, *Digital Signal Processing. Principles, Algorithms, and Applications*. New Jersey: Prentice-Hall, 3rd ed., 1996.

[136] P. Laguna, R. Jané, O. Meste, P. W. Poon, P. Caminal, H. Rix, and N. V. Thakor, "Adaptive filter for event-related bioelectric signals using an impulse correlated reference input: Comparison with signal averaging techniques," *IEEE Trans. Biomed. Eng.*, vol. 39, no. 10, pp. 1032–1044, 1992.

[137] P. Laguna, R. Jané, E. Masgrau, and P. Caminal, "The adaptive linear combiner with a periodic-impulse reference input as a linear comb filter," *Signal Proc.*, vol. 48, pp. 193–203, 1996.

[138] F. H. Y. Chan, F. K. Lam, P. W. F. Poon, and W. Qiu, "Detection of brainstem auditory evoked potentials by adaptive filtering," *Med. Biol. Eng. & Comput.*, vol. 33, pp. 69–75, 1995.

[139] V. Parsa and P. A. Parker, "Multireference adaptive noise cancellation applied to somatosensory evoked potentials," *IEEE Trans. Biomed. Eng.*, vol. 41, no. 8, pp. 792–800, 1994.

[140] J. M. Moser and J. I. Aunon, "Classification and detection of single evoked brain potentials using time–frequency amplitude features," *IEEE Trans. Biomed. Eng.*, vol. 33, pp. 1096–1106, 1986.

[141] S. Olmos, P. Laguna, and L. Sörnmo, "Block adaptive filters with deterministic reference inputs for event-related signals: BLMS and BRLS," *IEEE Trans. Signal Proc.*, vol. 50, pp. 1102–1112, 2002.

[142] N. V. Thakor, "Adaptive filtering of evoked potentials," *IEEE Trans. Biomed. Eng.*, vol. 34, pp. 6–12, 1987.

[143] C. E. Davila, A. J. Welch, and H. G. Rylander III, "Adaptive estimation of single evoked potentials," in *Proc. Conf. IEEE Eng. Med. Biol. Soc. (EMBS)*, pp. 406–408, IEEE, 1986.

[144] G. P. Madhavan, H. deBruin, and A. R. M. Upton, "Improvements to adaptive noise cancellation," in *Proc. Conf. IEEE Eng. Med. Biol. Soc. (EMBS)*, pp. 482–486, IEEE, 1986.

[145] O. Svensson, "Tracking of changes in latency and amplitude of the evoked potentials by using adaptive LMS filters and exponential averagers," *IEEE Trans. Biomed. Eng.*, vol. 40, pp. 1074–1078, 1993.

[146] G. P. Madhavan, "Minimal repetition evoked potentials by modified adaptive line enhancement," *IEEE Trans. Biomed. Eng.*, vol. 39, pp. 760–764, 1992.

[147] S. Olmos and P. Laguna, "Steady-state MSE convergence of LMS adaptive filters with deterministic reference inputs with application to biomedical signals," *IEEE Trans. Signal Proc.*, vol. 48, pp. 2229–2241, 2000.

[148] P. Laguna, G. B. Moody, J. García, A. L. Goldberger, and R. G. Mark, "Analysis of the ST-T complex using the K-L transform: Adaptive monitoring and alternans detection," *Med. Biol. Eng. & Comput.*, vol. 37, pp. 175–189, 1999.

[149] V. J. Samar, A. J. Bopardikar, R. Rao, and K. P. Swartz, "Wavelet analysis of neuroelectric waveforms: A conceptual tutorial," *Brain and Language*, vol. 66, pp. 7–60, 1999.

[150] G. Strang and T. Nguyen, *Wavelets and Filter Banks*. Wellesley: Wellesley - Cambridge Press, 1997.

[151] C. S. Burrus, R. A. Gopinath, and H. Guo, *Introduction to Wavelets and Wavelet Transforms: A Primer*. New Jersey: Prentice-Hall, 1998.

[152] S. Mallat, *A Wavelet Tour of Signal Processing*. San Diego: Academic Press, 1998.

[153] A. Aldroubi and M. Unser (eds.), *Wavelets in Medicine and Biology*. Boca Raton: CRC Press, 1996.

[154] M. Akay (ed.), *Time Frequency and Wavelets in Biomedical Signal Processing*. New York: IEEE Press, 1996.

[155] P. S. Addison, *The Illustrated Wavelet Transform Handbook*. Bristol: Institute of Physics Publ., 2002.

[156] A. Haar, "Zur theorie der orthogonalen funktionensysteme," *Matematische Annalen*, vol. 69, pp. 331–371, 1910.

[157] I. Daubechies, "Orthonormal bases of compactly supported wavelets," *Comm. Pure and Appl. Math.*, vol. 41, pp. 909–996, 1988.

[158] T. Demiralp, A. Ademoglu, M. Schürmann, C. Başar-Eroglu, and E. Başar, "Detection of P300 waves in single trials by the wavelet transform (WT)," *Brain and Language*, vol. 66, pp. 108–128, 1999.

[159] E. A. Bartnik, K. J. Blinowska, and P. J. Durka, "Single evoked potential reconstruction by means of the wavelet transform," *Biol. Cybern.*, vol. 67, pp. 175–181, 1992.

[160] N. V. Thakor, G. Xin-Rong, S. Yi-Chun, and D. Hanley, "Multiresolution wavelet analysis of evoked potentials," *IEEE Trans. Biomed. Eng.*, vol. 40, no. 11, pp. 1085–1092, 1993.

[161] V. J. Samar, K. P. Swartz, and M. R. Raghuveer, "Multiresolution analysis of event-related potentials by wavelet decomposition," *Brain and Cognition*, vol. 27, pp. 298–438, 1995.

[162] J. Raz, L. Dickerson, and B. Turetsky, "A wavelet packet model of evoked potentials," *Brain and Language*, vol. 66, pp. 61–88, 1999.

[163] D. L. Donoho, "De-noising by soft-thresholding," *IEEE Trans. Inform. Theory*, vol. 41, pp. 613–627, 1995.

[164] D. L. Donoho and I. M. Johnstone, "Adapting to unknown smoothness via wavelet shrinkage," *J. Am. Stat. Assoc.*, vol. 90, pp. 1200–1224, 1995.

[165] R. Quian Quiroga, "Obtaining single stimulus evoked potentials with wavelet denoising," *Physica D: Nonlinear Phenomena*, vol. 145, pp. 278–292, 2002.

[166] R. Quian Quiroga and H. García, "Single-trial event-related potentials with wavelet denoising," *Clin. Neurophysiol.*, vol. 114, pp. 376–390, 2003.

[167] A. Janušauskas, L. Sörnmo, O. Svensson, and B. Engdahl, "Detection of transient-evoked otoacoustic emissions and the design of time windows," *IEEE Trans. Biomed. Eng.*, vol. 49, pp. 132–139, 2002.

[168] A. Ademoglu, E. Micheli-Tzanakou, and Y. Istefanopulos, "Analysis of pattern reversal visual evoked potentials (PRVEP's) by spline wavelets," *IEEE Trans. Biomed. Eng.*, vol. 44, pp. 881–890, 1997.

[169] A. Effern, K. Lehnertz, T. Grunwald, G. Fernández, P. David, and C. E. Elger, "Time adaptive denoising of single trial event-related potentials in the wavelet domain," *Psychophysiology*, vol. 37, pp. 859–865, 2000.

[170] E. Laciar, R. Jané, and D. H. Brooks, "Improved alignment method for noisy high-resolution ECG and Holter records using multiscale cross-correlation," *IEEE Trans. Biomed. Eng.*, vol. 50, pp. 344–353, 2003.

[171] O. Bertrand, J. Bohorquez, and J. Pernier, "Time–frequency digital filtering based on an invertible wavelet transform: An application to evoked potentials," *IEEE Trans. Biomed. Eng.*, vol. 41, pp. 77–88, 1994.

[172] L. M. Lim, M. Akay, and J. A. Daubenspeck, "Identifying respiratory-related evoked potentials," *IEEE Eng. Med. Biol. Mag.*, vol. 14, pp. 174–178, 1995.

[173] T. Demiralp, J. Yordanova, V. Kolev, A. Ademoglu, M. Devrim, and V. J. Samar, "Time–frequency analysis of single-sweep event-related potentials by means of fast wavelet transform," *Brain and Language*, vol. 66, pp. 129–145, 1999.

[174] L. J. Trejo and M. J. Shensa, "Feature extraction of event-related potentials using wavelets: An application to human performance monitoring," *Brain and Language*, vol. 66, pp. 89–107, 1999.

[175] J. W. Huang, Y. Lu., A. Nayak, and R. J. Roy, "Depth of anesthesia estimation and control," *IEEE Trans. Biomed. Eng.*, vol. 46, pp. 71–81, 1999.

[176] H. Heinrich, H. Dickhaus, A. Rothenberger, V. Heinrich, and G. H. Moll, "Single-sweep analysis of event-related potentials by wavelet networks—methodological basis and clinical application," *IEEE Trans. Biomed. Eng.*, vol. 46, pp. 867–879, 1999.

[177] A. Effern, K. Lehnertz, T. Schreiber, T. Grunwald, P. David, and C. E. Elger, "Nonlinear denoising of transient signals with application to event-related potentials," *Physica D: Nonlinear Phenomena*, vol. 140, pp. 257–266, 2000.

[178] U. Hoppe, S. Weiss, R. W. Stewart, and U. Eysholdt, "An automatic sequential recognition method for cortical auditory evoked potentials," *IEEE Trans. Biomed. Eng.*, vol. 48, pp. 154–164, 2001.

[179] R. Quian Quiroga and E. L. J. M. van Luijtelaar, "Habituation and sensitization in rat auditory evoked potentials: A single-trial analysis with wavelet denoising," *Int. J. Psychophysiology*, vol. 43, pp. 141–153, 2002.

[180] S. Turner, P. Picton, and J. Campbell, "Extraction of short-latency evoked potentials using a combination of wavelets and evolutionary algorithms," *Med. Eng. & Physics*, vol. 25, pp. 407–412, 2003.

[181] S. M. Kay, *Fundamentals of Statistical Signal Processing. Detection Theory*. New Jersey: Prentice-Hall, 1998.

Problems

4.1 In a BAEP investigation, the EPs are assumed to be modeled by (4.4) and related assumptions on statistical properties. The SNR of the first potential $\mathbf{x}_1$ is defined by

$$\text{SNR} = 10 \cdot \log \frac{\mathbf{s}^T \mathbf{s}}{E[\mathbf{v}_1^T \mathbf{v}_1]}$$

and is assumed to be equal to –5 dB. All other EPs in the ensemble have identical SNRs. How many EPs need to be averaged, using (4.12), in order to increase the SNR to 10 dB?

4.2 The difference between two subaverages $\hat{s}_{a_0}(n)$ and $\hat{s}_{a_1}(n)$ is denoted

$$\Delta\hat{s}_a(n) = \hat{s}_{a_0}(n) - \hat{s}_{a_1}(n).$$

The two subaverages have been obtained by splitting the ensemble in a suitable way.

    a. Explain why it is of interest to study the quantity $\Delta\hat{s}_a(n)$ during the acquisition of EPs.

    b. Show that the variance of $\Delta\hat{s}_a(n)$ is equal to $4\sigma_v^2/M$ by making use of the common assumptions associated with ensemble averaging.

4.3 Determine the impulse response of the exponential averager in (4.35). The answer should be expressed as a function of the weight factor α and the length N of the EP. Assume that $\hat{s}_{e,0}(n) = 0$, and recall that all EPs are concatenated. Sketch the impulse response.

4.4 Computation of the ensemble variance estimate $\hat{\sigma}_v^2(n)$ in (4.17) has the disadvantage of requiring that the entire ensemble must be available before $\hat{\sigma}_v^2(n)$ can be computed. However, it may by desirable to monitor how the ensemble variance evolves as the number of EPs increases. Derive an approximate estimator which recursively computes the estimate of $\hat{\sigma}_{v,M}^2(n)$. It can be assumed that the ensemble average $\hat{s}_{a,M}(n)$ has been stabilized to such a degree that it can be approximated by its preceding estimate $\hat{s}_{a,M-1}(n)$.

4.5 Determine the mean of the exponential averager $\hat{s}_{e,M}(n)$ when $\hat{s}_{e,0}(n) = x_1(n)$. Discuss the fact that $E[\hat{s}_{e,M}(n)]$ is unbiased, whereas it is asymptotically unbiased when $\hat{s}_{e,0}(n) = 0$.

4.6   a. The exponential averager is usually initialized by either $\hat{s}_{e,0}(n) = 0$ or $\hat{s}_{e,0}(n) = x_1(n)$. However, both these initializations suffer from certain disadvantages. What are these disadvantages?

   b. Find that value of α of the exponential averager which makes the variance of $\hat{s}_{e,M}(n)$ equal to the variance of the ensemble averager $\hat{s}_{a,M}(n)$.

4.7 Derive the expression of the variance $V[\hat{s}_{e,M}(n)]$ in (4.38).

4.8 Determine the width of the frequency lobes of the comb filter at the -3 dB point corresponding to:

   a. the ensemble averager as a function of M and N, and

   b. the exponential averager as a function of α.

   In both these cases, it is assumed that the poles are well-separated such that the influence of neighboring poles can be neglected. Compare the role of M and α of the respective estimators.

4.9 Determine a closed-form expression for the -3 dB bandwidth of the peaks in the exponential averager, expressed in terms of the parameters α and N.

4.10 In addition to determining the magnitude function of the ensemble averager, cf. page 202, it is also of interest to determine its phase function.

   a. Derive an expression for the phase function of the ensemble averager and plot it.

   b. Discuss how the phase function influences the repetitive signal and the noise, respectively.

4.11 An anesthetized patient is periodically stimulated by short sound pulses to continuously monitor the BAEP. The resulting waveforms first stabilize at an amplitude of 0.6 μV in peak IV, but then suddenly decrease to an amplitude of 0.2 μV.

   a. For ensemble averaging, determine the delay in terms of the number of stimuli until the amplitude (in the mean) has dropped below 0.3 μV?

   b. Repeat the exercise in (a) for exponential averaging.

4.12 The ensemble average $\hat{s}_{a,M}(n)$ is often used to estimate the signal $s(n)$ in the observation model $x_i(n) = s(n) + v_i(n)$, where $v_i(n)$ is zero-mean noise with variance σ_v^2. The ensemble average can be computed recursively using the following expression

$$\hat{s}_{a,M}(n) = \hat{s}_{a,M-1}(n) + g_M(x_M(n) - \hat{s}_{a,M-1}(n)),$$

where $g_M = 1/M$. Analogously, the weighted average $\hat{s}_{w,M}(n)$ can be computed recursively but using another expression of g_M. Determine g_M for weighted averaging under the assumption that the noise variance is $\sigma_{v_i}^2$, and then determine the recursion for $\sigma_{v_i}^2 \equiv \sigma_v^2$.

4.13 In the interval preceding the stimulus elicited at time $n = 0$, we want to estimate the variance σ_v^2 from the background EEG signal, e.g., for later use in the computation of the weighted average. It is assumed that the samples $x(-N), \ldots, x(-1)$, are modeled as uncorrelated, Gaussian noise with mean m_v and variance σ_v^2. Determine the ML estimator of σ_v^2.

4.14 Determine the expression with which the variance of the weighted average $V[\hat{s}_{w,M}(n)]$ can be recursively computed from $V[\hat{s}_{w,M-1}(n)]$. The variance $V[\hat{s}_{w,M}(n)]$ is given in (4.68).

4.15 Determine $E[\hat{s}_w(n)]$ and $V[\hat{s}_w(n)]$ for weighted averaging under the assumption that the signal amplitude varies and the noise variance remains fixed for all EPs. Comment on bias and consistency.

4.16 Derive the optimal weights $\hat{w}_i$ of the weighted average that minimizes the following MSE criterion,

$$E\left[\left(s(n) - \sum_{i=1}^{M} w_i x_i(n)\right)^2\right].$$

Each EP is described by $x_i(n) = s(n) + v_i(n)$, where $s(n)$ is deterministic and $v_i(n)$ is random with variance $\sigma_{v_i}^2$.

   a. Determine the optimal weights, and comment on their dependence on the signal and noise.

   b. Show that the optimal weights approach those in (4.67) when the constraint

$$\sum_{i=1}^{M} w_i = 1$$

   is introduced; this constraint assures that the ensemble average is unbiased.

4.17 Two cases of weighted averaging have been described in the text—either varying signal amplitude or varying noise variance. In this problem, a third case is examined where both amplitude and noise variance are allowed to vary. Find the optimal weight vector for this case.

4.18 Weighted averaging requires that the noise variance of each EP be estimated. Although the estimator in (4.73) is adequate for certain applications such as BAEP and SEP, it is less suitable for VEPs where the SNR is relatively good. Suggest a variance estimator for the latter case which draws upon the better SNR.

4.19 The weights required in weighted averaging can be adaptively estimated by taking advantage of the assumption that signal and noise are uncorrelated [53]. The estimation is based on the adaptive linear combiner, shown in Figure 3.13, but now with the primary input (i.e., the upper branch of the block diagram) given by the ensemble average $\hat{s}_a(n)$ and the M reference inputs given by $x_i(n) = s(n) + v_i(n)$ for $i = 1, \ldots, M$.

    a. Assuming a steady-state situation, show that the LMS algorithm converges in the mean to the optimum weight vector $\mathbf{w}^\circ(n)$, cf. (3.55),

$$\mathbf{w}^\circ(n) = \frac{s^2(n)}{1 + \displaystyle\sum_{i=1}^{M} \frac{s^2(n)}{\sigma_i^2}} \begin{bmatrix} \dfrac{1}{\sigma_1^2} & \dfrac{1}{\sigma_2^2} & \cdots & \dfrac{1}{\sigma_M^2} \end{bmatrix}^T .$$

    b. Unfortunately, the weight vector that results in (a) is time-varying through $s(n)$ despite the fact that the noise is assumed to be stationary. As a result, the weight vector obtained by the LMS algorithm is biased. By introducing the constraint

$$\mathbf{w}^T\mathbf{1} = 1, \tag{4.376}$$

which assures that the estimate is unbiased, a constrained LMS algorithm can be developed which minimizes the MSE,

$$\mathcal{E}_\mathbf{w} = E\left[\left(\hat{s}_a(n) - \mathbf{w}^T\mathbf{x}(n) \right)^2 \right] - \lambda(\mathbf{w}^T\mathbf{1} - 1),$$

where the constraint is multiplied by the Lagrange multiplier λ. Derive the constrained LMS algorithm.

4.20 Another estimate of the normalization constant in (4.78) is given by

$$\widehat{\mathbf{a}^T\mathbf{a}} = \mathrm{tr}(\mathbf{X}^T\mathbf{X}).$$

Explain why this estimate is less suitable than the one given in the text.

4.21 The ML estimator of an signal corrupted by stationary (i.e., $\sigma_{v_i}^2 \equiv \sigma_v^2$), Laplacian noise is the ensemble median. Determine the ML estimator—the *weighted median*—when the Laplacian noise has a variance which varies from potential to potential.

4.22 Show that the recursive, robust averager with outlier rejection in (4.104), whose influence function is given by the sgn function, tends to converge to the median.

4.23 Determine an approximate expression for the –3 dB cut-off frequency F_c of the lowpass filter in Figure 4.20(b), assuming that the latency shifts τ are uniformly distributed. In other words, determine that $\Omega_c \ (= 2\pi F_c)$ for which

$$P_\tau(\Omega_c) = \frac{\sin \frac{1}{2}\Omega_c T}{\frac{1}{2}\Omega_c T} = \frac{1}{\sqrt{2}}.$$

4.24 For discrete-time jitter, show that the characteristic function $P_\theta(e^{j\omega})$ of the "discretized" Gaussian PDF,

$$p_\theta(\theta) = \frac{1}{\sqrt{2\pi\sigma_\theta^2}} \, e^{-\frac{\theta^2}{2\sigma_\theta^2}},$$

where θ is an integer, is given by

$$P_\theta(e^{j\omega}) = \sum_{n=-\infty}^{\infty} e^{-\sigma_\theta^2(\omega - 2\pi n)^2/2}.$$

For Gaussian jitter, the resulting effect is lowpass filtering with a cut-off frequency ω_c of

$$\omega_c = \frac{0.83}{\sigma_\theta}.$$

4.25 Latency estimation based on cross-correlation of $x(n)$ of length N and a deterministic waveform $s(n)$ of length $M < N$ may be formulated as

$$\hat{\tau} = \arg \max_{n_0 \in [0, N-M]} \sum_{n=n_0}^{n_0+M-1} x(n)s(n-n_0),$$

where $\hat{\tau}$ is the estimated latency, i.e., the argument which maximizes the above cross-correlation. Suggest two different techniques for latency estimation with better time resolution than that offered by the sampling interval of the original signal.

4.26 Derive a filter $h(n)$ of length N that maximizes the SNR at $n = N - 1$ for the model

$$x(n) = s(n) + v(n), \quad n = 0, 1, \ldots, N - 1,$$

where $s(n)$ is deterministic and $v(n)$ is stationary, colored, zero-mean, Gaussian noise with correlation matrix $\mathbf{R}_v$.

4.27 Derive the ML estimator of the delay θ_i when the EP is corrupted by stationary, colored, zero-mean, Gaussian noise with correlation matrix $\mathbf{R}_v$. Assuming that the signal length N is much larger than the correlation time d for the noise $v(n)$ and $r_v(k) = 0$ for $|k| \geq d$ (the so-called *asymptotic Gaussian PDF* assumption), it can be shown that the inverse of the noise correlation matrix $\mathbf{R}_v^{-1}$ is given by [181, p. 33]

$$\mathbf{R}_v^{-1} = \sum_{i=0}^{N-1} \frac{1}{\lambda_i} \boldsymbol{\varphi}_i \boldsymbol{\varphi}_i^T,$$

where λ_i is an eigenvalue of $\mathbf{R}_v$ for which the corresponding eigenvectors are given by the discrete Fourier transform vector,

$$\boldsymbol{\varphi}_i = \frac{1}{\sqrt{N}} [\, 1 \quad e^{j2\pi f_i} \quad e^{j4\pi f_i} \quad \cdots \quad e^{j2\pi(N-1)f_i} \,]^T.$$

Make use of this result in the derivation of the ML estimator.

4.28 A multichannel variant of the Woody method may be used in which each channel $x_i(l)$ is first processed by its corresponding matched filter $h_i(l)$, followed by a weighted summation of the filter outputs which is used for time delay estimation, i.e.,

$$y(n) = \sum_{i=1}^{P} \beta_i \left(\sum_{l=0}^{K} h_i(l) x_i(n - l) \right),$$

where P denotes the number of channels. Discuss, in general terms, how to choose the channel weights β_i.

4.29 The inverse of the correlation matrix $\mathbf{R}_x$ in (4.149) is required for weighting of the samples in the averaged EPs with the ML estimate of the ensemble correlation. The correlation matrix $\mathbf{R}_x$ can be expressed as

$$\mathbf{R}_x = (1 - \rho(n))\mathbf{I} + \rho(n)\mathbf{1}\mathbf{1}^T,$$

where, for simplicity, it is assumed that the power of the observed signal is normalized to unity, i.e., $\sigma_s^2(n) + \sigma_v^2 = 1$. Find the inverse $\mathbf{R}_x^{-1}$ expressed in terms of $\rho(n)$ and the number M of EPs. *Hint:* Use the matrix inversion lemma in (A.31).

4.30 In weighting an averaged EP with the ensemble correlation, we have assumed that $s(n)$ is random. An alternative approach is to assume that $s(n)$ is deterministic (once the ensemble has been fixed).

a. Show that the weight minimizing the MSE criterion is given by

$$w(n) = \frac{s^2(n)}{s^2(n) + \dfrac{\sigma_v^2}{M}}.$$

b. Propose estimators of $s(n)$ and σ_v^2 in order to determine the weight $w(n)$ in (a).

4.31 When performing single-trial analysis, it may be of interest to minimize the following MSE criterion:

$$\mathcal{E}_{\mathbf{w}} = E\left[||\mathbf{x}_i - \mathbf{\Phi}\mathbf{w}_i||^2\right].$$

In the text, it was tacitly assumed that the obtained solution in (4.202) corresponded to the minimum of the MSE. Show that this solution really corresponds to the minimum.

4.32 An estimate of the signal correlation matrix $\mathbf{R}_s$ is required for implementation of the a posteriori FIR Wiener filter in (4.184). One approach to develop an estimator is based on the model $\mathbf{x}_i = \mathbf{s} + \mathbf{v}_i$, where it is assumed that $\mathbf{s}$ is stationary and $\mathbf{v}_i$ is uncorrelated from EP to EP.

a. Suggest an estimator which involves the summation of all cross-products $\mathbf{x}_i\mathbf{x}_j^T$ for $i, j = 1, \ldots, M$, while excluding $i = j$.

b. For this estimator, evaluate its behavior in terms of mean and variance.

4.33 Rather than minimizing the squared error over all possible realizations, as done in (4.199), we can minimize the "instantaneous" error

$$\mathcal{E}(\mathbf{w}_i) = ||\mathbf{x}_i - \mathbf{\Phi}\mathbf{w}_i||^2$$

for one particular realization of $\mathbf{x}_i$. Proceeding in a way similar to the minimization of (4.199), the solution to this problem is found to be

$$\hat{\mathbf{w}}_i = \mathbf{\Phi}^T\mathbf{x}_i,$$

which is identical to the right-hand side of (4.203).

4.34 When adaptive estimation is performed with the instantaneous LMS algorithm for a complete set of basis functions, i.e., $K = N$, it can be shown that the mean of the weights is asymptotically unbiased,

$$\lim_{n \to \infty} E[\mathbf{w}(n)] = \mathbf{w}.$$

This result can be obtained from addressing the following three issues.

a. Show that the $\mathbf{F}_m(n)$ factor in (4.283) reduces to

$$\mathbf{F}_m(n) = \mathbf{I} - \mu \sum_{j=m}^{n} \boldsymbol{\varphi}_s(j)\boldsymbol{\varphi}_s^T(j).$$

b. Show that the recursion in (4.282) for time iN can be expressed as

$$\mathbf{w}(iN) = (1 - \mu)^i \mathbf{w}(0) + \mu \sum_{l=0}^{i-1}(1 - \mu)^i \sum_{m=0}^{N-1} x(lN + m)\boldsymbol{\varphi}_s(m).$$

c. Show that the expression in (b) is unbiased as $n \to \infty$.

4.35 Multiresolution signal analysis with the wavelet transform is efficiently implemented with an analysis filter bank where the detail coefficients at scale j are computed by

$$d_j(k) = h_\psi(-n) * c_{j+1}(n)|_{n=2k}$$
$$c_j(k) = h_\varphi(-n) * c_{j+1}(n)|_{n=2k}$$

and involve filtering and decimation by a factor of two. The recursion is commonly initiated by $c_J(n) = x(n)$, where $N = 2^J$ and N is the signal length, and thus it is inviting to view the analysis filter bank in terms of filtering of $x(n)$.

Since decimation is a time-varying operation, each state of the cascaded filter in Figure 4.43 can be implemented without decimation by interpolating the filter impulse response of the previous stage, so that time resolution is maintained from scale to scale. This algorithm is known as *algoritme à trous*, i.e., insertion of zeros ("holes") in the signal [152]. Find the filter transfer function $D_j(e^{j\omega})$ of the detail coefficients $d_j(n)$.

4.36 Starting from the refinement equation in (4.323), derive the iterative cascade algorithm in (4.363) which is used to produce the scaling and wavelet functions.

Chapter 5

The Electromyogram

The EMG signal reflects the electrical activity of skeletal muscles and contains information about the structure and function of muscles which make different parts of the body move. The EMG signal conveys information about the controller function of the central and peripheral nervous systems on the muscles. As such, the EMG signal provides a highly useful characterization of the neuromuscular system since many pathological processes, whether arising in the nervous system or the muscle, are manifested by alterations in the signal properties.

Although it has been known for centuries that muscles generate electricity, it is only since the 1960s that electromyography has come into widespread clinical use. During the last few decades, the accuracy of EMG signal interpretation has been considerably improved thanks to advances in recording technology and computer-based signal analysis and interpretation. Invasive electrodes are now available which can record localized activity of individual muscle fibers, as well as arrays of noninvasive electrodes which provide a two-dimensional spatial characterization of the electrical potential distribution. With the development of novel signal processing techniques, it has become possible to look deeper into the electrophysiology of muscles, for example, by decomposing the EMG signal into different components reflecting the activity of individual muscle units. Advances in engineering have extended electromyography beyond the traditional diagnostic applications to also include applications in diverse areas such as ergonomics, exercise physiology, rehabilitation, movement analysis, biofeedback, and myoelectric control of prosthesis.

This chapter begins with a brief description of the electrical activity of muscles and its manifestations in EMG signal recordings, acquired invasively or noninvasively. Then, a number of applications are listed which incorporate analysis of myoelectric information (Section 5.1). The remaining sections of this chapter focus on issues of central importance to EMG signal processing, namely, amplitude estimation, muscle fiber conduction velocity estimation, and decomposition of the intramuscular EMG. In most cases, the signal processing methods are derived from a phenomenological model of the EMG signal to which optimal estimation techniques are applied. Sections 5.2–5.4 deal with methods suitable for use in surface EMG analysis, while the model and the methods in Sections 5.5 and 5.6 are intended for analysis of intramuscular EMGs.

Readers wishing to learn more about electromyography and its applications should consult the excellent textbook by Merletti and Parker [1].

5.1 The Electrical Activity of Muscles

5.1.1 Action Potentials and Motor Units

The contraction of muscle tissue makes it possible to move different parts of the body such as the eyes and limbs, as well as to move fluid within the body. Depending on its purpose, the muscle may be categorized as either *skeletal, smooth,* or *cardiac.* Skeletal muscle is attached to the skeleton and facilitates movement and position of the body, whereas smooth muscle is found within the intestines and blood vessels. Skeletal muscle is the type of interest in this chapter. Cardiac muscle builds the heart walls and produces the contraction of the heart, creating a heartbeat; the function of the heart is described separately in Chapter 6.

In skeletal muscle, contraction is controlled by electrical impulses, i.e., action potentials, which propagate between the central and peripheral nervous systems and the muscles. The action potentials are transmitted down the axons of the motor neurons, originating in the brain or the spinal cord, to the muscle fibers. Each motor neuron is connected to muscle fibers through a specialized synapse called the *neuromuscular junction* which allows the action potentials to stimulate contraction. Taken together, a motor neuron and the fibers to which it connects (innervates) comprise a *motor unit* and represent a functional unit of contraction. Depending on the purpose of the muscle, a single motor unit may comprise just a few muscle fibers or more than a thousand muscle fibers [1, 2]. Muscles which control fine movements, for example, of an eye or a finger, have fewer muscle fibers per motor unit than muscles which control gross movements, for example, activated during running and jumping.

The contraction of a muscle fiber is initiated when neuronal action potentials reach the neuromuscular junction and fire action potentials which spread along the excitable membranes of the muscle fiber. A *motor unit action potential* (MUAP) results from spatial and temporal summation of

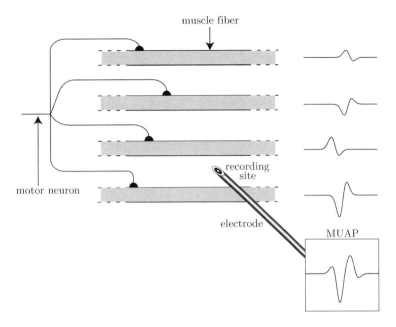

Figure 5.1: The generation of a MUAP of a single motor unit with four muscle fibers. The recorded MUAP results from the summation of the four single fiber action potentials, whose amplitudes decrease as the fibers become more distant from the electrode (in this case a concentric needle).

individual action potentials as they spread through the different muscle fibers of a single motor unit, see Figure 5.1. The EMG signal results, in turn, from summation of the different MUAPs which are sufficiently near the recording electrode. The number of MUAPs within the pick-up (detection) area of the electrode depends on the selected type of electrode (described in Section 5.1.2) and is typically larger than one because fibers of different motor units are interspersed throughout the entire muscle. In fact, a muscle cross-section of a few square millimeters may contain fibers belonging to as many as 50 motor units.

Motor unit recruitment is a fundamental muscular process in which the force exerted by muscle contraction is controlled by the central nervous system through spatial and temporal recruitment of motor units. Spatial recruitment means that force is increased by recruiting more motor units, whereas temporal recruitment means that force is increased by firing of action potentials at faster rates. Although both types of recruitment can occur at the same time, spatial recruitment dominates from lower levels of muscle contraction until most motor units have been recruited. At high levels of muscle contraction, temporal recruitment dominates and drives the motor

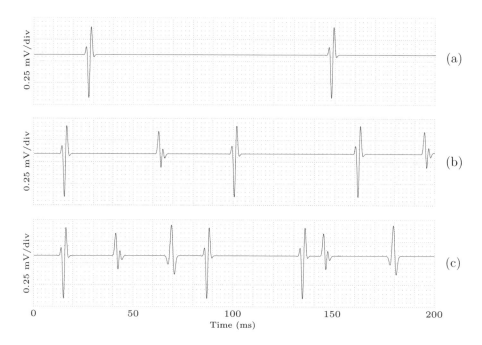

Figure 5.2: Normal motor unit recruitment pattern observed in the intramuscular EMG. (a) A single motor unit is firing at 8 Hz at a low level of muscle contraction. (b) Recruitment of a second motor unit results in a gradual increase in strength and firing rate and is (c) further increased by the recruitment of a third motor unit. Each motor unit has its own particular MUAP morphology and firing rate.

units with firing rates at about 50 Hz and faster. A high firing rate implies that individual MUAP waveforms no longer can be discerned due to temporal superimposition, and the resulting EMG signal exhibits a noise-like, random appearance, referred to as an interference pattern.

Motor units are usually recruited in order of their size, starting with the smallest motor units, with larger units progressively recruited for increasing strength of muscle contraction. As a result, initial activation is weak and followed by a smooth increase in strength through the successive addition of larger and stronger motor units. Figure 5.2 illustrates motor unit recruitment at different levels of muscle strength in a normal subject, involving up to three different motor units.

Examination of motor unit recruitment at low levels of muscle contraction is important for diagnosing disorders related to the nervous system (neuropathy) and muscle tissue (myopathy). The recruitment pattern is usually evaluated in terms of the firing rate of the first motor unit when the second motor unit is recruited. The recruitment rate is easily discerned due to the presence of few active motor units. A neuropathic recruitment pattern is characterized by a firing rate of the first motor unit which exceeds 20 Hz before the second motor unit appears. On the other hand, a myopathic muscle is characterized by a recruitment pattern in which several motor units are active already at minimal voluntary contraction in order to produce the necessary force; thus, a myopathic recruitment pattern is associated with several MUAPs of different shapes.

The information in an intramuscular EMG signal can be grouped into the two basic categories, namely, *morphology* and *firing pattern*—a most useful decomposition to be pursued when we later develop models of, and processing algorithms for, the EMG signal. Some important properties of each category are now summarized.

Morphology. A morphologic description of the MUAP waveform is essential when performing the standard clinical examination and includes the following parameters derived from the *intramuscular* EMG [3].

- *Amplitude* is the peak-to-peak amplitude of the MUAP and ranges from 0.25 to 5 mV for a normal subject. The amplitude is determined by the number of active muscle fibers within the immediate vicinity of the electrode. In general, large amplitudes are associated with neuropathies and small amplitudes with myopathies.

- The *number of phases* of a MUAP waveform reflects the degree of misalignment of action potentials propagating in different single fibers. While the action potential of a single fiber has only one or two phases, the recording electrode detects the summation of action potentials of many fibers which, due to varying degrees of misalignment, can result in a MUAP with multiple phases. *Polyphasic* MUAPs have more than four phases, and are mostly observed in neuropathic and myopathic conditions.

- *Duration* is defined by the onset and end of the MUAP waveform, usually taken as the first and the last time instant when the signal deviates from the baseline level by a certain fixed amplitude. The duration depends on the number of muscle fibers within the motor unit and increases as the number of fibers increases. The normal MUAP duration is 2–10 ms.

Motor unit action potential morphology is exemplified in Figure 5.3 for signals recorded by a needle electrode inserted into the muscle; signals recorded with other types of electrodes may exhibit different MUAP morphology with, for example, altered duration.

In a normal subject, MUAP morphology usually remains stable from discharge to discharge [4]. Unstable morphology may, however, arise when the

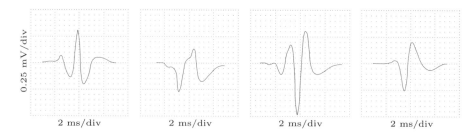

Figure 5.3: Examples of MUAPs with varying amplitude, number of phases, and duration. The EMG signals were recorded using a concentric needle electrode.

neuromuscular junction is impaired, causing certain nerve action potentials to be blocked from transmission. Consequently, not all muscle fibers of the motor unit are activated. In such cases, MUAP amplitude may actually double from one discharge to next and, sometimes, may also be accompanied by changes in the number of phases. Gradual changes in MUAP morphology are observed when the biochemical properties of a motor unit change, for example, occurring during muscle fatigue, or when the relative location of the electrode versus the motor unit changes.

Firing pattern. A muscle contraction is sustained through repeated activation of the motor units, with each motor unit being associated with a particular *train of MUAPs* (Figure 5.2). As the number of active motor units increases, the recording electrode detects an increasing number of simultaneous MUAP trains, and individual MUAP trains become more difficult to discern from each another. The timing with which MUAPs are repeated defines the firing pattern and conveys information on how the central nervous system controls the motor unit. The firing pattern is normally regular (rhythmic) with relatively small variability of successive *interfiring intervals*, i.e., the times between successive MUAPs of the same origin, during contraction. For regular firing patterns, it is often adequate to characterize the pattern by its average firing rate, defined as the inverse of the average length of the interfiring intervals. During voluntary muscle control, a motor unit normally begins firing at a rate of 4–5 Hz, although lower rates may be observed due to spontaneous activity beyond voluntary control, and increases its rate as force increases.

The average firing rate for normal muscle can be exemplified by those of the biceps brachii where the motor units have a rate of 7–12 Hz at the lowest level of muscle contraction; this level is commonly expressed as a percentage of *maximal voluntary contraction* (MVC), which defines a subject-specific scale of contraction. With increasing *isometric*[1] force, the average firing rate of the biceps brachii increases to reach a maximum of approximately 20 Hz at 100% MVC [5, 6]. When assessing motor unit abnormalities in general, it is necessary to take into account that different muscles have their own specific firing rates, as well as rates at which additional motor units are recruited.

While the average firing rate is an essential parameter for analysis of intramuscular EMGs, it is unsuitable for characterizing irregular firing patterns observed during dynamic contractions with rapid changes in force level. The same limitation also applies to the characterization of firing patterns associated with impaired neuromuscular transmission, where the interfiring intervals exhibit increased variability. Hence, alternative rhythm parameters are warranted which more appropriately reflect the transient properties of a firing pattern.

5.1.2 Recording of Myoelectric Signals

Myoelectric activity is measured invasively using a needle electrode or non-invasively by placing a surface electrode on the skin overlying the muscle. The needle EMG is a standard clinical tool used for diagnostic purposes, whereas the surface EMG is analyzed and processed in a variety of applications including prosthesis control, ergonomics, movement and gait analysis, and sports medicine.

The *needle EMG* is measured by inserting a needle electrode through the skin directly into the muscle. This technique provides a high-resolution, localized description of the muscle's electrical activity, albeit relatively painful for the patient. The pick-up area of the needle electrode depends on its specific design and may include as few as one or two muscle fibers. The *monopolar* needle electrode requires a reference electrode placed away from the needle insertion point at an electrically neutral site such as over a bone. The *concentric* needle electrode avoids the reference electrode by referencing the active surface of the electrode to the cannula of the needle (exemplified by the electrode in Figure 5.1). Both types of needle electrodes are used in clinical recordings of the EMG signal. By placing several needle electrodes at different locations, a more detailed description of a large muscle's electrical activity can be obtained.

The *surface EMG* (often abbreviated to sEMG) reflects the gross activity produced by a large number of motor units. Its spatial resolution is more limited than that of the needle EMG, and the high-frequency content

---

[1]A muscle contraction is said to be isometric when performed against resistance but without movement. The length of the muscle remains almost unchanged during isometric contraction.

of a MUAP is smoothed. The placement of surface electrodes depends on the muscle of interest and involves factors such as muscle fiber orientation, anatomical landmarks, and minimization of electrical cross-talk from other muscles.[2] The surface EMG is primarily used when the time of activation and the amplitude of the signal contain the desired information, for example, in connection with studies of motor behavior or myoelectric prosthesis control. The surface EMG does not generally allow the detection of individual MUAPs, although MUAP trains may be detected at low levels of muscle contraction [2]. A further development of the surface electrode technique is the linear electrode array, designed to provide a spatial description of the myoelectric activity. The multichannel EMG signal—resulting from several electrodes equidistantly placed along the muscle's direction—makes it possible to study the generation and extinction of action potentials and to estimate the velocity by which action potentials propagate in the muscle fiber (conduction velocity) [7].

The surface EMG can be recorded at lower sampling rates than the needle EMG since the intervening tissue between the motor units and the surface electrode acts as a lowpass filter of the electrical signal. The surface EMG has most of its spectral power below 400–500 Hz, implying that a sampling rate of 1 kHz or higher is required [8, 9]. For the needle EMG, the sampling rate should be chosen such that different MUAP waveforms, which may contain frequencies in the range up to 10 kHz, are accurately reproduced; therefore, a sampling rate of 50 kHz is often used.

The recording of an EMG is associated with different types of noise and artifacts which, to various degrees, hamper signal quality [9]. In the surface EMG, *electrode motion artifact* is caused by relative movement between the skin and the electrode and by deformation of the skin under the electrode: cf. the discussion on ECG noise and artifacts on page 441. Since this type of artifact is low-frequency in nature with most of its spectral components below 20 Hz, it may be reduced by highpass filtering without significantly altering the spectral content of the EMG signal [10, 11].

Similar to the situations when an EEG or ECG signal is recorded (see page 76), insufficient shielding of the EMG electrode cable makes it susceptible to electromagnetic fields caused by currents flowing in nearby powerlines or electrical devices. As a result, the recorded EMG signal will contain *powerline interference* at 50/60 Hz. Filtering techniques suitable for removing such a narrowband noise component are described in detail in Section 7.2 in the context of ECG signal processing. However, these filtering techniques are equally applicable to the processing of EMG signals, see also [12].

---

[2]The presence of cross-talk makes it more difficult to identify the origin of the electrical signal when two or more muscles, being in close proximity to each other, are active simultaneously.

Electrocardiographic activity may contaminate the surface EMG signal when electrodes are positioned on the trunk and neck so that the characteristic rhythmic ECG pattern is superimposed on the EMG signal. Since the spectral characteristics of the EMG and ECG signals overlap, straightforward application of linear, time-invariant filtering leads to loss of desirable signal information, thus calling for more advanced noise cancellation techniques [9, 13–15].

5.1.3 EMG Applications

Diagnostic EMG. The needle EMG is the standard clinical recording technique used for diagnosing neuromuscular pathology. When, for example, a patient consults a doctor for muscle weakness, she or he is examined by recording the needle EMG during contraction of specific muscles. The morphology of individual MUAP waveforms provides essential clinical information about the muscle's ability to respond to the central nervous system. This information may help to detect abnormal activity that can occur in conditions such as inflammation of muscles, damage to nerves in the arms and legs, pinched nerves, and muscular dystrophy. The needle EMG is also studied in connection with nerve injury and may be used to determine if the injury heals and returns to normal with full reinnervation of the muscle, for example, by examining changes in motor unit recruitment over a certain time span.

The diagnostic EMG includes examination of spontaneous motor activity which may occur during muscle relaxation. In normal conditions, the muscle is electrically silent at relaxation; however, abnormal spontaneous waveforms and waveform patterns may be generated which are associated with involuntary muscular movement and spasms.

Kinesiology. Kinesiology is the study of body movement and is aimed at understanding the processes that control movement. The surface EMG is essential to several aspects of kinesiology, including the study of motor control strategies, mechanics of muscle contraction, and gait [16]. As an example, the EMG pattern recorded during gait is characterized by successive bursts which reflect intervals of muscle activation (see Figure 5.4). Delineation of the times for each burst's onset and end, accompanied by an analysis of the resulting onset/end timing pattern, is an essential task for clinical assessment of various movement disorders. The EMG pattern analysis is facilitated by signal processing techniques which automate the delineation of bursts and characterize the onset/end timing pattern.

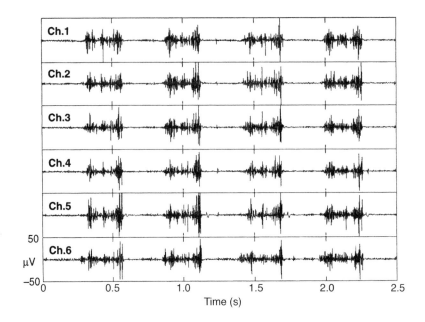

Figure 5.4: Six-channel surface EMG recorded during walking at natural speed. The electrodes were positioned on the leg over the tibialis anterior muscle. (Reprinted from Frigo and Shiavi [16] with permission.)

Ergonomics. The amplitude of the surface EMG signal provides a valuable, quantitative measure of muscle load, often used to assess physical load during work. In ergonomic analysis, the EMG signal is often recorded during light, repetitive work with particular interest to assess the activity of specific muscles at certain work positions. When combined with other ergonomic information, the outcome of the myoelectric signal analysis may ultimately help to avoid work-related disorders, design better workplaces, and improve productivity.

The study of muscle fatigue is central in ergonomics due to the natural wish to avoid fatigue in work situations. A useful definition of muscle fatigue is the condition when a subject no longer is able to maintain a required force [17]. From a signal processing perspective, a fatiguing muscle contraction is manifested by a gradual increase in amplitude of the surface EMG signal and significant changes in its power spectrum. Methods which quantify these two signal properties are described in Sections 5.2 and 5.3, respectively.

Prosthesis control. Myoelectric control of battery-powered prostheses is used by individuals with amputations or congenitally deficient upper limbs.

The control signal is derived with surface electrodes placed over muscles or muscle groups under voluntary control within the residual limb, i.e., the remaining portion of the limb. The control signal is fed to the prosthesis, where its characteristics are analyzed and translated such that the intended function can be performed. Depending on the type of prosthesis, the control information ranges from simple on/off commands generated by a single muscle to complex multifunction commands by a group of muscles. The single-muscle controller is usually based on the EMG amplitude so that muscle contractions of different strengths, as reflected by different amplitudes, can differentiate between hand closing and opening or elbow flexion and extension. In practice, such controllers can reliably perform only three functions [18]. By contrast, a multifunction prosthesis combines the use of several electrodes over different muscle groups with advanced signal processing algorithms in order to increase the amount of information that can be extracted about the active muscle state. The multifunction prosthesis achieves better accuracy of the user's intent by, for example, analyzing transient signal patterns occurring at the onset of rapid contractions, using wavelets or some other time–frequency technique. It is important to realize that the development of algorithms for prosthetic control is associated with a real-time constraint, requiring the response time to not exceed 300 ms; otherwise the user will perceive an unacceptable delay in operation. A recent review of schemes for control of powered upper limb prostheses can be found in [18], see also, e.g., [19–27].

5.2 Amplitude Estimation in the Surface EMG

5.2.1 Signal Model and ML Estimation

The amplitude of the surface EMG is a fundamental quantity which increases monotonically with the force developed in the muscle (Figure 5.5). The amplitude is frequently studied in both clinical routine and scientific studies, for example, in relation to muscle fatigue and muscle coordination. Another important application of EMG amplitude is as control input to a myoelectric prosthesis. Since the operation of the prosthesis requires amplitude estimates of high accuracy, optimal estimation techniques should be employed [18].

Since the surface EMG represents a stochastic signal, its *amplitude* is given by the *standard deviation* of the observed signal or by a similar dispersion estimate. A time-varying estimate of the EMG amplitude can be determined by successively processing the samples in a window sliding through the signal. The derivation of an amplitude estimator can take its starting point in a phenomenological model of the surface EMG in which the observed signal is modeled as the output of a linear, time-invariant filter $h(n)$

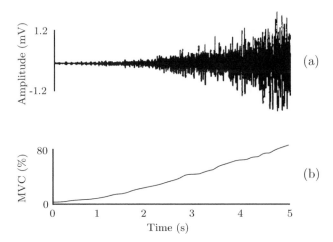

Figure 5.5: (a) The EMG signal and (b) the force curve during force-varying isometric muscle contraction, expressed as a percentage of maximal voluntary contraction. (Reprinted from Moritani et al. [28] with permission.)

fed by random noise [29], i.e., the well-known linear, filtered-noise model introduced in Section 3.1.3. The signal model is defined by

$$\mathbf{x} = \mathbf{Hv}, \tag{5.1}$$

where

$$\mathbf{x} = \begin{bmatrix} x(0) \\ x(1) \\ \vdots \\ x(N-1) \end{bmatrix}, \quad \mathbf{v} = \begin{bmatrix} v(0) \\ v(1) \\ \vdots \\ v(N-1) \end{bmatrix}. \tag{5.2}$$

The noise $\mathbf{v}$ is assumed to be white, Gaussian with variance σ_v^2. The matrix $\mathbf{H}$ defines a causal, linear filtering operation from the impulse response $h(n)$ and is lower triangular and Toeplitz,

$$\mathbf{H} = \begin{bmatrix} h(0) & 0 & 0 & \cdots & 0 \\ h(1) & h(0) & 0 & \cdots & 0 \\ h(2) & h(1) & h(0) & \cdots & 0 \\ \vdots & \vdots & \vdots & \ddots & \vdots \\ h(N-1) & h(N-2) & h(N-3) & \cdots & h(0) \end{bmatrix}. \tag{5.3}$$

Evidently, the signal model in (5.1) accounts neither for the influence of the firing pattern nor for the influence of individual MUAPs on the surface

EMG, but assumes that the spectral properties of the surface EMG signal are suitably shaped by the filter $h(n)$. Since the filtered signal $\mathbf{x}$ exhibits an initial transient without physiological counterpart, the transient should be omitted from subsequent analysis. The exact number of samples to be omitted depends on the properties of $h(n)$.

In order to assure that changes in the estimate of σ_v, occurring from one window to another, are not attributed to changes in filter gain, $h(n)$ is always subjected to normalization such that the variance σ_x^2 of the observed signal,

$$\sigma_x^2 = r_x(0) = \frac{1}{2\pi} \int_{-\pi}^{\pi} S_x(e^{\jmath\omega}) d\omega,$$

$$= \frac{1}{2\pi} \int_{-\pi}^{\pi} |H(e^{\jmath\omega})|^2 \sigma_v^2 d\omega, \tag{5.4}$$

is identical to σ_v^2; here, the Fourier transform of $h(n)$ is denoted $H(e^{\jmath\omega})$. Hence, we must require that

$$\frac{1}{2\pi} \int_{-\pi}^{\pi} |H(e^{\jmath\omega})|^2 d\omega = 1. \tag{5.5}$$

Although $h(n)$ is initially treated as being a priori known, it is later estimated from the observed signal.

With the assumption that the input noise $\mathbf{v}$ is Gaussian, the observed signal $\mathbf{x}$ is completely characterized by the PDF

$$p_v(\mathbf{x}) = \frac{1}{(2\pi)^{\frac{N}{2}} |\mathbf{R}_x|^{\frac{1}{2}}} \exp\left[-\frac{1}{2}\mathbf{x}^T \mathbf{R}_x^{-1} \mathbf{x}\right]. \tag{5.6}$$

Since the correlation matrix can be expressed as

$$\mathbf{R}_x = E\left[\mathbf{x}\mathbf{x}^T\right]$$
$$= \mathbf{H} E\left[\mathbf{v}\mathbf{v}^T\right] \mathbf{H}^T$$
$$= \sigma_v^2 \mathbf{H}\mathbf{H}^T, \tag{5.7}$$

its inverse equals

$$\mathbf{R}_x^{-1} = \frac{1}{\sigma_v^2} (\mathbf{H}^{-1})^T \mathbf{H}^{-1}, \tag{5.8}$$

and, accordingly, the PDF of $\mathbf{x}$ can be written as

$$p_v(\mathbf{x}; \sigma_v) = \frac{1}{(2\pi)^{\frac{N}{2}} |\sigma_v^2 \mathbf{H}\mathbf{H}^T|^{\frac{1}{2}}} \exp\left[-\frac{1}{2\sigma_v^2} (\mathbf{H}^{-1}\mathbf{x})^T (\mathbf{H}^{-1}\mathbf{x})\right]. \tag{5.9}$$

It should be noted that the Gaussian signal assumption has been found adequate in various surface EMG applications, including the present one on amplitude estimation [30, 31] (although inadequate in certain situations, see Section 5.2.2).

We will now derive the ML estimator of the standard deviation σ_v, determined from maximization of the logarithm of the likelihood function (cf. page 217),

$$\hat{\sigma}_v = \arg\max_{\sigma_v} \ln p_v(\mathbf{x}; \sigma_v). \tag{5.10}$$

Taking the logarithm of $p_v(\mathbf{x}; \sigma_v)$ in (5.9) and differentiating with respect to σ_v yields

$$\frac{\partial \ln p_v(\mathbf{x}; \sigma_v)}{\partial \sigma_v} = \frac{\partial}{\partial \sigma_v}\left[-\frac{N \ln 2\pi}{2} - N \ln \sigma_v - \frac{\ln|\mathbf{H}\mathbf{H}^T|}{2} - \frac{(\mathbf{H}^{-1}\mathbf{x})^T(\mathbf{H}^{-1}\mathbf{x})}{2\sigma_v^2} \right]$$

$$= -\frac{N}{\sigma_v} + \frac{1}{\sigma_v^3}(\mathbf{H}^{-1}\mathbf{x})^T(\mathbf{H}^{-1}\mathbf{x}), \tag{5.11}$$

where we have made use of the result on determinants given in (A.26). Setting the result in (5.11) equal to zero, we obtain the following expression,

$$\hat{\sigma}_v^2 = \frac{1}{N}(\mathbf{H}^{-1}\mathbf{x})^T(\mathbf{H}^{-1}\mathbf{x}), \tag{5.12}$$

which, considering that σ_v is the desired quantity, becomes

$$\hat{\sigma}_v = \sqrt{\frac{1}{N}(\mathbf{H}^{-1}\mathbf{x})^T(\mathbf{H}^{-1}\mathbf{x})}. \tag{5.13}$$

Filtering of the observed signal $\mathbf{x}$ with $\mathbf{H}^{-1}$,

$$\mathbf{y} = \mathbf{H}^{-1}\mathbf{x}, \tag{5.14}$$

has the interpretation of decorrelating ("whitening") the samples of $\mathbf{x}$ because the correlation matrix of $\mathbf{y}$ is diagonal,

$$\mathbf{R}_y = E\left[\mathbf{y}\mathbf{y}^T\right] = E\left[\mathbf{H}^{-1}\mathbf{x}\mathbf{x}^T(\mathbf{H}^{-1})^T\right]$$

$$= E\left[\mathbf{H}^{-1}\mathbf{H}\mathbf{v}\mathbf{v}^T\mathbf{H}^T(\mathbf{H}^{-1})^T\right]$$

$$= \sigma_v^2\mathbf{I}. \tag{5.15}$$

Since $\mathbf{H}$ is lower triangular and Toeplitz, it can be shown that its inverse is also lower triangular and Toeplitz and, therefore, corresponds to a causal operation. The columns of $\mathbf{H}^{-1}$ correspond to another linear, time-invariant filter whose impulse response is shifted downwards as the column index increases, in the same way as $h(n)$ was shifted downwards in (5.3) [32].

Before the ML estimator in (5.13) can be used in practice, we must first determine the matrix $\mathbf{H}$ from the observed signal. Since the power spectrum of the surface EMG usually exhibits a few pronounced peaks, the impulse response $h(n)$ is often modeled by an all-pole filter of low order, and thus the observed signal is modeled by a low-order AR process [29, 33–37]. From Section 3.4 we know that the coefficients of an all-pole filter are obtained as the solution of the normal equations in (3.126) and that the related whitening filter is always stable due to its FIR structure. The matrix $\mathbf{H}$ can be determined from an initial calibration phase in which the subject is instructed to perform a constant-force, isometric contraction [9].[3] Following the calibration phase, which typically lasts for a few seconds, the subsequent parts of the EMG recording are processed with the determined whitening filter. This filter remains appropriate to employ as long as the spectral properties of the recording remain unaltered. The estimate $\hat{\mathbf{H}}$ is substituted into (5.13) to yield the amplitude estimate

$$\hat{\sigma}_v = \sqrt{\frac{1}{N}(\hat{\mathbf{H}}^{-1}\mathbf{x})^T(\hat{\mathbf{H}}^{-1}\mathbf{x})}. \tag{5.16}$$

It should be noted that this estimator assumes that $\mathbf{x}$ is zero-mean; if this is not the case, the amplitude will be overestimated. Hence, it is essential to estimate and subtract the DC level from $\mathbf{x}$ prior to computation of $\hat{\sigma}_v$. Figure 5.6 summarizes the main blocks of the signal model and the ML amplitude estimator, respectively.

Although whitening of the observed signal is inherent to the ML estimator, it is essential to realize that the amplitude estimator that results from the assumption of a white observed signal,

$$\mathbf{H} = \mathbf{I}, \tag{5.17}$$

is the *standard technique* for analyzing the surface EMG in the time domain. Since the impulse response is given by $h(n) = \delta(n)$, the root mean-square (RMS) value of the observed signal,

$$\hat{\sigma}_v = \sqrt{\frac{1}{N}\mathbf{x}^T\mathbf{x}} = \sqrt{\frac{1}{N}\sum_{n=0}^{N-1} x^2(n)}, \tag{5.18}$$

is the amplitude estimate.

Since the amplitude estimate is influenced by a number of physiological factors, it is difficult to make a direct comparison of amplitude values

---

[3]During the calibration phase, the variance of the whitened signal directly results from solving the set of normal equations in (3.126), and, accordingly, the desired EMG amplitude estimate is the square-root of σ_e^2 in (3.128).

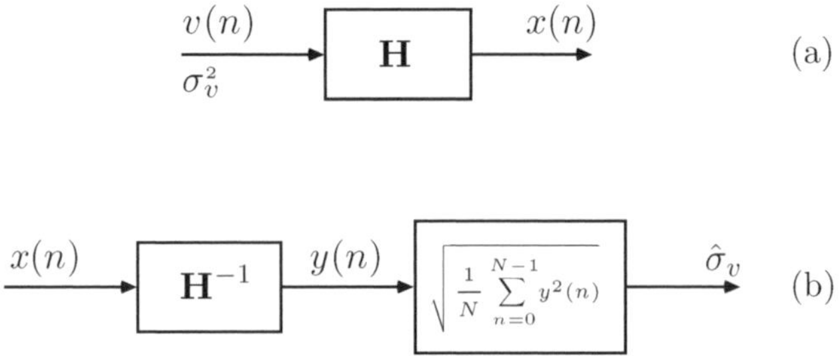

Figure 5.6: Amplitude estimation in EMG signal processing. (a) The linear, filtered-noise signal model and (b) the related ML estimator of the amplitude σ_v.

between subjects. This problem is usually handled by normalizing the amplitude estimate $\hat{\sigma}_v$ relative to the amplitude estimate at MVC. The normalized amplitude estimate reflects the degree of muscular activation and is expressed as a percentage of MVC.

5.2.2 Modifications of the ML Amplitude Estimator

Sliding window. In many situations, the EMG amplitude is changing during muscle contraction, and, therefore, the single-amplitude estimator in (5.16) must be replaced by another which is capable of producing a time-varying estimate. The common approach to track changes in EMG amplitude is through repeated estimation of σ_v in a sliding window, i.e., the same processing technique as was employed when computing the Hjorth parameter $\mathcal{H}_0(n)$ in (3.105). The estimator in Figure 5.6(b) can be easily extended to produce a time-varying estimate from the whitened signal $y(n)$, resulting from filtering of $x(n)$ using the first column of the whitening matrix $\mathbf{H}^{-1}$ as the impulse response. The time-varying amplitude estimate $\hat{\sigma}_v(n)$ is obtained by

$$\hat{\sigma}_v(n) = \sqrt{\frac{1}{N} \sum_{m=n-N+1}^{n} y^2(m)}. \qquad (5.19)$$

For a short window length N, the estimator $\hat{\sigma}_v(n)$ tracks rapid changes in amplitude at the expense of estimates with larger variance, whereas longer windows result in slower tracking but with smoother amplitude estimates.

The length of the sliding window may be selected with reference to the application of interest and the degree with which the amplitude is expected to vary. For example, the window length has been adapted to the walking speed during human locomotion [38], see also [39, 40]. Another approach is to adapt the window length to the local characteristics of the EMG signal so that a short window is used during rapid changes in amplitude, and vice versa. The function which maps signal characteristics into window lengths may involve the signal's amplitude and its first derivative [41, 42], as well as its second derivative [43].

The performance of the RMS amplitude estimator, when combined with the sliding window technique, is illustrated by the simulation example in Figure 5.7. The observed signal $x(n)$ was generated by passing white, Gaussian noise $v(n)$ through a linear, time-invariant filter $h(n)$, designed to produce the power spectrum displayed in Figure 5.7(b) [44].[4] The standard deviation of the noise was increased by a factor of two halfway into the observation interval in order to mimic a sudden change in EMG amplitude. The RMS amplitude estimate was computed either directly from the observed signal (Figure 5.7(c)) or from the whitened signal (Figure 5.7(e)). Assuming that the observed signal derives from an AR model, the coefficients of the whitening filter were estimated using Burg's method. It can be concluded from this simulation example that the whitening operation leads to amplitude estimates with variance considerably lower than estimates obtained directly from the observed signal, see Figures 5.7(d) and (f), respectively.

Laplacian PDF. In situations when the surface EMG is recorded at low contraction levels, the number of MUAPs is relatively sparse so that gaps in time between different MUAPs are occurring more frequently. As a result, the PDF which models the signal amplitude should be more sharply peaked at values around zero than what the above-mentioned Gaussian PDF accounts for; at higher contraction levels, the amplitude histogram has been found to become increasingly Gaussian [31, 46]. The Laplacian PDF has been suggested as a model of the EMG amplitude when recorded at low contraction levels [31]. Assuming that the samples $x(n)$ are statistically independent, Laplacian random variables, their joint PDF is given by

$$p_v(\mathbf{x}; \sigma_v) = \prod_{n=0}^{N-1} p_v(x(n); \sigma_v)$$

$$= \prod_{n=0}^{N-1} \frac{1}{\sqrt{2\sigma_v^2}} \exp\left[-\sqrt{\frac{2}{\sigma_v^2}} |x(n)|\right], \qquad (5.20)$$

from which it is straightforward to derive the ML estimator of σ_v. Performing the same operations as was done in (5.10) and (5.11), we obtain the following

---

[4]This simulation model has been found useful for the purpose of comparing algorithmic performance during voluntary, isometric contractions [45], where the idea was to study the influence of whitening on different amplitude estimates.

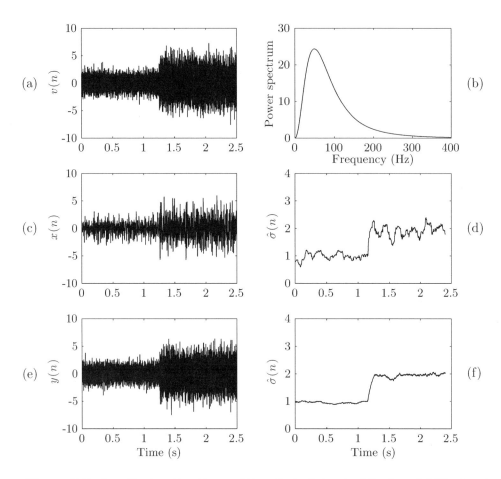

Figure 5.7: Amplitude estimation and the effect of signal whitening. (a) The input white, Gaussian noise whose variance increases by a factor of two at second 1.25, (b) the power spectrum of the filtered-noise signal, (c) the observed signal, (d) the amplitude computed by the estimator in (5.18), (e) the whitened signal, and (f) the amplitude computed by the estimator in (5.16). The sampling rate was set to 5 kHz, and the length of the sliding window was 100 ms. The parameters of the whitening filter were estimated from the first 500 ms of the observed signal.

estimator,

$$\hat{\sigma}_v = \frac{\sqrt{2}}{N} \sum_{n=0}^{N-1} |x(n)|, \tag{5.21}$$

commonly referred to as the *average rectified value* (ARV). Together with the RMS estimator in (5.18), the ARV estimator represents the most popular amplitude estimator in EMG analysis.

In contrast to the multivariate Gaussian PDF in (5.6) which accounts for correlated samples, the multivariate Laplacian PDF unfortunately has no manageable expression which lends itself to the derivation of a closed-form ML estimator. As a result, the implications of a Laplacian model in EMG analysis have, so far, been restricted to the case in (5.20) with statistically independent samples. While the whitening filter $\mathbf{H}^{-1}$ is inherent to the ML estimator when samples are Gaussian, the whitening operation has been found advantageous when used in conjunction with the ARV estimator. The improvement in performance achieved by the RMS estimator with signal prewhitening (Figure 5.7) is paralleled by a similar improvement when the ARV estimator makes use of a prewhitened signal [47].

Despite the fact that signals characterized by Gaussian and Laplacian PDFs are distinctly different (cf. the signals displayed in Figure 4.17), the difference in performance of the RMS and ARV estimators is practically negligible. This finding has been established experimentally in several EMG applications as well as in simulation studies [31, 48, 49].

Adaptive whitening. The use of a fixed whitening filter $\mathbf{H}^{-1}$, determined from a calibration phase, assumes that the spectral properties of the EMG signal remain the same throughout the recording. However, the spectral properties change with the task performed as different motor units are recruited within the muscle, thus implying that a fixed whitening filter cannot fulfil its original purpose. As long as the spectral changes are relatively slow, we may employ an adaptive technique to estimate the parameters of the whitening filter, for example, using the LMS or the GAL algorithm described in Section 3.6.5 which are both based on the assumption of an underlying AR signal model.

Another reason for considering an adaptive approach to signal whitening is the fact that the SNR changes with contraction level. Since a large number of motor units are active during stronger contractions, the muscular activity dominates relative to the noise of nonmuscular origin so that the SNR of the EMG signal is high. However, at weaker contractions (about 10% MVC and less) the presence of nonmuscular noise can no longer be neglected, as was done in the model in (5.1), but an additive, random noise term $\mathbf{w}$ should be included in the model,

$$\mathbf{x} = \mathbf{H}\mathbf{v} + \mathbf{w}. \tag{5.22}$$

Accordingly, the observed signal $\mathbf{x}$ can no longer be treated as deriving entirely from an AR process, calling for adaptive estimation techniques which produce an estimate of the signal part $\mathbf{H}\mathbf{v}$ in the presence of additive noise. Such adaptive techniques have been developed from the theory of Wiener filters, cf. Section 4.4 [50].

5.2.3   Multiple Electrode Sites

A major modification of the above ML amplitude estimator is to make it incorporate spatial information from multiple electrodes positioned about the muscle of interest. While the single electrode setup only views the adjacent area, the introduction of spatial sampling provides a more complete description of a muscle's electrical activity and can therefore be expected to produce improved amplitude estimates.

Figure 5.8(a) presents a multichannel model of the surface EMG when multiple electrode sites are employed. In this model, each of the M observed signals, denoted $x_1(n), \ldots, x_M(n)$, are assumed to exhibit not only temporal correlation, but also varying degrees of *spatial correlation* with the other $M - 1$ observed signals [49, 51, 52]. Spatial correlation is introduced by the $M \times M$ "mixing" matrix $\mathbf{S}$ that linearly combines the mutually uncorrelated input noise processes $v_1(n), \ldots, v_M(n)$ and is assumed to be independent of time. Each noise process $v_m(n)$ is assumed to be white, Gaussian with channel-independent, identical variances σ_v^2. *Temporal correlation* is introduced in each of the signals that results from mixing by $\mathbf{S}$ through the use of a channel-dependent filter whose impulse response is $h_m(n)$. Hence, the observed signals $x_m(n)$ are Gaussian because the Gaussian input noise $v_m(n)$ is modified by linear transformations. Since the EMG amplitude σ_v is common to all M channels, its estimation can be based on more information than what is available in the single-channel situation.

The whitening operation of the single-channel ML estimator in (5.13) turns out to be equally crucial to the multichannel ML estimator, although whitening should now be performed simultaneously in time and space. However, the derivation of the optimal multichannel ML estimator is exceedingly complicated, and, therefore, we restrict ourselves to describe the following three-step estimation procedure [52]:

1. Temporal whitening using a channel-dependent filter $\mathbf{H}_m^{-1}$, determined in the same way as the estimate $\hat{\mathbf{H}}^{-1}$ in (5.16), followed by

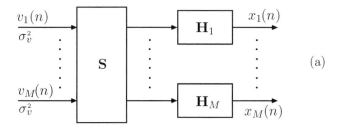

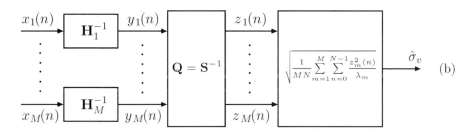

Figure 5.8: (a) A multichannel, linear, filtered-noise model of the surface EMG signal with identical input noise variances σ_v^2 and (b) the related ML amplitude estimator of σ_v.

2. spatial whitening which removes the channel-to-channel correlation introduced by the mixing matrix **S**, and, finally,

3. the multichannel amplitude estimate $\hat{\sigma}_v$ obtained as a weighted combination of the RMS amplitude estimates of the individual channels,

see the block diagram in Figure 5.8(b). This three-step estimation procedure is particularly plausible to investigate since the two whitening operations are performed in reverse to those of the signal model in Figure 5.8(a) which color the input noise. The order by which whitening is performed is uncritical since the correlation in space and time are assumed to be uncoupled.

Similar to the single-channel case, each of the M channels $x_m(n)$ is whitened in time using the filter $h_m(n)$ so that another signal $y_m(n)$ results whose variance is equal to σ_v^2. For the multichannel case, it is advantageous to make use of the vector representation of the whitened samples at time n,

$$\mathbf{y}(n) = \begin{bmatrix} y_1(n) \\ y_2(n) \\ \vdots \\ y_M(n) \end{bmatrix}, \quad n = 0, 1, \ldots, N-1, \tag{5.23}$$

a representation earlier introduced in (3.32). The correlation matrix of $\mathbf{y}(n)$ can be expressed as

$$\mathbf{R}_y = E\left[\mathbf{y}(n)\mathbf{y}(n)^T\right] = \sigma_v^2 \mathbf{A}_y, \tag{5.24}$$

where the matrix $\mathbf{A}_y$ describes spatial correlation. Since whitening of $x_m(n)$ does not alter the Gaussian signal property, the vector $\mathbf{y}(n)$ is equally characterized by a Gaussian PDF which, for each time n, is given by

$$p_v(\mathbf{y}(n); \sigma_v) = \frac{1}{(2\pi\sigma_v^2)^{\frac{M}{2}}|\mathbf{A}_y|^{\frac{1}{2}}} \exp\left[-\frac{1}{2\sigma_v^2}\mathbf{y}^T(n)\mathbf{A}_y^{-1}\mathbf{y}(n)\right], \tag{5.25}$$

where the unknown parameter σ_v has been separated from $\mathbf{R}_y$ by the introduction of $\mathbf{A}_y$. In order to proceed, the eigenvalue decomposition of a symmetric matrix is used to express $\mathbf{A}_y$ (see (A.37)),

$$\mathbf{A}_y = \mathbf{Q}\mathbf{\Lambda}\mathbf{Q}^T, \tag{5.26}$$

where the columns of the matrix $\mathbf{Q}$ are defined by the eigenvectors of $\mathbf{A}_y$, and $\mathbf{\Lambda}$ is a diagonal matrix defined by the eigenvalues

$$\mathbf{\Lambda} = \mathrm{diag}(\lambda_1, \lambda_2, \ldots, \lambda_M). \tag{5.27}$$

Introducing the linear transformation,

$$\mathbf{z}(n) = \mathbf{Q}^T\mathbf{y}(n), \tag{5.28}$$

and making use of the fact that

$$\mathbf{A}_y^{-1} = \mathbf{Q}\mathbf{\Lambda}^{-1}\mathbf{Q}^T, \tag{5.29}$$

we can write $p_v(\mathbf{y}(n); \sigma_v)$ in (5.25) as

$$\begin{aligned}
p_v(\mathbf{y}(n); \sigma_v) &= \frac{1}{(2\pi\sigma_v^2)^{\frac{M}{2}}|\mathbf{A}_y|^{\frac{1}{2}}} \exp\left[-\frac{1}{2\sigma_v^2}\mathbf{z}^T(n)\mathbf{\Lambda}_y^{-1}\mathbf{z}(n)\right] \\
&= \frac{1}{(2\pi\sigma_v^2)^{\frac{M}{2}}|\mathbf{A}_y|^{\frac{1}{2}}} \exp\left[-\frac{1}{2\sigma_v^2}\sum_{m=1}^{M}\frac{z_m^2(n)}{\lambda_m}\right].
\end{aligned} \tag{5.30}$$

Since this PDF can be expressed as a product of univariate, Gaussian PDFs of $z_1(n), \ldots, z_M(n)$, it is evident that linearly transforming $\mathbf{y}(n)$ with the eigenvector matrix $\mathbf{Q}$ is synonymous with removing the spatial correlation that exists between different channels.

For the Gaussian PDF in (5.30), it is easily shown that the ML estimator of σ_v is defined by the following eigenvalue-weighted RMS value,

$$\hat{\sigma}_v(n) = \sqrt{\frac{1}{M} \sum_{m=1}^{M} \frac{z_m^2(n)}{\lambda_m}}. \tag{5.31}$$

The weighting should be interpreted as a normalization of $z_m(n)$ with its variance since

$$E\left[z_m^2(n)\right] = \sigma_v^2 \lambda_m, \tag{5.32}$$

which stems from the fact that

$$\begin{aligned}
E\left[\mathbf{z}(n)\mathbf{z}(n)^T\right] &= \mathbf{Q}^T E\left[\mathbf{y}(n)\mathbf{y}(n)^T\right] \mathbf{Q} \\
&= \sigma_v^2 \mathbf{\Lambda}.
\end{aligned} \tag{5.33}$$

In order to develop the multichannel ML estimator, we are interested in the Gaussian PDF of all available data, i.e., $p_v(\mathbf{y}(0), \ldots, \mathbf{y}(N-1); \sigma_v)$, which, due to the whitening operation, can be expressed as a product,

$$p_v(\mathbf{y}(0), \ldots, \mathbf{y}(N-1); \sigma_v) = \prod_{n=0}^{N-1} p_v(\mathbf{y}(n); \sigma_v). \tag{5.34}$$

As a result, the multichannel ML estimator of the amplitude σ_v is given by the following expression,

$$\hat{\sigma}_v = \sqrt{\frac{1}{MN} \sum_{m=1}^{M} \sum_{n=0}^{N-1} \frac{z_m^2(n)}{\lambda_m}}. \tag{5.35}$$

Before the three-step estimator can be used in practice, the spatial correlation matrix $\mathbf{R}_y$ has to be determined from the temporally whitened signals using the estimator

$$\hat{\mathbf{R}}_y = \frac{1}{N} \sum_{n=0}^{N-1} \mathbf{y}(n)\mathbf{y}^T(n). \tag{5.36}$$

This estimator is reasonable to use when it can be assumed that each of the channels is stationary in time during the observation interval. Decomposition of $\hat{\mathbf{R}}_y$ produces an estimate of the rotation matrix $\mathbf{Q}$ because $\mathbf{R}_y$ and $\mathbf{A}_y$ only differ by a scalar factor, and, consequently, the eigenvalues λ_m of $\mathbf{A}_y$

are proportional to the eigenvalues λ'_m of $\mathbf{R}_y$. Since the trace of a matrix equals the sum of the eigenvalues, we have

$$\text{tr}(\mathbf{R}_y) = M\sigma_v^2 = \sigma_v^2 \sum_{m=1}^{M} \lambda_m, \tag{5.37}$$

and thus

$$\sum_{m=1}^{M} \lambda_m = M, \tag{5.38}$$

which, when combined with the fact that λ_m and λ'_m are proportional, leads to

$$\lambda_m = \frac{\lambda'_m M}{\displaystyle\sum_{m=1}^{M} \lambda'_m}. \tag{5.39}$$

It should be noted that the eigenvalue decomposition of $\hat{\mathbf{R}}_y$ will be ill-conditioned when two channels are almost perfectly correlated, i.e., when the channels are obtained from very closely spaced electrode sites. In such cases, certain eigenvalues are almost zero, due to $\hat{\mathbf{R}}_y$ being rank-deficient, and create numerical problems when calculating $\hat{\sigma}_v$ in (5.35) since $z_m^2(n)$ is divided by a value close to zero.[5]

Similar to the single-channel ML estimator, a sliding window approach may be employed to track changes in the amplitude of a multichannel EMG recording. It is desirable to update the correlation matrix $\mathbf{R}_y$ in each new window position and to perform the related eigenvalue decomposition so that spatial whitening remains meaningful over time.

The performance of the three-step, multichannel estimator has been compared to that of the single-channel estimator in terms of SNR of the amplitude estimate, using surface EMGs recorded during nonfatiguing, constant-force, isometric contractions [52]. The multichannel estimator was found to produce a considerably higher SNR which improved as the number of electrode sites increased. Simplified structures of the multichannel estimator, omitting the filter for temporal whitening, have also been studied and found to produce improved SNRs [33, 49, 51, 53, 54].

---

[5]A remedy to this problem may be to use a low-rank approximation of R_y based on the singular value decomposition (SVD), see page 639 in Appendix A.

5.3 Spectral Analysis of the Surface EMG

While amplitude characterization conveys important information on the surface EMG signal, expressed in terms of RMS and/or ARV values, additional information can be extracted by means of power spectral analysis. Such analysis has found particular significance in the study of muscle fatigue since, under voluntary, isometric contractions, fatigue is manifested as a shift of the power spectrum toward lower frequencies [55]; a phenomenon commonly referred to as *spectral compression*. Not only is amplitude estimation insusceptible to such a "slowing" behavior of the signal, but time domain analysis is generally less suitable for its quantification. The fatigue-related spectral shift to lower frequencies is mainly attributed to a decrease in muscle fiber conduction velocity, although other physiological factors also have an influence. Spectral compression is demonstrated by Figure 5.9 with a set of EMG signals recorded from a healthy subject who performed voluntary, isometric contraction, sustained during 100 s.

Spectral compression may be better understood by considering a simplistic model in which the (continuous-time, random) signal $x_1(t)$ is assumed to derive from a single motor unit [56]. Slightly later at the time τ, and at increased fatigue, the same motor unit produces another signal $x_2(t)$ related to $x_1(t)$ by[6]

$$x_2(t) = x_1(\nu t - \tau),\qquad(5.40)$$

where the scaling parameter ν models a slower conduction velocity through widening of $x_1(t)$, i.e., for $\nu < 1$. In terms of power spectra, the relationship in (5.40) corresponds to

$$S_{x_2}(\Omega) = \frac{1}{\nu}S_{x_1}\left(\frac{\Omega}{\nu}\right),\qquad(5.41)$$

which thus establishes that the original power spectrum $S_{x_1}(\Omega)$ is subjected to compression (scaling), as well as an increase in power.

As illustrated by the power spectra in Figure 5.9, the surface EMG signal is usually associated with a rather unimodal spectral shape, implying that parameters which describe the dominant frequency are relevant to apply. The *mean frequency* (MNF) and the *median frequency* (MDF) are two spectral parameters which have become exceedingly popular in EMG analysis. Assuming that the observed signal is discrete-time with power spectrum

---

[6]As the reader already may have noted, the continuous-time signal description is preferred when scaling in time is of interest since the scaling operation is more easily accommodated with such a description.

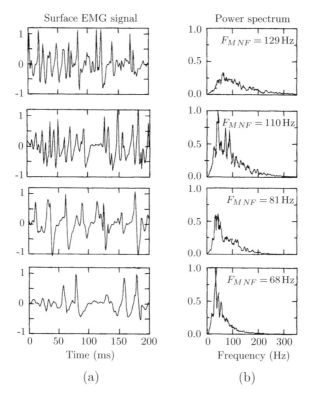

Figure 5.9: (a) The surface EMG signal recorded during voluntary, isometric contraction at 60% MVC of a healthy tibialis anterior muscle, sustained for 100 s. The displayed signals are from second 1, 30, 60, and 90 (top to bottom). (b) The corresponding power spectra were obtained by averaging and smoothing the periodograms of five contiguous signal segments such as those displayed in (a). (Reprinted from Merletti et al. [56] with permission.)

$S_x(e^{j\omega})$, the mean frequency is defined as the normalized, one-sided, first-order spectral moment,

$$
\omega_{\text{MNF}} = \frac{\displaystyle\int_0^{\pi} \omega S_x(e^{j\omega})\,d\omega}{\displaystyle\int_0^{\pi} S_x(e^{j\omega})\,d\omega}.
\tag{5.42}
$$

This definition bears close resemblance with the definition of the Hjorth mobility parameter $\mathcal{H}_1$ in (3.97), defined as the square-root of the normalized, two-sided, first-order spectral moment. However, while the integration interval in (5.42) is one-sided and only includes positive frequencies, the two-sided definition in (3.95) leads to odd-numbered moments, e.g., for $n = 1$, that are identical to zero due to the symmetry of a power spectrum. In comparison,

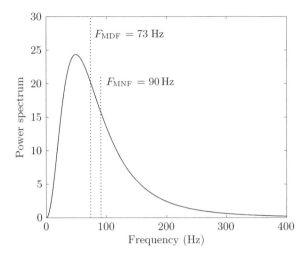

Figure 5.10: An example of an EMG power spectrum and the values of the spectral parameters F_{MDF} and F_{MNF}.

$\mathcal{H}_1$ and ω_{MNF} both describe the dominant frequency of a signal, although the value of $\mathcal{H}_1$ is slightly larger than that of ω_{MNF}; both parameters are useful for characterizing spectral compression since they change in exactly the same way during compression (Problems 5.7 and 5.9). It should be noted that higher-order spectral moments have also been considered for obtaining a more detailed shape characterization of the EMG power spectrum than what is offered by the dominant frequency [57, 58].

The other popular spectral parameter in EMG analysis is the median frequency ω_{MDF}, defined as that particular frequency which divides the total area under $S_x(e^{\jmath\omega})$ into two parts of equal size,

$$\int_0^{\omega_{\mathrm{MDF}}} S_x(e^{\jmath\omega})d\omega = \int_{\omega_{\mathrm{MDF}}}^{\pi} S_x(e^{\jmath\omega})d\omega. \qquad (5.43)$$

The median frequency ω_{MDF} is identical to ω_{MNF} when the positive part of $S_x(e^{\jmath\omega})$ is symmetric with respect to a certain frequency, whereas ω_{MDF} is associated with lower values than ω_{MNF} when $S_x(e^{\jmath\omega})$ comes with a high-frequency tail, see Figure 5.10. The median frequency ω_{MDF} has the advantage of being more robust than ω_{MNF} when the signal has very low SNRs, whereas it exhibits a larger variance at high SNRs [59–61].

Another way of writing the definition in (5.43) is

$$\int_0^{\omega_{\mathrm{MDF}}} S_x(e^{\jmath\omega})d\omega = \frac{1}{2}\int_0^{\pi} S_x(e^{\jmath\omega})d\omega, \qquad (5.44)$$

a form which lends itself to the generalization of the median frequency into a "percentile" frequency. Rather than determining the median frequency, which corresponds to the factor $\frac{1}{2}$ on the right-hand side of (5.44), a percentile frequency can be determined for any choice of factor contained within the interval $[0, 1]$ [62].

Similar to the estimation of EMG amplitude, spectral parameters are commonly computed using the sliding window approach in order to monitor, for example, muscle fatigue [63]. Returning to the example in Figure 5.9, the mean frequency is monitored and found to decrease by almost a factor of two, i.e., from 129 to 68 Hz, as muscle fatigue becomes increasingly pronounced during the 100 s of sustained voluntary contraction. Besides characterization of spectral compression during muscle fatigue, ω_{MDF} and ω_{MNF} have been employed for several other purposes such as the investigation of back and neck pain and age-induced muscle changes.

From a methodological point of view, spectral analysis of the surface EMG signal largely parallels that of the EEG signal, earlier presented in Chapter 3, and involves nonparametric (Fourier) as well as parametric spectral estimation techniques [45, 64]. In a similar way, parametric spectral analysis has been synonymous with AR modeling and has, in addition to being part of signal whitening in EMG amplitude estimation, been employed for prosthesis control [20, 36, 65, 66]. In the latter application, the AR parameters have been used to define a set of features for discrimination between different limb functions. Time–frequency analysis (see Section 3.6) has also found its way into the toolbox of techniques for studying the surface EMG and has been found particularly suitable for characterization of muscle contraction during dynamic conditions [58, 67–72].

We conclude this section by pointing out that additional aspects on spectral properties of the EMG signal are provided in Section 5.5, where the relationship between a signal model of the intramuscular EMG and its power spectrum is derived. In that model, it is assumed that the power spectrum of individual MUAP waveforms and the statistics of the firing pattern are known. Another mathematical model of the EMG power spectrum has been developed with which the spectrum can be explained in terms of physiological and geometrical parameters accounting for the size of the muscle fibers, the conduction velocity, the number of fibers per motor unit, the electrode-to-muscle distance, and the electrode configuration; see [73] for a description of this model and [74] for a recent refinement with a considerably more detailed account of the spatial filtering properties.

5.4 Conduction Velocity Estimation

Since certain muscle disorders are associated with a reduction in conduction velocity of the muscle fiber, it is important to estimate this parameter from either intramuscular or surface EMG recordings [7, 75]. Conduction velocity is studied using a recording setup that involves two electrode locations a certain distance d apart. The time it takes for the action potentials to propagate along the muscle fiber is reflected by the delay between the two recorded signals $x_1(n)$ and $x_2(n)$. Thus, by estimating the time delay θ between the signals, the conduction velocity ν is obtained from

$$\nu = \frac{d}{\theta}. \tag{5.45}$$

Conduction velocity estimation from the intramuscular EMG is made difficult since the shape of a motor unit waveform is changing between the two recording locations due to factors such as different conduction velocities of the active motor units and fiber orientation with respect to the electrodes [7]. Consequently, the estimation of θ requires more sophisticated techniques than merely determining the time distance between related peaks of the two signals. The presence of noise is another factor which makes it necessary to develop more sophisticated techniques than simple peak-picking in order to produce reliable estimates of θ. The same type of problem is associated with the surface EMG, even though the signal instead reflects the summation of many firing motor units and thus an "average" conduction velocity estimate is obtained for the muscle fiber.

Figure 5.11 displays six simultaneously recorded surface EMG signals, using a linear array of electrodes with identical inter-electrode distances (10 mm) positioned along the direction of the muscle fiber. The time delay between different channels can be clearly discerned by the eye and increases, as one would expect, for electrodes farther apart. Figure 5.11 also illustrates the fact that one channel is not a pure delay of another, but the wave shape changes quite considerably when viewed over the six channels; some of the changes in shape may be attributed to noise.[7]

The multichannel EMG signal in Figure 5.11 suggests various approaches to conduction velocity estimation such as analysis in pairs of the channels or simultaneous analysis of all channels. Below, we will introduce a simple, phenomenological model of the time delay between two EMG signals and

---

[7]The linear array of electrodes should be located so that all electrodes are placed away from the innervation zone of the muscle in the same direction. If not, channels 2, 3, and higher will not be delayed in relation to channel 1, but from the channel being closest to the innervation zone; as a result, the estimation of the conduction velocity ν becomes much more difficult to accomplish.

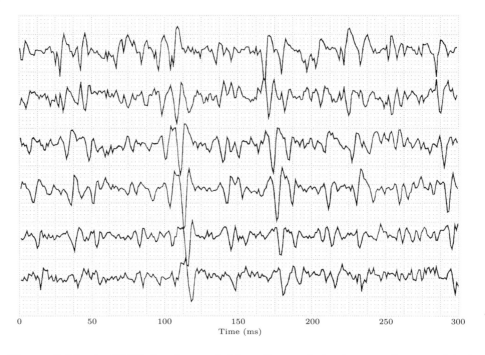

Figure 5.11: An EMG signal recorded by a linear electrode array from a subject performing a low-level (10% MVC), voluntary, isometric contraction of the biceps brachii muscle of the upper arm. Note the increasing time delay of the signals when reading the plot from top to bottom. (Reprinted from Farina et al. [7] with permission.)

derive the corresponding ML estimator (Section 5.4.1). In a next step, the model is extended to account for multichannel signals recorded with a linear electrode array, and again the corresponding ML estimator is derived (Section 5.4.2).

Before proceeding with the actual estimation problem, it is of vital importance to realize that the input to the conduction velocity estimator is a sampled signal with limited temporal resolution. In fact, it can be easily shown that time delay estimation requires subsample resolution when the surface EMG is acquired with the typical sampling rate of 1 kHz, i.e., with a sampling interval of 1 ms [7]. Considering that conduction velocity ranges from 2 to 5 m/s and the distance between two electrodes may be 5 mm, the actual time delay between two signals is within the range from 1 to 2.5 ms. Hence, it is clear that the time delay estimate has insufficient resolution when the samples have a crude resolution of 1 ms. A number of estimation techniques are fortunately available which are not limited by the temporal resolution of the original signal; one of these techniques is presented below.

5.4.1 Two-Channel Time Delay Estimation

Estimation of the time delay θ can, in its simplest form, be formulated as a two-channel problem with the following statistical signal model as the starting point,

$$x_1(n) = s(n) + v_1(n), \qquad (5.46)$$
$$x_2(n) = s(n - \theta) + v_2(n), \qquad (5.47)$$

where $n = 0, 1, \ldots, N - 1$. The observed signal $x_1(n)$ is assumed to be composed of a fixed, deterministic but unknown signal $s(n)$ and additive, white, Gaussian noise $v_1(n)$ with variance σ_v^2. The same assumption applies to the second channel $x_2(n)$, except that $s(n)$ is delayed by the unknown time θ. It assumed that θ is constrained such that $s(n)$ is always completely contained in the observation interval (cf. the model in (4.120)). Although it is natural to consider θ to be integer-valued in (5.47), this restriction will be loosened once the ML estimator of θ has been derived. Furthermore, the noise components $v_1(n)$ and $v_2(n)$ are assumed to be uncorrelated with each other—an assumption which may be questioned when the two electrode locations are close.

The ML estimator of θ is derived by maximizing the PDF which characterizes the available observations. Since $x_1(n)$ does not depend on θ, and $v_1(n)$ and $v_2(n)$ are uncorrelated, the joint PDF $p_v(\mathbf{x}_1, \mathbf{x}_2; \theta, \mathbf{s})$ of the model can be expressed as

$$p_v(\mathbf{x}_1, \mathbf{x}_2; \theta, \mathbf{s}) = p_v(\mathbf{x}_1; \mathbf{s}) p_v(\mathbf{x}_2; \theta, \mathbf{s}), \qquad (5.48)$$

where $\mathbf{x}_1$ and $\mathbf{x}_2$ are the vector representations of $x_1(n)$ and $x_2(n)$, respectively. From before we know that the ML estimator results from the following maximization,

$$[\hat{\theta}, \hat{\mathbf{s}}] = \arg\max_{\theta, \mathbf{s}} \ln p_v(\mathbf{x}_1, \mathbf{x}_2; \theta, \mathbf{s}). \qquad (5.49)$$

Making use of the assumption that $p_v(\mathbf{x}_1, \mathbf{x}_2; \theta, \mathbf{s})$ is Gaussian, we have

$$p_v(\mathbf{x}_1, \mathbf{x}_2; \theta, \mathbf{s})$$
$$= \prod_{n=0}^{N-1} \frac{1}{2\pi\sigma_v^2} \exp\left[-\frac{(x_1(n) - s(n))^2}{2\sigma_v^2}\right] \exp\left[-\frac{(x_2(n) - s(n - \theta))^2}{2\sigma_v^2}\right].$$
$$(5.50)$$

Taking the logarithm and grouping factors independent of θ or $\mathbf{s}$, we obtain

$$\ln p_v(\mathbf{x}_1, \mathbf{x}_2; \theta, \mathbf{s}) = \text{constant} +$$
$$\frac{1}{2\sigma_v^2} \sum_{n=0}^{N-1} \left(2x_1(n)s(n) + 2x_2(n)s(n - \theta) - s^2(n) - s^2(n - \theta)\right). \quad (5.51)$$

Maximization of the log-likelihood function in (5.51) is done by first differentiating it with respect to the continuous-valued parameter $s(n)$ for a given value of θ,

$$\frac{\partial \ln p_v(\mathbf{x}_1, \mathbf{x}_2; \theta, \mathbf{s})}{\partial s(n)} = 2x_1(n) + 2x_2(n + \theta) - 2s(n) - 2s(n), \qquad (5.52)$$

which, when set to zero, becomes

$$\hat{s}(n; \theta) = \frac{1}{2}\left(x_1(n) + x_2(n + \theta)\right). \qquad (5.53)$$

Inserting $\hat{s}(n; \theta)$ into the log-likelihood function in (5.51) and maximizing with respect to the other parameter θ, we obtain

$$\hat{\theta} = \arg\max_{\theta}\left(\frac{1}{2}\sum_{n=0}^{N-1}\left(x_1(n)x_2(n + \theta) + x_2(n)x_1(n - \theta)\right) - \frac{1}{4}E_x\right), \qquad (5.54)$$

where E_x denotes the total energy of the observed signals,

$$E_x = \sum_{n=0}^{N-1}\left(x_1^2(n - \theta) + x_2^2(n + \theta)\right). \qquad (5.55)$$

Making use of the assumption that $s(n)$ is completely contained in the observation interval, the ML estimator of θ can be written as

$$\hat{\theta} = \arg\max_{\theta}\left(\sum_{n=0}^{N-1} x_2(n)x_1(n - \theta)\right). \qquad (5.56)$$

Hence, the time delay estimate $\hat{\theta}$ is given by that integer value which maximizes the cross-correlation between $x_1(n)$ and $x_2(n)$.

In order to bypass the limited temporal resolution imposed by the sampling rate, we can interpolate $x_1(n)$ and $x_2(n)$ to a sampling rate which is sufficiently high. However, half of the computations required for such interpolation can be circumvented by instead interpolating the cross-correlation function

$$y(\theta) = \sum_{n=0}^{N-1} x_2(n)x_1(n - \theta) \qquad (5.57)$$

before the location of its maximum value is determined [45]. In practice, it is usually sufficient to only interpolate $y(\theta)$ in a short interval centered around its main peak so that the location may be determined from the ML estimator defined in (5.56). Since the cross-correlation function often exhibits smooth behavior, the peak location can be found from fitting a second-order polynomial that passes through the peak sample and its two surrounding samples (parabolic interpolation).

Another approach to improve the temporal resolution of θ involves the frequency domain expression of the cross-correlation operation in (5.57). By virtue of Parseval's theorem, the ML estimator in (5.56) can be written as

$$\hat{\theta} = \arg\max_{\theta} \left(\frac{1}{2\pi} \int_{-\pi}^{\pi} X_2(e^{\jmath\omega}) X_1^*(e^{\jmath\omega}) e^{\jmath\theta\omega} d\omega \right), \qquad (5.58)$$

where $X_1(e^{\jmath\omega})$ and $X_2(e^{\jmath\omega})$ denote the Fourier transform of $x_1(n)$ and $x_2(n)$, respectively, which are continuous-valued functions of ω. The crucial observation to be made from (5.58) is that θ no longer needs to be constrained to integer values but can be treated as continuous-valued. Thus, $\hat{\theta}$ can either be determined by evaluating the integral in (5.58) on a sufficiently fine frequency grid or by adopting a gradient-based optimization technique which finds the maximum of the integral [76]; the details of the gradient-based technique are developed in Problem 5.11.

5.4.2 Multichannel Time Delay Estimation

The accuracy of the conduction velocity estimate can be expected to improve when the time delay estimation is based on a multichannel EMG recording since more information is available on how the signal propagates along the muscle fiber. Therefore, it is desirable to generalize the two-channel model in (5.46) and (5.47) so that it accounts for the signal delay from one electrode location to another. For the case when the multichannel recording is acquired by a linear electrode array positioned away from the innervation zone, it may be assumed that the time delay between adjacent channels is fixed,

$$\theta_m = (m-1)\theta, \quad m = 1, \ldots, M, \qquad (5.59)$$

where M denotes the number of channels [77]. Thus, the estimation problem becomes one of finding a single time delay θ rather than $M-1$ different ones. With (5.59), the multichannel signal model is defined by

$$x_m(n) = s(n - (m-1)\theta) + w_m(n), \quad m = 1, \ldots, M, \qquad (5.60)$$

where assumptions on signal and noise properties are the same as those of the two-channel model; in fact, the two models are identical when $M = 2$ in (5.60).

Since the noise in a channel is assumed to be uncorrelated with the noise in the other channels, the multivariate Gaussian PDF can be factorized as

$$p_v(\mathbf{x}_1, \ldots, \mathbf{x}_M; \theta, \mathbf{s}) = \prod_{m=1}^{M} p_v(\mathbf{x}_m; \theta, \mathbf{s}). \qquad (5.61)$$

Taking the logarithm of $p_v(\mathbf{x}_1, \ldots, \mathbf{x}_M; \theta, \mathbf{s})$, we obtain the following quadratic expression of the log-likelihood function,

$$\ln p_v(\mathbf{x}_1, \ldots, \mathbf{x}_M; \theta, \mathbf{s}) = \text{constant} - \frac{1}{2\sigma_v^2} \sum_{m=1}^{M} \sum_{n=0}^{N-1} (x_m(n) - s(n - (m-1)\theta))^2,$$
(5.62)

which is to be maximized with respect to θ and $\mathbf{s}$ in order to produce the desired ML estimator. Proceeding in a way similar to the two-channel case, differentiation of the log-likelihood function with respect to $s(n)$ yields

$$\frac{\partial \ln p_v(\mathbf{x}_1, \ldots, \mathbf{x}_M; \theta, \mathbf{s})}{\partial s(n)} = \sum_{m=1}^{M} 2x_m(n + (m-1)\theta) - 2Ms(n), \qquad (5.63)$$

which, when set to zero, is given by the average of the M channels resynchronized in time,

$$\hat{s}(n, \theta) = \frac{1}{M} \sum_{m=1}^{M} x_m(n + (m-1)\theta). \qquad (5.64)$$

Replacing $s(n)$ in (5.62) with $\hat{s}(n, \theta)$, the log-likelihood function can be expressed as

$$\ln p_v(\mathbf{x}_1, \ldots, \mathbf{x}_M; \theta, \mathbf{s}) = \text{constant}$$
$$- \frac{1}{2\sigma_v^2} \sum_{m=1}^{M} \sum_{n=0}^{N-1} \left(x_m(n) - \frac{1}{M} \sum_{l=1}^{M} x_l(n + (l-m)\theta) \right)^2. \qquad (5.65)$$

Omitting the factors that do not depend on θ, the multichannel ML estimator of θ is given by

$$\hat{\theta} = \arg\max_{\theta} \left(\sum_{m=1}^{M} \sum_{n=0}^{N-1} \left(\frac{2}{M} \sum_{l=1}^{M} x_m(n) x_l(n + (l-m)\theta) \right. \right.$$
$$\left. \left. - \frac{1}{M^2} \left(\sum_{l=1}^{M} x_l(n + (l-m)\theta) \right)^2 \right) \right), \qquad (5.66)$$

which, following several algebraic manipulations (see Problem 5.12), can be simplified to

$$\hat{\theta} = \arg\max_{\theta} \left(\sum_{m=1}^{M-1} \sum_{l=m+1}^{M} \sum_{n=0}^{N-1} x_m(n) x_l(n + (l-m)\theta) \right). \qquad (5.67)$$

Thus, the multichannel estimator averages all possible combinations of pairwise cross-correlation functions and then finds the location of the maximum of the averaged function—this location is the ML estimate of θ. For $M = 2$, it is straightforward to show that the ML estimator in (5.67) is identical to the estimator in (5.56).

Since the multichannel estimate suffers equally from limited temporal resolution, it is necessary to apply a technique which improves the resolution. Although being more complicated than in the two-channel case, the frequency domain counterpart of (5.67),

$$\hat{\theta} = \arg \max_{\theta} \left(\sum_{m=1}^{M-1} \sum_{l=m+1}^{M} \frac{1}{2\pi} \int_{-\pi}^{\pi} X_m(e^{j\omega}) X_l^*(e^{j\omega}) e^{j(m-l)\theta\omega} \right), \qquad (5.68)$$

can be used in combination with, for example, a gradient-based optimization technique [77].

The performance of multichannel estimation is illustrated by Figure 5.12, where conduction velocity is estimated from a surface EMG recording using either two or six channels. The results are based on the signals displayed in Figure 5.11, although considerably longer segments than the ones displayed are processed. In this example, improved performance is manifested by a considerably lower variance of the conduction velocity estimates when six channels are used.

The signal model for multichannel time delay estimation can be refined so that the constraint in (5.59) of a linear time delay is removed [78]. Another refinement is to account for changes in waveform shape as the signal propagates along the muscle fiber, for example, as modeled by a timescale parameter [79]. For either of these model refinements, the ML estimator requires that a multidimensional optimization problem be solved, which is very time-consuming. As a result, effort has been expended on developing computationally efficient, suboptimal estimators of conduction velocity.

5.5   Modeling the Intramuscular EMG

So far, we have described a number of phenomenological models of the surface EMG which have helped us develop methods for estimating amplitude and conduction velocity. In this section, the modeling aspect remains salient to the presentation, although the scope shifts to describing a signal model of the intramuscular EMG where information on firing pattern and MUAP waveform shape is taken into consideration. The modeling approach provides us with a better understanding of how various parameters with electrophysiological significance relate to signal amplitude and power spectrum, see

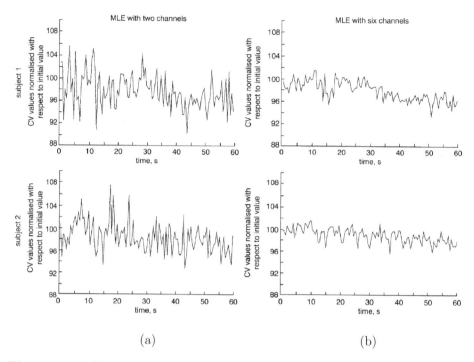

Figure 5.12: Conduction velocity estimation based on EMG signals recorded with a linear electrode array. The ML estimates were obtained for two different subjects using (a) two channels (channels 3 and 4 of the data partially displayed in Figure 5.11) and (b) six channels. The conduction velocity estimates are normalized with respect to the first value and are presented as percentages. The length of the sliding window was 250 ms. (Reprinted from Farina et al. [7] with permission.)

Sections 5.5.2 and 5.5.3, respectively. The model presented below is largely based on the publications by De Luca and coworkers; the interested reader is referred to [2, 6] and references therein.

5.5.1   A Signal Model of the MUAP Train

The electrical activity of a single motor unit is characterized by the time pattern with which action potentials are repeatedly fired to activate the muscle fibers and sustain contraction. The firing pattern can be given a statistical description in the form of a sum of unit impulse functions occurring at random times t_k,

$$d_{\mathrm{E}}(t) = \sum_{k=1}^{M} \delta(t - t_k), \tag{5.69}$$

where M is the total number of MUAPs assumed to occur in a train. The signal representation $d_E(t)$ is sometimes also referred to as an event series and will be revisited in Chapter 8 for the purpose of analyzing heart rate patterns. In order to characterize the firing pattern statistically, it is often preferable to consider the intervals r_k between successive firing times rather than the firing times themselves,

$$r_k = t_k - t_{k-1}, \quad k = 1, \ldots, M, \tag{5.70}$$

i.e., a series of interfiring intervals; for convenience, it is often assumed that $t_0 = 0$. Several experimental studies have concluded that the dependence between interfiring intervals is very weak, suggesting that it is appropriate to assume that the intervals are statistically independent. Hence, the joint PDF $p_r(r_1, \ldots, r_M)$ of the interfiring intervals can be expressed as a product of individual PDFs,

$$p_r(r_1, \ldots, r_M) = \prod_{k=1}^{M} p_{r_k}(r_k). \tag{5.71}$$

While nonstationary behavior of the interfiring intervals may be handled with this model, a structure must be assigned to $p_{r_k}(r_k)$ that describes the time-varying properties [80]. The structure of the PDF may be generalized to explicitly account for the influence of various physiological factors on the interfiring interval pattern, for example, the level of force and the type of muscle fiber. Since such aspects on the PDF structure cannot be introduced without substantial effort, we assume that all interfiring intervals are identically distributed,

$$p_{r_k}(r_k) \equiv p_r(r_k), \quad k = 1, \ldots, M. \tag{5.72}$$

One approach to characterize the properties of interfiring intervals is by the *average firing rate* λ_r, defined as the inverse of the mean length of the intervals r_k,

$$\lambda_r \stackrel{\text{def}}{=} \frac{1}{\displaystyle\int_{-\infty}^{\infty} r p_r(r) dr}. \tag{5.73}$$

This definition of λ_r may be expressed in terms of the mean number of interfiring intervals and the length of the observation interval, denoted $E[M]$ and T, respectively. Since $E[M]$ is given by T divided by the mean interval length,

$$E[M] = \frac{T}{\displaystyle\int_{-\infty}^{\infty} r p_r(r) dr}, \tag{5.74}$$

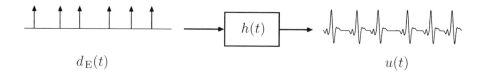

$d_{\mathrm{E}}(t)$ $u(t)$

Figure 5.13: The generation of a MUAP train using the filtered-impulse signal model.

the average firing rate is

$$\lambda_r = \frac{E[M]}{T}. \tag{5.75}$$

With M interfiring intervals contained in the observation interval $[0, T]$, the average firing rate can be approximated by

$$\lambda_r \approx \frac{M}{T}. \tag{5.76}$$

The MUAP waveform results from the superposition of several muscle fiber action potentials of a single motor unit, as illustrated by Figure 5.1. Its shape is largely determined by the geometry of the fibers of each motor unit with respect to the electrode and by the properties of the muscle tissue, see Section 5.1.1. The mathematical representation of a MUAP is here synonymous with an impulse response $h(t)$ whose shape should be chosen to mimic that of a MUAP waveform. The MUAP train is modeled by the output of the linear, time-invariant filter $h(t)$ when a train of unit impulse function $d_{\mathrm{E}}(t)$ is the input,

$$u(t) = d_{\mathrm{E}}(t) * h(t) = \sum_{k=1}^{M} h(t - t_k). \tag{5.77}$$

In this model, the PDF of the interfiring intervals should incorporate restrictions on their minimum length in order to assure that successive waveforms do not overlap in time since a new firing cannot occur until the previous one has terminated. A schematic illustration of the filtered-impulse signal model is presented in Figure 5.13.

Unless the intramuscular EMG is recorded during very weak muscle contractions, multiple motor units are simultaneously activated to produce the necessary muscle force. Since the electrode detects the combined contributions of multiple recruited motor units, the intramuscular EMG signal is better modeled as the summation of multiple MUAP trains $u_1, \ldots, u_L(t)$,

$$x(t) = \sum_{l=1}^{L} u_l(t) + v(t)$$

$$= \sum_{l=1}^{L} \sum_{k=1}^{M_l} h_l(t - t_{l,k}) + v(t). \tag{5.78}$$

Each train $u_l(t)$ results from by the filtered-impulse model in (5.77), associated with a particular MUAP waveform $h_l(t)$ and a firing pattern defined by the random times $t_{l,1}, \ldots, t_{l,M_l}$. The multiple MUAP train model in (5.78) is extended to include the additive term $v(t)$ which accounts for nonmuscular noise produced by, e.g., electronic amplifiers, electrode-wire movements, and other bioelectrical sources.

During certain conditions, a synchronization behavior can be observed between MUAP trains in which the firing times of one MUAP train tend to coincide with those of other motor units. Even in the absence of synchronization, the firing times of two MUAP waveforms in different trains may occasionally coincide. Analysis of the intramuscular EMG signal must therefore deal with the fact that MUAP waveforms do not not always occur as isolated events, but can be part of a composite of two or several superimposed MUAP waveforms, especially when the EMG signal is recorded at moderate to high MVCs. The problem of resolving superimposed waveforms is discussed in Section 5.6, and a method is described whose structure is inspired by the multiple MUAP train model in (5.78).

Several MUAP trains ($L = 25$) have been simulated (Figure 5.14(a)) and summed to produce the synthesized EMG signal displayed in Figure 5.14(b). The shape of the MUAP waveforms is fixed within each train but differs from train to train. The shape of the impulse response $h_l(t)$ is defined by the coefficients of a linear combination of basis functions (the so-called Hermite functions) [81–83]. The PDF of the interfiring intervals $p_r(r_k)$ is assumed to be Gaussian with mean and standard deviation equal to 50 and 5 ms, respectively.

The filtered-impulse model described above assumes that the MUAP shape is fixed—a reasonable assumption as long as the electrode position remains fixed and the active muscle fibers remain the same. However, the local pick-up area of an intramuscular electrode implies that even a slight electrode movement can significantly alter the MUAP shape. When this observation is combined with the observation that the MUAP shape may change during sustained contraction, cf. (5.40), it is evident that the shape is more accurately modeled by a time-varying impulse response $h(t, t_k)$ whose shape depends on the firing time t_k.

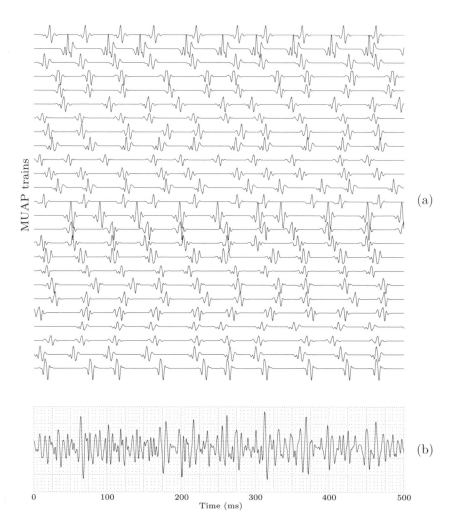

Figure 5.14: (a) Simulation of different MUAP trains with waveform shapes that differ from train to train. (b) The resulting EMG signal is obtained as the summation of all MUAP trains, described by (5.78).

Before we conclude the presentation of the signal model, it is worth pointing out that summation of a large number of MUAP trains implies that the signal amplitude approaches, by virtue of the law of large numbers, a Gaussian distribution ("large" is usually interpreted as at least 15). For such cases, the EMG signal is equally well modeled by the filtered-noise model used for amplitude estimation in the surface EMG. Unfortunately, the interesting connection between the EMG signal and the MUAP shape and firing pattern is lost with the filtered-noise model.

5.5.2 MUAP Train Amplitude

We will now examine the amplitude properties of a *single* MUAP train $u(t)$ as characterized by the MUAP shape $h(t)$ and the average firing rate λ_r. Since $u(t)$ is modeled as a random process, we need to consider the expected value of $u(t)$ when rectified or squared. The MUAP waveforms are nonoverlapping, and the PDF of each firing time t_k is assumed to be uniform over the observation interval $[0, T]$,

$$p_t(t_k) = \begin{cases} \dfrac{1}{T}, & 0 \le t_k \le T; \\ 0, & \text{otherwise.} \end{cases} \tag{5.79}$$

The expected value of the rectified MUAP train is given by

$$E\left[|u(t)|\right] = E\left[\left|\sum_{k=1}^{M} h(t - t_k)\right|\right], \tag{5.80}$$

which, when considering that MUAPs are nonoverlapping, can be written as

$$E\left[|u(t)|\right] = M \int_{-\infty}^{\infty} |h(t - t_k)| p_t(t_k) dt_k$$
$$= \lambda_r \int_{0}^{T} |h(t - t_k)| dt_k, \tag{5.81}$$

where the last step follows from setting M/T to the average firing rate λ_r, cf. (5.76). Since $h(t)$ has finite duration and is completely contained in the observation interval, the expression in (5.81) simplifies to

$$E\left[|u(t)|\right] = \lambda_r \int_{-\infty}^{\infty} |h(t)| dt. \tag{5.82}$$

We note that EMG amplitude estimation, described in Section 5.2 and used as a measure of muscle strength, may be related to the MUAP train model, since the expected value of the MUAP train is proportional to the firing rate λ_r and thus to strength.

The expected value of the squared MUAP train $u(t)$ is given by

$$E\left[u^2(t)\right] = E\left[\left(\sum_{k=1}^{M} h(t - t_k)\right)^2\right], \tag{5.83}$$

which, since MUAPs are nonoverlapping, can be written as

$$E\left[u^2(t)\right] = E\left[\sum_{k=1}^{M} h^2(t - t_k)\right]. \tag{5.84}$$

Proceeding in the same way as before, we obtain

$$E\left[u^2(t)\right] = \lambda_r \int_{-\infty}^{\infty} h^2(t)dt. \tag{5.85}$$

5.5.3   MUAP Train Power Spectrum

The power spectrum of a single MUAP train $u(t)$ can be derived by invoking the following well-known expression,

$$S_u(\Omega) = |H(\Omega)|^2 S_{d_E}(\Omega), \tag{5.86}$$

which relates the power spectrum $S_{d_E}(\Omega)$ of the input signal $d_E(t)$, defined in (5.69), to the power spectrum $S_u(\Omega)$ of the output signal $u(t)$ that results from filtering with $h(t)$. The derivation of $S_u(\Omega)$ mostly revolves around finding a closed-form expression of the autocorrelation function of $d_E(t)$ and requires the availability of certain statistical quantities:

- The PDF $p_t(t_k)$ of the firing time t_k, again assumed to be uniform within the interval $[0, T]$;

- the conditional PDF $m(\tau)$ which describes the probability of a firing time at $t_k + \tau$, conditioned on that a previous firing time occurred at t_k ($\tau > 0$); and

- the PDF $p_r(r_k)$ of the interfiring interval r_k which is also required, but can be determined from knowledge of $m(\tau)$.

Using the definition of $d_E(t)$ in (5.69), the autocorrelation function can be written as

$$r_{d_E}(\tau) = E\left[d_E(t)d_E(t - \tau)\right]$$
$$= E\left[\sum_{k=1}^{M}\sum_{j=1}^{M} \delta(t - t_k)\delta(t - t_j - \tau)\right]. \tag{5.87}$$

The above assumptions on PDFs lead to the expected value that can be calculated as

$$r_{d_E}(\tau) = \int_0^T \int_0^T \sum_{k=1}^M \delta(t - t_k)\delta(t - t_j - \tau)\frac{1}{T}m(t_k - t_j)dt_j dt_k$$

$$= \frac{M}{T}\int_0^T \int_0^T \delta(t - t_k)\delta(t - t_j - \tau)m(t_k - t_j)dt_j dt_k. \qquad (5.88)$$

Integration with respect to t_k yields

$$r_{d_E}(\tau) = \lambda_r \int_0^T \delta(t - t_j - \tau)m(t - t_j)dt_j$$

$$= \lambda_r m(\tau), \quad \tau > 0. \qquad (5.89)$$

Since $m(\tau)$ is only defined for positive values of τ, the autocorrelation function is easily extended to negative values due to the symmetry property of an autocorrelation function,

$$r_{d_E}(\tau) = \lambda_r m(-\tau), \quad \tau < 0. \qquad (5.90)$$

For $\tau = 0$, we have

$$r_{d_E}(0) = E\left[d_E(t)d_E(t)\right]$$

$$= E\left[\sum_{k=1}^M \sum_{j=1}^M \delta(t - t_k)\delta(t - t_j)\right]$$

$$= \frac{M}{T}\int_0^T \delta(t - t_k)\delta(t - t_k)dt_k$$

$$= \lambda_r \delta(\tau), \qquad (5.91)$$

which, when combined with the results in (5.89) and (5.90), becomes

$$r_{d_E}(\tau) = \lambda_r(\delta(\tau) + m(\tau) + m(-\tau)). \qquad (5.92)$$

In order to arrive at the desired result, we need to express the conditional PDF $m(\tau)$ in terms of $p_r(r)$. Conditioned on the fact that a previous firing time occurred at t_k, the PDF of the firing time at $t_k + \tau$ is given by the sum of the PDF of the spike at $t_k + \tau$ being the first spike, plus the PDF of the spike at $t_k + \tau$ being the second spike, and so on, which when combined becomes

$$m(\tau) = p_r(\tau) + \int_0^\infty p_r(\tau')p_r(\tau - \tau')d\tau'$$

$$+ \int_0^\infty \int_0^\infty p_r(\tau')p_r(\tau'')p_r(\tau - \tau' - \tau'')d\tau' d\tau'' + \dots. \qquad (5.93)$$

Alternatively, $m(\tau)$ can be calculated recursively using

$$m(\tau) = p_r(\tau) + \int_0^\infty p_r(\tau')m(\tau - \tau')d\tau'. \tag{5.94}$$

Combining the Fourier transform of this recursion with the Fourier transform of $r_{d_E}(\tau)$ in (5.92), the power spectrum of $d_E(t)$ can be expressed as

$$\begin{aligned}
S_{d_E}(\Omega) &= \lambda_r(1 + M(\Omega) + M^*(\Omega)) \\
&= \lambda_r\left(1 + \frac{P_r(\Omega)}{1 - P_r(\Omega)} + \frac{P_r^*(\Omega)}{1 - P_r^*(\Omega)}\right) \\
&= \frac{\lambda_r(1 - |P_r(\Omega)|^2)}{1 - 2\Re\{P_r(\Omega)\} + |P_r(\Omega)|^2}.
\end{aligned} \tag{5.95}$$

Consequently, the power spectrum of a single MUAP train $u(t)$ is described by the following expression [6, 84],

$$S_u(\Omega) = |H(\Omega)|^2 \frac{\lambda_r(1 - |P_r(\Omega)|^2)}{1 - 2\Re\{P_r(\Omega)\} + |P_r(\Omega)|^2}. \tag{5.96}$$

A deeper understanding of this expression can be obtained by assigning a specific PDF to characterize the interfiring intervals. Based on the results from several experimental studies that interfiring interval histograms are approximately modeled by a Gaussian PDF [85], the power spectrum $S_u(\Omega)$ is evaluated for interfiring intervals characterized by

$$p_r(r) = \frac{1}{2\pi\sigma_r^2} \exp\left[-\frac{(r - \frac{1}{\lambda_r})^2}{2\sigma_r^2}\right], \tag{5.97}$$

where the interval r has average length $1/\lambda_r$ and variance σ_r^2. The characteristic function of $p_r(r)$ is

$$P_r(\Omega) = \exp\left[-\frac{j\Omega}{\lambda_r}\right]\exp\left[-\frac{\sigma_r^2\Omega^2}{2}\right], \tag{5.98}$$

which gives

$$S_u(\Omega) = |H(\Omega)|^2 \frac{\lambda_r(1 - e^{-\sigma_r^2\Omega^2})}{1 + e^{-\sigma_r^2\Omega^2} - 2\cos\left(\frac{\Omega}{\lambda_r}\right)e^{-\frac{\sigma_r^2\Omega^2}{2}}}. \tag{5.99}$$

Figure 5.15 displays the three spectral quantities of main interest, namely, $S_{d_E}(\Omega), H(\Omega)$, and $S_u(\Omega)$, with the firing pattern $d_E(t)$ defined by $\lambda_r = 10$ Hz and $\sigma_r = 7$ ms and the MUAP waveform $h(t)$ with a shape shown in the diagram inset. The power spectrum $S_{d_E}(\Omega)$ is manifested by the fundamental

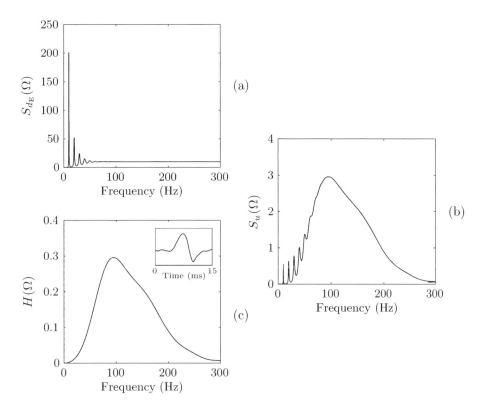

Figure 5.15: (a) The power spectrum $S_{d_\mathrm{E}}(\Omega)$ of the firing pattern for Gaussian interfiring intervals $\sigma_r = 7$ ms and an average firing rate of $\lambda_r = 10$ Hz. (b) The spectrum $H(\Omega)$ of the MUAP waveform $h(t)$ shown in the inset and (c) the power spectrum $S_u(\Omega)$ of the MUAP train.

frequency at λ_r and related harmonics with decreasing power; for the specific values of λ_r and σ_v, three harmonics can be discerned from $S_{d_\mathrm{E}}(\Omega)$). An important observation to be made from $S_{d_\mathrm{E}}(\Omega)$ is that the peaks become increasingly distinct when the standard deviation σ_r decreases in relation to λ_r, and vice versa. Examining the overall shape of the power spectrum $S_u(\Omega)$, it is evident that the frequency components of the firing pattern are mainly present at lower frequencies, i.e., below about 40 Hz, whereas the shape of $h(t)$ determines $S_u(\Omega)$ at higher frequencies.

Since the model of the intramuscular EMG signal is defined as a summation of multiple MUAP trains, the power spectrum $S_x(\Omega)$ of $x(t)$ in (5.78) not only involves the power spectra $S_{u_l}(\Omega)$ of individual MUAP trains, but also the cross-power spectra between trains whose firing patterns are correlated. A general expression of the power spectrum $S_x(\Omega)$ is obtained from the autocorrelation function of $x(t)$ which is

$$E\left[x(t)x(t-\tau)\right] = E\left[\left(\sum_{l=1}^{L} u_l(t) + v(t)\right)\left(\sum_{k=1}^{L} u_k(t-\tau) + v(t-\tau)\right)\right]$$

$$= \sum_{l=1}^{L} r_{u_l}(\tau) + \sum_{l=1}^{L}\sum_{\substack{k=1\\k\neq l}}^{L} r_{u_l,u_k}(\tau) + r_v(\tau), \qquad (5.100)$$

where the function $r_{u_l,u_k}(\tau)$ describes the cross-correlation between $u_l(t)$ and $u_k(t)$. Taking the Fourier transform of $E\left[x(t)x(t-\tau)\right]$, the resulting power spectrum of $x(t)$ is given by [2, 86, 87]

$$S_x(\Omega) = \sum_{l=1}^{L} S_{u_l}(\Omega) + \sum_{l=1}^{L}\sum_{\substack{k=1\\k\neq l}}^{L} S_{u_l,u_k}(\Omega) + S_v(\Omega), \qquad (5.101)$$

where $S_{u_l}(\Omega)$ and $S_v(\Omega)$ denote the power spectrum of $u_l(t)$ and $v(t)$, respectively, and $S_{u_l,u_k}(\Omega)$ denotes the cross-power spectrum of $u_l(t)$ and $u_k(t)$.

For uncorrelated MUAP trains, we conclude from (5.101) that the power spectra of each train will simply add to form the total power spectrum $S_x(\Omega)$. This behavior can be observed during weak contractions, i.e., 5–10% MVC, when few motor units are recruited, and leads to the peak corresponding to the average firing rate λ_r that can be discerned from the power spectrum estimated from the EMG [2]. An estimate of λ_r can thus be determined from the estimated spectrum by selecting the largest peak below 40 Hz. At higher levels of MVC, the spectral peak becomes increasingly indistinct as additional motor units are recruited with average firing rates that may differ quite considerably, causing the spectral peak to vanish. For situations when MUAP trains exhibit strong cross-correlation with respect to firing patterns (motor unit synchronization), the low-frequency components of the power spectrum are influenced although the exact behavior depends on the degree and structure of the cross-correlation.

Figure 5.16 presents 25 different MUAP trains, their summation into a simulated EMG signal, and the power spectrum estimated from the digitized EMG signal rather than from the theoretical expression (the average firing rate λ_r is either 20 or 40 Hz). The MUAP shape was defined by the coefficients of a linear combination of basis functions (Hermite basis functions) [82], and the shape was held fixed in each of the trains. In order to reduce variance, the periodograms of 100 realizations of the digitized signal were averaged, and the resulting power spectrum is presented at the bottom of Figure 5.16. In this simulation example, the average firing rate is easily estimated from the power spectrum where the corresponding peak in the low-frequency region is clearly visible. For the lower average firing rate of $\lambda_r = 20$ Hz, it is possible to discern the first harmonic at 40 Hz.

5.6 Intramuscular EMG Signal Decomposition

The intramuscular EMG signal contains important information concerning the motor control system not immediately quantifiable from the measured signal, but which needs to be disentangled with advanced signal processing and pattern recognition. Since myoelectric activity recorded during contraction by a needle electrode is a composite of concurrently active motor units, the purpose of *EMG signal decomposition* is to resolve the recorded signal into its constituent MUAP trains. The decomposition procedure involves several steps which together accomplish detection and identification of individual MUAP waveforms belonging to different MUAP trains, see Figure 5.17 [88]. The main steps of EMG signal decomposition are discussed in this section and accompanied by a description of select algorithms.

While the multiple MUAP train model in (5.78) only accounts for basic properties of the intramuscular EMG signal, it nevertheless offers a useful description of how the observed signal relates to individual motor units, expressed in MUAP waveforms and firing patterns. This model will be considered below when we develop the specific steps of the decomposition procedure; however, its notation is adopted already in order to facilitate the definition of signal decomposition. Assuming that the EMG is a discrete-time signal, denoted $x(n)$, the purpose of the signal decomposition can be expressed as

$$x(n) \to \left(h_1(n), \{\theta_{1,j}\}_{j=1}^{M_1} \right), \ldots, \left(h_L(n), \{\theta_{L,j}\}_{j=1}^{M_L} \right), \tag{5.102}$$

where L denotes the number of active motor units; the lth train is characterized by M_l different MUAPs with shape $h_l(n)$ and the firing pattern $\theta_{l,1}, \ldots, \theta_{l,M_l}$. The signal decomposition is illustrated by the simulation example in Figure 5.18, where the observed signal $x(n)$ results from summation of five different MUAP trains. The decomposition process leads to the information displayed in the frame—in this case perfectly recovered from $x(n)$ since the number of trains, MUAP waveforms, and firing patterns, are identical with the underlying MUAP trains. In real life, the MUAP waveforms belonging to a particular train vary slightly in shape so that a representative MUAP waveform has to be selected for display.

The main processing steps of intramuscular EMG signal decomposition are listed below:

MUAP detection. The first processing step performs detection of MUAPs in the EMG signal, and a series of firing times is produced. The detec-

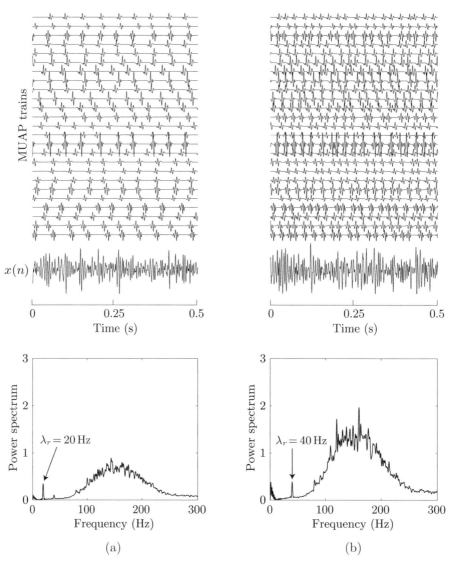

Figure 5.16: Power spectrum of simulated EMG signals with the average firing rate equal to (a) $\lambda_r = 20$ Hz and (b) $\lambda_r = 40$ Hz. For each rate, the individual MUAP trains, the resulting simulated EMG signal, and the power spectrum are shown from top to bottom. Each power spectrum resulted from averaging of the periodograms obtained from 100 realizations.

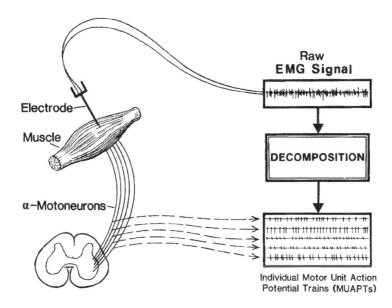

Raw
EMG Signal

DECOMPOSITION

Electrode

Muscle

α–Motoneurons

Individual Motor Unit Action
Potential Trains (MUAPTs)

Figure 5.17: Schematic illustration of the recording and decomposition of an intramuscular EMG signal into MUAP trains. (Reprinted from De Luca et al. [89] with permission.)

tor structure usually involves bandpass filtering to accentuate MUAP shape and improve separation in time between successive MUAPs. A MUAP is detected when the filtered signal exceeds a certain threshold value. The problem of detecting signals in noise is discussed at length in Section 7.4 and presents the rationale for using such a detector structure.

MUAP feature extraction. In order to group (cluster) MUAP waveforms with similar shape, the waveforms are first characterized by a set of features arranged into a feature vector (Section 5.6.1).

MUAP clustering. The objective of clustering is to determine the number of active motor units by grouping the detected MUAPs into different clusters so that all members of a cluster have similar shape (Section 5.6.1). Once the MUAPs have been clustered, the firing times of the different MUAP trains can be determined.

Resolution of superimposed MUAPs. The signal decomposition is rendered difficult by the fact that MUAPs belonging to different trains may occur at the same time, thus resulting in superposition of wave-

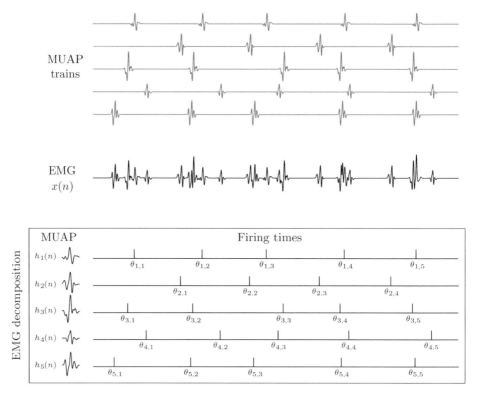

Figure 5.18: Intramuscular EMG signal decomposition. Five MUAP trains (which cannot be observed individually) are summed into the observed EMG signal. The desired outcome of EMG decomposition (framed) consists of the MUAP waveforms and the firing times of each MUAP train.

forms in the EMG signal. Signal processing techniques can be used to resolve superimposed MUAP waveforms (Section 5.6.2).

Under ideal conditions, each MUAP waveform has a unique shape due to the unique geometrical distribution of the muscle fibers of the motor unit relative the electrode. Unfortunately, the design of a decomposition procedure cannot assume that MUAP shapes remain fixed throughout the recording, but various issues related to signal nonstationarity should be taken into account to attain robust performance. These issues are exemplified by significant changes in MUAP shape caused by small movements of the needle electrode and the gradual change in shape of a MUAP related to changes in contraction level. Several approaches to EMG signal decomposition have been presented where these issues are addressed to various extents; the interested reader is referred to the abundant literature available on this topic, see, e.g., [90–102].

5.6.1 MUAP Feature Extraction and Clustering

Clustering refers to a self-organizing process which arranges feature vectors into clusters so that the members of a cluster are closer to each other than to the centers of other clusters. Each new feature vector is either assigned to the cluster being closest in the feature space or appointed as a representative of a new cluster. A clustering algorithm may be designed to group feature vectors based on prior knowledge of the number of clusters. However, such knowledge is not available in the EMG application, so the number of clusters is only determined once the clustering process has been fully completed. Rather than presenting an overview of the numerous algorithms for clustering, we will describe a frequently employed algorithm—the *leader–follower clustering* algorithm—which serves as an excellent introduction to the central ideas of clustering [103]. This algorithm has not only been employed for analysis of MUAP waveforms [91, 93, 104–106], but it has also found its way into many other areas of relevance for this textbook, for example, in automated analysis of epileptic spikes in EEG signals and arrhythmias in ECG signals.

The input feature vector $\mathbf{p}_i$ may be defined in different ways to represent the shape of a MUAP waveform. A straightforward representation is the one defined by the time domain samples $x_i(n)$,

$$\mathbf{p}_i = \begin{bmatrix} x_i(0) \\ x_i(1) \\ \vdots \\ x_i(N-1) \end{bmatrix}, \tag{5.103}$$

where i denotes waveform detection index. Often, the bandpass filtered samples of the MUAP detector are instead used to obtain a representation being less sensitive to noise. Implicit with the time domain representation is the selection of N samples, to be appropriately aligned (centered) around the detected MUAP waveform. Since the feature vector in (5.103) usually suffers from a dimensionality problem due to the many samples included, a basis function representation is often preferable since waveform shape can be represented with much fewer coefficients,

$$\mathbf{p}_i = \begin{bmatrix} w_{i,1} \\ w_{i,2} \\ \vdots \\ w_{i,K} \end{bmatrix}, \tag{5.104}$$

see Section 4.5. The Karhunen–Loève basis functions are of particular interest since the signal energy is concentrated into a few coefficients, offering

a robust description of the signal. The coefficients of the Fourier transform
and the wavelet series expansion have also been considered for clustering of
MUAP waveforms [93, 94, 99, 100]. Similar to the time domain representation in (5.103), it is essential that MUAPs are appropriately aligned to
assure that the coefficients are representative. Yet another approach is to
define the feature vector in terms of heuristic features such as amplitude,
width, and shape of the waveform, providing an intuitively attractive but
less robust description [95, 106, 107].

The leader–follower clustering algorithm is a self-learning technique which
clusters the set of input feature vectors $\{\mathbf{p}_1, \mathbf{p}_2, \ldots\}$ sequentially; no prior
knowledge of the number of clusters is required. The following three com-
ponents are central to the algorithm.

- The lth cluster is characterized by its *center* and *spread* in the feature
  space, defined by the mean vector $\boldsymbol{\mu}_l$ and the covariance matrix $\mathbf{C}_l$,
  respectively.

- The distance measure $d(\mathbf{p}_i, \boldsymbol{\mu}_l)$ quantifies *similarity* between the feature vector $\mathbf{p}_i$ and the mean $\boldsymbol{\mu}_l$ of the lth cluster. The distance determines whether $\mathbf{p}_i$ belongs to an existing cluster or should be the first
  member of a new cluster.

- Since the cluster parameters $\boldsymbol{\mu}_l$ and $\mathbf{C}_l$ are unknown, they must be
  subjected to estimation from the input feature vectors as the clustering
  process progresses.

Similarity between the feature vector $\mathbf{p}_i$ and the cluster mean $\boldsymbol{\mu}_l$ is here
defined by the squared *Mahalanobi's distance*,

$$d^2(\mathbf{p}_i, \boldsymbol{\mu}_l) = (\mathbf{p}_i - \boldsymbol{\mu}_l)^T \mathbf{C}_l^{-1} (\mathbf{p}_i - \boldsymbol{\mu}_l), \tag{5.105}$$

which can be understood as a weighted Euclidean distance where the weights
are chosen so as to normalize the errors in relation to their variance. The
Mahalanobi distance represents a natural choice for data characterized by a
multivariate Gaussian PDF since such data tend to cluster about the mean
$\boldsymbol{\mu}_l$, spreading in a cloud of ellipsoid shape whose principal axes are the
eigenvectors of $\mathbf{C}_l$, cf. (3.7).

The leader–follower clustering algorithm initializes the first cluster center
($L = 1$) by equating it with the initial feature vector,

$$\boldsymbol{\mu}_1 = \mathbf{p}_1, \tag{5.106}$$

and using a preset spread of the first cluster,

$$\mathbf{C}_1 = \kappa \mathbf{I}, \tag{5.107}$$

where the design parameter κ describes the uncertainty associated with the cluster. Proceeding to the next feature vector and generalizing the description of the algorithm, the nearest cluster l in the feature space is determined from

$$l = \arg \min_{j=1,\ldots,L} d^2(\mathbf{p}_i, \boldsymbol{\mu}_j). \tag{5.108}$$

The feature vector $\mathbf{p}_i$ is assigned to cluster l if the distance is less than the threshold η, i.e., $d^2(\mathbf{p}_i, \boldsymbol{\mu}_l) \leq \eta$. Otherwise, a new cluster is initialized in the same way as the first cluster was in (5.106) and (5.107), and the number of clusters is incremented by one, i.e., $L = L + 1$.

When a feature vector $\mathbf{p}_i$ has been assigned to a cluster l, the information about the cluster is updated so that its mean vector is modified to include a fraction of $\mathbf{p}_i$ through exponential averaging (Section 4.3.3),

$$\boldsymbol{\mu}_{l,k} = (1 - \alpha)\boldsymbol{\mu}_{l,k-1} + \alpha\mathbf{p}_i, \tag{5.109}$$

where the additional index k denotes the current number of members of cluster l, and α is the update rate of the cluster center. Recalling the definition of a covariance matrix,

$$\mathbf{C}_l = E\left[(\mathbf{p} - \boldsymbol{\mu}_l)(\mathbf{p} - \boldsymbol{\mu}_l)^T\right],$$

the covariance matrix is updated using the same type of recursive expression as in (5.109),

$$\mathbf{C}_{l,k} = (1 - \alpha)\mathbf{C}_{l,k-1} + \alpha(\mathbf{p}_i - \boldsymbol{\mu}_l)(\mathbf{p}_i - \boldsymbol{\mu}_l)^T. \tag{5.110}$$

Rather than having to invert $\mathbf{C}_{l,k}$ every time the squared Mahalanobi distance in (5.105) is computed, the inverse matrix can be updated directly thanks to the matrix inversion lemma in (A.31) allowing us to express $\mathbf{C}_{l,k}^{-1}$ in terms of $\mathbf{C}_{l,k-1}^{-1}$,

$$\mathbf{C}_{l,k}^{-1} = \frac{1}{(1-\alpha)}\mathbf{C}_{l,k-1}^{-1} - \frac{\dfrac{1}{(1-\alpha)}}{\dfrac{1-\alpha}{\alpha} + \mathbf{e}_l^T\mathbf{C}_{l,k-1}^{-1}\mathbf{e}_l}\mathbf{C}_{l,k-1}^{-1}\mathbf{e}_l\mathbf{e}_l^T\mathbf{C}_{l,k-1}^{-1}, \tag{5.111}$$

where

$$\mathbf{e}_l = \mathbf{p}_i - \boldsymbol{\mu}_l. \tag{5.112}$$

The clustering process is repeated until all input feature vectors have been processed.

The threshold η is a design parameter which determines the total number of clusters. A small value of η leads to a large number of small clusters, whereas a large value leads to a small number of large clusters. It is often advantageous to relate the threshold setting to the SNR of the MUAP waveform in order to reflect the uncertainty associated with the feature vector; a lower threshold may be used at higher SNRs, and vice versa. Evidently, a signal-dependent threshold requires that an estimate of the SNR can be determined.

The center and spread of a cluster is adapted to the input feature vectors at a rate defined by α in the exponential averager. This technique tracks a gradually changing shape of the MUAP waveform, while also providing an improved SNR of $\boldsymbol{\mu}_l$. Too large a value of α may, however, cause a cluster's center to drift and get nearer to the center of another cluster so that two similar clusters result. Too small a value of α leads to the very first feature vector $\mathbf{x}_1$ being overemphasized.

The leader–follower clustering algorithm is particularly suitable for on-line applications where clustering needs to be performed as the data becomes available. The total number of clusters depends on the order of data presentation and may be different if, for example, the data was presented in reverse order. The above algorithm does not include any mechanism for merging similar clusters, but may have to be supplemented with this once clustering is completed.

Intramuscular EMG signal decomposition based on a variant of the leader–follower clustering algorithm is illustrated by Figure 5.19. The input feature vector was defined by the discrete Fourier transform (DFT) coefficients of the bandpass filtered EMG signal displayed in Figure 5.19(b). Filtering was used to enhance temporal separation by making the MUAPs more spike-like. A total of nine clusters were produced by the algorithm, suggesting that nine different motor units were simultaneously active. The number of clusters increases with contraction level, and reaches a limit at 30–40% MVC above which the signal becomes difficult to analyze [93]. For the case presented in Figure 5.19, the feature vectors only included information related to the MUAP waveform, however, the vector may be augmented with information on the firing pattern in order to improve clustering performance.

Figure 5.20 presents another example of an intramuscular EMG signal where the MUAP waveforms have been clustered, but now with the firing pattern of each active motor unit determined and presented. The relatively regular pattern of interfiring intervals is disrupted by occasional gaps caused by unresolved MUAP superpositions.

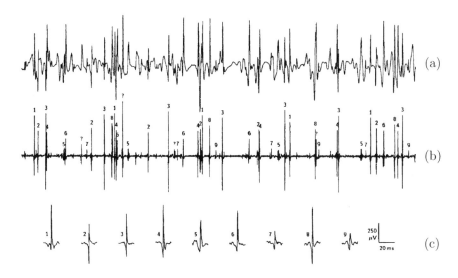

Figure 5.19: Clustering of MUAP waveforms. (a) The EMG signal recorded from biceps brachii using a needle electrode, (b) the bandpass filtered EMG, and (c) the resulting MUAP clusters, in this case equal to nine. (Reprinted from McGill et al. [93] with permission.)

5.6.2   Resolution of Superimposed MUAP Waveforms

The superposition of more than one MUAP waveform in the intramuscular EMG signal leads to difficulties in determining the correct firing patterns of the motor units, and calls for a technique which resolves a superimposed waveform into its constituent MUAPs. The success in resolving superimposed waveforms depends on how the shapes of the individual waveforms combine. Three categories of superimposed MUAPs may be defined, namely, those which are *completely*, *partially*, and *destructively* superimposed [88], see Figure 5.21. Completely and partially superimposed MUAPs may be distinguished from individual MUAPs by their increased amplitude or width, see Figure 5.21(b)–(c). Unfortunately, a newly recruited motor unit may produce a MUAP whose amplitude and width are similar to the superimposed MUAP, implying that superimposed MUAPs are not easily identified. Destructively superimposed MUAPs are particularly problematic since their amplitude may be so low that they are actually missed by the MUAP detector, see Figure 5.21(d).

The method for resolving superimposed MUAPs may be activated when a new MUAP shape has been found that does not fit into any of the existing MUAP clusters. Before initializing a new cluster, the new MUAP is matched to combinations of the previously detected and clustered MUAPs

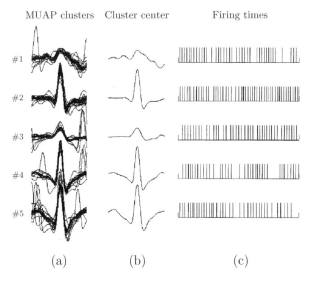

MUAP clusters　　Cluster center　　　　Firing times

#1

#2

#3

#4

#5

(a)　　　　　　(b)　　　　　　(c)

Figure 5.20: Clustering of MUAP waveforms and determination of firing patterns. (a) Five clusters with MUAP waveforms, (b) the cluster centers computed from the individual MUAP waveforms in (a), and (c) the related patterns of firing times. (Reprinted from Stashuk and Qu [108] with permission.)

in order to find out if it represents a superposition of two or more MUAPs. This approach is based on the assumption that the constituent MUAPs are available—something which cannot be completely guaranteed even though the likelihood of superimposed MUAPs is small when the EMG signal is recorded at a few percent MVC.

We will now describe a model-based technique for resolving superimposed MUAPs whose structure is inspired by the intramuscular EMG model in (5.78). Using the least-squares (LS) criterion, our task is to determine the amplitudes a_k and firing times θ_k of the M known waveforms $h_k(n)$ which yield the best fit to the superimposed MUAP waveform $x(n)$ [92, 109],

$$\mathcal{J}(\mathbf{p}) = \mathcal{J}(\mathbf{a}, \boldsymbol{\theta}) = \sum_{n=0}^{N-1} \left(x(n) - \sum_{k=1}^{M} a_k h_k(n - \theta_k) \right)^2, \qquad (5.113)$$

where $M > 1$. The amplitudes a_k are introduced to account for minor changes in MUAP amplitude and, typically, deviate in amplitude with less than 15–20% from previous MUAPs.[8] For convenience, we use vector nota-

---

[8]The inclusion of the amplitudes a_k may lead to a better description of the superimposed MUAPs, while also introducing additional degrees of freedom which make convergence to the minimum error of $\mathcal{J}(\mathbf{p})$ more difficult to achieve.

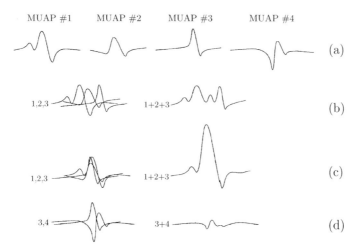

MUAP #1   MUAP #2   MUAP #3   MUAP #4

(a)

1,2,3   1+2+3   (b)

1,2,3   1+2+3   (c)

3,4   3+4   (d)

Figure 5.21: Different types of MUAP waveform superposition. (a) Four MUAP waveforms, and composite waveforms which are either (b) partially, (c) completely, or (d) destructively superimposed. (Reprinted from Etawil and Stashuk [98] with permission.)

tion of the amplitudes and firing times,

$$\mathbf{a} = \begin{bmatrix} a_1 \\ a_2 \\ \vdots \\ a_M \end{bmatrix}, \quad \boldsymbol{\theta} = \begin{bmatrix} \theta_1 \\ \theta_2 \\ \vdots \\ \theta_M \end{bmatrix}, \tag{5.114}$$

with the total parameter vector

$$\mathbf{p} = \begin{bmatrix} \mathbf{a} \\ \boldsymbol{\theta} \end{bmatrix}. \tag{5.115}$$

Since the number of constituent MUAPs is unknown in practice, the minimization of the error $\mathcal{J}(\mathbf{p})$ in (5.113) has to be repeated for different values of M. Denoting the total number of MUAPs with L ($M \leq L$), the error $\mathcal{J}(\mathbf{p})$ should be evaluated for all possible subsets of M different MUAPs. For $M = 2$, the following MUAP combinations are analyzed,

$$(h_i(n), h_j(n)), \quad i, j = 1, \ldots, L \text{ for } i \neq j, \tag{5.116}$$

for $M = 3$,

$$(h_i(n), h_j(n), h_k(n)), \quad i, j, k = 1, \ldots, L \text{ for } i \neq j, j \neq k, i \neq k, \tag{5.117}$$

and so on. The number of constituent MUAPs is determined when $\mathcal{J}(\mathbf{p})$ reaches a sufficiently small error; the error usually corresponds to that the residual signal is mostly noise. In practice, the number of constituent MUAPs is small, i.e., $M = 2, 3,$ or 4, while L can be much larger. Consequently, the number of combinations to be analyzed is less than the maximal number of MUAP combinations, which is $(L! - L + 1)$.

The minimization of $\mathcal{J}(\mathbf{p})$ is done by differentiation with respect to $\mathbf{a}$ and $\boldsymbol{\theta}$.[9] First, we differentiate $\mathcal{J}(\mathbf{p})$ with respect to $\mathbf{p}$ and set the result to zero, yielding the following set of normal equations,

$$\sum_{n=0}^{N-1} x(n) h_k(n - \theta_k) = \sum_{i=1}^{M} a_i \sum_{n=0}^{N-1} h_i(n - \theta_i) h_k(n - \theta_k) \qquad (5.118)$$

$$\sum_{n=0}^{N-1} x(n) h_k'(n - \theta_k) = \sum_{i=1}^{M} a_i \sum_{n=0}^{N-1} h_i(n - \theta_i) h_k'(n - \theta_k) \qquad (5.119)$$

for $k = 1, \ldots, M$. The derivative waveform $h_k'(n)$ is obtained from

$$h_k'(n) = \left. \frac{dh_k(t)}{dt} \right|_{t=nT_s}, \qquad (5.120)$$

where $h_k(t)$ may be obtained from interpolation of $h_k(n)$; T_s denotes the length of the sampling interval. By introducing the following two correlation quantities,

$$r_{xh_k}(\theta_k) = \sum_{n=0}^{N-1} x(n) h_k(n - \theta_k) \qquad (5.121)$$

and

$$r_{h_i h_k}(\theta_k - \theta_i) = \sum_{n=0}^{N-1} h_i(n - \theta_i) h_k(n - \theta_k), \qquad (5.122)$$

and the related notations,

$$\mathbf{r}_{xh}(\boldsymbol{\theta}) = \begin{bmatrix} r_{xh_1}(\theta_1) \\ r_{xh_2}(\theta_2) \\ \vdots \\ r_{xh_M}(\theta_M) \end{bmatrix}, \quad \mathbf{r}_{hh_k}(\boldsymbol{\theta}) = \begin{bmatrix} r_{h_1 h_k}(\theta_k - \theta_1) \\ r_{h_2 h_k}(\theta_k - \theta_2) \\ \vdots \\ r_{h_M h_k}(\theta_k - \theta_M) \end{bmatrix}, \qquad (5.123)$$

---

[9]Another approach for finding the solution to the least-squares problem in (5.113) is presented in [110] where differentiation is avoided. The firing times are estimated at a time resolution finer than the sampling interval.

$$\mathbf{R}_{hh}(\boldsymbol{\theta}) = \begin{bmatrix} \mathbf{r}_{hh_1}(\boldsymbol{\theta}) & \mathbf{r}_{hh_2}(\boldsymbol{\theta}) & \cdots & \mathbf{r}_{hh_M}(\boldsymbol{\theta}) \end{bmatrix}, \tag{5.124}$$

the normal equations in (5.118) and (5.119) can be compactly expressed as

$$\begin{bmatrix} \mathbf{R}_{hh}^T(\boldsymbol{\theta}) \\ \mathbf{R}_{hh'}^T(\boldsymbol{\theta}) \end{bmatrix} \mathbf{a} = \begin{bmatrix} \mathbf{r}_{xh}(\boldsymbol{\theta}) \\ \mathbf{r}_{xh'}(\boldsymbol{\theta}) \end{bmatrix}. \tag{5.125}$$

Solving for $\mathbf{a}$, we obtain an estimator of $\mathbf{a}$, being a function of $\boldsymbol{\theta}$, which can be inserted into (5.113) to produce the LS estimator of the integer-valued parameter $\boldsymbol{\theta}$,

$$\hat{\boldsymbol{\theta}} = \arg\min_{\boldsymbol{\theta}} \mathcal{J}(\hat{\mathbf{a}}(\boldsymbol{\theta}), \boldsymbol{\theta})$$

$$= \arg\min_{\boldsymbol{\theta}} \left(\sum_{n=0}^{N-1} \left(x(n) - \sum_{k=1}^{M} \hat{a}_k(\boldsymbol{\theta}) h_k(n - \theta_k) \right)^2 \right). \tag{5.126}$$

The amplitude estimate is obtained from $\hat{\mathbf{a}} = \mathbf{a}(\hat{\boldsymbol{\theta}})$.

The resulting estimation procedure is computationally demanding since it involves the two coupled equations in (5.125) and (5.126), which must be evaluated for all possible combinations of $\theta_1, \ldots, \theta_M$. By considering a subset of all values, the procedure becomes computationally more feasible, and may be used to find a good guess of the initial values for use in a gradient-based search for the optimal solution, see below. For example, the error in (5.126) may be evaluated on a grid of θ_k which only includes every second or third sample.

The computational demands associated with the LS estimator can be substantially reduced if gradient-based minimization is employed, since it may produce the optimal estimates in relatively few iterations. Our approach to gradient-based minimization is to use the well-known *Newton's method* which can be derived from the Taylor series expansion of $\mathcal{J}(\mathbf{p})$,

$$\mathcal{J}(\mathbf{p} + \Delta\mathbf{p}) = \mathcal{J}(\mathbf{p}) + [\nabla_{\mathbf{p}}\mathcal{J}(\mathbf{p})]^T \Delta\mathbf{p} + \frac{1}{2!}(\Delta\mathbf{p})^T [\nabla_{\mathbf{p}}\nabla_{\mathbf{p}}^T\mathcal{J}(\mathbf{p})]^T (\Delta\mathbf{p}) + \cdots \tag{5.127}$$

Neglecting higher-order terms of the expansion, the change in $\mathcal{J}(\mathbf{p})$ due to the change $\Delta\mathbf{p}$ is described by

$$\Delta\mathcal{J}(\mathbf{p}) = \mathcal{J}(\mathbf{p} + \Delta\mathbf{p}) - \mathcal{J}(\mathbf{p})$$

$$= [\nabla_{\mathbf{p}}\mathcal{J}(\mathbf{p})]^T \Delta\mathbf{p} + \frac{1}{2!}(\Delta\mathbf{p})^T [\nabla_{\mathbf{p}}\nabla_{\mathbf{p}}^T\mathcal{J}(\mathbf{p})]^T (\Delta\mathbf{p}). \tag{5.128}$$

Differentiating this expression with respect to $\Delta\mathbf{p}$, we find that the expression $\Delta\mathcal{J}(\mathbf{p})$ which describes the change has its minimum when

$$\Delta\mathbf{p} = - \left(\nabla_{\mathbf{p}}\nabla_{\mathbf{p}}^T\mathcal{J}(\mathbf{p}) \right)^{-1} [\nabla_{\mathbf{p}}\mathcal{J}(\mathbf{p})]. \tag{5.129}$$

Using the method of steepest descent (see Section 3.2.5), the parameter vector $\mathbf{p}_{(q)}$ is iteratively updated with the gradient information in order to bring the next estimate $\mathbf{p}_{(q+1)}$ closer to the minimum,

$$\mathbf{p}_{(q+1)} = \mathbf{p}_{(q)} + \alpha \Delta \mathbf{p}_{(q)}, \tag{5.130}$$

where q is the iteration index and α the step size. Insertion of (5.129) leads to the expression which defines Newton's method,

$$\mathbf{p}_{(q+1)} = \mathbf{p}_{(q)} - \alpha \left(\nabla_{\mathbf{p}} \nabla_{\mathbf{p}}^T \mathcal{J}(\mathbf{p}) \right)^{-1} \left[\nabla_{\mathbf{p}} \mathcal{J}(\mathbf{p}_{(q)}) \right]. \tag{5.131}$$

For our specific structure of $\mathcal{J}(\mathbf{p})$ in (5.113), the gradient is

$$\nabla_{\mathbf{p}} \mathcal{J}(\mathbf{p}) = \begin{bmatrix} \mathbf{R}_{hh}^T(\boldsymbol{\theta})\mathbf{a} - \mathbf{r}_{xh}(\boldsymbol{\theta}) \\ -\mathbf{A}\mathbf{R}_{hh'}^T(\boldsymbol{\theta})\mathbf{a} + \mathbf{A}\mathbf{r}_{xh'}(\boldsymbol{\theta})) \end{bmatrix}, \tag{5.132}$$

where

$$\mathbf{A} = \mathrm{diag}(a_1, a_2, \ldots, a_M). \tag{5.133}$$

The matrix with second derivatives (the *Hessian* matrix) reorients the negative gradient vector $\nabla_{\mathbf{p}} \mathcal{J}(\mathbf{p}_{(q)})$ toward the minimum since the gradient does not necessarily point to the minimum when the error surface is nonquadratic. The Hessian matrix is defined by

$$\nabla_{\mathbf{p}} \nabla_{\mathbf{p}}^T \mathcal{J}(\mathbf{p}) = \begin{bmatrix} \nabla_{\mathbf{a}} \nabla_{\mathbf{a}}^T \mathcal{J}(\mathbf{p}) & \vdots & \nabla_{\mathbf{a}} \nabla_{\theta}^T \mathcal{J}(\mathbf{p}) \\ \cdots\cdots\cdots\cdots\cdots\cdots\cdots\cdots\cdots \\ \nabla_{\theta} \nabla_{\mathbf{a}}^T \mathcal{J}(\mathbf{p}) & \vdots & \nabla_{\theta} \nabla_{\theta}^T \mathcal{J}(\mathbf{p}) \end{bmatrix}, \tag{5.134}$$

which after evaluation becomes

$$\begin{bmatrix} \mathbf{R}_{hh}^T(\boldsymbol{\theta}) & \vdots & -\mathbf{R}_{h'h}^T(\boldsymbol{\theta})\mathbf{A} \\ & \vdots & \mathrm{diag}(\mathbf{r}_{xh'}(\boldsymbol{\theta})) \\ & \vdots & -\mathrm{diag}(\mathbf{r}_{h'h}(\mathbf{0}))\mathbf{A} \\ \cdots\cdots\cdots\cdots\cdots\cdots\cdots\cdots\cdots \\ -\mathbf{R}_{hh'}^T(\boldsymbol{\theta})\mathbf{A} & \vdots & -\mathrm{diag}(\mathbf{r}_{xh''}(\boldsymbol{\theta}))\mathbf{A} \\ \mathrm{diag}(\mathbf{r}_{xh'}(\boldsymbol{\theta})) & \vdots & \mathbf{A}\mathbf{R}_{h'h'}^T(\boldsymbol{\theta})\mathbf{A} \\ -\mathrm{diag}(\mathbf{r}_{hh'}(\mathbf{0}))\mathbf{A} & \vdots & \mathbf{A}\,\mathrm{diag}(\mathbf{r}_{hh''}(\mathbf{0}))\mathbf{A} \end{bmatrix}. \tag{5.135}$$

It is worth commenting on certain aspects of Newton's method related to initialization, integer-valued parameters, and convergence properties. The minimization may be initialized by choosing the a_k values equal to one and

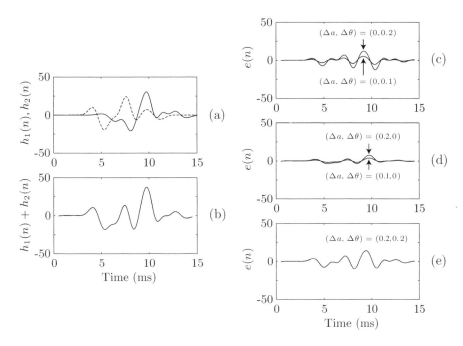

Figure 5.22: The effect of errors in amplitude a and time θ when two MUAPs are superimposed. (a) The two individual MUAPs and (b) the resulting superimposed MUAP. The residual signal is displayed when (c) the error in amplitude between the two waveforms differs by either 10 or 20% but no time error, (d) the error in firing time differs by either 0.1 or 0.2 ms but no amplitude error, and (e) with errors in both amplitude and time.

the θ_k values such that the $h_k(n)$'s are well-aligned to the superimposed waveform $x(n)$; the alignment may be accomplished using, for example, matched filtering techniques.

Since the parameter vector $\mathbf{p}$ is treated as continuous-valued in Newton's method, the part of $\mathbf{p}$ which relates to $\boldsymbol{\theta}$ must in each iteration be rounded to the nearest integer. The rounding is required in each iteration since the correlation functions $r_{xh_k}(\theta_k)$ and $r_{h_ih_k}(\theta_k - \theta_i)$ need to be computed.

It has proven advantageous to split the minimization procedure into two stages to achieve better convergence. In the first stage, only the firing times are adjusted whereas, in the next stage, amplitudes and firing times are simultaneously adjusted [109]. This approach is motivated by the observation that the gradient with respect to amplitude is more sensitive to errors in time than is the gradient with respect to time to errors in amplitude. Hence, the minimization is more difficult to perform with respect to time than amplitude. This observation is demonstrated by the example in Figure 5.22 where two superimposed MUAP waveforms are considered. The effect of errors in

a and θ is described by the residual signal

$$e(n) = x(n) - a_1 h_1(n - \theta_1) - a_2 h_2(n - \theta_2).$$

It is noted from this example that a time shift as small as 0.2 ms, i.e., a fraction of the waveform width, causes a much larger residual signal than does a 20% error in amplitude, see Figures 5.22(c) and (d), respectively. The combined effect of errors in both amplitude and time is presented in Figure 5.22(e).

We conclude this section by mentioning that suboptimal approaches have been presented to the problem of resolving superimposed MUAPs which avoid the heavy computations of LS estimation. Rather than jointly dealing with the constituent MUAPs, as is done in LS estimation, the "peel-off" approach resolves superimposed MUAPs by sequentially matching and subtracting individual MUAPs; in that way less computations are required [88, 90, 99, 100]. The first MUAP to be peeled off from the superimposed waveform may be chosen as the one which produces the smallest residual signal, provided that these two waveforms have been first appropriately aligned in time. The procedure continues with peeling off the next MUAP, and so on, until the amplitude of the residual signal has become sufficiently small.

Bibliography

[1] R. Merletti and P. A. Parker (eds.), *Electromyography. Physiology, Engineering, and Noninvasive Applications.* Piscataway, NJ: IEEE Press Series in Biomedical Engineering, 2004.

[2] J. V. Basmajian and C. J. De Luca, *Muscles Alive. Their Functions Revealed by Electromyography.* Baltimore: Williams & Wilkins, 1985.

[3] K.-Å. Henneberg, "Principles of electromyography," in *The Biomedical Engineering Handbook* (J. D. Bronzino, ed.), ch. 14, pp. 14.1–14.11, Boca Raton: CRC Press, 2000.

[4] D. C. Preston and B. E. Shapiro, *Electromyography and Neuromuscular Disorders. Clinical-Electrophysiologic Correlations.* Boston: Butterworth-Heinemann, 1998.

[5] H. P. Clamann, "Activity of single motor units during isometric tension," *Neurology*, vol. 20, pp. 254–260, 1970.

[6] C. J. De Luca, "Physiology and mathematics of myoelectric signals," *IEEE Trans. Biomed. Eng.*, vol. 26, pp. 313–325, 1979.

[7] D. Farina, R. Merletti, and C. Disselhorst-Klug, "Multi-channel techniques for information extraction from the surface EMG," in *Electromyography. Physiology, Engineering, and Noninvasive Applications* (R. Merletti and P. A. Parker, eds.), ch. 7, pp. 169–203, Piscataway, NJ: IEEE Press Series in Biomedical Engineering, 2004.

[8] R. Merletti, "Standards for reporting EMG data," *J. Electromyogr. Kinesiol.*, vol. 6, pp. III–IV, 1996.

[9] E. A. Clancy, E. L. Morin, and R. Merletti, "Sampling, noise-reduction and amplitude estimation issues in surface electromyography," *J. Electromyogr. Kinesiol.*, vol. 12, pp. 1–16, 2002.

[10] V. R. Zschorlig, "Digital filtering of EMG signals," *Clin. Neurophysiol.*, vol. 29, pp. 81–86, 1989.

[11] S. Conforto, T. D'Alessio, and S. Pignatelli, "Optimal rejection of movement artefacts from myoelectric signals by means of a wavelet filtering procedure," *J. Electromyogr. Kinesiol.*, vol. 9, pp. 47–57, 1999.

[12] R. V. Baratta, M. Solomonow, B.-H. Zhou, and M. Zhu, "Methods to reduce the variability of EMG power spectrum estimates," *J. Electromyogr. Kinesiol.*, vol. 8, pp. 279–285, 1998.

[13] P. Akkiraju and D. C. Reddy, "Adaptive cancellation technique in processing myoelectric activity of respiratory muscles," *IEEE Trans. Biomed. Eng.*, vol. 39, pp. 652–655, 1992.

[14] M. S. Redfern, R. E. Hughes, and D. B. Chaffin, "High-pass filtering to remove electrocardiographic interference from torso EMG recordings," *J. Clin. Biomech.*, vol. 8, pp. 44–48, 1993.

[15] Y. Deng, W. Wolf, R. Schnell, and U. Appel, "New aspects to event-synchronous cancellation of ECG interference: An application of the method in diaphragmatic EMG signals," *IEEE Trans. Biomed. Eng.*, vol. 47, pp. 1177–1184, 2000.

[16] C. Frigo and R. Shiavi, "Applications in movement and gait analysis," in *Electromyography. Physiology, Engineering, and Noninvasive Applications* (R. Merletti and P. A. Parker, eds.), ch. 15, pp. 381–401, Piscataway, NJ: IEEE Press Series in Biomedical Engineering, 2004.

[17] G. M. Hägg, B. Melin, and R. Kadefors, "Applications in ergonomics," in *Electromyography. Physiology, Engineering, and Noninvasive Applications* (R. Merletti and P. A. Parker, eds.), ch. 13, pp. 343–363, Piscataway, NJ: IEEE Press Series in Biomedical Engineering, 2004.

[18] P. A. Parker, K. B. Englehart, and B. S. Hudgins, "Control of powered upper limb prostheses," in *Electromyography. Physiology, Engineering, and Noninvasive Applications* (R. Merletti and P. A. Parker, eds.), ch. 18, pp. 453–475, Piscataway, NJ: IEEE Press Series in Biomedical Engineering, 2004.

[19] P. Herberts, "Myoelectric signals in control of prostheses," *Acta Ortop. Scand. (suppl)*, vol. 40, p. 124, 1969.

[20] D. Graupe, J. Magnussen, and A. Beex, "A microprocessor system for multifunctional control of upper-limb prostheses via myoelectric signal identification," *IEEE Trans. Autom. Control*, vol. 23, pp. 538–544, 1978.

[21] S. Jacobsen, D. F. Knutti, R. T. Johnson, and H. H. Sears, "Development of the Utah artificial arm," *IEEE Trans. Biomed. Eng.*, vol. 29, pp. 249–269, 1982.

[22] P. A. Parker and R. N. Scott, "Myoelectric control of prostheses," *CRC Crit. Rev. Biomed. Eng.*, vol. 13, pp. 283–310, 1986.

[23] S. Lee and G. Saridis, "The control of a prosthetic arm by EMG pattern recognition," *IEEE Trans. Autom. Control*, vol. 29, pp. 290–302, 1984.

[24] B. S. Hudgins, P. A. Parker, and R. N. Scott, "A new strategy for multifunction myoelectric control," *IEEE Trans. Biomed. Eng.*, vol. 40, pp. 82–94, 1993.

[25] E. Park and S. G. Meek, "Fatigue compensation of the electromyographic signal for prosthetic control and force estimation," *IEEE Trans. Biomed. Eng.*, vol. 40, pp. 1019–1023, 1993.

[26] P. J. Kyberd, M. Evans, and S. te Winkel, "An intelligent anthropomorphic hand, with automatic grasp," *Robotica*, vol. 16, pp. 531–536, 1998.

[27] K. B. Englehart and B. S. Hudgins, "A robust, real-time control scheme for multifunction myoelectric control," *IEEE Trans. Biomed. Eng.*, vol. 50, pp. 848–854, 2003.

[28] T. Moritani, D. Stegeman, and R. Merletti, "Basic physiology and biophysics of EMG signal generation," in *Electromyography. Physiology, Engineering, and Noninvasive Applications* (R. Merletti and P. A. Parker, eds.), ch. 1, pp. 1–25, Piscataway, NJ: IEEE Press Series in Biomedical Engineering, 2004.

[29] E. A. Clancy and N. Hogan, "Single site electromyograph amplitude estimation," *IEEE Trans. Biomed. Eng.*, vol. 41, pp. 159–167, 1994.

[30] P. A. Parker, J. A. Stuller, and R. N. Scott, "Signal processing for the multistate myoelectric channel," *Proc. IEEE*, vol. 65, pp. 662–673, 1977.

[31] E. A. Clancy and N. Hogan, "Probability density of the surface electromyogram and its relation to amplitude detectors," *IEEE Trans. Biomed. Eng.*, vol. 46, pp. 730–739, 1999.

[32] M. Hayes, *Statistical Digital Signal Proccessing and Modeling*. New York: John Wiley & Sons, 1996.

[33] M. I. Harba and P. A. Lynn, "Optimizing the acquisition and processing of surface electromyographic signals," *J. Biomed. Eng.*, vol. 3, pp. 100–106, 1981.

[34] G. F. Inbar and A. E. Noujaim, "On surface EMG spectral characterization and its application to diagnostic classification," *IEEE Trans. Biomed. Eng.*, vol. 31, pp. 597–604, 1984.

[35] O. Paiss and G. F. Inbar, "Autoregressive modeling surface EMG and its spectrum with application to fatigue," *IEEE Trans. Biomed. Eng.*, vol. 34, pp. 761–770, 1987.

[36] R. J. Triolo, D. H. Nash, and G. D. Moskowitz, "The identification of time series models of lower extremity EMG for the control of prostheses using Box-Jenkins criteria," *IEEE Trans. Biomed. Eng.*, vol. 35, pp. 584–594, 1988.

[37] T. Kiryu, C. J. De Luca, and Y. Saitoh, "AR modeling of myoelectric interference signals during a ramp contraction," *IEEE Trans. Biomed. Eng.*, vol. 41, pp. 1031–1038, 1994.

[38] C. Hershler and M. Milner, "An optimality criterion for processing electromyographic (EMG) signals relating to human locomotion," *IEEE Trans. Biomed. Eng.*, vol. 25, pp. 413–420, 1978.

[39] H. Miyano, T. Masuda, and T. Sadoyama, "A note on the time constant in low-pass filtering of rectified surface EMG," *IEEE Trans. Biomed. Eng.*, vol. 27, pp. 274–278, 1980.

[40] F. Q. Xiong and E. Shwedyk, "Some aspects of nonstationary myoelectric signal processing," *IEEE Trans. Biomed. Eng.*, vol. 34, pp. 166–172, 1987.

[41] T. D'Alessio, "Some results on the optimization of a digital processor for surface EMG signals," *Electroencephal. Clin. Neurophysiol.*, vol. 24, pp. 625–643, 1984.

[42] E. Park and S. G. Meek, "Adaptive filtering of the electromyographic signal for prosthetic control and force estimation," *IEEE Trans. Biomed. Eng.*, vol. 42, pp. 1048–1052, 1995.

[43] E. A. Clancy, "Electromyogram amplitude estimation with adaptive smoothing window length," *IEEE Trans. Biomed. Eng.*, vol. 46, pp. 717–729, 1999.

[44] E. Shwedyk, R. Balasubramanian, and R. N. Scott, "A non-stationary model for the electromyogram," *IEEE Trans. Biomed. Eng.*, vol. 24, pp. 417–424, 1977.

[45] D. Farina and R. Merletti, "Comparison of algorithms for estimation of EMG variables during voluntary isometric contractions," *J. Electromyogr. Kinesiol.*, vol. 10, pp. 337–349, 2000.

[46] M. Bilodeau, M. Cincera, A. B. Arsenault, and D. Gravel, "Normality and stationarity of EMG signals of elbow flexor muscles during ramp and step isometric contractions," *J. Electromyogr. Kinesiol.*, vol. 7, pp. 87–96, 1997.

[47] Y. St-Amant, D. Rancourt, and E. A. Clancy, "Influence of smoothing window length on electromyogram amplitude estimates," *IEEE Trans. Biomed. Eng.*, vol. 45, pp. 795–799, 1998.

[48] J. G. Kreifeldt and S. Yao, "A signal-to-noise investigation of nonlinear electromyographic processors," *IEEE Trans. Biomed. Eng.*, vol. 21, pp. 298–308, 1974.

[49] N. Hogan and R. W. Mann, "Myoelectric signal processing: Optimal estimation applied to electromyography—Part II: Experimental demonstration of optimal myoprocessor performance," *IEEE Trans. Biomed. Eng.*, vol. 27, pp. 396–410, 1980.

[50] E. A. Clancy and K. A. Farry, "Adaptive whitening of the electromyogram to improve amplitude estimation," *IEEE Trans. Biomed. Eng.*, vol. 47, pp. 709–719, 2000.

[51] N. Hogan and R. W. Mann, "Myoelectric signal processing: Optimal estimation applied to electromyography—Part I: Derivation of the optimal myoprocessor," *IEEE Trans. Biomed. Eng.*, vol. 27, pp. 382–395, 1980.

[52] E. A. Clancy and N. Hogan, "Multiple site electromyograph amplitude estimation," *IEEE Trans. Biomed. Eng.*, vol. 41, pp. 203–211, 1995.

[53] W. R. Murray and W. D. Rolph, "An optimal real-time digital processor for the electrical activity of muscle," *Med. Instrum.*, vol. 19, pp. 77–82, 1985.

[54] S. Thusneyapan and G. I. Zahalak, "A practical electrode-array myoprocessor for surface electromyography," *IEEE Trans. Biomed. Eng.*, vol. 36, pp. 295–299, 1989.

[55] L. H. Lindström, R. I. Magnusson, and I. Petersén, "Muscular fatigue and action potential conduction velocity changes studies with frequency analysis of EMG signals," *Electromyography*, vol. 10, pp. 341–356, 1970.

[56] R. Merletti, A. Rainoldi, and D. Farina, "Myoelectric manifestations of muscle fatigue," in *Electromyography. Physiology, Engineering, and Noninvasive Applications* (R. Merletti and P. A. Parker, eds.), ch. 9, pp. 233–258, Piscataway, NJ: IEEE Press Series in Biomedical Engineering, 2004.

[57] R. Merletti, A. Gulisashvili, and L. R. Lo Conte, "Estimation of shape characteristics of surface muscle signal spectra from time domain data," *IEEE Trans. Biomed. Eng.*, vol. 42, pp. 759–776, 1995.

[58] S. Karlsson, J. Yu, and M. Akay, "Time–frequency analysis of myoelectric signals during dynamic contractions: A comparative study," *IEEE Trans. Biomed. Eng.*, vol. 47, pp. 228–238, 2000.

[59] F. B. Stulen and C. J. De Luca, "Frequency parameters of the myoelectric signal as a measure of muscle conduction velocity," *IEEE Trans. Biomed. Eng.*, vol. 28, pp. 515–523, 1981.

[60] A. L. Hof, "Errors in frequency parameters of EMG power spectra," *IEEE Trans. Biomed. Eng.*, vol. 38, pp. 1077–1088, 1991.

[61] M. A. Mañanas, R. Jané, J. A. Fiz, J. Morera, and P. Caminal, "Influence of estimators of spectral density on the analysis of electromyographic and vibromyographic signals," *Med. Biol. Eng. & Comput.*, vol. 40, pp. 90–98, 2002.

[62] M. M. Lowery, C. L. Vaughan, P. J. Nolan, and M. J. O'Malley, "Spectral compression of the electromyographic signal due to decreasing muscle fiber conduction velocity," *IEEE Trans. Rehab. Eng.*, vol. 8, pp. 353–361, 2000.

[63] R. Merletti, D. Biey, M. Biey, G. Prato, and A. Orusa, "On-line monitoring of the median frequency of the surface EMG power spectrum," *IEEE Trans. Biomed. Eng.*, vol. 32, pp. 1–7, 1985.

[64] E. A. Clancy, D. Farina, and G. Filligoi, "Single-channel techniques for information extraction from the surface EMG signal," in *Electromyography. Physiology, Engineering, and Noninvasive Applications* (R. Merletti and P. A. Parker, eds.), ch. 6, pp. 133–168, Piscataway, NJ: IEEE Press Series in Biomedical Engineering, 2004.

[65] P. C. Doerschuk, D. E. Gustafson, and A. S. Willsky, "Upper extremity limb function discrimination using EMG signal analysis," *IEEE Trans. Biomed. Eng.*, vol. 30, pp. 18–28, 1983.

[66] R. J. Triolo and G. D. Moskowitz, "The experimental demonstration of a multichannel time-series myoprocessor: System testing and evaluation," *IEEE Trans. Biomed. Eng.*, vol. 36, pp. 1018–1027, 1989.

[67] J. Dûchene, D. Devedeux, S. Mansour, and C. Marque, "Analyzing uterine EMG: Tracking instantaneous burst frequency," *IEEE Eng. Med. Biol. Mag.*, vol. 14, pp. 125–132, 1995.

[68] M. Knaflitz and P. Bonato, "Time–frequency methods applied to muscle fatigue assessment during dynamic contractions," *J. Electromyogr. Kinesiol.*, vol. 9, pp. 337–350, 1999.

[69] S. Karlsson and B. Gerdle, "Mean frequency and signal amplitude of the surface EMG of the quadriceps muscles increase with increasing torque—A study using the continuous wavelet transform," *J. Electromyogr. Kinesiol.*, vol. 11, pp. 131–140, 2001.

[70] D. MacIsaac, P. A. Parker, and R. N. Scott, "The short-time Fourier transform and muscle fatigue assessment in dynamic contractions," *J. Electromyogr. Kinesiol.*, vol. 11, pp. 439–449, 2001.

[71] P. Bonato, S. H. Roy, M. Knaflitz, and C. J. De Luca, "Time–frequency parameters of the surface myoelectric signal for assessing muscle fatigue during cyclic dynamic contractions," *IEEE Trans. Biomed. Eng.*, vol. 48, pp. 745–753, 2001.

[72] N. Östlund, J. Yu, and J. S. Karlsson, "Improved maximum frequency estimation with application to instantaneous mean frequency estimation of surface electromyography," *IEEE Trans. Biomed. Eng.*, vol. 51, pp. 1541–1546, 2004.

[73] L. H. Lindström and R. I. Magnusson, "Interpretation of myoelectric power spectra: A model and its application," *Proc. IEEE*, vol. 65, pp. 653–662, 1977.

[74] D. Farina and R. Merletti, "A novel approach for precise simulation of the EMG signal detected by surface electrodes," *IEEE Trans. Biomed. Eng.*, vol. 48, pp. 637–646, 2001.

[75] L. Arendt-Nielsen and M. Zwarts, "Measurement of muscle fiber conduction velocity in humans: Techniques and applications," *J. Clin. Neurophysiol.*, vol. 6, pp. 173–190, 1989.

[76] K. C. McGill and L. J. Dorfman, "High-resolution alignment of sampled waveforms," *IEEE Trans. Biomed. Eng.*, vol. 31, pp. 462–468, 1984.

[77] D. Farina, W. Muhammad, E. Fortunato, O. Meste, R. Merletti, and H. Rix, "Estimation of single motor unit conduction velocity from surface electromyogram signals detected with linear electrode arrays," *Med. Biol. Eng. & Comput.*, vol. 39, pp. 225–236, 2001.

[78] W. Muhammad, O. Meste, and H. Rix, "Comparison of single and multiple time delay estimators: Application to muscle fiber conduction velocity estimation," *Signal Proc.*, vol. 82, pp. 925–940, 2002.

[79] W. Muhammad, O. Meste, H. Rix, and D. Farina, "A pseudo joint estimation of time delay and scale factor for M-wave analysis," *IEEE Trans. Biomed. Eng.*, vol. 50, pp. 459–468, 2003.

[80] K. Englehart and P. A. Parker, "Single motor unit myoelectric signal analysis with nonstationary data," *IEEE Trans. Biomed. Eng.*, vol. 41, pp. 168–180, 1994.

[81] L. R. Lo Conte, R. Merletti, and G. V. Sandri, "Hermite expansions of compact support waveforms: Applications to myoelectric signals," *IEEE Trans. Biomed. Eng.*, vol. 41, no. 12, pp. 1147–1159, 1994.

[82] D. Farina, A. Crosetti, and R. Merletti, "A model for the generation of synthetic intramuscular EMG signals to test decomposition algorithms," *IEEE Trans. Biomed. Eng.*, vol. 48, pp. 66–77, 2001.

[83] A. Merlo, D. Farina, and R. Merletti, "A fast and reliable technique for muscle activity detection from surface EMG signals," *IEEE Trans. Biomed. Eng.*, vol. 50, pp. 316–323, 2003.

[84] P. J. Lago and N. B. Jones, "Effect of motor unit firing time statistics on e.m.g. spectra," *Med. Biol. Eng. & Comput.*, vol. 5, pp. 648–655, 1977.

[85] D. F. Stegeman, R. Merletti, and H. J. Hermens, "EMG modeling and simulation," in *Electromyography. Physiology, Engineering, and Noninvasive Applications* (R. Merletti and P. A. Parker, eds.), ch. 8, pp. 205–231, Piscataway, NJ: IEEE Press Series in Biomedical Engineering, 2004.

[86] C. J. De Luca and E. J. van Dyk, "Derivation of some parameters of myoelectric signals recorded during sustained constant force isometric contractions," *Biophys. J.*, vol. 15, pp. 1167–1180, 1975.

[87] J. L. Weytjens and D. van Steenberghe, "The effects of motor unit synchronization on the power spectrum of the electromyogram," *Biol. Cybern.*, vol. 51, pp. 71–77, 1984.

[88] D. W. Stashuk, D. Farina, and K. Søgaard, "Decomposition of intramuscular EMG signals," in *Electromyography. Physiology, Engineering, and Noninvasive Applications* (R. Merletti and P. A. Parker, eds.), ch. 3, pp. 47–80, Piscataway, NJ: IEEE Press Series in Biomedical Engineering, 2004.

[89] C. J. De Luca, R. S. LeFever, M. P. McCue, and A. P. Xenakis, "Behaviour of human motor units in different muscle during linear-varying contractions," *J. Physiol. (London)*, vol. 329, pp. 113–128, 1982.

[90] R. S. LeFever and C. J. De Luca, "A procedure for decomposing the myoelectric signal into its constituent action potentials—Part I: Technique, theory and implementation," *IEEE Trans. Biomed. Eng.*, vol. 29, pp. 149–157, 1982.

[91] R. S. LeFever, A. P. Xenakis, and C. J. De Luca, "A procedure for decomposing the myoelectric signal into its constituent action potentials—Part II: Execution and test for accuracy," *IEEE Trans. Biomed. Eng.*, vol. 29, pp. 158–164, 1982.

[92] A. Gerber, R. M. Studer, R. J. P. De Figueiredo, and G. S. Moschytz, "A new framework and computer program for quantitative EMG signal analysis," *IEEE Trans. Biomed. Eng.*, vol. 31, pp. 857–863, 1984.

[93] K. C. McGill, K. L. Cummins, and L. J. Dorfman, "A procedure for decomposing the myoelectric signal into its constituent action potentials—Part II: Execution and test for accuracy," *IEEE Trans. Biomed. Eng.*, vol. 32, pp. 470–477, 1985.

[94] D. W. Stashuk and H. De Bruin, "Automatic decomposition of selective needle-detected myoelectric signals," *IEEE Trans. Biomed. Eng.*, vol. 35, pp. 1–10, 1988.

[95] G. H. Loudon, N. B. Jones, and A. S. Sehmi, "New signal processing techniques for the decomposition of EMG signals," *Med. Biol. Eng. & Comput.*, vol. 30, pp. 591–599, 1992.

[96] D. W. Stashuk and R. K. Naphran, "Probabilistic inference-based classification applied to myoelectric signal decomposition," *IEEE Trans. Biomed. Eng.*, vol. 39, pp. 346–355, 1992.

[97] C. J. De Luca, "Precision decomposition of EMG signals," *Meth. Clin. Neurophysiol.*, vol. 4, pp. 1–28, 1993.

[98] H. A. Y. Etawil and D. W. Stashuk, "Resolving superimposed motor unit action potentials," *Med. Biol. Eng. & Comput.*, vol. 34, pp. 33–40, 1996.

[99] J. Fang, G. C. Agarwal, and B. T. Shahani, "Decomposition of multiunit electromyographic signals," *IEEE Trans. Biomed. Eng.*, vol. 46, pp. 685–697, 1999.

[100] I. C. Christodoulou and C. S. Pattichis, "Unsupervised pattern recognition for the classification of EMG signals," *IEEE Trans. Biomed. Eng.*, vol. 46, pp. 169–178, 1999.

[101] R. Gut and G. S. Moschytz, "High-precision EMG signal decomposition using communication techniques," *IEEE Trans. Signal Proc.*, vol. 48, pp. 2487–2494, 2000.

[102] D. Farina, R. Colombo, R. Merletti, and H. Baare Olsen, "Evaluation of intramuscular EMG signal decomposition algorithms," *J. Electromyogr. Kinesiol.*, vol. 11, pp. 175–187, 2001.

[103] R. O. Duda, P. E. Hart, and D. G. Stork, *Pattern Classification.* New York: Wiley–Interscience, 2nd ed., 2001.

[104] R. L. Joynt, R. F. Erlandson, S. J. Wu, and C. M. Wang, "Electromyography interference pattern decomposition," *Arch. Phys. Med. Rehabil.*, vol. 472, pp. 567–572, 1991.

[105] S. D. Nandedkar, P. E. Barkhaus, and A. Charles, "Multi-motor unit action potential analysis (MMA)," *Muscle Nerve*, vol. 18, pp. 1155–1166, 1995.

[106] E. Stålberg, B. Falck, M. Sonoo, S. Stålberg, and M. Åström, "Multi-MUP EMG analysis—A two year experience in daily clinical work," *Electroencephal. Clin. Neurophysiol.*, vol. 97, pp. 145–154, 1995.

[107] C. S. Pattichis, C. N. Schizas, and L. T. Middleton, "Neural network models in EMG diagnosis," *IEEE Trans. Biomed. Eng.*, vol. 42, pp. 486–496, 1995.

[108] D. W. Stashuk and Y. Qu, "Adaptive motor unit action potential clustering using shape and temporal information," *Med. Biol. Eng. & Comput.*, vol. 34, pp. 41–49, 1996.

[109] R. J. P. De Figueiredo and A. Gerber, "Separation of superimposed signals by a cross-correlation method," *IEEE Trans. Acoust. Speech Sig. Proc.*, vol. 31, pp. 1084–1089, 1983.

[110] K. C. McGill, "Optimal resolution of superimposed action potentials," *IEEE Trans. Biomed. Eng.*, vol. 49, pp. 640–650, 2002.

[111] J. Vredenbregt and G. Rau, "Surface electromyography in relation to force, muscle length and endurance," in *New Developments in EMG and Clinical Neurophysiology, Vol. 1* (J. E. Desmedt, ed.), pp. 607–622, Basel: Karger, 1973.

Problems

5.1 In Section 5.2, the ML procedure for estimating the standard deviation σ_v was found to include a whitening operation of the observed signal $x(n)$. Another way to motivate this operation is to study the MSE of the estimator $\hat{\sigma}_v^2$,

$$\text{MSE}\left(\hat{\sigma}_v^2\right) = E\left[\left(\hat{\sigma}_v^2 - \sigma_v^2\right)^2\right],$$

where

$$\hat{\sigma}_v^2 = \frac{1}{N}\sum_{n=0}^{N-1} y^2(n).$$

In order to show that the MSE is minimized when the filtered signal $y(n)$ becomes white, we divide the derivation into two steps.

   a. Derive an expression for the MSE of $\hat{\sigma}_v^2$. *Hint:* When $x(n)$ is a zero-mean, Gaussian process with variance σ_x^2, we have that

$$E\left[x^2(n)x^2(n-k)\right] = \sigma_x^4 + 2r_x^2(k).$$

   Also, it can be assumed that significant correlation lags are much shorter than the observation interval N.

   b. Minimize the MSE using a Lagrange multiplier in order to include the constraint that the power of $y(n)$ must not change with the whitening operation, cf. (5.5). *Hint:* Express the cost function to be minimized in the frequency domain using Parseval's theorem.

5.2 Experimental studies have shown that the EMG amplitude (σ) is nonlinearly related to muscle force ($\mathcal{F}$) through

$$\sigma = g(\mathcal{F}) = k\mathcal{F}^a,$$

where a and k are positive-valued constants. Hence, $\mathcal{F}$ can be written as [111]

$$\mathcal{F} = g^{-1}(\sigma) = \left(\frac{\sigma}{k}\right)^{\frac{1}{a}}.$$

Show that the ML estimator of $\mathcal{F}$ is

$$\hat{\mathcal{F}} = g^{-1}\left(\sqrt{\frac{1}{N}(\mathbf{H}^{-1}\mathbf{x})^T(\mathbf{H}^{-1}\mathbf{x})}\right).$$

5.3 Assuming that the whitened signal $\mathbf{y} = \mathbf{H}^{-1}\mathbf{x}$ is Gaussian with standard deviation related to force by $\sigma = k\mathcal{F}^a$, the ML estimator of $\mathcal{F}$ can be expressed as (see Problem 5.2 and [51]),

$$\hat{\mathcal{F}} = \left(\frac{\hat{\sigma}}{k}\right)^{\frac{1}{a}} = \left(\frac{1}{k^2 N}\sum_{n=0}^{N-1} y^2(n)\right)^{\frac{1}{2a}}.$$

The sum $\xi = \sum_{n=0}^{N-1} y^2(n)$ is a *chi-squared* random variable whose PDF is defined by

$$p(\xi,\sigma) = \begin{cases} \dfrac{\xi^{(N/2-1)}e^{\frac{-\xi}{2\sigma^2}}}{(2\sigma^2)^{N/2}\Gamma(N/2)} & \xi \geq 0 \\ 0 & \xi < 0, \end{cases}$$

where the Gamma function $\Gamma(x)$ is defined by

$$\Gamma(x) = \int_0^\infty t^{x-1}e^{-t}dt.$$

a. Show that $\hat{\mathcal{F}}$ is biased.

b. The performance of $\hat{\mathcal{F}}$ can be studied in terms of the "signal-to-noise ratio," defined as the ratio between the energy of the expected value of $\hat{\mathcal{F}}$ and its variance,

$$\text{SNR} = \frac{E^2\left[\hat{\mathcal{F}}\right]}{E\left[\left(\hat{\mathcal{F}} - E\left[\hat{\mathcal{F}}\right]\right)^2\right]}.$$

Show that the SNR is independent of $\mathcal{F}$. Comment on the behavior of the SNR as the number of observations N increases, under the assumption that a linear relationship exists between amplitude and force (i.e, $a = 1$).

5.4 The multichannel ML amplitude estimator is given by (see (5.35))

$$\hat{\sigma} = \sqrt{\frac{1}{NM}\sum_{n=0}^{N-1}\sum_{m=1}^{M}\frac{z_m^2(n)}{\lambda_m}}.$$

It was pointed out that this estimator exhibits numerical problems when an eigenvalue λ_m is close to zero. In order to derive a more robust estimator, the following approximation is useful,

$$\frac{1}{N}\sum_{n=0}^{N-1} z_m^2(n) \approx \sigma_v^2\lambda_m,$$

which draws on (5.32). Using this result, modify the multichannel ML estimator so that it becomes more robust.

5.5 At weaker contractions the presence of nonmuscular noise can no longer be neglected as in (5.1), but an additive, random noise term should be included in the model. Accordingly, the observed signal can no longer be treated as deriving entirely from an AR process. In this problem, we develop an amplitude estimation technique in which the observed signal instead is modeled by

$$x(n) = s(n) + v(n).$$

The technique is developed in two steps consisting of whitening of the EMG signal component (part (a.) below), followed by reduction of the remaining noise component (part (b.) below) [50].

  a. Propose a filter $h(n)$ to whiten $s(n)$, being inversely related to the power spectrum $S_s(e^{j\omega})$. It can be assumed that the observed signal $x(n)$ at 0% MVC only contains the noise component $v(n)$.

  b. In order to reduce the noise that remains after using the filter developed in (a), the signal is subjected to Wiener filtering. Suggest means for determining the quantities which define the Wiener filter.

5.6 Show that the spectral parameters ω_{MNF} and ω_{MDF}, describing the mean and median frequency, respectively, change by the same factor when the power spectrum is subjected to scaling. As a result, the ratio $\omega_{\mathrm{MNF}}/\omega_{\mathrm{MDF}}$ is constant during spectral compression.

5.7 The spectral parameters ω_{MNF} and ω_{MDF} are influenced differently by noise. This aspect can be investigated by calculating the bias of the parameters, assuming that the observed signal is modeled by

$$x(n) = s(n) + v(n).$$

Here, the signal $s(n)$ has the power spectrum $S_s(e^{j\omega})$, and the noise $v(n)$ is white with variance σ_v^2; the two components are assumed to be uncorrelated. Show that the median frequency is less influenced by noise in terms of bias. *Hint:* In the derivations it can be assumed that the SNR is high so that

$$\frac{1}{\pi} \int_0^\pi S_s(e^{j\omega})d\omega \gg \sigma_v^2.$$

5.8 In order to quantify the degree of spectral compression, for example, observed during muscle fatigue, the spectral scaling factor ν in (cf. (5.41))

$$S_{x_2}(e^{J\omega}) = \frac{1}{\nu} S_{x_1}(e^{J\omega/\nu})$$

needs to be determined. This is often done by computing the median frequencies ω_{MDF,x_1} and ω_{MDF,x_2} of the discrete-time signal segments $x_1(n)$ and $x_2(n)$, respectively, and estimating the scale factor by

$$\hat{\nu} = \frac{\omega_{\text{MDF},x_2}}{\omega_{\text{MDF},x_1}}.$$

a. Show that the percentile frequencies ω_c, obtained for different values of the percentile c in

$$\int_0^{\omega_c} S_x(e^{J\omega})d\omega = c \int_0^{\pi} S_x(e^{J\omega})d\omega,$$

change by the same factor when the power spectrum is subjected to scaling. Recall that the percentile frequency is defined for $c = \frac{1}{2}$ in (5.44), i.e., the median frequency, see also Problem 5.6.

b. In view of the result in (a), the scale factor ν can be estimated for different values of the percentile factor c. Suggest an estimator which combines the ν-estimates, resulting for different values of c, so that a "global" estimate of ν is obtained.

5.9 An alternative to the mean frequency ω_{MNF} is the mobility parameter $\mathcal{H}_1$, defined as (see (3.97))

$$\mathcal{H}_1 = \sqrt{\frac{\int_{-\pi}^{\pi} \omega^2 S_x(e^{J\omega})d\omega}{\int_{-\pi}^{\pi} S_x(e^{J\omega})d\omega}},$$

which also can be used as an estimate of dominant frequency. Show that $\mathcal{H}_1$ changes in the same way as ω_{MDF} and ω_{MNF} when the power spectrum is subjected to compression.

5.10 An advantage of the mobility parameter $\mathcal{H}_1$ is that it can be easily computed in the time domain, see page 101. Show that same advantage applies to the mean frequency ω_{MNF} by developing a related time domain procedure. *Hint:* Make use of the fact that the one-sided power spectrum of a signal can be interpreted as the analytic signal, resulting from summation of the original signal and the Hilbert transform (see page 145).

5.11 The conduction velocity ν can be estimated from the time delay θ between the two EMG signals $x_1(n)$ and $x_2(n)$, recorded at locations separated by a distance d, using $\hat{\nu} = d/\hat{\theta}$. In this problem, we look at certain aspects related to two-channel conduction velocity estimation, assuming that the model defined by (5.46) and (5.47) is valid [7].

   a. Derive the maximum error when estimating the velocity for $\nu = 4$ m/s and $d = 10$ mm at a sampling rate of 1 kHz.

   b. Show that the frequency domain algorithm in (5.58) is equivalent to interpolation of the signal prior to time delay estimation.

   c. Derive an iterative, frequency domain algorithm which produces an estimate of the time delay θ between the two signals $x_1(n)$ and $x_2(n)$ [76]. *Hint:* Use Newton's algorithm defined by

$$\theta^{(i+1)} = \theta^{(i)} - \alpha \frac{\left.\frac{d\mathcal{J}(\theta)}{d\theta}\right|_{\theta=\theta^{(i)}}}{\left.\frac{d^2\mathcal{J}(\theta)}{d\theta^2}\right|_{\theta=\theta^{(i)}}},$$

   where $\mathcal{J}(\theta)$ is the function to minimized respect to θ.

5.12 Derive the multichannel ML amplitude estimator in (5.68) starting with the expression in (5.66).

5.13 Derive the multichannel version of the iterative algorithm in Problem 5.11 for the time delay estimation. The frequency domain expression in (5.68) defines the starting point for this algorithm.

5.14 The intramuscular EMG signal $x(t)$ can be model by linear summation of L different MUAP trains as detected by the recording electrode,

$$x(t) = \sum_{l=1}^{L} u_l(t).$$

It was shown in Section 5.5.3 that the power spectrum $S_x(\Omega)$ of $x(t)$ not only involves the power spectra $S_{u_l}(\Omega)$ of individual MUAP trains, but also the cross-power spectra between trains, see (5.101). In this problem, we determine $S_x(\Omega)$ for two special cases of interrelations between different MUAP trains, namely, when

   a. the firing times are uncorrelated, or

   b. the firing times are identical in all MUAP trains.

Chapter 6

The Electrocardiogram— A Brief Background

An electrocardiogram (ECG) describes the electrical activity of the heart recorded by electrodes placed on the body surface. The voltage variations measured by the electrodes are caused by the action potentials of the excitable cardiac cells as they make the cells contract. The resulting heartbeat in the ECG is manifested by a series of waves whose morphology and timing convey information which is used for diagnosing diseases that are reflected by disturbances of the heart's electrical activity. The time pattern that characterizes the occurrence of successive heartbeats is also very important.

The very first ECG recordings in man were made by Augustus Waller in the 1880s. The Dutch physiologist Willem Einthoven further developed the recording device in the early 20th century by making use of a string galvanometer which was sensitive enough to record electrical potentials on the body surface. He also defined sites for electrode placement on the arms and legs which remain in use today. The pioneering effort of Einthoven were rewarded with the Nobel Prize in Medicine in 1924. Since then, the ECG has undergone dramatic development and become an indispensable clinical tool in many different contexts. From having been a signal which was recorded at rest under favorable conditions, the ECG is today recorded in diverse clinical applications, often during strenuous or ambulatory conditions where signal processing algorithms are essential for extraction of reliable information [1]. The importance of the ECG has been further strengthened through the discoveries of subtle variability patterns which are present in rhythm or wave morphology on a beat-to-beat basis and cardiac micropotentials. Both these discoveries were made possible thanks to the availability of suitable signal processing techniques.

The number of electrodes attached to the body surface depends on the type of clinical information desired. It is usually sufficient to use a few electrodes when only heart rhythm is being studied, whereas ten electrodes are typically used when information on waveform morphology is required. Much denser spatial sampling of the body surface can be achieved by employing a recording technique known as "body surface potential mapping", in which an array of 100–200 electrodes is attached to the torso [2, 3]. The signals recorded with such an electrode array can be used to produce a sequence of "electrical images" which offer a more detailed view of the spatiotemporal potential distribution over the body surface [4]. For example, an electrical image may capture the presence of local large gradients in potential which would be missed when too sparse spatial sampling is used.

Although this textbook is concerned with the processing of noninvasive signals, it is important to be aware that the electrical activity of the heart may also be studied by invasively recorded signals (electrograms). The electrogram signal is recorded by internally implanted electrodes or an electrode catheter guided through the skin and through a blood vessel leading into the heart. Due to the proximity of the electrode to the heart, the electrogram gives a much more local description of the electrical activity of cardiac cells than does the ECG. The electrogram is studied in electrophysiological testing when the heart is artificially stimulated to help identify the region where an arrhythmia may develop. Another important use of the electrogram signal is to provide an implanted cardiac pacemaker with information on whether the heart's natural pacemaker is working properly or needs to be replaced by the artificial pacemaker. Generally, an artificial pacemaker corrects an abnormally slow heart rate by taking over the function of the natural pacemaker. An implantable cardioverter defibrillator is equally dependent on the electrogram signal, but is specifically designed to detect the presence of life-threatening arrhythmias which, for example, cause the heart to cease pumping blood. Once such a condition is detected, a high-energy electrical shock is given to terminate the arrhythmia [5].

This chapter contains a brief description of the heart's electrical activity. The generation of the ECG and techniques for its recording are described, and the manifestations in the ECG of certain important heart diseases are considered. Finally, an overview of the most common clinical ECG applications is presented.

6.1   Electrical Activity of the Heart

The heart is a muscular organ the size of a large fist whose primary function is to pump oxygen-rich blood throughout the body. Its anatomy is divided

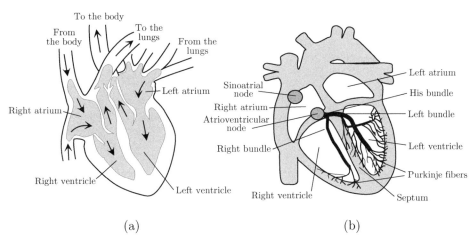

To the body

From the body

To the lungs

From the lungs

Right atrium

Left atrium

Right ventricle

Left ventricle

Sinoatrial node

Right atrium

Atrioventricular node

Right bundle

Right ventricle

Left atrium

His bundle

Left bundle

Left ventricle

Purkinje fibers

Septum

(a)   (b)

Figure 6.1: Schematic illustration of (a) the heart's anatomy (the arrows indicate the directions of the blood flow into and out of the heart) and (b) its electrical conduction system.

into two "mirrored" sides, left and right, which support different circulatory systems but which pump in a synchronized, rhythmic manner. Each side of the heart consists of two chambers, the *atrium* where the blood enters and the *ventricle* where the blood is forced into further circulation, see Figure 6.1(a). The two sides are divided by a muscular wall called the *septum*. The direction of blood flow is controlled by four different valves which are located between the atria and the ventricles (atrioventricular valves) and between the ventricles and the arteries (pulmonary and aortic valves).

The wall of the heart is called the *myocardium* and is primarily composed of muscle cells which produce mechanical force during contraction of the heart. The myocardium also contains specialized muscle cells which are connected into a network (conduction system) that allows an electrical impulse to rapidly spread throughout the heart. A cardiac cycle is created when such an impulse propagates through the conduction system. The electrical impulse is the event that triggers the mechanical force, and thus the electrical event precedes heart contraction.

The sequence of mechanical events that defines a cardiac cycle can be assumed to start in the right atrium where blood is collected from all the veins in the body except those of the lungs. When the right atrium is triggered to contract, it forces blood into the right ventricle. When the right ventricle has been filled, it contracts and forces blood into the lungs, where the excess carbon dioxide is replaced by oxygen. The pulmonary veins return the oxygenated blood to the left atrium which in turn empties into the left ventricle. In its capacity as a high-pressure pump, the left ventricle forces blood to all of the body organs and tissues (except the lungs) through the arterial vessels which evolve into capillaries and, finally, into the venous return system of the heart.

Each cardiac cycle is composed of two phases, activation and recovery, which are referred to in electrical terms as *depolarization* and *repolarization* and in mechanical terms as contraction and relaxation. Depolarization is manifested by a rapid change in the membrane potential of the cell (from −90 to 20 mV in approximately 1 ms) and constitutes the initial phase of the cardiac action potential, see Figure 1.2. The rapid change in voltage causes neighboring cells to depolarize, and, as a result, an electrical impulse spreads from cell to cell throughout the myocardium. Depolarization is immediately followed by repolarization during which the membrane potential of the cells gradually returns to its resting state.

The initialization of a cardiac cycle occurs in a mass of pacemaker cells with the ability to spontaneously fire an electrical impulse. These cells are collectively referred to as the *sinoatrial* (SA) *node* and are situated in the upper part of the right atrium. The electrical impulse (sometimes also called the electrical wavefront) then propagates through the conduction system so that atrial and ventricular contraction and relaxation can take place with the correct timing (Figure 6.1(b)). After electrical activation of the right and left atria, the impulse is collected and delayed at the *atrioventricular* (AV) *node* before it enters into the ventricles. The delay allows the atrial contraction to further increase the blood volume in the ventricles before ventricular contraction occurs. The delay in the AV node is caused by slower conduction of the impulse by the muscle tissue in this area. The impulse enters the wall between the two ventricles at the bundle of His; this is the only location that electrically connects the atria and the AV node/ventricles. The pathway is then divided into rapidly conducting bundles with branches to the left and right ventricles and then further into an extensive network of specialized conduction fibers called *Purkinje fibers*. The large size of the two ventricles requires the electrical impulse to propagate rapidly to initiate a unified contraction. The conduction velocity of the heart ranges from only 0.05 m/s at the AV node to as much as 4 m/s at the Purkinje fibers.

The SA node is the natural pacemaker of the heart which determines the rate of beating. Since the cells of the SA node have the fastest pacemaker rate, all other cells follow in synchrony; this property of beating on its own is called *automaticity*. The rate of the SA node is, however, not only determined by its inherent discharge rate ("clock frequency"), but also by external information which is mediated through the autonomic nervous system. The balance between the parasympathetic and sympathetic parts of the autonomic nervous system determines the heart rate so that an increase in parasympathetic activity decreases the heart rate, while an increase in sympathetic activity increases the heart rate. An upper limit is set on the discharge rate by the time during which a cell is electrically inactive, i.e., the refractory period, and ranges from 200–250 ms, which corresponds to a theoretical maximum heart rate of 240–300 beats/minute. However, a rhythm initiated by the SA node rarely exceeds a rate of 220 beats/minute, and this only applies to young individuals; the maximum heart rate decreases with age.

The SA node has precedence over other pacemaker cells in the normal heart. During certain conditions, however, another mass of cells, referred to as an ectopic focus, may take precedence over the SA node. An ectopic focus, which can be located in the atria or in the ventricles, determines the heart rate when the discharge rate of the SA node falls below a certain level. The intrinsic rate of atrial cells is about 50–60 times per minute and is higher than the ventricular cells which discharge at a rate between 20 and 40 times per minute.

The electrical activity of the heart can be characterized by measurements acquired from the cellular level as well as from the body surface. We will restrict ourselves to electrocardiographic measurements, although the study of cardiac action potentials is essential for a deeper understanding of the mechanisms behind various cardiac disorders such as arrhythmias. The ECG describes the different electrical phases of a cardiac cycle and represents a summation in time and space of the action potentials generated by millions of cardiac cells. Thus, rather than directly reflecting changes in membrane potential across the cells, the ECG provides a measure of the electrical currents generated in the extracellular fluid by these potential changes. The waveforms produced during depolarization and repolarization deviate from a baseline level which corresponds to the resting state of the cells. The depolarization waves are generally steeper and more peaked than those related to the subsequent repolarization which are smooth and rounded. These characteristics are illustrated in Figure 6.2 where the action potentials associated with different regions of the heart are depicted; the timing relationship between the different action potentials and the ECG measured on the body surface is also illustrated.

6.2 Generation and Recording of an ECG

6.2.1 Depolarization and Repolarization

Of the millions of individual cells in the heart that depolarize during a cardiac cycle, only groups of cells in the myocardium depolarize at any given instant. Each group of cells which is simultaneously depolarizing may be represented

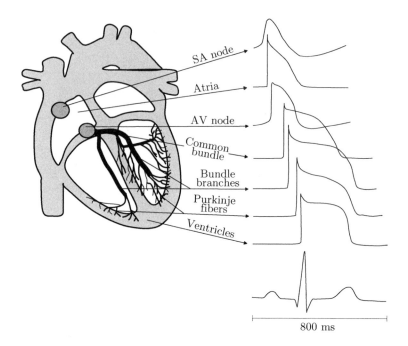

Figure 6.2: The morphology and timing of action potentials from different regions of the heart and the related cardiac cycle of the ECG as measured on the body surface.

as an equivalent current dipole source to which a vector is associated, describing the dipole's time-varying position, orientation, and magnitude. The related vectors of all these groups can be summed to give a "dominant" vector which describes the main direction of the electrical impulse, see Figure 6.3.

The dipole/vector representation has proven very useful for understanding how the waves of the ECG are generated as recorded by electrodes attached to the body surface.[1] Figure 6.4 illustrates the sequence of vectors associated with different phases of depolarization and repolarization and the related ECG wave, in this example viewed by an exploring electrode which is positioned on the chest (known as lead V_5, see below). This electrode position primarily reflects the activity of the left ventricle, although the activity of the atria and the right ventricle can also be observed.

---

[1]It has been shown that the ECG is proportional to the projection of the dominant vector onto a *lead vector* whose direction is defined by the positions of the heart and the electrode(s) [6]. The lead vector modulus decreases as the electrodes retreat from the heart due to increased impedance and, consequently, with a lowered ECG amplitude as a result. It is often helpful to consider the ECG as being proportional to the projection of the dominant vector onto the lead vector.

dominant vector

Figure 6.3: The vector associated with each group of cells in the myocardium, during both depolarization and repolarization, can be summed into a dominant vector describing the main direction of the electrical impulse.

Before a new heartbeat is initiated by the SA node, all cardiac cells are at rest, which is reflected by a horizontal line (the isoelectric line) in the ECG which forms its baseline, see Figure 6.4(a). During atrial depolarization, the dominant vector is directed downwards toward the AV node. As a result, an atrial wave with positive polarity is generated in the ECG recorded at the position of the exploring electrode, see Figure 6.4(b). The amplitude of the resulting wave is low because the muscle mass of the atria that produces the electrical wavefront is relatively small.

Once depolarization of the atria has been completed, the ECG returns to the isoelectric line where it remains until the ventricles become depolarized, see Figure 6.4(c). Depolarization of the AV node and the His bundle starts toward the end of the atrial wave but does not produce any visible ECG waves because of the small muscle masses.

The waves associated with ventricular depolarization are much larger than the atrial wave since the ventricles have a much larger muscle mass. Ventricular depolarization begins in the wall between the ventricles (septum) in such a way that the associated vector is directed away from the exploring electrode, see Figure 6.4(d); hence, the related ECG wave has negative polarity. Due to the high conduction velocity of the cells in this part of the heart, the negative wave has a short duration. During continued ventricular depolarization, the dominant direction of the vector gradually turns toward the exploring electrode. This behavior is related to the fact that the wall of the left ventricle is three times thicker than that of the right ventricle and consequently takes longer to depolarize, see Figure 6.4(e)–(g).

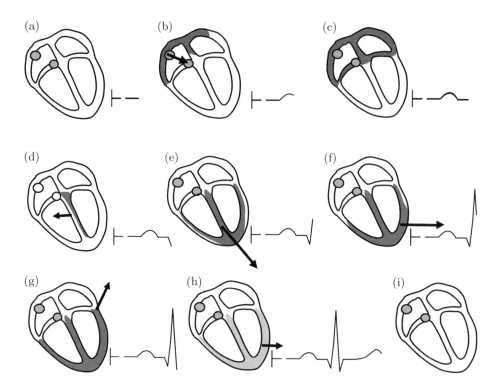

Figure 6.4: The ECG recorded in by an electrode positioned at the location of the symbol ⊢. (a) All cardiac cells at rest, (b) atrial depolarization, (c) the electrical impulse passing through the AV node, (d)–(g) ventricular depolarization, (h) ventricular repolarization, and (i) all cardiac cells again at rest.

Depolarization terminates with the dominant vector pointing away from the electrode, and thus a wave with negative polarity is produced in the ECG, see Figure 6.4(g).

Once ventricular depolarization has been completed, the ECG returns to the isoelectric line where it remains until ventricular repolarization occurs. During ventricular repolarization, a similar sequence of dominant vectors to those during ventricular depolarization appears, and a wave with positive polarity is produced, see Figure 6.4(h). Since atrial repolarization is concurrent with ventricular depolarization, the related atrial repolarization wave is masked by the ventricular waves which have much larger amplitudes.

It is important to realize that the polarity and morphology of individual waves are strongly dependent on where the electrodes are positioned on the body. For some positions, a wave may actually be completely absent because the wavefront is propagating perpendicularly to the electrode. Furthermore, wave amplitude depends on the distance between the heart and the electrode.

6.2.2 ECG Recording Techniques

The electrical activity of the heart is measured on the body surface by attaching a set of electrodes to the skin. The electrodes are positioned so that the spatiotemporal variations of the cardiac electrical field are sufficiently well-reflected. For an ECG recording, the difference in voltage between a pair of electrodes is referred to as a *lead*. The ECG is typically recorded with a multiple-lead configuration which includes unipolar or bipolar leads, or both. A so-called *unipolar lead* reflects the voltage variation of a single electrode and is measured in relation to a reference electrode which is positioned so that the voltage remains almost constant throughout the cardiac cycle; the reference is commonly called the "central terminal". A *bipolar lead* reflects the voltage difference between two electrodes, e.g., between the left and right arm.

The electrode wires are connected to a differential amplifier which is specially designed for bioelectrical signals. An amplifier with high gain and large dynamic range is needed since the ECG ranges from a few microvolts to about 1 V in magnitude; although the individual waves have a maximal magnitude of only a few millivolts, a wandering baseline in the ECG due to variations in electrode-skin impedance may reach 1 V. The amplifier bandwidth is commonly between 0.05 and 100–500 Hz, where the upper limit depends on the ECG application of interest. Insulation must also be taken into consideration to assure patient safety and implies that everything on the patient side is battery powered.

A number of lead systems exist today with standardized electrode positions of which we will describe the two that have received the most attention, namely, the standard 12-lead ECG and the orthogonal lead system producing a *vectorcardiogram* (VCG). We will briefly consider techniques for synthesizing one lead system from the other. In practice, the preferred lead system is not necessarily chosen on the basis of maximized information content, but the choice is more often guided by various clinical issues and practical considerations.

Standard 12-lead ECG. The standard 12-lead ECG is the most widely used lead system in clinical routine and is defined by a combination of three different lead configurations: the bipolar limb leads, the augmented unipolar limb leads, and the unipolar precordial leads (Figure 6.5). The 12-lead ECG is recorded by placing 10 electrodes at standardized positions on the body surface.

The three *bipolar limb leads* are denoted I, II, and III and are obtained by measuring the voltage difference between the left arm, right arm, and left leg in the following combinations:

$$I = V_{LA} - V_{RA}, \tag{6.1}$$

$$II = V_{LL} - V_{RA}, \tag{6.2}$$

$$III = V_{LL} - V_{LA}, \tag{6.3}$$

where V_{LA}, V_{RA}, and V_{LL} denote the voltage recorded on the left arm, right arm, and left leg, respectively. Since these three electrode positions can be viewed as the corners of an equiangular triangle ("Einthoven's triangle") with the heart at its center, the resulting limb leads describe the cardiac electrical activity in three different directions of the frontal plane, see Figure 6.6(a); each direction is thus separated by an angle of 60°. It is not necessary to record lead III since it can be computed from the leads I and II by the relation $III = II - I$.

The *augmented unipolar limb leads* $(aVF, aVL$ and $aVR)$ were introduced to fill the 60° gaps in the directions of the bipolar limb leads. These leads use the same electrodes as the bipolar limb leads, but are defined as voltage differences between one corner of the triangle and the average of the remaining two corners:

$$aVR = V_{RA} - \frac{V_{LA} + V_{LL}}{2}, \tag{6.4}$$

$$aVL = V_{LA} - \frac{V_{RA} + V_{LL}}{2}, \tag{6.5}$$

$$aVF = V_{LL} - \frac{V_{LA} + V_{RA}}{2}, \tag{6.6}$$

see Figure 6.6(b). Hence, the augmented limb leads describe directions which are shifted 30° from those of the bipolar limb leads: the gap between I and II is filled by $-aVR$, between II and III by aVF, and between III and I by aVL (Figure 6.7(a)). The augmented limb leads are considered to be unipolar because one electrode is exploring while the average of the other two serves as the reference electrode, cf. the expressions (6.4)–(6.6). Similar to lead III, the augmented limb leads do not have to be recorded but can be easily computed from leads I and II.

The *precordial leads* are positioned in succession on the front and left side of the chest in order to provide a more detailed view of the heart than do the limb leads (Figure 6.6(c)). The six precordial leads, by convention labeled $V_1, ..., V_6$, are unipolar and related to a central terminal which is defined by

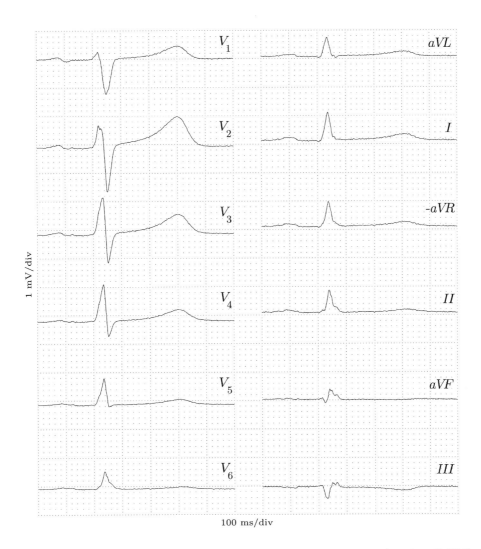

Figure 6.5: The standard 12-lead ECG with bipolar limb leads (I, II, and III), augmented unipolar limb leads (aVF, aVL, and $-aVR$), and unipolar precordial leads ($V_1, ..., V_6$). The ECG was recorded from a healthy subject.

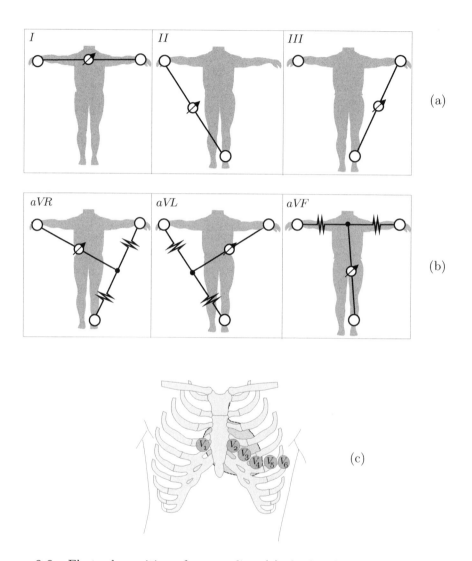

Figure 6.6: Electrode positions for recording (a) the bipolar limb leads I, II, and III (together these three leads define Einthoven's triangle), (b) the augmented unipolar limb leads aVR, aVL, and aVF (the output signal is measured between the two resistors), and (c) the precordial leads $V_1, \ldots, V_6$.

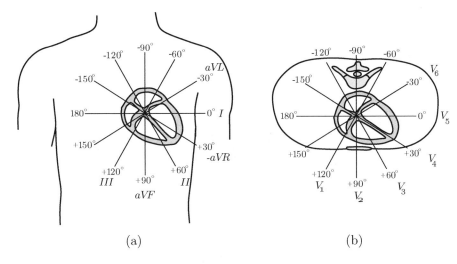

(a) (b)

Figure 6.7: (a) The directions of the bipolar limb leads and the augmented limb leads in the frontal plane. Note that the negative polarity of lead aVR is commonly used since the six limb leads then reflect directions from $-30°$ to $+120°$ which are $30°$ apart. (b) The precordial leads are approximately $30°$ apart in the transversal plane.

the average of the voltages measured on the right and left arms and the left leg,

$$V_{WCT} = \frac{V_{LA} + V_{RA} + V_{LL}}{3}. \tag{6.7}$$

Based on theoretical considerations, this definition has been found advantageous since V_{WCT} remains almost constant throughout the entire cardiac cycle. The abbreviation WCT stands for Wilson central terminal to honor the physician who introduced the definition given in (6.7). Leads V_1 and V_2 primarily reflect the activity of the right ventricle. Leads V_3 and V_4 primarily view the front of the left ventricle (anterior wall), while its side (lateral wall) is viewed by V_5 and V_6 (Figure 6.7(b)).

In general, the ECG waveforms of the six limb leads have a relatively low amplitude and tend to be more noisy than the precordial leads since the electrodes are positioned on the extremities at larger distances from the heart. Thus, the signal-to-noise ratios of the limb leads are often lower than those observed in the precordial leads.

Orthogonal leads. An orthogonal lead system is attractive since it reflects the electrical activity in the three perpendicular directions X, Y, and Z. For such a lead system, the ECG interpretation is not confined to findings in in-

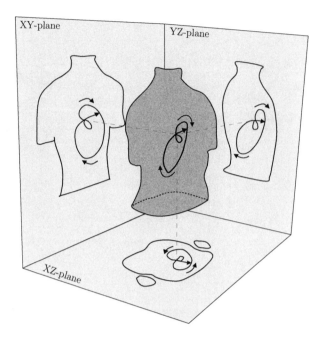

Figure 6.8: A vectorcardiographic loop and its projection onto the three orthogonal planes. The two arrows outside each loop indicate the direction in which the loop evolves.

dividual leads, but additional information is acquired through the visualization of a three-dimensional loop together with its projection onto the XY-, XZ-, and YZ-planes, see Figure 6.8. Since a loop is traced out by the tip of the vector which describes the dominant direction of the electrical wavefront during the cardiac cycle, this particular type of recording is referred to as a vectorcardiogram. While the VCG provides a time-varying description of how the magnitude and direction of the dominant vector change with time, it does not provide an anatomically exact description of where the wavefront is propagating in the myocardium.

Pairs of electrodes which are positioned along mutually perpendicular lines on the body surface may, at first glance, appear to produce leads which are orthogonal.[2] Based on mathematical modeling as well as on experimental results, however, it has been found that additional electrodes are required in order to account for the geometry of the human torso. The corrected orthogonal leads, known as the *Frank lead system* after its inventor [7], are obtained as linear combinations of seven electrodes positioned on the chest,

---

[2]The word "orthogonal" is used here to signify the different, perpendicular spatial directions and should not be confused with its usage in the mathematical sense as on page 255.

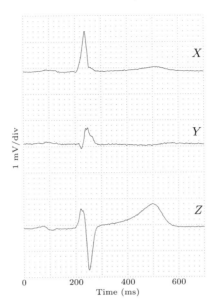

Figure 6.9: The orthogonal vectorcardiographic leads X, Y, and Z; these leads were recorded from the same subject as the ECG leads shown in Figure 6.5.

back, neck, and left foot. The resulting leads X, Y, and Z view the heart from the left side, from below, and from the front. The waves observed in leads X, Y, and Z have a fairly close resemblance to those of V_5, aVF, and V_2 (cf. Figures 6.9 and 6.5). However, analysis of the orthogonal leads on an individual basis does not significantly add to the diagnostic information gathered from the analysis of V_5, aVF, and V_2. The benefit of using VCG leads reveals itself rather when the loop, or its projections on the planes, is analyzed with respect to spatial quantities such as loop morphology, direction of rotation, and area. These quantities, which do not have any counterpart in the analysis of individual leads, have been used to improve ECG-based diagnosis of, e.g., myocardial infarction.

Although the information contained in VCG leads has been found useful in certain applications, the 12-lead ECG remains by far the preferred lead system in the clinical routine due to the existence of well-established criteria for its interpretation.

Synthesized leads. The concept of synthesized leads was first introduced to handle situations in which VCG leads were available but where analysis of the standard 12-lead ECG was desired [8]. Such lead synthesis may be employed to extend the clinical value of a database containing digitized VCG signals. Another situation arises when the attachment of certain electrodes of the 12-lead ECG is inconvenient or even impossible, such as during open-heart surgery when some of the electrodes cannot be attached. With synthesized leads, the clinician can still monitor changes that may occur in some of the ECG leads. In practice, synthesis of the 12-lead ECG from the VCG has not gained wide acceptance due to diagnostic inaccuracies which have been noted between the true and synthesized leads. However, the 12-lead ECG synthesized from another configuration known as the EASI leads, requiring only four electrodes, has received more attention [9, 10].

The opposite idea, of synthesizing the orthogonal leads of the VCG from the standard 12-lead ECG, has been implemented with greater success [11, 12], almost certainly because more leads are available. The addition of synthesized VCG leads to computer programs which originally relied on 12-lead ECG recordings has contributed to improved detection of certain electrocardiographic abnormalities. The synthesized VCG leads are sometimes referred to as the *12-lead vectorcardiogram* in order to acknowledge the origin of the lead system.

Both these types of lead synthesis rely on the basic assumption that the voltage at any point on the body surface at any instant in time n can be approximated by a linear combination of the L recorded leads $x_1(n), \ldots, x_L(n)$. Hence, the M synthesized leads $y_1(n), \ldots, y_M(n)$ result from the following simple, linear relationship,

$$
\begin{bmatrix} y_1(n) \\ y_2(n) \\ \vdots \\ y_M(n) \end{bmatrix} = \mathbf{T}_{M \times L} \begin{bmatrix} x_1(n) \\ x_2(n) \\ \vdots \\ x_L(n) \end{bmatrix}, \tag{6.8}
$$

where the synthesis matrix $\mathbf{T}_{M \times L}$ has the dimensions $M \times L$. The original matrix $\mathbf{T}_{12 \times 3}$ was derived from geometrical considerations of the heart's electrical field [13]. In determining $\mathbf{T}_{3 \times 12}$ for VCG lead synthesis, both heuristic and statistical approaches have been investigated [12]. So far, the best agreement with the original Frank leads, expressed in diagnostic terms, has been achieved by using an inverse of the original matrix $\mathbf{T}_{12 \times 3}$ [11, 14], see also [15]. The calculation of this particular choice of $\mathbf{T}_{3 \times 12}$ represents an overdetermined least-squares problem (i.e., where the number of equations is larger than the number of unknown variables) whose solution is briefly described in Appendix A.

6.2.3   ECG Waves and Time Intervals

We will now describe some important ECG wave characteristics, central to the development of signal processing algorithms, along with the wave-naming

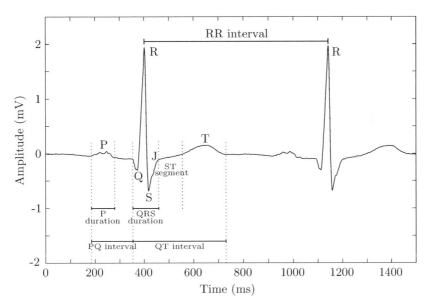

Figure 6.10: Wave definitions of the cardiac cycle and important wave durations and intervals. The J point defines the point in time when the QRS complex curves into the ST segment.

convention. Atrial depolarization is reflected by the P wave, and ventricular depolarization is reflected by the QRS complex, whereas the T wave reflects ventricular repolarization, see Figure 6.10. Atrial repolarization cannot usually be discerned from the ECG since it coincides with the much larger QRS complex. The amplitude of a wave is measured with reference to the ECG baseline level, commonly defined by the isoelectric line which immediately precedes the QRS complex. The duration of a wave is defined by the two time instants at which the wave either deviates significantly from the baseline or crosses it.

The **P wave** reflects the sequential depolarization of the right and left atria. In most leads, the P wave has positive polarity and a smooth, monophasic morphology. Its amplitude is normally less than 300 μV, and its duration is less than 120 ms. An absent P wave may, for example, suggest that the rhythm has its origin in the ventricles, i.e., a ventricular ectopic focus has taken precedence over the SA node causing atrial depolarization to coincide with ventricular depolarization.

The spectral characteristic of a normal P wave is usually considered to be low-frequency, below 10–15 Hz (Figure 6.11). However, the application of ensemble averaging techniques to produce a noise-reduced ECG has helped demonstrate that much higher frequency components of the P wave exist;

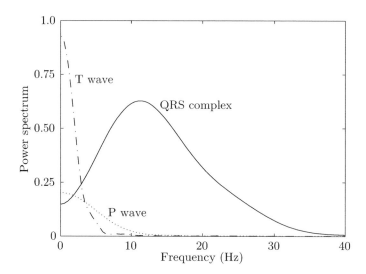

Figure 6.11: Power spectrum of the P wave, QRS complex, and T wave. The diagram serves primarily as a rough guide to where the spectral components are located; large variations exist between beats of different lead, origin, and subjects.

such components have been found useful for predicting the occurrence of certain arrhythmias of atrial origin.

It is sometimes problematic to determine the time instants that define the onset and end of a P wave because of a low amplitude and smooth morphology. As a result, the analysis of individual P waves is excluded from certain ECG applications where the presence of noise is considerable.

The **QRS complex** reflects depolarization of the right and left ventricles which in the normal heart lasts for about 70–110 ms. The first negative deflection of the QRS complex is denoted the *Q wave*, and the first positive is denoted the *R wave*, while the negative deflection subsequent to the R wave is denoted the *S wave* (Figure 6.10). Although the QRS complex may be composed of less than three individual waves, it is nevertheless referred to as a QRS complex. The morphology of the QRS complex is highly variable and depends on the origin of the heartbeat: the QRS duration of an ectopic beat may extend up to 250 ms, and is sometimes composed of more than three waves.

Since the QRS complex has the largest amplitude of the ECG waveforms, sometimes reaching 2–3 mV, it is the waveform of the ECG which is first identified in any type of computer-based analysis. The algorithm that performs the search is termed the QRS detector and produces the "landmark information" required to further analyze the ECG characteristics, see Section 7.4.

Due to its steep slopes, the frequency content of the QRS complex is considerably higher than that of the other ECG waves and is mostly concentrated in the interval 10–50 Hz (Figure 6.11). Similar to the P wave, ensemble averaging of the QRS complex has, in certain ECG recordings, uncovered high-frequency components which have been found to convey diagnostic information. In particular, the presence of late potentials in the terminal portion of the QRS complex has received considerable attention; see page 447 for further details.

The **ST segment** is not really a wave, but represents the interval during which the ventricles remain in an active, depolarized state. The ST segment begins at the end of the S wave (the J point) from where it proceeds nearly horizontally until it curves into the T wave (Figure 6.10). Changes in the ST segment, which make it either more elevated, depressed, or more steeply sloped, often indicate various underlying cardiac conditions.

The **T wave** reflects ventricular repolarization and extends about 300 ms after the QRS complex. The position of the T wave is strongly dependent on heart rate, becoming narrower and closer to the QRS complex at rapid rates; this "contraction" property does not apply to the P wave or the QRS complex. The normal T wave has a smooth, rounded morphology which, in most leads, is associated with a single positive peak.

The T wave is sometimes followed by another slow wave (the U wave) whose origin is unclear but is probably ventricular after-repolarization. At rapid heart rates, the P wave merges with the T wave, causing the T wave end point to become fuzzy as well as the P wave onset. As a result, it becomes extremely difficult to determine the T wave end point because of the gradual transition from wave to baseline.

The **RR interval** represents the length of a ventricular cardiac cycle, measured between two successive R waves, and serves as an indicator of ventricular rate. The RR interval is the fundamental rhythm quantity in any type of ECG interpretation and is used to characterize different arrhythmias as well as to study heart rate variability.

The **PQ interval** is the time interval from the onset of atrial depolarization to the onset of ventricular depolarization. Accordingly, the PQ interval reflects the time required for the electrical impulse to propagate from the SA node to the ventricles. The length of the PQ interval is weakly dependent on heart rate.

The **QT interval** represents the time from the onset of ventricular depolarization to the completion of ventricular repolarization. This interval normally varies with heart rate and becomes shorter at more rapid rates. It is therefore customary to correct the QT interval for heart rate—using nonlinear [16] or, better, linear techniques [17]—so that the corrected QT interval allows an assessment that is roughly independent of heart rate. Prolongation of the QT interval has been observed in various cardiac disorders associated with increased risk of sudden death.

6.3 Heart Rhythms

The rhythm of the normal heart is controlled by the electrical impulses formed within the SA node and produces a heart rate between 50 and 100 beats/minute during rest. A deviation from or a disturbance of the normal sinus rhythm is called *arrhythmia*. An arrhythmia may come about when depolarization is initiated by other pacemaker cells of the heart than those of the SA node, thus altering the *formation* of the electrical impulses. Another mechanism which produces arrhythmia is when the *conduction* of the electrical impulses is altered [18, 19].

Problems of impulse formation arise when an ectopic focus below the SA node exhibits accelerated automaticity. Alternatively, decelerated automaticity of the SA node makes way for an ectopic focus to take over control. In contrast to a sinus rhythm, an ectopic rhythm lacks modulation by the parasympathetic and sympathetic components of the autonomic nervous system.

Problems of impulse conduction are related to an area of the heart where the conduction of the cells is partially or totally blocked. A partial block delays propagation of the electrical impulse (e.g., causing prolongation of the PR interval), whereas a total block causes conduction failure and an accompanying decrease in heart rate.

The *reentry* phenomenon is another impulse conduction problem that sustains arrhythmia and is caused by changes in the refractory period and propagation speed of the heart. When neighboring areas of the myocardium have different refractory periods, the electrical impulse may depolarize an area which is receptive, while another area remains inactive since the cells are still refractory from the previous cardiac cycle. Once the inactive area has recovered, the impulse may be accepted and the area can serve as a pathway back to the area which was initially depolarized. Reentry occurs if the latter area has had time to recover and can be depolarized again and continues until the cells become unreceptive. The resulting movement of the electrical impulse is often circular and is said to occur in a reentry circuit.

The classification of cardiac arrhythmia involves the site of its origin. The following list describes different relationships between the atria and ventricles among different kinds of arrhythmia [18].

1. The atrial and ventricular rhythms are associated and have the same rate, and the rhythm may originate in the atria or the ventricles, respectively.

2. The atrial and ventricular rhythms are associated, but the atrial rhythm is faster than the ventricular rate (the rhythm originates in the atria).

3. The atrial and ventricular rhythms are associated, but the ventricular rate is faster than the atrial rate (the rhythm originates in the ventricles).

4. The atrial and ventricular rhythms are independent, and the atrial and ventricular rates are either the same or one is faster than the other.

An arrhythmia is also classified with respect to its rate: *bradyarrhythmia* has a slow heart rate of less than 60 beats/minute, whereas *tachyarrhythmia* has a rapid rate of more than 100 beats/minute. Hence, "atrial tachyarrhythmia" refers to an arrhythmia which is initiated in the atria and has a rate that exceeds 100 beats/minute. It should be noted that not all arrhythmias are beyond these rate limits. An arrhythmia which suddenly begins and ends is referred to as *paroxysmal*; otherwise, it is said to be *persistent/permanent*.

The significance of an arrhythmia differs widely and may be benign, symptomatic, life threatening, or fatal. Its consequence depends not only on its manifestation, but also on the presence of abnormal structural conditions of the heart, see Section 6.4.

6.3.1 Sinus Rhythm

The normal sinus rhythm originates from the SA node and has a rate between 50 and 100 beats/minute at rest (Figure 6.12(a)). The rhythm is called sinus *bradycardia* when the rate is below the lower limit and sinus *tachycardia* when it is above the upper limit. At rest, the heart rate is essentially regular but not totally so, even if external perturbations in the form of physical or mental stress are absent. These small variations in heart rate are caused by continual variation of the balance between the two components of the autonomic nervous system which influence the firing rate of the SA node: increased parasympathetic activity decreases the rate, whereas increased sympathetic activity increases the rate.

The dynamics of spontaneous *heart rate variability* serve as an indicator of how heart rate, respiration, blood pressure, and temperature are controlled by the body [20–22]. Decreased heart rate variability may be associated with significant underlying cardiac or autonomic abnormalities. For

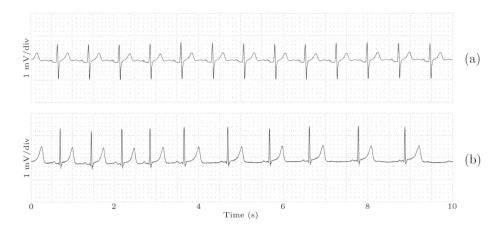

Figure 6.12: (a) Normal sinus rhythm and (b) respiratory sinus arrhythmia.

example, decreased variability has been associated with increased mortality after myocardial infarction, caused by a malignant ventricular arrhythmia [23, 24]. Heart rate variability has been increasingly studied in recent years for the purpose of diagnosing and predicting cardiovascular diseases. As a result, tailored signal processing algorithms that can handle, for example, the presence of ectopic beats that disrupt the sinus rhythm are much in demand. Chapter 8 contains a comprehensive description of some of the algorithms used for the analysis of heart rate variability.

In a normal subject, the variability in heart rate is related to the phases of respiration so that the rate increases with inspiration and decreases with expiration (Figure 6.12(b)). When pronounced, this type of sinus rhythm is called *respiratory sinus arrhythmia* and is normal. Figure 6.13 illustrates heart rate variability in terms of RR intervals for a normal subject who has been instructed to breath deeply at different, predetermined respiratory rates. Each of the four respiratory rates was maintained for one minute. In this example, the variability is as large as 25–30 beats/minute during the course of just a few heartbeats. It is obvious that the variability increases as the respiratory rate decreases and the breaths become increasingly deeper.

Important information on autonomic function may be inferred from an ECG by studying changes in the sinus rhythm in response to different types of provocation. The so-called *orthostatic test* represents a provocation in which the subject is tilted from lying to a standing position using a tilt table. Other types of provocation are used in the *deep breathing test,* where the subject is instructed to hyperventilate (cf. Figure 6.13(a)), and the *Valsalva test,* where the subject is forcibly trying to exhale with both the nose and mouth closed [25].

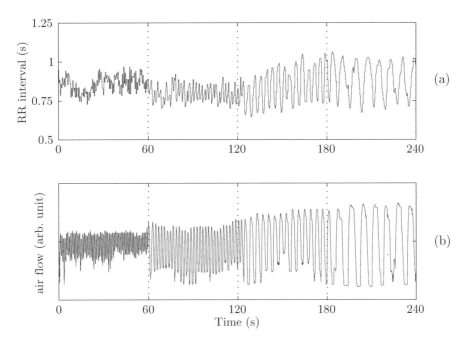

Figure 6.13: (a) Variation in RR intervals during controlled respiration with deep breaths. (b) The corresponding variations in air flow. The healthy subject was instructed to maintain the same respiratory rate for one minute for each respiratory rate.

6.3.2 Premature Beats

The normal sinus rhythm is sometimes interrupted by a beat occurring before the expected time of the next sinus beat and is therefore referred to as a *premature beat*; the terms "ectopic beat" and "extrasystole" are frequently used synonyms. The ectopic focus producing the premature beat may be located in any part of the heart other than the SA node. Depending on the location of the focus, the premature beat may have either normal or abnormal morphology and may be either preceded by a P wave or not.

A premature beat is either called a *supraventricular premature beat* (SVPB) if its origin is above the ventricles, i.e., in the atria or the AV node, or a *ventricular premature beat* (VPB) if its origin is in the ventricles. The presence of a VPB almost always prevents the occurrence of the next sinus beat. Although the SA node discharges on schedule, the impulse cannot propagate to the ventricles because the tissue has been made refractory by the premature beat. The pause that results between the VPB and the next sinus beat is called the *compensatory pause*. The most common kind of SVPB is associated with an abnormal P wave morphology and a QRS complex morphology resembling that of a normal sinus beat. The related compensatory pause is so that the interval between the two sinus beats that enclose the SVPB is less than the length of two normal RR intervals (Figure 6.14(a)).

A VPB may originate from any area beyond the point where the common bundle has branched into the left and right bundle branches. Since the electrical impulse of the ventricular ectopic focus does not follow the normal conduction pathways, the produced QRS complex is abnormally prolonged and has a morphology which deviates considerably from that of a sinus beat—it is often much larger and bizarre-looking. A VPB usually inhibits the next sinus beat and introduces a compensatory pause which is twice the length of the normal RR interval (Figure 6.14(b)). When several different ectopic foci exist in the ventricles, the resulting premature beats have widely different QRS complex morphologies and are called multiform VPBs. When a premature beat follows every normal sinus beat, the rhythm pattern is called *bigeminy* (Figure 6.14(c)); similarly, when a premature beat follows every other normal sinus beat, the rhythm is called *trigeminy* (Figure 6.14(d)).

When the heart rate of the SA node is slow and the VPB is extremely premature, the next sinus beat can still occur on time because the AV node and ventricles are no longer refractory. Such a VPB is referred to as "interpolated" and is not associated with a compensatory pause, but is enclosed by two short RR intervals (Figure 6.14(e)).

Isolated premature beats are commonly found in normal subjects. Abundant premature beats may, however, be a manifestation of an underlying cardiac disease.

6.3.3 Atrial Arrhythmias

Various rhythm disturbances originate from one or multiple ectopic foci in the atria. The resulting arrhythmias are characterized either by the presence of abnormal P waves or a complete lack of distinct P waves. Abnormal P waves appear when an ectopic focus is located far away from the SA node, i.e., closer to the AV node, since the electrical impulse then propagates in a direction which is opposite to the normal one; a normal P wave thus becomes negative. If the focus is near the AV node and the ventricles, depolarization of the ventricles commences at about the same time as does depolarization of the atria. In consequence, the occurrence of the P wave coincides with that of the QRS complex and cannot be discerned in the ECG.

Atrial tachycardia is an arrhythmia produced by increased automaticity in the pacemaking cells of one or multiple foci within the atria. Most of the electrical impulses are conducted to the ventricles, leading to a heart rate from 140 to 220 beats/minute. Due to the rapid rate, P waves are often masked by the T wave or even the QRS complex of the previous heart cycle.

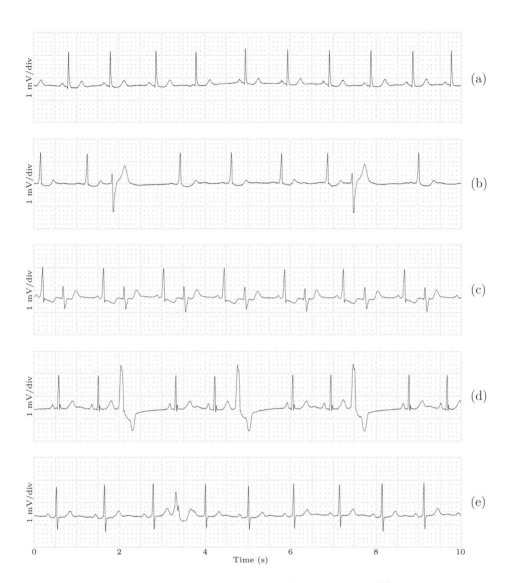

Figure 6.14: Examples of ectopic beats with different origins. (a) A supraventricular premature beat with a small, negative P wave (occurring after the third sinus beat). (b) A premature ventricular beat followed by a compensatory pause. The sum of the two RR intervals adjacent to the ectopic beat is approximately equal to twice the normal RR interval. (c) Bigeminy, (d) trigeminy, and (e) an interpolated ventricular premature beat (occurring after the third sinus beat).

Atrial flutter and *atrial fibrillation* are two tachyarrhythmias in which the atria are unsynchronized with the ventricles and beat at a rate which is much faster than that of the ventricles. Both these arrhythmias are caused by the continuing reentry of an electrical impulse in the atria. This is manifested in the ECG by an undulating baseline which replaces the P waves, and, consequently, the isoelectric line is no longer well-defined. The rapid and irregular rate of atrial flutter/fibrillation causes the blood to flow more slowly than normal through the atria, increasing the likelihood that a blood clot will be produced. If the clot is pumped out of the heart, a stroke may result if it reaches the brain, but a clot may also cause damage to other parts of the body.

Atrial flutter is the more organized arrhythmia of the two and is characterized by the atria beating regularly at a rate of usually around 300 beats/minute. Not all electrical impulses reach the ventricles but are blocked by a refractory AV node, which thus protects the ventricles from too rapid activation. When only every second impulse is conducted to the ventricles, a 2:1 block is said to exist, producing a ventricular rate of around 150 beats/minute, and so on. In the ECG, flutter waves are referred to as F waves and have an appearance which resembles that of a sawtooth (Figure 6.15(a)).

Atrial fibrillation is a very rapid, chaotic rhythm (400–700 beats/minute) which makes the atria quiver and the ventricles beat irregularly. The arrhythmia is produced by reentry within multiple circuits in the atria, giving rise to impulses that bombard the AV node. Only some of the impulses get through the AV node and then produce a ventricular rate which is highly irregular and often also rapid. In the ECG, fibrillation waves are referred to as f waves and have an irregular, multiform appearance (Figure 6.15(b)).

6.3.4   Ventricular Arrhythmias

The most common ventricular tachyarrhythmias result from the reentry mechanism and include *ventricular tachycardia, ventricular flutter*, and *ventricular fibrillation*. The mechanisms behind these three arrhythmias are analogous to the three above-mentioned atrial tachyarrhythmias, i.e., they are reentrant ventricular tachyarrhythmias; however, their manifestations in the ECG are completely different.

Ventricular tachycardia occurs at a rate over 120 beats/minute and consists of beats with a morphology similar to that of premature beats, i.e., increased QRS width and large amplitude. The P waves are often lost because wide QRS complexes or T waves are constantly occurring so that one ventricular cycle immediately succeeds the other. In Figure 1.7, an ECG was presented which contains an episode of ventricular tachycardia with

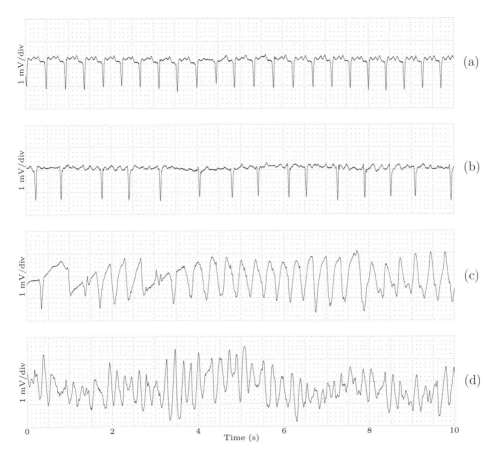

Figure 6.15: Examples of atrial and ventricular tachyarrhythmias. (a) Atrial flutter, (b) atrial fibrillation, (c) ventricular flutter, and (d) ventricular fibrillation.

three beats. This is the minimum number of consecutive beats required to be defined as arrhythmia.

Ventricular flutter is a rapid organized rhythm without any discernible QRS complexes or T waves. It is similar to atrial flutter but has a much larger amplitude which fluctuates considerably over time. Ventricular flutter can lead to ventricular fibrillation, which is a totally disorganized rhythm during which the ventricles cease to depolarize in an orderly fashion. As a result, a heart undergoing ventricular fibrillation cannot deliver oxygenated blood to the brain. Ventricular fibrillation leads to cardiac arrest, cessation of respiration, loss of consciousness, and, if no immediate treatment is given, it is almost invariably fatal. Ventricular flutter and fibrillation produce ECGs of similar appearance, although the latter rhythm is associated with a more chaotic signal (Figures 6.15(c) and (d)).

6.3.5   Conduction Blocks

The propagation of an electrical impulse can be disturbed by a block along its normal conduction pathway. The block causes depolarization and repolarization to become abnormal, disturbing the function of the heart. One conduction block is related to the AV node and makes the electrical connection between the atria and ventricles abnormal to various degrees. The severity of the AV block is graded from minor, when all impulses are conducted with delay, through moderate, when some impulses do not reach the ventricles, to complete, when no impulses are conducted. A complete AV block is manifested by P waves and QRS complexes at two different, independent rates; the P waves are produced by the SA node, while the QRS complexes have their origin in a ventricular ectopic focus. Many other types of conduction blocks may occur, e.g., in the left or right bundle branches [18].

6.4   Heartbeat Morphologies

A wide variety of abnormal beat morphologies can be seen in an ECG which reflect abnormal structural conditions of the heart, such as enlargement of the atria or ventricles (hypertrophy) and inflammation of the sac-like covering of the heart (pericarditis). Abnormal morphologies are also typical of several arrhythmias. In this section, two important structural conditions are considered which are caused either by insufficient blood supply to the myocardium (myocardial ischemia) or by death of the tissue of the myocardial wall (myocardial infarction). During the acute phase of an ischemic episode or myocardial infarction, the beat morphology sometimes undergoes dynamic changes which are so rapid that they occur almost from one beat to the next.

6.4.1   Myocardial Ischemia

Myocardial ischemia is a condition where the blood flow to the cells of the heart is restricted, causing a lack of oxygenated blood. Myocardial ischemia arises when one or more coronary arteries has become narrowed and the demand for oxygenated blood to the heart muscle increases due to exercise or mental stress. A temporary reduction in flow usually causes chest pain or discomfort known as *angina pectoris*. However, myocardial ischemia is sometimes completely unrelated to chest pain and is then called silent ischemia. Since ischemia is associated with electrical instability of the heart, it may cause life-threatening ventricular tachyarrhythmias such as ventricular fibrillation.

Myocardial ischemia is manifested in an ECG as a change in morphology of the ST segment and T wave, jointly referred to as an ST–T change. While the normal ST segment starts at the isoelectric line and curves smoothly upwards into the T wave, the ischemic ST segment is instead horizontal or slopes downwards and may start well below the isoelectric line. An ST segment which drops below the isoelectric line is referred to as an ST depression. An ischemic T wave is often more flat than a normal T wave and may exhibit biphasic morphology or negative polarity.

Tiny beat-to-beat alternations in T wave morphology are related to myocardial ischemia as a presage of malignant ventricular arrhythmias that often lead to sudden cardiac death [26–29]. The morphologic alternations follow a flip-flop pattern (i.e., ABABA...) in which every other T wave has the same morphology. Since *T wave alternans* is a phenomenon in the microvolt range, it cannot be perceived by the naked eye from a standard ECG print-out, but requires signal processing techniques for its detection and quantification [30].

Subjects with suspected angina pectoris, who may also have had episodes of silent ischemia, are referred to a hospital to perform an exercise stress test. From this test, a diagnosis can be made by the physician once the ST–T reaction, as provoked by an exercise stress test, has been examined; a similar test can be used when T wave alternans is of interest. Another common test is ambulatory ECG monitoring, where the subject carries a portable recording device for 24 hours or more so that episodes of myocardial ischemia can be identified. Both these tests are discussed in more detail below.

6.4.2 Myocardial Infarction

Myocardial infarction causes death (necrosis) of some heart cells due to sudden and sustained loss of blood supply. The loss is caused by a complete blockage of a coronary artery and typically results from the process of arteriosclerosis. In that process, an artery becomes hardened by a build-up of plaque which will eventually rupture so that a blood clot forms and blocks the artery. Similar to silent ischemia, an infarction is associated with electrical instability of the heart and dramatically increases the risk of ventricular fibrillation.

Most patients today survive myocardial infarction thanks to a number of efficient treatment options. Although the heart is unable to pump as much blood as before infarction, there is usually sufficient viable heart muscle left to perform the required work load, making recovery almost complete.

The infarcted area is electrically inert and disturbs the normal propagation pathways of the electrical impulse. As a result, the dominant vector which describes the main direction of the impulse is altered, and waves are

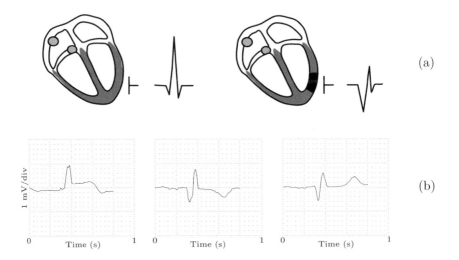

Figure 6.16: (a) The normal QRS complex (left panel) and the QRS complex after myocardial infarction (right panel). The dark area represents the infarcted area. (b) Evolution of a myocardial infarction in lead aVF. The ECG was recorded upon arrival at the emergency room (left panel), after 10 days (middle panel), and after 6 months (right panel).

produced which differ significantly from the normal ECG. Figure 6.16(a) illustrates morphologic changes caused by myocardial infarction: in lead aVF, the amplitude of the R wave is reduced, and a much larger, pathological Q wave is observed.

During the healing period, when the necrotic heart muscle completes the scarring process, the QRST complex undergoes a gradual change in morphology. Figure 6.16(b) illustrates the changes that occur during a time span of 6 months following infarction. In this example, the initial ST–T segment elevation has disappeared after 10 days, and a large Q wave has appeared which remains after 6 months.

6.5 Noise and Artifacts

An important reason behind the success of computer-based ECG analysis is the capability to improve poor signal quality by means of signal processing algorithms. This result has been achieved thanks to good knowledge of not only signal properties but also noise properties. It is therefore important to become familiarized with the most common types of noise and artifacts in the ECG before addressing methods in the next chapter which compensate for their presence. Below follows a list of common noncardiac noise sources of which the first three are of technical origin whereas the fourth is of physiological origin. Even parts of the cardiac activity can sometimes be viewed as a source of noise when detecting QRS complexes, see Section 7.4.

Baseline wander is an extraneous, low-frequency activity in the ECG (Figure 6.17(a)) which may interfere with the signal analysis, rendering the clinical interpretation inaccurate and misleading. For example, ECG measurements defined with reference to the isoelectric line cannot be computed because the isoelectric line is no longer well-defined. Baseline wander, which is often exercise-induced, may result from a variety of noise sources including perspiration, respiration, body movements, and poor electrode contact. The magnitude of the undesired wander may exceed the amplitude of the QRS complex by several times. Its spectral content is usually confined to an interval well below 1 Hz, but it may contain higher frequencies during strenuous exercise. Signal processing techniques for the removal of baseline wander are presented in detail in Section 7.1.

Electrode motion artifacts are mainly caused by skin stretching which alters the impedance of the skin around the electrode. Motion artifacts resemble the signal characteristics of baseline wander, but are more problematic to combat since their spectral content considerably overlaps that of the PQRST complex. They occur mainly in the range from 1 to 10 Hz [31, 32]. In the ECG, these artifacts are manifested as large-amplitude waveforms which are sometimes mistaken for QRS complexes (Figure 6.17(b)). Electrode motion artifacts are particularly troublesome in the context of ambulatory ECG monitoring where they constitute the main source of falsely detected heartbeats.

Powerline interference (50/60 Hz) is caused by improper grounding of the ECG equipment and interference from nearby equipment [33]. Such interference can be removed in many situations by means of linear or nonlinear filtering, see Section 7.2.

The electrical activity of skeletal muscles during periods of contraction causes *electromyographic noise* (EMG noise), commonly seen in ECGs recorded during ambulatory monitoring or exercise. The main characteristics of such noise have already been presented on page 74 in connection with artifact rejection in EEG signal processing (different muscles are, however, active in producing the noise which corrupts the ECG signal), see also Section 5.1. Electromyographic noise can either be intermittent in nature, e.g., due to a sudden body movement (Figure 6.17(c)), or have more stationary noise properties. The frequency components of EMG considerably overlap those of the QRS complex while also extending into higher frequencies. As a result, difficulties in removing EMG noise from the EEG signal without introducing distortion are unfortunately also present in ECG signal processing. Some approaches that deal with EMG noise are briefly presented in

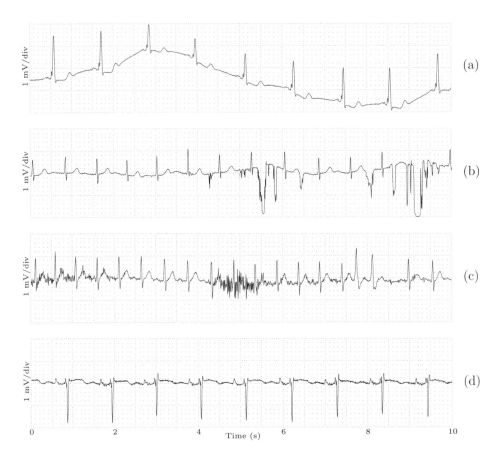

Figure 6.17: Different types of noise and artifacts in the ECG. (a) Baseline wander, (b) electrode motion artifacts, (c) electromyographic noise, and (d) respiration-induced modulation of the QRS amplitude.

Section 7.3. The influence of EMG noise can also be reduced by ensemble averaging when the recurrent property of the heartbeats can be exploited.

Respiratory activity influences electrocardiographic measurements not only through heart rate but also through beat morphology. Such beat-to-beat variations in morphology are caused by chest movements, changes in the position of the heart, and changes in lung conductivity [34, 35]. During the respiratory cycle, the vector describing the dominant direction of the electrical wave propagation changes so that variations in beat morphology arise. Figure 6.17(d) presents an ECG with a pronounced variation in QRS amplitude being induced by respiration; in this example, the period length of a breath is approximately 5 s, suggesting that the subject is breathing at a rate of 12 breaths/minute. Although variations in QRS amplitude represent an undesirable signal characteristic, it may be exploited for estimation of the respiratory frequency [36–39].

6.6 Clinical Applications

Since its invention, the usage of the ECG has become so greatly diversified that it is today inappropriate to refer to the ECG as being just one clinical test. In order to give the reader an overview, the most common clinical applications of the ECG are briefly described, while a comprehensive description can be found elsewhere [1]. Each of these applications has become intimately associated with computer processing of the ECG, a fact that has contributed to improving test accuracy.

6.6.1 Resting ECG

The resting ECG is one of the most widely used diagnostic tests in clinical routines of all kinds and is called for when a wide range of diseases are suspected, not necessarily of cardiac origin. The standard 12-lead ECG is recorded for 10 s in conditions which are favorable from a signal quality perspective since the patient is at rest in the supine position. The brief recording time limits the significance of the test to heart problems of a permanent nature, while transiently occurring arrhythmias must be investigated by other tests such as the ambulatory ECG (see below).

Resting ECGs are today conventionally interpreted by computer, and this involves software that derives a set of measurements describing waveform morphology and rhythm. Due to the good signal quality of resting ECGs, P wave information can be part of the analysis even if the P waves are small; such information is a prerequisite for correct classification of atrial arrhythmias. Although some basic signal processing is needed when analyzing a resting ECG (see Chapter 7), the big challenge has been to develop software that satisfactorily implements all the different criteria that are applied by the human interpreter to the ECG. While the diagnostic accuracy of a modern system is very good, the ECG print-out is usually checked by a physician to assure that the diagnosis is correct.

Once a resting ECG has been diagnosed, the signal is stored in a database for retrieval, if necessary, at a later date. Database storage facilitates *serial ECG analysis* in which two or more successive ECG recordings from the same patient are compared to reveal possible changes related, for example, to myocardial infarction. Unfortunately, interrecording changes in the ECG caused by nonphysiological factors, such as different electrode placement or positional changes of the heart, degrade the reliability of serial analysis.

6.6.2 Intensive Care Monitoring

A patient who has suffered myocardial infarction or undergone heart surgery is placed in an intensive care unit (ICU) or coronary care unit (CCU). Such a unit has a number of beds wired to a central computer so that the ECG of each individual patient can be continuously monitored. The objective of ECG monitoring is primarily to detect life-threatening arrhythmias, such as ventricular fibrillation, at a very early stage. It is also important to detect episodes of acute myocardial ischemia by monitoring changes in the ST segment. Of the five ECG applications discussed in this section, intensive care monitoring is the only one which is critically dependent on real-time signal processing: a serious event such as cardiac arrest must be detected within a few seconds so that the staff can be alerted to immediately begin emergency life-saving procedures.

The ECG signal is recorded under conditions associated with considerable amounts of noise and artifacts caused by, for example, muscle activity and changes in body position. Poor signal quality produces an increased number of false alarms and reduced diagnostic performance. As a result, the intensive care staff must deal with increased distraction and workload, sometimes leading to critical cardiac events being overlooked. Since the patient's stay at the ICU/CCU may last for more than a week, it is important to address the signal quality issue. With the current trend of continuously monitoring the 12-lead ECG rather than a subset of leads, this issue is even more crucial since more leads are liable to result in more problems.

6.6.3 Ambulatory Monitoring

Ambulatory ECG monitoring is used to identify patients with transient symptoms, e.g., palpitations, light-headedness, or syncope, which are indicative of arrhythmias. Another group of patients are those at high risk of sudden death after infarction. Ambulatory monitoring is also used in patients who are on antiarrhythmic drugs and whose reaction to the therapy needs to be assessed. During 24 hours or more of normal daily activities, the patient carries a solid state recording device that stores the ECG. A 3-lead configuration is often used because the 12-lead ECG is impractical in these recording conditions. The patient is instructed to note activities and symptoms in a diary, used later to facilitate assessment of the ECG. The connection between symptoms and arrhythmia can be made even more precise by asking the patient to press the event button of the device whenever a symptom occurs. The ambulatory ECG recording technique is also called *Holter monitoring* after its inventor Norman Holter who introduced the first portable (analog) device to record an ECG in the late 1950s [40].

The ambulatory recording technique has since then also been successfully utilized in other applications such as the monitoring of EEG (see page 43) or blood pressure.

Digital storage of a 24-hour, three-lead ambulatory ECG requires considerable amount of memory. For many years, it was far from feasible to manufacture portable recorders with sufficient memory, and, therefore, data compression algorithms became a hot research topic in ECG signal processing; Section 7.6 describes the major principles behind such algorithms. Although current memory technology can easily accommodate 24 hours or more of uncompressed ECG, data compression remains an important aspect since it is needed, for example, when storing large numbers of ambulatory ECG recordings in a database available for later retrieval and scrutiny.

Once the patient has returned the device to the hospital, the recorded ECG is analyzed by computer with respect to the occurrence of arrhythmias. The results are then assessed by a physician or technician to make sure that artifacts, sometimes numerous, have not jeopardized the analysis and introduced false arrhythmias. For several years, much emphasis was placed on developing algorithms for classification of beat morphologies since it was believed that the VPB count per hour represented an important risk factor in sudden death. Although it was later shown that this belief was unfounded, VPB detection remains an essential part of the analysis of ambulatory ECGs. For example, it is necessary to deal with the presence of ectopic beats when an ECG is analyzed with respect to heart rate variability. The latter type of analysis in Holter monitoring has shown great promise in predicting mortality rates in patients after myocardial infarction. Another important use of Holter monitoring is for the detection of silent ischemia where changes in the ST segment are analyzed.

Diagnosis of atrial arrhythmias using the ambulatory ECG is rendered difficult by the fact that P waves are frequently masked by noise and artifacts. As a result of this, it is extremely difficult to design algorithms for P wave detection that give a reliable diagnosis.

6.6.4 Stress Test

Exercise stress testing is a method of investigating the ability of the heart to cope with physical work. When the body works harder, the demand for oxygen increases, and the heart needs to pump more blood. With the stress test it is possible to assess if the blood supply to the arteries that supply the heart is sufficient; for example, the test can be used to diagnose patients with angina pectoris who may have suffered from undiagnosed episodes of silent ischemia.

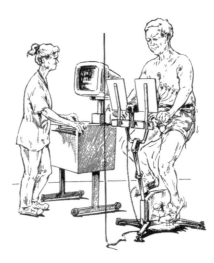

Figure 6.18: Exercise stress testing using an ergometer bicycle.

Exercise usually starts at a low workload, and the load is thereafter increased progressively; either a bicycle (Figure 6.18) or a treadmill is used. During exercise, the standard 12-lead ECG is recorded and subjected to real-time signal processing to provide the physician with a reliable ECG. The processed signal can be monitored on a screen together with trends that describe the time evolution of different ECG measurements, notably changes in the ST segment. The stress test is terminated when the patient experiences fatigue, when symptoms like chest pain and shortness of breath prevent further exercise, or when abnormal ECG changes appear. The overall response to exercise is assessed in terms of maximum workload, maximum heart rate, ECG changes, blood pressure, and respiratory rate. In addition, the recovery period subsequent to exercise is assessed to determine if the ECG returns to its initial appearance before exercise.

Figures 6.19(a) and (b) illustrate the ECG reaction observed in a normal subject and a patient suffering from myocardial ischemia, respectively. In the normal ECG the ST depression that develops during exercise largely disappears after 4 minutes' recovery. On the other hand, for the ischemic patient the ST depression endures and the ST segment actually starts to slope slightly downwards after 4 minutes' recovery. It should be noted that the much shortened QT interval observed at peak exercise is primarily explained by the higher heart rate.

Considering the difficult circumstances under which the exercise ECG is recorded, the production of accurate measurements by signal processing has attracted much research attention [41–43]. Especially at high workloads,

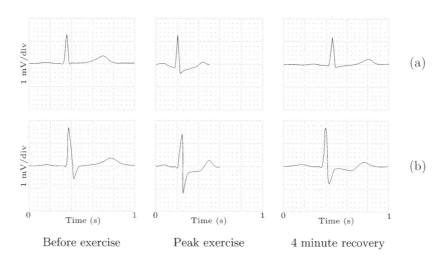

Figure 6.19: ECG reaction during exercise stress test for (a) a healthy subject and (b) a patient suffering from myocardial ischemia. The beats were noise-reduced using exponential averaging. Note that the beat at peak exercise is shorter because of the higher heart rate.

baseline wander and EMG noise seriously degrade the signal quality and preclude direct measurements from the ECG. Ensemble averaging is conventionally implemented in commercial systems for stress testing in order to produce a noise-reduced ECG. Exponential averaging or related recursive algorithms are commonly used since it is essential to track the exercise-induced changes in the ST segment (see Section 4.3.3). Although VPBs are of little interest *per se* in stress testing, their presence must nonetheless be determined in order to exclude them from averaging since that technique only involves sinus beats.

6.6.5 High-Resolution ECG

For many years the interpretation of resting ECGs was based on measurements derived from waves whose amplitude were at least several tens of microvolts; waves with smaller amplitudes were ignored since these were almost always caused by noise. This limitation was, however, removed with the advent of the high-resolution ECG with which it became possible to detect signals on the order of 1 μV thanks to signal averaging techniques (therefore, these signals are sometimes denoted "micropotentials"). The high-resolution ECG has helped unlock novel information and has demonstrated that signal processing for the purpose of noise reduction is a clinically viable technique. The acquisition procedure is usually the same as for the resting ECG, except that the signal is recorded over an extended time period so that a sufficiently low noise level is attained, i.e., sufficiently many heartbeats must be available for averaging.

In contrast to the averaging of evoked potentials, where information is available on when the external stimulus is elicited, the time reference ("fiducial point") must be determined from each individual heartbeat before ensemble averaging can be performed. The fiducial point must be accurate, otherwise low-amplitude, high-frequency components of the ECG will be distorted by smearing (cf. Section 4.3.6 on the effects of latency shifts). The high-resolution ECG rests on the assumption that the signal to be estimated has a fixed beat-to-beat morphology, whereas signal averaging during exercise must be able to track slow changes in morphology. Since the high-resolution ECG is often expected to contain high-frequency components, the sampling rate is at least 1 kHz (a lower sampling rate is sufficient in the other, above-mentioned ECG applications).

Several subintervals of the cardiac cycle have received special attention in high-resolution ECG analysis, and low-level signals have been considered in connection with

- the bundle of His which depolarizes during the PR segment, i.e., an interval which in the resting ECG is considered silent [44, 45],

- the terminal part of the QRS complex and the ST segment where *late potentials* may be present [46–48],

- intra-QRS potentials [49–51], and

- the P wave [52–54].

Of these four applications, the analysis of late potentials has received the most widespread clinical attention. Late potentials may be found in patients with myocardial infarction where ventricular depolarization can terminate many milliseconds after the end of the QRS complex (Figure 6.20). This prolongation is due to delayed and fragmented depolarization of the cells in the myocardium which surround the dead region (scarred tissue) caused by infarction; the conduction capability of the bordering cells is severely impaired by infarction. Many studies have demonstrated the importance of late potentials when, for example, identifying postinfarct patients at high risk of future life-threatening arrhythmias [55].

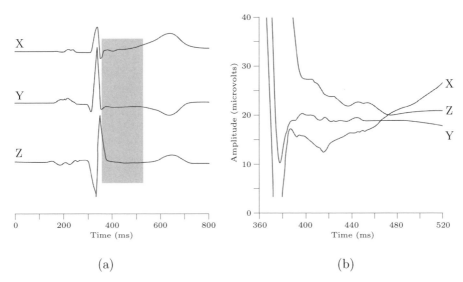

Figure 6.20: (a) The high-resolution ECG obtained by signal averaging the orthogonal X, Y, and Z leads. (b) The terminal part of the QRS complex and the ST segment, i.e., the interval shaded gray in (a), is magnified 10 times in amplitude to better display the small undulations known as late potentials.

Bibliography

[1] P. W. Macfarlane and T. D. W. Lawrie (eds.), *Comprehensive Electrocardiology. Theory and Practice in Health and Disease. (Vols. 1, 2, 3)*. New York: Pergamon Press, 1989.

[2] R. L. Lux, "Mapping techniques," in *Comprehensive Electrocardiology. Theory and Practice in Health and Disease (Vol. 2)* (P. W. Macfarlane and T. D. W. Lawrie, eds.), ch. 26, pp. 1001–1014, New York: Pergamon Press, 1989.

[3] L. De Ambroggi, E. Musso, and B. Taccardi, "Body-surface mapping," in *Comprehensive Electrocardiology. Theory and Practice in Health and Disease (Vol. 2)* (P. W. Macfarlane and T. D. W. Lawrie, eds.), ch. 27, pp. 1015–1049, New York: Pergamon Press, 1989.

[4] D. H. Brooks and R. S. MacLeod, "Electrical imaging of the heart," *IEEE Signal Proc. Mag.*, vol. 14, pp. 24–42, 1997.

[5] J. M. Jenkins and S. A. Caswell, "Detection algorithms in implantable cardioverter defibrillators," *Proc. IEEE*, vol. 84, pp. 428–445, 1996.

[6] J. Malmivuo and R. Plonsey, *Bioelectromagnetism*. Oxford: Oxford University Press, 1995.

[7] E. Frank, "An accurate, clinically practical system for spatial vectorcardiography," *Circulation*, vol. 13, pp. 737–749, 1956.

[8] G. E. Dower, "A lead synthesizer for the Frank system to simulate the standard 12-lead electrocardiogram," *J. Electrocardiol.*, vol. 1, pp. 101–116, 1968.

[9] G. E. Dower, A. Yakush, S. B. Nazzal, R. V. Jutzy, and C. E. Ruiz, "Deriving the 12-lead electrocardiogram from four (EASI) electrodes," *J. Electrocardiol.*, vol. 21 (Suppl.), pp. S182–187, 1988.

[10] B. J. Drew, M. G. Adams, M. M. Pelter, S. F. Wung, and M. A. Caldwell, "Comparison of standard and derived 12-lead electrocardiograms for diagnosis of coronary angioplasty-induced myocardial ischemia," *Am. J. Cardiol.*, vol. 79, pp. 639–644, 1997.

[11] L. Edenbrandt and O. Pahlm, "Vectorcardiogram synthesized from a 12-lead ECG: Superiority of the inverse Dower matrix," *J. Electrocardiol.*, vol. 21, pp. 361–367, 1988.

[12] P. W. Macfarlane, L. Edenbrandt, and O. Pahlm, *12-lead Vectorcardiography.* Oxford: Butterworth-Heinemann, 1994.

[13] G. E. Dower, H. B. Machado, and J. A. Osborne, "On deriving the electrocardiogram from vectorcardiographic leads," *Clin. Cardiol.*, vol. 3, pp. 87–95, 1980.

[14] P. Rubel, I. Benhadid, and J. Fayn, "Quantitative assessment of eight different methods for synthesizing Frank VCGs from simultaneously recorded standard ECG leads," *J. Electrocardiol.*, vol. 24 (Suppl), pp. 197–202, 1991.

[15] J. A. Kors, G. van Herpen, A. C. Sittig, and J. H. van Bemmel, "Reconstruction of the Frank vectorcardiogram from standard electrocardiographic leads: Diagnostic comparison of different methods," *Eur. Heart J.*, vol. 11, pp. 1083–1092, 1990.

[16] H. C. Bazett, "An analysis of the time relations of electrocardiograms," *Heart*, vol. 7, pp. 353–370, 1920.

[17] M. Hodges, D. Salerno, and D. Erlien, "Bazett's QT correction reviewed. Evidence that a linear QT correction for heart rate is better," *J. Am. Coll. Cardiol.*, vol. 1, p. 694, 1983.

[18] G. S. Wagner, *Marriott's Practical Electrocardiography.* Baltimore: Lippincott Williams & Wilkins, 10th ed., 2001.

[19] A. Bayes de Luna, *Clinical Electrocardiography: A Textbook.* Mount Kisco: Futura Publ. Co., 2nd ed., 1998.

[20] M. Malik and A. J. Camm (eds.), *Heart Rate Variability.* Armonk: Futura Publ., 1995.

[21] Task Force of The European Society of Cardiology and The North Americam Society for Pacing and Electrophysiology, "Heart rate variability: Standards of measurement, physiological interpretation, and clinical use," *Circulation*, vol. 93, pp. 1043–1065, 1996.

[22] S. Akselrod, D. Gordon, F. A. Ubel, D. C. Shannon, A. C. Barger, and R. J. Cohen, "Power spectrum analysis of heart rate fluctuations. A quantitative probe of beat-to-beat cardiovascular control," *Science*, vol. 213, pp. 220–222, 1981.

[23] M. M. Wolf, G. A. Varigos, D. Hunt, and J. G. Sloman, "Sinus arrhythmia in acute myocardial infarction," *Med. J. Australia*, vol. 2, pp. 52–53, 1978.

[24] R. E. Kleiger, J. P. Miller, J. T. Bigger, and A. J. Moss, "Decreased heart rate variability and its association with increased mortality after myocardial infarction," *Am. J. Cardiol.*, vol. 59, pp. 256–262, 1987.

[25] D. Andresen, T. Brüggemann, S. Behrens, and C. Ehlers, "Heart rate response to provocative maneuvers," in *Heart Rate Variability* (M. Malik and A. J. Camm, eds.), ch. 21, pp. 267–274, Armonk: Futura Publ., 1995.

[26] D. R. Adam, J. M. Smith, S. Akselrod, S. Nyberg, A. O. Powell, and R. J. Cohen, "Fluctuations in T-wave morphology and susceptibility to ventricular fibrillation," *J. Electrocardiol.*, vol. 17, pp. 209–218, 1984.

[27] J. M. Smith, E. A. Clancy, R. Valeri, J. N. Ruskin, and R. J. Cohen, "Electrical alternans and cardiac electrical instability," *Circulation*, vol. 77, pp. 110–121, 1988.

[28] M. J. Janse and A. L. Wit, "Electrophysiological mechanism of ventricular arrhythmias resulting from myocardial ischemia and infarction," *Physiol. Rev.*, vol. 69, pp. 1049–1169, 1989.

[29] M. A. Murdah, W. J. McKenna, and A. J. Camm, "Repolarization alternans: Techniques, mechanisms, and cardiac vulnerability," *Pacing Clin. Electrophysiol.*, vol. 20, pp. 2641–2657, 1997.

[30] J. P. Martínez and S. Olmos, "Methodological principles of T wave alternans analysis: A unified framework," *IEEE Trans. Biomed. Eng.*, vol. 52, pp. 599–613, 2005.

[31] H. Tam and J. G. Webster, "Minimizing electrode motion artifact by skin abrasion," *IEEE Trans. Biomed. Eng.*, vol. 24, pp. 134–139, 1977.

[32] D. P. Burbank and J. G. Webster, "Reducing skin potential motion artifact by skin abrasion," *Med. Biol. Eng. & Comput.*, vol. 16, pp. 31–38, 1978.

[33] J. C. Huhta and J. G. Webster, "60-Hz interference in electrocardiography," *IEEE Trans. Biomed. Eng.*, vol. 43, pp. 91–101, 1973.

[34] J. T. Flaherty, S. Blumenschein, A. W. Alexander, R. D. Gentzler, T. M. Gallie, J. P. Boineau, and M. S. Spach, "Influence of respiration on recording cardiac potentials," *Am. J. Cardiol.*, vol. 20, pp. 21–28, 1967.

[35] H. Riekkinen and P. Rautaharju, "Body position, electrode level and respiration effects on the Frank lead electrocardiogram," *Circulation*, vol. 53, pp. 40–45, 1976.

[36] R. C. Wang and T. W. Calvert, "A model to predict respiration from the vectorcardiogram," *Ann. Biomed. Eng.*, vol. 2, pp. 47–57, 1974.

[37] F. Pinciroli, R. Rossi, L. Vergani, P. Carnevali, S. Mantero, and O. Parigi, "Remarks and experiments on the construction of respiratory waveforms from electrocardiographic tracings," *Comput. Biomed. Res.*, vol. 19, pp. 391–409, 1985.

[38] G. B. Moody, R. G. Mark, A. Zoccola, and S. Mantero, "Derivation of respiratory signals from multi-lead ECG's," in *Proc. Computers in Cardiology*, pp. 113–116, IEEE Computer Society Press, 1985.

[39] S. Leanderson, P. Laguna, and L. Sörnmo, "Estimation of respiration frequency using spatial information from the VCG," *Med. Eng. & Physics*, vol. 25, pp. 501–507, 2003.

[40] N. J. Holter, "New method for heart studies: Continuous electrocardiography of active subjects over long period is now practical," *Science*, vol. 134, p. 1214, 1961.

[41] M. L. Simoons, H. B. K. Boom, and E. Smallenburg, "On-line processing of orthogonal exercise electrocardiograms," *Comput. Biomed. Res.*, vol. 8, pp. 105–117, 1975.

[42] O. Pahlm and L. Sörnmo, "Data processing of exercise ECGs," *IEEE Trans. Biomed. Eng.*, vol. 34, pp. 158–165, 1987.

[43] V. X. Afonso, W. J. Tompkins, T. Q. Nguyen, K. Michler, and L. Shen, "Comparing stress ECG enhancement algorithms," *IEEE Eng. Med. Biol. Mag.*, vol. 15, pp. 37–44, 1996.

[44] E. J. Berbari, R. Lazzara, P. Samet, and B. J. Scherlag, "Noninvasive techniques for detection of electrical activity during the P-R segment," *Circulation*, vol. 148, pp. 1005–1013, 1973.

[45] N. C. Flowers, R. C. Hand, P. C. Orander, C. B. Miller, M. O. Walden, and L. G. Horan, "Surface recording of electrical activity from the region of the bundle of His," *Am. J. Cardiol.*, vol. 33, pp. 384–389, 1974.

[46] M. B. Simson, "Use of signals in the terminal QRS complex to identify patients with ventricular tachycardia after myocardial infarction," *Circulation*, vol. 64, pp. 235–242, 1981.

[47] E. J. Berbari, "High resolution electrocardiography," *CRC Crit. Rev.*, vol. 16, pp. 67–103, 1988.

[48] E. J. Berbari and P. Lander, "Principles of noise reduction," in *High-Resolution Electrocardiography* (N. El-Sherif and G. Turitto, eds.), pp. 51–66, Armonk: Futura Publ. Comp., 1992.

[49] S. Abboud, "Subtle alterations in the high-frequency QRS potentials during myocardial ischemia in dogs," *Comput. Biomed. Res.*, vol. 20, pp. 384–395, 1987.

[50] S. Abboud, "High-frequency electrocardiogram analysis of the entire QRS in the diagnosis and assessment of coronary artery disease," *Prog. Cardiovasc. Dis.*, vol. 35, pp. 311–328, 1993.

[51] P. Lander, P. Gomis, R. Goyal, E. J. Berbari, P. Caminal, R. Lazzara, and J. S. Steinberg, "Improved predictive value for arrhythmic events using the signal-averaged electrocardiogram," *Circulation*, vol. 95, pp. 1386–1393, 1997.

[52] M. Fukunami, T. Yamada, M. Ohmori, K. Kumagai, K. Umemoto, A. Sakai, N. Kondoh, T. Minamino, and N. Hoki, "Detection of patients at risk for paroxysmal atrial fibrillation during sinus rhythm by P wave-triggered signal-averaged electrocardiogram," *Circulation*, vol. 83, pp. 162–169, 1991.

[53] J. S. Steinberg, S. Zelenkofske, S. C. Wong, M. Gelernt, R. Sciacca, and E. Menchavez, "Value of the P-wave signal-averaged ECG for predicting atrial fibrillation after cardiac surgery," *Circulation*, vol. 88, pp. 2618–2622, 1992.

[54] F. A. Ehlert, D. Korenstein, and J. S. Steinberg, "Evaluation of P wave signal-averaged electrocardiographic filtering and analysis methods," *Am. Heart. J.*, vol. 134, pp. 985–993, 1997.

[55] M. B. Simson and P. W. Macfarlane, "The signal-averaged electrocardiogram," in *Comprehensive Electrocardiology. Theory and Practice in Health and Disease (Vol. 2)* (P. W. Macfarlane and T. D. W. Lawrie, eds.), ch. 33, pp. 1199–1218, New York: Pergamon Press, 1989.

Chapter 7

ECG Signal Processing

Electrocardiographic analysis was one of the very first areas in medicine where computer processing was introduced [1–3]. Early work mostly dealt with the development of decision tree logic for ECG interpretation, mimicking the rules a cardiologist would apply. It soon became quite evident, however, that the outcome of computer interpretation was critically dependent on the accuracy of the measurements. As a result, the role of signal processing has become increasingly important in producing accurate measurements, especially when analyzing ECGs recorded under ambulatory or strenuous conditions. In addition, theoretical advances in signal processing have contributed significantly to a new understanding of the ECG signal and, in particular, its dynamic properties.

So far, no system offers a "universal" type of ECG signal analysis, but systems are designed to process signals recorded under particular conditions. It is, therefore, customary to speak of systems for resting ECG interpretation, stress testing, ambulatory ECG monitoring, intensive care monitoring, and so on. Common to all these systems is a set of algorithms which condition the signal with respect to different types of noise, extract basic ECG measurements of wave amplitudes and durations, and compress the data for efficient storage or transmission. The block diagram in Figure 7.1 presents this set of signal processing algorithms, i.e., filtering for noise reduction, QRS detection, wave delineation, and data compression; their respective descriptions define the scope of the present chapter. While these algorithms are frequently implemented to operate in sequential order, information on the occurrence time of a QRS complex, as produced by the QRS detector, is sometimes incorporated into the other algorithms to improve performance. The complexity of each algorithm varies from application to application so that, for example, noise filtering performed in ambulatory monitoring is much more sophisticated than that required in resting ECG analysis.

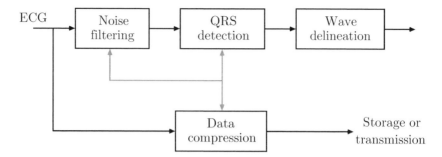

Figure 7.1: Algorithms for basic ECG signal processing. The timing information produced by the QRS detector may be fed to the blocks for noise filtering and data compression (indicated by gray arrows) to improve their respective performance. The output of the upper branch is the conditioned ECG signal and related temporal information, including the occurrence time of each heartbeat and the onset and end of each wave.

Before describing the details of the algorithms for basic ECG signal processing, we will provide a brief introduction on their use and the analysis which often follows.

ECG filtering. Considerable attention has been paid to the design of filters, which may have linear or nonlinear structures, for the removal of baseline wander and powerline interference, see Sections 7.1 and 7.2, respectively. Both these types of disturbance imply the design of a narrowband filter. Removal of noise due to muscle activity represents another important filtering problem being much more difficult to handle because of the substantial spectral overlap between the ECG and muscle noise (Section 7.3). This circumstance is identical to the situation where the EEG signal is disturbed by muscle noise, see page 74. In contrast to the EEG, muscle noise present in the ECG can be reduced whenever it is appropriate to employ techniques that benefit from the fact that the ECG is a recurrent signal. For example, ensemble averaging techniques used for noise reduction of evoked potentials, described in Chapter 4, can be successfully applied to time-aligned heartbeats for reduction of muscle noise.

The filtering techniques described in Sections 7.1–7.3 are primarily used for preprocessing of the signal and have as such been implemented in a wide variety of systems for ECG analysis. It should be remembered, however, that filtering of the ECG, as with any other type of biomedical signal, is contextual and should be performed only when the desired information remains undistorted. This important insight may be exemplified by filtering for the removal of powerline interference. Such filtering is suitable in a system for the analysis of heart rate variability, whereas it is inappropriate in a system for the analysis of late potentials, as late potentials spectrally overlap the interference.

A major concern when filtering out noise is the degree to which the QRS complexes influence the output of the filter. The QRS complex acts, in fact, as an unwanted, large-amplitude impulse input to the filter. Since linear, time-invariant filters are generally more sensitive to the presence of such impulses, filters with a nonlinear structure may be preferable. In order to assure that a filter does not introduce unacceptable distortion, its performance should be assessed by means of simulated signals so that distortion can be exactly quantified.

QRS detection. The presence of a QRS complex and its occurrence time is basic information required in all types of ECG signal processing. The design of a QRS detector, described in Section 7.4, is of crucial importance since poor detection performance may propagate to subsequent processing steps and, consequently, may limit the overall performance of the system. Beats which remain undetected constitute a more severe error than do false detections; the former type of error can be difficult to correct at a later stage in the chain of processing algorithms, while, hopefully, false detections can be eliminated by, for example, an algorithm for classification of QRS morphologies.

Once the QRS complex has been detected, the T wave can be analyzed since ventricular repolarization always follows depolarization. Conversely, the P wave does not lend itself as easily to analysis since atrial and ventricular rhythms may be independent of each other. In the vast majority of cases, however, atrial and ventricular rhythms are associated so that P wave detection may be based on a backward search in time beginning at the QRS complex and ending at the end of the preceding T wave. The success rate of the P wave detector is strongly dependent on the noise level of the ECG. As a result, comprehensive rhythm interpretation, which assumes the availability of P wave information, is precluded from those applications where the ECG signal is relatively noisy. Further aspects of the problem of detecting P waves can be found in, e.g., [4–8].

Wave delineation. Since essential diagnostic information is contained in the wave amplitudes and durations of a heartbeat, cf. Figure 6.10, wave delineation represents an important step in ECG signal processing (Section 7.5). The design of such delineation algorithms continues to receive attention with the all-embracing goal of elegantly handling the fact that the signal amplitude is low at the wave boundaries and often obscured by noise waves. The lack of universally acknowledged rules for finding the onset and end of ECG waves is another factor that makes the design process challenging. Once the onset and end of a wave have been determined, its duration and peak amplitude can be readily computed.

Data compression. The ECG signal exhibits a certain amount of redundancy, as manifested by correlation between adjacent samples, the recurrence of heartbeats with similar morphology, and the relative resemblance between different leads. Considerable savings can be achieved in terms of storage capacity and transmission time by exploiting the different types of redundancy so that each sample can be represented by fewer bits than in the original signal. The use of data compression is, however, only acceptable as long as the desired diagnostic information is preserved in the reconstructed signal. The major approaches to designing methods for ECG data compression are presented in Section 7.6, together with various considerations on how to evaluate performance.

Further analysis. The above aspects of basic ECG signal processing are usually accompanied by further analysis dealing with morphology and rhythm. *Feature extraction* is performed for the purpose of characterizing the morphology of a QRS complex. Although the durations and amplitudes that result from wave delineation contain important diagnostic information, additional features are required to reliably group beats with similar morphology into the same cluster. One approach to feature extraction is to derive a set of "heuristic" features which, e.g., describe the area, polarity, and slopes of the waves. Another, more robust, approach is to make use of the coefficients that result from the correlation of each beat with either a set of orthonormal basis functions or a set of QRS templates, being either predefined or created dynamically during the analysis.

Based on the set of extracted features, *clustering of QRS morphologies* can be performed. In its simplest form, clustering may be used to single out beats that deviate from the predominant morphology, which is usually that belonging to the normal sinus beat. Once this is done, beats belonging to the "sinus cluster" can be subjected, for example, to ensemble averaging or heart rate variability analysis. In other situations, there is reason to study the entire range of beat clusters. Since clustering does not assign a label with a physiological meaning to a beat, it may be necessary to classify the beats according to their cardiac origin. The steps of feature extraction and clustering were briefly described in Section 5.6.1 within the context of electromyographic analysis and clustering of motor unit action potentials.

Rhythm analysis is, for natural reasons, based on the pattern of RR intervals, but must also embrace morphologic information since most arrhythmias are manifested by a joint deviation in rhythm and morphology. The scope of rhythm analysis is strongly application-dependent. For example, rhythm analysis in a system for resting ECG interpretation is limited by the very short duration of the recording and thus only deals with less complicated, persistent arrhythmias, whereas a system for continuous arrhythmia monitoring is designed to detect life-threatening, transient arrhythmias such as ventricular fibrillation.

In addition to these types of analysis, a wide variety of specialized signal processing algorithms have been developed over the years [9]. A small selection of such algorithms include those for noise reduction in stress testing, detection of ST–T segment changes in ischemia monitoring, characterization of heart rate variability (Chapter 8), detection and characterization of "unorganized" arrhythmias such as atrial and ventricular fibrillation, serial comparison of ECG/VCG recordings, interpretation of pacemaker performance ("pacemaker ECG"), detection of late potentials, and dynamic analysis of the repolarization phase including detection of T wave alternans.

7.1 Baseline Wander

Removal of baseline wander is required in order to minimize changes in beat morphology which do not have cardiac origin. This is especially important when subtle changes in the "low-frequency" ST–T segment are analyzed for the diagnosis of ischemia, which may be observed, for example, during the course of a stress test [10, 11]. The frequency content of baseline wander is usually in the range below 0.5 Hz. However, increased movement of the body during the latter stages of a stress test further increases the frequency content of baseline wander, see Figure 7.2. Patients unable to perform a traditional treadmill or ergometer stress test may still be able to perform a stress test by either sitting, running an ergometer by hand, or using a special rowing device. In such cases, baseline wander related to motion of the arms severely distorts the ECG signal. The bandwidth of such baseline wander is considerably larger than that caused by respiratory activity and perspiration.

We will below describe the two major techniques employed for the removal of baseline wander from the ECG, namely, linear filtering and polynomial fitting. Linear filtering can be further divided into filtering based on time-invariant or time-variant structures.

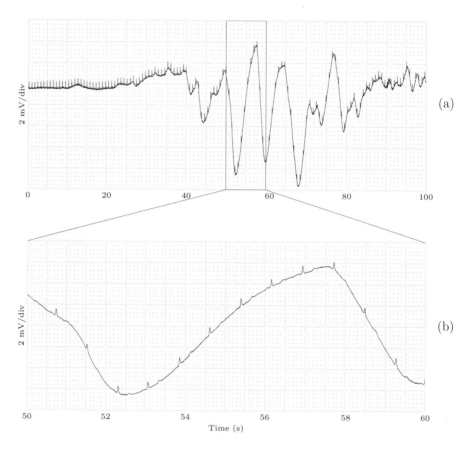

Figure 7.2: (a) Electrocardiographic baseline wander due to sudden body movements. The amplitude of the baseline wander is considerably larger than that of the QRS complexes. (b) A close-up in time (10×) of the ECG signal framed in (a).

7.1.1 Linear, Time-Invariant Filtering

The design of a linear, time-invariant, highpass filter involves several considerations, of which the most crucial are the choice of filter cut-off frequency and phase response characteristic. The cut-off frequency should obviously be chosen so that the clinical information in the ECG signal remains undistorted while as much as possible of the baseline wander is removed. Hence, it is essential to find the lowest frequency component of the ECG spectrum. In general, the slowest heart rate is considered to define this particular frequency component; the PQRST waveform is attributed to higher frequencies. During bradycardia the heart rate may drop to approximately 40 beats/minute, implying that the lowest frequency contained in the ECG is approximately 0.67 Hz [12]. Since the heart rate is not perfectly regular

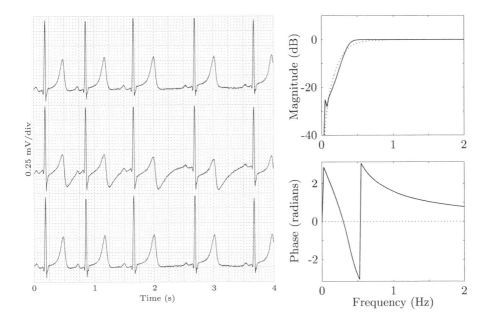

Figure 7.3: Linear, time-invariant, highpass filtering and the effect of a nonlinear phase response. The original ECG (top panel) was processed with two filters having almost identical magnitude functions, but with either a highly nonlinear phase (solid line) or a zero-phase (dotted line). The signal produced by the filter with the nonlinear phase contains severely distorted ST–T segments (middle panel), while the zero-phase filter introduces virtually no distortion at all (bottom panel). The filter cut-off frequency was $F_c = 0.5$ Hz, and the sampling rate was $F_s = 250$ Hz.

but always fluctuates from one beat to the next, it is necessary to choose a slightly lower cut-off frequency, approximately $F_c = 0.5$ Hz. If too high a cut-off frequency is employed, the output of the highpass filter contains an unwanted, oscillatory component which is strongly correlated to the heart rate.

The other crucial design consideration is related to the properties of the phase response and, consequently, the choice of filter structure. Linear phase filtering is highly desirable in order to prevent phase distortion from altering various wave properties of the cardiac cycle such as the duration of the QRS complex, the ST–T segment level, or the end point of the T wave, see Figure 7.3. It is well-known that FIR filters can have an exact linear phase response, provided that the impulse response is either symmetric or antisymmetric [13, Ch. 4]. On the other hand, IIR filters introduce signal distortion due to the nonlinear phase response.

If we assume that the filter cut-off frequency is $F_c = 0.5$ Hz, and the signal is sampled at a rate of $F_s = 250$ Hz, the corresponding normalized cut-off frequency becomes

$$f_c = \frac{F_c}{F_s} = 0.002, \tag{7.1}$$

which thus establishes that baseline wander removal must be treated as a narrowband filtering problem, i.e., only a fraction of the signal spectrum should be attenuated. Although a linear phase FIR, highpass filter can be designed in numerous ways, the result is invariably a filter with a very long impulse response. A straightforward approach to the design of a filter is to choose the ideal highpass filter as a starting point,

$$H(e^{j\omega}) = \begin{cases} 0, & 0 \le |\omega| \le \omega_c; \\ 1, & \omega_c < |\omega| < \pi, \end{cases} \tag{7.2}$$

where $\omega_c = 2\pi f_c$. Since the corresponding impulse response has an infinite length,

$$h(n) = \frac{1}{2\pi} \int_{\omega_c}^{\pi} 1 \cdot e^{j\omega n} d\omega + \frac{1}{2\pi} \int_{-\pi}^{-\omega_c} 1 \cdot e^{j\omega n} d\omega$$

$$= \begin{cases} 1 - \dfrac{\omega_c}{\pi}, & n = 0; \\ -\dfrac{\sin(\omega_c n)}{\pi n}, & n = \pm 1, \pm 2, \ldots, \end{cases} \tag{7.3}$$

truncation can be done by multiplying $h(n)$ by a rectangular window function, defined by

$$w(n) = \begin{cases} 1, & |n| = 0, \ldots, L; \\ 0, & \text{otherwise}, \end{cases} \tag{7.4}$$

or by another window function if more appropriate, see [14, Ch. 8]. Such an FIR filter should have an order $2L + 1$ of approximately 1150 to achieve a reasonable trade-off between stopband attenuation (at least 20 dB) and the width of the transition band, see Table 7.1. Although the symmetry of the impulse response can be exploited to reduce the number of multiplications in the filter, a considerable number of multiplications is nevertheless required. The use of a filter with lower order, for example, 400, hardly provides any attenuation at all in the stopband, see Figure 7.4.[1]

---

[1]The well-informed filter designer may, at this point, argue that more advanced design techniques will produce shorter filters than does the windowing method. While this is certainly true, the use of any such advanced technique will not sufficiently reduce filter complexity.

Table 7.1: Filter lengths required to achieve a certain stopband attenuation using the window method, in this case with a Hamming window. The cut-off frequency of the highpass filter is 0.5 Hz, the stopband interval is 0–0.3 Hz, and the sampling rate is 250 Hz.

Minimum stopband attenuation (dB)	Required filter length, $2L+1$
20	1142
30	1564
40	1884

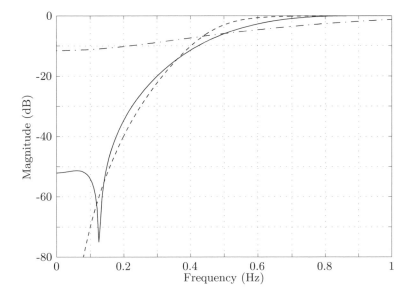

Figure 7.4: Frequency response for highpass filters designed with the window method, using a Hamming window and a filter length of 400 (dash–dotted line) and 1142 (solid line). The frequency response of the forward-backward, fifth-order Butterworth filter is also displayed (dashed line). The cut-off frequency F_c was 0.5 Hz, and the sampling rate was 250 Hz.

A number of techniques exist with which the above problem of filter complexity can be dramatically reduced, while preserving the linear phase property. These techniques include

- forward-backward IIR filtering,

- insertion of zeros into an FIR filter, and

- sampling rate alteration.

All three techniques are briefly described in the following.

Forward-backward IIR filtering.   An IIR filter meets the magnitude specifications more easily with a much lower filter order due to the freedom of positioning its poles. However, this property is accompanied by a non-linear phase response. The use of forward-backward filtering remedies this disadvantage since the overall result is filtering with a zero-phase transfer function. Implementation of such a filtering scheme involves three steps: processing of the input signal $x(n)$ with an IIR filter $h(n)$, time reversal of the filter output, and repeated processing with $h(n)$, followed by time reversal of the doubly filtered signal to produce the output signal $s(n)$, or, equivalently,

$$z_1(n) = h(n) * x(n), \tag{7.5}$$

$$z_2(n) = h(n) * z_1(-n), \tag{7.6}$$

$$s(n) = z_2(-n). \tag{7.7}$$

The overall effect of this scheme is established by determining the input–output relationship in the frequency domain. Using the discrete-time Fourier transform of a real-valued signal $x(n)$,

$$X(e^{\jmath\omega}) = \sum_{n=-\infty}^{\infty} x(n)e^{-\jmath\omega n}, \tag{7.8}$$

and the transform property

$$x(-n) \overset{F}{\longleftrightarrow} X^*(e^{\jmath\omega}),$$

(7.5)–(7.7) can be combined and written as

$$S(e^{\jmath\omega}) = Z_2^*(e^{\jmath\omega}) = H^*(e^{\jmath\omega})Z_1(e^{\jmath\omega})$$
$$= H^*(e^{\jmath\omega})H(e^{\jmath\omega})X(e^{\jmath\omega})$$
$$= |H(e^{\jmath\omega})|^2 X(e^{\jmath\omega}). \tag{7.9}$$

Thus, $x(n)$ is processed with a filter whose magnitude function is $|H(e^{\jmath\omega})|^2$ and phase function is zero, albeit that $h(n)$ itself has a nonlinear phase response. The order of the overall filter is twice that of $h(n)$.

   In order to exemplify forward-backward IIR filtering, the overall frequency response for a fifth-order Butterworth filter is shown in Figure 7.4 for $f_c = 0.5/250$. It is evident that the frequency response is close to that of the

FIR filter with an order of 1142, but with better attenuation of frequencies close to zero. The number of multiplications required for forward-backward IIR filtering is dramatically lower than that of straightforward FIR filtering; the number of multiplications required in each of these approaches is the subject of Problem 7.2.

Forward-backward IIR filtering is primarily a scheme for off-line processing since the requirement of causality has to be relaxed when a time-reversed signal is processed [15, 16]. If one is willing to accept a relatively short time delay, this type of filtering can also be implemented in "almost" real time by processing successive, overlapping signal segments [17]. The delay can be reduced by cleverly choosing the initial conditions of the forward-backward filters such that the initial transients at both ends of the filtered signal are minimized [18].

The application of forward-backward filtering to baseline wander removal becomes increasingly difficult at higher sampling rates, i.e., for 1000 Hz and higher, since the poles of the filter move closer and closer to the unit circle, or even outside, and the filter thus becomes unstable. Another potential disadvantage of forward-backward filtering is that this technique is not easily extended to handle time-variant filtering in which the cut-off frequency of the highpass filter varies in time, see below.

Insertion of zeros into an FIR filter. Insertion of zeros into a finite impulse response $h_0(n)$, designed for a much lower sampling rate F_{s_0}, is a cheap way to reduce filter complexity [19]. The insertion of $D - 1$ zeros in between every sample in $h_0(n)$,

$$h(n) = \begin{cases} h_0(n/D), & n = 0, \pm D, \pm 2D, \ldots; \\ 0, & \text{otherwise}, \end{cases} \tag{7.10}$$

has the effect of a D-fold repetition of the corresponding filter transfer function $H_0(e^{j\omega})$: a result easily established by calculating the discrete-time Fourier transform of $h(n)$,

$$H(e^{j\omega}) = \sum_{n=-\infty}^{\infty} h(n)e^{-j\omega n}$$

$$= \sum_{n=-\infty}^{\infty} h_0(n)e^{-j\omega n D}$$

$$= H_0(e^{j\omega D}). \tag{7.11}$$

The effect of zero insertion is illustrated in Figure 7.5, where the magnitude function $|H_0(e^{j\omega})|$ of an ideal highpass filter is shown together with the

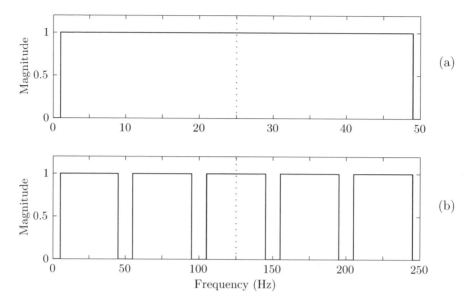

Figure 7.5: The effect of inserting zeros into the impulse response. (a) The original magnitude function and (b) the magnitude function after insertion of four zeros between each sample, i.e., $D = 5$. The stopbands occur at 50 and 100 Hz for a sampling rate of $F_s = 250$ Hz and $F_{s_0} = 50$ Hz (half the sampling rate is indicated by the dotted line).

function $|H(e^{j\omega})|$ resulting from setting $D = 5$. The D-fold repetition of the filter transfer function attenuates not only the desired baseline wander, but also signal frequencies present at multiples of the original sampling rate F_{s_0} of $h_0(n)$. The shape of the frequency response increasingly resembles the shape of a comb as the number of inserted zeros increases, and, therefore, such filters are commonly referred to as *comb filters*. In certain situations, the repetition property can be explored for attenuating powerline interference with the baseline filter; this requires that the powerline frequency be equal to the ratio of the final sampling rate F_s to D or be a multiple of that ratio [19]. However, it is essential to realize that this type of multiple stopband filtering can severely distort the diagnostic information, for example, that of the QRS complex.

Sampling rate alteration. Filter complexity can be drastically reduced by the introduction of sampling rate alteration in which filtering of baseline wander is performed on a signal sampled at a much lower rate than the original ECG. Sampling rate alteration involves the two steps of

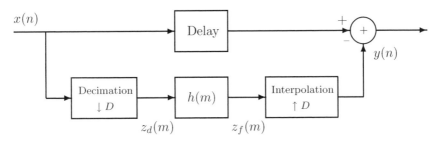

Figure 7.6: Baseline removal by means of linear highpass filtering. The baseline wander is estimated by an FIR lowpass filter $h(m)$, using sampling rate decimation and interpolation by a factor of D. The resulting baseline estimate $y(n)$ is subtracted from the observed signal $x(n)$, assuming that $x(n)$ has been appropriately delayed. Note that lowpass filtering is included in the blocks representing decimation and interpolation.

1. *decimation* of the original signal to a lower sampling rate better suited to filtering, and

2. *interpolation* of the processed signal back to its original sampling rate.

Since decimation removes the high-frequency content of the signal, the previously mentioned highpass filtering technique for baseline wander removal has to be replaced by a lowpass filter which instead outputs an estimate of the baseline wander. Once the baseline estimate has been interpolated to the original sampling rate, it can be subtracted from the original signal which then, in effect, causes the signal to be highpass filtered. The block diagram in Figure 7.6 shows the different steps involved in filtering based on decimation and interpolation.

Decimation by an integer factor D requires that the original signal $x(n)$ be first bandlimited to $|\omega| \leq \pi/D$ to avoid aliasing [13, Ch. 10]. Consequently, the decimation process must involve two different steps, namely, lowpass filtering using a linear phase filter $h_a(n)$ (where "a" denotes that the filter is required for alteration of the sampling rate),

$$z(n) = h_a(n) * x(n), \tag{7.12}$$

followed by a factor-of-D downsampling,[2]

$$z_d(m) = z(mD). \tag{7.13}$$

---

[2]Strictly speaking, sampling rate alteration is not a time-invariant procedure, but, for all practical purposes, it can here be considered as such since the spectral content of the signal component to be extracted is very low-frequency.

Ideally, the specifications of $h_a(n)$ should be such that frequencies in the range $\pi/D \leq |\omega| \leq \pi$ are eliminated,

$$H_a(e^{j\omega}) = \begin{cases} 1, & |\omega| < \pi/D; \\ 0, & \pi/D \leq |\omega| \leq \pi. \end{cases} \tag{7.14}$$

Since the frequency content of baseline wander is typically far below π/D, the definition of the transition band of $h_a(n)$ does not have to be nearly as strict as suggested by (7.14). Instead, the cut-off frequency of $h_a(n)$ can be chosen well below π/D, thus implying that low-order FIR filters are appropriate for decimation.

Once $x(n)$ has been decimated to a lower sampling rate, the design specifications of the lowpass filter $h(m)$ are much less demanding since the normalized cut-off frequency f_c is now D times higher than the original one given in (7.1),

$$f_c = \frac{F_c}{F_s}D = 0.002D. \tag{7.15}$$

The design of $h(m)$ can be based on the previously mentioned windowing method or, better, by considering some criterion-based technique which produces a filter with linear phase [20].

The output of the filter $h(m)$ constitutes the estimated baseline wander which, prior to being subtracted from $x(n)$, must be interpolated in order to have the original sampling rate, see the block diagram in Figure 7.6. The interpolation process is initialized by insertion of zeros between successive samples in the output signal $z_f(m)$,

$$z_u(n) = \begin{cases} z_f(n/D), & n = 0, \pm D, \pm 2D, \ldots; \\ 0, & \text{otherwise.} \end{cases} \tag{7.16}$$

As already pointed out in (7.11), this operation causes periodic repetition of the spectrum for a downsampled signal, and, consequently, $z_u(n)$ must be lowpass filtered to eliminate undesired spectral components. The lowpass filter $h_a(n)$, previously used for decimation, is also used for interpolation since the alteration factors D are identical (although the interpolation filter should have an additional gain factor D in order to assure that the power of the baseline wander estimate is correct).

From a computational point of view, it is useful to observe that $h_a(n)$ only needs to produce an output for every m^{th} sample in the decimation process, cf. the convolution in (7.12). Furthermore, filtering of $z_u(n)$ for interpolation is much simplified by the fact that $(m-1)$ out of m samples are equal to zero, thus making most filtering multiplications unnecessary. Both these properties can easily be profited from when $h_a(n)$ is assigned

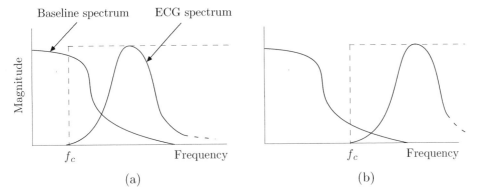

Figure 7.7: Illustration displaying schematic spectra of the ECG (here represented by its fundamental frequency which reflects heart rate) and baseline wander at a (a) low heart rate and (b) high heart rate. The magnitude function of the highpass filter (dashed line) has a cut-off frequency f_c which increases at higher heart rates.

an FIR structure. Further details on how to design systems for sampling rate alteration can be found in [13, Ch. 10]. For example, large alterations in sampling rate are more efficiently implemented using several, successive stages of decimation; interpolation is then implemented analogously [21].

7.1.2   Linear, Time-Variant Filtering

The cut-off frequency of the filter for baseline wander removal was in the previous section related to the minimal heart rate in order to avoid distortion of the signal. Unfortunately, filtering based on such a choice of cut-off frequency cannot sufficiently remove baseline wander that may occur, for example, during the latter stages of a stress test. Since the heart rate increases as the workload of the ergometer increases, it may be advantageous to couple the cut-off frequency to the prevailing heart rate, rather than to the lowest possible heart rate, to further improve baseline removal. Figure 7.7 illustrates changes in the ECG spectrum which occur as the heart rate increases.

The notion "prevailing heart rate" can be represented in several ways, of which the instantaneous RR interval length estimate $r(n)$ is a simple but useful way, and, consequently, heart rate is inversely proportional to $r(n)$. It is assumed that two successive heartbeats occur at the times θ_i and θ_{i+1}; the corresponding RR interval is then given by

$$r_{i+1} = \theta_{i+1} - \theta_i. \tag{7.17}$$

At the occurrence times of the beats, the instantaneous RR interval estimate is defined as $r(\theta_i) = r_i$ and $r(\theta_{i+1}) = r_{i+1}$, whereas linear interpolation may, for example, be used to define the interior values of the interval $[\theta_i, \theta_{i+1}]$ at different time instants,

$$r(n) = r_i + \frac{r_{i+1} - r_i}{\theta_{i+1} - \theta_i}(n - \theta_i), \quad n = \theta_i, \ldots, \theta_{i+1}. \tag{7.18}$$

Since the time-varying cut-off frequency $f_c(n)$ is related to heart rate, it is desirable to make it inversely proportional to the instantaneous RR interval estimate $r(n)$,

$$f_c(n) \sim \frac{1}{r(n)}. \tag{7.19}$$

The time-varying cut-off frequency $f_c(n)$ is used to design a lowpass filter at every time instant n, for example, integrated with the filter structure in Figure 7.6. Considering the ideal lowpass filter, a time-varying impulse response $h_I(k, n)$ can be derived from the inverse DTFT of its frequency response,

$$
\begin{aligned}
h_I(k, n) &= \frac{1}{2\pi} \int_{-\omega_c(n)}^{\omega_c(n)} 1 \cdot e^{j\omega k} d\omega \\
&= \begin{cases} \dfrac{2\pi f_c(n)}{\pi}, & k = 0; \\[2ex] \dfrac{\sin(2\pi f_c(n) \cdot k)}{\pi k}, & k \neq 0, \end{cases}
\end{aligned} \tag{7.20}
$$

where k denotes time within the impulse response, and n denotes the time at which the filter should be applied; $\omega_c(n) = 2\pi f_c(n)$ denotes the variable cut-off radian frequency. In practice, an upper limit must be imposed on $f_c(n)$ in order to avoid waveform morphology being distorted during a very short RR interval. The problem of designing several lowpass filters, each having a slightly different cut-off frequency, can be reduced to the design of one single, prototype lowpass filter subjected to a simple transformation to produce the other filters; such a transformation of filter coefficients is discussed in Problem 7.5.

Linear filtering based on filters with variable cut-off frequency was initially suggested for off-line processing of ECG signals [22] and later extended for use in on-line processing [23]. Other approaches to linear, time-variant filtering have also been described based on adaptive, LMS techniques [7, 24].

Baseline wander removal relying on linear, time-invariant or time-variant FIR filtering is illustrated by the example in Figures 7.8(a)–(c), where the baseline wander changes relatively fast at a heart rate of approximately 120 beats/minute. It is evident from Figure 7.8(c) that the time-variant filter, with its higher cut-off frequency, performs better than does the time-invariant filter, where the fixed cut-off frequency is related to the lowest possible heart rate.

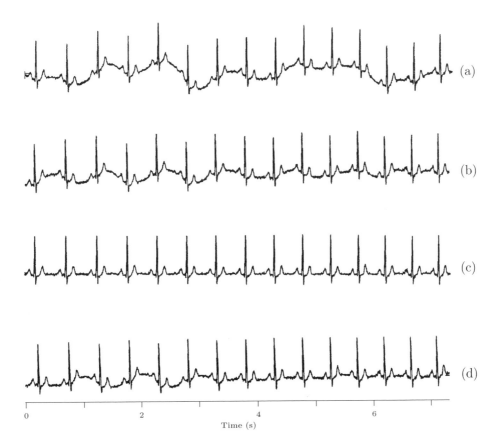

Figure 7.8: Comparison of three methods for baseline wander removal at a heart rate of approximately 120 beats/minute. (a) The original ECG signal and the signals resulting from (b) time-invariant filtering, (c) heart rate dependent filtering, and (d) cubic spline fitting (see page 470).

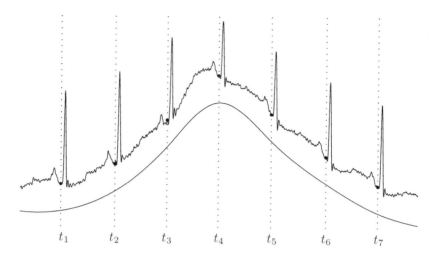

Figure 7.9: Polynomial fitting for baseline wander removal based on a series of knots positioned within the PQ intervals. To facilitate interpretation, the resulting baseline estimate is plotted with an offset from the signal.

7.1.3 Polynomial Fitting

An alternative to baseline wander removal with linear filtering techniques is to fit a polynomial to representative samples ("knots") of the ECG, with one knot being defined for each beat. The knots are chosen from the "silent", isoelectric line which, in most heart rhythms, is best represented by the PQ interval. The polynomial estimating the baseline is fitted by requiring it to pass through each of the knots smoothly, see Figure 7.9. In contrast to linear filtering, baseline wander removal by polynomial fitting requires that the QRS complexes first be detected and that the corresponding PQ intervals be accurately determined.

Using a first-order polynomial, successive knots are simply connected by straight lines [25]. However, the resulting baseline estimate tracks variations rather poorly, and its derivatives at the knots are discontinuous. By using higher-order polynomials, the likelihood of producing an accurate baseline estimate increases, although it is associated with increased computational complexity. Instead of letting the order increase as the number of knots increases, third-order polynomial fitting to successive triplets of knots represents a popular approach [26–29]. This technique is commonly referred to as *cubic spline* baseline estimation and is described in more detail in the following.

Since the cubic spline technique has its starting point in a Taylor series expansion, we will present its development within a continuous-time framework; discretization of the technique is then straightforward. The knots of successive beats located at the times t_i are denoted,

$$x(t_i), \quad i = 0, 1, 2, \ldots, \tag{7.21}$$

and are assumed to have been determined by some suitable method. The baseline estimate $y(t)$ is computed for the interval $[t_i, t_{i+1}]$ by incorporating the three knots $x(t_i)$, $x(t_{i+1})$, and $x(t_{i+2})$ into the Taylor series expanded around t_i,

$$y_\infty(t) = \sum_{l=0}^{\infty} \frac{(t - t_i)^l}{l!} \, y_\infty^{(l)}(t_i), \tag{7.22}$$

which, for a third-order polynomial description, is truncated to

$$y(t) = y(t_i) + (t - t_i)y^{(1)}(t_i) + \frac{(t - t_i)^2}{2} y^{(2)}(t_i) + \frac{(t - t_i)^3}{6} y^{(3)}(t_i), \tag{7.23}$$

where $y^{(l)}(t)$ denotes the lth derivative of $y(t)$, which is identical to $y_\infty^{(l)}(t)$ for $l = 1$, 2, and 3. Furthermore, the series expansion for the first derivative $y^{(1)}(t)$,

$$y^{(1)}(t) = y^{(1)}(t_i) + (t - t_i)y^{(2)}(t_i) + \frac{(t - t_i)^2}{2} y^{(3)}(t_i), \tag{7.24}$$

is considered since it constrains the degree of polynomial change that may take place at each knot. The goal is now to find appropriate values of $y(t_i)$, $y^{(1)}(t_i)$, $y^{(2)}(t_i)$, and $y^{(3)}(t_i)$ which will allow us to compute $y(t)$ from (7.23).

As already pointed out, it is natural to require that $y(t)$ passes through the knot $x(t_i)$,

$$y(t_i) = x(t_i). \tag{7.25}$$

We may also approximate the first derivative $y^{(1)}(t_i)$ at t_i by the slope between $x(t_{i+1})$ and $x(t_i)$,[3]

$$y^{(1)}(t_i) \approx \frac{x(t_{i+1}) - x(t_i)}{t_{i+1} - t_i}. \tag{7.26}$$

In order to find the remaining two variables $y^{(2)}(t_i)$ and $y^{(3)}(t_i)$ in (7.23), the Taylor series for $y(t)$ and $y^{(1)}(t)$ is studied for $t = t_{i+1}$,

$$y(t_{i+1}) = y(t_i) + y^{(1)}(t_i)\Delta t_{i_1} + y^{(2)}(t_i)\frac{\Delta t_{i_1}^2}{2} + y^{(3)}(t_i)\frac{\Delta t_{i_1}^3}{6} \tag{7.27}$$

---

[3]This equation could just as well make use of t_{i-1} instead of t_i; however, this was not suggested in the original derivation of the cubic spline technique in [26].

and

$$y^{(1)}(t_{i+1}) = y^{(1)}(t_i) + y^{(2)}(t_i)\Delta t_{i_1} + y^{(3)}(t_i)\frac{\Delta t_{i_1}^2}{2}, \qquad (7.28)$$

where

$$\Delta t_{i_j} = t_{i+j} - t_i.$$

The left-hand side of the two equations in (7.27) and (7.28) can be determined in a way analogous to (7.25) and (7.26),

$$y(t_{i+1}) = x(t_{i+1}), \qquad (7.29)$$

$$y^{(1)}(t_{i+1}) \approx \frac{x(t_{i+2}) - x(t_i)}{t_{i+2} - t_i} = \frac{x(t_{i+2}) - x(t_i)}{\Delta t_{i_2}}. \qquad (7.30)$$

With the values of $y(t_{i+1})$ and $y^{(1)}(t_{i+1})$ inserted into (7.27) and (7.28), the following solution for $y^{(2)}(t_i)$ and $y^{(3)}(t_i)$ is obtained,

$$y^{(2)}(t_i) = \frac{6(y(t_{i+1}) - y(t_i))}{\Delta t_{i_1}^2}$$

$$- \frac{2(2y^{(1)}(t_i) + (y(t_{i+2}) - y(t_i))/\Delta t_{i_2})}{\Delta t_{i_1}}, \qquad (7.31)$$

$$y^{(3)}(t_i) = -\frac{12(y(t_{i+1}) - y(t_i))}{\Delta t_{i_1}^3}$$

$$+ \frac{6(y^{(1)}(t_i) + (y(t_{i+2}) - y(t_i))/\Delta t_{i_2})}{\Delta t_{i_1}^2}, \qquad (7.32)$$

where $y(t_{i+2})$ is identical to the knot $x(t_{i+2})$.

The baseline estimate $y(t)$ in (7.23) is now completely specified and can be computed within the interval $[t_i, t_{i+1}]$. Since the ECG is assumed to be a discrete-time signal, $y(t)$ is, of course, computed equidistantly in time for $t_n = nT$ and subsequently subtracted from the ECG signal at samples $x(n) = x(t_n)$. This procedure is then repeated for the next interval $[t_{i+1}, t_{i+2}]$ using the knots $x(t_{i+1})$, $x(t_{i+2})$, and $x(t_{i+3})$, and so on. Problem 7.7 deals with efficient computation of the baseline estimate, making use of a recursive procedure.

The performance of the cubic spline technique is critically dependent on the accuracy of the knot determination. While the PQ interval is relatively easy to find in ECGs recorded during resting conditions, this interval may be exceedingly difficult to delimit in recordings with muscle noise or when certain types of chaotic arrhythmias are present, such as ventricular tachycardia [30, 31]. In the latter situation, the PQ interval is no longer well-defined, and, therefore, the cubic spline technique is inapplicable.

The cubic spline approach results in linear filtering with a time-variable cut-off frequency in the sense that the baseline estimate better tracks rapid baseline wander when a fast heart rate is encountered. This behavior is explained by the fact that more knots become available at faster heart rates. On the other hand, polynomial fitting performs poorly when the available knots are too far apart, a property illustrated by the example in Figure 7.8(d). The problem of too few knots has been addressed by defining additional knots within each beat, depending on the zero-crossing pattern of the signal [32].

7.2 Powerline Interference (50/60 Hz)

Electromagnetic fields caused by a powerline represent a common noise source in the ECG, as well as to any other bioelectrical signal recorded from the body surface. Such noise is characterized by 50 or 60 Hz sinusoidal interference, possibly accompanied by a number of harmonics. Such narrowband noise renders the analysis and interpretation of the ECG more difficult, since the delineation of low-amplitude waveforms becomes unreliable and spurious waveforms may be introduced [33]. Although various precautions can be taken to reduce the effect of powerline interference, for example, by selecting a recording location with few surrounding electrical devices or by appropriately shielding and grounding the location, it may still be necessary to perform signal processing to remove such interference [34, 35]. Several techniques have been presented for this purpose, ranging from straightforward linear, bandstop filtering to more advanced techniques which handle variations in powerline frequency and suppress the influence of transients manifested by the occurrence of QRS complexes.

7.2.1 Linear Filtering

A very simple approach to the reduction of powerline interference is to consider a filter defined by a complex-conjugated pair of zeros that lie on the unit circle at the interfering frequency ω_0,

$$z_{1,2} = e^{\pm j\omega_0}.$$

Such a second-order FIR filter has the transfer function

$$
\begin{aligned}
H(z) &= (1 - z_1 z^{-1})(1 - z_2 z^{-1}) \\
&= 1 - 2\cos(\omega_0)z^{-1} + z^{-2}.
\end{aligned}
\tag{7.33}
$$

Since this filter has a notch with a relatively large bandwidth, it will attenuate not only the powerline frequency but also the ECG waveforms with frequencies close to ω_0. It is, therefore, necessary to modify the filter in (7.33) so that the notch becomes more selective, for example, by introducing a pair of complex-conjugated poles positioned at the same angle as the zeros $z_{1,2}$ but at a radius r,

$$p_{1,2} = re^{\pm j\omega_0}, \tag{7.34}$$

where $0 < r < 1$. Thus, the transfer function of the resulting IIR filter is given by

$$\begin{aligned} H(z) &= \frac{(1 - z_1 z^{-1})(1 - z_2 z^{-1})}{(1 - p_1 z^{-1})(1 - p_2 z^{-1})} \\ &= \frac{1 - 2\cos(\omega_0)z^{-1} + z^{-2}}{1 - 2r\cos(\omega_0)z^{-1} + r^2 z^{-2}}. \end{aligned} \tag{7.35}$$

The notch bandwidth is determined by the pole radius r and is reduced as r approaches the unit circle. Figure 7.10 shows the impulse response and the magnitude function for two different values of the radius, $r = 0.75$ and 0.95. From Figure 7.10 it is obvious that the bandwidth decreases at the expense of increased transient response time of the filter. The practical implication of this observation is that a transient present in the signal causes a ringing artifact in the output signal. For causal filtering, such filter ringing will occur after the transient, thus mimicking the low-amplitude cardiac activity that sometimes occurs in the terminal part of the QRS complex, i.e., late potentials. Figure 7.11 shows one heartbeat without any contamination by powerline interference and the ringing artifact that results from processing with the IIR filter given in (7.35) using $r = 0.97$. This example clearly demonstrates that uncritical use of linear, time-invariant filtering can have a devastating effect on the ECG signal, significantly modifying its diagnostic content.

More sophisticated linear filters than the above second-order IIR filters can be designed for the removal of powerline interference, for example, by increasing the filter order to obtain a narrower notch or by employing filter design criteria involving both time and frequency properties. Since increased frequency resolution is always obtained to the detriment of decreased time resolution, it is impossible to design a linear, time-invariant filter which only removes powerline interference while not introducing a certain amount of ringing.

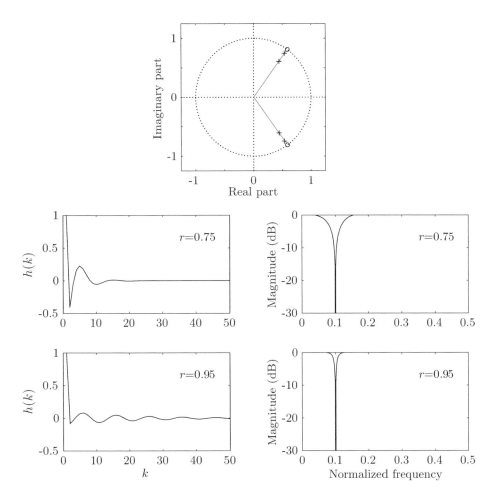

Figure 7.10: Pole-zero diagram for two second-order IIR filters whose zeros are identically positioned but whose poles are at a radius r of either 0.75 or 0.95. The impulse response $h(k)$ and the corresponding magnitude function are shown in the left and right panels, respectively.

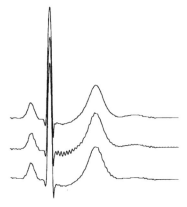

Figure 7.11: Filtering of powerline interference. The original ECG signal without powerline interference (top), the output signal from the second-order IIR filter defined in (7.35) for $r = 0.97$ (middle), and the nonlinear filter defined in (7.41) with transient suppression using $\alpha = 10$ μV (bottom). Note that the ringing caused by IIR filtering masquerades as late potentials. (Reprinted from Hamilton [36] with permission.)

7.2.2   Nonlinear Filtering

From the above observation, it is obviously desirable to develop a method for removal of powerline interference which is less sensitive to transients. We will now describe a nonlinear filter based on the idea of subtracting a sinusoid, generated internally by the filter, from the observed signal [36–39]. The amplitude of the internal sinusoid is adapted to the powerline interference present in the observed signal $x(n)$. The adaptation process is the key to making the filter less sensitive to transients and avoiding related filter ringing. The internal sinusoid is generated by

$$v(n) = w_0 \sin(\omega_0 n).$$

Taking into account the fact that the amplitude w_0, in practice, is unknown and changing with time, it is preferable to generate the sinusoid recursively allowing us to update $v(n)$ at every sample so that amplitude changes can be tracked. The sinusoid can be generated by an oscillator defined by a pair of complex-conjugated poles located on the unit circle at frequency ω_0. The transfer function for the oscillator is

$$H(z) = \frac{V(z)}{U(z)} = \frac{1}{1 - 2\cos\omega_0 z^{-1} + z^{-2}}, \tag{7.36}$$

and, accordingly, the sinusoid is generated by the following difference equation,

$$v(n) = 2 \cos \omega_0 v(n-1) - v(n-2) + u(n), \qquad (7.37)$$

using the initial conditions $v(-1) = v(-2) = 0$. The input signal $u(n)$ is given by

$$u(n) = \delta(n), \qquad (7.38)$$

where $\delta(n)$ is the unit impulse function.

An error function $e(n)$ is now introduced that indicates how well $v(n)$ predicts the powerline interference contained in the signal $x(n)$,

$$e(n) = x(n) - v(n). \qquad (7.39)$$

Since this error definition suffers from a dependence on the DC level of $x(n)$, it must be modified so that it becomes insensitive to the DC level, for example, by computing the first difference of $e(n)$,

$$\begin{aligned} e'(n) &= e(n) - e(n-1) \\ &= x(n) - x(n-1) - (v(n) - v(n-1)). \end{aligned} \qquad (7.40)$$

We can, of course, use other types of filter to efficiently remove the DC level while retaining the sinusoidal interference; however, the first difference filter is extremely simple to implement. Depending on the sign of $e'(n)$, the current value of $v(n)$ is either updated by a fixed positive or negative increment α or kept constant to produce a new estimate $\hat{v}(n)$ of the powerline interference. The update equation is given by

$$\hat{v}(n) = v(n) + \alpha \operatorname{sgn}(e'(n)), \qquad (7.41)$$

where the sgn function has been defined earlier in (4.99). The output signal $y(n)$ of the nonlinear filter results from subtraction of $\hat{v}(n)$ from $x(n)$,

$$y(n) = x(n) - \hat{v}(n). \qquad (7.42)$$

The nonlinear equation in (7.41) implements the transient suppression property of the filter since changes in amplitude are limited by the increment α. We note that too small a value of α causes the filter to poorly track changes in amplitude of the powerline interference, whereas too large a value of α causes the filter to introduce extra noise in $y(n)$ because of the large step alterations which will occur in $\hat{v}(n)$.

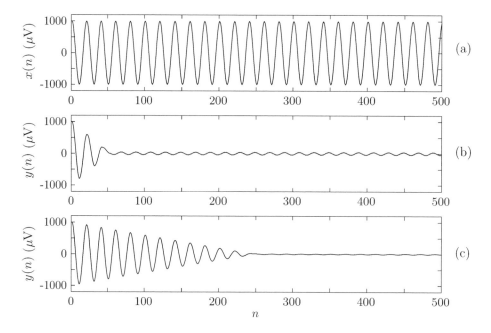

Figure 7.12: Convergence properties of the nonlinear 50/60 Hz filter, illustrated by processing (a) a purely sinusoidal signal $x(n)$. The output signal $y(n)$ of the nonlinear filter is computed for a fixed increment of either (b) $\alpha = 10 \ \mu\text{V}$ or (c) $\alpha = 2 \ \mu\text{V}$.

Before the next sample at time $n+1$ is processed, $v(n)$ is replaced by its estimate in (7.41),

$$v(n) = \hat{v}(n), \tag{7.43}$$

and then used in the recursion (7.37) to generate $v(n+1)$, and so on.

The performance of the nonlinear filter is exemplified in Figure 7.11. In contrast to the above-mentioned IIR notch filter, no ringing artifact can be discerned after the QRS complex when this filter is applied. In the ECG signal, adaptation of the internal sinusoid primarily takes place during the isoelectric line and the T wave; the duration of the QRS complex is short enough not to significantly influence the interference estimate $\hat{v}(n)$ [38]. Figure 7.12 illustrates the convergence properties of the nonlinear filter: too large a value of α will produce an output signal in which a substantial part of the sinusoid remains.

The frequency characteristics of this filter are not easily analyzed due to its nonlinear structure. By replacing the nonlinear update equation in (7.41) with a linear one,

$$\hat{v}(n) = v(n) + \alpha e'(n), \qquad (7.44)$$

the resulting filter can be identified as an IIR notch filter with the transfer function [40]

$$H(z) = \frac{(1-\alpha)(1 - 2\cos\omega_0 z^{-1} + z^{-2})}{1 - (\alpha + 2(1-\alpha)\cos\omega_0)z^{-1} + (1-\alpha)z^{-2}}. \qquad (7.45)$$

This filter has a pair of zeros on the unit circle at ω_0 and a pair of poles located inside the unit circle. The update parameter α in (7.44) determines the pole position and, thus, the bandwidth of the IIR filter. Clearly, the transient suppression property is lost when (7.41) is replaced by the linear update equation in (7.44), and thus, QRS-related ringing artifacts will occur in the filtered signal. It may be noted from the transfer function in (7.45) that the poles are not located at the angles $\pm\omega_0$, but rather at positions modified by the increment α.

7.2.3 Estimation–Subtraction

Another approach to the removal of powerline interference is to estimate the amplitude and phase of the interfering sinusoid in an isoelectric segment, followed by subtraction of the estimated sinusoid within the entire heartbeat [41–43]. Since it is only of interest to estimate the properties of the interference, the isoelectric segment can be made even more "silent" by appropriate use of bandpass filtering centered around the powerline frequency. The position of this segment can be defined by the PQ interval (Figure 7.13), cf. the determination of knots in the cubic spline method for baseline removal, or with reference to certain detection criteria [44, 45]. The performance of the estimation–subtraction method deteriorates when parts of the P or the Q wave are included in the segment; the interference is then overestimated and causes an increase, rather than a decrease, in the 50/60 Hz content of the output signal.

The sinusoid is subtracted not only from the interval in which it is estimated, but also from the remaining parts of the heartbeat [42]. A new estimate is determined for the subsequent beat and subtracted, and so on. Such a beat-to-beat oriented procedure implies, however, that sudden shifts in amplitude may occur in the output signal at the boundaries of successive beats.

In order to proceed in more detail with the estimation–subtraction approach, our goal is to find the amplitude w and phase ϕ,

$$v(n) = w\sqrt{\frac{2}{N}}\sin(\omega_0 n + \phi), \quad n = 0, \ldots, N-1, \qquad (7.46)$$

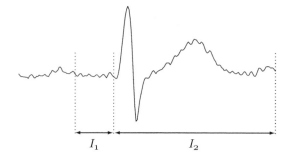

Figure 7.13: A heartbeat disturbed by 50-Hz powerline interference. The samples of the first interval, denoted I_1, are used for estimating the amplitude and phase of the interference, after which a sinusoid, defined by the estimated amplitude and phase, is subtracted from the samples of both intervals, i.e., I_1 and I_2. The procedure is repeated for the next heartbeat, and so on.

that produce the best fit to the signal $x(n)$, indexed by $n = 0, \ldots, N-1$ in the isoelectric segment, i.e., the interval I_1 in Figure 7.13. The factor $\sqrt{2/N}$ is included in (7.46) for normalization purposes. The frequency ω_0 is assumed to be known and is restricted to be harmonically related to the fundamental frequency $2\pi/N$ such that

$$\omega_0 = 2\pi \frac{k_0}{N}. \tag{7.47}$$

The observation interval has a length N which is chosen such that an integer number of periods of the sinusoidal interference is contained in the interval. The fit of $v(n)$ to $x(n)$ turns out to be much more easily solved when (7.46) is rewritten so that the two unknown parameters w and ϕ enter $v(n)$ in a linear way,

$$v(n) = w_1 \sqrt{\frac{2}{N}} \cos(\omega_0 n) + w_2 \sqrt{\frac{2}{N}} \sin(\omega_0 n), \tag{7.48}$$

where

$$w_1 = w \sin\phi,$$
$$w_2 = w \cos\phi.$$

Based on techniques described in Section 4.5 for analysis of evoked potentials, an estimator of w_1 and w_2 can be developed by noting that $v(n)$ is a linear combination of two orthonormal basis functions. In vector–matrix notation, (7.48) can be expressed as

$$\mathbf{v} = \mathbf{\Phi w}, \tag{7.49}$$

where

$$\mathbf{v} = \begin{bmatrix} v(0) \\ v(1) \\ \vdots \\ v(N-1) \end{bmatrix}$$

and

$$\mathbf{w} = \begin{bmatrix} w_1 \\ w_2 \end{bmatrix},$$

and the matrix $\boldsymbol{\Phi}$ is orthogonal,

$$\boldsymbol{\Phi} = \begin{bmatrix} \boldsymbol{\varphi_1} & \boldsymbol{\varphi_2} \end{bmatrix}$$

$$= \sqrt{\frac{2}{N}} \begin{bmatrix} 1 & 0 \\ \cos \omega_0 & \sin \omega_0 \\ \vdots & \vdots \\ \cos \omega_0 (N-1) & \sin \omega_0 (N-1) \end{bmatrix}. \tag{7.50}$$

The weight vector can be determined from minimization of the MSE between the observed signal $\mathbf{x}$ and the sinusoidal model defined by $\boldsymbol{\Phi}\mathbf{w}$,

$$E\left[\|\mathbf{x} - \boldsymbol{\Phi}\mathbf{w}\|^2\right].$$

The related estimator has already been presented in (4.203) and is given by

$$\hat{\mathbf{w}} = \boldsymbol{\Phi}^T \mathbf{x}. \tag{7.51}$$

Hence, the weight estimates are determined by the inner products between $\mathbf{x}$ and the cosine and sine basis functions $\boldsymbol{\varphi_1}$ and $\boldsymbol{\varphi_2}$, respectively. The powerline interference estimate $\hat{\mathbf{v}}$ is then computed from (cf. (4.206))

$$\hat{\mathbf{v}} = \boldsymbol{\Phi}\boldsymbol{\Phi}^T \mathbf{x} \tag{7.52}$$

and subtracted from the observed signal $\mathbf{x}$ within the interval from which $\hat{\mathbf{v}}$ was estimated,

$$\mathbf{y} = \mathbf{x} - \hat{\mathbf{v}}. \tag{7.53}$$

Alternatively, the subtraction can be expressed as

$$\mathbf{y} = \left(\mathbf{I} - \boldsymbol{\Phi}\boldsymbol{\Phi}^T\right)\mathbf{x}$$
$$= \mathbf{H}\mathbf{x}, \tag{7.54}$$

where each row of the matrix $\mathbf{H} = \mathbf{I} - \boldsymbol{\Phi}\boldsymbol{\Phi}^T$ is an impulse response which describes how the input samples are filtered to produce the output sample at a certain time instant. For the above sine–cosine basis functions, the matrix $\mathbf{H}$ has the following form,

$$
\mathbf{H} = -\frac{2}{N}
\begin{bmatrix}
1 - \frac{N}{2} & \cos\omega_0 & \cos 2\omega_0 & \cdots & \cos\omega_0(N-1) \\
\cos\omega_0 & 1 - \frac{N}{2} & \cos\omega_0 & \cdots & \cos\omega_0(N-2) \\
\cos 2\omega_0 & \cos\omega_0 & 1 - \frac{N}{2} & \cdots & \vdots \\
\vdots & \vdots & \vdots & \ddots & \cos\omega_0 \\
\cos(N-1)\omega_0 & \cos(N-2)\omega_0 & \cdots & \cos\omega_0 & 1 - \frac{N}{2}
\end{bmatrix}.
$$

$$(7.55)$$

This particular matrix is highly structured, since it is not only symmetric and Toeplitz, but each row is a circularly shifted version of the row above and is, therefore, referred to as a *circulant matrix*. Calculation of the magnitude response for each of the N linear, time-invariant filters in (7.55) produces identical results since a circular shift of a finite length sequence in the frequency domain corresponds to multiplying its DFT by a phase shift [46]. The magnitude response is displayed in Figure 7.14 and has, as one would expect, bandstop characteristics. As the interval length N, over which the sinusoid is fitted, increases, the stopband becomes increasingly narrow and the passband becomes increasingly flat; the oscillatory phenomenon at both sides of the stopband will, however, persist as N increases (Gibbs phenomenon). The phase response of each of the filters is nonlinear, except the one with a symmetric impulse response having the value $-\frac{2}{N}(1 - \frac{N}{2})$ at its midpoint (N being odd).

   Hence, we can interpret the estimation–subtraction technique in terms of linear filtering which allows us to compare its frequency response to those of other linear filtering methods. The simplicity of the estimation–subtraction technique, as defined by (7.52) and (7.53), relies on the assumption that the two basis functions are orthonormal. It is, however, possible to develop an estimation–subtraction technique for any value of ω_0, not necessarily harmonically related to $1/N$, but at the expense of a more complex procedure.

   Finally, the estimated sinusoid $\hat{\mathbf{v}} = \boldsymbol{\Phi}\boldsymbol{\Phi}^T\mathbf{x}$ is extended in time so that it can be subtracted from the samples located outside the isoelectric segment, i.e., the interval I_2 in Figure 7.13. In doing this, the estimation–subtraction technique can no longer be interpreted as a linear filtering operation since it has become completely insusceptible to the input signal in the interval I_2.

   We conclude the description of methods for cancellation of powerline interference by mentioning that the estimation–subtraction technique may be modified to adaptively compute estimates of the weights w_1 and w_2. Such a modification may be based on the LMS algorithm for which the reference

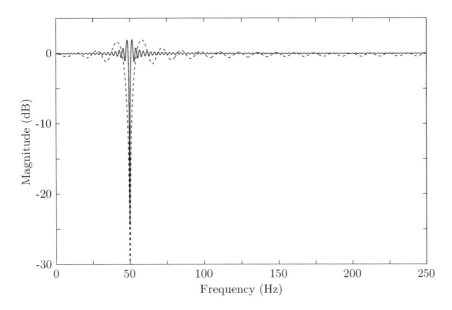

Figure 7.14: Cancellation of powerline interference using the estimation–subtraction technique. The magnitude function of the filter is displayed for two different interval lengths, $N = 40$ (dashed line) and 200 (solid line), assuming a sampling rate of 500 Hz.

input is either generated internally or taken directly from a wall outlet, of course, after appropriate insulation and attenuation [47, Ch. 12], [48, 49]. The weights w_1 and w_2, for each new instant in time, are modified so that the MSE between the powerline frequency and the observed signal is minimized, see the block diagram in Figure 7.15.

It has been shown that the adaptive estimation–subtraction approach presented in Figure 7.15, for a fixed frequency ω_0, is characterized by a linear, time-invariant, second-order IIR filter, the transfer function of which is equal to [47, Ch. 12],

$$H(z) = \frac{1 - 2\cos\omega_0 z^{-1} + z^{-2}}{1 - 2(1 - \mu)\cos\omega_0 z^{-1} + (1 - 2\mu)z^{-2}}, \tag{7.56}$$

where μ denotes the step size parameter of the LMS algorithm. Due to the adaptive structure of the filter, we have the capability of tracking variations in ω_0 and ϕ, provided that the reference comes from an external outlet that reflects these variations. However, the adaptive filter will unfortunately introduce a ringing artifact at the end of the QRS complex due to its linear structure.

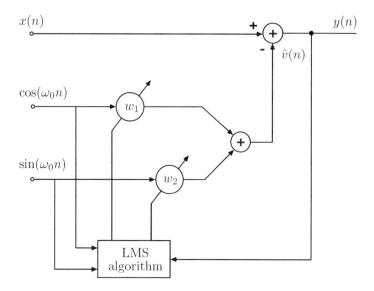

Figure 7.15: Adaptive removal of 50/60 Hz powerline interference based on the LMS algorithm. The ECG signal $x(n)$ is fed to the upper branch which constitutes the primary input to the algorithm.

7.3   Muscle Noise Filtering

The presence of muscle noise represents a major problem in many ECG applications, especially in recordings acquired during exercise, since low-amplitude waveforms may become completely obscured. Muscle noise is, in contrast to baseline wander and 50/60 Hz interference, not removed by narrowband filtering, but presents a much more difficult filtering problem since the spectral content of muscle activity considerably overlaps that of the PQRST complex. Since the ECG is a repetitive signal, techniques can be used to reduce muscle noise in a way similar to the processing of evoked potentials. Successful noise reduction by ensemble averaging is, however, restricted to one particular QRS morphology at a time and requires that several beats be available. Hence, there is still a need to develop signal processing techniques which can reduce the influence of muscle noise.

One approach to dealing with this problem is offered by time-varying lowpass filtering using a filter with a variable frequency response [50, 51]. For example, a filter with a Gaussian impulse response has been suggested for this purpose as the filter's bandwidth is easily changed from one sample to another through a function $\beta(n)$ which defines the width of the Gaussian,

$$h(k, n) \sim e^{-\beta(n)k^2}. \tag{7.57}$$

The width function $\beta(n)$ is designed to reflect *local* signal properties so that smooth segments of the ECG are subjected to considerable lowpass filtering, whereas the QRS interval, with its much steeper slopes, largely remains unfiltered. By making $\beta(n)$ proportional to the derivative of the signal, slow signal changes produce small values of $\beta(n)$, thus making the Gaussian impulse response to decay more slowly to zero so as to produce greater noise suppression, and vice versa. Details of designing the width function $\beta(n)$, truncating $h(k, n)$ in (7.57), and the resulting performance on ECG signals can be found in [50], see also Problem 7.14. The idea of adapting the cut-off frequency of a linear lowpass filter to the slopes of the ECG has also been explored for other types of filters [52].

It is evident that the Gaussian filtering technique is related to the time-varying lowpass filter earlier described for the removal of baseline wander, with the major difference being that the baseline filter is adapted to the pre-vailing heart rate rather than to the morphologic properties of the signal. It may also be of interest to point out that the noise reduction technique based on a truncated series expansion of basis functions may also be considered; this technique can be interpreted in terms of time-varying lowpass filtering, see Section 4.5.4.

The overall value of muscle noise reduction, resulting from any of the above-mentioned time-varying filter techniques, must be judged in relation to the distortion introduced in the PQRST complex. In fact, the Achilles' heel of such filtering is that its time-varying properties may introduce artificial waves: a filter that provides considerable smoothing of the low-frequency ECG segments outside the QRS complex, including the P and T waves, while essentially no filtering at all is done within the QRS complex, is likely to exhibit undesirable effects during the transitional periods. Such distortion may be acceptable when the filter output is used as input to other signal processing steps; however, the distortion renders the filtered signal unsuitable for diagnostic interpretation of the ECG because of the artificial waves that may have been introduced.

Although a host of additional techniques have been proposed for muscle noise reduction, see, e.g., [53–57], no single method has gained wide acceptance for use in clinical routine. As a result, the muscle noise problem remains largely unsolved, similar to the muscle noise problem in EEG signal processing.

7.4 QRS Detection

A QRS detector must be able to detect a large number of different QRS morphologies in order to be clinically useful and able to follow sudden or gradual changes of the prevailing QRS morphology. Furthermore, the detector must not lock onto certain types of rhythm, but treat the next possible event as if it could occur at almost any time after the most recently detected beat.

Several detector-critical types of noise and artifacts exist depending on the ECG application of interest. The noise may be highly transient in nature or be of a more persistent nature, as exemplified by the presence of powerline interference. In the case of an ECG recording with episodes containing excessive noise, it may be necessary to exclude such episodes from further analysis.

Most QRS detectors described in the literature have been developed from ad hoc reasoning and experimental insight. The detectors can, in general terms, be described by the block diagram presented in Figure 7.16 [58]. Within such a detector structure, the purpose of the preprocessor is to enhance the QRS complexes while suppressing noise and artifacts; the preprocessor is usually implemented as a linear filter followed by a nonlinear transformation. The output of the preprocessor is then fed to a decision rule for detection. The purpose of each processing block is summarized below.

a. The *linear filter* is designed to have bandpass characteristics such that the essential spectral content of the QRS complex is preserved, while unwanted ECG components such as the P and the T waves are suppressed. The center frequency of the filter varies from 10 to 25 Hz and the bandwidth from 5 to 10 Hz. In contrast to other types of ECG filtering, waveform distortion is not a critical issue in QRS detection. The focus is instead on improving the SNR to achieve good detector performance.

b. The *nonlinear transformation* further enhances the QRS complex in relation to the background noise as well as transforming each QRS complex into a single positive peak better suited for threshold detection. The transformation may consist of a memoryless operation, such as rectification or squaring of the bandpass filtered signal, or a more complex transformation with memory. Not all preprocessors employ nonlinear transformations, but the filtered signal is instead fed directly to the decision rule.

c. The *decision rule* takes the output of the preprocessor and performs a test on whether a QRS complex is present or not. The decision rule can be implemented as a simple amplitude threshold procedure, but may also include additional tests, for example, on reasonable waveform duration, to assure better immunity against various types of noise.

As indicated by its name, the QRS detector is designed to detect heartbeats, while rarely producing occurrence times of the QRS complexes with

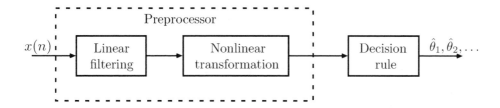

Figure 7.16: Block diagram of a commonly used QRS detector structure. The input is the ECG signal, and the output $\hat{\theta}_1, \hat{\theta}_2, \ldots$ is a series of occurrence times of the detected QRS complexes.

high temporal resolution. Hence, it may be necessary to improve the resolution using an algorithm that performs time alignment of the detected beats. Such alignment reduces the problem of smearing which may occur when computing the ensemble average of several beats; cf. the methods for estimation of latency shifts on page 229.

7.4.1 Signal and Noise Problems

The problem of conditioning the ECG with respect to noise has already been addressed. However, the term "noise" acquires a somewhat different meaning when considered from a QRS detection point of view because the P and T waves, although being part of the same cardiac cycle as the QRS complex, must now be treated as noise. Hence, noise may have physiological as well as technical origins. Signal and noise problems in QRS detection can be classified into two main categories, namely, morphologic changes (including amplitude changes) and the occurrence of noise. These categories can be further subclassified:

I. Changes in QRS morphology:

    a. of physiological origin, or

    b. due to technical problems.

II. Occurrence of noise:

    a. with large P or T waves,

    b. of myoelectric origin, or

    c. due to transient artifacts (mainly related to electrode problems).

Changes in QRS morphology are demonstrated by the ECGs shown in Figures 7.17(a)–(c). In Figure 7.17(a), every third QRS complex is quite different in morphology from the others, although roughly equal in amplitude; this rhythm is known as trigeminy since every second normal beat is followed by a VPB. In Figure 7.17(b), two morphologies occur of which one is markedly lower in amplitude than the other and has biphasic morphology; this rhythm is known as bigeminy. Both these examples contain VPBs and, therefore, represent signal problems of physiological origin. Figure 7.17(c) demonstrates a technically mediated, rather drastic variation in QRS amplitude.

Various common types of noise are illustrated in Figures 7.17(d)–(g). Figures 7.17(d) and (e) present situations where the P and T waves could be misinterpreted as QRS complexes. Tall, sharp, and uniphasic P waves precede the rather small and biphasic QRS complexes in Figure 7.17(d), whereas the QRS complexes are very low in amplitude relative to the T waves in Figure 7.17(e). Figure 7.17(f) shows a short burst of noise, probably of muscular origin, where some of the noise wave shapes resemble the QRS complex, making them rather difficult to deal with. Figure 7.17(g) shows an example of an even more delicate type of artifact, probably due to electrode problems. The shape and amplitude of such artifacts are also fairly representative of QRS complexes, and only the time relationships and the complete absence of T waves reveal to the trained eye that these artifacts are not QRS complexes.

7.4.2   QRS Detection as an Estimation Problem

The rationale for using the QRS detector structure in Figure 7.16 can be appreciated from a theoretical viewpoint by investigating a number of statistical models of the ECG signal. In doing so, we will rely on ML estimation techniques in order to derive the detector structure that corresponds to the model of interest.[4]

Unknown occurrence time.   As a starting point, we will reconsider the signal model introduced on page 230 for derivation of Woody's method, developed for the alignment of evoked responses with varying latency. Here, it is assumed that a QRS complex with known morphology $s(n)$ occurs at an unknown time θ so that it is completely contained in the observation interval of length N. The noise $v(n)$ is again modeled by a stationary, white,

---

[4]Since we have tried to pursue an "estimation viewpoint" in this textbook, we will not consider QRS detection in terms of classical detection theory, as presented, for example, in [59]. Instead, the detection theory viewpoint is briefly explored in Problem 7.16, where the central concepts of hypothesis testing and likelihood ratio are introduced. For the signal models of interest here, it should be emphasized that both viewpoints lead to essentially the same detector structure.

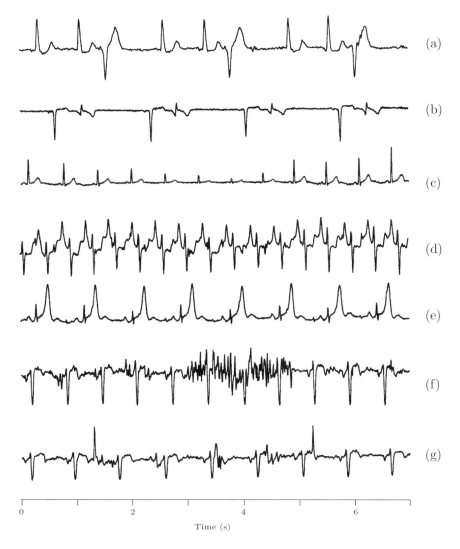

Figure 7.17: Various common types of noise in the ECG of physiological and technical origins. The specific details of each ECG are discussed in the text.

Gaussian process with variance σ_v^2. In summary, the observation model is given by

$$x(n) = \begin{cases} v(n), & 0 \le n \le \theta - 1; \\ s(n - \theta) + v(n), & \theta \le n \le \theta + D - 1 \\ v(n), & \theta + D \le n \le N - 1, \end{cases} \qquad (7.58)$$

where the integer D denotes the duration of $s(n)$. The ML estimate of the occurrence time θ is determined by the value that maximizes the log-likelihood function,

$$\hat{\theta} = \arg \max_{\theta} \ln p(\mathbf{x}; \theta), \qquad (7.59)$$

where $\mathbf{x} = \begin{bmatrix} x(0) & x(1) & \cdots & x(N-1) \end{bmatrix}^T$. Proceeding in the same way as we did on page 231, maximization of the log-likelihood function is equivalent to finding the location of the peak amplitude in the signal $y(\theta)$,

$$\hat{\theta} = \arg \max_{\theta} y(\theta), \qquad (7.60)$$

where $y(\theta)$ denotes the output of the matched filter $h(n)$,

$$y(\theta) = \sum_{n=\theta}^{\theta+D-1} x(n)h(\theta - n). \qquad (7.61)$$

The impulse response of the matched filter was found to be equivalent to a time-reversed replica of $s(n)$. We remind the reader that $h(n)$ can be modified so that it is matched to stationary, colored noise rather than to white noise, see page 236.

For the model in (7.58), a QRS complex is always assumed to be present in the observation interval. Such an assumption is, however, not very realistic since the interval may be empty, and, as a result, the maximum amplitude occurring at $\hat{\theta}$ will correspond to a falsely detected event. It is, therefore, necessary to include a threshold test to determine whether or not the potential event at $\hat{\theta}$ has a sufficiently large amplitude. A QRS complex is only detected if the output of the matched filter at $\hat{\theta}$ exceeds the threshold η,

$$y(\hat{\theta}) > \eta. \qquad (7.62)$$

Following a "refractory period" immediately after an event has been detected, the threshold procedure is repeated for the subsequent observation interval, and a new event is detected when η is exceeded. Hence, a series of occurrence times $\hat{\theta}_1, \hat{\theta}_2, \ldots$ of the detected QRS complexes is produced. The threshold η, which in (7.62) is assumed to be fixed, is in practice made dependent on the amplitude of the most recently detected QRS complexes in order to track changes in QRS amplitude. Aspects of threshold adaptation, as well as other aspects on the decision rule, are further described below.

Unknown occurrence time and amplitude. A major limitation of the model in (7.58) is the assumption of a QRS complex having a fixed amplitude. A natural extension of the model is, therefore, to assume that the occurrence time θ and the amplitude a are unknown parameters which are jointly subjected to ML estimation. Thus, the observed signal $x(n)$ is instead modeled by

$$x(n) = as(n - \theta) + v(n). \tag{7.63}$$

Before continuing with the estimator derivation, it is helpful to recall that since $s(n)$ is completely contained in the observation interval, its energy is equal to E_s for all possible values of θ,

$$E_s = \sum_{n=\theta}^{\theta+D-1} s^2(n - \theta) = \sum_{n=0}^{N-1} s^2(n).$$

Our goal is to choose θ and a so that the log-likelihood function is maximized,

$$[\hat{\theta}, \hat{a}] = \arg\max_{\theta,a} \ln p_v(\mathbf{x}; \theta, a). \tag{7.64}$$

Although the estimate of a may be of less interest for the final output of the detector, it is still important to investigate how the very presence of this unknown parameter influences the overall structure of the ML estimator. For the Gaussian, white noise assumption, the joint PDF of the observed signal is given by

$$
p_v(\mathbf{x}; a, \theta) = \prod_{n=0}^{\theta-1} \frac{1}{\sqrt{2\pi\sigma_v^2}} \exp\left[-\frac{x^2(n)}{2\sigma_v^2}\right]
$$
$$
\cdot \prod_{n=\theta}^{\theta+D-1} \frac{1}{\sqrt{2\pi\sigma_v^2}} \exp\left[-\frac{(x(n) - as(n - \theta))^2}{2\sigma_v^2}\right] \tag{7.65}
$$
$$
\cdot \prod_{n=\theta+D}^{N-1} \frac{1}{\sqrt{2\pi\sigma_v^2}} \exp\left[-\frac{x^2(n)}{2\sigma_v^2}\right].
$$

Hence, the corresponding log-likelihood function is

$$
\ln p_v(\mathbf{x}; a, \theta) = -\frac{N}{2}\ln(2\pi\sigma_v^2) - \frac{1}{2\sigma_v^2}\sum_{n=0}^{N-1} x^2(n)
$$
$$
+ \frac{1}{2\sigma_v^2}\sum_{n=\theta}^{\theta+D-1} \left(2ax(n)s(n - \theta) - a^2 s^2(n - \theta)\right)
$$
$$
= \text{constant} + \frac{1}{2\sigma_v^2}\left(2ay(\theta) - a^2 E_s\right). \tag{7.66}
$$

Maximization of the log-likelihood function in (7.66) may be performed in two steps. The first step is performed by differentiation with respect to the continuous-valued parameter a, resulting in an estimator of a being a function of θ which is used to determine an estimator of the integer-valued parameter θ. Differentiation with respect to a yields

$$\frac{\partial \ln p_v(\mathbf{x}; \theta, a)}{\partial a} = \frac{1}{2\sigma_v^2}(2y(\theta) - 2aE_s), \qquad (7.67)$$

which, when set to zero, becomes

$$\hat{a}(\theta) = \frac{1}{E_s}y(\theta) = \bar{y}(\theta). \qquad (7.68)$$

The estimator of a is thus equal to the energy-normalized output $\bar{y}(\theta)$ of the matched filter. The next step is to insert $\hat{a}(\theta)$ into (7.66), yielding the following ML estimator of θ,

$$\hat{\theta} = \arg\max_{\theta} \left[\frac{E_s}{2\sigma_v^2}\bar{y}^2(\theta) \right]. \qquad (7.69)$$

The multiplicative factor $E_s/2\sigma_v^2$ is, in practice, unknown but can fortunately be omitted since it does not influence the estimation of θ. Squaring the output signal of the matched filter implies that the estimator treats waveforms with positive or negative amplitude in the same way. If required, an estimate of a can be computed by simply inserting the resulting $\hat{\theta}$ into (7.68).

Again, the detection of QRS complexes may be based on a threshold test, although now modified so that the squared, filtered signal $\bar{y}^2(\theta)$ is compared to a threshold η,

$$\bar{y}^2(\hat{\theta}) > \eta. \qquad (7.70)$$

For the signal model in (7.63), we have thus shown that $\hat{\theta}$ should be chosen to be the instant in time at which the largest amplitude occurs in the squared output of the matched filter. Hence, these different processing steps may serve as the rationale for considering a detector structure in which both linear filtering and nonlinear transformation are embraced, as illustrated in Figure 7.16.

It may be instructive to examine how the structure of the ML estimator is modified when the absolute value of the amplitude a is constrained to a certain interval bounded by the known parameters a_1 and a_2 [60],

$$0 < a_1 \leq |a| \leq a_2. \qquad (7.71)$$

Since the log-likelihood function is quadratic in a, amplitude estimates produced by (7.68) which fall outside the allowed intervals should be set to the closest boundary value of the interval in order to maximize the log-likelihood function. Thus, the amplitude estimator is similar to that in (7.68) but now with a clipping operation included,

$$\hat{a}(\theta) = \begin{cases} a_1 \, \text{sgn}(\bar{y}(\theta)), & |\bar{y}(\theta)| < a_1; \\ \bar{y}(\theta), & a_1 \leq |\bar{y}(\theta)| \leq a_2; \\ a_2 \, \text{sgn}(\bar{y}(\theta)), & a_2 < |\bar{y}(\theta)|. \end{cases} \tag{7.72}$$

By inserting this expression for $\hat{a}(\theta)$ into the log-likelihood function

$$\ln p_v(\mathbf{x}; \hat{a}(\theta), \theta) \sim \frac{1}{2\sigma_v^2} (2\hat{a}(\theta)y(\theta) - \hat{a}^2(\theta)E_s), \tag{7.73}$$

we obtain the ML estimator of θ,

$$\hat{\theta} = \arg \max_{\theta} \left[\frac{E_s}{2\sigma_v^2} f(\bar{y}(\theta)) \right], \tag{7.74}$$

where the function $f(\cdot)$ is defined by

$$f(\bar{y}(\theta)) = \begin{cases} 2a_1|\bar{y}(\theta)| - a_1^2, & |\bar{y}(\theta)| < a_1; \\ \bar{y}^2(\theta), & a_1 \leq |\bar{y}(\theta)| \leq a_2; \\ 2a_2|\bar{y}(\theta)| - a_2^2, & a_2 < |\bar{y}(\theta)|. \end{cases} \tag{7.75}$$

This function constitutes a memoryless, nonlinear transformation in which the filtered signal $\bar{y}(\theta)$ is rectified whenever the amplitude estimate falls outside the allowed interval, defined in (7.71), whereas the signal is squared when the amplitude is within the limits.

For the special case when the amplitude a is assumed to be a priori known except its polarity,

$$a = \pm a_0, \tag{7.76}$$

the nonlinear transformation becomes a rectifier,

$$f(\bar{y}(\theta)) = 2a_0|\bar{y}(\theta)| - a_0^2. \tag{7.77}$$

On the other hand, complete elimination of the amplitude constraint yields the squarer in (7.69),

$$f(y(\theta)) = \bar{y}^2(\theta). \tag{7.78}$$

Hence, it can be demonstrated that the two most common types of nonlinear transformations—rectifier and squarer—are intimately related to the model in (7.63) with unknown occurrence time and amplitude of $s(n)$.

Unknown occurrence time, amplitude, and duration. Yet another generalization of the observation model is to account for the fact that QRS complexes exhibit considerable variation in duration. While the normal QRS duration is around 120 ms, beats originating from foci in the ventricles may be twice as wide, or even more. For the continuous-time case, variations in duration of a waveform $s_c(t)$ can be easily modeled by the scaling parameter β,

$$s_c\left(\frac{t}{\beta}\right), \quad \beta > 0,$$

where the duration thus increases as β becomes larger. A discrete-time waveform $s(n)$ can be obtained from $s_c(t)$ by periodic sampling,

$$s(n) = s_c\left(\frac{nT}{\beta}\right), \quad n = 0, 1, \ldots,$$

where the ratio T/β can be interpreted as the sampling period. As a result, different QRS durations can be modeled by sampling $s_c(t)$ at slightly different sampling periods. This can, for example, be achieved by considering uniformly spaced values of the duration, obtained according to the following expression,

$$\beta_l = \beta_{\min} + l \cdot \Delta\beta, \quad l = 0, 1, \ldots, l_{\max}, \tag{7.79}$$

where $\beta_{\min}$ corresponds to the shortest duration of physiological interest, and $\Delta\beta$ is the duration step size. The resulting discrete-time waveform is indexed by the duration parameter l such that

$$s_c\left(\frac{nT}{\beta_l}\right) \rightarrow s(n, l), \quad n = 0, 1, \ldots.$$

Discretization of β does not necessarily have to be uniform as suggested by (7.79); another approach is to use a dyadic sequence $\{\beta_l = 2^l, l = 0, 1, 2, \ldots\}$ resulting in rather coarse discretization of the duration. It should be pointed out that the continuous-time waveform $s_c(t)$ is rarely available a priori, but the discrete-time waveform $s(n)$, assumed to be known in the model (7.58), is instead available. However, this limitation can be overcome by appropriate use of interpolation and decimation in order to increase the time resolution of $s(n)$ so that waveforms $s(n, l)$ with different durations can be determined.

An example of a QRS waveform with different durations is presented in Figure 7.18. From the most narrow waveform, the remaining waveforms are obtained by increasing the original sampling rate with rational factors

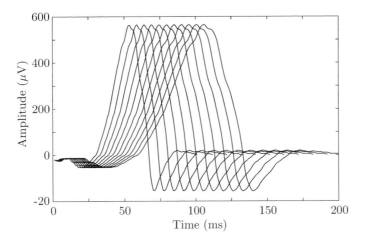

Figure 7.18: An example of QRS waveforms with identical morphology but different durations. The duration is increased in steps of 10% until twice the initial duration is reached.

11/10, 12/10, and so on. The increase in sampling rate is implemented by first performing interpolation by the numerator factor and then decimating the interpolated signal by the denominator factor. For example, an increase in sampling rate of a factor of 1.1 is obtained by interpolation by a factor of 11 followed by decimation by a factor of 10.

The procedure of finding an estimate of the occurrence time θ is similar to that in (7.69), except that we now have to include the estimation of l. As before, we can first determine an estimate of the amplitude a as a function of θ and l and then perform the maximization with respect to θ and l. The resulting ML estimator is

$$\hat{\theta} = \arg\max_{\theta} \left(\frac{1}{2\sigma_v^2} \max_l \left[E_s(l)\bar{y}^2(\theta, l) \right] \right), \tag{7.80}$$

where the energy of $s(n)$ is now a function of l,

$$E_s(l) = \sum_{n=\theta}^{\theta+D-1} s^2(n - \theta, l), \tag{7.81}$$

and can, therefore, no longer be omitted from the maximization with respect to θ and l. The output signal $\bar{y}(\theta, l)$ can be viewed as the output of a bank of $l_{\max}$ different filters, with each filter being matched to a waveform with a

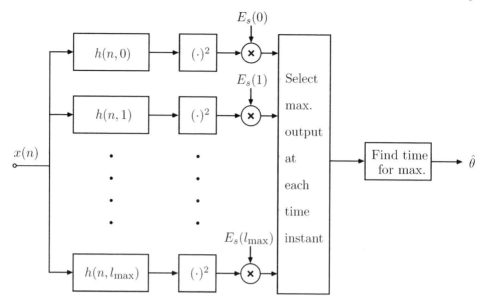

Figure 7.19: Maximum likelihood estimation of the occurrence time θ when the amplitude and duration of the waveform are unknown. Each filter $h(n, l)$ is matched to a certain duration defined by the parameter l. The output signal at $\hat{\theta}$ is then fed to the decision rule of the QRS detector.

certain duration. The filter output is obtained from

$$\bar{y}(\theta, l) = \frac{1}{E_s(l)} \sum_{n=\theta}^{\theta+D-1} x(n)s(n - \theta, l). \tag{7.82}$$

The block diagram in Figure 7.19 presents the ML estimator in (7.80) which produces $\hat{\theta}$ and $\hat{l}$.

Another, much simpler approach to model duration is to consider $s(n)$ as being composed of two identical waveforms $q(n)$, of which one is shifted l samples in time and has the opposite sign,

$$s(n, l) = q(n) - q(n - l), \tag{7.83}$$

where $l = l_{\min}, \ldots, l_{\max}$. The waveform $s(n, l)$ is assumed to be completely contained in the observation interval for all durations. To determine the ML estimator based on the earlier assumptions on noise, it is helpful to first introduce some notations related to the model in (7.83),

$$\bar{y}(\theta, l) = \bar{y}_q(\theta, l) - \bar{y}_q(\theta - l, l), \tag{7.84}$$

where

$$\bar{y}_q(\theta, l) = \frac{1}{E_s(l)} \sum_{n=\theta}^{\theta+D-1} x(n)q(n-\theta). \tag{7.85}$$

The energy of $s(n)$ can be expressed as

$$E_s(l) = 2E_q(1 - \rho_q(l)), \tag{7.86}$$

where the function $\rho_q(l)$ denotes the energy-normalized autocorrelation function of $q(n)$,

$$\rho_q(l) = \frac{1}{E_q} \sum_{n=0}^{D-1} q(n)q(n-l), \tag{7.87}$$

and E_q denotes the energy of $q(n)$. Using these notations, we can express the ML estimator in (7.80) as

$$\hat{\theta} = \arg \max_{\theta} \left(\frac{E_q}{\sigma_v^2} \max_{l} \left[(1 - \rho_q(l)) \left(\bar{y}_q(\theta) - \bar{y}_q(\theta - l) \right)^2 \right] \right). \tag{7.88}$$

Although the estimator in (7.88) can be implemented along the same lines as the one in (7.80), it is interesting to observe that local extreme values of the filtered signal $\bar{y}_q(\theta)$ enter the maximization in (7.88). It is evident that the term $(\bar{y}_q(\theta) - \bar{y}_q(\theta - l))^2$ is maximized when $\bar{y}_q(\theta)$ and $\bar{y}_q(\theta - l)$ are selected from among local extreme values with opposite sign, i.e., positive maximum and negative minimum, and separated by a distance within the interval $[l_{\min}, l_{\max}]$. An approximate, but computationally much more efficient, implementation of the ML estimator in (7.88) is, therefore, to determine the local extreme values of the filtered signal and to use these values for the estimation of θ.

Independently of the above approximate ML technique, the idea of using local extreme values as a basis for QRS detection is well-established and is sometimes referred to as a *peak-and-valley picking strategy*. In fact, several QRS detectors have been presented which take into account the properties of adjacent pairs of local extreme values with opposite signs [61–64]; for such pairs to be considered an event, the distance between two extreme values must be within certain limits to qualify as a heartbeat. The peak-picking procedure is also central to certain methods of ECG data compression which will be described in Section 7.6.

7.4.3 Detector Preprocessing

Linear filtering. We will now describe certain linear, time-invariant filters which may be used for QRS detection. The earliest attempts to condition the ECG signal employed differentiation in order to emphasize segments of the signal with rapid transients, i.e., the QRS complex [65–67]. In discrete-time, differentiation can be approximated by a filter $H(z)$ that produces the difference between successive samples,

$$H(z) = 1 - z^{-1}. \tag{7.89}$$

Such a differencing filter may perhaps be an acceptable choice when analyzing resting ECGs; however, it accentuates high-frequency noise and is, therefore, inappropriate in situations with moderate or low SNRs.

A better approach is to combine differentiation with lowpass filtering so that noise activity above a certain cut-off frequency $\omega_c = 2\pi f_c$ is attenuated [68, 69]. The frequency response of the ideal *lowpass differentiator* is given by

$$H(e^{j\omega}) = \begin{cases} j\omega, & |\omega| \leq \omega_c; \\ 0, & \omega_c < |\omega| < \pi, \end{cases} \tag{7.90}$$

and the corresponding impulse response is

$$h(n) = \frac{1}{2\pi} \int_{-\omega_c}^{\omega_c} j\omega e^{j\omega n} d\omega$$

$$= \begin{cases} 0, & n = 0; \\ \dfrac{1}{\pi n}\left(\omega_c \cos(\omega_c n) - \dfrac{1}{n}\sin(\omega_c n)\right), & n \neq 0. \end{cases} \tag{7.91}$$

Before the filter is used in practice, its infinite impulse response must be truncated using windowing or, better, by determining the coefficients of an FIR filter so that the error between its magnitude function and $H(e^{j\omega})$ in (7.90) is minimized in the MSE sense [69].

The large variability in signal and noise properties of the ECG implies that the requirements on frequency response have to be rather loose, and, as a result, simple structured filters can be applied. One family of such filters is defined by [60]

$$H(z) = \left(1 - z^{-L_1}\right)\left(1 + z^{-1}\right)^{L_2}, \tag{7.92}$$

where L_1 and L_2 are two integer-valued parameters. The corresponding frequency response is given by

$$H(e^{j\omega}) = j2^{L_2+1} e^{-j\omega(L_1+L_2)/2} \sin\left(\frac{\omega L_1}{2}\right) \cos^{L_2}\left(\frac{\omega}{2}\right). \tag{7.93}$$

The first part, $(1 - z^{-L_1})$, forms the difference between the input signal and the delayed input, whereas the second part, $(1 + z^{-1})^{L_2}$, is a lowpass filter

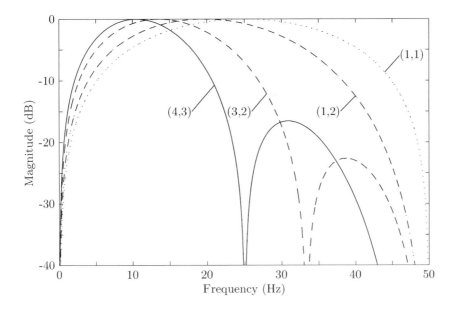

Figure 7.20: The magnitude function of the filter in (7.92), defined by the two integer parameters L_1 and L_2, displayed for the combinations (1,1), (1,2), (3,2), and (4,3). Each magnitude function has been normalized so that its maximum gain corresponds to 0 dB. The sampling rate is assumed to be 100 Hz.

whose bandwidth decreases as L_2 increases. Filters belonging to the family in (7.92) can be implemented without multipliers, thus only requiring addition and subtraction. Consequently, these filters are attractive for systems which analyze long-term ECG recordings. For example, good detection performance has been reported for $(L_1, L_2) = (1, 2)$, resulting in a bandpass filter with a rather large bandwidth and a center frequency of 20 Hz, assuming a sampling rate of $F_s = 100$ Hz, see Figure 7.20 [70]. The filter defined by $(L_1, L_2) = (1, 1)$ has been employed in [71–73], and a filter defined by $(L_1, L_2) = (1, 3)$ has been employed in [74]. The use of a filter with a lower center frequency, such as that of $(L_1, L_2) = (4, 3)$, increases the number of false detections due to large-amplitude T waves. The filter $(L_1, L_2) = (5, 4)$ may be a suitable choice for a higher sampling rate of 250 Hz, resulting in a filter with a center frequency of 20 Hz [75].

The idea of designing a filter matched to a certain waveform $s(n)$, as suggested by the ML estimator of an unknown occurrence time θ, is not feasible in practice due to the widely different QRS morphologies, as well as different noise characteristics. Instead, the notion of an "optimal" linear filter may be realized for a given filter structure by optimizing detector performance with respect to the filter parameters, for example, by selecting appropriate values of L_1 and L_2 in (7.92). Such an optimization approach was pursued in [76] where the center frequency and bandwidth of a second-order, Butterworth bandpass filter were selected in order to produce the largest SNR. Based on a large database with QRS complexes, it was found that a bandpass filter with a center frequency of 17 Hz, and a relatively small bandwidth, yielded the highest SNR.

Perhaps the most straightforward approach to designing a matched filter has not yet been mentioned, namely, to identify the impulse response with the detected QRS complexes (tacitly assuming that the noise is white). By initially using a filter with a fixed impulse response, the impulse response may subsequently be updated as new beats are detected using exponential averaging, or any other recursive technique [77]. An advantage of this approach is that the filter is better matched to the QRS morphology of individual ECGs and, therefore, can be expected to yield better performance than does a fixed filter. On the other hand, the degradation in performance that results when beats with deviating morphologies are encountered may be unacceptable since such beats can have special importance from a clinical viewpoint.

Although certain variability in QRS morphology is accounted for by allowing different amplitudes and durations, the filter $h(n)$ remains mismatched to a range of morphologies. Without going into details of the resulting ML estimator, it is interesting to point out that different QRS morphologies $s(n)$ can be modeled through linear combinations of a set of K basis functions (cf. Section 4.5),

$$s(n) = \sum_{i=1}^{K} a_i \varphi_i(n), \qquad (7.94)$$

where the amplitude a_i of each basis function has to be estimated. The corresponding ML estimator includes a bank of K filters in which each subfilter is matched to a certain basis function $\varphi_i(n)$ [78–80].

Nonlinear transformations.   The common objective in nonlinear transformation is to produce a single, positive-valued peak for each QRS complex, which allows the use of peak detection or a one-sided detection threshold. Similar to the case with linear filtering, the transformation should be designed so that it produces a signal in which QRS complexes are enhanced relative to the background of P and T waves, noise, and artifacts.

The QRS detector previously derived with ML estimation techniques includes a squarer as nonlinear transformation. Alas, the squarer introduces additional peaks and valleys in the output signal which may cause spurious

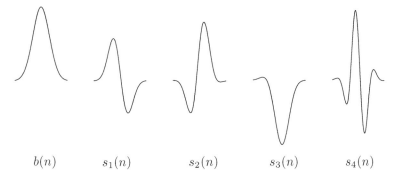

$$b(n) \qquad s_1(n) \qquad s_2(n) \qquad s_3(n) \qquad s_4(n)$$

Figure 7.21: Various waveforms $s_1(n), \ldots, s_4(n)$ obtained using the signal model in (7.96). The waveforms all have envelopes identical to the leftmost Gaussian waveform $b(n)$.

events. It is, therefore, advisable to smooth $y^2(n)$ through linear filtering,

$$z(n) = \sum_{k=n-L+1}^{n} y^2(k)h_s(n-k), \qquad (7.95)$$

where $h_s(k)$ denotes a lowpass FIR filter whose length is L. The smoothing of the signal should be such that only large-amplitude peaks of sufficient duration, i.e., the QRS complexes, are preserved in $z(n)$. Simple structured smoothing filters have been used, for example, those defined by a rectangular [74, 81] or a triangular impulse response [82, 83].

Another approach to the design of a nonlinear transformation is based on a model in which the QRS complex is described by the deterministic, positive-valued, lowpass signal $b(n)$, modulated by a cosine function defined by the modulation frequency ω_m and the phase angle ϕ,

$$s(n) = b(n)\cos(\omega_m n + \phi). \qquad (7.96)$$

The lowpass signal $b(n)$ is commonly referred to as the *envelope* of $s(n)$. By varying the two parameters ω_m and ϕ, a wide variety of waveform morphologies can be modeled which resemble different QRS complexes. Figure 7.21 displays a number of waveforms for which the common envelope $b(n)$ is defined by the Gaussian function.

It is of great interest to develop a technique with which the envelope $b(n)$ can be extracted from $s(n)$ without any prior knowledge of ω_m or ϕ. In order to solve this problem, we start by expressing $s(n)$ in the frequency domain,

$$S(e^{j\omega}) = \frac{1}{2}\left(B\left(e^{j(\omega-\omega_m-\phi)}\right) + B\left(e^{j(\omega+\omega_m+\phi)}\right)\right), \qquad (7.97)$$

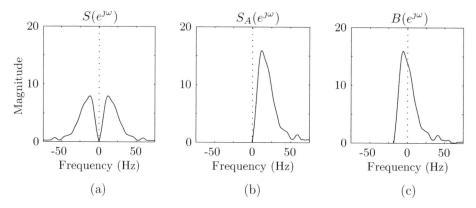

Figure 7.22: Spectral operations required to obtain the envelope of $s(n)$. (a) Spectrum of the original signal $s(n)$, (b) spectrum after cancellation of negative frequencies using the Hilbert transform, and (c) spectrum of the frequency-shifted signal.

where $B(e^{\jmath\omega})$ and $S(e^{\jmath\omega})$ denote the DTFTs of $b(n)$ and $s(n)$, respectively. The envelope $B(e^{\jmath\omega})$ can be obtained by introducing a technique which cancels out negative frequencies and shifts the remaining spectrum to the origin, as illustrated in Figure 7.22. Such a technique is well-known from the representation and demodulation of bandpass signals in the area of communication [84].

In order to cancel out frequencies in the interval $-\pi \leq \omega < 0$, we introduce the function $S_A(e^{\jmath\omega})$, defined by

$$S_A(e^{\jmath\omega}) = S(e^{\jmath\omega}) + \jmath H(e^{\jmath\omega})S(e^{\jmath\omega})$$
$$= S(e^{\jmath\omega}) + \jmath \check{S}(e^{\jmath\omega}), \qquad (7.98)$$

where $H(e^{\jmath\omega})$ is a linear, time-invariant filter whose transfer function is defined by

$$H(e^{\jmath\omega}) = \begin{cases} -\jmath, & 0 \leq \omega < \pi; \\ \jmath, & -\pi \leq \omega < 0. \end{cases} \qquad (7.99)$$

This filter is known as the *Hilbert transformer* and has a unit magnitude frequency response and a phase response equal to $-\pi/2$ for $0 < \omega < \pi$ and $\pi/2$ for $-\pi < \omega < 0$ [46]. The output of the Hilbert transformer is thus a $90°$ phase-shifted version of $s(n)$ which, in the following, is denoted $\check{s}(n)$ and is thus the Hilbert transform. The one-sided spectrum $S_A(e^{\jmath\omega})$ that results from the operation in (7.98) is expressed as

$$S_A(e^{\jmath\omega}) = \begin{cases} 2S(e^{\jmath\omega}), & 0 \le \omega < \pi; \\ 0, & -\pi \le \omega < 0, \end{cases}$$

$$= \begin{cases} B\left(e^{\jmath(\omega-\omega_m)}\right), & 0 \le \omega < \pi; \\ 0, & -\pi \le \omega < 0. \end{cases} \tag{7.100}$$

The time domain signal $s_A(n)$, known as the analytic signal, represents a frequency-shifted version of the envelope $b(n)$,

$$s_A(n) = b(n)e^{\jmath\omega_m n}.$$

By computing the absolute value of $s_A(n)$, we obtain the positive-valued envelope $b(n)$ without knowledge of ω_m or ϕ, since

$$b(n) = |s_A(n)|$$
$$= \sqrt{s^2(n) + \check{s}^2(n)}, \tag{7.101}$$

where the last equality in (7.101) is due to the complex-valued definition in (7.98) which, in the time domain, equals $s_A(n) = s(n) + \jmath\check{s}(n)$. The Hilbert transform and the envelope for a number of QRS complexes with different morphologies are presented in Figure 7.23.

The impulse response of the Hilbert transformer in (7.99) is

$$h(n) = \begin{cases} \dfrac{2}{\pi}\dfrac{\sin^2(\pi n/2)}{n}, & n \ne 0; \\ 0, & n = 0. \end{cases} \tag{7.102}$$

Since this impulse response is infinite and noncausal, it must be approximated before the envelope can be computed. In its simplest version, the approximation involves truncation and appropriate time shifting in order to make the filter causal.

When the squaring and square-root operations of the envelope computation are undesirable for implementational reasons, for example, in the context of a pacemaker where strict demands are put on low power consumption, the Euclidean distance in (7.101) can be approximated by the "city block" distance,

$$\sqrt{s^2(n) + \check{s}^2(n)} \approx |s(n)| + |\check{s}(n)|.$$

By truncating the impulse response of the Hilbert transformer to the shortest possible length, an approximate envelope $\hat{b}(n)$ can be computed from

$$\hat{b}(n) = |s(n)| + \frac{2}{\pi}|s(n+1) - s(n-1)|. \tag{7.103}$$

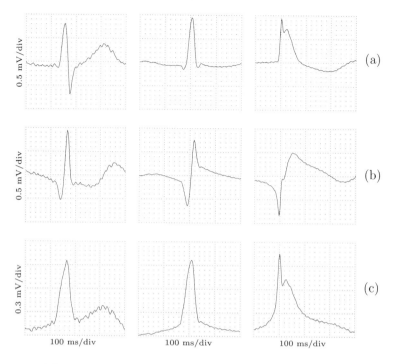

Figure 7.23: (a) An example of different QRS complex morphologies and their corresponding (b) Hilbert transforms and (c) envelope functions.

This approximation is associated with a certain amount of ripple which can be eliminated with lowpass filtering before the envelope is suitable for threshold detection [85], see also [86–88].

In addition to the above-mentioned nonlinear transformations, several others have been described which exploit the degree of changes that are characteristic of the ECG signal, see, e.g., [89–92].

7.4.4 Decision Rules

A decision rule must be applied to the output of the preprocessor to determine whether or not a QRS complex has occurred. The decision rule in (7.62) or (7.70) was synonymous with a test where the preprocessed signal was compared to a fixed threshold η. However, it is highly desirable to incorporate adaptivity into the threshold because QRS amplitude and morphology may change drastically during the course of just a few seconds (Figure 7.17); the detection of low-amplitude QRS complexes with a fixed threshold inevitably implies accepting several false detections. Although the focus here is on amplitude-related decision rules, it is certainly possible to introduce additional rules which relate to other signal properties, such as the duration of a waveform.

The *interval-dependent* QRS detection threshold is updated once for each new detection at θ_i and is held fixed during the subsequent interval until the threshold is exceeded and a new QRS is detected. A popular structure of the interval-dependent threshold $\eta_I(n)$ is the one based on the exponentially updated peak amplitude $\tilde{z}_{e,i}$ of the previously detected QRS complexes,

$$\eta_I(n) = \mu\tilde{z}_{e,i}, \quad n = \theta_i, \theta_i + 1, \ldots, \tag{7.104}$$

with

$$\tilde{z}_{e,i} = \tilde{z}_{e,i-1} + \alpha\left(z(\theta_i) - \tilde{z}_{e,i-1}\right), \quad i \geq 1, \tag{7.105}$$

where $\tilde{z}_{e,i}$ is the exponential average, and $z(\theta_i)$ represents the amplitude in the preprocessed signal of the most recently detected QRS complex at time θ_i. The parameter μ in (7.104) determines the fraction of the amplitude $\tilde{z}_{e,i}$ to be used in the threshold computation; typically, μ is chosen within the interval 0.5–0.7. The parameter α defines the speed with which the amplitude threshold can change, cf. the exponential averager defined in (4.35). The recursion in (7.105) may be initialized by setting $\tilde{z}_{e,0}$ to a fixed value. Alternative techniques for tracking the QRS amplitude include the mean or median value of the most recently detected beats [93].

The QRS detector using an interval-dependent threshold can be extended to become a *time-dependent* threshold for the purpose of improving the rejection of large-amplitude T waves, while still allowing low-amplitude ectopic beats to be detected. Subsequent to detection of a QRS complex at θ_i, the time-dependent threshold $\eta(n)$ may be assigned the following structure:

$$\eta(n) = \begin{cases} \eta_{\max}, & n = \theta_i + 1, \ldots, \theta_i + D_0; \\ g(n - \theta_i - D_0 - 1), & n = \theta_i + D_0 + 1, \ldots, \theta_i + D_1; \\ \mu\tilde{z}_{e,i}, & n = \theta_i + D_1 + 1, \ldots, \end{cases} \tag{7.106}$$

where $\eta_{\max}$ is a constant, and $g(n)$ is a function defined such that it decreases over the transition interval $[\theta_i + D_0 + 1, \theta_i + D_1]$ until it reaches the interval-dependent threshold $\eta_I(n)$,

$$\eta_{\max} = g(0) > g(1) > \ldots > g(D_1 - D_0 + 1) = \mu\tilde{z}_{e,i}. \tag{7.107}$$

The most common choice of (7.106) involves the *eye-closing period* during which nothing is detected until $\theta_i + D_1$, i.e., $\eta_{\max} = \infty$ and $D_0 = D_1 - 1$. The length of the eye-closing period is chosen within the interval 160–200 ms, motivated by the existence of an absolute refractory period during which the heart is unresponsive to electrical stimuli.

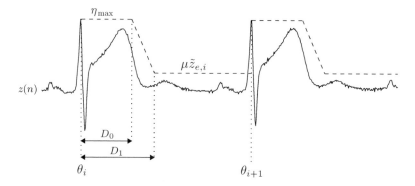

Figure 7.24: Time-dependent thresholding in QRS detection. Following each detected QRS, the threshold is initially set to a fixed level η_{max} during D_0 samples, after which it decreases linearly until a lower level $\mu \tilde{z}_{e,i}$ is reached after D_1 samples. In this case, the original ECG signal itself is used as the threshold signal $z(n)$.

The obvious risk of missing premature ventricular ectopic beats necessitates choosing a value of D_1 which is not too large. On the other hand, a very short eye-closing interval will increase the number of false detections due not only to T waves, but also to very wide ectopic beats. A compromise between these conflicting demands may be to use a finite value of η_{max} in (7.106), but still with $D_0 = D_1 - 1$ [94]. Another possible choice of $\eta(n)$ is given by $\eta_{max} = \infty$ and with $g(n)$ chosen as a linearly decreasing function in the interval $[\theta_i + D_0 + 1, \theta_i + D_1]$ [95]. While the eye-closing period D_0 should be independent of heart rate, the transition period D_1 may be related to the length of the average RR interval such that $g(n)$ decreases faster as the heart beats faster. From an electrophysiological viewpoint, $g(n)$ corresponds to the relative refractory period, following the absolute refractory period, during which the cardiac cells can respond to a stronger than normal stimulus, but with abnormally slow conduction. Thresholding based on a finite value of η_{max}, combined with linearly decreasing $g(n)$, is illustrated in Figure 7.24.

The signal fed to the decision rule is typically processed in a sequential fashion, implying that QRS complexes are detected in temporal order despite the fact that the entire recording may be available from a storage device. A detector which, to a certain degree, incorporates both past and future signal properties is one that first delimits an observation interval and then detects QRS complexes in their order of magnitude rather than in temporal order [60]. Owing to this "noncausal" property, eye-closing periods of equal length may be applied both before and after each detected beat. The detection threshold could be adapted with respect to the properties of the QRS complexes which delimit the interval. This approach allows the detector to find the QRS complexes even when a sudden decrease in amplitude occurs.

Various add-on techniques may be used to improve detection performance. Since the spectral content of muscle noise overlaps that of the QRS complex, such noise causes the performance to deteriorate. To cope with this situation, noise measurements are of great value and can, in combination with information on QRS amplitude, be used to better adjust the level of the detection threshold [74, 81, 93, 96]. When including such measurements, it is crucial to make sure that the measurement interval is positioned so that it only contains noise.

The above described decision rules do not impose any constraints on heart rhythm, except that a certain time must elapse between two successive beats. In fact, an inherent property of any QRS detector is its restrictive use of information available from the pattern of preceding RR intervals; even though a regular rhythm has prevailed for a long time, the next QRS complex to be detected must be treated as if it could occur at almost any time in the observation interval. Still, certain basic information on rhythm may be employed to control a *look-back detection mode* to avoid low-amplitude beats that are being missed [97, 98]. The occurrence of an RR interval of approximately twice the length of the average interval length may, in cases of a stable sinus rhythm, be explained by a missed low-amplitude ectopic beat. The processing of a prolonged RR interval in look-back mode by using a lower detection threshold may result in detection of the beat which was initially missed.

Although we have presented approaches to single-lead QRS detection, detection based on multilead ECG recordings is preferable since this is associated with a substantial improvement in performance. Noise and artifacts tend to occur independently in different leads, so improved immunity can be achieved with a multilead approach. Detection of ectopic beats will, in addition, be more reliable since ectopic beats of low amplitude in one lead are usually larger in another. Multilead QRS detectors may incorporate either a single decision function, for example, the sum of the preprocessed signals of the different leads [88, 90, 99], or decision logic based on the outcome of QRS detection in individual leads [79, 100].

7.4.5 Performance Evaluation

Before a QRS detector can be implemented in a clinical setting, suitable parameter values must be determined, and the performance for the chosen set of parameter values must be evaluated. It may be tempting to choose the parameter values which were found useful during algorithm development rather than the values that would result from a separate performance optimization. In doing this, however, one runs the serious risk of choosing values that are too attuned to the training data, but not necessarily well-suited to subsequent data.

A large number of parameter values must usually be fixed within each detector structure. Joint optimization of all parameters with respect to a suitable performance measure may imply a massive amount of computation which may be unrealistic. A natural way to cope with this problem is to optimize only those parameters that have the most profound effects on detector performance. Other values can be fixed on physiological grounds or determined by various ad hoc decisions.

Detector performance is commonly measured in terms of

- P_D, the probability of a true beat being detected, and

- P_F, the probability of a false beat being detected.

The probability of a missed beat P_M is related to the probability of detection through $P_D = 1 - P_M$. Although these probabilities may be calculated theoretically for certain statistical models of the ECG signal and noise, it is, in practice, much more interesting to estimate them from the performance that results from a database of ECGs with a large variety of QRS morphologies and noise types. In this case, the estimation is based on ratios that include the number of correctly detected QRS complexes N_D, the number of false alarms N_F, and the number of missed beats N_M. The probability of false detection can be estimated from

$$\hat{P}_F = \frac{N_F}{N_D + N_F}, \tag{7.108}$$

where the denominator includes the term N_F to assure that $\hat{P}_F$ is always between zero and one. The probability of detection is estimated from

$$\hat{P}_D = \frac{N_D}{N_D + N_M}. \tag{7.109}$$

Since each probability is determined for each of the ECG recordings in the database, it is customary to compute a "gross" average of the estimates in order to reflect the overall performance of the QRS detector.[5]

The numbers N_D, N_F, and N_M can only be computed once the database has been subjected to manual annotation. Such annotation is typically a laborious process, involving one or several skilled ECG readers, and leads to

---

[5]In clinically oriented literature, different terminology is used to describe detector performance, in which P_D is referred to as *sensitivity*, P_F is referred to as 1–*positive predictivity*, and N_T, N_F, and N_M are referred to as the number of *true positives, false positives*, and *false negatives*, respectively.

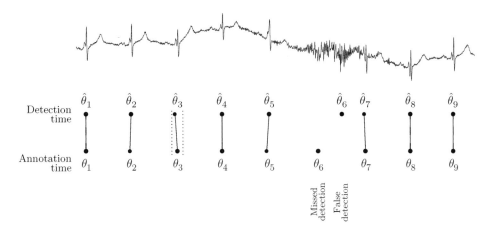

Figure 7.25: Comparison of the QRS detector output to the manual annotations, in this case indicating that all beats are correctly detected except the beat occurring at θ_6 which is missed. A noise wave is falsely detected at $\hat{\theta}_6$. The matching window, with length $\Delta\theta$, is displayed for the beat at θ_3 but is, of course, equally applicable to the other beats.

every QRS complex being assigned its correct occurrence time θ_i. A beat is said to have been detected when the difference between the estimated occurrence time $\hat{\theta}_j$ and the annotation time θ_i is within a certain matching window defined by $\Delta\theta$,

$$|\hat{\theta}_j - \theta_i| \leq \Delta\theta.$$

A false detection is produced when $\hat{\theta}_j$ is located at a distance larger than $\Delta\theta$ from any θ_i, and a beat is considered to have been missed when no detection occurs closer than $\Delta\theta$ to θ_i. The process of comparing the detector output with the annotated QRS complexes is illustrated in Figure 7.25.

To study the behavior of the QRS detector for different parameter values, the estimate $\hat{P}_D$ can be displayed versus $\hat{P}_F$ in a *receiver operating characteristic* (ROC), see Figure 7.26. From such a diagram, we may choose suitable parameter values for the detector in order to achieve an acceptable trade-off between the two counterbalancing measures $\hat{P}_D$ and $\hat{P}_F$; the chosen trade-off will differ from application to application [70, 101]. The ROC offers, among other things, a very practicable means of investigating and comparing the performance of different QRS detectors with respect to their robustness against noise and artifacts.

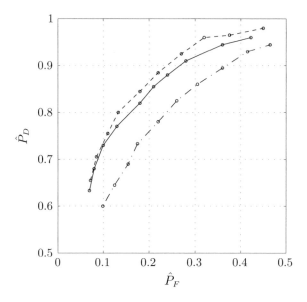

Figure 7.26: Receiver operating characteristics (ROCs) of a QRS detector using three different frequency responses of its linear filter. Each symbol corresponds to a certain value of an amplitude detection threshold. The performance was determined on a set of ECG recordings with very low SNRs [58].

7.5   Wave Delineation

A method of wave delineation determines the boundaries of each wave within the PQRST complex so that, with the resulting time instants, wave duration can be computed. Once the wave has been delineated, other measures characterizing the wave, such as amplitude and morphology, can be easily computed. Such a method must also be able to detect when a certain wave is absent; this situation is commonly encountered since, for example, only the R wave or the S wave is present in certain leads or pathologies. Although delineation is primarily applied to beats originating from the sinus node, it may be applied to any type of beat to produce measurements for use in automated beat classification.

The classical definition of a wave boundary is the time instant at which the wave crosses a certain amplitude threshold level. Unfortunately, this definition is not well-suited for the common situation when the ECG contains baseline wander, and, therefore, this definition is rarely applied in practice. Instead, many methods for wave delineation exploit the change in slope that occurs at a boundary to avoid the problems due to low-frequency noise. Hence, the first derivative of the signal is calculated and analyzed

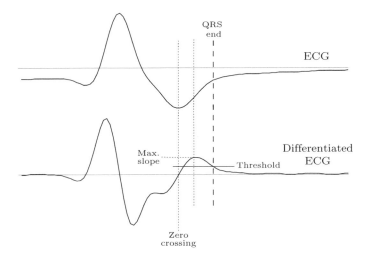

Figure 7.27: Determination of the QRS end using slope information. The QRS end is the time at which the differentiated signal crosses a threshold after the maximum slope has occurred. The threshold level is usually expressed as a percentage of the maximum slope.

with respect to zero crossings and extreme values.[6] This type of delineation is illustrated by Figure 7.27 where the aim is to find the end of the S wave; the other wave boundaries of the PQRST complex can be found in a similar way. In this example, the search for the end point starts when the steepest upslope of the S wave occurs and continues until the derivative of the signal falls below a certain threshold value. The time instant at which the level is crossed defines the QRS end. Since the above search procedure is based on the assumption that each of the different waves is present, it is necessary to first establish which waves are absent to ensure meaningful delineation. Such wave detection is usually done by analyzing the pattern of successive peak amplitudes and interpeak distances of the differentiated signal in an interval positioned around the QRS complex.

The threshold level which determines the position of a wave boundary may be fixed and chosen with reference to a slope value that is representative of the boundary to be determined [102–104]. Alternatively, the threshold may be related to signal morphology so that its level is set to a certain percentage of the maximum slope [100, 105]. The latter type of thresholding is

---

[6]We remind the reader that a zero crossing of the derivative signal coming from below corresponds to a minimum of the original signal, while that from above corresponds to a maximum. The maximum of the derivative signal corresponds to the point of the original signal with the steepest upslope, while the minimum corresponds to the steepest downslope.

more suggestive of a cardiologist's approach to delineation since the boundaries of a large-amplitude wave with steep slopes and a low-amplitude wave with less steep slopes will occur at about the same position; this is not the case when fixed thresholding is applied.

In noisy ECG signals, wave delineation from the differentiated signal performs poorly since an already low signal amplitude at the wave boundary is disturbed by noise. The performance can, to a certain degree, be improved by combining signal differentiation with lowpass filtering to attenuate high-frequency noise, cf. (7.90). The cut-off frequency of the lowpass filter may be fixed or, better, adapted to the spectral content of the wave to be delineated [69]. For example, delineation of the QRS complex should be based on a filter with a higher cut-off frequency than the filter used to find the end of the T wave, reflecting the fact that the T wave contains much less high-frequency components, see Figure 7.28. Furthermore, wave delineation can be made more robust to noise by replacing thresholding with template matching [50]. While thresholding relates to highly local signal behavior, and in that way becomes vulnerable to noise, the matching of a template waveform to the lowpass differentiated signal through correlation makes use of more information in the signal.

The threshold levels, or the shapes of the waveform templates, should be chosen such that the resulting delineation agrees with those obtained by cardiological expertise. Following training of the delineation method to obtain suitable parameter values, its performance should be evaluated on a database with P, QRS, and T wave boundaries having been manually annotated; such databases are today publicly available [106, 107]. Delineation performance is described in terms of the mean and standard deviation of the error between the boundaries produced by the method and the experts [108, 109]. It is important to realize that a zero value of the standard deviation can never be attained since a certain dispersion will always exist even among experts. However, a method's performance is judged as satisfactory when the dispersion is approximately on the same order as that among experts [110].

Wave delineation is especially problematic when determining the end of the T wave, which is often characterized by a very gradual transition to the isoelectric line of the ECG, see, for example, the T wave in Figure 7.28. In fact, its delineation is problematic even among cardiologists, and differences between cardiologists may occasionally approach as much as 100 ms [110]. Despite these difficulties, the end of the T wave is an extremely important boundary, required to compute the length of the QT interval, i.e., the total duration of ventricular depolarization and repolarization, cf. page 429. Due to the importance of this measurement, several special techniques have been developed for the purpose of robustly determining the T wave end, see, e.g., [111–116]. Multiresolution signal analysis of the ECG using the

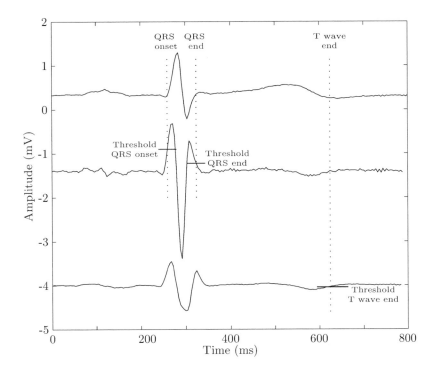

Figure 7.28: Wave delineation based on lowpass differentiated signals. The original signal (top) is differentiated and lowpass filtered to yield the QRS onset and end (middle) and the T wave end (bottom). The threshold level which determines each wave boundary is indicated by the horizontal line. A higher cut-off frequency of the lowpass differentiator was used for QRS complex delineation than for the T wave.

dyadic wavelet transform in which the signal is analyzed at different time resolutions, has proven to be particularly well-suited for T wave delineation. By first determining a robust, but preliminary, boundary position from a smooth approximation of the original signal, the position can be refined by analyzing the properties of better approximations in an interval positioned around the preliminary boundary [62, 63, 117, 118]. The wavelet-based approach can, with an appropriate choice of wavelet function, be viewed as a filter bank of lowpass differentiators with varying cut-off frequencies. Evaluating the performance of the methods based on either lowpass differentiation or wavelet analysis, the latter method has been found to produce T wave ends in better agreement with those produced by cardiologists [118].

7.6 Data Compression

Since a wide range of clinical examinations involve the recording of ECG signals, huge amounts of data are produced not only for immediate scrutiny, but also for storage in a patient database for future retrieval and review. It is well-known that the availability of one or several previous ECG recordings improves diagnostic accuracy of various cardiac disorders, including myocardial infarction. Today, such serial ECG comparison encompasses short-duration recordings acquired during rest, but may in the future encompass long signals, for example, acquired during stress testing or ambulatory monitoring. Although hard disk technology has undergone dramatic improvements in recent years, increased disk size is parallelled by the ever-increasing wish of physicians to store more information. In particular, the inclusion of additional ECG leads, the use of higher sampling rates and finer amplitude resolution, the inclusion of other, noncardiac signals such as blood pressure and respiration, and so on, lead to rapidly increasing demands on disk size. It is evident that efficient methods of data compression will be required for a long time to come.

Another important driving force behind the development of methods for data compression is the transmission of ECG signals across public telephone networks, cellular networks, intrahospital networks, and wireless communication systems ("telemetry"). Such data transmission may be initiated from an ambulance or a patient's home to the hospital and has, among other things, been found to be valuable for early diagnosis of an infarct. The transmission of uncompressed data is today too slow, making it incompatible with the real-time demand that often accompanies such ECG applications.

Any signal can be subjected to data compression as long as it contains a certain amount of redundancy—a fact that applies to virtually all signals of practical interest and to the ECG signal in particular. The notion of redundancy has already been touched upon when the samples of a signal were considered to be correlated; thus, decorrelation would represent a potential approach to designing a method for data compression. The overall goal is to represent a signal as accurately as possible using the fewest number of bits, by applying either *lossless* compression, in which the compressed/reconstructed signal is an exact replica of the original signal, or *lossy* compression, in which the reconstructed signal is allowed to differ from the original signal. With lossy compression, a certain amount of distortion has to be accepted in the reconstructed signal, although the distortion must remain small enough not to modify or jeopardize the diagnostic content of the ECG.

Data compression is today a well-established area of technology, resting on a solid theoretical basis, and has found its way into a wide range of applications, from voice communication to image/video processing. Instead of presenting the general theory on data compression, as has already been done in a number of textbooks [119–121], we will describe the major approaches to ECG data compression which, to various degrees, have been developed with reference to the specific characteristics of the ECG signal. Electrocardiographic data compression must take into account the fact that both small- and large-amplitude waveforms are present in the signal, carrying important diagnostic information, while the isoelectric line contains negligible information. Preferably, the design process should also take into account the fact that the signal contains recurrent heartbeats, often with similar morphology, and that the signal is, almost invariably, a multilead recording.

The outcome of data compression is critically dependent on the sampling rate and the number of bits used to represent each sample of the original signal. For example, a signal acquired at a low sampling rate contains less redundancy than one acquired at a high rate; as a result, the compression ratio, defined as the bit size of the original signal divided by the bit size of the compressed signal, is lower for a signal acquired at a lower sampling rate. Other factors that influence the outcome of data compression are the signal bandwidth, the number of leads, and the noise level. For example, a signal sampled at a rate of 500 Hz but bandlimited to 50 Hz is associated with a better compression ratio than is a signal bandlimited to the Nyquist frequency of 250 Hz. Consequently, it is imperative that any comparison of performance for different compression methods is based on identical values of the system parameters.

In the presentation below, methods for data compression are categorized according to the following three main types of data redundancy found in ECG recordings.

- *Intersample* or, equivalently, *intrabeat* redundancy is exploited by employing either direct or transform-based methods as described in Sections 7.6.2 and 7.6.3, respectively.

- *Interbeat* redundancy is manifested, within each lead, by successive, similar-looking heartbeats. As a result, their occurrence times must be determined by a QRS detector before interbeat redundancy can be exploited (Section 7.6.4).

- *Interlead* redundancy is due to the fundamental fact that a heartbeat is "viewed" concurrently in different leads. Therefore, waveforms exhibit varying degrees of interlead correlation which depend on the distance between electrodes on the body surface (Section 7.6.5).

It should be noted that many methods of data compression have been designed to solely deal with the first type of redundancy. However, methods

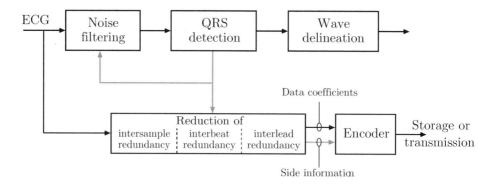

Figure 7.29: Data compression of ECG signals. The output of the block performing redundancy reduction is a sequence of data coefficients. The output may also include side information which, for example, describes the set of basis functions used for computing the data coefficients. The encoder translates the input into an efficiently coded bit stream.

which deal with all three types combined are becoming increasingly common. The block diagram in Figure 7.29 presents the two main steps in data compression. In the first step, the redundancy of the original signal is reduced so that a more compact signal representation is obtained. The output data is then fed to an encoder whose purpose is to produce an efficiently coded bit stream suitable for storage or transmission. Although this section is primarily focused on methods of redundancy reduction, data compression performance cannot be properly evaluated unless the performance of the encoder is also taken into account.

The above-mentioned measure, *compression ratio*, is frequently used to describe a method's performance. Unfortunately, this measure does not provide sufficient detail on the performance when lossy data compression is used since it does not reflect the distortion of the reconstructed signal; thus, an excellent compression ratio may be achieved at the expense of a severely distorted signal. A crucial aspect of ECG data compression is, therefore, to define complementary performance measures which reflect the accuracy with which the diagnostic information in the original ECG signal is preserved, see Section 7.6.7. It is essential to point out that the amount of distortion acceptable differs from application to application. For example, more distortion may be accepted if all the detailed signal analysis is done prior to data compression, while the reconstructed signal is only used for overall visual review [122].

Another important aspect of performance evaluation is the choice of ECG database. Since the performance of a method depends on the noise level, the evaluation should be based on data representative of the application in

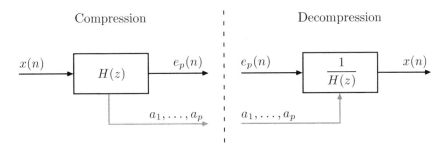

Figure 7.30: Block diagram of lossless data compression based on linear prediction. The output of the filter with transfer function $H(z)$ is the sequence of prediction errors $e_p(n)$. The predictor coefficients $a_1, \ldots, a_p$ may have to be included as side information for reconstruction.

question. The amount of ectopic beats and arrhythmias are other factors which, to various degrees, influence the outcome of an evaluation.

7.6.1 Lossless Compression

The use of lossless data compression seems very well-suited for applications where the demands on preserving diagnostic details are high. At an early stage, such compression was also applied to ECG signals, but proved to be relatively inefficient in achieving high compression ratios. Therefore, we will not delve into lossless compression techniques except to briefly mention linear prediction, which has been considered in the context of ECG signal processing [123–127]. With this technique, intersample redundancy is reduced by predicting the current sample $x(n)$ by a linear combination of the p previous samples,

$$\hat{x}_p(n) = -a_1 x(n-1) - \cdots - a_p x(n-p), \qquad (7.110)$$

so that the prediction error

$$e_p(n) = x(n) - \hat{x}_p(n) \qquad (7.111)$$

only needs to be considered. Since $e_p(n)$ has a magnitude which is typically much smaller than that of $x(n)$, fewer bits are required for its representation. Hence, the compressed data is represented by the sequence of prediction errors $e_p(n)$, possibly in combination with the predictor coefficients $a_1, \ldots, a_p$ as side information when these coefficients are not a priori known, see Figure 7.30.

In its simplest form, $x(n)$ may be predicted by the preceding sample $x(n-1)$ so that the predictor computes the first difference,

$$e_1(n) = x(n) - x(n-1), \qquad (7.112)$$

where $a_1 = -1$. Hence, the transfer function of the prediction error filter producing $e_1(n)$ is equal to $H(z) = 1 - z^{-1}$. Another simple structured predictor is the one which predicts $x(n)$ by extending the straight line defined by $x(n-1)$ and $x(n-2)$ so that

$$e_2(n) = x(n) - 2x(n-1) + x(n-2)$$
$$= e_1(n) - e_1(n-1), \tag{7.113}$$

where $(a_1, a_2) = (-2, 1)$; obviously, this predictor computes the second difference of the signal. Both these predictors have been applied to the compression of ECG data [123].

A systematic approach to determining the predictor coefficients is through minimization of the MSE

$$E\left[(x(n) - \hat{x}_p(n))^2\right],$$

assuming that $x(n)$ is a zero-mean stationary process characterized by its correlation function $r_x(k)$. The set of linear equations which yields the optimal MSE coefficients is well-known from AR-based spectral analysis and is presented in Section 3.4. Using, for example, the autocorrelation/covariance estimation method, the predictor coefficients can be determined by solving the normal equations given in (3.126).

The output signals of the two simple structured predictors, together with the output signal of a third-order MSE predictor, are displayed in Figure 7.31 for an ECG signal sampled at a rate of 200 Hz. While all three predictors remove the P and T waves, the third-order predictor is, in this particular example, markedly better at predicting the QRS complexes. Predictors with orders higher than three have been found to offer only minor improvements in performance when the sampling rate is 200 Hz [124].

The original signal $x(n)$ can be reconstructed from the prediction errors $e_p(n)$ through inverse filtering with $1/H(z)$ (Figure 7.30). To make reconstruction successful, however, the zeros of $H(z)$ must be located inside or, possibly, on the unit circle; if not, the inverse filter becomes unstable since its poles are located outside the unit circle. Of the three methods described in Section 3.4, Burg's method is preferred since it always produces a function $H(z)$ whose zeros are inside the unit circle.

Although linear prediction represents a lossless compression technique, the end result may nonetheless be lossy since $e_p(n)$ has to be rounded off to a number of bits which, in general, is lower than the number of bits used for the internal computation of $H(z)$. This is often the situation for MSE predictors since, in contrast to the two simple structured predictors above, their coefficients are almost invariably nonintegers.

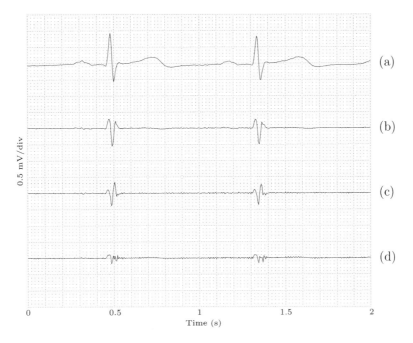

Figure 7.31: Reducing intersample redundancy with linear prediction. (a) The ECG signal, sampled at a rate of 200 Hz, was processed by (b) the first difference predictor, (c) the second difference predictor, and (d) a third-order predictor whose coefficients $(a_1, a_2, a_3) = (-2.15, 1.86, -0.63)$ were chosen so as to minimize the MSE.

7.6.2 Lossy Compression—Direct Methods

Direct methods operate in the time domain by extracting a set of K "significant" samples $x(n_k)$ from the original signal $x(n)$ such that

$$(n, x(n)), n = 0, \ldots, N-1 \quad \rightarrow \quad (n_k, x(n_k)), k = 0, \ldots, K-1,$$

where $K < N$. The resulting subset of K samples is retained for data compression, while the other samples are discarded. Reconstruction of the samples between the significant samples is achieved by interpolation using the following general expression:

$$\tilde{x}(n) = \begin{cases} x(n), & n = n_0, \ldots, n_{K-1}; \\ f_{n_0, n_1}(n), & n = n_0 + 1, \ldots, n_1 - 1; \\ \vdots & \vdots \\ f_{n_{K-2}, n_{K-1}}(n), & n = n_{K-2} + 1, \ldots, n_{K-1} - 1. \end{cases} \tag{7.114}$$

The first and last significant samples of the signal $x(n)$ are usually chosen to be $n_0 = 0$ and $n_{K-1} = N - 1$, respectively. The interpolating function

$f_{n_{k-1},n_k}(n)$ usually has a polynomial form of low order, approximating the signal with zero- or first-order polynomials, i.e., by a sequence of plateaus or straight lines. First-order (linear) interpolation has become especially popular since the signal can be completely reconstructed from the set of significant samples $x(n_k)$. The reconstructed signal has by some been labeled a "polygon," and, therefore, the significant samples are sometimes referred to as "vertices". Although more advanced interpolating functions can be used, e.g., rational or trigonometric functions, additional parameters need to be stored as side information to reconstruct the signal. As a result, improvements in performance may still be lost due to the additional cost of representing the interpolating function.

The selection of significant samples can be viewed as an "intelligent" sub-sampling of the signal in which the isoelectric segments are approximated by a small number of samples, whereas the QRS complex is much more densely sampled so that the essential information contained in the ECG is preserved. A simplistic approach would be to select the significant samples from among the turning points of the signal, i.e., its peaks and valleys; however, the error between the original and reconstructed signal may at times be quite considerable. Therefore, the selection of significant samples is usually based on a criterion assuring that the reconstruction error remains within a certain tolerance. The selection process can be performed sequentially so that the next significant sample is selected with reference to the properties of preceding signal properties. Alternatively, a larger block of samples can be processed at the same time so that significant samples are selected with reference to the enclosing signal properties. While the block-based approach can be expected to yield better performance, it is less suitable for real-time processing.

The performance of direct methods is particularly influenced by the noise level of the ECG, since the number of significant samples required to meet the maximal error tolerance increases as the noise level increases. Accordingly, poorer compression ratios are achieved at high noise levels. While direct methods work satisfactorily when processing ECGs acquired during resting conditions, the very idea of selecting significant samples can be questioned in noisy recordings.

In the following, we will describe two methods, called AZTEC and SAPA, which belong to the family of direct methods. In addition to these methods, which have both been extensively studied, several other direct methods have been presented in the literature [128].

AZTEC.   For many years, the *amplitude zone time epoch coding* (AZTEC) method represented a popular approach to ECG data compression due to its very modest computational requirements [129, 130]. With AZTEC, the original signal is converted into a sequence of plateaus and slopes which, following reconstruction, may be useful in certain types of automated ECG analysis, but hardly for diagnostic interpretation.

The definition of a plateau is based on two sequences, $x_{\min}(n)$ and $x_{\max}(n)$, which describe the extreme values of the signal from the starting time n_{k-1} and onwards:

$$x_{\min}(n) = \min\{x(n_{k-1}), x(n_{k-1}+1), \ldots, x(n)\}, \qquad (7.115)$$
$$x_{\max}(n) = \max\{x(n_{k-1}), x(n_{k-1}+1), \ldots, x(n)\}. \qquad (7.116)$$

A plateau extends in time as long as the difference between the maximum and minimum values does not exceed a certain preset error tolerance ε,

$$x_{\max}(n) - x_{\min}(n) \leq \varepsilon. \qquad (7.117)$$

The last sample n for which (7.117) holds true is denoted n_k. Rather than retaining $x(n_k)$ for storage or transmission, it is replaced by the average of the maximum and minimum values, considered to be representative of the plateau's amplitude. Since the plateau representation is inadequate for waveforms with steep slopes, AZTEC also includes a procedure for retaining slopes whenever $(n_k - n_{k-1})$ is less than a certain distance. In such cases, the sample $x(n_k)$ is retained in place of the two-sample average. The reconstructed signal $\tilde{x}(n)$ results from expanding the series of plateaus and slopes by

$$\tilde{x}(n) = \begin{cases} \dfrac{x_{\min}(n_k) + x_{\max}(n_k)}{2}, & \text{for a plateau;} \\ x(n_{k-1}) + (n - n_{k-1}) \cdot \dfrac{x(n_k) - x(n_{k-1})}{n_k - n_{k-1}}, & \text{for a slope,} \end{cases}$$
$$(7.118)$$

where $n = n_{k-1}, \ldots, n_k$, and $k = 0, \ldots, K-1$.

Figure 7.32 illustrates the performance of AZTEC for different values of the error tolerance ε. For large values, it is evident that the reconstructed signal exhibits a disturbing, nonphysiological staircase appearance. Although the discontinuities can be smoothed with lowpass filtering, amplitude distortion is inevitable and leads to the wave amplitudes being underestimated.

A number of modifications of the AZTEC method have been suggested in order to improve signal reconstruction. For example, the tolerance ε can be made time-dependent so that the isoelectric line is associated with a larger value of ε than are intervals with diagnostic information [131]. Another approach to dealing with the two types of intervals is to let AZTEC operate only on isoelectric segments while another algorithm is applied to intervals

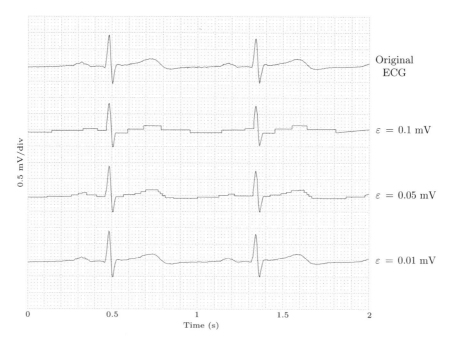

Figure 7.32: Reconstruction using the AZTEC method. The original ECG signal is reconstructed for three different values of the error tolerance ε. Slopes are stored if the interval $(n_k - n_{k-1})$ is less than 10 ms. Of the 500 samples in the original ECG, 42, 66, and 245 significant samples are retained for $\varepsilon = 0.1, 0.05$, and 0.01, respectively.

with high-frequency content. One such algorithm is the turning-point algorithm which compares pairs of samples, always retaining the second sample unless the first sample represents a turning point, i.e., a sample where the slope of the signal is changing [132], see also [133, 134]. Thus, the turning-point algorithm always downsamples the signal by a factor of two while still preserving the peak amplitudes of the different waves. The combined data compression method is designed to produce plateaus whenever their lengths exceed a certain minimum length and, otherwise, turning points [135, 136].

SAPA. Another well-known method for data compression is the *scan-along polygonal approximation* (SAPA) [137]. The leading principle behind this method is that the signal is represented by consecutive straight lines, thus avoiding the plateau representation of AZTEC. Assuming that a significant sample (vertex) has been found at n_{k-1}, the next vertex at n_k is the sample that is furthest away from n_{k-1} for which the error between $x(n)$ and the straight line reconstruction $\tilde{x}(n)$ remains within the error tolerance ε,

$$|x(n) - \tilde{x}(n)| < \varepsilon, \quad n = n_{k-1}, \ldots, n_k. \tag{7.119}$$

The straight line $\tilde{x}(n)$ is defined by the two enclosing samples $x(n_{k-1})$ and $x(n_k)$,

$$\tilde{x}(n) = x(n_{k-1}) + (n - n_{k-1}) \cdot \frac{x(n_k) - x(n_{k-1})}{n_k - n_{k-1}}, \quad n = n_{k-1}, \ldots, n_k. \tag{7.120}$$

Linear interpolation is then repeated by starting at vertex n_k and continuing until $x(n)$ is completely processed. The result is a sequence of vertices $x(n_0), \ldots, x(n_{K-1})$ for which the value of K depends on the properties of the analyzed signal.

Determination of the next vertex at n_k is facilitated by the introduction of a "slope" function $g(n, \varepsilon)$ which involves ε,

$$g(n, \varepsilon) = \frac{x(n) + \varepsilon - x(n_{k-1})}{n - n_{k-1}}, \quad n = n_{k-1}, \ldots. \tag{7.121}$$

The next vertex may be found as soon as the maximum value of all slopes at the lower tolerance $-\varepsilon$ exceeds the minimum value of all slopes at the upper tolerance ε,

$$\max_m g(m, -\varepsilon) > \min_m g(m, \varepsilon), \tag{7.122}$$

where the search for extreme values is assumed to start at $m = n_{k-1}$. Unfortunately, the test in (7.122) produces a reconstruction error whose magnitude may be as large as 2ε; therefore, the test is replaced by two other slope tests which require that the slope $g(n, 0)$ of the straight line between $x(n_{k-1})$ and $x(n)$ either falls below the maximum value of all slopes at $-\varepsilon$,

$$g(n, 0) < \max_m g(m, -\varepsilon), \tag{7.123}$$

or exceeds the minimum value of all slopes at ε,

$$g(n, 0) > \min_m g(m, \varepsilon). \tag{7.124}$$

The newly found vertex is defined by the sample that immediately precedes the sample violating any of these two tests. The above procedure is then repeated by starting from n_k to find the next vertex, and so on, see Figure 7.33. The performance of the SAPA method is illustrated in Figure 7.34 for three different values of the error tolerance ε. It can be observed from

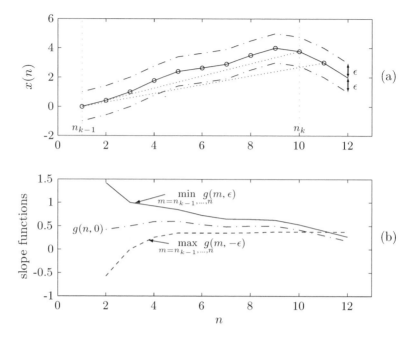

Figure 7.33: Determination of a vertex using the SAPA method. (a) With an existing vertex at $n_{k-1} = 1$, the next vertex occurs at $n = 10$ since the straight line from n_{k-1} to $n = 11$ is partially outside the corridor defined by the error tolerance ε. All straight lines up to $n = 10$ are within the tolerance. (b) In this example, the test in (7.123) determines the next vertex since $g(n, 0)$ is smaller than the maximum value of all slopes at $-\varepsilon$ at $n = 11$.

this example that the Q wave disappears in the reconstructed ECG when ε becomes too large.

The SAPA method assures that the magnitude of the reconstruction error in (7.119) is always less than ε. However, it may occasionally fail to find the very last sample which remains within the error tolerance and, therefore, a search terminating too early may lead to less efficient compression. This minor deficiency can be addressed by including additional tests on the slope [137].[7] Improved compliance with the error tolerance does not, however, imply better reconstruction of the signal; in fact, the additional tests were found to smooth out small Q waves, and, as a result, these tests have never been considered.

---

[7]The SAPA method with additional tests on the slope is originally named SAPA–3, while the method involving the tests in (7.123)–(7.124) is named SAPA–2. The SAPA–1 method is defined by the single test in (7.122).

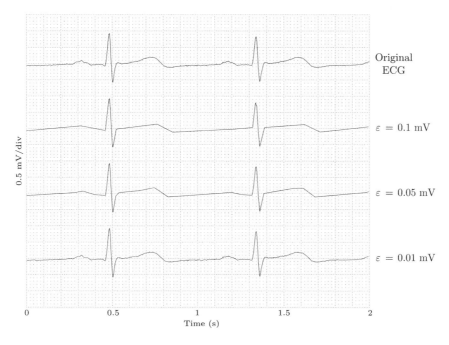

Figure 7.34: Reconstruction using the SAPA method. The original ECG signal is reconstructed for different error tolerances ε. Of the 500 samples in the original ECG, 21, 33, and 187 significant samples are retained for $\varepsilon = 0.1, 0.05$, and 0.01, respectively.

Similar to SAPA, the *fan method* draws straight lines between significant samples selected so that the signal reconstruction is always within a certain error tolerance. This method was presented already in the 1960s for use in telemetry of ECG data [138] and was later claimed to be identical to the SAPA method described above [139]. The performance of the fan method has been studied in considerable detail [140, 141], as has the efficient implementation of the method in a microprocessor for real-time operation [142–144]. Other methods closely related to the SAPA/fan method have also been presented [145–147].

A salient property of the SAPA/fan method is its sequential selection of vertices, implying that the resulting vertices are not necessarily associated with the minimal reconstruction error. By instead employing a block-based optimization criterion in which, for example, one cardiac cycle is processed at a time, it is possible to find the vertices that produce the minimal reconstruction error, for example, in the root mean-square (RMS) error sense [148–150]. It can be shown that the solution to this optimization problem is identical to finding the shortest path from one point to another in a graph, subject to the constraint that the path can only intersect a certain number of vertices; the details of this algorithm can be found in [148]. At a fixed compression ratio, the block-based optimization approach produces a lower reconstruction error than does the SAPA/fan method—a performance improvement which is accompanied by increased processing time.

7.6.3 Lossy Compression—Transform-based Methods

Transform-based data compression assumes that a compact signal representation exists in terms of the coefficients of a truncated orthonormal expansion. The idea of representing a signal $\mathbf{x} = \begin{bmatrix} x(0) & x(1) & \cdots & x(N-1) \end{bmatrix}^T$ by a set of orthonormal basis functions $\boldsymbol{\varphi}_k$ is already familiar to us since it was used for single-trial analysis of EPs in Section 4.5: an estimate of the signal is obtained from truncation of the complete series expansion so that only K out of the N terms are included. In the context of data compression, the transform-based approach is closely related since the coefficients $w_1, \ldots, w_K$ are retained for storage or transmission, hopefully providing adequate signal reconstruction, while the remaining $(N - K)$ coefficients, being near zero, are discarded. The coefficients w_k are obtained by correlating $\mathbf{x}$ with each of the basis functions, i.e., the inner product $w_k = \boldsymbol{\varphi}_k^T \mathbf{x}$, cf. (4.195). Hence, the subset of K coefficients constitutes the information to be compressed and from which the signal is later reconstructed. If the basis functions are a priori unknown, the set of coefficients must be supplemented with the samples of the required basis functions. Figure 7.35 illustrates the property of a transform domain, i.e., the domain defined by $\boldsymbol{\varphi}_k$, which offers a more compact representation than that of the time domain samples, packing the energy into a few coefficients w_k. Following data compression, the reconstructed signal $\tilde{\mathbf{x}}_K$ is obtained from

$$\tilde{\mathbf{x}}_K = \sum_{k=1}^{K} w_k \boldsymbol{\varphi}_k. \tag{7.125}$$

In contrast to most direct methods, transform-based methods require that the ECG first be *partitioned* into a series of successive blocks, where each block is subsequently subjected to data compression. The signal may be partitioned so that each block contains one heartbeat, and, therefore, QRS detection must always precede such compression methods. Each block is positioned around the QRS complex, starting at a fixed distance before the QRS which includes the P wave and extending beyond the end of the T wave to the beginning of the next beat. Since the heart rate is not constant, the distance by which the block extends after the QRS complex is adapted to the prevailing heart rate. Hence, the resulting blocks vary in length,

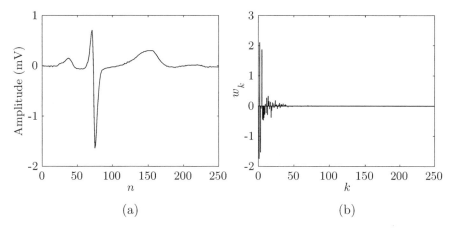

Figure 7.35: Transform-based data compression. (a) The original ECG and (b) the corresponding coefficients in the transform domain (here defined by Karhunen–Loève basis functions). In the transform domain, the signal energy is concentrated to a few coefficients with low index, implying that the signal is well-reconstructed by a much truncated series expansion of basis functions. The sample index n is here used, rather than time, to underline the fact that the transform coefficients are equal in number to the time domain samples.

introducing a potential problem in transform-based compression where a fixed block length is assumed. This problem may be solved by padding too short blocks with a suitable sample value, whereas too long blocks can be truncated to the desired length. It should be noted that partitioning of the ECG is bound to fail when certain chaotic rhythms are encountered, most notably ventricular fibrillation during which no QRS complexes are present.

A fixed number of basis functions are often considered for data compression, with the value of K being chosen from considerations concerning overall performance expressed in terms of compression ratio and reconstruction error. While serving as an important guideline to the choice of K, such an approach may occasionally produce an unacceptable representation of certain beat morphologies. Since the loss of morphologic detail causes incorrect interpretation of the ECG, the choice of K can be adapted for every beat to the properties of the reconstruction error $(\mathbf{x} - \tilde{\mathbf{x}}_K)$ [151]. For example, the value of K may be chosen such that the RMS value of the reconstruction error does not exceed the error tolerance ε or, more demanding, that none of the reconstruction errors of the entire block exceeds ε. It is evident that the value of K sometimes becomes much larger than the value suggested based on considerations on overall performance; however, it sometimes also becomes smaller. By letting K be variable, we can fully control the quality of the reconstructed signal, while also being forced to increase the amount

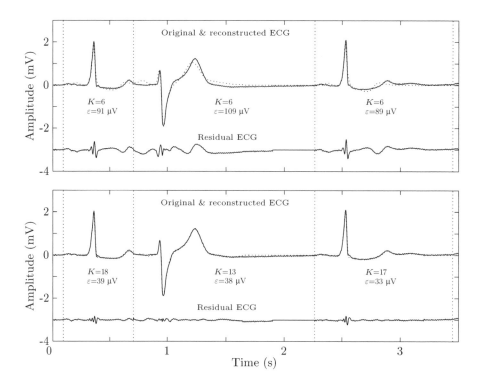

Figure 7.36: Quality control and data compression using a transform-based method. (a) A fixed number of basis functions ($K = 6$) are used for signal reconstruction. (b) The number of basis function is chosen so that the RMS error, denoted ε, between the original and reconstructed signal is always below 40 μV. The basis functions were defined by the KL transform.

of side information since we must keep track of the value of K for every data block. If the basis functions are a priori unknown, a larger number of basis functions must also be part of the side information. Figure 7.36 illustrates signal reconstruction for a fixed number of basis functions and a number determined by an RMS-based quality control criterion. In this example, the indicated error tolerance is attained by using different numbers of basis functions for each of the three displayed beats.

The Karhunen–Loève transform. The most important question to address is, of course, which set of basis functions to choose for data compression. We know from Section 4.5.3 that the Karhunen–Loève (KL) expansion is optimal in that it minimizes the MSE of approximation, and, therefore, the KL basis functions have become popular [122, 151–154]. Unlike the development of the KL expansion in Section 4.5.3, no distinction is made here between signal and noise, but our aim is instead to find the set of φ_k that makes $\tilde{\mathbf{x}}_K$ resemble $\mathbf{x}$ as closely as possible. Since this aim is identical to setting $\mathbf{x} = \mathbf{s}$, the derivation of the matrix equation in (4.235) holds without modification, and, thus, the basis functions are obtained as eigenvectors of the correlation matrix $\mathbf{R}_x$. The coefficient vector $\mathbf{w}$ of the *Karhunen–Loève transform* (KLT) of $\mathbf{x}$ is defined by

$$\mathbf{w} = \mathbf{\Phi}^T \mathbf{x}, \tag{7.126}$$

where the columns of $\mathbf{\Phi}$ contain the KL basis functions, and the coefficient vector $\mathbf{w}$ defines the transform domain.

The performance of the KLT may be described by the index $\mathcal{R}$, defined in (4.243), which reflects how well the original signal is approximated by the basis functions. While this index describes the performance on the chosen ensemble of data as an average, it does not provide information on the reconstruction error in individual beats. Therefore, it is appropriate to include a criterion for quality control when the number of basis function K is chosen, see Figure 7.36.

The calculation of $\mathbf{R}_x$ can be based on different types of data sets. The basis functions are labeled *universal*, when the data set originates from a large number of patients, or *subject-specific*, when the data originates from a single recording. While it is rarely necessary to store or transmit universal basis functions, subject-specific functions need to be part of the side information. Still, subject-specific basis functions offer superior energy concentration of the signal because these functions are better tailored to the data, provided that the ECG contains few beat morphologies. Figure 7.37 illustrates the latter observation by presenting the reconstructed signal for both types of basis functions. It can be seen in this figure that two subject-specific basis functions produce a much lower reconstruction error than do eight universal functions.

When calculating either type of basis function, it is important that the beats are well-aligned in time before estimating the correlation matrix $\mathbf{R}_x$. If they are not, the data ensemble exhibits artificial morphologic variability which makes the basis function representation less efficient. The occurrence times $\hat{\theta}_i$, produced by the QRS detector, can be further improved by aligning the ensemble of beats with respect to the QRS complex interval using, for example, Woody's method.

Noise reduction of the reconstructed signal is an interesting side effect when KL basis functions are employed for data compression—a property presaged by the results presented earlier in the context of single-trial EP analysis. This property is attributed to the fact that the most significant basis functions represent most of the signal energy, whereas noise is mostly represented by the basis functions excluded through truncation.

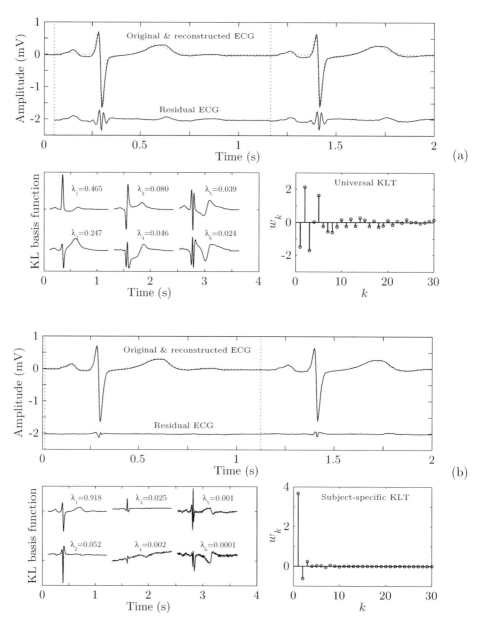

Figure 7.37: Transform-based data compression using the KL basis functions derived from either (a) a huge database including thousands of ECGs from different subjects or (b) subject-specific data. The basis functions φ_k and associated eigenvalues λ_k are presented, as are the 30 largest coefficients of the original ECG's KLT. The ECGs are reconstructed with $K = 8$ and 2 for universal and subject-specific basis functions, respectively.

A limitation of the KL basis functions comes to light when compressing ECGs with considerable changes in heart rate and, consequently, changes in the position of the T wave. Such ECG changes are observed during the course of a stress test. Since the basis functions account for the T wave occurrence at a fixed distance from the QRS complex, the basis functions become ill-suited for representing beats whose T waves occur earlier or later than this interval. As a result, additional basis functions are required to achieve the desired reconstruction error, thus leading to a reduction in performance.

Wavelets and wavelet packets. Although the KLT is optimal in the MSE sense, a certain amount of side information is required, especially for subject-specific basis functions, implying the possibility that other sets of basis functions may produce better performance. Moreover, the KLT is a signal-dependent transform, and, therefore, no algorithm exists which offers fast implementation.

The discrete wavelet transform (DWT), described in Section 4.7.2, has been found useful for compression of ECG signals since the information is concentrated into a fairly small number of coefficients [155–163]. Similar to the KLT approach, coefficients of large magnitude are retained first for compression because they convey most of the signal energy: cf. Section 4.7.6 which describes techniques for signal denoising. However, wavelet coefficients often exhibit a temporal relationship across scales which may be analyzed to facilitate the selection of coefficients.

The *discrete wavelet packet transform* (DWPT) represents a powerful generalization of the DWT [164]. While the DWT successively decomposes the scaling coefficients $c_j(k)$ which define the approximation signals at different scales, see Figure 4.44, the DWPT successively decomposes *both* the scaling coefficients $c_j(k)$ and the wavelet coefficients $d_j(k)$ which define the detail signals. As a result, the DWPT produces N coefficients at each scale, whereas the DWT produces a total of N coefficients. From the different scales of the wavelet packet decomposition, a total of K, out of N, coefficients are selected to represent the signal in the transform domain.

Like the DWT, the DWPT can be implemented by the filter bank with highpass and lowpass filters shown in Figure 4.43. However, decomposition of both scaling and wavelet coefficients means that the output of each branch of the filter bank is split into lowpass and highpass filters. Therefore, the dyadic tree structure of the DWT is replaced by a binary structure with one two-channel filter bank at the first stage, two two-channel filter banks at the second stage, four two-channel filter banks at the third stage, and so on.

The calculation of the DWPT is illustrated in Figure 7.38(a) for a signal of length $N = 8$; in this particular example, the eight transform coefficients

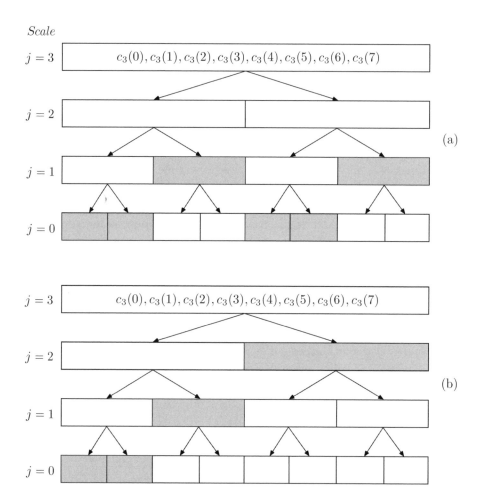

Figure 7.38: (a) The discrete wavelet packet transform (DWPT) and one particular choice of transform coefficients (indicated by shaded boxes); note that both scaling and wavelet coefficients are successively decomposed by this transform. (b) The DWPT and another choice of transform coefficients identical to those produced by the discrete wavelet transform. No truncation of coefficients was performed in these two examples.

were selected from scales $j = 1$ and 0. Figure 7.38(b) presents another signal decomposition for which the coefficients were selected exactly so that the DWT results. With the possibility of selecting coefficients in various ways, the DWPT offers a flexible signal representation which allows adaptation to individual signals. The selection of DWPT coefficients is often based on an information measure, like entropy, which concentrates as much information in as few coefficients as possible [165]. Another approach is to select the coefficients so that a distortion measure does not exceed a certain error tolerance [166], see below.

Interestingly, it has been shown that the DWPT is a good approximation of the KLT while, at the same time, circumventing the disadvantage of the KLT as a signal-dependent transform [164]. Rather than coding the total basis functions, as the KLT does, the DWPT only requires that the binary tree structure be coded.

Other transforms. It should be mentioned that many signal-independent transforms have, over the years, been considered for ECG data compression. These transforms include the discrete Fourier transform [125, 167, 168], the Walsh transform [169–172], the discrete cosine transform [173–175], the discrete Legendre transform [176], the Hermite transform [177], and the optimally warped transform [178]. While these transforms exhibit performances inferior to the KLT and DWPT transforms, some of them have the advantage of being efficiently calculated thanks to the existence of a fast algorithm such as the FFT implements the discrete Fourier transform.

7.6.4 Interbeat Redundancy

The above compression methods are designed to reduce intersample redundancy of the ECG, while not dealing with the fact that successive beats often have almost identical morphology. A simplistic approach to dealing with interbeat redundancy is to use the previous beat to predict the next beat. To proceed, we assume that the ith beat $x_i(n)$ starts at a fixed time Δ before $\hat{\theta}_i$ and lasts for N samples,

$$x_i(n) = x(n + \hat{\theta}'_i), \quad n = 0, \dots, N - 1, \tag{7.127}$$

where

$$\hat{\theta}'_i = \hat{\theta}_i - \Delta. \tag{7.128}$$

The distance Δ is chosen to be large enough to make sure that the onset of the beat, i.e., the P wave, is included. The error $e_i(n)$ when predicting $x_i(n)$

by $\hat{x}_i(n)$ is then given by

$$e_i(n) = x_i(n) - \hat{x}_i(n), \quad n = 0, \ldots, N-1, \tag{7.129}$$

where $\hat{x}_i(n)$ is defined by the previous beat [179],

$$\hat{x}_i(n) = x_{i-1}(n) = x(n + \hat{\theta}'_{i-1}), \quad n = 0, \ldots, N-1. \tag{7.130}$$

The prediction is initialized by

$$\hat{x}_1(n) = 0, \quad n = 0, \ldots, N-1, \tag{7.131}$$

which means that $e_1(n)$ is identical to the first beat $x_1(n)$. By repeating the prediction for all beats, a signal is produced whose magnitude is much smaller than the original one, thus requiring fewer bits for its representation; the resulting signal is commonly referred to as the *residual ECG*.

For a perfectly periodic heart rhythm, where all interval lengths are identical to N samples, it is easily realized that the original signal can be perfectly reconstructed by

$$x_i(n) = e_i(n) + \hat{x}_i(n), \tag{7.132}$$

from which we then obtain $\hat{x}_{i+1}(n) = x_i(n)$, and so on. Again, the initialization in (7.131) is used. Since heart rate always varies, care must exercised in situations when the end of the beat to be predicted overlaps with the onset of the next. If this happens, the prediction interval must end earlier, immediately preceding the onset of the next beat. When $\hat{x}_i(n)$ is too short to allow the prediction of $x_i(n)$, it may be padded with zeros so that the original samples at the end of the beat are retained for compression.

A drawback of the simple "previous-beat" predictor in (7.130) is its vulnerability to noise, a property which can be improved by instead using a predictor based on averaging of the J most recent beats [180–182],

$$\hat{x}_i(n) = \frac{1}{J} \sum_{j=1}^{J} x(n + \hat{\theta}'_{i-j}), \quad n = 0, \ldots, N-1. \tag{7.133}$$

When this particular predictor is employed, the prediction is often referred to as *average beat subtraction*.[8] Ensemble averaging can, of course, be replaced

---

[8] Average beat subtraction has been found useful in other applications as well, such as for the purpose of extracting the f waves of the ECG which reflect atrial activity during atrial fibrillation (Section 6.3.3). Due to the fact that atrial and ventricular activity are uncoupled, subtraction of the average QRST complex will produce a residual ECG which essentially contains only the fibrillatory f waves [183–185].

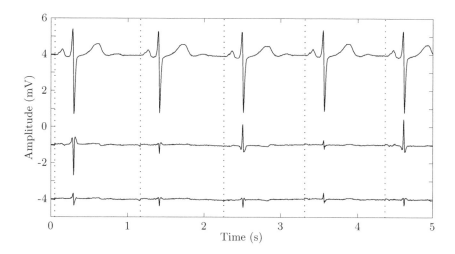

Figure 7.39: The influence of time alignment on the residual ECG. The original ECG (top), the residual ECG resulting from the use of θ_i produced by the QRS detector (middle), and θ_i improved by performing energy minimization between $x_i(n)$ and $\hat{x}_i(n)$ (bottom). The residuals are considerably smaller after time alignment.

by any of the techniques presented for noise reduction in Section 4.3, for example, by exponential averaging which allows faster adaptation to the most recent beats. Using average beat subtraction, interval length considerations similar to the above should be applied to achieve appropriate reconstruction. An extra long average beat, for example, of 2 s length, facilitates the prediction process, on condition that the average beat is only updated during the relevant interval of the beats $x_i(n)$ [180].

A fundamental assumption of the beat subtraction approach is that the beats used to compute $\hat{x}_i(n)$ exhibit similar morphology. To ensure this, it is necessary to first categorize the beats according to their respective morphology so that several average beats can be initialized [180, 186, 187], see also [188]. A straightforward approach to such beat categorization (clustering) would be to consider the energy of the prediction error $\sum_n e_i^2(n)$ in a beat interval: a new average beat is initialized if the energy exceeds a certain threshold, unless the current beat matches an already existing average beat category. Methods which cluster different beat morphologies are briefly considered in Section 8.5.

It is essential that the estimation of θ_i be accurate so that it relates to same fiducial point in all beats having similar morphology. It is usually necessary to improve the occurrence time of the QRS detector by optimally aligning the average beat with the beat to be predicted so that, for example, the energy of the prediction error is minimized. Figure 7.39 illustrates the importance of time alignment by displaying the residual ECG when no further alignment is done after QRS detection and when the error energy is minimized. With time alignment, it is obvious from Figure 7.39 that large prediction errors, costing several bits to represent, are considerably reduced. An issue intimately related to time alignment is the choice of sampling rate, since with too low a sampling rate, the residual ECG will contain large prediction errors [180].

The above predictors for reducing interbeat redundancy can be further generalized to a predictor which also accounts for intersample correlation— an approach suggested in [127] and later developed in detail in [189, 190]. In this approach, intersample correlation within the prediction interval is modeled by an AR model of order p_0 with regression parameters $a_1, \ldots, a_{p_0}$, whereas intersample correlation within the $(i - j)^{\text{th}}$ beat is modeled by an AR model of order p_j with parameters $a_{j,1}, \ldots, a_{j,p_j}$. Combining these, the long-term predictor is defined by [190]

$$\hat{x}_i(n) = \sum_{k=1}^{p_0} a_k x(n + \hat{\theta}'_i - k) +$$

$$\sum_{j=1}^{J} \sum_{k=0}^{p_j-1} a_{j,k} x(n + \hat{\theta}'_{i-j} - k), \quad n = 0, \ldots, N - 1. \qquad (7.134)$$

The parameters of the *long-term predictor* can be obtained by minimizing the prediction error, using techniques similar to those presented in Section 3.4 for AR power spectral analysis [190]; the details of the predictor parameter estimation problem are worked out in Problem 7.24.

It is interesting to observe that the previous-beat predictor, defined in (7.130), results from choosing the parameters values

$$p_0 = 0, \quad J = 1, \quad p_1 = 1, \quad a_{1,0} = 1, \quad a_{2,0} = \cdots = a_{J,0} = 0,$$

whereas the average beat predictor in (7.133) results from $p_0 = 0$ and the J most recent beats using

$$p_1 = \cdots = p_J = 1, \quad a_{1,0} = \cdots = a_{J,0} = 1/J.$$

Finally, we note that the above time domain techniques for reducing interbeat redundancies can be equally applied in the transform domain, as defined by the coefficients $w_1, \ldots, w_K$.

7.6.5   Interlead Redundancy

Since considerable correlation exists between different ECG leads, data compression of multilead ECGs would benefit from exploring interlead redundancy rather than just applying the previously described methods to one lead at a time. Direct methods for single-lead data compression have turned out to be not easily extended to multilead compression, although a few adaptations have been presented, for example, of the AZTEC method [191].

With transform-based methods, interlead correlation may be dealt with in two steps, namely,

1. a transformation which concentrates the signal energy spread over the available L leads into a few leads, followed by

2. compression of each transformed lead using a single-lead technique.

Since the first step is exactly what the KLT is designed to do, its original definition in (7.126) is modified to suit the case of interlead correlation. Defining the $L \times 1$ lead vector $\mathbf{x}(n)$ as

$$\mathbf{x}(n) = \begin{bmatrix} x_1(n) \\ x_2(n) \\ \vdots \\ x_L(n) \end{bmatrix}, \qquad (7.135)$$

an $L \times 1$ transformed lead vector $\mathbf{w}(n)$ is obtained by

$$\mathbf{w}(n) = \mathbf{\Phi}^T \mathbf{x}(n), \quad n = 0, \ldots, N-1, \qquad (7.136)$$

where the columns of the matrix $\mathbf{\Phi}$ are now defined by the eigenvectors of an $L \times L$ matrix $\mathbf{R_x}$ describing the correlation between leads. This matrix is estimated by

$$\hat{\mathbf{R}}_{\mathbf{x}} = \frac{1}{N_t} \sum_{n=0}^{N_t-1} \mathbf{x}(n)\mathbf{x}^T(n), \qquad (7.137)$$

where N_t denotes the total number of samples of several beats. Figure 7.40 illustrates the transformation in (7.136) when applied to the standard 12-lead ECG (recall that only 8 leads are unique for this lead system, whereas the remaining 4 are obtained as linear combinations). Using the samples of the displayed signal to estimate $\mathbf{R_x}$, the energy of the original leads is redistributed so that only three out of the eight transformed leads contain significant energy; the remaining five leads mostly account for noise. Since the KLT is orthonormal, we can easily reconstruct the original signal from the transformed leads using

$$\mathbf{x}(n) = \left(\mathbf{\Phi}^T \right)^{-1} \mathbf{w}(n) = \mathbf{\Phi}\mathbf{w}(n). \qquad (7.138)$$

Following concentration of the signal energy using (7.136), different approaches to data compression may be applied to the transformed leads, of

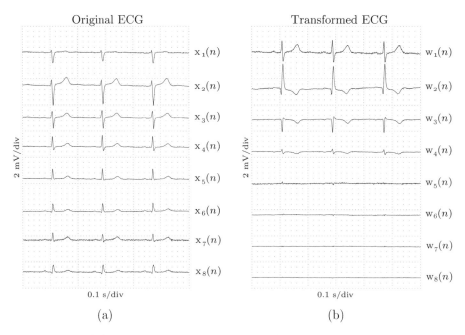

Figure 7.40: (a) The standard 12-lead ECG ($V_1, \ldots, V_6, I$, and II from top to bottom) and (b) its KL transform, obtained using (7.136), which concentrates the signal energy to only three of the leads.

which the simplest one is to only retain those leads whose energy exceeds a certain limit. Each retained lead is then compressed using any of the direct or transform-based methods described above. If a more faithful reconstruction of the ECG is required, leads with less energy can be retained, although they will be subjected to more drastic compression than the other leads [192].

A unified approach, which jointly deals with intersample and interlead redundancy, is to pile up all the segmented leads $\mathbf{x}_{i,1}, \ldots, \mathbf{x}_{i,L}$ into a single $LN \times 1$ vector (Figure 7.41),

$$\mathbf{x}'_i = \begin{bmatrix} \mathbf{x}_{i,1} \\ \mathbf{x}_{i,2} \\ \vdots \\ \mathbf{x}_{i,L} \end{bmatrix}, \tag{7.139}$$

where $\mathbf{x}_{i,l}$ denotes an $N \times 1$ vector containing the ith beat of the lth lead. The vector $\mathbf{x}'_i$ is then subjected to compression by any of the transform-based methods described above [153, 193]. Applying the KLT, the piled vector approach provides a more efficient signal representation than does the two-step approach, although the calculation of basis functions through

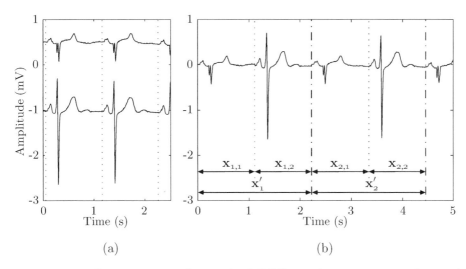

Figure 7.41: Concatenation of a two-lead ECG signal containing two beats. (a) Each beat of the two leads is concatenated ("piled up") into (b) one single vector.

diagonalization of the $LN \times LN$ correlation matrix is much more costly, in terms of computational measures, than for the $L \times L$ matrix in (7.137). Figure 7.42 shows the different processing blocks of transform-based data compression which together account for all three types of redundancy.

Finally, we note that the long-term predictor in (7.134) can be modified to also incorporate information from different leads [194]. This is done by augmenting the sum in (7.134) so that it not only accounts for interbeat correlation, but also for interlead correlation. Hence, the ith beat of the lth lead is predicted by

$$\hat{x}_{i,l}(n) = \sum_{k=1}^{p_0} a_{k,l} x_l(n-k) +$$

$$\sum_{q=1}^{L} \sum_{j=1}^{J} \sum_{k=0}^{p_j - 1} a_{j,k,q} x_q(n + \hat{\theta}_{i-j} - k), \quad n = 0, \dots, N-1, \quad (7.140)$$

where the interlead correlation properties are assumed to remain the same from beat to beat.

7.6.6 Quantization and Coding

The design of a system for ECG data compression must involve considerations on how to quantize and code the data resulting from redundancy

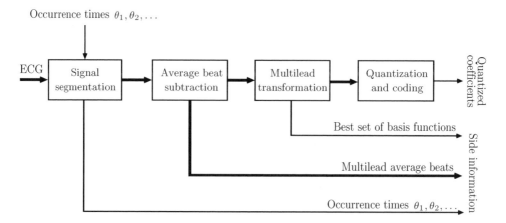

Figure 7.42: The processing steps of transform-based data compression where account is taken of all three types of redundancy, namely, intersample, interbeat, and interlead redundancy. Heavier arrows signify multilead processing of the data.

reduction. While it would seem natural to preserve the original bit resolution of the significant samples retained when a direct method is applied, it is not equally evident how many bits should be allocated to represent the truncated transform coefficients. It is appropriate to quantize the coefficients so that more bits are allocated to coefficients with large magnitude, being important for accurate reconstruction, than to coefficients with small magnitude. The process of mapping the transform coefficients into another alphabet ("codebook") is referred to as *quantization*.

Quantization is said to be scalar when the coefficients are quantized on an individual basis, using either a uniform or nonuniform grid [119]. The uniform quantizer is optimal for amplitudes which obey a uniform PDF and is often employed in practice because its design only requires determination of the quantization step size and the number of quantization levels; "optimality" here refers to finding the particular quantizer that minimizes the distortion in the MSE sense for a given number of quantization levels. When the PDF is nonuniform, the steps of the optimal quantizer are nonuniform and are defined by several parameters. While the distortion associated with nonuniform quantizers is less than that of uniform ones, the side information needed for coding a nonuniform codebook is larger—an effect which may be burdensome if the quantizer needs to be periodically updated to track nonstationary characteristics of the ECG signal.

Quantization can also be done on a vector basis in which several data samples are quantized at a time [120]. Vector quantization has theoretically been shown to produce a compression performance superior to scalar quantization and has, as a result, been studied for compression of ECG signals [191, 195–198]. However, if the coefficients are uncorrelated, as is the case with coefficients of the KLT, cf. (4.242), vector quantization does not improve the performance [120].

7.6.7 Performance Evaluation

The *compression ratio* $\mathcal{P}_{\mathrm{CR}}$ is a crucial measure when evaluating the performance of data compression methods and is defined as

$$\mathcal{P}_{\mathrm{CR}} = \frac{\#\text{bits to represent } x(n)}{\#\text{bits to represent } \tilde{x}(n)}. \tag{7.141}$$

Another measure is the *bit rate* $\mathcal{P}_{\mathrm{BR}}$, defined as the average number of bits required per second to represent the ECG, and is, in contrast to $\mathcal{P}_{\mathrm{CR}}$, independent of sampling rate and word length. However, the definitive compression performance can only be evaluated when any of these two measures are used in combination with another measure reflecting distortion of the reconstructed signal.

The *percentage root mean-square difference* (PRD) is a frequently employed distortion measure which quantifies the error between the original signal $x(n)$ and the reconstructed $\tilde{x}(n)$,

$$\mathcal{P}_{\mathrm{PRD}} = \sqrt{\frac{\displaystyle\sum_{n=0}^{N-1}(x(n) - \tilde{x}(n))^2}{\displaystyle\sum_{n=0}^{N-1} x^2(n)}} \cdot 100, \tag{7.142}$$

where it is assumed that the mean value of $x(n)$ has been subtracted prior to data compression. The measure $\mathcal{P}_{\mathrm{PRD}}$ has become popular because of its computational simplicity and the ease with which distortion can be compared from one signal to another. However, $\mathcal{P}_{\mathrm{PRD}}$ has certain flaws which make it unsuitable for performance evaluation. For example, less distortion would result from artificially adding baseline wander to the ECG since a lower $\mathcal{P}_{\mathrm{PRD}}$ would then result. Furthermore, compression of ECGs with large-amplitude QRS complexes results in less distortion than does compression of an ECG with small-amplitude QRS complexes, even if the squared error $(x(n) - \tilde{x}(n))^2$ is identical in both cases. These disadvantages can, to a certain degree, be mitigated by replacing the energy normalization in $\mathcal{P}_{\mathrm{PRD}}$ with a fixed normalization so that the modified measure, denoted $\mathcal{P}_{\mathrm{RMS}}$,

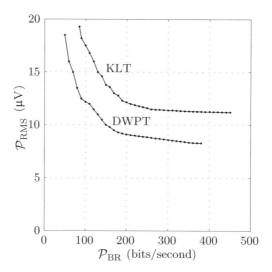

Figure 7.43: Rate distortion curves for data compression based on the KLT and the DWPT, where account is taken of the required side information when calculating $\mathcal{P}_{\mathrm{BR}}$. The results were obtained from 10 minutes of ECG data, selected from the MIT–BIH database (the diagram is adapted from [193]).

describes the error in absolute terms,

$$\mathcal{P}_{\mathrm{RMS}} = \sqrt{\frac{1}{N}\sum_{n=0}^{N-1}(x(n) - \tilde{x}(n))^2}. \tag{7.143}$$

Such a description is somewhat more suggestive of diagnostic ECG interpretation where criteria are expressed in terms of millivolt wave amplitudes rather than in percentages of signal energy.

Performance is often presented as a *rate distortion curve* where signal distortion measurements are displayed as a function of the bit rate $\mathcal{P}_{\mathrm{BR}}$. Such curves are shown in Figure 7.43 for two different transform-based compression methods, the KLT and DWPT, with $\mathcal{P}_{\mathrm{RMS}}$ as the chosen distortion measure. With this type of curve, the operating point of a compression method can be easily defined, specifying the bit rate at which acceptable distortion of the reconstructed signal is achieved.

By requiring the distortion to be low, e.g., a $\mathcal{P}_{\mathrm{PRD}}$ of only 1% or a $\mathcal{P}_{\mathrm{RMS}}$ of only 10 μV, it is tempting to believe that the diagnostic information in the reconstructed signal is preserved. However, both these distortion measures suffer from an inability to reflect loss of diagnostic information; instead, all samples are treated equally whether located in the QRS complex or in the uninformative isoelectric segment. While the loss of a tiny Q wave in the reconstructed signal essentially goes unreflected in $\mathcal{P}_{\mathrm{PRD}}$ or $\mathcal{P}_{\mathrm{RMS}}$, the absence of a Q wave represents an essential loss from a diagnostic point of view when, for example, diagnosing myocardial infarction.

The *weighted diagnostic distortion* (WDD) measure $\mathcal{P}_{\mathrm{WDD}}$ is one of the very few mathematically defined measures which addresses the limitations of distortion measures based on the error between samples of the original and reconstructed signal [199, 200]. The measure $\mathcal{P}_{\mathrm{WDD}}$ is composite since it involves various wave parameters essential to ECG interpretation, especially wave amplitudes and durations of the PQRST complex. Assuming that measurements of the kth ECG parameter have been obtained from the original and reconstructed signals, denoted β_k and $\tilde{\beta}_k$, respectively, a normalized error $\Delta\beta_k$ can be defined,

$$\Delta\beta_k = \frac{|\beta_k - \tilde{\beta}_k|}{\max(|\beta_k|, |\tilde{\beta}_k|)}, \qquad (7.144)$$

which is constrained to the interval $0 < \Delta\beta_k \leq 1$; it is tacitly assumed that any meaningful ECG measurement has a nonzero value. When several beats are available for measurement, the resulting values of $\Delta\beta_k$ are averaged before further processing is done.

For a set of P different parameters on amplitude and duration, the WDD measure is defined as [199]

$$\mathcal{P}_{\mathrm{WDD}} = \frac{\displaystyle\sum_{k=1}^{P} \alpha_k (\Delta\beta_k)^2}{\displaystyle\sum_{k=1}^{P} \alpha_k} \cdot 100, \qquad (7.145)$$

where the coefficients α_k make it possible to weight the parameter measurement errors $\Delta\beta_k$ in relation to their overall significance. Such weighting can be used to emphasize measurements of particular significance, such as ST segment measurements in ischemia monitoring. When all measurements are considered to be equally significant, we have $\alpha_1 = \cdots = \alpha_P = 1$.

A prerequisite for making use of $\mathcal{P}_{\mathrm{WDD}}$ is the availability of an algorithm that computes the desired set of diagnostic measurements. Once available, the algorithm must produce accurate and reproducible measurements so as to avoid $\mathcal{P}_{\mathrm{WDD}}$ reflecting a poorly performing measurement algorithm rather than distortion of the reconstructed signal.

We conclude this section by presenting two examples which illustrate the fundamental differences between $\mathcal{P}_{\mathrm{PRD}}$ and $\mathcal{P}_{\mathrm{WDD}}$ in characterizing signal distortion, see Figure 7.44. In this example, $\mathcal{P}_{\mathrm{WDD}}$ includes measurements

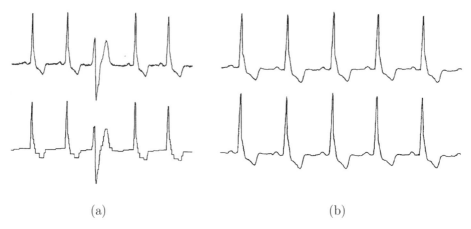

(a)   (b)

Figure 7.44: Distortion characterized by $\mathcal{P}_{\mathrm{PRD}}$ and $\mathcal{P}_{\mathrm{WDD}}$. (a) An example of poor-quality signal reconstruction where the original ECG (top) has been compressed by AZTEC (bottom); the corresponding values of $\mathcal{P}_{\mathrm{PRD}}$ and $\mathcal{P}_{\mathrm{WDD}}$ are 10.6 and 30.1%, respectively. (b) An example of good-quality signal reconstruction where the original ECG (top) has been compressed by an algorithm based on long-term prediction (bottom); the corresponding values of $\mathcal{P}_{\mathrm{PRD}}$ and $\mathcal{P}_{\mathrm{WDD}}$ are 15.4 and 3.5%, respectively. (Reprinted from Zigel et al. [199] with permission.)

on the RR interval, QT interval, P wave duration and amplitude, QRS duration and amplitudes, T wave amplitude, and ST segment slope [199]. Figure 7.44(a) presents a reconstructed signal which, following application of AZTEC, contains unacceptable distortion manifested by a staircase appearance. While $\mathcal{P}_{\mathrm{PRD}}$ is not alarmingly high for this signal, $\mathcal{P}_{\mathrm{WDD}}$ clearly indicates that the reconstructed signal is useless for diagnostic purposes. Figure 7.44(b) presents a reconstructed signal which closely resembles the original ECG, an observation well-reflected by the WDD measure which is equal to 3.5%; on the other hand, $\mathcal{P}_{\mathrm{PRD}}$ is as high as 15.4% and, therefore, incorrectly indicates that the reconstructed signal is severely distorted. In Figure 7.44(b), data compression was performed using a method based on long-term prediction and an error control algorithm ("analysis-by-synthesis") which compressed the signal so that $\mathcal{P}_{\mathrm{WDD}}$ never exceeded a certain limit [200].

Bibliography

[1] L. Taback, E. Marden, H. L. Mason, and H. V. Pipberger, "Digital recording of electrocardiographic data for analysis by a digital computer," *IRE Trans. Med. Electron.*, vol. 6, pp. 167–171, 1959.

[2] H. V. Pipberger, R. J. Arms, and F. W. Stallmann, "Automatic screening of normal and abnormal electrocardiograms by means of a digital electronic computer," *Proc. Soc. Exp. Biol. Med.*, vol. 106, pp. 130–132, 1961.

[3] C. A. Caceres, C. A. Steinberg, S. Abraham, W. J. Carbery, J. M. McBride, and W. E. Tollens, "Computer extraction of electrocardiographic parameters," *Circulation*, vol. 25, pp. 356–362, 1962.

[4] S. J. Hengevald and J. H. van Bemmel, "Computer detection of P waves," *Comput. Biomed. Res.*, vol. 9, pp. 125–132, 1976.

[5] F. Gritzali, G. Frangakis, and G. Papakonstantinou, "Detection of the P and T waves in an ECG," *Comput. Biomed. Res.*, vol. 22, pp. 83–91, 1989.

[6] D. A. Coast, R. M. Stern, G. G. Cano, and S. A. Briller, "An approach to cardiac arrhythmia analysis using hidden Markov models," *IEEE Trans. Biomed. Eng.*, vol. 37, pp. 826–836, 1990.

[7] N. V. Thakor and Z. Yi-Sheng, "Applications of adaptive filtering to ECG analysis: Noise cancellation and arrhythmia detection," *IEEE Trans. Biomed. Eng.*, vol. 38, pp. 785–794, 1991.

[8] B. R. S. Reddy, P. E. Elko, D. W. Christenson, and G. I. Rowlandson, "Detection of P waves in resting ECG: A preliminary study," in *Proc. Computers in Cardiology*, pp. 87–90, IEEE Computer Society Press, 1992.

[9] P. W. Macfarlane and T. D. W. Lawrie (eds.), *Comprehensive Electrocardiology. Theory and Practice in Health and Disease. (Vol. 2)*. New York: Pergamon Press, 1989.

[10] A. S. Berson, D. B. Geselowitz, and H. V. Pipberger, "Low-frequency requirements for electrocardiographic recording of ST segment," *Am. J. Cardiol.*, vol. 60, pp. 939–940, 1987.

[11] D. Tayler and R. Vincent, "Signal distortion in the electrocardiogram due to inadequate phase response," *IEEE Trans. Biomed. Eng.*, vol. 30, pp. 352–356, 1983.

[12] J. J. Bailey, A. S. Berson, A. Garson, L. G. Horan, P. W. Macfarlane, D. W. Mortara, and C. Zywietz, "Recommendations for the standardization and specifications in automated electrocardiogrpahy: Bandwidth and signal processing," *Circulation*, vol. 81, pp. 730–739, 1990.

[13] S. K. Mitra, *Digital Signal Processing: A Computer-based Approach*. New York: McGraw–Hill, 1998.

[14] J. G. Proakis and D. G. Manolakis, *Digital Signal Processing. Principles, Algorithms, and Applications*. New Jersey: Prentice-Hall, 3rd ed., 1996.

[15] R. L. Longini, J. P. Giolma, C. Wall, and R. F. Quick, "Filtering without phase shift," *IEEE Trans. Biomed. Eng.*, vol. 22, pp. 432–433, 1975.

[16] E. W. Pottala, J. J. Bailey, M. R. Horton, and J. R. Gradwohl, "Suppression of baseline wander in the ECG using a bilinearly transformed, null-phase filter," *J. Electrocardiol.*, vol. 22 (Suppl.), pp. 243–247, 1989.

[17] J. J. Kormylo and V. K. Jain, "Two-pass recursive digital filter with zero phase shift," *IEEE Trans. Acoust. Speech Sig. Proc.*, vol. 22, pp. 384–387, 1974.

[18] F. Gustafsson, "Determining the initial states in forward-backward filtering," *IEEE Trans. Signal Proc.*, vol. 44, pp. 988–992, 1996.

[19] J. A. van Alsté and T. S. Schilder, "Removal of base-line wander and power-line interference from the ECG by an efficient FIR filter with a reduced number of taps," *IEEE Trans. Biomed. Eng.*, vol. 32, no. 12, pp. 1052–1060, 1985.

[20] T. W. Parks and C. S. Burrus, *Digital Filter Design*. New York: Wiley–Interscience, 1987.

[21] J. P. Marques de Sá, "Digital FIR filtering for removal of ECG baseline wander," *J. Clin. Eng.*, vol. 7, pp. 235–239, 1982.

[22] J. A. van Alsté, W. van Eck, and O. E. Herrman, "ECG baseline wander reduction using linear phase filters," *Comput. Biomed. Res.*, vol. 19, pp. 417–427, 1986.

[23] L. Sörnmo, "Time-variable digital filtering of ECG baseline wander," *Med. Biol. Eng. & Comput.*, vol. 31, pp. 503–508, 1993.

[24] R. Jané, P. Laguna, N. V. Thakor, and P. Caminal, "Adaptive baseline wander removal in the ECG: Comparative analysis with cubic spline technique," in *Proc. Computers in Cardiology*, pp. 143–146, IEEE Computer Society Press, 1992.

[25] P. W. Macfarlane, J. Peden, G. Lennox, M. P. Watts, and T. D. V. Lawrie, "The Glasgow system," in *Trends in Computer-Processed Electrocardiograms* (J. H. van Bemmel and J. L. Willems, eds.), pp. 143–150, Amsterdam, North-Holland, 1977.

[26] C. R. Meyer and H. N. Keiser, "Electrocardiogram baseline noise estimation and removal using cubic splines and state-space computation techniques," *Comput. Biomed. Res.*, vol. 10, pp. 459–470, 1977.

[27] F. Bartoli, S. Cerutti, and E. Gatti, "Digital filtering and regression algorithms for an accurate detection of the baseline in ECG signals," *Med. Inform.*, vol. 8, pp. 71–82, 1983.

[28] H. T. Le, C. van Arsdel, A. M. Macowski, E. W. Pottala, and J. J. Bailey, "Automated analysis of rodent three-channel electrocardiograms and vectorcardiograms," *IEEE Trans. Biomed. Eng.*, vol. 22, pp. 43–50, 1985.

[29] F. Badilini, A. J. Moss, and E. L. Titlebaum, "Cubic spline baseline estimation in ambulatory ECG recordings for the measurement of ST segment displacements," in *Proc. Conf. IEEE Eng. Med. Biol. Soc. (EMBS)*, pp. 584–585, IEEE, 1991.

[30] J. Froning, V. F. Froelicher, and M. D. Olson, "Application and limitations of continuous baseline estimation and removal using cubic-spline technique during exercise ECG testing," in *Proc. Computers in Cardiology*, pp. 537–540, IEEE Computer Society Press, 1987.

[31] J. Froning, M. D. Olson, and V. F. Froelicher, "Problems and limitations of ECG baseline estimation and removal using cubic spline technique during exercise ECG testing: Recommendations for proper implementation," *J. Electrocardiol.*, vol. 21 (Suppl.), pp. 149–157, 1988.

[32] M. Raifel and S. Ron, "Estimation of slowly changing components of physiological signals," *IEEE Trans. Biomed. Eng.*, vol. 44, pp. 215–220, 1997.

[33] J. C. Huhta and J. G. Webster, "60-Hz interference in electrocardiography," *IEEE Trans. Biomed. Eng.*, vol. 43, pp. 91–101, 1973.

[34] E. Cramer, C. D. McManus, and D. Neubert, "Estimation and removal of power line interference in the electrocardiogram: A comparison of digital approaches," *Comput. Biomed. Res.*, vol. 20, pp. 12–28, 1987.

[35] C. D. McManus, D. Neubert, and E. Cramer, "Characterization and elimination of AC noise in the electrocardiogram: A comparison of digital filtering methods," *Comput. Biomed. Res.*, vol. 26, pp. 48–67, 1993.

[36] P. S. Hamilton, "A comparison of adaptive and nonadaptive filters for the reduction of powerline interference in the ECG," *IEEE Trans. Biomed. Eng.*, vol. 43, pp. 105–109, 1996.

[37] D. Mortara, "Digital filters for ECG signals," in *Proc. Computers in Cardiology*, pp. 511–514, IEEE Computer Society Press, 1977.

[38] M. L. Ahlstrom and W. J. Tompkins, "Digital filters for real-time ECG signal processing using microprocessors," *IEEE Trans. Biomed. Eng.*, vol. 32, pp. 708–713, 1985.

[39] S.-C. Pei and C.-C. Tseng, "Elimination of AC interference in electrocardiogram using IIR notch filter with transient suppression," *IEEE Trans. Biomed. Eng.*, vol. 42, pp. 1128–1132, 1995.

[40] J. R. Glover, Jr., "Comments on digital filters for real-time ECG processing using microprocessors," *IEEE Trans. Biomed. Eng.*, vol. 34, pp. 962–963, 1987.

[41] H. J. Scheer, "Line frequency rejection for biomedical application," *IEEE Trans. Biomed. Eng.*, vol. 37, pp. 68–69, 1987.

[42] Y. Z. Ider and H. Köymen, "A new technique for line interference monitoring and reduction in biopotential amplifiers," *IEEE Trans. Biomed. Eng.*, vol. 37, pp. 624–631, 1990.

[43] Y. Z. Ider, M. C. Şaki, and H. A. Güçer, "Removal of power line interference monitoring in signal-averaged electrocardiography systems," *IEEE Trans. Biomed. Eng.*, vol. 42, pp. 731–735, 1995.

[44] C. Levkov, G. Michov, R. Ivanov, and I. K. Daskalov, "Subtraction of 50 Hz interference from the electrocardiogram," *Med. Biol. Eng. & Comput.*, vol. 22, pp. 371–373, 1984.

[45] I. A. Dotsinsky and I. K. Daskalov, "Accuracy of 50 Hz interference subtraction from an electrocardiogram," *Med. Biol. Eng. & Comput.*, vol. 34, pp. 489–494, 1996.

[46] A. V. Oppenheim, R. W. Schafer, and J. R. Buck, *Discrete-Time Signal Processing*. New Jersey: Prentice-Hall, 2nd ed., 1999.

[47] B. Widrow and S. D. Stearns, *Adaptive Signal Proccessing*. New Jersey: Prentice-Hall, 1985.

[48] J. R. Glover, Jr., "Adaptive noise cancelling applied to sinusoidal interferences," *IEEE Acoust. Speech Signal Proc.*, vol. 25, pp. 484–491, 1977.

[49] M. Ferdjallah and R. Barr, "Adaptive digital notch filter design on the unit circle for the removal of powerline noise from biomedical signals," *IEEE Trans. Biomed. Eng.*, vol. 41, pp. 529–536, 1994.

[50] J. L. Talmon, J. A. Kors, and J. H. van Bemmel, "Adaptive Gaussian filtering in routine ECG/VCG analysis," *IEEE Trans. Acoust. Speech Sig. Proc.*, vol. 34, pp. 527–534, 1986.

[51] E. K. Hodson, D. R. Thayer, and C. Franklin, "Adaptive Gaussian filtering and local frequency estimates using local curvature analysis," *IEEE Trans. Acoust. Speech Sig. Proc.*, vol. 29, pp. 854–859, 1981.

[52] V. de Pinto, "Filters for the reduction of baseline wander and muscle artifact in the ECG," *J. Electrocardiol.*, vol. 25 (Suppl.), pp. 40–48, 1991.

[53] D. W. Mortara, "Source consistency filtering," *J. Electrocardiol.*, vol. 25, pp. 200–206, 1993.

[54] W. Philips, "Adaptive noise removal from biomedical signals using warped polynomials," *IEEE Trans. Biomed. Eng.*, vol. 43, pp. 480–492, 1996.

[55] V. X. Afonso, W. J. Tompkins, T. Q. Nguyen, K. Michler, and L. Shen, "Comparing stress ECG enhancement algorithms," *IEEE Eng. Med. Biol. Mag.*, vol. 15, pp. 37–44, 1996.

[56] W. Kaiser and M. Findeis, "Novel signal processing methods for exercise ECG," *Int. J. Bioelectromagn.*, vol. 2, 2000. (http://www.tut.fi/ijbem/).

[57] J. S. Paul, M. R. Reddy, and V. J. Kumar, "A transform domain SVD filter for suppression of muscle noise artefacts in exercise ECG's," *IEEE Trans. Biomed. Eng.*, vol. 47, pp. 654–663, 2000.

[58] O. Pahlm and L. Sörnmo, "Software QRS detection in ambulatory monitoring—a review," *Med. Biol. Eng. & Comput.*, vol. 22, pp. 289–297, 1984.

[59] S. M. Kay, *Fundamentals of Statistical Signal Processing. Detection Theory.* New Jersey: Prentice-Hall, 1998.

[60] P. O. Börjesson, O. Pahlm, L. Sörnmo, and M. E. Nygårds, "Adaptive QRS detection based on maximum-a-posteriori estimation," *IEEE Trans. Biomed. Eng.*, vol. 29, pp. 341–351, 1982.

[61] S. Azevedo and R. L. Longini, "Abdominal-lead fetal electrocardiographic R-wave enhancement for heart rate determination," *IEEE Trans. Biomed. Eng.*, vol. 27, pp. 255–260, 1980.

[62] C. Li, C. Zheng, and C. Tai, "Detection of ECG characteristic points using the wavelet transform," *IEEE Trans. Biomed. Eng.*, vol. 42, pp. 21–28, 1995.

[63] J. S. Sahambi, S. N. Tandon, and R. K. P. Bhatt, "Using wavelet transforms for ECG characterization: An on-line digital signal processing system," *IEEE Eng. Med. Biol. Mag.*, vol. 16, pp. 77–83, 1997.

[64] S. Kadambe, R. Murray, and G. F. Boudreaux-Bartels, "Wavelet transform-based QRS complex detector," *IEEE Trans. Biomed. Eng.*, vol. 46, pp. 838–848, 1999.

[65] T. A. Pryor, R. Russel, A. Budkin, and W. G. Price, "Electrocardiographic interpretation by computer," *Comput. Biomed. Res.*, vol. 2, pp. 538–548, 1969.

[66] W. P. Holsinger, K. M. Kempner, and M. H. Miller, "A QRS preprocessor based on digital differentiation," *IEEE Trans. Biomed. Eng.*, vol. 18, pp. 212–217, 1971.

[67] G. Belforte, R. D. Mori, and F. Ferraris, "A contribution to the automatic processing of electrocardiograms using syntactic methods," *IEEE Trans. Biomed. Eng.*, vol. 26, pp. 125–136, 1979.

[68] S. Usui and I. Amidror, "Digital low-pass differentiation for biological signal processing," *IEEE Trans. Biomed. Eng.*, vol. 29, no. 10, pp. 686–693, 1982.

[69] P. Laguna, N. V. Thakor, P. Caminal, and R. Jané, "Low-pass differentiators for biological signals with known spectra: Application to ECG signal processing," *IEEE Trans. Biomed. Eng.*, vol. 37, pp. 420–424, 1990.

[70] L. Sörnmo, O. Pahlm, and M. E. Nygårds, "Adaptive QRS detection: A study of performance," *IEEE Trans. Biomed. Eng.*, vol. 32, pp. 392–401, 1985.

[71] M. E. Nygårds and J. Hulting, "An automated system for ECG monitoring," *Comput. Biomed. Res.*, vol. 12, pp. 181–202, 1979.

[72] T. Fancott and D. Wong, "A minicomputer system for direct high-speed analysis of cardiac arrhythmia in 24h ambulatory ECG tape recordings," *IEEE Trans. Biomed. Eng.*, vol. 27, pp. 685–693, 1980.

[73] G. M. Friesen, T. C. Jannett, M. A. Jadallah, S. L. Yates, S. R. Quint, and H. T. Nagle, "A comparison of the noise sensitivity of nine QRS detection algorithms," *IEEE Trans. Biomed. Eng.*, vol. 37, pp. 85–98, 1990.

[74] A. Ligtenberg and M. Kunt, "A robust-digital QRS detection algorithm for arrhythmia monitoring," *Comp. Biomed. Res.*, vol. 16, pp. 273–286, 1983.

[75] W. A. H. Engelse and C. Zeelenberg, "A single-scan algorithm for QRS detection and feature extraction," in *Proc. Computers in Cardiology*, pp. 37–42, IEEE Computer Society Press, 1979.

[76] N. V. Thakor, J. G. Webster, and W. J. Tompkins, "Estimation of QRS complex power spectrum for design of a QRS filter," *IEEE Trans. Biomed. Eng.*, vol. 31, no. 11, pp. 702–706, 1984.

[77] A. Cohen and D. Landsberg, "Adaptive real time wavelet detection," *IEEE Trans. Biomed. Eng.*, vol. 30, pp. 332–340, 1983.

[78] D. T. Kaplan, "Simultaneous QRS detection and feature extraction using simple matched filter basis functions," in *Proc. Computers in Cardiology*, pp. 503–506, IEEE Computer Society Press, 1990.

[79] V. X. Afonso, W. J. Tompkins, T. Q. Nguyen, and L. Shen, "ECG beat detection using filter banks," *IEEE Trans. Biomed. Eng.*, vol. 46, pp. 192–202, 1999.

[80] M. Åström, S. Olmos, and L. Sörnmo, "Wavelet-based detection in cardiac pacemakers," in *Proc. Conf. IEEE Eng. Med. Biol. Soc. (EMBS)*, pp. 2121–2124, IEEE, 2001.

[81] J. Pan and W. J. Tompkins, "A real-time QRS detection algorithm," *IEEE Trans. Biomed. Eng.*, vol. 32, pp. 230–236, 1985.

[82] M. Kunt, H. Rey, and A. Ligtenberg, "Preprocessing of electrocardiograms by digital techniques," *Signal Proc.*, vol. 4, pp. 215–222, 1982.

[83] I. S. N. Murthy and M. R. Rangaraj, "New concepts for PVC detection," *IEEE Trans. Biomed. Eng.*, vol. 26, pp. 409–416, 1979.

[84] J. G. Proakis, *Digital Communications*. New Jersey: Prentice-Hall, 4th ed., 2001.

[85] M. E. Nygårds and L. Sörnmo, "Delineation of the QRS complex using the envelope of the ECG," *Med. Biol. Eng. & Comput.*, vol. 21, pp. 538–547, 1983.

[86] R. A. Balda, G. Diller, E. Deardorff, J. Done, and P. Hsieh, "The HP ECG analysis program," in *Trends in Computer-processed electrocardiograms* (J. H. van Bemmel and J. L. Willems, eds.), pp. 49–56, Amsterdam: North-Holland, 1977.

[87] N. V. Thakor, J. G. Webster, and W. J. Tompkins, "Design, implementation and evaluation of a microcomputer-based portable arrhythmia monitor," *Med. Biol. Eng. & Comput.*, vol. 22, pp. 151–159, 1984.

[88] F. E. M. Brekelmans and C. D. R. de Vaal, "A QRS detection scheme for multichannel ECG analysis," in *Proc. Computers in Cardiology*, pp. 437–440, IEEE Computer Society Press, 1981.

[89] M. Okada, "A digital filter for the QRS complex detection," *IEEE Trans. Biomed. Eng.*, vol. 26, pp. 700–703, 1979.

[90] F. Gritzali, "Towards a generalized scheme for QRS detection in ECG waveforms," *Signal Proc.*, vol. 15, pp. 183–192, 1988.

[91]   S. Suppappola and Y. Sun, "Nonlinear transforms of the ECG signals for digital QRS detection: A quantitative analysis," *IEEE Trans. Biomed. Eng.*, vol. 41, pp. 397–400, 1994.

[92]   Q. Xue, Y. H. Hu, and W. J. Tompkins, "Neural-network-based adaptive matched filter for QRS detection," *IEEE Trans. Biomed. Eng.*, vol. 39, pp. 317–329, 1992.

[93]   P. S. Hamilton and W. J. Tompkins, "Quantitative investigation of QRS detection rules using the MIT/BIH arrhythmia database," *IEEE Trans. Biomed. Eng.*, vol. 33, pp. 1157–1165, 1986.

[94]   P. M. Shah, J. M. Arnold, N. A. Haberern, D. T. Bliss, K. M. McClelland, and W. B. Clarke, "Automatic real time arrhythmia monitoring in the intensive care unit," *Am. J. Cardiol.*, vol. 39, pp. 701–708, 1977.

[95]   R. Dillman, N. Judell, and S. Kuo, "Replacement of AZTEC by correlation for more accurate VPB detection," in *Proc. Computers in Cardiology*, pp. 29–32, IEEE Computer Society Press, 1978.

[96]   K. Akazawa, K. Motoda, A. Sasamori, T. Ishizawa, and E. Harasawa, "Adaptive threshold QRS detection algorithm for ambulatory ECG," in *Proc. Computers in Cardiology*, pp. 445–448, IEEE Computer Society Press, 1991.

[97]   O. Pahlm, P. O. Börjesson, K. Johansson, B. Jonson, K. Pettersson, L. Sörnmo, and O. Werner, "Efficient data compression and arrhythmia detection for long term ECG's," in *Proc. Computers in Cardiology*, pp. 395–396, IEEE Computer Society Press, 1978.

[98]   C. N. Mead, K. W. Clark, S. J. Potter, S. M. Moore, and L. J. Thomas, Jr., "Development and evaluation of a new QRS detector/delineator," in *Proc. Computers in Cardiology*, pp. 251–254, IEEE Computer Society Press, 1979.

[99]   K.-J. Falk, J.-E. Angelhed, and T. I. Bjurö, "Real-time processing of multiple-lead exercise electrocardiogram," *Med. Progr. Technol.*, vol. 8, pp. 159–174, 1982.

[100] P. Laguna, R. Jané, and P. Caminal, "Automatic detection of wave boundaries in multilead ECG signals: Validation with the CSE database," *Comput. Biomed. Res.*, vol. 27, pp. 45–60, 1994.

[101] N. V. Thakor, J. G. Webster, and W. J. Tompkins, "Optimal QRS detector," *Med. Biol. Eng. & Comput.*, vol. 21, pp. 343–350, 1983.

[102] A. Algra, H. Le Brun, and C. Zeelenberg, "An algorithm for computer measurement of QT intervals in the 24 hour ECG," in *Proc. Computers in Cardiology*, pp. 117–119, IEEE Computer Society Press, 1987.

[103] I. K. Daskalov, I. A. Dotsinsky, and I. I. Christov, "Developments in ECG acquisition, preprocessing, parameter measurement, and recording," *IEEE Eng. Med. Biol. Mag.*, vol. 17, pp. 50–58, 1998.

[104] I. K. Daskalov and I. I. Christov, "Electrocardiogram signal preprocessing for automatic detection of QRS boundaries," *Med. Eng. & Physics*, vol. 21, pp. 37–44, 1999.

[105] G. Speranza, G. Nollo, F. Ravelli, and R. Antolini, "Beat-to-beat measurement and analysis of the R-T interval in 24 hour ECG Holter recordings," *Med. Biol. Eng. & Comput.*, vol. 31, pp. 487–494, September 1993.

[106] J. L. Willems, P. Arnaud, J. H. van Bemmel, P. J. Bourdillon, R. Degani, B. Denis, I. Graham, F. M. A. Harms, P. W. Macfarlane, G. Mazzocca, J. Meyer, and C. Zywietz, "A reference data base for multi-lead electrocardiographic computer measurement programs," *J. Am. Coll. Cardiol.*, vol. 10, pp. 1313–1321, 1987.

[107] P. Laguna, R. G. Mark, A. L. Goldberger, and G. B. Moody, "A database for evaluation of algorithms for measurement of QT and other waveform intervals in the ECG," in *Proc. Computers in Cardiology*, pp. 673–676, IEEE Press, 1997.

[108] J. L. Willems, P. Arnaud, J. H. van Bemmel, and et al., "Assessment of the performance of electrocardiographic computer program with the use of a reference data base," *Circulation*, vol. 71, pp. 523–534, 1985.

[109] R. Jané, A. Blasi, J. García, and P. Laguna, "Evaluation of an automatic detector of waveform limits in Holter ECGs with the QT database," in *Proc. Computers in Cardiology*, pp. 295–298, IEEE Press, 1997.

[110] The CSE Working Party, "Recommendations for measurement standards in quantitative electrocardiography," *Eur. Heart J.*, vol. 6, pp. 815–825, 1985.

[111] J. H. van Bemmel, C. Zywietz, and J. A. Kors, "Signal analysis for ecg interpretation," *Methods Inf. Med.*, vol. 29, pp. 317–329, 1990.

[112] N. B. McLaughlin, R. W. Campbell, and A. Murray, "Comparison of automatic QT measurement techniques in the normal 12 lead electrocardiogram," *Br. Heart J.*, vol. 74, pp. 84–89, 1995.

[113] Q. Xue and S. Reddy, "Algorithms for computerized QT analysis," *J. Electrocardiol.*, vol. 30, pp. 181–186, 1998.

[114] B. Acar, G. Yi, K. Hnatkova, and M. Malik, "Spatial, temporal and wavefront direction characteristics of 12-lead T-wave morphology," *Med. Biol. Eng. & Comput.*, vol. 37, pp. 574–584, 1999.

[115] J. A. Vila, G. Yi, J. M. R. Presedo, M. Fernandez-Delgado, S. Barro, and M. Malik, "A new approach for TU complex characterization," *IEEE Trans. Biomed. Eng.*, vol. 47, pp. 764–772, 2000.

[116] K. Lund, H. Nygaard, and P. A. Kirstein, "Weighing the QT intervals with the slope or the amplitude of the T wave," *Ann. Noninv. Electrocardiol.*, vol. 7, pp. 4–9, 2002.

[117] M. Bahoura, M. Hassani, and M. Hubin, "DSP implementation of wavelet transform for real time ECG waveforms detection and heart rate analysis," *Comput. Meth. Progr. Biomed.*, vol. 52, pp. 35–44, 1997.

[118] J. P. Martínez, R. Almeida, S. Olmos, A. P. Rocha, and P. Laguna, "A wavelet-based ECG delineator: Evaluation on standard databases," *IEEE Trans. Biomed. Eng.*, vol. 51, pp. 570–581, 2004.

[119] N. S. Jayant and P. Noll, *Digital Coding of Waveforms*. New Jersey: Prentice-Hall, 1984.

[120] A. Gersho and R. M. Gray, *Vector Quantization and Signal Compression*. Dordrecht/Norwell, MA: Kluwer Academic Publ., 1992.

[121] D. Salomon, *Data Compression: The Complete Reference*. New York, NY: Springer-Verlag, 2000.

[122] R. Degani, G. Bortolan, and R. Murolo, "Karhunen-Loéve coding of ECG signals," in *Proc. Computers in Cardiology*, pp. 395–398, IEEE Computer Society Press, 1990.

[123] O. Pahlm, P. O. Börjesson, and O. Werner, "Compact digital storage of ECG's," *Comput. Prog. Biomed.*, vol. 9, pp. 293–300, 1979.

[124] U. E. Ruttimann and H. V. Pipberger, "Compression of the ECG by prediction or interpolation and entropy coding," *IEEE Trans. Biomed. Eng.*, vol. 26, pp. 613–623, 1979.

[125] M. Shridhar and M. F. Stevens, "Analysis of ECG data, for data compression," *Int. J. Bio-med. Comput.*, vol. 10, pp. 113–128, 1979.

[126] P. O. Börjesson, G. Einarsson, and O. Pahlm, "Comments on compression of the ECG by predication or interpolation and entropy coding," *IEEE Trans. Biomed. Eng.*, vol. 27, pp. 674–675, 1980.

[127] A. S. Krishnakumar, J. L. Karpowicz, N. Belic, D. H. Singer, and J. M. Jenkins, "Microprocessor-based data compression scheme for enhanced digital transmission of Holter recordings," in *Proc. Computers in Cardiology*, pp. 435–437, IEEE Computer Society Press, 1980.

[128] S. M. S. Jalaleddine, C. G. Hutchens, R. D. Strattan, and W. A. Coberly, "ECG data compresssion techniques: A unified approach," *IEEE Trans. Biomed. Eng.*, vol. 37, pp. 329–343, 1990.

[129] J. R. Cox, F. M. Nolle, H. A. Fozzard, and G. C. Oliver, "AZTEC, a preprocessing program for real time ECG rhythm analysis," *IEEE Trans. Biomed. Eng.*, vol. 15, pp. 128–129, 1968.

[130] J. R. Cox, H. A. Fozzard, and R. M. Arthur, "Digital analysis of the electroencephalogram, the blood pressure wave, and the electrocardiogram," *Proc. IEEE*, vol. 60, pp. 1137–1164, 1972.

[131] B. Furht and A. Perez, "An adaptive real-time ECG compression algorithm with variable threshold," *IEEE Trans. Biomed. Eng.*, vol. 35, pp. 489–494, 1988.

[132] W. C. Mueller, "Arrhythmia detection program for an ambulatory ECG monitor," *Biomed. Sci. Instrum.*, vol. 14, pp. 81–85, 1978.

[133] G. B. Moody, K. Soroushian, and R. G. Mark, "ECG data compression for tapeless ambulatory monitors," in *Proc. Computers in Cardiology*, pp. 467–470, IEEE Computer Society Press, 1987.

[134] G. B. Moody, R. G. Mark, and A. L. Goldberger, "Evaluation of the 'TRIM' ECG data compressor," in *Proc. Computers in Cardiology*, pp. 167–170, IEEE Computer Society Press, 1988.

[135] W. J. Tompkins and J. G. Webster (eds.), *Design of Microcomputer-based Medical Instrumentation*. New Jersey: Prentice-Hall, 1981.

[136] J. P. Abenstein and W. J. Tompkins, "A new data reduction algorithm for real-time ECG analysis," *IEEE Trans. Biomed. Eng.*, vol. 29, pp. 43–48, 1982.

[137] M. Ishijima, S.-B. Shin, G. H. Hostetter, and J. Sklansky, "Scan-along polygonal approximation for data compression of electrocardiograms," *IEEE Trans. Biomed. Eng.*, vol. 30, pp. 723–729, 1983.

[138] L. W. Gardenshire, "Data redundancy reduction for biomedical telemetry," in *Biomedical Telemetry* (C. A. Caceres, ed.), ch. 11, pp. 255–298, San Diego: Academic Press, 1965.

[139] R. C. Barr, S. M. Blanchard, and D. A. DiPersio, "SAPA–2 is the fan," *IEEE Trans. Biomed. Eng.*, vol. 32, p. 337, 1985.

[140] S. M. Blanchard and R. C. Barr, "Comparison of methods for adaptive sampling of cardiac electrograms and electrocardiograms," *Med. Biol. Eng. & Comput.*, vol. 23, pp. 377–386, 1985.

[141] D. A. DiPersio and R. C. Barr, "Evaluation of the fan method of adaptive sampling on human electrocardiograms," *Med. Biol. Eng. & Comput.*, vol. 23, pp. 401–410, 1985.

[142] L. N. Bohs and R. C. Barr, "Prototype for real-time adaptive sampling using the fan algorithm," *Med. Biol. Eng. & Comput.*, vol. 26, pp. 574–583, 1988.

[143] L. N. Bohs and R. C. Barr, "Real-time adaptive sampling with the fan algorithm," *Med. Biol. Eng. & Comput.*, vol. 26, pp. 565–573, 1988.

[144] X. B. Huang, M. J. English, and R. Vincent, "Fast ECG data compression algorithms suitable for microprocessor system," *J. Biomed. Eng.*, vol. 14, pp. 64–68, 1992.

[145] S. C. Tai, "SLOPE—a real-time ECG data compressor," *Med. Biol. Eng. & Comput.*, vol. 29, pp. 175–179, 1991.

[146] S. C. Tai, "ECG data compression by corner detection," *Med. Biol. Eng. & Comput.*, vol. 30, pp. 584–590, 1992.

[147] S. C. Tai, "AZTDIS—a two phase real-time ECG data compressor," *J. Biomed. Eng.*, vol. 15, pp. 510–515, 1993.

[148] D. Haugland, J. Heber, and J. Husøy, "Optimisation algorithms for ECG data compression," *Med. Biol. Eng. & Comput.*, vol. 35, pp. 420–424, 1997.

[149] S. O. Aase, R. Nygaard, J. H. Husøy, and D. Haugland, "Optimized time and frequency domain methods for ECG signal compression," *Applied Signal Proc.*, vol. 5, pp. 210–225, 1998.

[150] R. Nygaard, G. Melnikov, and A. K. Katsaggelos, "A rate distortion optimal ECG coding algorithm," *IEEE Trans. Biomed. Eng.*, vol. 48, pp. 28–40, 2001.

[151] T. Blanchett, G. C. Kember, and G. A. Fenton, "KLT-based quality controlled compression of single-lead ECG," *IEEE Trans. Biomed. Eng.*, vol. 45, pp. 942–945, 1998.

[152] N. Ahmed, P. J. Milne, and S. G. Harris, "Electrocardiographic data compression via orthogonal transforms," *IEEE Trans. Biomed. Eng.*, vol. 22, pp. 484–487, 1975.

[153] M. E. Womble, J. S. Halliday, S. K. Mitter, M. C. Lancaster, and J. H. Triebwasser, "Data compression for storing and transmitting ECG's/VCG's," *Proc. IEEE*, vol. 65, pp. 702–706, 1977.

[154] S. Olmos, M. Millán, J. García, and P. Laguna, "ECG data compression with the Karhunen-Loève transform," in *Proc. Computers in Cardiology*, pp. 253–256, IEEE Press, 1996.

[155] J. Crowe, N. Gibson, M. Woolfson, and M. Somekh, "Wavelet transform as a potential tool for ECG analysis and compression," *J. Biomed. Eng.*, vol. 14, pp. 268–272, 1992.

[156] N. V. Thakor, Y. Sun, H. Rix, and P. Caminal, "MULTIWAVE: A multiresolution wavelet-based ECG data compression," *IEICE Trans. Inform. Sys.*, vol. 76, pp. 1462–1469, 1993.

[157] J. Chen, S. Itoh, and T. Hashimoto, "ECG data compression by using wavelet transform," *IEICE Trans. Inform. Sys.*, vol. 76, pp. 1454–1461, 1993.

[158] M. L. Hilton, "Wavelet and wavelet packet compression of electrocardiograms," *IEEE Trans. Biomed. Eng.*, vol. 44, pp. 394–402, 1997.

[159] J. Chen and S. Itoh, "A wavelet transform-based ECG compression method guaranteeing desired signal quality," *IEEE Trans. Biomed. Eng.*, vol. 45, pp. 1414–1419, 1998.

[160] A. G. Ramakrishnan and S. Saha, "ECG coding by wavelet-based linear prediction," *IEEE Trans. Biomed. Eng.*, vol. 44, pp. 1253–1261, 1997.

[161] Z. Lu, D. Y. Kim, and W. Pearlman, "Wavelet compression of ECG signals by the set partitioning in hierarchial trees algorithm," *IEEE Trans. Biomed. Eng.*, vol. 47, pp. 849–856, 2000.

[162] R. S. H. Istepanian and A. A. Petrosian, "Optimal zone wavelet-based ECG data compression for a mobile telecardiology system," *IEEE Trans. Info. Tech. Biomed.*, vol. 4, pp. 200–211, 2000.

[163] R. S. H. Istepanian, L. I. Hadjileontiadis, and S. M. Panas, "ECG data compression using wavelets and higher order statistics methods," *IEEE Trans. Info. Tech. Biomed.*, vol. 5, pp. 108–115, 2001.

[164] M. V. Wickerhauser, *Adapted Wavelet Analysis from Theory to Software*. Piscataway, NJ, USA: IEEE Press, 1994.

[165] B. Bradie, "Wavelet packet-based compression of single lead ECG," *IEEE Trans. Biomed. Eng.*, vol. 43, pp. 493–501, 1996.

[166] K. Nagarajan, E. Kresch, S. S. Rao, and Y. Kresh, "Constrained ECG compression using best adapted wavelet packet basis," *IEEE Trans. Signal Proc.*, vol. 3, pp. 273–275, 1996.

[167] B. R. S. Reddy and I. S. N. Murthy, "ECG data compression using Fourier descriptors," *IEEE Trans. Biomed. Eng.*, vol. 33, pp. 428–434, 1986.

[168] H. A. M. Al-Nashash, "ECG data compression using adaptive Fourier coefficients estimation," *Med. Eng. & Physics*, vol. 16, pp. 62–66, 1994.

[169] W. S. Kuklinski, "Fast Walsh transform data-compression algorithm: ECG applications," *Med. Biol. Eng. & Comput.*, vol. 21, pp. 465–472, 1983.

[170] G. P. Frangakis, G. Papakonstantinou, and S. G. Tzafestas, "A fast Walsh transform-based data compression multi-microprocessor system: Application to ECG signals," *Math. Comput. Simul.*, vol. 27, pp. 491–502, 1985.

[171] T. A. De Perez, M. C. Stefanelli, and F. D'Alvano, "ECG data data compression via exponential quantization of the Walsh spectrum," *J. Clin. Eng.*, vol. 12, pp. 373–378, 1987.

[172] E. Berti, F. Chiaraluce, N. E. Evans, and J. J. McKee, "Reduction of Walsh-transformed electrocardiograms by double logarithmic coding," *IEEE Trans. Biomed. Eng.*, vol. 47, pp. 1543–1547, 2000.

[173] V. A. Allen and J. Belina, "ECG data compression using the discrete cosine transform," in *Proc. Computers in Cardiology*, pp. 687–690, IEEE Computer Society Press, 1992.

[174] H. Lee and K. M. Buckley, "ECG data compression using cut and align beats approach and 2-D transforms," *IEEE Trans. Biomed. Eng.*, vol. 46, pp. 556–564, 1999.

[175] L. V. Batista, E. U. Melcher, and L. C. Carvalho, "Compression of ECG signals by optimized quantization of discrete cosine transform coefficients," *Med. Eng. & Physics*, vol. 23, pp. 127–134, 2001.

[176] W. Philips and G. De Jonghe, "Data compression of ECG's by high-degree polynomial approximation," *IEEE Trans. Biomed. Eng.*, vol. 39, pp. 330–337, 1992.

[177] R. Jané, S. Olmos, P. Laguna, and P. Caminal, "Adaptive Hermite models for ECG data compression: Performance and evaluation with automatic wave detection," in *Proc. Computers in Cardiology*, pp. 389–392, IEEE Computer Society Press, 1993.

[178] W. Philips, "ECG data compression with time-warped polynomials," *IEEE Trans. Biomed. Eng.*, vol. 43, pp. 480–492, 1996.

[179] D. L. Cohn and J. L. Melsa, "Efficient digitization methods for electrocardiograms," in *Proc. Joint Aut. Control Conf.*, pp. 1386–1391, IEEE Computer Society Press, 1977.

[180] P. S. Hamilton and W. J. Tompkins, "Compression of the ambulatory ECG by average beat subtraction and residual differencing," *IEEE Trans. Biomed. Eng.*, vol. 38, pp. 253–259, 1991.

[181] P. S. Hamilton and W. J. Tompkins, "Theoretical and experimental rate distortion performance in compression of ambulatory ECG's," *IEEE Trans. Biomed. Eng.*, vol. 38, pp. 260–266, 1991.

[182] P. S. Hamilton, "Adaptive compression of the ambulatory electrocardiogram," *Biomed. Instr. & Techn.*, vol. 27, pp. 56–63, 1993.

[183] J. Slocum, E. Byrom, L. McCarthy, A. Sahakian, and S. Swiryn, "Computer detection of atrioventricular dissociation from surface electrocardiograms during wide QRS complex tachycardia," *Circulation*, vol. 72, pp. 1028–1036, 1985.

[184] S. Shkurovich, A. Sahakian, and S. Swiryn, "Detection of atrial activity from high-voltage leads of implantable ventricular defibrillators using a cancellation technique," *IEEE Trans. Biomed. Eng.*, vol. 45, no. 2, pp. 229–234, 1998.

[185] M. Stridh and L. Sörnmo, "Spatiotemporal QRST cancellation techniques for analysis of atrial fibrillation," *IEEE Trans. Biomed. Eng.*, vol. 48, no. 1, pp. 105–111, 2001.

[186] T. Uchiyama, K. Akazawa, and A. Sasamori, "Data compression of ambulatory ECG by using multi-template matching and residual coding," *IEICE Trans. Inform. Sys.*, vol. 76, pp. 1419–1424, 1993.

[187] C. Paggetti, M. Lusini, M. Varanini, A. Taddei, and C. Marchesi, "A multichannel template based data compression algorithm," in *Proc. Computers in Cardiology*, pp. 629–632, IEEE Computer Society Press, 1994.

[188] G. D. Barlas and E. S. Skordalakis, "A novel family of compression algorithms for ECG and other semiperiodical, one-dimensional, biomedical signals," *IEEE Trans. Biomed. Eng.*, vol. 43, pp. 820–828, 1996.

[189] P.-W. Hsia, "Electrocardiographics data compression using preceding consecutive QRS information," in *Proc. Computers in Cardiology*, pp. 465–468, IEEE Computer Society Press, 1988.

[190] G. Nave and A. Cohen, "ECG compression using long-term prediction," *IEEE Trans. Biomed. Eng.*, vol. 40, pp. 877–885, 1993.

[191] C. P. Mammen and B. Ramamurthi, "Vector quantization of multichannel ECG," *IEEE Trans. Biomed. Eng.*, vol. 37, pp. 821–825, 1990.

[192] A. E. Çetin, H. Köymen, and M. C. Aydin, "Multichannel ECG data compression by multirate signal processing and transform domain coding techniques," *IEEE Trans. Biomed. Eng.*, vol. 40, pp. 495–499, 1993.

[193] S. Olmos and P. Laguna, "Multi-lead ECG data compression with orthogonal expansions: KLT and wavelet packets," in *Proc. Computers in Cardiology*, pp. 539–542, IEEE Press, 1999.

[194] A. Cohen and Y. Zigel, "Compression of multichannel ECG through multichannel long-term prediction," *IEEE Eng. Med. Biol. Mag.*, vol. 17, pp. 109–115, 1998.

[195] K. Anant, F. Dowla, and G. Rodrigue, "Vector quantization of ECG wavelet coefficients," *IEEE Trans. Signal Proc. Lett.*, vol. 2, pp. 129–131, 1995.

[196] B. Wang and G. Yuan, "Compression of ECG data by vector quantization," *IEEE Eng. Med. Biol. Mag.*, vol. 16, pp. 23–26, 1997.

[197] J. L. Cárdenas-Barrera and J. V. Lorenzo-Ginori, "Mean-shape vector quantizer for ECG signal compression," *IEEE Trans. Biomed. Eng.*, vol. 46, pp. 62–70, 1999.

[198] S.-G. Miaou and H.-L. Yen, "Quality driven gold washing adaptive vector quantization and its application to ECG data compression," *IEEE Trans. Biomed. Eng.*, vol. 47, pp. 209–218, 2000.

[199] Y. Zigel, A. Cohen, and A. Katz, "The weighted diagnostic distortion (WDD) measure for ECG signal compression," *IEEE Trans. Biomed. Eng.*, vol. 47, pp. 1422–1430, 2000.

[200] Y. Zigel, A. Cohen, and A. Katz, "ECG signal compression using analysis by synthesis coding," *IEEE Trans. Biomed. Eng.*, vol. 47, pp. 1308–1316, 2000.

[201] P. Jarske, Y. Nuevo, and S. K. Mitra, "A simple approach to the design of linear phase FIR digital filters with variable characteristics," *Signal Proc.*, vol. 14, pp. 313–326, 1988.

[202] R. A. Haddad and T. W. Parsons, *Digital Signal Processing. Theory, Applications and Hardware.* New York: Computer Science Press, 1991.

[203] O. Rompelman and H. H. Ros, "Coherent averaging technique: A tutorial review. Part 2: Trigger jitter, overlapping responses and non-periodic stimulation," *J. Biomed. Eng.*, vol. 8, pp. 30–35, 1986.

[204] S. Olmos and P. Laguna, "A clinical distortion index for ECG data compression performance evaluation," in *Proc. Conf. Eur. Soc. Eng. Med. (ESEM)*, pp. 381–382, 1999.

Problems

7.1 Explain the sentence, "In case of too high a cut-off frequency, the output of the highpass filter contains an unwanted, oscillatory component strongly correlated with the prevailing heart rate", using, for example, the continuous-time Fourier series as a starting point,

$$x(t) = \frac{a_0}{2} + \sum_{k=1}^{\infty} (a_k \cos(k\Omega_0 t) + b_k \sin(k\Omega_0 t)),$$

where Ω_0 is the repetition rate (i.e., the "heart rate").

7.2 Determine the number of multiplications required for implementing

    a. an FIR filter with a symmetric impulse response of length N,

    b. forward/backward filtering using an Nth order IIR filter, and

    c. sampling rate decimation and interpolation (the decimator and interpolator are of order M).

7.3 Develop an approach by which forward/backward filtering can be implemented for use in real-time processing [17]. Comment on the resulting time delay of the output signal.

7.4 A second-order IIR filter is defined by a pair of conjugate poles. Describe how the poles change when the sampling rate is increased for a fixed value of the cut-off frequency, and comment on potential problems that may arise.

7.5 The design of a time-varying lowpass filter for baseline removal (see page 468) may involve a tunable filter and a rule which relates the cut-off frequency of the filter to the observed signal.

    a. The design of a tunable lowpass filter is based on the impulse response of the ideal lowpass filter with cut-off frequency at ω_{c_0},

$$h_0(n) = \frac{\sin(\omega_{c_0} n)}{\pi n}.$$

Show that the impulse response $h(n)$ of a linear, time-invariant filter with cut-off frequency ω_c can be expressed in terms of $h_0(n)$ by the following expression [201],

$$h(n) = \begin{cases} c(0)\omega_c, & n = 0; \\ c(n)\sin(\omega_c n), & n = \pm 1, \pm 2, \ldots, \pm N, \end{cases}$$

where the sequence $c(n)$ is to be determined. Explain the advantage with this approach.

b. With the impulse response $h(n)$ available, propose a rule how the cut-off frequency ω_c can be adapted to the properties of the ECG signal.

7.6 The properties of an interpolator can be analyzed in terms of its impulse response and related magnitude function. In this problem, it is assumed that the input signal, acquired at a sampling rate of $1/T$, is to be linearly interpolated by a factor of L. The signal processing is illustrated by the block diagram below for $L = 3$.

a. Determine the transfer function for the linear interpolator, and interpret the result in terms of filtering.

b. Determine the -3 dB cut-off frequency as a function of the interpolation order L.

c. Determine the corresponding cut-off frequency F_c when the sampling interval is $T = 1$ s. Comment on the consequences of an altered sampling interval T, which is the case when the interval between successive heartbeats, i.e., $T_i = t_{i+1} - t_i$, is varying.

7.7 The cubic spline method for estimation of baseline wander can be made efficient by introducing a state-space formulation [26].

a. From the Taylor expansion in (7.23), determine the matrix $\mathbf{A}$ which relates

$$\begin{bmatrix} y(t) \\ y^{(1)}(t) \\ \vdots \end{bmatrix} = \mathbf{A} \begin{bmatrix} y(t_0) \\ y^{(1)}(t_0) \\ \vdots \end{bmatrix}.$$

b. Determine $\mathbf{A}$ when t is equal to the sampling interval $T = 1$. Describe how the resulting algorithm can be used to estimate the baseline wander at any sample.

7.8 Find the transfer function in (7.45) for the 50/60 Hz powerline interference filter when the update equation of the nonlinear filter is linear, i.e., defined by (7.44) [40].

7.9 Derive the bandwidth of the second-order FIR and IIR filters for removal of 50/60 Hz powerline interference.

7.10 Show that the LMS algorithm in Figure 7.15 for adaptive removal of 50/60 Hz powerline interference is identical to a first-order filter with two coefficients.

7.11 Show that the matrix $\mathbf{H}$ in (7.55) for removal of 50/60 Hz powerline interference, using the estimation–subtraction technique, has the form

$$\mathbf{H} = -\frac{2}{N} \begin{bmatrix} 1 - \frac{N}{2} & \cos \omega_0 & \cos 2\omega_0 & \cdots & \cos \omega_0 (N-1) \\ \cos \omega_0 & 1 - \frac{N}{2} & \cos \omega_0 & \cdots & \cos \omega_0 (N-2) \\ \cos 2\omega_0 & \cos \omega_0 & 1 - \frac{N}{2} & \cdots & \vdots \\ \vdots & \vdots & \vdots & \ddots & \cos \omega_0 \\ \cos(N-1)\omega_0 & \cos(N-2)\omega_0 & \cdots & \cos \omega_0 & 1 - \frac{N}{2} \end{bmatrix}.$$

7.12 It could be argued that the rows of the matrix $\mathbf{H}$ in (7.55) constitute impulse responses which are asymmetric, implying filters with nonlinear phase. Show that such an argument is incorrect, i.e., all rows of $\mathbf{H}$ correspond to filters with linear phase.

7.13 Determine the zeros and poles of (7.56).

7.14 The design of a filter for EMG attenuation can be made time-varying by adjusting the cut-off frequency with respect to the intervals of the heartbeat; a higher cut-off frequency is used for the QRS interval than for the P and T wave intervals. One approach to designing such a filter is to estimate the noise level ϵ of the observed signal and to use the obtained ϵ as the maximal acceptable error tolerance between the observed and filtered signals [50]. This approach preserves high-frequency components of the original signal at the same level as low-frequency components. The time-varying, Gaussian impulse response has been suggested for this purpose, defined by

$$h(k, n) = \left(\frac{\beta(n)}{\pi} \right)^{1/2} e^{-\beta(n)k^2},$$

where $\beta(n)$ controls the cut-off frequency of the filter and is normalized such that

$$\sum_{k=-\infty}^{\infty} h(k, n) = 1.$$

a. From a Taylor series expansion of the observed signal $x(n)$, evaluated around n, determine a closed-form expression of the filter output $y(n)$ such that $y(n) = x(n) + \epsilon$.

b. From the closed-form expression in part (a.) it is evident that ϵ depends on $\beta(n)$. Propose a procedure to estimate $\beta(n)$ from $x(n)$ so that the difference between $x(n)$ and the filtered signal is always guaranteed to be lower than ϵ. *Hint:* Make use of the result

$$\int_{-\infty}^{\infty} t^{2m} e^{-\beta(n)t^2} dt = \frac{1 \cdot 3 \cdot 5 \cdots (2m-1)}{2^m} \left(\frac{\pi}{\beta(n)^{2m+1}} \right)^{1/2}.$$

7.15 In designing time-varying filters for EMG attenuation, the result in Section 4.5.4—stating that a truncated orthonormal series expansion of a signal can be interpreted in terms of a linear, time-variant filtering—can be exploited. For each segmented PQRST complex of N samples, denoted $\mathbf{x}_i$, the EMG-attenuated signal $\mathbf{y}_i$ results from orthogonal transformation and truncation (cf. (4.206)),

$$\mathbf{y}_i = \Phi_s \Phi_s^T \mathbf{x}_i,$$

where Φ_s is given in (4.204). The K columns of Φ_s ($K < N$) contain the basis functions which are used to estimate the signal.

Propose a rule to determine the number of basis functions K so that the difference between $\mathbf{x}_i$ and $\mathbf{y}_i$ is always guaranteed to be lower than a certain tolerance ϵ.

7.16 QRS detection: The purpose of hypothesis testing is to select the most likely hypothesis assuming the model

$$\mathcal{H}_0 : x(n) = v(n), \quad n = 0, \ldots, N-1;$$
$$\mathcal{H}_1 : x(n) = s(n) + v(n), \quad n = 0, \ldots, N-1,$$

where $s(n)$ is a deterministic, known waveform, and $v(n)$ is white, Gaussian noise with variance σ_v^2.

Define a likelihood ratio

$$L(\mathbf{x}) = \frac{p(\mathbf{x}|\mathcal{H}_1)}{p(\mathbf{x}|\mathcal{H}_0)},$$

then the following test can be used to decide $\mathcal{H}_1$,

$$L(\mathbf{x}) > \eta.$$

The threshold can be used to attain a certain performance (and possibly related to a priori probabilities for the two hypotheses). Conclude that this detector is identical to the estimation-based approach presented in the text.

7.17 The binomial–Hermite family of filters is defined by the following transfer function [202],

$$H(z) = \left(1 - z^{-1}\right)^K \left(1 + z^{-1}\right)^{L-K}$$

and is, evidently, closely related to the filters introduced in (7.92) for use in QRS detection.

    a. Find the frequency response of the binomial–Hermite filters.

    b. For which values of K and L will the filter have lowpass, bandpass, and highpass characteristics?

7.18 Morphologic variability of the QRS complex has been found to convey clinically valuable information. One approach to quantify such variability is to study the ensemble variance using the estimator in (4.17), provided that the QRS complexes have been first time-aligned so that an ensemble of signals can be created.

    a. It is assumed that a QRS complex is modeled by

$$x_i(n) = s(n) + v_i(n), \quad i = 1, \ldots, M,$$

    where $s(n)$ is fixed from beat-to-beat, whereas $v_i(n)$ is a zero-mean component which *accounts for morphologic variability*. Show that the ensemble variance can be estimated by

$$\hat{\sigma}^2(n) = \frac{1}{M} \sum_{i=1}^{M} v_i^2(n).$$

    b. When the ensemble of QRS complexes are misaligned, the observed signal is modeled by

$$x_i(t) = s(t - \tau_i) + v_i(t).$$

    For this model, show that an estimator of the ensemble variance is

$$\hat{\sigma}^2(n) = \frac{1}{M} \sum_{i=1}^{M} v_i^2(n) + s'^2(n)\sigma_\tau^2,$$

where $s'(n)$ denotes the derivative of $s(n)$. It is assumed that morphologic variability and misalignment are statistically independent and that $s(t - \tau_i)$ can be approximated, for small values of τ_i around t, with a first-order Taylor series expansion. *Hint:* Make the derivation in continuous-time and convert the result to discrete-time.

c. In part (b.) it was shown that $\hat{\sigma}^2(n)$ is essentially composed of two components related to morphologic variability and misalignment. Here, the purpose is to find that sampling rate F_s which assures that the influence of misalignment due to sampling is sufficiently small. When τ_i has a uniform PDF, it has been shown that the standard deviation σ_τ is related to the sampling rate F_s by [203]

$$\sigma_\tau = \frac{1}{2\sqrt{3}F_s}.$$

The power of $\hat{\sigma}(n)$,

$$\mathcal{P} = \frac{1}{N} \sum_{n=0}^{N-1} \hat{\sigma}^2(n),$$

is used as a measure of morphologic variability; the power of $s(n)$, $s'(n)$, and $v_i(n)$, denoted $\mathcal{P}_s$, $\mathcal{P}_{s'}$, and $\mathcal{P}_v$, respectively, is defined analogously. In this example, it is assumed that

$$\mathcal{P}_s = 1, \quad \frac{\mathcal{P}_s}{\mathcal{P}_v} = 10^{-3}, \quad \mathcal{P}_{s'} = 1.9 \times 10^4.$$

Determine the sampling rate F_s which assures that the misalignment contribution in $\mathcal{P}$ is lower than 10%.

7.19 Late potentials represent a high-frequency activity which occurs in the terminal part of the QRS complex (Section 6.6.5) and requires ensemble averaging for their detection. Since late potentials are obscured by the waves of the normal QRST complex, the averaged signal is usually highpass filtered before detection is performed.

a. Comment on possible problems with such linear filtering considering the location of late potentials relative to the QRS complex.

b. With bidirectional filtering the signal is filtered in the forward direction until the middle of the QRS complex is reached and then filtered backwards from the end of the T wave until the middle of the QRS complex is again reached. Explain why this type of filtering mitigates problems associated with conventional causal filtering.

7.20 Noise reduction by ensemble averaging requires that the QRS complexes first be aligned in time. Since alignment is limited by the sampling resolution it is of interest to study the effects on the ensemble average (Section 4.3.6). It is assumed that the sampling rate is $F_s = 2000$ Hz and that the time jitter is uniformly distributed in the sampling interval.

    a. Which is the standard deviation of the misalignment error?

    b. Which is the maximum frequency that can be studied, assuming that the –3 dB cut-off frequency is acceptable?

    c. What is the more restrictive factor of high-frequency estimation in this case: sampling or ensemble averaging?

7.21 A simplified decision rule for vertex determination in the SAPA data compression algorithm is the so-called SAPA–1 algorithm. With this strategy, the net vertex to n_0 is determined as the sample immediately preceding the first n sample that satisfies

$$\min_{n_0 > m > n} g(m, \varepsilon) > \max_{n_0 > m > n} g(m, -\varepsilon).$$

Show that the reconstruction error has a tolerance of 2ε with this strategy.

7.22 In ECG data compression, average beat subtraction is a technique which reduces the amplitude of a signal (see page 534). For this technique to be efficient, it is important that the average beat $\hat{s}(t)$ is well-aligned in time with the current beat $s(t)$; otherwise, the residual signal will include an additional component due to misalignment. Assuming that the observed signal is modeled by $x_i(t) = s(t) + v_i(t)$ and that $\hat{s}(t)$ is taken as the previous beat $x_{i-1}(t)$ (cf. 7.130), the residual ECG can be written as

$$y_i(t) = x_i(t) - \hat{s}_i(t) = s(t) + v_i(t) - s(t - \tau_i) + v_{i-1}(t).$$

The alignment error τ_i is uniformly distributed in the interval $[-0.5T, 0.5T]$, where T denotes the length of the sampling interval. The noise $v_i(t)$ is assumed to be uncorrelated from beat to beat.

    Determine the power spectrum of the residual signal, and comment on how the spectrum of $s(t)$ is modified by poor time alignment.

7.23 Amplitude quantization of the signal $x(n)$ should be considered when dealing with data compression. The error introduced by an analog-to-digital converter is given by

$$e_q(n) = x(n) - Q[x(n)] = x(n) - x_q(n),$$

where $Q[\cdot]$ denotes a function that describes the quantizer. Average beat subtraction reduces the dynamic range of $x(n)$ so that fewer bits are needed for representing the residual ECG, while still yielding a quantization error $e_q(n)$ which is acceptable. Various strategies can be pursued for residual estimation and quantization [180], of which two are developed below. It is assumed that previous ECG beats are identical, i.e., $s_i(n) = s(n)$, and perfectly aligned. The noise $v_i(n)$ is additive and stationary across beats, with zero-mean and variance σ_v^2. Furthermore, it is assumed that the quantization error $e_q(n)$ is uncorrelated with $s(n)$ and with variance $\sigma_e^2 = \Delta^2/12$, where Δ denotes the quantization step. The noise at the averager output is assumed to be negligible (the number of averaged beats is large enough) except for the roundoff error.

a. The input signal $x(n)$ is coarsely quantized into $x_q(n)$ with the quantizer Q_c, and average beat subtraction is based on $x_q(n)$ (upper block diagram below). Determine the power of the residual signal.

b. The input signal $x(n)$ is finely quantized into $x_q(n)$ with the quantizer Q_f, while average beat subtraction is based on a coarsely quantized residual signal (lower block diagram below). Determine the power of the residual signal, and compare the result with that obtained for the other quantization strategy.

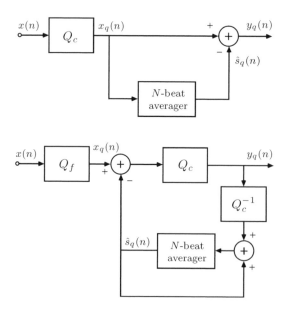

7.24 Interbeat redundancy can be addressed by using a long-term prediction strategy. This strategy has been applied by restricting the long-term prediction to only include the previous beat at $\hat{\theta}_{i-1}$ and without AR modeling of the running beat. Thus, a long-term prediction of the current beat $x_i(n)$ is obtained by

$$\hat{x}_i(n) = \sum_{k=-p}^{p} a_k x(n + \hat{\theta}_{i-1} - k), \quad n = -N_1, \ldots, N_2 - 1.$$

Derive the coefficients $\{a_k\}_{k=-p}^{p}$ which minimize the least-squares error

$$\mathcal{E}_i = \sum_{n=-N_1}^{N_2-1} e_i^2(n) = \sum_{n=-N_1}^{N_2-1} \left(x(n + \hat{\theta}_i) - \hat{x}_i(n) \right)^2,$$

where $x(n)$, $n = \hat{\theta}_i - N_1, \ldots, \hat{\theta}_i + N_2 - 1$, denotes the current beat.

7.25 Evaluation of data compression techniques using a performance measure defined by a mathematical norm, such as $\mathcal{P}_{RMS}$ in (7.143), runs the risk of misjudging performance so that $\mathcal{P}_{RMS}$ is larger than one would expect. This behavior can be observed when a noisy signal is compressed by a transform-based method such as the Karhunen–Loève transform; the reconstructed signal is essentially noise-free, while the observed signal noise and, therefore, an unreasonably large value of $\mathcal{P}_{RMS}$ results.

Propose a new performance measure for transform-based compression which does not suffer from the above-mentioned problem. *Hint:* The proposal can be based on the model $\mathbf{x} = \mathbf{s} + \mathbf{v}$, where the noise $\mathbf{v}$ is assumed to be white with variance σ_v^2. Since the white property implies that the noise is evenly spread in the transform domain, we can, for a low truncation value K ($<< N$), neglect the noise component in the reconstructed signal,

$$\hat{\mathbf{x}} = \hat{\mathbf{s}} + \hat{\mathbf{v}} \approx \hat{\mathbf{s}}.$$

7.26 For each beat, the performance measure $\mathcal{P}_{WDD}$ is defined as the weighted, quadratic norm of the normalized error $\Delta\beta_k$ of the kth measurement, see (7.145). Since the normalization factor is beat-dependent, it is influenced by beat-to-beat changes of a measurement, for example, as observed in QRS amplitude. Another property is that the normalization factor is relative, implying that it may exhibit widely different magnitudes which depend on the specific measurement; the magnitude may range from the low amplitude of a P wave to the much larger of a QRS complex.

Another approach to developing a performance measure is to start by obtaining several measurements of a certain parameter from different human annotators. From these measurements, an estimate of the inter-annotator variability can be computed which can be considered as the expected variability of a certain parameter. Based on this information, propose a performance measure which is less sensitive to beat-to-beat variations as well as to widely differing magnitudes [204].

Chapter 8

ECG Signal Processing: Heart Rate Variability

Heart rate variability (HRV) has, in recent years, received widespread research interest since the state of the autonomic nervous system, and related diseases, can be investigated noninvasively using relatively basic signal processing techniques. Despite the seeming simplicity of deriving the series of RR intervals from the ECG signal and defining related measures of dispersion, it is nonetheless essential to assure that HRV is analyzed accurately. Several definitions of signals for representing the heart rhythm have been suggested which characterize variability either in terms of successive RR intervals or instantaneous heart rate.[1] In particular, spectral analysis of heart rhythm signals has received considerable attention since oscillations embedded in the rhythm, for example, due to respiratory activity or variations in blood pressure, can be quantified from the corresponding peaks in the estimated power spectrum. The oscillations are characterized by low-frequency components which typically are located in the interval below 0.5 Hz. These low-frequency components will remain in the ECG signal despite the inclusion of baseline wander removal based on, for example, a highpass filter with a cut-off frequency at 0.5 Hz. Since HRV is solely characterized by the pattern of heartbeat occurrence times, and baseline wander is related to the morphology of the ECG signal, baseline wander removal does not alter the HRV information.

The present chapter is dedicated to the analysis of HRV. Following an overview in Sections 8.1 and 8.2 of the demands on data acquisition for HRV analysis, the conditioning of RR intervals, and simple time domain measures, the most important signals for representing the heart rhythm are presented

---

[1]The term "heart rate variability" here signifies fluctuations in both RR intervals and instantaneous heart rate [1].

in Section 8.3. These signals form the basis for HRV measures which are more advanced than the time domain measures and which quantify the correlation that usually exists between different RR intervals. Spectral analysis of a heart rhythm signal may, in addition to the straightforward application of Fourier-based analysis, be implemented by techniques which directly account for the fact that the heart rhythm derives from an unevenly sampled signal (Section 8.4). Section 8.6 contains a description of methods developed for the correction of the RR interval series when ectopic beats are present; such correction requires that sinus beats and ectopic beats first be separated into different clusters with respect to their morphologies (Section 8.5). Finally, Section 8.7 discusses briefly the interaction between heart rate and other physiological signals, and how such interaction can be mathematically modeled. We remind the reader that a brief physiological background to HRV is found on page 431.

8.1 Acquisition and RR Interval Conditioning

The analysis of HRV is based on the series of occurrence times $\theta_0, \ldots, \theta_M$, originally produced by the QRS detector, but usually refined by an algorithm for time alignment.[2] Since an important purpose of HRV analysis is to investigate the influence of autonomic activity on the sinoatrial node, the onset of the P wave is actually a more appropriate fiducial point of the heartbeat than a fiducial point related to the QRS complex. However, a fiducial point related to the P wave is extremely difficult to determine with sufficient accuracy since the P wave has a low amplitude; sometimes the P wave is completely missing. Therefore, the fiducial point is commonly related to the QRS complex—evidently under the assumption that the QRS fiducial point has been determined by a reliable technique. The use of the RR intervals instead of the PP intervals has been generally accepted because the PR interval can be considered as relatively fixed, and thus, the RR intervals reflect the activity of the sinoatrial node.

The requirements on data acquisition are primarily concerned with the sampling rate by which the ECG signal should be acquired; subtle beat-to-beat variations in rhythm will be lost if the signal is too coarsely discretized in time. The sampling rate commonly used in resting ECG analysis, i.e., 250–500 Hz, is sufficient for most types of HRV analysis. However, certain clinical HRV investigations have been based on Holter recordings, digitized at a considerably lower sampling rate, i.e., 100–125 Hz. Methodological studies have shown that such low sampling rates are inappropriate and cause,

---

[2]For pedagogical reasons, the indexing of events in this chapter starts from zero rather than from one as otherwise assumed.

when spectral analysis of the heart rhythm signal is of interest, exaggeration of the high-frequency components of the power spectrum [2]. Based on a simple statistical model of the RR interval series, the error introduced in the power spectrum for different sampling rates of the ECG signal is discussed in Problem 8.1.

While HRV analysis is essentially unproblematic when analyzing recordings acquired during rest, artifacts are usually present in Holter recordings which pose some serious limitations on the analysis. For example, noise bursts may cause the QRS detector to produce false detections as well as to miss low-amplitude QRS complexes, thus resulting in an RR interval series with invalid intervals. Hence, the exclusion of non-normal RR intervals represents an important step in conditioning the series in order to make HRV analysis more reliable. The resulting interval series is commonly referred to as the normal-to-normal intervals (NN intervals). Since manual editing of a 24-hour Holter recording, containing approximately 100 000 RR intervals, is extremely laborious, automated exclusion procedures have been developed in order to accomplish rejection of artifacts. A simple approach is to apply an exclusion criterion by which an RR interval is considered abnormal if it deviates more than 20% from the mean length of the preceding RR intervals [3]. Such an approach is based on the assumption that the physiological mechanisms controlling the heart during sinus rhythm do not abruptly change the heart rate. Other, more complex decision criteria for exclusion of non-normal intervals have been presented in which the shape of the distribution of beat-to-beat differences in interval length is investigated [4, 5]. However, more complex approaches have not necessarily been found to result in a more valid RR interval series, but may rather produce the opposite result [5].

One particular aspect of artifact rejection is to handle the presence of ectopic beats which interrupt fluctuations in heart rate modulated by changes in autonomic balance. Since an ectopic beat interrupts the sinus rhythm by its premature occurrence, i.e., prior to the time when the next normal beat is expected to occur, it is necessary to correct for both the preceding, shorter than normal RR interval and the subsequent, longer than normal, compensatory pause before HRV analysis can be adequately performed (Section 8.6).

It should be emphasized that different techniques for HRV characterization are, to varying degrees, sensitive to the presence of artifacts in the RR interval series. Simple time domain measures, derived from the RR interval distribution, are less sensitive to the presence of artifacts than measures derived from the power spectrum. The "brute force" approach, based on minute-length ECG segments completely free of artifacts, may, in fact, be the sole approach when power spectral analysis of HRV is required.

8.2   Time Domain Measures

Clinical studies of HRV have frequently been synonymous with the use of simple time domain measures such as the standard deviation of the RR intervals. Although a variety of heart rhythm representations may be used, the series of RR intervals is the preferred starting point for the design of time domain measures. An important consideration is whether the measure should reflect long- or short-term HRV so as to convey information primarily related to parasympathetic or sympathetic activity. This consideration is particularly important when Holter recordings are analyzed and may be addressed by calculating the time domain measures from successive segments of shorter lengths or from the entire ECG recording [1]. Figure 8.1 illustrates the large variability in heart rate that may be observed during one day and night.

A straightforward way to quantify HRV is to calculate the standard deviation of the available NN intervals,

$$\text{SDNN} = \sqrt{\frac{1}{M-1} \sum_{k=1}^{M} (r_k - T_I)^2}, \tag{8.1}$$

where r_k denotes the kth NN interval, see (7.17), and T_I denotes the mean length of the M intervals r_k. For long-term recordings, SDNN only provides a rough characterization of HRV since the mean heart rate changes considerably from the active parts of the day to sleep during the night. Another commonly used measure is the standard deviation of the average length of NN intervals in 5-minute segments, abbreviated to SDANN, which, due to the 5-minute averages, primarily reflects very slow, circadian variations in heart rate. Since the resulting values of SDNN and SDANN depend, to a certain degree, on the length of the ECG recording, such information must also be taken into account to make a comparison of results meaningful.

Since both SDNN and SDANN reflect long-term variability in heart rate, additional dispersion measures have been suggested which reflect short-term variability through analysis of the difference between successive NN intervals. The effect of the difference operation is to accentuate the high-frequency content of the NN interval series. Hence, the standard deviation of successive NN interval differences is a frequently used dispersion measure in clinical studies; this measure is commonly referred to as the root mean-square of successive differences (rMSSD). The proportion of intervals differing more than a certain limit value from the preceding interval represents another measure which reflects short-term variability. Since the limit value is typically set to 50 ms, this HRV measure is referred to as "pNN50". Comparing

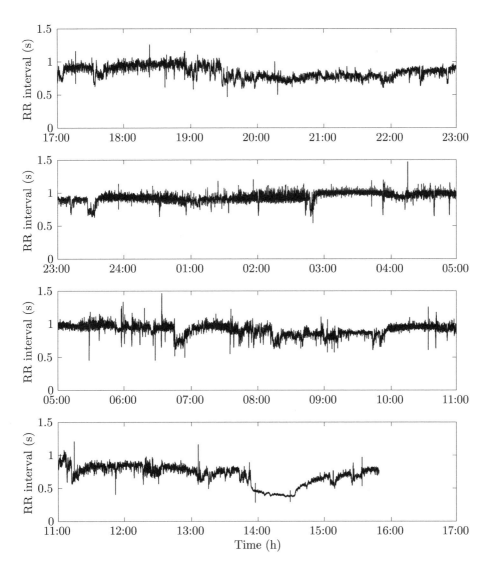

Figure 8.1: The series of RR intervals observed during almost one day and night. During this time span, the RR intervals vary in length from about 0.3 to 1.5 s.

Table 8.1: Common time domain measures for characterization of HRV, using the abbreviations which have become generally accepted in clinical studies. The table is a shortened version of the one presented in a task force paper on HRV [1].

Measure	Definition
SDNN	Standard deviation of all NN intervals.
SDANN	Standard deviation of the averages of NN intervals in all 5-minute segments of the entire ECG recording.
rMSSD	Root mean-square of successive differences of adjacent NN intervals.
pNN50	Percentage of pairs of adjacent NN intervals differing by more than 50 ms.
TINN	Triangular interpolation index. The base of a triangle fitted to the RR interval histogram (see text and Figure 8.2).

rMSSD and pNN50, it is evident that rMSSD provides a more detailed description of short-term variability, whereas pNN50 is much less vulnerable than rMSSD to artifacts that may be present in the RR interval series [6]. A list of common time domain measures, and their respective definitions, is presented in Table 8.1.

Time domain measures reflecting long-term variability in heart rate have, in certain patient groups, such as alcoholics and diabetics suffering from neuropathy, offered better performance in detecting abnormalities in autonomic function than short-term measures. In a similar fashion, the prediction of mortality in patients who have suffered from an earlier myocardial infarction has improved when the very slow variability in heart rate is investigated. Both these results were found in studies which were based on the analysis of Holter recordings [7].

Another group of time domain measures that deserves mentioning is that which is derived from the geometrical properties of the RR interval histogram [8]. The main idea behind such measures is the observation that the histogram often contains a dominant peak which can be well-characterized in terms of some simple geometrical shape such as a triangle. After finding the best fit of the triangle to the dominant peak, for example, expressed in the least-squares sense, a robust measure of the variability in heart rate is given by the width of the triangle base; this measure is referred to as the triangular interpolation index (TINN) [9], see Figure 8.2.

One important motivation for developing histogram-based HRV methods is their relative robustness to inclusion of non-NN intervals caused by ectopic beats or artifacts; such intervals often tend to fall outside the dominant

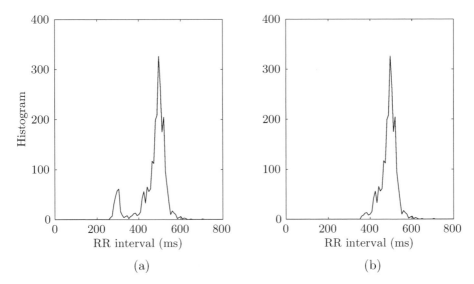

Figure 8.2: (a) An RR interval histogram with two dominant peaks, one reflecting the predominant RR interval lengths and the other reflecting the very short intervals due to falsely detected T waves and whose lengths are about 300 ms. (b) The RR interval histogram after removal of the T wave intervals using a criterion based on the triangular index.

peak of the interval histogram. However, the application of these methods is limited to Holter recordings, preferably of 24 hour duration or more, in order to get a sufficient number of intervals to construct a reliable histogram. Since a 24-hour recording contains periods from day activities as well as from rest during night-time, the histogram may become bimodal and sometimes multimodal. As a result, the use of triangular methods is no longer suitable since they tend to overestimate the variability in heart rate [8].

The use of histogram-based methods is not restricted to the study of variability in normal sinus rhythm, but has also been found valuable for studying mechanisms behind certain arrhythmias. For example, the shape of the RR interval histogram has been analyzed during atrial fibrillation to better understand the random behavior of electrical impulses that occasionally propagate through the atrioventricular node and activate the ventricles [10, 11].

8.3 Heart Rhythm Representations

The purpose of a heart rhythm representation is to produce a signal which accurately reflects variations in heart rhythm and which lends itself to different types of HRV analysis. We have already touched upon the representation issue when mentioning that the heart rhythm can be represented in terms of either *interval* or *rate* (the latter entity defined by the inverse of the RR intervals). Other representations have also been put forward which take their starting point in the series of occurrence times of the QRS complexes ("event series") rather than in the series of successive RR intervals. The distinction between these two types of series is important from a conceptual viewpoint, although the latter series is easily derived from the former. An overview of the different approaches to represent the heart rhythm is provided in this section.

The heart rhythm signal is based on the times at which the QRS complexes occur and, consequently, on a process that is "sampled" at unevenly spaced time instants. As a result, it is highly desirable to regularize the sampling rate of the heart rhythm signal in order to make the signal compatible with the multitude of analysis methods which require an evenly sampled signal. It is crucial not to confuse the sampling rate inherent to the heart with the one used for digitizing the ECG signal. The latter sampling rate determines the resolution of the QRS occurrence times θ_k and is, in the following, assumed to be sufficiently high so that it can be replaced by its continuous-time counterpart t_k [2]. It should be noted that while the ECG is typically sampled at a rate of 500–1000 Hz, the evenly sampled heart rhythm signal has a much lower sampling rate of only a few hertz. This lower rate is not only sufficient to completely characterize HRV, but it also results in a substantial reduction of the number of samples required to perform the analysis.

The performance requirement mentioned above of a heart rhythm representation that "well reflects variations in heart rhythm" is, unfortunately, not easily expressed in exact terms. However, by developing a mathematical model of HRV, it is possible to not only model the autonomic nervous influence on the heart rate, but also to provide a tool which helps to indicate which representation exhibits the better performance. One such model is the integral pulse frequency modulation (IPFM) model which has gained wide popularity in the field of HRV analysis.

8.3.1   The Integral Pulse Frequency Modulation Model

The IPFM model is used to generate an event series, such as the series of heartbeat occurrences, and assumes the existence of a continuous-time input signal with a particular physiological interpretation. Figure 8.3 presents a block diagram of the IPFM model and illustrates the signals as they may appear in different steps of the model. In this model, the input signal is integrated until a threshold R is reached at which an event is generated at

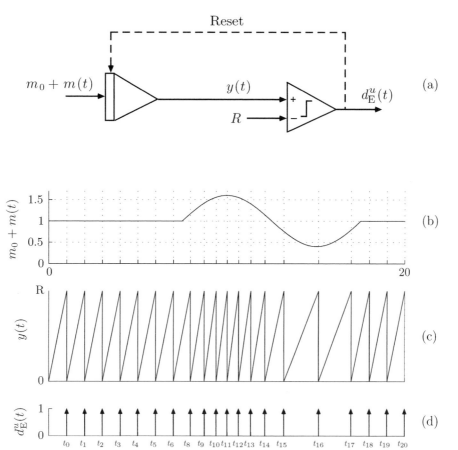

Figure 8.3: (a) The integral pulse frequency modulation (IPFM) model, (b) the input function $m_0 + m(t)$ which modulates the variation in interval length, (c) the output $y(t)$ of the integrator assuming a threshold level at R, and (d) the resulting event series $d_{\mathrm{E}}^u(t)$ at occurrence times $t_0, t_1, \dots, t_M$.

time t_k. The integrator is then reset to zero, the procedure is repeated, and so on. The threshold R defines the mean interval length between successive events. The input signal, being positive-valued, is the sum of two quantities, namely, a DC level m_0 and a modulating function $m(t)$ whose DC component is equal to zero and whose amplitude is bounded such that $|m(t)| \ll m_0$ to assure that the input signal always remains positive. Assuming that the IPFM model is valid, our objective is to design a method which can retrieve information on $m(t)$ from the observed series of event times t_k, represented by the signal

$$d_{\mathrm{E}}^u(t) = \sum_{k=0}^{M} \delta(t - t_k), \qquad (8.2)$$

where the superscript "u" denotes that the events occur unevenly in time (and later also denoting uneven sampling). A definition of the unit impulse function $\delta(t)$ and the impulse-train sampling in (8.2) can be found in most textbooks covering the fundamentals on signals and systems, see, e.g., [12].

In physiological terms, the output signal of the integrator in Figure 8.3 can be viewed as the charging of the membrane potential of a sinoatrial pacemaker cell [13]. The membrane potential increases until a certain threshold is exceeded and then triggers off an action potential which, when combined with the effect of many other action potentials, initiates a new heartbeat. The input to the integrator consists of m_0, which defines the mean heart rate, and the modulating signal $m(t)$, which describes the variations in heart rate as modulated by the autonomic activity on the sinoatrial node. In general, $m(t)$ is bandlimited such that spectral components above 0.4–0.5 Hz can be neglected during resting conditions. The assumption $|m(t)| \ll m_0$ is included in order to assure that the HRV is small when compared to the mean heart rate.

In mathematical terms, the event series is defined by the following equation which is central to the IPFM model,

$$\int_{t_{k-1}}^{t_k} (m_0 + m(\tau))d\tau = R, \quad k = 1, \ldots, M. \qquad (8.3)$$

The modulating function $m(t)$ determines the variation in interval length between two successive events occurring at t_{k-1} and t_k. Without any modulation, i.e., for $m(t) \equiv 0$, the resulting event series is perfectly regular and has a constant interval length equal to R/m_0; the corresponding unmodulated mean repetition frequency $F_I = 1/T_I$ is given by

$$F_I = \frac{m_0}{R}. \qquad (8.4)$$

The constant m_0 is usually set to one, implying that the inversely-related threshold R specifies the mean repetition frequency F_I in units of hertz; also, R is identical to the mean RR interval length T_I,

$$R = \frac{1}{F_I} = T_I. \tag{8.5}$$

Hence, the "heart rate" of the IPFM model is equal to 60 events/minute when T_I is chosen to be 1 s.

Assuming that the initial event occurs at $t_0 = 0$, the integral in (8.3) can alternatively be expressed as

$$\int_0^{t_k} (1 + m(\tau))d\tau = kT_I, \quad k = 0, \dots, M, \tag{8.6}$$

where k is an integer that indexes the kth event. Furthermore, rather than having the IPFM model defined for only those time instants t_k when the threshold T_I is exceeded, it can be generalized to a continuous-time function by introducing the following definition [14],

$$\int_0^t (1 + m(\tau))d\tau = \kappa(t)T_I. \tag{8.7}$$

Here, integration up to a certain time t is proportional to a continuous-valued *indexing function* $\kappa(t)$, whose value at t_k is identical to the integer-valued event index k, i.e., $\kappa(t_k) = k$. The generalization of the IPFM model in (8.7) will later make it possible to develop a heart rhythm representation known as the *heart timing* signal.

The behavior of the modulating function $m(t)$ conveys essential information on the HRV. No prior knowledge of $m(t)$ is, however, available, and, therefore, $m(t)$ has to be estimated from the occurrence times of the observed event series. When evaluating the performance of various methods developed for heart rhythm representations, it is commonly assumed that $m(t)$ is defined as a sum of P sinusoids with amplitudes m_p and frequencies F_p,

$$m(t) = \sum_{p=1}^{P} m_p \sin(2\pi F_p t). \tag{8.8}$$

The multiple sinusoid model may account for HRV caused by respiration, changes in blood pressure, and other physiological factors. The amplitudes $m_1, \dots, m_P$ in (8.8) are commonly assumed to have a value much smaller than one. Naturally, the modulating function $m(t)$ can be assigned other structures than the one suggested in (8.8). For example, it may be defined by

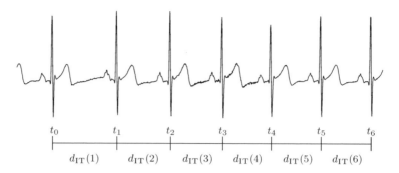

Figure 8.4: Definition of the interval tachogram $d_{IT}(k)$, which is identical to the series of RR intervals. Note that indexing of $d_{IT}(k)$ starts at $k = 1$.

a bandlimited AR process whose spectral peaks correspond to the frequencies $F_1, \ldots, F_P$, introduced in the multiple sinusoid model in (8.8) [14, 15].

The IPFM model is an important tool for simulation studies and to better understand the mechanisms behind HRV [14, 16–21]. Nonetheless, it should be remembered that this model does not provide an exact description of sinoatrial activity; more sophisticated models may be of interest to consider [22–25]. It should also be pointed out that the IPFM model is, by no means, restricted to the study of HRV, but has been found equally useful in other biomedical applications where an event series is observed, for example, in the area of neurophysiology [26–28].

8.3.2   Interval Series Representions

A frequently used heart rhythm representation is the *interval tachogram* $d_{IT}(k)$ in which the events, occurring at $t_0, \ldots, t_M$, are transformed into a discrete-time signal consisting of the successive intervals, i.e., the RR intervals,

$$d_{IT}(k) = t_k - t_{k-1}, \quad k = 1, \ldots, M, \tag{8.9}$$

see Figure 8.4. Hence, the interval tachogram is the heart rhythm representation upon which the simple time domain measures rest (Section 8.2), and it has been extensively used in the literature on HRV analysis, see, e.g., [29–31] (note that the interval tachogram $d_{IT}(k)$ is identical to r_k in (8.1)). The *inverse interval tachogram* $d_{IIT}(k)$ is the "companion" representation to the interval tachogram and is defined by

$$d_{IIT}(k) = \frac{1}{t_k - t_{k-1}}, \quad k = 1, \ldots, M, \tag{8.10}$$

which thus reflects instantaneous heart rate.

A major drawback when using either $d_{\mathrm{IT}}(k)$ or $d_{\mathrm{IIT}}(k)$ is that both these signals are indexed by an interval number rather than by a sample number as is commonly the case with the discrete-time signal, evenly sampled in time. Consequently, power spectral analysis of these two signals cannot be expressed in units of "cycles per second" (Hertz), but can be expressed by the far less attractive unit of "cycles per interval".

Transformation of the tachogram signals into evenly sampled time domain signals is essential not only for obtaining a spectral description in hertz, but also for more advanced variability analysis when the heart rate is cross-correlated with other physiological time domain signals such as blood pressure and respiration. These signals are often sampled at time instants which differ from those of the beat occurrences. Yet another motivation is provided by the situation in which heart rhythm response is studied in relation to certain types of stimulus, such as when solving a mental task or rising from a recumbent to standing position. In such situations, time-synchronized averaging of heart rhythm signals from several successive stimuli may be necessary to establish a reliable response, and, therefore, a signal evenly sampled in time is required.

In contrast to the above tachogram representations, the *interval function* $d_{\mathrm{IF}}(t)$ is defined on a continuous-time basis such that the QRS complex, occurring at time t_k, is represented by a unit impulse function $\delta(t - t_k)$ scaled by the length of the preceding RR interval [32, 33],

$$
\begin{aligned}
d_{\mathrm{IF}}^u(t) &= \sum_{k=1}^{M} (t_k - t_{k-1})\, \delta(t - t_k) \\
&= \sum_{k=1}^{M} d_{\mathrm{IF}}(t)\delta(t - t_k).
\end{aligned}
\tag{8.11}
$$

Similar to $d_{\mathrm{IIT}}(k)$ in (8.10), the *inverse interval function* $d_{\mathrm{IIF}}(t)$ is inversely related to the length of the RR interval,

$$
\begin{aligned}
d_{\mathrm{IIF}}^u(t) &= \sum_{k=1}^{M} \left(\frac{1}{t_k - t_{k-1}} \right) \delta(t - t_k) \\
&= \sum_{k=1}^{M} d_{\mathrm{IIF}}(t)\delta(t - t_k),
\end{aligned}
\tag{8.12}
$$

and reflects instantaneous heart rate [34]. The heart rhythm representations based on the tachogram or the interval function are exemplified in Figure 8.5.

Since both $d_{\mathrm{IF}}^u(t)$ and $d_{\mathrm{IIF}}^u(t)$ represent unevenly sampled signals, it is desirable to resample these functions to become evenly spaced. Resampling

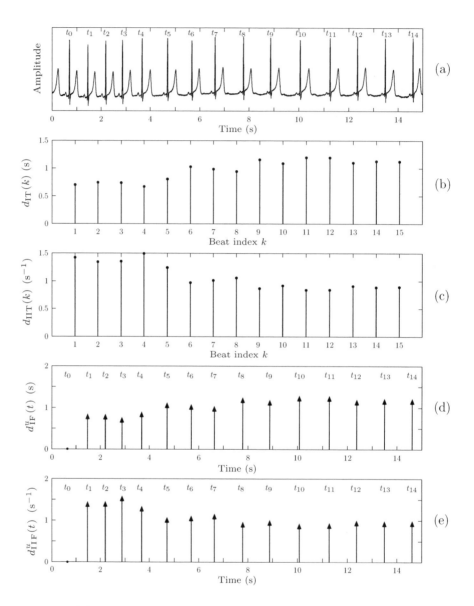

Figure 8.5: (a) An ECG signal with the beat occurrence times t_k. The heart rhythm is represented by (b) the interval tachogram $d_{IT}(k)$, (c) the inverse interval tachogram $d_{IIT}(k)$, (d) the interval function $d_{IF}^u(t)$, and (e) the inverse interval function $d_{IIF}^u(t)$. The functions displayed in (d) and (e) are unevenly sampled.

can be implemented by employing interpolation between the existing samples, thereby resulting in a signal denoted $d^i(t)$, where the superscript "i" denotes interpolated. Regular sampling of the interpolated signal is then performed at the desired rate. The simplest approach to interpolation is to hold the interval value at t_k until the next occurrence time t_{k+1}, and so on. Such a technique is referred to as zero-order, or sample-and-hold, interpolation. The function that results from zero-order interpolation of the inverse interval function $d^i_{\text{IIF}}(t)$ can be expressed as

$$d^i_{\text{IIF}}(t) = \frac{1}{t_1} u(t - t_1) + \sum_{k=2}^{M} \left(\frac{1}{t_k - t_{k-1}} - \frac{1}{t_{k-1} - t_{k-2}} \right) u(t - t_k), \quad (8.13)$$

and its shape is exemplified in Figure 8.6(b). In (8.13), we have assumed that the first beat occurs at $t_0 = 0$. The unit step function, denoted $u(t)$, is defined by

$$u(t) = \begin{cases} 1, & t \geq 0; \\ 0, & t < 0. \end{cases} \quad (8.14)$$

It is obvious from Figure 8.6(b) that a short RR interval, such as the one occurring at about 5 s, causes a disproportionately long, delayed interval with a large value in the interpolated signal $d^i_{\text{IIF}}(t)$. This undesirable effect can be mitigated by introducing a minor modification to (8.13); by shifting the RR intervals one step such that the interval $(t_{k+1} - t_k)$ is instead used to scale $u(t)$ at time t_k, the interpolated function in (8.13) becomes

$$d^i_{\text{IIFs}}(t) = \frac{1}{t_1} + \sum_{k=1}^{M-1} \left(\frac{1}{t_{k+1} - t_k} - \frac{1}{t_k - t_{k-1}} \right) u(t - t_k). \quad (8.15)$$

Figure 8.6(c) shows that the use of $d^i_{\text{IIFs}}(t)$, instead of $d^i_{\text{IIF}}(t)$, leads to an instantaneous heart rate with better tracking properties and has therefore been found to be more suitable for power spectral analysis [18, 35]. The advantage of using $d^i_{\text{IIFs}}(t)$ over $d^i_{\text{IIF}}(t)$ has also been supported by the use of the IPFM model, with results showing that $m(t)$ is better estimated from $d^i_{\text{IIFs}}(t)$ [35].

It is well-known from the design of digital-to-analog converters that the staircase signal, resulting from the sample-and-hold operation, contains high-frequency components. Hence, it is necessary to bandlimit the signal before resampling in order to avoid aliasing. Although the bandlimiting operation, strictly speaking, calls for the design and use of a continuous-time lowpass filter, there are filtering approaches which, fortunately, can be implemented digitally.

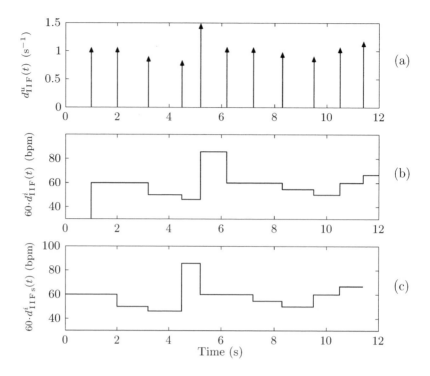

Figure 8.6: (a) The inverse interval function $d_{\text{IIF}}^u(t)$, (b) the corresponding zero-order interpolation using (8.13), and (c) the zero-order interpolation using (8.15) with a delay of one RR interval. The functions displayed in (b) and (c) have been multiplied by a factor of 60 to allow for interpretation in terms of "beats per minute" (bpm). The first beat is assumed to occur at $t_0 = 0$.

Other, more sophisticated interpolation techniques than the zero-order interpolation may also be considered, e.g., involving polynomial fitting. Since the signal is unevenly sampled, the interpolation operation can be interpreted in terms of time-varying filtering, an aspect detailed in Problem 8.7.

8.3.3　Event Series Representation

The heart rhythm representations previously described are redundant since the information on occurrence times t_k and scale factors $(t_k - t_{k-1})$ are closely knit together. This observation may serve as an important motivation for instead considering an event series representation of the heart rhythm, defined by

$$d_{\text{E}}^u(t) = \sum_{k=0}^{M} \delta(t - t_k). \tag{8.16}$$

This expression is identical to the one in (8.2), introduced in connection with the IPFM model, but has a different interpretation since t_k is now estimated from the heartbeats, whereas t_k in (8.2) was produced by the IPFM model. The very low-frequency components of $d_E^u(t)$ contain the information which completely characterizes the variability in heart rate, whereas the high-frequency components can be discarded from further analysis. Therefore, it has been suggested that a useful heart rhythm representation results from lowpass filtering of $d_E^u(t)$ using a linear, time-invariant filter $h(t)$ whose cut-off frequency is chosen well below the mean heart rate [33, 36, 37], see also [38]. The output signal $d_{LE}(t)$ of $h(t)$ is obtained from the following convolution,

$$d_{LE}(t) = \int_{-\infty}^{\infty} h(t-\tau)d_E^u(\tau)d\tau$$
$$= \sum_{k=0}^{M} h(t-t_k). \qquad (8.17)$$

Hence, $d_{LE}(t)$ is computed for any value of t by simply summing the values of the impulse response $h(t)$ at $(M+1)$ different points in time $(t-t_k)$, see Figure 8.7.

Considering an ideal, continuous-time, lowpass filter, the impulse response $h(t)$ is defined by a sinc function,

$$h(t) = \frac{\sin(2\pi F_c t)}{\pi t}, \quad -\infty < t < \infty, \qquad (8.18)$$

where the cut-off frequency is denoted F_c. The cut-off frequency is usually chosen within the interval of 0.4–0.5 Hz in order to comply with heart rates being typical at rest. Alternatively, F_c can be related to the prevailing heart rate which sometimes allows the use of a higher cut-off frequency and, consequently, the analysis of frequency components which describe faster fluctuations in heart rate.

Since the tails of the sinc function in (8.18) drop off to zero, it is possible to truncate the number of terms in the sum of (8.17) once the distance between t and t_k has exceeded a certain value. The accuracy of the lowpass filtered signal $d_{LE}(t)$ decreases when M is small or when output values close to the interval end points t_0 or t_M are to be computed. Another property of the lowpass filtered event series is its noncausal computation; however, this property does not impose any serious limitation as long as the ECG signal is subjected to off-line analysis which is usually the case.

When the event series is assumed to be produced by the IPFM model and a sinusoidal modulating function $m(t)$ with frequency F_1, an ideal lowpass filter can be employed to extract F_1, provided that $F_1 < F_c$. As a

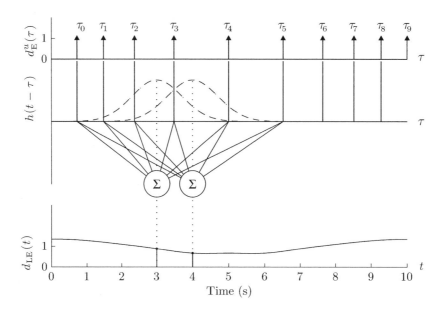

Figure 8.7: The transformation of an event series $d_E^u(t)$ into a lowpass filtered event series $d_{\mathrm{LE}}(t)$ using a filter with impulse response $h(t)$. The computational procedure for obtaining output samples at $t = 3$ and 4 is indicated by the summation networks. Note that output samples can be computed at any time.

result, the lowpass filtered event series yields an estimate of the modulating function $m(t)$ [13],

$$\hat{m}(t) = d_{\mathrm{LE}}(t). \tag{8.19}$$

The case with $m(t)$ being a sinusoid with frequency F_1 is further considered on page 596–597 when the power spectrum of the resulting event series is derived.

Since $d_{\mathrm{LE}}(t)$ in practice is computed by digital techniques, we briefly summarize its discrete-time version which is based on the occurrence times θ_k, estimated from an ECG signal $x(n)$ digitized at the sampling rate F_x. The sampling rate of $d_{\mathrm{LE}}(t)$, denoted F_d, is, of course, chosen to be much lower than that of $x(n)$ and, without much loss of generality, such that the sampling interval $1/F_d$ is an integer multiple L of the interval $1/F_x$, i.e., $1/F_d = L/F_x$. Assuming that the impulse response of the discrete-time lowpass filter $h(n)$ is sampled at the rate F_x, the discrete-time version of $d_{\mathrm{LE}}(t)$ is obtained by

$$d_{\mathrm{LE}}(n) = \sum_{k=0}^{M} h(nL - \theta_k). \tag{8.20}$$

For the special case when $h(n)$ is a truncated version of an ideal lowpass filter with a cut-off frequency that exactly matches the Nyquist frequency, i.e., $F_c = F_d/2$, an efficient algorithm can be derived for the computation of the evenly spaced sample values of $d_{\mathrm{LE}}(t)$ [36]; the algorithm, known as the *French–Holden algorithm*, is discussed in Problem 8.5. This particular choice of F_c may not, however, provide sufficient attenuation at or above the Nyquist frequency, and, as a result, aliasing distortion may be introduced in $d_{\mathrm{LE}}(n)$.

Another efficient algorithm has been presented in which the signal is oversampled by choosing a sampling rate being a factor of two larger than the maximal frequency of interest, i.e., $F_c = F_d/4$. Such a choice provides a much better attenuation of frequency components near the Nyquist frequency and thus avoids aliasing [39]. Since straightforward truncation of the impulse response of the ideal lowpass filter does not produce very good attenuation of the stopband, the use of, for example, windowing techniques is warranted to improve the filter design [40]; see also page 461. Like the earlier mentioned problem of baseline wander removal, lowpass filtering of the event series is synonymous with narrowband filtering which may be efficiently implemented using a multirate filter structure.

8.3.4 Heart Timing Representation

The heart timing signal is, in contrast to the previous heart rhythm representations, based on the IPFM model and is aimed explicitly at estimating the modulating function $m(t)$ [14]. The heart timing signal $d_{\mathrm{HT}}^u(t)$ is an unevenly sampled signal defined as the deviation of the event time t_k from the expected occurrence time, related to the mean RR interval length kT_I, which in mathematical terms is

$$d_{\mathrm{HT}}^u(t) = \sum_{k=0}^{M}(kT_I - t_k)\delta(t - t_k)$$

$$= \sum_{k=0}^{M} d_{\mathrm{HT}}(t)\delta(t - t_k), \tag{8.21}$$

see Figure 8.8. To understand how $d_{\mathrm{HT}}^u(t)$ is related to the IPFM model, we rewrite the model equation in (8.6) for a particular time t_k such that

$$\int_0^{t_k} m(\tau)d\tau = kT_I - t_k$$

$$= d_{\mathrm{HT}}(t_k). \tag{8.22}$$

Hence, $d_{\mathrm{HT}}(t_k)$ and $m(t)$ are linearly related to each other through integration of $m(t)$ until t_k. In order to compute $d_{\mathrm{HT}}^u(t)$, an estimate of T_I is first

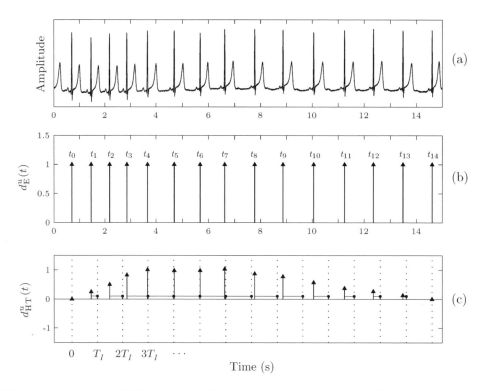

Figure 8.8: (a) An ECG signal and (b) the related event series $d_{\mathrm{E}}^u(t)$ which displays the beat occurrence times t_k. (c) The heart timing signal $d_{\mathrm{HT}}^u(t)$ is defined as the deviation of the event time t_k from the expected occurrence time kT_I (dotted vertical lines). The deviation's magnitude is indicated by the horizontal bar, as well as by the amplitude of the arrow. The time base of $d_{\mathrm{HT}}^u(t)$ is shifted so that its origin is $t = t_0 = 0$.

required from the available data. This parameter can be obtained by simply dividing the occurrence time of the last event with the number of events,

$$\hat{T}_I = \frac{t_M - t_0}{M} = \frac{t_M}{M},\tag{8.23}$$

where we have assumed that $t_0 = 0$. Thus, $d_{\mathrm{HT}}(t)$ depends on where the interval $[t_0, t_M]$ is positioned within the ECG recording. It should be noted that the end point values are such that $d_{\mathrm{HT}}(t_0) = d_{\mathrm{HT}}(t_M) = 0$, see Figure 8.8(c).

The rationale for using the heart timing signal becomes evident when the Fourier transform of its generalization to continuous-time, denoted $d_{\mathrm{HT}}(t)$, is determined. To do this, we make use of the generalized IPFM model in

(8.7), by which the heart timing signal can be expressed as

$$d_{\mathrm{HT}}(t) = \int_0^t m(\tau)d\tau$$

$$= \int_{-\infty}^t m(\tau)d\tau. \tag{8.24}$$

Here, the integration interval has been extended to $-\infty$ due to the nonrestrictive assumption that $m(t)$ is a causal function, i.e., equal to zero for $t < 0$. The Fourier transform of (8.24) is given by [12]

$$D_{\mathrm{HT}}(\Omega) = \int_{-\infty}^{\infty} d_{\mathrm{HT}}(t)e^{-\jmath\Omega t}dt$$

$$= \frac{M(\Omega)}{\jmath\Omega} + \pi M(0)\delta(\Omega)$$

$$= \frac{M(\Omega)}{\jmath\Omega}, \tag{8.25}$$

where $D_{\mathrm{HT}}(\Omega)$ and $M(\Omega)$ denote the Fourier transform of $d_{\mathrm{HT}}(t)$ and $m(t)$, respectively, and $\Omega = 2\pi F$. The term $\pi M(0)\delta(\Omega)$ is identical to zero since $m(t)$ was assumed to have a DC component equal to zero.

Consequently, an estimate of the power spectrum $S_m(\Omega)$ of $m(t)$ can be obtained by multiplying $D_{\mathrm{HT}}(\Omega)$, calculated from the event times $t_0, \ldots, t_M$ in the observation interval, by $\jmath\Omega$,

$$\hat{S}_m(\Omega) = \frac{1}{(M+1)T_I}|\hat{M}(\Omega)|^2$$

$$= \frac{1}{(M+1)T_I}|\Omega\hat{D}_{\mathrm{HT}}(\Omega)|^2. \tag{8.26}$$

The multiplicative factor $1/((M+1)T_I)$ is included to account for the total time interval with events. Once the spectrum of $d_{\mathrm{HT}}(t)$ has been computed, it is straightforward to estimate the spectrum of $m(t)$ (further details on the spectral computations follow below). The modulating function $m(t)$ is assumed to be bandlimited to a maximal frequency lower than half the mean heart rate $1/(2T_I)$. As a result, $d_{\mathrm{HT}}(t)$ will also be bandlimited, being the integral of $m(t)$, and can therefore be fully retrieved from the time instants t_k.

The agreement between $m(t)$ and the different heart rhythm representations, except $d_{\mathrm{LE}}(t)$, is illustrated in Figure 8.9, assuming that $m(t)$ is sinusoidal. Figures 8.9(a)–(c) show the signals at different stages of the IPFM model, namely, the input signal $m(t)$, the output signal of the integrator $\kappa(t)$, and the resulting event series $d_{\mathrm{E}}^u(t)$. In order to interpret $d_{\mathrm{IT}}(k)$

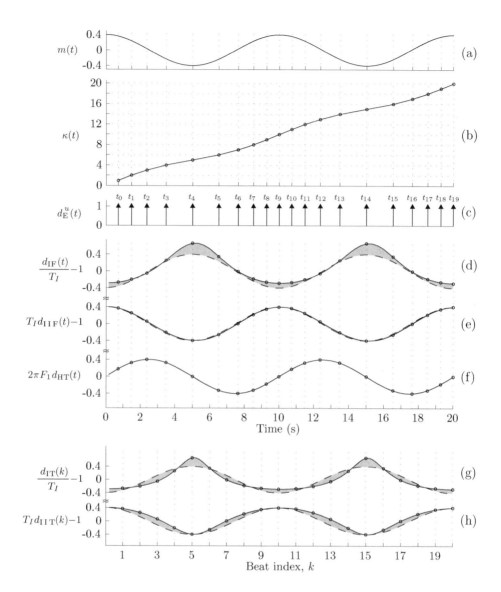

Figure 8.9: Different heart rhythm representations for an event series generated by the IPFM model assuming that $m(t) = 0.4\cos(2\pi F_1 t)$, $F_1 = 0.1$ Hz, and $T_I = 1$ s. (a) The modulating function, (b) the indexing function, (c) the resulting event series, (d) the interval function, (e) the inverse interval function, (f) the heart timing signal, (g) the interval tachogram, and (h) the inverse interval tachogram. Values between the event times were obtained by interpolation [14]. For comparison, $m(t)$ (dashed line) is superimposed in (e) and (h); the inverse of $m(t)$ (dashed line) is superimposed in (d) and (g) since $d_{\mathrm{IF}}(t)$ and $d_{\mathrm{IT}}(k)$ are estimates of $1/m(t)$. In (f), $m(t)$ is shifted by $\pi/2$ in order to account for the $\jmath$ factor which relates $d_{\mathrm{HT}}(t)$ to $m(t)$ in (8.25).

and $d_{\mathrm{IF}}(t)$ as estimates of $m(t)$, we scale these signals with the mean heart rate $1/T_I$ and subtract the mean, which, after scaling, is equal to one (Figures 8.9(d) and (g)). Similarly, $d_{\mathrm{IIT}}(k)$ and $d_{\mathrm{IIF}}(t)$ are scaled with T_I and the mean subtracted (Figures 8.9(e) and (h)).

Figure 8.9(f) demonstrates, as expected, that $d_{\mathrm{HT}}(t)$ is the preferred representation for recovering $m(t)$, although the inverse interval function $d_{\mathrm{IIF}}(t)$ comes close. Another observation is that the representations inversely related to the interval length are better in estimating $m(t)$ than those proportional to the interval length. This observation can be explained by (8.6), in which the term $(1 + m(t))/T_I$ can be interpreted as the instantaneous heart rate which, when integrated over time, gives the beat index k. Since $d_{\mathrm{IIF}}(t)$ reflects instantaneous heart rate, $d_{\mathrm{IIF}}(t)T_I - 1$ can be interpreted as an estimate of $m(t)$, thus explaining the better performance than what is achieved by $d_{\mathrm{IF}}(t)$.

Finally, it seems appropriate to point out that although $d_{\mathrm{HT}}(t)$ exhibits superior performance within the context of IPFM modeling than the other representations, model-based studies do not fully account for HRV observed in humans. Hence, the performance improvement achieved by using $d_{\mathrm{HT}}(t)$ remains to be demonstrated from a clinical point of view and is possibly embedded in the IPFM modeling error.

8.4 Spectral Analysis of Heart Rate Variability

The spontaneous variability in heart rate found in healthy subjects during rest usually exhibits an oscillatory behavior. Such variability is influenced by respiratory activity as well as by feedback mechanisms of the systems for regulation of temperature and blood pressure. The different systems oscillate spontaneously at rest with characteristic frequencies in different intervals: a thermoregulatory peak in the interval below 0.05 Hz, a peak related to blood pressure at about 0.1 Hz, and a peak related to respiration in an interval ranging from 0.2 to 0.4 Hz, see Figure 8.10.

By quantifying the power of the different spectral components, information may be inferred on various pathologies related to cardiac autonomic function. Unfortunately, the oscillations are sometimes poorly pronounced, especially those reflecting changes in thermoregulation and blood pressure, and, therefore, it may be difficult to identify the peaks of the estimated spectrum. This problem is commonly alleviated by instead quantifying the power of low- and high-frequency components in the two intervals 0.04–0.15 Hz and 0.15–0.40 Hz. The spectral power measured in these two intervals is closely associated with autonomic balance; an increase in sympathetic activity is related to an increase of the low-frequency power, whereas an in-

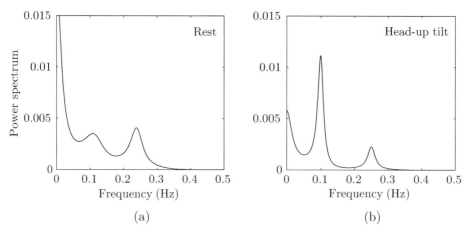

Figure 8.10: Power spectrum of a heart rate signal acquired from a normal subject during (a) resting conditions and (b) a $90°$ head-up tilt. The power spectra were obtained by fitting a 7th-order AR model to the signal. The head-up tilt increases sympathetic activity as reflected by the increased peak at 0.1 Hz. The peak at 0.25 Hz can be attributed to respiration as controlled by parasympathetic activity.

crease in parasympathetic activity is primarily related to an increase of the high-frequency power.[3] Hence, the ratio between these two spectral power measures serves as an index of autonomic balance and has, as such, been extensively used in clinical HRV studies [41–43].

Stationarity is naturally an important consideration when a heart rate signal is subjected to spectral analysis. Although various stationarity tests of a signal have been proposed which, for example, test for deviations from the assumption of a constant mean ("trend shifts"), these tests have rarely found their way into clinical use. Instead, practical tests have been applied on the presence of ectopic beats since such beats clearly violate stationarity; if included for spectral analysis, a false increase in spectral power results which is distributed over the entire frequency interval.

A crucial insight when investigating the spectral content of a heart rate signal is that frequencies above half the mean heart rate cannot be analyzed because the sampling rate is intrinsically defined by the time instants when the beats occur. In reality, the highest frequency has to be somewhat lower than half the mean heart rate since the heart rate may fluctuate considerably so that the length of the longest RR interval bounds the highest frequency [33, 44]. Nevertheless, fruitless attempts have been made to ana-

---

[3]These relations have been established through experiments in which, for example, a substance has been injected which is known to block either the sympathetic or the parasympathetic activity.

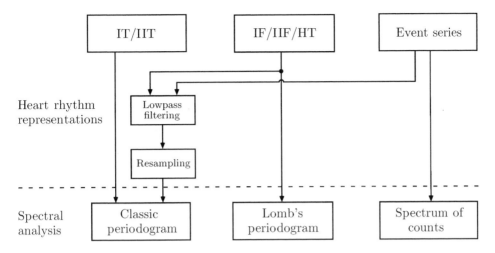

Figure 8.11: Interrelationships between different heart rhythm representations and techniques for spectral analysis. The classical periodogram is based on an evenly sampled signal, whereas the other two spectral techniques assume that the samples are unevenly spaced. Note that lowpass filtering and interpolation are represented by a single block since interpolation can be interpreted as a lowpass filtering operation.

lyze frequencies above half the Nyquist frequency despite the fact that this interval only contains aliased frequency components and, therefore, does not carry meaningful information.

The interrelationships between different heart rhythm representations and techniques for spectral analysis are presented in the block diagram of Figure 8.11. The two tachogram signals $d_{\mathrm{IT}}(k)$ and $d_{\mathrm{IIT}}(k)$ can be analyzed using either classical periodogram-based methods or model-based, parametric methods such as AR modeling; both approaches have been described within the context of EEG signal processing. These two approaches to spectral analysis are relatively straightforward to apply and are not further described. Studies making use of AR modeling in HRV analysis can, for example, be found in [15, 31, 45], and its possibilities and limitations are discussed in [46, 47].

The interval functions $d_{\mathrm{IF}}^{u}(t)$ and $d_{\mathrm{IIF}}^{u}(t)$ or the heart timing $d_{\mathrm{HT}}^{u}(t)$ may be interpolated and resampled at evenly spaced times and then processed with the same methods as those used for the tachogram signals. However, since these signals are unevenly sampled, we may alternatively consider spectral techniques designed to directly handle such sampling. *Lomb's periodogram* is one such technique which, based on the least-squares criterion, produces a nonparametric estimate of the power spectrum; this periodogram is identical to the classical definition when an evenly sampled signal is analyzed. Finally, the event series may be spectrally analyzed after lowpass filtering and resampling. Alternatively, the event series may be inserted directly into the definition of the Fourier transform and evaluated to yield the *spectrum of counts*.

8.4.1   Direct Estimation from Unevenly Spaced Samples

We will start the presentation of different spectral techniques by taking a closer look at the Fourier transform of a general, unevenly sampled signal $d^u(t)$. This signal is conveniently represented by the product of the sampling function $d_E^u(t)$, defined by a train of unit impulse functions positioned at the event times t_k, and the continuous-time signal $d(t)$ to be sampled,

$$d^u(t) = d(t)d_E^u(t), \tag{8.27}$$

where

$$d_E^u(t) = \sum_{k=-\infty}^{\infty} \delta(t - t_k). \tag{8.28}$$

Here, the event series is extended to include both negative and positive values of the index k. The signal $d(t)$ may be defined by any of the presented heart rhythm signals, i.e., $d_{IF}(t), d_{IIF}(t), d_{HT}(t)$, or $d_{LE}(t)$. Since multiplication in the time domain corresponds to convolution in the frequency domain, the Fourier transform of the product $d(t)d_E^u(t)$ in (8.27) is

$$D^u(\Omega) = \frac{1}{2\pi} \int_{-\infty}^{\infty} D(\xi)D_E^u(\Omega - \xi)d\xi$$
$$= D(\Omega) * D_E^u(\Omega), \tag{8.29}$$

where

$$D_E^u(\Omega) = \sum_{k=-\infty}^{\infty} e^{-j\Omega t_k}. \tag{8.30}$$

Thus, the spectrum $D^u(\Omega)$ is related to a version of $D(\Omega)$ which is modified by convolution with $D_E^u(\Omega)$. Based on a finite-length observation interval with $M+1$ events, the power spectrum of $d^u(t)$ is estimated by the following expression,

$$\hat{S}_{d^u}(\Omega) = \frac{1}{(M+1)}|\hat{D}^u(\Omega)|^2$$

$$= \frac{1}{(M+1)}\left|\sum_{k=0}^{M} d^u(t_k)e^{-j\Omega t_k}\right|^2. \tag{8.31}$$

The Fourier transform $\hat{D}^u(\Omega)$ is obtained from the available series of event times $t_0, \ldots, t_M$, which may be viewed as windowing of $d^u(t)$ using a rectangular window. It may be worthwhile to call the reader's attention to the fact that the Fourier transform $D^u(\Omega)$ results from an unevenly sampled, continuous-time signal represented by delta functions and is not a periodic function as the Fourier transform of a discrete-time signal. However, the frequency interval of interest when studying HRV is still limited upwards by half the mean heart rate.

In general, it is difficult to describe the effect of $D_E^u(\Omega)$ on the original spectrum $D(\Omega)$ in (8.29). Certain insight may be gained from the special case when the event times t_k are transformed into evenly spaced samples by interpolation and resampling. Assuming that the event times are integer multiples of the interval length T_I, i.e., $t_k = kT_I$, the convolution in (8.29) becomes

$$D^e(\Omega) = D^i(\Omega) * \left(\sum_{k=-\infty}^{\infty} e^{-\jmath\Omega k T_I} \right), \qquad (8.32)$$

where $D^i(\Omega)$ is the spectrum of the interpolated signal, and $D^e(\Omega)$ is the spectrum of the evenly resampled signal. Poisson's formula can be used to express the sum of complex exponentials in (8.32) as a train of equidistantly spaced impulse functions,

$$\sum_{k=-\infty}^{\infty} e^{-\jmath k \Omega T_I} = \frac{1}{T_I} \sum_{k=-\infty}^{\infty} \delta\left(\Omega - k\Omega_I\right), \qquad (8.33)$$

where $\Omega_I = 2\pi/T_I$. Inserting (8.33) in (8.32), we obtain the well-known result from sampling theory which states that the spectrum $D^e(\Omega)$ of the sampled signal is a repetition of the spectrum $D^i(\Omega)$ of the original signal,

$$D^e(\Omega) = \frac{1}{T_I} \sum_{k=-\infty}^{\infty} D^i\left(\Omega - k\Omega_I\right). \qquad (8.34)$$

When the spectrum $D^i(\Omega)$ is bandlimited, such that its highest frequency component does not exceed half the repetition rate of t_k, i.e., $F_I = 1/T_I$, the sampled signal will not be distorted by aliasing.

8.4.2 The Spectrum of Counts

The spectrum of the event series $d_E^u(t)$ deserves special mentioning since it has been widely studied in the literature, commonly referred to as the spectrum of counts. From the definition of $d_E^u(t)$ in (8.28), it is straightforward to calculate its Fourier transform,

$$D_{\mathrm{E}}^u(\Omega) = \int_{-\infty}^{\infty} d_{\mathrm{E}}^u(t) e^{-\jmath\Omega t} dt$$

$$= \sum_{k=-\infty}^{\infty} e^{-\jmath\Omega t_k}, \tag{8.35}$$

by simply inserting the values of the observed event times in the sum. The corresponding power spectrum is obtained by

$$\hat{S}_{d_{\mathrm{E}}^u}(\Omega) = \frac{1}{(M+1)} |\hat{D}_{\mathrm{E}}^u(\Omega)|^2$$

$$= \frac{1}{(M+1)} \left[\left(\sum_{k=0}^{M} \cos(\Omega t_k) \right)^2 + \left(\sum_{k=0}^{M} \sin(\Omega t_k) \right)^2 \right]. \tag{8.36}$$

From (8.35) it is evident that the spectrum of counts is identical to the term $D_{\mathrm{E}}^u(\Omega)$ in (8.29) by which the spectrum $D(\Omega)$ is convolved. For the event series representation, the spectrum $D(\Omega)$ is equal to $\delta(\Omega)$ since $d(t)$ is a constant function with unit amplitude.

Considerable insight on the properties of the spectrum of counts can be obtained when the event times t_k are represented as deviations from the mean heart rate, i.e., described by the heart timing signal $d_{\mathrm{HT}}(t)$ [14]. For this particular representation, we can derive an analytic expression of the spectrum $D_{\mathrm{E}}^u(\Omega)$ in terms of the modulating function $m(t)$ which defines HRV in the IPFM model. Using the definition of $d_{\mathrm{HT}}(t)$ in (8.22), we can express t_k as

$$t_k = kT_I - d_{\mathrm{HT}}(t_k) \tag{8.37}$$

and, consequently, the event series as

$$d_{\mathrm{E}}^u(t) = \sum_{k=-\infty}^{\infty} \delta(t - kT_I + d_{\mathrm{HT}}(t_k)). \tag{8.38}$$

A key step in the derivation of $D_{\mathrm{E}}^u(\Omega)$ is to use a technique by which t_k in (8.38) can be completely eliminated from the impulse functions, thereby making further manipulations tractable. It can be shown that for any function $g(t)$ with a single first-order zero at $t = \tau$, i.e., $g(\tau) = 0$, $g(t \neq \tau) \neq 0$, and $\partial g(t)/\partial t|_{t=\tau} \neq 0$, the time-shifted impulse function can be written as [48]

$$\delta(t - \tau) = \left| \frac{\partial g(t)}{\partial t} \right| \delta(g(t)), \tag{8.39}$$

where the right-hand side is independent of the shift τ.

Inspired by the appearance of the impulse functions in (8.38), we define the function $g(t)$ as

$$g(t) = t - kT_I + d_{\mathrm{HT}}(t), \tag{8.40}$$

which can be shown to satisfy the above requirements at $\tau = t_k$. Insertion of this particular choice of $g(t)$ into (8.39) and setting $\tau = t_k$ yield

$$\delta(t - t_k) = \left| 1 + \frac{\partial d_{\mathrm{HT}}(t)}{\partial t} \right| \delta(t - kT_I + d_{\mathrm{HT}}(t)). \tag{8.41}$$

Since $d_{\mathrm{HT}}(t)$ is related to $m(t)$ through integration, cf. (8.24), we have that

$$\frac{\partial d_{\mathrm{HT}}(t)}{\partial t} = m(t), \tag{8.42}$$

which, together with the property of HRV being small in comparison with the mean heart rate (i.e., $|m(t)| \ll 1$), enables us to express $\delta(t - t_k)$ as

$$\delta(t - t_k) = (1 + m(t))\delta(t - kT_I + d_{\mathrm{HT}}(t)). \tag{8.43}$$

Insertion of this result into the definition of the event series yields

$$d_{\mathrm{E}}^u(t) = (1 + m(t)) \sum_{k=-\infty}^{\infty} \delta(t - kT_I + d_{\mathrm{HT}}(t)), \tag{8.44}$$

which can be rewritten in a more suitable format using Poisson's formula,

$$
\begin{aligned}
d_{\mathrm{E}}^u(t) &= \frac{1 + m(t)}{T_I} \left[\sum_{k=-\infty}^{\infty} e^{j\frac{2\pi k}{T_I}(t + d_{\mathrm{HT}}(t))} \right] \\
&= \frac{1 + m(t)}{T_I} \left[1 + \sum_{k=1}^{\infty} 2\cos\left(\frac{2\pi k}{T_I}(t + d_{\mathrm{HT}}(t)) \right) \right].
\end{aligned} \tag{8.45}
$$

The Fourier transform of this expression is equal to

$$D_{\mathrm{E}}^u(\Omega) = \left(\frac{\delta(\Omega) + M(\Omega)}{T_I} \right) * \left[\delta(\Omega) + \sum_{k=1}^{\infty} D_{\mathrm{HT}_k}(\Omega) \right], \tag{8.46}$$

where the term $D_{\mathrm{HT}_k}(\Omega)$ denotes the Fourier transform of a frequency modulated (FM) function $d_{\mathrm{HT}}(t)$ whose "carrier frequency" is located at k/T_I,

$$D_{\mathrm{HT}_k}(\Omega) = \mathcal{FT}\left\{ 2\cos\left(\frac{2\pi k}{T_I}(t + d_{\mathrm{HT}}(t)) \right) \right\}. \tag{8.47}$$

Recalling that we are only interested in spectral components below half the mean heart rate, $d_{\mathrm{HT}}(t)$ is bandlimited to $1/(2T_I)$, and assuming that $d_{\mathrm{HT}}(t) < T_I$, only frequencies represented by the first term $D_{\mathrm{HT}_1}(\Omega)$ are located within the interval of interest [44]. Consequently, the expression of $D_{\mathrm{E}}^u(\Omega)$ in (8.46) can be well-approximated by

$$D_{\mathrm{E}}^u(\Omega) \approx \frac{1}{T_I} \left(\delta(\Omega) + M(\Omega) + D_{\mathrm{HT}_1}(\Omega) + M(\Omega) * D_{\mathrm{HT}_1}(\Omega) \right). \quad (8.48)$$

From this expression of the spectrum of counts it can be concluded that, apart from the desired term $M(\Omega)$, three unwanted terms exist in (8.48) related to the DC component and the term $D_{\mathrm{HT}_1}(\Omega)$. At a first glance, the DC component does not seem to represent a problem since it is outside the frequency interval of interest. However, the DC component turns out to be more problematic since the DC power leaks to adjacent frequencies in the very low-frequency interval of the estimated spectrum; cf. the leakage effect of the periodogram described on page 94. In contrast to the usual situation where the DC component is subtracted prior to spectral analysis in order to avoid this effect, such subtraction is obviously not meaningful when an event series is subjected to spectral analysis. It can be concluded from (8.48) that, using suitable lowpass filtering to compute $d_{\mathrm{LE}}(t)$, an estimate of $m(t)$ is obtained.

For the case when $m(t) = m_1 \sin(2\pi F_1 t)$, the spectrum of the resulting event series $d_{\mathrm{E}}^u(t)$ can be determined analytically [26]. In addition to the expected frequency peaks at F_I (the mean heart rate) and F_1, it can be shown that the spectrum also contains a number of spurious peaks centered around F_I at distances which are integer multiples of F_1, see Figure 8.12. The amplitudes of the spurious peaks depend on the degree of modulation; a large value of m_1 causes the spurious peaks to interfere significantly with the peak at F_1.

Returning to the Fourier transform of a general, unevenly spaced heart rate signal $d^u(t)$ in (8.29), we recall that the spectrum $D^u(\Omega)$ is the convolution between $D(\Omega)$ and $D_{\mathrm{E}}^u(\Omega)$. Using the approximation of $D_{\mathrm{E}}^u(\Omega)$ in (8.48), we obtain the following result,

$$D^u(\Omega) = D(\Omega) * D_{\mathrm{E}}^u(\Omega)$$

$$\approx \frac{1}{T_I} (D(\Omega) + D(\Omega) * M(\Omega)), \quad |\Omega| < \frac{1}{2T_I}, \quad (8.49)$$

where the second step results from neglect of the FM terms since these components are mainly located outside the frequency band of interest. Hence, the term $D(\Omega) * M(\Omega)$ in (8.49) provides the explanation as to why $D^u(\Omega)$, i.e., the spectrum of an unevenly sampled signal, differs from the desired spectrum $D(\Omega)$.

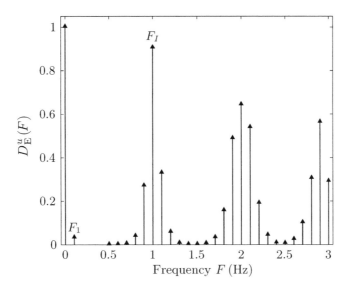

Figure 8.12: The spectrum of the event series generated by the IPFM model with the modulating function $m(t)$ chosen to be a sinusoid with frequency $F_1 = 0.1$ Hz. The heart rate is reflected by the peak at $F_I = 1$ Hz.

8.4.3 Lomb's Periodogram

Lomb's periodogram [49, 50] is useful for estimating the power spectrum directly from an unevenly sampled signal and constitutes an alternative to the classical periodogram combined with interpolation and resampling. Similar to the classical periodogram, Lomb's periodogram is a nonparametric estimation technique which does not make any assumptions on the genesis of the analyzed signal. The main idea behind this approach is the definition of a spectrum that results from minimization of the squared error between $d(t_k)$ and a sinusoidal model signal $s(t_k; \Omega)$,

$$\mathcal{E} = \sum_{k=0}^{M} (d(t_k) - s(t_k; \Omega))^2, \tag{8.50}$$

where

$$s(t_k; \Omega) = a_1 \cos(\Omega t_k) + a_2 \sin(\Omega t_k). \tag{8.51}$$

The energy of $s(t_k; \Omega)$ is a measure which reflects the degree by which a certain frequency Ω is contained in $d(t_k)$. Once estimates of the amplitudes

a_1 and a_2 have been computed, the energy is obtained by

$$\sum_{k=0}^{M} \hat{s}^2(t_k; \Omega) = \sum_{k=0}^{M} (\hat{a}_1(\Omega) \cos(\Omega t_k) + \hat{a}_2(\Omega) \sin(\Omega t_k))^2, \qquad (8.52)$$

where the dependence on Ω for the two estimates $\hat{a}_1(\Omega)$ and $\hat{a}_2(\Omega)$ is explicitly indicated.

Instead of solving the least-squares (LS) problem in (8.50) by straightforward minimization of $\mathcal{E}$ with respect to a_1 and a_2, we recast the problem in matrix notation and derive the general LS solution and the minimum error. The LS problem can be expressed as

$$\mathcal{E}(\mathbf{a}) = (\mathbf{d} - \mathbf{H}\mathbf{a})^T (\mathbf{d} - \mathbf{H}\mathbf{a}), \qquad (8.53)$$

where

$$\mathbf{H} = \begin{bmatrix} \mathbf{h}_1 & \mathbf{h}_2 \end{bmatrix} = \begin{bmatrix} \cos(\Omega t_0) & \sin(\Omega t_0) \\ \cos(\Omega t_1) & \sin(\Omega t_1) \\ \vdots & \vdots \\ \cos(\Omega t_M) & \sin(\Omega t_M) \end{bmatrix} \qquad (8.54)$$

and

$$\mathbf{d} = \begin{bmatrix} d(t_0) & d(t_1) & \cdots & d(t_M) \end{bmatrix}^T, \qquad (8.55)$$

$$\mathbf{a} = \begin{bmatrix} a_1 & a_2 \end{bmatrix}^T. \qquad (8.56)$$

The model signal is thus $\mathbf{s} = \mathbf{H}\mathbf{a}$. Minimization of $\mathcal{E}(\mathbf{a})$ is accomplished by calculating the gradient with respect to $\mathbf{a}$ (see Appendix A for differentiation of vectors and matrices),

$$\nabla_{\mathbf{a}} \mathcal{E}(\mathbf{a}) = -2\mathbf{H}^T (\mathbf{d} - \mathbf{H}\mathbf{a}). \qquad (8.57)$$

By setting this gradient equal to zero, the LS estimator is obtained as

$$\hat{\mathbf{a}} = (\mathbf{H}^T \mathbf{H})^{-1} \mathbf{H}^T \mathbf{d}. \qquad (8.58)$$

The matrix $(\mathbf{H}^T \mathbf{H})^{-1}$ is invertible since $\mathbf{H}$ is assumed to have full rank—an assumption which is fulfilled for the specific choice of $\mathbf{H}$ in (8.54). The minimum LS error is found by rewriting (8.53) such that

$$\begin{aligned} \mathcal{E}(\mathbf{a}) &= (\mathbf{d} - \mathbf{H}\mathbf{a})^T (\mathbf{d} - \mathbf{H}\mathbf{a}) \\ &= \mathbf{d}^T (\mathbf{d} - \mathbf{H}\mathbf{a}) - (\mathbf{H}\mathbf{a})^T (\mathbf{d} - \mathbf{H}\mathbf{a}) \\ &= \mathbf{d}^T \mathbf{d} - \mathbf{d}^T \mathbf{H}\mathbf{a} - \mathbf{a}^T \mathbf{H}^T (\mathbf{d} - \mathbf{H}\mathbf{a}) \end{aligned} \qquad (8.59)$$

and by making use of the fact that the last term is equal to the gradient in (8.57), equal to zero in order to assure optimality. Hence, the minimum LS error is

$$\mathcal{E}_{\min} = \mathbf{d}^T\mathbf{d} - \mathbf{d}^T\mathbf{H}\hat{\mathbf{a}}. \tag{8.60}$$

While it is evident that the first term in (8.60) represents the energy of $\mathbf{d}$, it is probably not equally evident in what way the other term, $\mathbf{d}^T\mathbf{H}\hat{\mathbf{a}}$, should be interpreted. To shed some light on this, we introduce the rather general assumptions of additive noise, $\mathbf{d} = \mathbf{s}+\mathbf{v}$, and $\mathbf{v}^T\hat{\mathbf{s}} = 0$; for a stochastic model, the latter assumption states that signal and noise are uncorrelated. In this case, the model signal $\mathbf{s}$ is the component to be estimated, containing the frequency Ω, whereas $\mathbf{v}$ represents the remaining signal components which are labeled as "noise". Then, we may interpret the term $\mathbf{d}^T\mathbf{H}\hat{\mathbf{a}}$ as an estimate of the energy of $\mathbf{s}$ since

$$\begin{aligned}\mathbf{d}^T\mathbf{H}\hat{\mathbf{a}} &= (\mathbf{s} + \mathbf{v})^T\hat{\mathbf{s}} \\ &= \mathbf{s}^T\hat{\mathbf{s}}.\end{aligned} \tag{8.61}$$

Since this energy interpretation agrees with the spectral measure suggested in (8.52), it is not surprising that Lomb's periodogram is defined as

$$\hat{S}_{d^u}(\Omega) \stackrel{\text{def}}{=} \frac{1}{M+1}\mathbf{d}^T\mathbf{H}\hat{\mathbf{a}}, \tag{8.62}$$

where both $\hat{\mathbf{a}}$ and $\mathbf{H}$ depend on Ω. As we will see later, this definition is also attractive because it will reduce to the classical periodogram when the analyzed signal is evenly sampled. Insertion of the LS estimate $\hat{\mathbf{a}}$ into (8.62) yields an expression of Lomb's periodogram in terms of $\mathbf{H}$ and $\mathbf{d}$,

$$\hat{S}_{d^u}(\Omega) = \frac{1}{M+1}\mathbf{d}^T \begin{bmatrix} \mathbf{h}_1 & \mathbf{h}_2 \end{bmatrix} \begin{bmatrix} \mathbf{h}_1^T\mathbf{h}_1 & \mathbf{h}_2^T\mathbf{h}_1 \\ \mathbf{h}_1^T\mathbf{h}_2 & \mathbf{h}_2^T\mathbf{h}_2 \end{bmatrix}^{-1} \begin{bmatrix} \mathbf{h}_1^T \\ \mathbf{h}_2^T \end{bmatrix} \mathbf{d}. \tag{8.63}$$

From a computational point of view, however, it would be highly desirable if the expression in (8.63) could be simplified such that the cross-terms

$$\mathbf{h}_1^T\mathbf{h}_2 = \mathbf{h}_2^T\mathbf{h}_1 = \sum_{k=1}^{M}\cos(\Omega t_k)\sin(\Omega t_k) \tag{8.64}$$

of the inverted matrix could be made equal to zero (in general, the cross-terms are nonzero since the unevenly sampled sine and cosine functions are not orthogonal). Therefore, the question is whether some technique exists by which Lomb's periodogram can be modified to become

$$\hat{S}_{d^u}(\Omega) \stackrel{?}{=} \frac{1}{M+1}\left(\frac{(\mathbf{h}_1^T\mathbf{d})^2}{\mathbf{h}_1^T\mathbf{h}_1} + \frac{(\mathbf{h}_2^T\mathbf{d})^2}{\mathbf{h}_2^T\mathbf{h}_2}\right). \tag{8.65}$$

In order to answer this question positively, Lomb came up with the idea in his original paper [49] to introduce a delay τ in the model signal,

$$\mathbf{H}_\tau = \begin{bmatrix} \mathbf{h}_{1,\tau} & \mathbf{h}_{2,\tau} \end{bmatrix} = \begin{bmatrix} \cos(\Omega(t_0 - \tau)) & \sin(\Omega(t_0 - \tau)) \\ \cos(\Omega(t_1 - \tau)) & \sin(\Omega(t_1 - \tau)) \\ \vdots & \vdots \\ \cos(\Omega(t_M - \tau)) & \sin(\Omega(t_M - \tau)) \end{bmatrix}, \qquad (8.66)$$

and choose τ such that

$$\mathbf{h}_{1,\tau}^T \mathbf{h}_{2,\tau} = \sum_{k=0}^{M} \cos(\Omega(t_k - \tau)) \sin(\Omega(t_k - \tau)) = 0. \qquad (8.67)$$

With the help of certain trigonometric identities, it can be shown (see Problem 8.9) that the value of τ which makes $\mathbf{h}_1^T \mathbf{h}_2$ equal to zero is given by

$$\tau = \frac{1}{2\Omega} \arctan \left(\frac{\displaystyle\sum_{k=0}^{M} \sin(2\Omega t_k)}{\displaystyle\sum_{k=0}^{M} \cos(2\Omega t_k)} \right). \qquad (8.68)$$

Another reason for the introduction of τ is to make Lomb's periodogram translation invariant in time [50]. This crucial property implies that identical periodograms are produced irrespective of where the observed samples are located in time; such translation invariance is not achieved with the matrix $\mathbf{H}$ initially proposed in (8.54).

Thus, Lomb's periodogram in (8.62) is, after modification with the delay parameter τ, given by

$$\hat{S}_{d^u}(\Omega) = \frac{1}{M+1} \left(\frac{(\mathbf{h}_{1,\tau}^T \mathbf{d})^2}{\mathbf{h}_{1,\tau}^T \mathbf{h}_{1,\tau}} + \frac{(\mathbf{h}_{2,\tau}^T \mathbf{d})^2}{\mathbf{h}_{2,\tau}^T \mathbf{h}_{2,\tau}} \right)$$

$$= \frac{1}{M+1} \left[\frac{\left(\displaystyle\sum_{k=0}^{M} d(t_k) \cos(\Omega(t_k - \tau)) \right)^2}{\displaystyle\sum_{k=0}^{M} \cos^2(\Omega(t_k - \tau))} + \frac{\left(\displaystyle\sum_{k=0}^{M} d(t_k) \sin(\Omega(t_k - \tau)) \right)^2}{\displaystyle\sum_{k=0}^{M} \sin^2(\Omega(t_k - \tau))} \right]. $$
$$(8.69)$$

In contrast to the classical periodogram, Lomb's periodogram is not a periodic function, but may convey information on frequencies which span slightly above the Nyquist frequency. These higher frequencies may be analyzed when several samples are much closer in time than the "average" sampling interval. However, such analysis should be exercised with a great deal of caution and not be interpreted such that *any* frequency above the Nyquist frequency is meaningful to study, because aliasing may be present. Since Lomb's periodogram is associated with a considerable amount of computation, a fast algorithm has been developed, similar to the FFT algorithm [51, 52].

Lomb's periodogram reduces to the classical periodogram when the event times t_k are evenly sampled with the sampling interval T_I, i.e., $t_k = kT_I$, at the Nyquist rate or higher. Since $\tau = 0$, we have that

$$\sum_{k=0}^{M} \cos^2(\Omega k T_I) = \sum_{k=0}^{M} \sin^2(\Omega k T_I) = \frac{M+1}{2}, \qquad (8.70)$$

which, when inserted into (8.69), yields the following well-known expression of a Fourier power spectrum,

$$\hat{S}_{d^u}(\Omega) = \frac{1}{M+1} \left(\sum_{k=0}^{M} d(kT_I) \cdot \sqrt{\frac{2}{M+1}} \cos(\Omega k T_I) \right)^2$$

$$+ \frac{1}{M+1} \left(\sum_{k=0}^{M} d(kT_I) \cdot \sqrt{\frac{2}{M+1}} \sin(\Omega k T_I) \right)^2. \qquad (8.71)$$

Figure 8.13 presents an example where an event series, generated by the IPFM model, is analyzed with different spectral estimation techniques. The input to the IPFM model is a two-tone signal with modulation frequencies at 0.1 and 0.25 Hz. In this particular example, the best agreement with the original spectrum is obtained by the classical periodogram based on the heart timing signal $d_{\text{HT}}^i(t)$ which follows from interpolation of $d_{\text{HT}}^u(t)$ and resampling (the interpolation was based on cubic splines and resampling was done at a rate of 2 Hz [14]). Lomb's periodogram, based on either the interval function $d_{\text{IF}}^u(t)$ or the inverse interval function $d_{\text{IIF}}^u(t)$, contains a number of low-amplitude spurious peaks. The same observation applies to the classical periodogram, based on interpolated and resampled versions of $d_{\text{IF}}^u(t)$ or $d_{\text{IIF}}^u(t)$, although the high-frequency components (both expected and spurious ones) are more attenuated due to the lowpass effect of interpolation.

From the example in Figure 8.13, we may conclude that differences in performance between Lomb's periodogram and the classical periodogram are relatively insignificant. Therefore, other aspects such as those related to the amount of computations, e.g., due to interpolation and resampling, and the

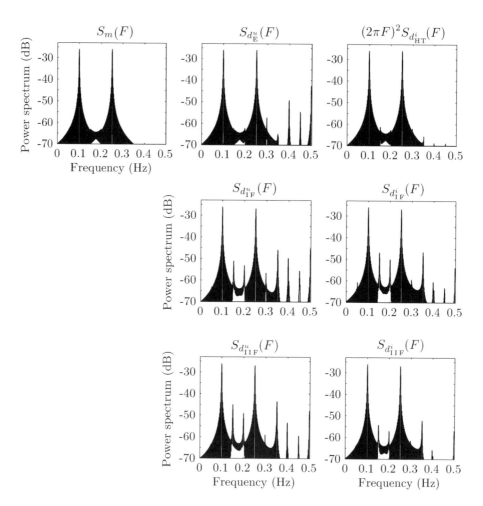

Figure 8.13: Spectral analysis of data generated by the IPFM model for the two-tone case with modulation frequencies at 0.1 and 0.25 Hz [14]. The original two-tone spectrum $S_m(F)$, the related spectrum of counts of $d_E^u(t)$, the classical periodogram based on $d_{HT}^i(t)$, Lomb's periodogram of $d_{IF}^u(t)$, the classical periodogram of $d_{IF}^i(t)$, Lomb's periodogram of $d_{IIF}^u(t)$, and the classical periodogram of $d_{IIF}^i(t)$. The mean RR interval length T_I is equal to 1 s, and the mean value was subtracted from $d_{IF}(t)$ and $d_{IIF}(t)$ before the corresponding spectrum was computed.

handling of gaps due to ectopic beats should be considered when selecting a method for spectral analysis [53, 54].

We conclude this section on spectral analysis by presenting power spectra of sinus rhythm and atrial fibrillation computed by Lomb's method. While the former rhythm is mainstream to this chapter, the latter rhythm is not considered in traditional HRV analysis due to the fact that the sinus node is no longer in control of atrial activation, see page 434. Still, it is instructive to compare the outcome of spectral analysis for these two rhythms, see Figure 8.14. While the power spectrum of sinus rhythm exhibits a pronounced peak at about 0.2 Hz, corresponding to respiration, no such peak can be discerned from the power spectrum of atrial fibrillation since respiration no longer modulates heart rate. Furthermore, atrial fibrillation has a considerably larger variability than sinus rhythm which spectrally is manifested by a larger area under the spectrum, especially at higher frequencies, see Figure 8.14(b); the flatter power spectrum of atrial fibrillation indicates that this rhythm contains less structured information than sinus rhythm.

8.5 Clustering of Beat Morphologies

Analysis of HRV requires that sinus beats be labeled as such before the sinus rhythm can be analyzed. Such labeling is typically accomplished by clustering heartbeat morphologies in exactly the same way that motor unit action potentials (MUAPs) are clustered for the purpose of decomposing intramuscular EMG signals, see the discussion in Section 5.6.1. In contrast to MUAP clustering, where each cluster is equally important, it is only the cluster containing the sinus beats which matters in HRV analysis. Although no prior knowledge is available on which of the clusters contains the sinus beats, the "sinus" cluster can usually be identified as the cluster with the largest number of beats. Therefore, in HRV analysis it is not necessary to find out if, for example, a beat has a P wave and a QRS duration of about 120 ms or less, two properties characteristic of a sinus beat (such beat classification is, however, required in certain clinical applications such as resting ECG analysis and ambulatory monitoring where it is of interest to assign a label to each cluster with a certain type of ectopic beat).

As described in Section 5.6.1, clustering is based on a set of features which describe waveform morphology and, possibly, also rhythm properties. The time domain representation, given in (5.103), has frequently been employed in ECG signal processing, especially when combined with the cross-correlation coefficient as a measure of pattern similarity [55–61], defined by

$$d^2(\mathbf{p}_i, \boldsymbol{\mu}_l) = \frac{\mathbf{p}_i^T \boldsymbol{\mu}_l}{\|\mathbf{p}_i\|_2 \|\boldsymbol{\mu}_l\|_2}. \tag{8.72}$$

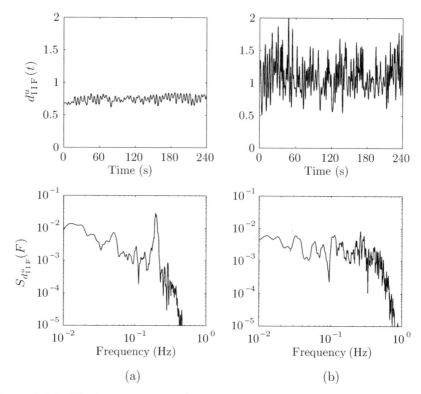

Figure 8.14: The inverse interval function and the corresponding power spectrum for (a) sinus rhythm and (b) atrial fibrillation. Note that the power spectra are displayed with log-log scales.

The vector $\mathbf{p}_i$ contains the QRS samples of the current beat, and $\boldsymbol{\mu}_l$ is the mean of the cluster (i.e., the "template" beat). The ECG signal is usually bandpass filtered before it is clustered so that the influence of baseline wander and EMG noise is reduced.

Using the cross-correlation coefficient as a measure of similarity, it is easily shown that clustering becomes invariant to changes in QRS amplitude. Such a situation is exemplified by Figure 7.17(c) where the sinus beats, exhibiting a drastic, short-term change, would all be assigned to the same cluster. Amplitude invariance is an acceptable property in HRV analysis where the information in demand is restricted to the timing of sinus beats. However, amplitude invariance is undesirable when the purpose is to average the sinus beats of a cluster for noise reduction as, for example, required in high-resolution ECG analysis (Section 6.6.5); averaging of similar-shaped beats with widely differing QRS amplitudes produces a nonrepresentative ensemble average.

The basis function representation in (5.104) has also been considered for feature extraction when clustering heartbeats, often expressed in terms of the Karhunen–Loève or the Hermite functions [62–72]. In such cases, the Mahalanobi distance, defined in (5.105), is preferable as a measure of pattern similarity.

Improved accuracy of the occurrence time t_k is intimately related to the clustering process because the current beat $\mathbf{p}_i$ can be optimally aligned in time to $\boldsymbol{\mu}_l$ when similarity is measured. The availability of morphologic information through $\boldsymbol{\mu}_l$ may be used to improve the accuracy of t_k, originally determined by the QRS detector which operates at a lower temporal resolution (and determined without considering the morphology of previous beats). When clustering is based on the cross-correlation coefficient, the samples of $\mathbf{p}_i$ are correlated to the mean of the cluster $\boldsymbol{\mu}_l$ and shifted in time until the highest cross-correlation value is obtained; the resulting value is used for cluster assignment. The procedure for aligning two waveforms is actually well-known from latency estimation of evoked potentials and is described in detail in Section 4.3.7. Once clustering is finished, the occurrence times of beats contained in the sinus cluster can be further time improved using Woody's method.

8.6   Dealing with Ectopic Beats

The presence of ectopic beats perturbs the impulse pattern initiated by the SA node and implies that the RR intervals adjacent to an ectopic beat cannot be used for HRV analysis, see Figure 8.15. In such cases, autonomic modulation of the SA node is temporarily lost, and, instead, an ectopic focus prematurely initiates the next beat. The location of the ectopic focus gives rise to different types of RR interval perturbation; a beat of ventricular origin inhibits the next sinus beat so that a compensatory pause is introduced after the ectopic beat (Figure 6.14(b)), whereas a beat of supraventricular origin discharges the SA node ahead of schedule and causes the following sinus beat to also occur ahead of schedule (Figure 6.14(a)). Another perturbation is that related to an interpolated ectopic beat, manifested by two short RR intervals adjacent to the ectopic beat (Figures 6.14(e)). Perturbations in rhythm may also be due to missed or falsely detected beats; such errors are usually the result of incorrect decisions made by the QRS detector [73, 74].

Since ectopic beats occur in both normal subjects and patients with heart disease, their presence represents an important error source to be dealt with prior to spectral analysis of the heart rate signal. If not dealt with, the analysis of an RR interval series containing ectopic beats results in a power spec-

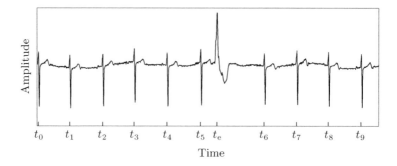

Figure 8.15: A premature ectopic beat occurring at time t_e followed by a compensatory pause. The occurrence times of the sinus beats are denoted $t_0, \ldots, t_9$.

trum with fictitious frequency components, manifested as a "white noise" level as illustrated by Figure 8.16. The increased spectral level is caused by the impulses produced by the two RR intervals adjacent to the ventricular ectopic beat.

From a signal processing viewpoint, we need, of course, to know if each beat is of ectopic origin or not before a correction technique can be applied; beat clustering provides this knowledge in most systems for ECG analysis (Section 8.5). Ventricular ectopic beats are relatively easy to algorithmically single out since their morphologies deviate considerably from that of the normal sinus beat; the same observation is also valid for many types of artifacts being falsely detected by the QRS detector. On the other hand, rhythm perturbations primarily manifested by changes in the RR interval pattern, such as those associated with supraventricular ectopic beats, tend to be more difficult to detect since interval-based criteria by necessity are less specific than those which also involve morphology.

A number of techniques have been developed which deal with the presence of ectopic beats with all techniques conforming to the restriction that *only* ECG segments with occasional ectopic beats should be processed. Segments containing frequent ectopic beats or, worse, runs of ectopic beats perturb the underlying sinus rhythm and must therefore be excluded from further analysis [75]. A straightforward approach to correction of an occasional ectopic beat is to delete the aberrant RR intervals from the interval tachogram $d_{IT}(k)$. However, interval deletion does not try to fill in the interval variation that should have been present, had no ectopic beat occurred, and, as a result, the "corrected" interval tachogram remains less suitable for further HRV analysis.

In this section, we describe three vastly different techniques which deal with the presence of ectopic beats by either modifying an existing processing

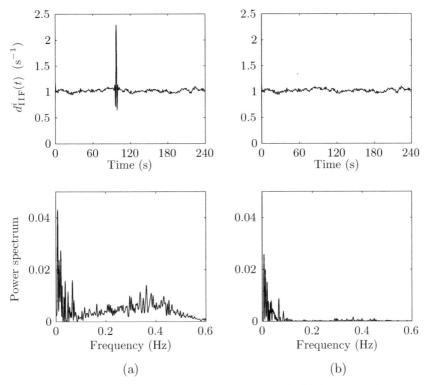

Figure 8.16: (a) The presence of an ectopic beat in $d_{\mathrm{IIF}}^i(t)$ (top panel) is associated with high-frequency components (bottom panel). (b) The corrected $d_{\mathrm{IIF}}^i(t)$ (top panel) produces a power spectrum with much less high-frequency content (bottom panel). The mean value of $d_{\mathrm{IIF}}^i(t)$ was subtracted before the power spectrum was computed.

block or inserting an additional processing step in the analysis, see the block diagram presented in Figure 8.17. The main idea behind each of these three techniques is

1. to modify the very definition of the heart timing signal $d_{\mathrm{HT}}(t_k)$,

2. to modify the estimator of the correlation function such that only the NN intervals are included (used in combination with $d_{\mathrm{IT}}(k)$, $d_{\mathrm{IIT}}(k)$, $d_{\mathrm{IF}}(t_k)$, or $d_{\mathrm{IIF}}(t_k)$), and

3. to interpolate over the gap caused by the ectopic beat in order to obtain values of the heart rate signal that align well with the adjacent NN intervals (used in combination with $d_{\mathrm{IF}}(t_k)$ or $d_{\mathrm{IIF}}(t_k)$).

A technique which may be used in combination with the lowpass filtered event series $d_{\mathrm{LE}}(t)$ is considered in Problem 8.11. Other techniques

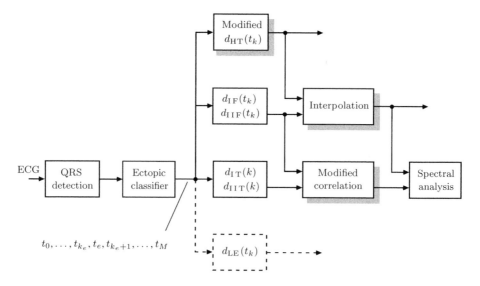

Figure 8.17: Three different techniques for correction of ectopic beats (marked by the shadowed boxes). Note that the correlation-based technique is developed for use in spectral analysis, while no such restriction applies to the other two techniques. Correction techniques for use with the lowpass filtered event series $d_{\mathrm{LE}}(t)$ are not included in the block diagram.

which deal with ectopic beats have been described in [76, 77], with the latter reference also including a comparison of performance.

In the presentation below, we will assume that the normal sinus beats have occurrence times $t_0, t_1, \ldots, t_M$ and that only one single ectopic beat occurs at time t_e; the time t_e is *not* included in the series $t_0, t_1, \ldots, t_M$. From an indexing point of view, we note that the sinus beat immediately preceding the ectopic occurs at t_{k_e}, and the one immediately following occurs at t_{k_e+1}. It is assumed that the ectopic beat does not occur at the edges of the signal but is always preceded by, and followed by, a number of sinus beats. Finally, the ectopic beat has been classified as such using some suitable clustering algorithm.

8.6.1   Correlation-based Correction

If the aim is to spectrally analyze the heart rate signal using the nonparametric approach described in Section 3.3, the correlation function estimate required for the periodogram in (3.79) can be modified to account for ectopic beats.[4] Here, the derivation of the modified estimator is based on the inter-

---

[4]A model-based, parametric approach to spectral analysis of data with missing observations can also be used, see [78].

val tachogram $d_{IT}(k)$, or its inverse. However, a similar modification can also be introduced when the interval function, or its inverse, is considered [79].

The interval tachogram $d_{IT}(k)$ reflects the sinus rhythm, here denoted $d_{SR}(k)$, except in intervals with an ectopic beat where the value of $d_{IT}(k)$ is considered missing. In mathematical terms, this property can be expressed by

$$d_{IT}(k) = o(k)d_{SR}(k), \qquad (8.73)$$

where $o(k)$ denotes a binary variable which is equal to one when an NN interval occurs, but otherwise zero. In order to proceed, we assume that $o(k)$ and $d_{SR}(k)$ are independent random variables characterized by their respective correlation functions $r_o(l)$ and $r_{d_{SR}}(l)$. Then, we can write

$$
\begin{aligned}
r_{d_{IT}}(l) &= E[d_{IT}(k)d_{IT}(k-l)] \\
&= E[o(k)d_{SR}(k)o(k-l)d_{SR}(k-l)] \\
&= E[o(k)o(k-l)] \cdot E[d_{SR}(k)d_{SR}(k-l)] \\
&= r_o(l)r_{d_{SR}}(l), \qquad (8.74)
\end{aligned}
$$

where the third equality results from the assumption of independence.

Hence, the result in (8.74) suggests that an estimate of the correlation function for the desired signal $d_{SR}(k)$ can be obtained by [80]

$$\hat{r}_{d_{SR}}(l) = \frac{\hat{r}_{d_{IT}}(l)}{\hat{r}_o(l)}, \qquad (8.75)$$

where

$$\hat{r}_{d_{IT}}(l) = \frac{1}{M_o(l)} \sum_{k=l}^{M_o(l)} d_{IT}(k)d_{IT}(k-l). \qquad (8.76)$$

The parameter $M_o(l)$ denotes the number of terms $d_{IT}(k)d_{IT}(k-l)$ that are nonzero; this number depends on the lag l. The correlation estimate of the binary variables $o(k)$ is obtained in the same way as $\hat{r}_{d_{IT}}(l)$, but is required to satisfy $\hat{r}_o(l) \neq 0$ in order to avoid division by zero in (8.75). For short lags, this requirement is usually fulfilled since only ECGs with occasional ectopic beats are subject to analysis. For large lags, $\hat{r}_o(l)$ may become equal to zero, implying that a truncated version of $\hat{r}_{d_{SR}}(l)$ should instead be used in the periodogram computation.

8.6.2 Interpolation-based Correction

Another possibility is to use interpolation in the interval function $d_{IF}(t_k)$, or its inverse $d_{IIF}(t_k)$, over the gap caused by the ectopic beat in order to insert

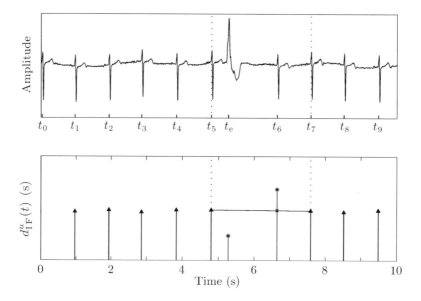

Figure 8.18: Linear interpolation of $d_{\mathrm{IF}}^u(t)$ to correct for the presence of an ectopic beat occurring at t_e. The two RR intervals altered by the ectopic beat are marked with asterisks. The interpolated value is positioned at t_6 and is marked with "×".

the samples required for producing an evenly sampled signal, see Figure 8.18. Since the ectopic beat is assumed to occur in the interval $t_{k_e} < t_e < t_{k_e+1}$, it is clear that interpolation must be based on samples up to $d_{\mathrm{IF}}(t_{k_e})$ and then from $d_{\mathrm{IF}}(t_{k_e+2})$ and onwards; the sample $d_{\mathrm{IF}}(t_{k_e+1})$ cannot be used since it involves the two times t_{k_e} and t_{k_e+1} that define the aberrant RR interval.

By means of linear interpolation, the interval function can be interpolated over the interval $t_{k_e} < t < t_{k_e+2}$ using the following expression,

$$d_{\mathrm{IF}}^i(t) = d_{\mathrm{IF}}(t_{k_e}) + \frac{d_{\mathrm{IF}}(t_{k_e+2}) - d_{\mathrm{IF}}(t_{k_e})}{t_{k_e+2} - t_{k_e}}\,(t - t_{k_e}), \quad t_{k_e} < t < t_{k_e+2},$$

$$(8.77)$$

where only two samples, i.e., $d_{\mathrm{IF}}(t_{k_e})$ and $d_{\mathrm{IF}}(t_{k_e+2})$, are required. Higher-order polynomial interpolation can also be applied involving additional samples of $d_{\mathrm{IF}}(t_k)$ from both sides of the ectopic beat.

Finally, the new samples that result from the interpolation in (8.77) are merged with the existing values of the interval function so as to define the corrected signal which is subjected to further analysis.

8.6.3 The Heart Timing Signal and Ectopic Beats

The definition of the heart timing signal in (8.22) can be modified to account for the presence of an ectopic beat occurring at t_e [81, 82], see also [83] for a similar IPFM-based approach. In the modified definition, the occurrence times subsequent to the ectopic beat are related to the time basis kT_I involving a parameter s such that

$$d_{\mathrm{HT}}(t_k) = \begin{cases} kT_I - t_k, & k = 1, \ldots, k_e; \\ (k + s)T_I - t_k, & k = k_e + 1, \ldots, M. \end{cases} \tag{8.78}$$

The parameter s can be viewed as the jump occurring when the integral in the IPFM model is reset, defined by

$$s = \frac{1}{T_I} \int_{t_{k_e}}^{t_{k_e}^b} (1 + m(\tau))d\tau, \tag{8.79}$$

where $t_{k_e}^b$ denotes the reset time at which the SA node has been "restarted" by the wave propagating from the ectopic focus. A value of s close to zero indicates that the event at t_e is probably caused by an artifact, whereas a value close to one probably indicates that the event is a premature ectopic beat followed by a compensatory pause. Recalling the generalized IPFM model from (8.7), the continuous-time heart timing signal is defined by

$$d_{\mathrm{HT}}(t) = \kappa(t)T_I - t,$$

where

$$\kappa(t) = \frac{1}{T_I} \int_0^t (1 + m(\tau))d\tau.$$

With the presence of one ectopic beat, the indexing function $\kappa(t)$ is, when sampled at the occurrence times t_k, given by

$$\kappa(t_k) = \begin{cases} k, & k \leq k_e; \\ k + s, & k \geq k_e + 1. \end{cases} \tag{8.80}$$

In order to use the modified definition of $d_{\mathrm{HT}}(t_k)$ in (8.78), we need to estimate the parameter s as well as to modify our previous estimator of the mean RR interval length T_I such that it now accounts for the presence of the ectopic beat (Figure 8.19). The estimator of s requires most of our attention since the estimation procedure consists of several steps, whereas the estimator of T_I, requiring that $\hat{s}$ be available, has a very simple structure.

Our starting point is the observation that the indexing function $\kappa(t)$ can be estimated from both the occurrence times preceding the ectopic beat and

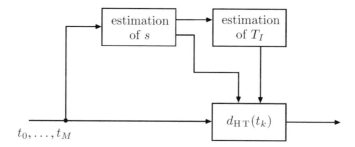

Figure 8.19: Correction of an ectopic beat using the modified heart timing signal in (8.78). The signal can be determined once the estimates of the jump parameter s and the mean RR interval length T_I are available.

the times following the ectopic beat, with the latter times associated with the offset s. Thus, two different estimators related to $\kappa(t)$ can be based on the samples of $(t, \kappa(t))$. The "forward estimator", denoted $\hat{\kappa}^f(t)$, is based on the occurrences at $(t_0, 0), \ldots, (t_{k_e}, k_e)$ and produces an estimate of $\kappa(t)$. The "backward estimator", denoted $\hat{\kappa}^b(t)$, is based on $(t_{k_e+1}, k_e+1), \ldots, (t_M, M)$ and produces, in contrast to $\hat{\kappa}^f(t)$, an estimate which is offset by s from $\kappa(t)$,

$$\kappa^b(t) = \kappa(t) - s. \tag{8.81}$$

Since the resulting indexing functions $\hat{\kappa}^f(t)$ and $\hat{\kappa}^b(t)$ would differ by an offset equal to the desired parameter s, it is possible to extrapolate these two functions forward and backward in time, respectively, to such an extent that they overlap and, thereby, make it possible to estimate s, see Figure 8.20.

Extrapolation of the indexing functions is done by first forwardly extending the series of occurrence times $t_0, \ldots, t_{k_e}$ with a new time $\hat{t}^f_{k_e+1}$ under the assumption that the sinus rhythm continues. In the same way, the series of occurrence times $t_{k_e+1}, \ldots, t_M$ is backwardly extended with $\hat{t}^b_{k_e}$ under the assumption that the sinus rhythm precedes t_{k_e+1}. For now, we will assume that the new occurrence times $\hat{t}^f_{k_e+1}$ and $\hat{t}^b_{k_e}$ are located such that the desired overlap exists, i.e., $\hat{t}^f_{k_e+1} > \hat{t}^b_{k_e}$. If not, the two series have to be further extended until this requirement is fulfilled. The computation of these two occurrence times is given by

$$\hat{t}^f_{k_e+1} = t_{k_e} + d^i_{\mathrm{IF}}(\hat{t}^f_{k_e+1}) \tag{8.82}$$

and

$$\hat{t}^b_{k_e} = t_{k_e+1} - d^i_{\mathrm{IF}}(t_{k_e+1}), \tag{8.83}$$

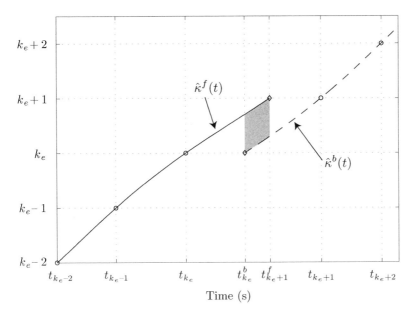

Figure 8.20: Forward extrapolation of the indexing function $\kappa^f(t)$, based on the occurrences at $(t_0, 0), \ldots, (t_{k_e}, k_e)$, and backward extrapolation of $\kappa^b(t)$, based on $(t_{k_e+1}, k_e + 1), \ldots, (t_M, M)$. The indexing functions are extrapolated until they overlap in time. An estimate of the jump parameter s is given by the shaded area divided by its length, cf. (8.88).

where $d_{\mathrm{IF}}^i(t)$ denotes the interpolated interval function. Before interpolation, the interval function is given by

$$d_{\mathrm{IF}}(t_k) = t_k - t_{k-1}, \quad k \neq k_e + 1, \tag{8.84}$$

where the intervals adjacent to the ectopic beat have been excluded from the computation of $d_{\mathrm{IF}}(t_k)$. From the expression in (8.84), the interpolated function $d_{\mathrm{IF}}^i(t)$ can be obtained. In (8.82), the value of $\hat{t}_{k_e+1}^f$ is obtained by solving the equation recursively.

Next, the two indexing functions $\hat{\kappa}^f(t)$ and $\hat{\kappa}^b(t)$ are both interpolated in the interval $\hat{t}_{k_e}^b \leq t \leq \hat{t}_{k_e+1}^f$ using a suitable interpolation function $g(\cdot)$ that makes use of available occurrence times,

$$\hat{\kappa}^f(t) = g((t_0, 0), \ldots, (t_{k_e}, k_e), (\hat{t}_{k_e+1}^f, k_e + 1)), \tag{8.85}$$

$$\hat{\kappa}^b(t) = g((\hat{t}_{k_e}^b, k_e), (t_{k_e+1}, k_e + 1), \ldots, (t_M, M)). \tag{8.86}$$

Although the occurrence times before and after the ectopic beat have been included in the interpolation functions for the sake of completeness, interpolation is typically based on only a few occurrence times on each side of the ectopic beat.

With the estimates of the two indexing functions available, we are now in a position to define an error criterion by which the value of s, which offsets $\hat{\kappa}^f(t)$ from $\hat{\kappa}^b(t)$, can be determined. Once again we adopt the LS criterion which, in the present context, is defined by

$$\mathcal{J}(s) = \int_{\hat{t}^b_{k_e}}^{\hat{t}^f_{k_e+1}} \left(\hat{\kappa}^f(t) - (\hat{\kappa}^b(t) + s) \right)^2 dt. \tag{8.87}$$

Differentiation of $\mathcal{J}(s)$ with respect to s and setting the result equal to zero yield the value of s that minimizes $\mathcal{J}(s)$,

$$\hat{s} = \frac{1}{\hat{t}^f_{k_e+1} - \hat{t}^b_{k_e}} \int_{\hat{t}^b_{k_e}}^{\hat{t}^f_{k_e+1}} \left(\hat{\kappa}^f(t) - \hat{\kappa}^b(t) \right) dt. \tag{8.88}$$

Hence, the estimator is equivalent to the area enclosed by the two indexing functions within the overlapping time interval and normalized by the length of the overlap interval. In practice, the integral in (8.88) is approximated by summation over a discrete set of times.

Once $\hat{s}$ is available, estimation of the mean RR interval length T_I can be done in a way similar to that in (8.23), except that the occurrence time of the last event, i.e., t_M, must be divided by a factor which accounts for the perturbation introduced by s,

$$\hat{T}_I = \frac{t_M}{M + \hat{s}}. \tag{8.89}$$

As before, the resulting heart timing signal is finally subjected to interpolation and resampling to become suitable for further HRV analysis.

It has been reported in [82] that the above correction technique for the heart timing signal avoids the artificial increase in low-frequency components (i.e., <0.05 Hz) which is accompanied by interpolation-based correction techniques. The difference in performance is increasingly pronounced when the degree of ectopy increases [84].

8.7   Interaction with Other Physiological Signals

The variability in heart rate is influenced by different physiological signals, of which respiration and blood pressure are the most dominant ones and whose presence may be reflected by different peaks in the HRV power spectrum (see page 589). While such spectral information is very valuable, a deeper understanding of the mechanisms which control the cardiovascular system can be achieved by employing multivariate (multichannel) signal models to characterize the mutual interaction between heart rate and other physiological signals. This characterization is referred to as a closed-loop identification problem since the system of interest must be identified during "operation", based on measurements of the different physiological signals. In this section, we conclude the HRV chapter by very briefly mentioning how the signal interaction can be modeled, while leaving the methods for estimating model parameters to the interested reader.

The *baroreceptor reflex* is an essential component to this interaction, being the control system of the body for rapidly dealing with changes in blood pressure. Baroreceptors are nerve cells which are specialized to sense changes in blood pressure. If an increase in blood pressure is sensed, the heart rate will, through a negative feedback loop, decrease to compensate; if, on the other hand, a decrease in blood pressure is sensed, the heart rate will increase. Accordingly, it is not entirely surprising that the variabilities observed in heart rate and blood pressure are highly correlated [85, 86]; this observation applies to both low- and high-frequency components in the intervals 0.04–0.15 and 0.15–0.40 Hz, respectively.

By analyzing the interaction between heart rate and blood pressure, valuable insight into the dynamics of the baroreceptor mechanisms can be obtained [87–90]. Such analysis starts by obtaining simultaneous measurements on heart rate and blood pressure; the latter measurement usually being synonymous to the systolic arterial blood pressure which is measured as the peak amplitude of the pressure signal, see Figure 8.21. Similar to heart rate, the systolic arterial pressure is sampled at uneven points in time, and, therefore, it is necessary to perform interpolation and resampling at even times before cross-analysis of the signals can be performed (unless the tachogram is employed as heart rhythm representation when the blood pressure measurements can be used directly).

Before describing a model of the interaction between different physiological signals, we will introduce the *cross-power spectrum* which is a general, nonparametric approach to characterize the correlation between two stationary processes $x(n)$ and $y(n)$. The cross-power spectrum is defined as the DTFT of the cross-correlation function $r_{xy}(k)$,

$$S_{xy}(e^{j\omega}) = \sum_{k=-\infty}^{\infty} r_{xy}(k)e^{-j\omega k}, \qquad (8.90)$$

where

$$r_{xy}(k) = E[x(n)y(n-k)]. \qquad (8.91)$$

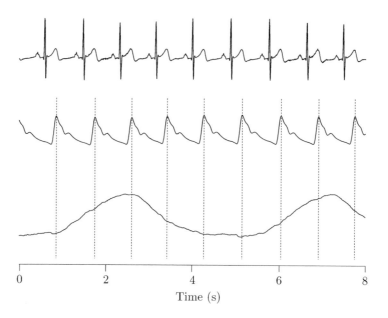

Figure 8.21: Simultaneous recording of the ECG, arterial blood pressure, and abdominal respiration (top to bottom). In each cardiac cycle, the systolic pressure is given by the value of the maximal amplitude (marked by dashed lines). The respiration signal is sampled at the same time instant when a tachogram representation is used, while it is preferable to sample at evenly spaced times when other representations are employed.

The cross-power spectrum $S_{xy}(e^{j\omega})$ can be interpreted as the correlation between $x(n)$ and $y(n)$ at a given frequency. The normalized cross-power spectrum is defined by

$$\Gamma_{xy}(e^{j\omega}) = \frac{S_{xy}(e^{j\omega})}{\sqrt{S_x(e^{j\omega})}\sqrt{S_y(e^{j\omega})}} \tag{8.92}$$

and is known as the *coherence function*; normalization is done with the square-root of the two power spectra. However, the *magnitude squared coherence*, given by

$$|\Gamma_{xy}(e^{j\omega})|^2 = \frac{|S_{xy}(e^{j\omega})|^2}{S_x(e^{j\omega})S_y(e^{j\omega})}, \tag{8.93}$$

is more often used in practice and has the attractive property of being normalized such that

$$0 \le |\Gamma_{xy}(e^{j\omega})|^2 \le 1. \tag{8.94}$$

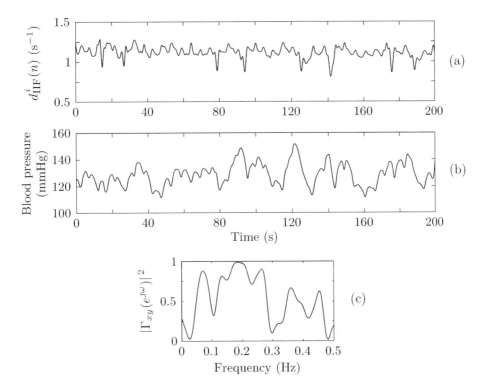

Figure 8.22: From the signals displayed in Figure 8.21, (a) the interpolated inverse interval function $x(n) = d^i_{\mathrm{IIF}}(n)$ of the heart rate (sampled at even times) and (b) the systolic arterial blood pressure $y(n)$, also obtained from interpolation, are used to compute (c) the magnitude squared coherence.

The magnitude squared coherence can be viewed as the frequency domain counterpart of the cross-correlation coefficient earlier used, for example, in the context of EP subaveraging, see (4.23). The magnitude squared coherence is exemplified in Figure 8.22 for two signals describing heart rate and systolic arterial pressure. From this example it is obvious that the two signals are strongly correlated around 0.2 Hz since the magnitude squared coherence is almost one; this frequency actually corresponds to the respiratory frequency which can be roughly estimated from the period length of the respiration signal displayed in Figure 8.21.

Autoregressive modeling is a useful, parametric approach for studying the interaction between blood pressure and heart rate. Since blood pressure variability not only influences HRV but is also influenced by HRV through feedback, a two-channel (bivariate) AR model has been found appropriate for describing how these signals oscillate around their respective mean values [91–93]. The bivariate AR model is defined by the following two coupled equations,

$$x_1(n) = -\sum_{k=1}^{p} a_{11,k} x_1(n-k) + \sum_{k=1}^{p} a_{12,k} x_2(n-k) + v_1(n) \qquad (8.95)$$

and

$$x_2(n) = -\sum_{k=1}^{p} a_{22,k} x_1(n-k) + \sum_{k=1}^{p} a_{21,k} x_2(n-k) + v_2(n), \qquad (8.96)$$

where $x_1(n)$ and $x_2(n)$ denote blood pressure and heart rate, respectively. It is assumed that the four subsystems, defined by the four parameter sets $\{a_{11,k}\}, \{a_{12,k}\}, \{a_{21,k}\}$, and $\{a_{22,k}\}$, have the same model order p. The input noise sources $v_1(n)$ and $v_2(n)$ are assumed to be white and uncorrelated with each other. Note that $x_1(n)$ and $x_2(n)$ account for variability but not for the absolute level since they are both zero-mean. In dealing with this model, it is convenient to combine the two equations into a matrix equation, see (3.24),

$$\mathbf{x}(n) = -\sum_{k=1}^{p} \mathbf{A}_k \mathbf{x}(n-k) + \mathbf{v}(n), \qquad (8.97)$$

where

$$\mathbf{A}_k = \begin{bmatrix} a_{11,k} & -a_{12,k} \\ -a_{21,k} & a_{22,k} \end{bmatrix}, \qquad (8.98)$$

and

$$\mathbf{x}(n) = \begin{bmatrix} x_1(n) \\ x_2(n) \end{bmatrix}, \quad \mathbf{v}(n) = \begin{bmatrix} v_1(n) \\ v_2(n) \end{bmatrix}. \qquad (8.99)$$

The expression in (8.97) thus clearly demonstrates the feedback structure of the bivariate signal model.

An alternative way of representing the model is in terms of four scalar transfer functions as illustrated by the block diagram in Figure 8.23. Two transfer functions $H_{ii}(z)$ relate $v_i(n)$ to $x_i(n)$ by

$$H_{ii}(z) = \frac{1}{1 + \sum_{k=1}^{p} a_{ii,k} z^{-k}}, \quad i = 1, 2, \qquad (8.100)$$

and two cross-transfer functions $G_{ij}(z)$ relate $x_i(n)$ to $x_j(n)$ by

$$G_{ij}(z) = \frac{\sum_{k=1}^{p} a_{ij,k} z^{-k}}{1 + \sum_{k=1}^{p} a_{ii,k} z^{-k}}, \quad i,j = 1,2, \ i \neq j. \qquad (8.101)$$

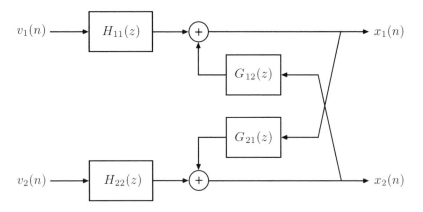

Figure 8.23: Block diagram of the bivariate AR model used for modeling of the interaction between blood pressure $x_1(n)$ and heart rate $x_2(n)$.

In physiological terms, the transfer function $G_{21}(z)$ represents the effect of variability in systolic arterial pressure on heart rate mediated through the autonomic nervous system (baroreceptor feedback dynamics), whereas $G_{12}(z)$ represents the mechanical effect of variability in heart rate on systolic arterial pressure [92]. With the availability of $G_{ij}(z)$, the gain and phase relationship between $x_i(n)$ and $x_j(n)$ can be calculated for different frequencies.

In order to determine the cross-spectral information related to the bivariate AR model, it is helpful to consider its input–output relation which is given by

$$\mathbf{X}(z) = \mathbf{H}(z)\mathbf{V}(z), \tag{8.102}$$

where $\mathbf{X}(z)$ and $\mathbf{V}(z)$ denote the z-transforms of the two vectors in (8.99). The multichannel transfer function $\mathbf{H}(z)$ is defined by

$$
\begin{aligned}
\mathbf{H}(z) &= \begin{bmatrix} H_{11}(z) & H_{12}(z) \\ H_{21}(z) & H_{22}(z) \end{bmatrix} \\
&= \begin{bmatrix} H_{11}(z) & G_{12}(z)H_{22}(z) \\ G_{21}(z)H_{11}(z) & H_{22}(z) \end{bmatrix},
\end{aligned} \tag{8.103}
$$

where the second step connects $H_{12}(z)$ and $H_{21}(z)$ to the transfer functions in (8.100) and (8.101). The cross-power spectral matrix contains all the spectral information of the model,

$$\mathbf{S}_x(e^{j\omega}) = \begin{bmatrix} S_{x_1x_1}(e^{j\omega}) & S_{x_1x_2}(e^{j\omega}) \\ S_{x_2x_1}(e^{j\omega}) & S_{x_2x_2}(e^{j\omega}) \end{bmatrix}, \tag{8.104}$$

and is calculated by the following expression [94, 95],

$$\mathbf{S}_x(e^{j\omega}) = \mathbf{H}^*(e^{j\omega})\mathbf{S}_v(e^{j\omega})\mathbf{H}^T(e^{j\omega}). \tag{8.105}$$

where $\mathbf{H}(e^{j\omega})$ is obtained by evaluating $\mathbf{H}(z)$ on the unit circle, i.e., $z = e^{j\omega}$, and the diagonal matrix $\mathbf{S}_v(e^{j\omega})$ describes the variance of the input noise,

$$\mathbf{S}_v(e^{j\omega}) = \begin{bmatrix} \sigma_{v_1}^2 & 0 \\ 0 & \sigma_{v_2}^2 \end{bmatrix}. \tag{8.106}$$

It should be noted that the scalar counterpart to (8.105) has already been introduced in (3.19).

Parameter estimation in a multichannel AR model can be approached by, for example, modifying the single-channel autocorrelation/covariance method, earlier discussed in Section 3.4, so that a multichannel version of the normal equations results [94, 95]. If, on the other hand, the model parameters are known to vary slowly over time, it may be preferable to employ a recursive estimation technique similar to the ones presented in Section 3.6.5, but modified for the multichannel case; for details on such estimation techniques, see [96].

Figure 8.24 presents an example of variability in heart rate and systolic arterial blood pressure observed during a tilting maneuver, causing syncope at the end of the recording. Based on the assumption that the data in Figure 8.24(a)–(b) can be adequately modeled by a bivariate AR model, the magnitude squared coherence is estimated using a time-varying, recursive approach [88, 97]. It is evident from Figure 8.24(c) that the coherence is particularly pronounced at frequencies around 0.4 Hz before the tilting maneuver, but then gradually decreases to much lower frequencies until the syncope occurs. The syncope is accompanied by a sudden drop in systolic blood pressure and a sudden decrease in heart rate, i.e., a prolongation of the RR intervals.

Finally, we note that the above matrix formulation of the bivariate AR model lends itself well to handle the inclusion of additional physiological signals. For example, a trivariate model should be considered for the interaction between heart rate, blood pressure, and respiration. In doing so, certain cross-transfer functions has to be omitted from the model, i.e., set to zero, since these are not physiologically meaningful. For example, it is well-known that respiration influences heart rate as reflected by respiratory sinus arrhythmia (see page 432), and that respiration has a mechanical effect on systolic blood pressure. However, neither heart rate nor blood pressure has a significant influence on respiration, and, therefore, the corresponding cross-transfer functions are omitted.

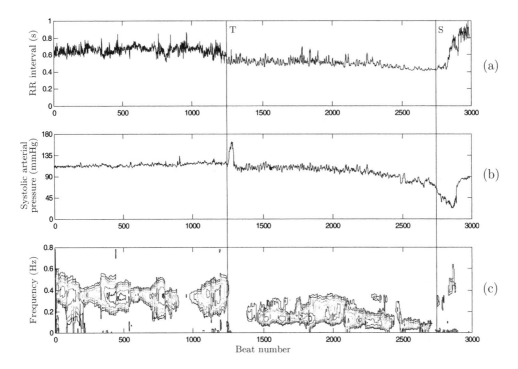

Figure 8.24: Interaction between heart rate and blood pressure during a tilting maneuver which causes syncope. The tachograms of (a) the RR intervals and (b) the systolic arterial pressure, and (c) the related time-varying magnitude squared coherence, displayed using the isocontour format. The letter "T" marks the time when the subject is tilted from a resting position, and "S" marks the time when syncope occurs. (Reprinted from Cerutti et al. [88] with permission.)

Bibliography

[1] Task Force of The European Society of Cardiology and The North Americam Society for Pacing and Electrophysiology, "Heart rate variability: Standards of measurement, physiological interpretation, and clinical use," *Circulation*, vol. 93, pp. 1043–1065, 1996.

[2] M. Merri, D. C. Farden, J. G. Mottley, and E. L. Titlebaum, "Sampling frequency of the electrocardiogram for spectral analysis of the heart rate variability," *IEEE Trans. Biomed. Eng.*, vol. 37, pp. 99–105, 1990.

[3] R. E. Kleiger, J. P. Miller, J. T. Bigger, and A. J. Moss, "Decreased heart rate variability and its association with increased mortality after myocardial infarction," *Am. J. Cardiol.*, vol. 59, pp. 256–262, 1987.

[4] G. G. Berntson, K. S. Quigley, J. F. Jang, and S. T. Boysen, "An approach to artifact identification: Application to heart period data," *Psychophysiology*, vol. 27, pp. 586–598, 1990.

[5] M. Malik, T. Cripps, T. Farrell, and A. J. Camm, "Prognostic value of heart rate variability after myocardial infarction. A comparison of different data processing methods," *Med. Biol. Eng. & Comput.*, vol. 27, pp. 603–611, 1989.

[6] M. A. García-González and R. Pallàs-Areny, "A novel robust index assess beat-to-beat variability in heart rate time series analysis," *IEEE Trans. Biomed. Eng.*, vol. 48, pp. 617–621, 2001.

[7] R. E. Kleiger, P. K. Stein, M. S. Bosner, and J. N. Rottman, "Time-domain measurements of heart rate variability," in *Heart Rate Variability* (M. Malik and A. J. Camm, eds.), ch. 3, pp. 33–45, Armonk: Futura Publ., 1995.

[8] M. Malik, T. Farrell, T. Cripps, and A. J. Camm, "Heart rate variability in relation to prognosis after myocardial infarction: Selection of optimal processing techniques," *Eur. Heart J.*, vol. 10, pp. 1060–1074, 1989.

[9] T. G. Farrell, Y. Bashir, T. Cripps, M. Malik, J. Poloniecki, E. D. Bennett, D. E. Ward, and A. J. Camm, "Risk stratification for arrhythmic events in postinfarction patients based on heart rate variability, ambulatory electrocardiographic variables, and the signal-averaged electrocardiogram," *J. Am. Coll. Cardiol.*, vol. 18, pp. 687–697, 1991.

[10] R. J. Cohen, R. D. Berger, and T. Dushane, "A quantitative model for the ventricular response during atrial fibrillation," *IEEE Trans. Biomed. Eng.*, vol. 30, pp. 769–780, 1983.

[11] N. Cai, M. Dohnal, and S. B. Olsson, "Methodological aspects of the use of heart rate stratified RR interval histograms in the analysis of atrioventricular conduction during atrial fibrillation," *Cardiovasc. Res.*, vol. 21, pp. 455–462, 1987.

[12] A. V. Oppenheim, A. S. Willsky, and S. H. Nawab, *Signals and Systems*. New Jersey: Prentice-Hall, 1997.

[13] B. W. Hyndman and R. K. Mohn, "A model of the cardiac pacemaker and its use in decoding the information content of cardiac intervals," *Automedica*, vol. 1, pp. 239–252, 1975.

[14] J. Mateo and P. Laguna, "Improved heart rate variability signal analysis from the beat occurrence times according to the IPFM model," *IEEE Trans. Biomed. Eng.*, vol. 47, pp. 997–1009, 2000.

[15] L. T. Mainardi, A. M. Bianchi, G. Baselli, and S. Cerutti, "Pole-tracking algorithms for the extraction of time variant heart rate variability spectral parameters," *IEEE Trans. Biomed. Eng.*, vol. 42, pp. 250–259, 1995.

[16] O. Rompelman, J. B. I. M. Snijders, and C. J. van Spronsen, "The measurement of heart rate variability spectra with the help of a personal computer," *IEEE Trans. Biomed. Eng.*, vol. 29, pp. 503–510, 1982.

[17] R. W. de Boer, J. M. Karemaker, and J. Strackee, "Spectrum of a series of point event, generated by the integral pulse frequency modulation model," *Med. Biol. Eng. & Comput.*, vol. 23, pp. 138–142, 1985.

[18] R. D. Berger, S. Akselrod, D. Gordon, and R. J. Cohen, "An efficient algorithm for spectral analysis of heart rate variability," *IEEE Trans. Biomed. Eng.*, vol. 33, pp. 900–904, 1986.

[19] P. Castiglioni, "Evaluation of heart rhythm variability by heart or heart period: Differences, pitfalls and help from logarithms," *Med. Biol. Eng. & Comput.*, vol. 33, pp. 323–330, 1995.

[20] B. J. TenVoorde, T. J. C. Faes, and O. Rompelman, "Spectra of data sampled at frequency modulated rates in application to cardiovascular signals: Part 1 Analytical derivation of the spectra," *Med. Biol. Eng. & Comput.*, vol. 32, pp. 63–70, 1994.

[21] I. P. Mitov, "Spectral analysis of heart rate variability using the integral pulse frequency modulation model," *Med. Biol. Eng. & Comput.*, vol. 39, pp. 348–354, 2001.

[22] G. B. Stanley, K. Poolla, and R. A. Siegel, "Threshold modeling of autonomic control of heart rate variability," *IEEE Trans. Biomed. Eng.*, vol. 47, pp. 1147–1153, 2000.

[23] S. R. Seydenejad and R. I. Kitney, "Time-varying threshold integral pulse frequency modulation," *IEEE Trans. Biomed. Eng.*, vol. 48, pp. 949–962, 2001.

[24] H. Chiu and T. Kao, "A mathematical model for autonomic control of heart rate variation," *IEEE Eng. Med. Biol. Mag.*, vol. 20, pp. 69–76, 2001.

[25] E. Pyetan and S. Akselrod, "Do the high-frequency indexes of HRV provide a faithful assessment of cardiac vagal tone? A critical theoretical evaluation," *IEEE Trans. Biomed. Eng.*, vol. 50, pp. 777–783, 2003.

[26] E. J. Bayly, "Spectral analysis of pulse frequency modulation in the nervous system," *IEEE Trans. Biomed. Eng.*, vol. 15, pp. 257–265, 1968.

[27] A. M. Bruckstein and Y. Y. Zeevi, "Analysis of 'integrate to threshold' neural coding schemes," *Biol. Cybern.*, vol. 34, pp. 63–79, 1979.

[28] G. Giunta and A. Neri, "Neural correlation based on the IPFM model," *IEEE Trans. Syst. Man Cybernetics*, vol. 20, pp. 262–268, 1990.

[29] B. McA. Sayers, "Analysis of heart rate variability," *Ergonomics*, vol. 16, pp. 17–32, 1973.

[30] S. Akselrod, D. Gordon, F. A. Ubel, D. C. Shannon, A. C. Barger, and R. J. Cohen, "Power spectrum analysis of heart rate fluctuations. A quantitative probe of beat-to-beat cardiovascular control," *Science*, vol. 213, pp. 220–222, 1981.

[31] F. Bartoli, G. Baselli, and S. Cerutti, "AR identification and spectral estimate applied to the R-R interval measurements," *Int. J. Biomed. Comput.*, vol. 16, pp. 201–205, 1985.

[32] H. Luczak and W. Laurig, "An analysis of heart rate variability," *Ergonomics*, vol. 16, no. 4, pp. 85–97, 1973.

[33] O. Rompelman, A. J. R. M. Coenen, and R. I. Kitney, "Measurement of heart-rate variability: Part 1—comparative study of heart-rate variability analysis methods," *Med. Biol. Eng. & Comput.*, vol. 15, pp. 239–252, 1977.

[34] B. F. Womack, "The analysis of respiratory sinus arrhythmia using spectral analysis methods," *IEEE Trans. Biomed. Eng.*, vol. 18, pp. 399–409, 1971.

[35] R. W. de Boer, J. M. Karemaker, and J. Strackee, "Description of heart rate variability data in accordance with a physiological model for the genesis of heart beats," *Psychophysiology*, vol. 22, pp. 147–155, 1985.

[36] A. S. French and A. V. Holden, "Alias-free sampling of neuronal spike trains," *Kybernetik*, vol. 8, pp. 165–175, 1971.

[37] A. J. R. M. Coenen, O. Rompelman, and R. I. Kitney, "Measurement of heart-rate variability: Part 2—hardware digital device for the assessment of heart-rate variability," *Med. Biol. Eng. & Comput.*, vol. 15, pp. 423–430, 1977.

[38] O. Rompelman, "Tutorial review on processing the cardiac event series: A signal analysis approach," *Automedica*, vol. 7, pp. 191–212, 1986.

[39] R. J. Peterka, A. C. Sanderson, and D. P. O'Leary, "Practical considerations in the implementation of the French-Holden algorithm for sampling of neuronal spike trains," *IEEE Trans. Biomed. Eng.*, vol. 25, pp. 192–195, 1978.

[40] S. R. Seydenejad and R. I. Kitney, "Real-time heart rate variability extraction using the Kaiser window," *IEEE Trans. Biomed. Eng.*, vol. 44, pp. 990–1005, 1997.

[41] M. V. Kamath and E. L. Fallen, "Power spectral analysis of HRV: A noninvasive signature of cardiac autonomic functions," *Crit. Rev. Biomed. Eng.*, vol. 21, pp. 245–311, 1993.

[42] S. Cerutti, A. Bianchi, and L. T. Mainardi, "Spectral analysis of the heart rate variability signal," in *Heart Rate Variability* (M. Malik and A. J. Camm, eds.), ch. 6, pp. 63–74, Armonk: Futura Publ., 1995.

[43] S. Akselrod, "Components of heart rate variability: Basic studies," in *Heart Rate Variability* (M. Malik and A. J. Camm, eds.), ch. 12, pp. 147–163, Armonk: Futura Publ., 1995.

[44] P. Laguna, G. B. Moody, and R. G. Mark, "Power spectral density of unevenly sampled data by least-square analysis: Performance and application to heart rate signals," *IEEE Trans. Biomed. Eng.*, vol. 45, pp. 698–715, 1998.

[45] G. Baselli, D. Bolis, S. Cerutti, and C. Freschi, "Autoregressive modeling and power spectral estimate of R-R interval time series in arrhythmic patients," *Comput. Biomed. Res.*, vol. 18, pp. 510–530, 1985.

[46] R. L. Burr and M. J. Cowan, "Autoregressive spectral models of heart rate variability," *J. Electrocardiol.*, vol. 25 (Suppl.), pp. 224–233, 1992.

[47] D. J. Christini, A. Kulkarni, S. Rao, E. Stutman, F. M. Bennett, J. M. Hausdorff, N. Oriol, and K. Lutchen, "Uncertainty of AR spectral estimates," in *Proc. Computers in Cardiology*, pp. 451–454, IEEE Computer Society Press, 1993.

[48] A. B. Carlson, *Communication Systems. An Introduction to Signal and Noise in Electrical Comunication.* New York: McGraw-Hill, 3rd ed., 1986.

[49] N. R. Lomb, "Least-squares frequency analysis of unequally spaced data," *Astrophys. Space Sci.*, vol. 39, pp. 447–462, 1976.

[50] J. D. Scargle, "Studies in astronomical time series analysis II. Statistical aspects of spectral analysis of unevenly spaced data," *Astrophys. J.*, vol. 263, pp. 835–853, 1982.

[51] W. H. Press and G. B. Rybicki, "Fast algorithm for spectral analysis of unevenly sampled data," *Astrophys. J.*, vol. 338, pp. 277–280, 1989.

[52] W. H. Press, S. A. Teukolsky, W. T. Vetterling, and B. P. Flannery, *Numerical Recipes in C: The Art of Scientific Computing.* New York: Cambridge Univ. Press, 2nd ed., 1992.

[53] G. B. Moody, "Spectral analysis of heart rate without resampling," in *Proc. Computers in Cardiology*, pp. 715–718, IEEE Computer Society Press, 1993.

[54] G. D. Clifford and L. Tarassenko, "Quantifying errors in spectral estimates of HRV due to beat replacement and resampling," *IEEE Trans. Biomed. Eng.*, vol. 52, pp. 630–638, 2005.

[55] G. J. Balm, "Crosscorrelation techniques applied to the electrocardiogram interpretation problem," *IEEE Trans. Biomed. Eng.*, vol. 14, pp. 258–262, 1967.

[56] C. L. Feldman, P. G. Amazeen, M. D. Klein, and B. Lown, "Computer detection of ectopic beats," *Comput. Biomed. Res.*, vol. 3, pp. 666–674, 1971.

[57] J. H. van Bemmel and S. J. Hengevald, "Clustering algorithm for QRS and ST–T waveform typing," *Comput. Biomed. Res.*, vol. 6, pp. 442–456, 1973.

[58] M. E. Nygårds and J. Hulting, "An automated system for ECG monitoring," *Comput. Biomed. Res.*, vol. 12, pp. 181–202, 1979.

[59] J. A. Kors, J. Talmon, and J. H. van Bemmel, "Multilead ECG analysis," *Comput. Biomed. Res.*, vol. 19, pp. 28–46, 1986.

[60] K. M. Strand, L. R. Smith, M. E. Turner, and J. A. Mantle, "A comparison of simple and template variable models for discrimination between normal and PVC waveforms," in *Proc. Computers in Cardiology*, pp. 21–26, IEEE Computer Society Press, 1980.

[61] S. H. Rappaport, L. Gillick, G. B. Moody, and R. G. Mark, "QRS morphology classification: Quantitative evaluation of different strategies," in *Proc. Computers in Cardiology*, pp. 33–38, IEEE Computer Society Press, 1982.

[62] T. Y. Young and W. H. Huggins, "On the representation of electrocardiograms," *IEEE Trans. Biomed. Eng.*, vol. 10, pp. 86–95, 1963.

[63] S. Karlsson, "Representation of ECG records," in *Dig. 7th Int. Conf. Med. & Biol. Eng.*, (Stockholm), p. 105, 1967.

[64] A. R. Hambley, R. L. Moruzzi, and C. L. Feldman, "The use of intrinsic components in an ECG filter," *IEEE Trans. Biomed. Eng.*, vol. 21, pp. 469–473, 1974.

[65] L. Sörnmo, P. O. Börjesson, M. E. Nygårds, and O. Pahlm, "A method for evaluation of QRS shape features using a mathematical model for the ECG," *IEEE Trans. Biomed. Eng.*, vol. 28, no. 10, pp. 713–717, 1981.

[66] G. Bortolan, R. Degani, and J. L. Willems, "Neural networks for ECG classification," in *Proc. Computers in Cardiology*, pp. 269–272, IEEE Computer Society Press, 1990.

[67] L. Senhadji, G. Carrault, J. J. Bellanger, and G. Passariello, "Comparing wavelet transforms for recognizing cardiac patterns," *IEEE Eng. Med. Biol. Mag.*, vol. 14, pp. 167–173, 1995.

[68] P. Laguna, R. Jané, S. Olmos, N. V. Thakor, H. Rix, and P. Caminal, "Adaptive estimation of QRS complex by the Hermite model for classification and ectopic beat detection," *Med. Biol. Eng. & Comput.*, vol. 34, pp. 58–68, 1996.

[69] Y. H. Hu, S. Palreddy, and W. J. Tompkins, "A patient-adaptable ECG beat classifier using a mixture of experts approach," *IEEE Trans. Biomed. Eng.*, vol. 44, pp. 891–900, 1997.

[70] M. Lagerholm, C. Peterson, G. Braccini, L. Edenbrandt, and L. Sörnmo, "Clustering ECG complexes using Hermite functions and self-organizing maps," *IEEE Trans. Biomed. Eng.*, vol. 47, pp. 838–848, 2000.

[71] S. Osowski and T. H. Linh, "ECG beat recognition using fuzzy hybrid neural network," *IEEE Trans. Biomed. Eng.*, vol. 48, pp. 1265–1271, 2001.

[72] T. H. Linh, S. Osowski, and M. Stodolski, "On-line heart beat recognition using Hermite polynomials and neuro-fuzzy network," *IEEE Trans. Instrum. Measure.*, vol. 52, pp. 1224–1231, 2003.

[73] M. N. Cheung, "Detection of recovery from errors in the cardiac interbeat intervals," *Psychophysiology*, vol. 18, pp. 341–346, 1981.

[74] M. V. Kamath and E. L. Fallen, "Correction of the heart rate variability signal for ectopics and missing beats," in *Heart Rate Variability* (M. Malik and A. J. Camm, eds.), ch. 6, pp. 75–86, Armonk: Futura Publ., 1995.

[75] G. A. Myers, G. J. Martin, N. M. Magid, P. S. Barnett, J. W. Schaad, J. S. Weiss, M. Lesch, and D. H. Singer, "Power spectral analysis of heart rate variability in sudden cardiac death: Comparison to other methods," *IEEE Trans. Biomed. Eng.*, vol. 33, pp. 1149–1156, 1986.

[76] L. J. M. Mulder, "Measurement and analysis methods of heart rate and respiration for use in applied environments," *Biol. Psychol.*, vol. 34, pp. 205–236, 1992.

[77] N. Lippman, K. M. Stein, and B. B. Lerman, "Comparison of methods for removal of ectopy in measurements of heart rate variability," *Am. J. Physiol.*, vol. 267, pp. H411–H418, 1994.

[78] P. J. Brockwell and R. A. Davis, *Time Series: Theory and Methods*. New York: Springer-Verlag, 2nd ed., 1991.

[79] P. Albrecht and R. J. Cohen, "Estimation of heart rate power spectrum bands from real world data: Dealing with ectopic beats and noise data," in *Proc. Computers in Cardiology*, pp. 311–314, IEEE Computer Society Press, 1988.

[80] E. Parzen, "On spectral analysis with missing observations and amplitude modulation," *Sankhya*, vol. 25, pp. 971–977, 1963.

[81] J. Mateo and P. Laguna, "Extension of the heart timing signal HRV analysis in the presence of ectopic beats," in *Proc. Computers in Cardiology*, pp. 813–816, IEEE Press, 2000.

[82] J. Mateo and P. Laguna, "Analysis of heart rate variability in the presence of ectopic beats using the heart timing signal," *IEEE Trans. Biomed. Eng.*, vol. 50, pp. 334–343, 2003.

[83] M. Brennan, M. Palaniswami, and P. Kamen, "A new model-based ectopic beat correction algorithm for heart rate variability," in *Proc. Conf. IEEE EMBS (CD-ROM)*, 2001.

[84] C. L. Birkett, M. G. Kienzle, and G. A. Myers, "Interpolation of ectopics increases low frequency power in heart rate variability spectra," in *Proc. Computers in Cardiology*, pp. 257–259, IEEE Computer Society Press, 1991.

[85] R. W. de Boer, J. M. Karemaker, and J. Strackee, "Relationships between short-term blood pressure fluctuations and heart rate variability in resting subjects," *Med. Biol. Eng. & Comput.*, vol. 23, pp. 352–364, 1985.

[86] G. Baselli, S. Cerutti, S. Civardi, D. Liberati, F. Lombardi, A. Malliani, and M. Pagani, "Spectral and cross-spectral analysis of heart rate and arterial blood pressure variability signals," *Comput. Biomed. Res.*, vol. 19, pp. 520–534, 1986.

[87] G. Baselli, A. Porta, and G. Ferrari, "Models for the analysis of cardiovascular variability signals," in *Heart Rate Variability* (M. Malik and A. J. Camm, eds.), ch. 11, pp. 135–145, Armonk: Futura Publ., 1995.

[88] S. Cerutti, G. Baselli, A. M. Bianchi, L. T. Mainardi, and A. Porta, "Analysis of the interactions between heart rate and blood pressure variabilities," in *Dynamic Electrocardiography* (M. Malik and A. J. Camm, eds.), ch. 18, pp. 170–179, New York: Blackwell Futura Publ., 2004.

[89] M. L. Appel, R. D. Berger, J. P. Saul, J. M. Smith, and R. J. Cohen, "Beat to beat variability in cardiovascular variables: Noise or music?," *J. Amer. Coll. Card.*, vol. 14, pp. 1139–1148, 1989.

[90] R. D. Berger, J. P. Saul, and R. J. Cohen, "Assessment of autonomic response by broad-band respiration," *IEEE Trans. Biomed. Eng.*, vol. 36, pp. 1061–1065, 1989.

[91] G. Baselli, S. Cerutti, S. Civardi, A. Malliani, and M. Pagani, "Cardiovascular variability signals: Towards the identification of a closed-loop model of the neural control mechanisms," *IEEE Trans. Biomed. Eng.*, vol. 35, pp. 1033–1046, 1988.

[92] R. Barbieri, A. M. Bianchi, J. K. Triedman, L. T. Mainardi, S. Cerutti, and J. P. Saul, "Model dependency of multivariate autoregressive spectral analysis," *IEEE Eng. Med. Biol. Mag.*, vol. 16, pp. 74–85, 1997.

[93] G. Nollo, A. Porta, L. Faes, M. Del Greco, M. Disertori, and F. Ravelli, "Causal linear parametric model for baroreflex gain assessment in patients with recent myocardial infarction," *Am. J. Physiol. (Heart Circ. Physiol.)*, vol. 280, pp. 1830–1839, 2001.

[94] S. L. Marple Jr., *Digital Spectral Analysis with Applications*. New Jersey: Prentice-Hall, 1987.

[95] S. M. Kay, *Modern Spectral Estimation. Theory and Application*. New Jersey: Prentice-Hall, 1988.

[96] S. Haykin, *Adaptive Filter Theory*. New Jersey: Prentice-Hall, 4th ed., 2002.

[97] L. T. Mainardi, A. M. Bianchi, R. Furlan, R. Barbieri, V. di Virgilio, A. Malliani, and S. Cerutti, "Multivariate time-variant identification of cardiovascular variability signals: A beat-to-beat spectral parameter estimation in vasovagal syncope," *IEEE Trans. Biomed. Eng.*, vol. 44, pp. 978–989, 1997.

[98] M. R. Spiegel, *Schaum's Mathematical Handbook of Formulas and Tables*. New York: McGraw–Hill, 1999.

Problems

8.1 The QRS detector outputs the discrete-valued occurrence times θ_k which provide the basis for HRV analysis. Due to the quantization introduced by the sampling process, θ_k will fluctuate around an underlying, continuous-valued time t_k by half the length of the sampling interval T. Determine how such quantization-related fluctuations influence the spectrum estimated from $d_{\mathrm{IT}}(k)$, $d_{\mathrm{IIT}}(k)$, $d^u{}_{\mathrm{E}}(t)$, $d_{\mathrm{IF}}(t)$, $d_{\mathrm{IIF}}(t)$, and $d_{\mathrm{HT}}(t)$.

8.2 The so-called *triangular index*, denoted I_{T}, is used in clinical studies of HRV in order to quantify RR interval dispersion from its histogram [1]. The triangular index is defined as the ratio between the area under the histogram, equal to the total number of RR intervals M, and the maximum value of the histogram,

$$I_{\mathrm{T}} = \frac{M}{\max\left(P(r)\right)},$$

where $P(r)$ denotes the RR interval histogram.

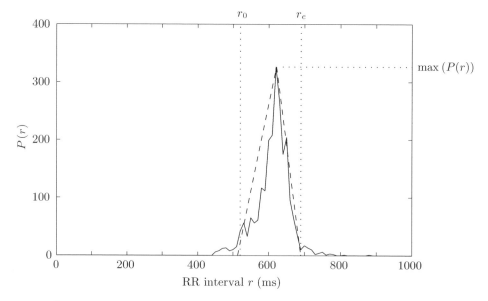

a. Comment on how different RR interval resolutions (i.e., bin sizes) of the histogram influence I_{T}.

b. The *triangular interpolation index*, denoted I_{TINN}, mitigates the problems of I_{T} by measuring the RR interval dispersion as the base of a triangle fitted to the histogram $P(r)$, defined by [1]

$$I_{\mathrm{TINN}} = r_e - r_o,$$

where r_o and r_e denote the onset and end, respectively, of the histogram (see figure). Note that the onset/end are *not* identical to the shortest/longest RR intervals but are to some robustly estimated interval lengths. Propose a procedure to estimate the onset and end required for computing I_{TINN}.

8.3 Many time domain HRV indices are calculated from the interval tachogram $d_{\mathrm{IT}}(k)$ such as the variance (i.e., SDNN squared). In a similar way, the variance of the inverse interval tachogram $d_{\mathrm{IIT}}(k)$ can be used to quantify variations in heart rate [19]. In order to study the behavior of the variance of these two heart rhythm representations, we assume that $d_{\mathrm{IT}}(k)$ is a random variable characterized by a uniform PDF with mean m_{IT},

$$p_{\mathrm{IT}}(x) = \begin{cases} \dfrac{1}{2A}, & m_{\mathrm{IT}} - A \leq x \leq m_{\mathrm{IT}} + A; \\ 0, & \text{otherwise.} \end{cases}$$

a. Determine the mean and variance of $d_{\mathrm{IT}}(k)$ and $d_{\mathrm{IIT}}(k)$. *Hint:* Start by expressing the probability distribution function of $d_{\mathrm{IIT}}(k)$,

$$P_{\mathrm{IIT}}(x) = \text{Probability}(d_{\mathrm{IIT}}(k) \leq x),$$

in terms of the probability distribution function of $d_{\mathrm{IT}}(k)$, and then determine the PDF of $d_{\mathrm{IIT}}(k)$.

b. Compare the variances of $d_{\mathrm{IT}}(k)$ and $d_{\mathrm{IIT}}(k)$ that result from the two sets of parameters, $(m_{\mathrm{IT}}, A) = (1.0, 0.2)$ and $(m_{\mathrm{IT}}, A) = (0.7, 0.1)$, where both parameters have the unit "seconds".

8.4 Show that the output of the IPFM model for a single sinusoidal input with modulating frequency F_1 is given by

$$d_{\mathrm{E}}(t) = \frac{1}{T_I} + \frac{m_1}{T_I} \cos(2\pi F_1 t)$$
$$+ \frac{2}{T_I} \sum_{k=1}^{\infty} \sum_{l=-\infty}^{\infty} \left(1 + \frac{lF_1 T_I}{k}\right) J_l\left(\frac{km_1}{F_1 T_I}\right) \cos\left(2\pi \left(\frac{k}{T_I} + lF_1\right)t\right).$$

The function $J_l\left(\frac{km_1}{F_1 T_I}\right)$ is a Bessel function of the first kind of order l; for a definition of this function, see, e.g., [98].

8.5 The computation of the lowpass filtered event series can be made more efficient for certain sampling rates.

a. Assuming that an ideal lowpass filter is used, show that sampling of the lowpass filtered event series at exactly the Nyquist sampling rate, i.e., $F_d = 2F_c$ with F_c being the highest frequency component of the signal, is given by [36]

$$d_{\mathrm{LE}}(n) = 2F_c \sum_{k=0}^{M} \frac{(-1)^{n+1} \sin(2\pi F_c t_k)}{\pi(n - 2F_c t_k)}.$$

Explain why this expression is computationally more efficient than the one valid for a general cut-off frequency F_c.

b. Show that the following expression can be derived for $d_{\mathrm{LE}}(n)$ when the sampling rate is twice that of the Nyquist rate, i.e., $F_d = 4F_c$ [39],

$$d_{\mathrm{LE}}(n) = \begin{cases} 2F_c \displaystyle\sum_{k=0}^{M} \frac{(-1)^{\frac{n+2}{2}} \sin(2\pi F_c t_k)}{\pi\left(\dfrac{n}{2} - 2F_c t_k\right)}, & n \text{ even}; \\[4ex] 2F_c \displaystyle\sum_{k=0}^{M} \frac{(-1)^{\frac{n+3}{2}} \cos(2\pi F_c t_k)}{\pi\left(\dfrac{n}{2} - 2F_c t_k\right)}, & n \text{ odd}. \end{cases}$$

8.6 A physiologist has questioned the validity of the IPFM model and the way it accounts for the influence of the autonomic nervous system on the heart rhythm. Instead, the double integrator is suggested as a more appropriate model (see below).

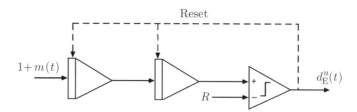

a. Supposing that the physiologist's suggestion is correct, how should the heart timing signal $d_{\mathrm{HT}}(t)$ be modified to produce an estimate of the power spectrum of $m(t)$?

b. How can the power spectrum be estimated from the new heart timing signal?

8.7 For various heart rhythm representations, interpolation is used to obtain a regularly sampled signal suitable for the Fourier-based estimation of the power spectrum. In this problem, linear interpolation is used in combination with the requirement that the frequency content below 0.4 Hz be essentially unaltered, i.e., the attenuation of these frequency components should be less than 3 dB. What is the maximal RR interval length acceptable for this property to hold true?

8.8 In the derivation of the spectrum of the event series signal, denoted $D_{\mathrm{E}}^u(\Omega)$, we made use of the relation

$$\delta(t - \tau) = \left|\frac{\partial g(t)}{\partial t}\right| \delta(g(t)),$$

valid when $g(t)$ is any function with a first-order zero at $t = \tau$, ($g(\tau) = 0$, $g(t \neq \tau) \neq 0$), and $\partial g(t)/\partial t|_{t=\tau} \neq 0$. Prove that this is true by making use of a Taylor series expansion of the function $g(t)$ around τ.

8.9 Derive the delay τ in (8.68) of Lomb's method.

8.10 Generalize the estimator of the mean RR interval length T_I in (8.89) to account for multiple ectopics.

8.11 Propose an approach for handling the presence of a single ectopic beat when the lowpass filtered event series $d_{\mathrm{LE}}(t)$ is considered.

8.12 For two stochastic signals $\mathbf{x}$ and $\mathbf{y}$, the cross-correlation coefficient ρ is defined as

$$\rho = \frac{E[\mathbf{x}^T\mathbf{y}]}{\sqrt{E[\mathbf{x}^T\mathbf{x}]}\sqrt{E[\mathbf{y}^T\mathbf{y}]}}.$$

Show that ρ can be expressed in terms of the coherence function $\Gamma_{xy}(e^{\jmath\omega})$ as

$$\rho = \frac{\frac{1}{2\pi}\int_{-\pi}^{\pi}\Gamma_{xy}(e^{\jmath\omega})\sqrt{S_x(e^{\jmath\omega})}\sqrt{S_y(e^{\jmath\omega})}d\omega}{\sqrt{E[\mathbf{x}^T\mathbf{x}]}\sqrt{E[\mathbf{y}^T\mathbf{y}]}},$$

where $S_x(e^{\jmath\omega})$ and $S_y(e^{\jmath\omega})$ denote the power spectra of $\mathbf{x}$ and $\mathbf{y}$, respectively.

8.13 Express the magnitude squared coherence $\Gamma_{x_1x_2}(e^{\jmath\omega})$ for the bivariate AR model as a function of $H_{ii}(e^{\jmath\omega})$ and $G_{ij}(e^{\jmath\omega})$ given in (8.100) and (8.101), respectively; see also the block diagram in Figure 8.23.

Appendix A

Review of Important Concepts

A.1 Matrix Fundamentals

A.1.1 Definitions

A matrix $\mathbf{A}$ is a rectangular array whose elements a_{ij} are arranged in m rows and n columns, referred to as an $m \times n$ matrix,

$$\mathbf{A} = \begin{bmatrix} a_{11} & a_{12} & \cdots & a_{1n} \\ a_{21} & a_{22} & \cdots & a_{2n} \\ \vdots & \vdots & & \vdots \\ a_{m1} & a_{m2} & \cdots & a_{mn} \end{bmatrix}. \tag{A.1}$$

The matrix is said to be *square* when $m = n$, a *row matrix* when $m = 1$, and a *column matrix* when $n = 1$. The column matrix has particular significance and is referred to as a *vector*.

It may be useful to *partition* a matrix into smaller *submatrices*. For example, an $m \times n$ matrix may be written as

$$\mathbf{A} = \begin{bmatrix} \mathbf{A}_{11} & \mathbf{A}_{12} \\ \mathbf{A}_{21} & \mathbf{A}_{22} \end{bmatrix}, \tag{A.2}$$

where $\mathbf{A}_{11}$, $\mathbf{A}_{12}$, $\mathbf{A}_{21}$, and $\mathbf{A}_{22}$ are $m_1 \times n_1$, $m_1 \times n_2$, $m_2 \times n_1$, and $m_2 \times n_2$ matrices, respectively, such that $m_1 + m_2 = m$ and $n_1 + n_2 = n$. By denoting the ith column of $\mathbf{A}$ as $\mathbf{a}_i$,

$$\mathbf{a}_i = \begin{bmatrix} a_{i1} \\ a_{i2} \\ \vdots \\ a_{in} \end{bmatrix}, \tag{A.3}$$

the matrix $\mathbf{A}$ can be written as a partitioned matrix

$$\mathbf{A} = \begin{bmatrix} \mathbf{a}_1 & \mathbf{a}_2 & \cdots & \mathbf{a}_n \end{bmatrix}. \tag{A.4}$$

The *transpose* of a matrix $\mathbf{A}$ is another matrix $\mathbf{A}^T$ whose rows and columns have been interchanged, and thus, if $\mathbf{A}$ is $m \times n$ then $\mathbf{A}^T$ is $n \times m$. The *Hermitian* transpose of a complex matrix $\mathbf{A}$ is another matrix $\mathbf{A}^H$ which has been transposed and with all elements complex-conjugated. The *inverse* of a square matrix $\mathbf{A}$ is a square matrix $\mathbf{A}^{-1}$ for which

$$\mathbf{A}\mathbf{A}^{-1} = \mathbf{A}^{-1}\mathbf{A} = \mathbf{I}, \tag{A.5}$$

where $\mathbf{I}$ is the *identity matrix* with ones in the principal diagonal and elsewhere zeros,

$$\mathbf{I} = \begin{bmatrix} 1 & 0 & \cdots & 0 \\ 0 & 1 & \cdots & 0 \\ \vdots & \vdots & & \vdots \\ 0 & 0 & \cdots & 1 \end{bmatrix}. \tag{A.6}$$

The *rank* of a matrix is defined as the number of linearly independent rows or columns, whichever is less.

The *reversal* of a vector

$$\mathbf{a} = \begin{bmatrix} a_1 \\ a_2 \\ \vdots \\ a_n \end{bmatrix} \tag{A.7}$$

is defined by

$$\tilde{\mathbf{a}} = \begin{bmatrix} a_n \\ a_{n-1} \\ \vdots \\ a_1 \end{bmatrix}, \tag{A.8}$$

where the tilde denotes the reversal operation. For the matrix $\mathbf{A}$ in (A.1), its reversal is given by

$$\tilde{\mathbf{A}} = \begin{bmatrix} a_{mn} & a_{m(n-1)} & \cdots & a_{m1} \\ a_{(m-1)n} & a_{(m-1)(n-1)} & \cdots & a_{(m-1)2} \\ \vdots & \vdots & & \vdots \\ a_{1n} & a_{1(n-1)} & \cdots & a_{11} \end{bmatrix}, \tag{A.9}$$

that is, the matrix $\mathbf{A}$ is reversed about both its vertical axis and its horizontal axis.

The inverse of a product of two matrices $\mathbf{A}$ and $\mathbf{B}$ is the product of the inverse matrices in reversed order,

$$(\mathbf{AB})^{-1} = \mathbf{B}^{-1}\mathbf{A}^{-1}. \tag{A.10}$$

Similarly, the transpose of the product of two matrices is given by

$$(\mathbf{AB})^T = \mathbf{B}^T\mathbf{A}^T. \tag{A.11}$$

A matrix is said to be *symmetric* if

$$\mathbf{A}^T = \mathbf{A}, \tag{A.12}$$

and *orthogonal* if

$$\mathbf{A}^T\mathbf{A} = \mathbf{I}, \tag{A.13}$$

or, equivalently,

$$\mathbf{A}^T = \mathbf{A}^{-1}. \tag{A.14}$$

A square matrix is called *diagonal* if all elements off the principal diagonal are zero,

$$\operatorname{diag}(a_{11}, a_{22}, \ldots, a_{nn}) = \begin{bmatrix} a_{11} & 0 & 0 & \cdots & 0 \\ 0 & a_{22} & 0 & \cdots & 0 \\ 0 & 0 & a_{33} & \cdots & 0 \\ \vdots & \vdots & \vdots & \ddots & \vdots \\ 0 & 0 & 0 & \cdots & a_{nn} \end{bmatrix}. \tag{A.15}$$

A *lower triangular matrix* has all of its elements above the principal diagonal equal to zero,

$$\begin{bmatrix} a_{11} & 0 & 0 & \cdots & 0 \\ a_{21} & a_{22} & 0 & \cdots & 0 \\ a_{31} & a_{32} & a_{33} & \cdots & 0 \\ \vdots & \vdots & \vdots & & \vdots \\ a_{n1} & a_{n2} & a_{n3} & \cdots & a_{nn} \end{bmatrix}. \tag{A.16}$$

An *upper triangular matrix* is defined as the transpose of a lower triangular matrix. A matrix is said to be *Toeplitz* if all of its elements along each diagonal have the same value,

$$
\begin{bmatrix}
a_0 & a_1 & a_2 & \cdots & a_{n-2} & a_{n-1} \\
a_{-1} & a_0 & a_1 & \cdots & a_{n-3} & a_{n-2} \\
a_{-2} & a_{-1} & a_0 & \cdots & a_{n-4} & a_{n-3} \\
\vdots & \vdots & \vdots & \ddots & \vdots & \vdots \\
a_{-n+2} & a_{-n+3} & a_{-n+4} & \cdots & a_0 & a_1 \\
a_{-n+1} & a_{-n+2} & a_{-n+3} & \cdots & a_{-1} & a_0
\end{bmatrix}.
\tag{A.17}
$$

The matrix is *symmetric Toeplitz* if $a_{-k} = a_k$.

The product of a transposed vector and another vector of the same dimension is a scalar a and is referred to as the *inner product* or the *scalar product*

$$
a = \mathbf{x}^T \mathbf{y} = \mathbf{y}^T \mathbf{x}.
\tag{A.18}
$$

The vectors are orthogonal if their scalar product is equal to zero,

$$
\mathbf{x}^T \mathbf{y} = 0.
\tag{A.19}
$$

The product of a vector and a transposed vector is a matrix $\mathbf{A}$ and is referred to as the *outer product*

$$
\mathbf{A} = \mathbf{x}\mathbf{y}^T.
\tag{A.20}
$$

Note that $\mathbf{x}\mathbf{y}^T \neq \mathbf{y}\mathbf{x}^T$ in general.

For a symmetric $n \times n$ matrix $\mathbf{A}$, a *quadratic form* Q is defined as

$$
Q = \mathbf{x}^T \mathbf{A} \mathbf{x} = \sum_{i=1}^{n} \sum_{j=1}^{n} a_{ij} x_i x_j.
\tag{A.21}
$$

A symmetric matrix $\mathbf{A}$ is said to be *positive definite* if

$$
\mathbf{x}^T \mathbf{A} \mathbf{x} > 0
\tag{A.22}
$$

for all $\mathbf{x} \neq \mathbf{0}$; the matrix $\mathbf{A}$ is *positive semidefinite* if the quadratic form Q is greater or equal to zero.

The sum of the diagonal elements of a square matrix $\mathbf{A}$ is called the *trace* of $\mathbf{A}$ and is

$$
\mathrm{tr}(\mathbf{A}) = \sum_{i=1}^{n} a_{ii}.
\tag{A.23}
$$

The trace of the sum of matrices is equal to the sum of the traces

$$
\mathrm{tr}(\mathbf{A} + \mathbf{B}) = \mathrm{tr}(\mathbf{A}) + \mathrm{tr}(\mathbf{B}),
\tag{A.24}
$$

and the trace of a matrix product is

$$\text{tr}(\mathbf{AB}) = \text{tr}(\mathbf{BA}). \tag{A.25}$$

The determinant for an $n \times n$ matrix $\mathbf{A}$ multiplied with a scalar a is

$$\det(a\mathbf{B}) = a^n \det(\mathbf{A}), \tag{A.26}$$

and for a matrix product it is

$$\det(\mathbf{AB}) = \det(\mathbf{BA}). \tag{A.27}$$

The determinant of a diagonal matrix $\mathbf{A}$ is

$$\det(\mathbf{A}) = \prod_{i=1}^{n} a_{ii}. \tag{A.28}$$

The determinant of an inverse matrix is

$$\det(\mathbf{A}^{-1}) = (\det(\mathbf{A}))^{-1}. \tag{A.29}$$

A useful relationship for the trace of the outer product of the vectors $\mathbf{x}$ and $\mathbf{y}$ is

$$\text{tr}(\mathbf{xy}^T) = \mathbf{y}^T\mathbf{x}. \tag{A.30}$$

The *matrix inversion lemma* states that

$$(\mathbf{A} + \mathbf{BCD})^{-1} = \mathbf{A}^{-1} - \mathbf{A}^{-1}\mathbf{B}(\mathbf{DA}^{-1}\mathbf{B} + \mathbf{C}^{-1})^{-1}\mathbf{DA}^{-1}, \tag{A.31}$$

where $\mathbf{A}$ is $n \times n$, $\mathbf{B}$ is $n \times m$, $\mathbf{C}$ is $m \times m$, and $\mathbf{D}$ is $m \times n$, with the assumption that the included inverse matrices exist.

The norm of a vector is a number that characterizes the magnitude of the vector. The *Euclidean norm* is one such measure which is defined by

$$\|\mathbf{x}\|_2 = \left(\sum_{i=1}^{n} |x_i|^2 \right)^{\frac{1}{2}} = \left(\mathbf{x}^T\mathbf{x} \right)^{\frac{1}{2}}. \tag{A.32}$$

Similarly, the norm of a matrix is a number that characterizes the magnitude of the matrix. Several matrix norm definitions exist, of which one of the most used is the *Frobenius norm*,

$$\|\mathbf{A}\|_F = \sqrt{\sum_{i=1}^{m} \sum_{j=1}^{n} |a_{ij}|^2} = \left(\text{tr} \left(\mathbf{AA}^T \right) \right)^{1/2}. \tag{A.33}$$

A.1.2   Matrix Decomposition

A square $n \times n$ matrix $\mathbf{A}$ has an *eigenvector* $\mathbf{v}$ which satisfies

$$\mathbf{A}\mathbf{v} = \lambda\mathbf{v} \tag{A.34}$$

for some scalar λ, also referred to as the *eigenvalue* corresponding $\mathbf{v}$. The eigenvectors are normalized to have unit length, i.e., $\mathbf{v}^T\mathbf{v} = 1$. The n eigenvalues of (A.34) are obtained as the roots of the *characteristic equation*

$$\det(\mathbf{A} - \lambda\mathbf{I}) = \mathbf{0}.$$

With these eigenvalues, the corresponding eigenvectors can be determined from

$$(\mathbf{A} - \lambda\mathbf{I})\mathbf{v} = \mathbf{0}.$$

When $\mathbf{A}$ is symmetric and positive definite (semidefinite), all eigenvalues are real-valued and positive (non-negative). The corresponding eigenvectors are orthonormal,

$$\mathbf{v}_i^T\mathbf{v}_j = \begin{cases} 1, & i = j; \\ 0, & i \neq j. \end{cases} \tag{A.35}$$

The relation in (A.34) can be expressed to include all n eigenvalues and eigenvectors

$$\mathbf{A}\begin{bmatrix} \mathbf{v}_1 & \mathbf{v}_1 & \cdots & \mathbf{v}_n \end{bmatrix} = \begin{bmatrix} \lambda_1\mathbf{v}_1 & \lambda_2\mathbf{v}_2 & \cdots & \lambda_n\mathbf{v}_n \end{bmatrix},$$

or

$$\mathbf{A}\mathbf{V} = \mathbf{V}\Lambda, \tag{A.36}$$

where $\mathbf{V} = \begin{bmatrix} \mathbf{v}_1 & \mathbf{v}_1 & \cdots & \mathbf{v}_n \end{bmatrix}$ and $\Lambda = \mathrm{diag}(\lambda_1, \lambda_2, \ldots, \lambda_n)$. Since the columns of $\mathbf{V}$ are orthonormal, i.e., $\mathbf{V}^T = \mathbf{V}^{-1}$, the matrix $\mathbf{A}$ can be decomposed into a weighted sum of rank-one matrices $\mathbf{v}_i\mathbf{v}_i^T$,

$$\mathbf{A} = \mathbf{V}\Lambda\mathbf{V}^T = \sum_{i=1}^{n} \lambda_i\mathbf{v}_i\mathbf{v}_i^T. \tag{A.37}$$

Thus, for a matrix defined by an outer product, i.e., $\mathbf{A} = \mathbf{a}\mathbf{a}^T$, it is evident from (A.37) that only one eigenvalue can be nonzero. The corresponding eigenvector $\mathbf{v}_1$ is, apart from a normalization factor, identical to the vector $\mathbf{a}$. The remaining eigenvectors must be selected such that these are orthogonal to $\mathbf{v}_1$.

Based on the expansion in (A.37), the determinant and the trace of a symmetric matrix $\mathbf{A}$ can be related to its eigenvalues,

$$\det(\mathbf{A}) = \det(\mathbf{V}\Lambda\mathbf{V}^T) = \det(\mathbf{V}^T\mathbf{V}\Lambda)$$
$$= \prod_{i=1}^{n} \lambda_i, \tag{A.38}$$

and

$$\text{tr}(\mathbf{A}) = \text{tr}(\mathbf{V}\Lambda\mathbf{V}^T) = \text{tr}(\mathbf{V}^T\mathbf{V}\Lambda)$$
$$= \sum_{i=1}^{n} \lambda_i, \tag{A.39}$$

respectively.

Another, more general type of matrix decomposition is the *singular value decomposition* (SVD) by which an $m \times n$ matrix $\mathbf{A}$ can be decomposed into two orthogonal matrices, an $m \times m$ matrix $\mathbf{U}$ and an $n \times n$ matrix $\mathbf{V}$ such that

$$\mathbf{A} = \mathbf{U}\Sigma\mathbf{V}^T, \tag{A.40}$$

where Σ is an $m \times n$ non-negative diagonal matrix defined by

$$\Sigma = \begin{bmatrix} \mathbf{S} & \mathbf{0} \\ \mathbf{0} & \mathbf{0} \end{bmatrix}, \tag{A.41}$$

and

$$\mathbf{S} = \text{diag}(\sigma_1, \sigma_2, \ldots, \sigma_l), \tag{A.42}$$

where $\sigma_1 \geq \sigma_2 \geq \ldots \geq \sigma_l > 0$ and $\sigma_{l+1} = \cdots = \sigma_p = 0$ denote the singular values of $\mathbf{A}$; the number of nonzero singular values l does not exceed $p = \min(m, n)$. Similar to eigendecomposition of a symmetric matrix in (A.37), the matrix $\mathbf{A}$ can be decomposed into a weighted sum of rank-one matrices $\mathbf{u}_i\mathbf{v}_i^T$,

$$\mathbf{A} = \sum_{i=1}^{l} \sigma_i \mathbf{u}_i\mathbf{v}_i^T. \tag{A.43}$$

The nonzero singular values of $\mathbf{A}$ are equal to the square-root of the positive eigenvalues of the matrices $\mathbf{A}^T\mathbf{A}$ and $\mathbf{A}\mathbf{A}^T$.

A.1.3   Matrix Optimization

When optimizing matrix equations the following vector differentiation rules are useful:

$$\nabla_{\mathbf{x}}(\mathbf{x}^T\mathbf{y}) = \mathbf{y}, \tag{A.44}$$

and

$$\nabla_{\mathbf{x}}(\mathbf{y}^T\mathbf{x}) = \mathbf{y}, \tag{A.45}$$

where the gradient $\nabla_{\mathbf{x}}$ of a vector function $f(\mathbf{x})$ is defined as

$$\nabla_{\mathbf{x}}f(\mathbf{x}) \stackrel{\text{def}}{=} \begin{bmatrix} \dfrac{\partial f}{\partial x_1} \\ \dfrac{\partial f}{\partial x_2} \\ \vdots \\ \dfrac{\partial f}{\partial x_n} \end{bmatrix}, \tag{A.46}$$

where $x_1, \ldots, x_n$ are the elements of $\mathbf{x}$. The gradient of a quadratic form is given by

$$\nabla_{\mathbf{x}}(\mathbf{x}^T\mathbf{A}\mathbf{x}) = 2\mathbf{A}\mathbf{x}, \tag{A.47}$$

where $\mathbf{A}$ is symmetric.

It may be necessary to optimize a vector function $f(\mathbf{x})$ subject to a number of constraints. For a set of L different constraints $g_l(\mathbf{x}) = 0$ with $l = 1, \ldots, L$, optimization with respect to $\mathbf{x}$ can be done in the following way.

1. First, define the *Lagrangian* function $\mathcal{L}(\mathbf{x}, \mu_1, \ldots, \mu_L)$,

$$\mathcal{L}(\mathbf{x}, \mu_1, \ldots, \mu_L) = f(\mathbf{x}) + \sum_{l=1}^{L} \mu_l g_l(\mathbf{x}), \tag{A.48}$$

where $\mu_1, \ldots, \mu_l$ are called Lagrange multipliers.

2. Then, solve the equation system

$$\begin{cases} \nabla_{\mathbf{x}}\mathcal{L}(\mathbf{x}, \mu_1, \ldots, \mu_L) = \nabla_{\mathbf{x}}\left(f(\mathbf{x}) + \displaystyle\sum_{l=1}^{L} \mu_l g_l(\mathbf{x}) \right) = \mathbf{0}, \\[2mm] \dfrac{\partial \mathcal{L}(\mathbf{x}, \mu_1, \ldots, \mu_L)}{\partial \mu_1} = g_1(\mathbf{x}) = 0, \\[1mm] \qquad\qquad\vdots \\[1mm] \dfrac{\partial \mathcal{L}(\mathbf{x}, \mu_1, \ldots, \mu_L)}{\partial \mu_L} = g_L(\mathbf{x}) = 0, \end{cases} \tag{A.49}$$

where the solutions to the L different equations

$$\frac{\partial \mathcal{L}(\mathbf{x}, \mu_1, \ldots, \mu_L)}{\partial \mu_l} = 0$$

are the specified constraints.

A.1.4 Linear Equations

The solution to a set of n linear equations in the m unknowns x_i, $i = 1, \ldots, m$,

$$a_{11}x_1 + a_{12}x_2 + \cdots + a_{1m}x_m = b_1,$$

$$a_{21}x_1 + a_{22}x_2 + \cdots + a_{2m}x_m = b_2,$$

$$\vdots$$

$$a_{n1}x_1 + a_{n2}x_2 + \cdots + a_{nm}x_m = b_n,$$

or, equivalently,

$$\mathbf{Ax} = \mathbf{b}, \tag{A.50}$$

depend on various factors such as the relative size of m and n and the rank of $\mathbf{A}$. When $\mathbf{A}$ is a square matrix, i.e., $m = n$, the solution is defined by

$$\mathbf{x} = \mathbf{A}^{-1}\mathbf{b}, \tag{A.51}$$

provided that the inverse matrix $\mathbf{A}^{-1}$ exists.

In many signal processing problems, the number of unknowns x_i is less than the number of linear equations, i.e., $m < n$, and the solution is said to be overdetermined. In such cases, the *least-squares solution* is usually considered and results from minimizing the error norm

$$\|\mathbf{Ax} - \mathbf{b}\|_2^2 \tag{A.52}$$

with respect to $\mathbf{x}$. When the matrix $\mathbf{A}$ has full rank, the least-squares solution is [1–3]

$$\mathbf{x} = \left(\mathbf{A}^T\mathbf{A}\right)^{-1}\mathbf{A}^T\mathbf{b}, \tag{A.53}$$

where the matrix $\left(\mathbf{A}^T\mathbf{A}\right)^{-1}\mathbf{A}^T$ is referred to as the pseudo-inverse of $\mathbf{A}$ for the overdetermined problem.

A.2   Discrete-Time Stochastic Processes

A stochastic process represents an ensemble of possible realizations of a process. Each realization of the stochastic process is called a time series, e.g., $x(0), \ldots, x(N-1)$, which in vector form can be written

$$
\mathbf{x} = \begin{bmatrix} x(0) \\ x(1) \\ \vdots \\ x(N-1) \end{bmatrix}.
\tag{A.54}
$$

It is assumed that $x(n)$ is real-valued, stochastic process.

A.2.1   Definitions

A stochastic process can be described using expected values or ensemble averages which are averages over all realizations. The mean or expected value of a function $f(x)$ of a random variable X, characterized by its PDF $p_X(x)$, is defined as

$$
E[f(x)] = \int_{-\infty}^{\infty} f(x) p_X(x) dx.
\tag{A.55}
$$

The mean value function of a stochastic process is defined by

$$
m_x(n) = E[x(n)]
\tag{A.56}
$$

and contains the averages of all possible outcomes for each individual sample. In vector representation (A.56) can be written

$$
\mathbf{m}_x = E[\mathbf{x}].
\tag{A.57}
$$

The variance function contains for each sample the ensemble average of the squared deviation from the mean value for that sample. The variance function is given by

$$
\sigma_x^2(n) = E\left[|x(n) - m_x(n)|^2\right],
\tag{A.58}
$$

or in vector form

$$
\sigma_x^2 = E\left[|\mathbf{x} - \mathbf{m}_x|^2\right].
\tag{A.59}
$$

The covariance function (also called the autocovariance function) describes the average joint deviation from the mean value for two samples n_1 and n_2 and is defined by

$$
c_x(n_1, n_2) = E[(x(n_1) - m_x(n_1))(x(n_2) - m_x(n_2))].
\tag{A.60}
$$

A positive covariance value indicates that the deviations from the mean value for these two samples, in average, have the same sign, while a negative value indicates that the deviations tend to have opposite sign. The covariance matrix $\mathbf{C}_x$ is defined by

$$
\begin{aligned}
\mathbf{C}_x &= E\left[(\mathbf{x} - \mathbf{m}_x)(\mathbf{x} - \mathbf{m}_x)^T\right] \\
&= \begin{bmatrix}
c_x(0,0) & c_x(0,1) & \cdots & c_x(0,N-1) \\
c_x(1,0) & c_x(1,1) & \cdots & c_x(1,N-1) \\
\vdots & \vdots & & \vdots \\
c_x(N-1,0) & c_x(N-1,1) & \cdots & c_x(N-1,N-1)
\end{bmatrix},
\end{aligned}
\tag{A.61}
$$

which is symmetric. The correlation function (also called the autocorrelation function) is defined by

$$
r_x(n_1, n_2) = E[x(n_1)x(n_2)],
\tag{A.62}
$$

which has an interpretation similar to that of the covariance function, although it does not reflect deviations from the mean value. The correlation matrix is defined by

$$
\begin{aligned}
\mathbf{R}_x &= E\left[\mathbf{x}\mathbf{x}^T\right] \\
&= \begin{bmatrix}
r_x(0,0) & r_x(0,1) & \cdots & r_x(0,N-1) \\
r_x(1,0) & r_x(1,1) & \cdots & r_x(1,N-1) \\
\vdots & \vdots & & \vdots \\
r_x(N-1,0) & r_x(N-1,1) & \cdots & r_x(N-1,N-1)
\end{bmatrix}.
\end{aligned}
\tag{A.63}
$$

A close relation exists between the covariance and the correlation matrices since

$$
\mathbf{C}_x = \mathbf{R}_x - \mathbf{m}_x \mathbf{m}_x^T,
\tag{A.64}
$$

and thus a zero-mean process has identical covariance and correlation matrices.

A cross-correlation function can be defined which describes the correlation properties between two different stochastic processes $x(n)$ and $y(n)$

$$
r_{xy}(n_1, n_2) = E[x(n_1)y(n_2)].
\tag{A.65}
$$

The corresponding cross-correlation matrix is defined by

$$
\mathbf{R}_{xy} = E\left[\mathbf{x}\mathbf{y}^T\right].
\tag{A.66}
$$

A.2.2   Stationarity

In the previous subsection the stochastic process was characterized by its first and second moments. It should be noted, however, that the first two moments do not provide a complete statistical description of a stochastic process. On the other hand, it is seldom possible to exactly determine the probability density function, and the process may be reasonably well-described by its first two moments.

A stochastic process is said to be strictly stationary if all of its moments are time-invariant. A less strict assumption is that its first two moments are time-invariant, and the process is called *wide-sense stationary*. A process is wide-sense stationary if the mean value function is a constant,

$$m_x(n) = m_x, \tag{A.67}$$

and the covariance and correlation functions depend only on the lag between the two samples,

$$c_x(n, n - k) = c_x(k) \tag{A.68}$$

and

$$r_x(n, n - k) = r_x(k). \tag{A.69}$$

The correlation function of a wide-sense stationary process is symmetric,

$$r_x(k) = r_x(-k). \tag{A.70}$$

For a zero lag, the correlation function is non-negative and equals the mean-square value of the process,

$$r_x(0) = E\left[|x(n)|^2\right] \geq 0. \tag{A.71}$$

Furthermore, the correlation function is bounded by the mean-square value,

$$|r_x(k)| \leq r_x(0). \tag{A.72}$$

The correlation matrix $\mathbf{R}_x$ of a wide-sense stationary process $x(n)$ is symmetric and Toeplitz,

$$\mathbf{R}_x = E[\mathbf{x}\mathbf{x}^T] = \begin{bmatrix} r_x(0) & r_x(-1) & \cdots & r_x(-N+1) \\ r_x(1) & r_x(0) & \cdots & r_x(-N+2) \\ \vdots & \vdots & & \vdots \\ r_x(N-1) & r_x(N-2) & \cdots & r_x(0) \end{bmatrix}, \tag{A.73}$$

and is positive semidefinite with non-negative, real-valued eigenvalues.

A.2.3 Ergodicity

Often different realizations of a signal are unavailable, and the mean value and the correlation function can be estimated using time averages instead of ensemble averages. If a large number of samples of a wide-sense stationary process are available, the mean value of a stochastic process can be estimated by using

$$\hat{m}_x(N) = \frac{1}{N} \sum_{n=0}^{N-1} x(n). \tag{A.74}$$

If

$$\lim_{N \to \infty} E\left[|\hat{m}_x(N) - m_x|^2\right] = 0, \tag{A.75}$$

the process $x(n)$ is said to be *ergodic in the mean*. Similarly, the correlation function can be estimated by

$$\hat{r}_x(k, N) = \frac{1}{N} \sum_{n=k}^{N-1} x(n)x(n-k). \tag{A.76}$$

If

$$\lim_{N \to \infty} E\left[|\hat{r}_x(k, N) - r_x(k)|^2\right] = 0, \tag{A.77}$$

the process is said to be *correlation ergodic*.

A.2.4 Bias and Consistency

An estimate $\hat{\theta}_N$ of the unknown parameter θ is referred to as being *unbiased* if

$$E\left[\hat{\theta}_N\right] = \theta, \tag{A.78}$$

or, otherwise, the difference

$$b(\theta) = E\left[\hat{\theta}_N\right] - \theta$$

is referred to as the *bias*; the parameter N denotes the number of observations used to compute an estimate of θ. The estimate is *asymptotically unbiased* if the bias approaches zero for an increasing number of observations,

$$\lim_{N \to \infty} E\left[\hat{\theta}_N\right] = \theta. \tag{A.79}$$

The estimate is said to be *consistent* if it is asymptotically unbiased and has a variance that approaches zero as the number of observations goes to infinity.

A.2.5   Power Spectrum

The discrete-time Fourier transform of the correlation function is called *power spectrum* or *power spectral density*

$$S_x(e^{j\omega}) = \sum_{k=-\infty}^{\infty} r_x(k)e^{-j\omega k}. \tag{A.80}$$

Inversely, the correlation function can be calculated from the power spectrum by

$$r_x(k) = \frac{1}{2\pi}\int_{-\pi}^{\pi} S_x(e^{j\omega})e^{j\omega k}d\omega. \tag{A.81}$$

The z-transform can be used instead of the discrete-time Fourier transform, and the power spectrum can be written

$$S_x(z) = \sum_{k=-\infty}^{\infty} r_x(k)z^{-k}. \tag{A.82}$$

For a wide-sense stationary process the power spectrum is symmetric,

$$S_x(e^{j\omega}) = S_x(e^{-j\omega}), \tag{A.83}$$

and non-negative,

$$S_x(e^{j\omega}) \geq 0. \tag{A.84}$$

The average power of a zero-mean, wide-sense stationary process can be written

$$E\left[|x(n)|^2\right] = \frac{1}{2\pi}\int_{-\pi}^{\pi} S_x(e^{j\omega})d\omega. \tag{A.85}$$

A.2.6   White Noise

A zero-mean process that has constant power spectrum is called a *white noise* process. A stationary, white noise process $v(n)$ is completely described by its second-order moment,

$$r_v(k) = \sigma_v^2\delta(k), \tag{A.86}$$

where σ_v^2 is the variance of the process. The correlation function shows that white noise is a sequence of uncorrelated random variables. If the process is a sequence of Gaussian random variables it is called white, Gaussian noise. The power spectrum of white noise is

$$S_v(e^{j\omega}) = \sigma_v^2. \tag{A.87}$$

A.2.7 Filtering of Stochastic Processes

Stochastic processes are often inputs to linear, time-invariant filters. The first and second-order moments of the output process $y(n)$ of a filter with impulse response $h(n)$ then relate to the input process $x(n)$ as follows:

$$m_y = E[y(n)] = m_x \sum_{k=-\infty}^{\infty} h(k), \qquad (A.88)$$

and

$$r_y(k) = r_x(k) * h(k) * h(-k). \qquad (A.89)$$

In the frequency domain these relations can be written

$$m_y = m_x H(e^{j0}), \qquad (A.90)$$

and

$$S_y(e^{j\omega}) = |H(e^{j\omega})|^2 S_x(e^{j\omega}). \qquad (A.91)$$

Similarly, using the z-transform,

$$S_y(z) = H(z)H(z^{-1})S_x(z). \qquad (A.92)$$

Filtering of white noise with variance σ_x^2 can then be expressed as

$$S_y(z) = H(z)H(z^{-1})\sigma_x^2. \qquad (A.93)$$

A white noise process $v(n)$ filtered with a filter $H(z)$ having a rational transfer function with q zeros and p poles of the form

$$H(z) = \frac{B_q(z)}{A_p(z)} = \frac{\displaystyle\sum_{k=0}^{q} b_q(k)z^{-k}}{1 + \displaystyle\sum_{k=1}^{p} a_q(k)z^{-k}} \qquad (A.94)$$

is called an *autoregressive moving average* (ARMA) process of order (p, q). The power spectrum of an ARMA process $x(n)$ can then be written

$$S_x(e^{j\omega}) = \frac{|B_q(e^{j\omega})|^2}{|A_p(e^{j\omega})|^2}\sigma_v^2. \qquad (A.95)$$

There are two special cases of an ARMA process and that is when $q = 0$ and when $p = 0$, respectively. When $q = 0$ the process is referred to as an *autoregressive* (AR) process of order p. The transfer function and power spectrum are given by

$$H(z) = \frac{b(0)}{1 + \sum_{k=1}^{p} a_q(k)z^{-k}} \tag{A.96}$$

and

$$S_x(e^{j\omega}) = \frac{|b(0)|^2}{|A_p(e^{j\omega})|^2}\sigma_v^2, \tag{A.97}$$

respectively. Similarly, when $p = 0$ the process is referred to as a *moving average* (MA) process of order q for which the transfer function and power spectrum are

$$H(z) = \sum_{k=0}^{q} b_q(k)z^{-k} \tag{A.98}$$

and

$$S_x(e^{j\omega}) = |B_q(e^{j\omega})|^2\sigma_v^2, \tag{A.99}$$

respectively.

Bibliography

[1] G. Strang, *Linear Algebra and Its Applications*. Philadelphia: Saunders, 3rd ed., 1988.

[2] G. H. Golub and C. F. van Loan, *Matrix Computations*. Baltimore: The Johns Hopkins University Press, 2nd ed., 1989.

[3] M. Hayes, *Statistical Digital Signal Proccessing and Modeling*. New York: John Wiley & Sons, 1996.

Appendix B

Symbols and Abbreviations

B.1 Mathematical Symbols

$\hat{}$	denotes estimator
$\check{}$	denotes approximate estimator
$\tilde{}$	denotes reversal of a vector or matrix
$\bar{}$	denotes mean value
$*$	convolution
$\xleftrightarrow{\mathcal{FT}}$	Fourier transform pair
$\lfloor \cdot \rfloor$	denotes integer part
$\vert \cdot \vert$	denotes absolute value
$\oint_C$	contour integral over C
$\arg\max\limits_{x} f(x)$	denotes the value of x that maximizes $f(x)$
$\mathbf{0}$	vector whose elements equal zero
$\mathbf{1}$	vector whose elements equal one
$\mathbf{1}_M$	vector whose M elements equal one
A	maximum amplitude of excitatory postsynaptic potentials
$\mathbf{A}_i$	$M \times M$ matrix describing temporal and spatial correlation
$A(z)$	denominator polynomial of an AR system transfer function
$A_p(z)$	denominator polynomial of a pth order AR system transfer function
$A_p(e^{\jmath\omega})$	discrete-time Fourier transform of denominator polynomial
$A_x(\tau, \nu)$	Ambiguity function of signal $x(t)$
a	amplitude factor,
	average time delay in excitatory postsynaptic potentials
$\mathbf{a}$	signal amplitude vector
a_i	feedback coefficient of a linear, time-invariant system,
	amplitude factor
$\mathbf{a}_p$	vector of feedback coefficient of a linear, time-invariant system,
	signal amplitude vector

α	gain factor in various recursive algorithms,
	weight coefficient
B	continuous-time signal bandwidth
	maximum amplitude of inhibitory postsynaptic potentials,
$B(z)$	numerator polynomial of the transfer function
$B(e^{\jmath\omega})$	discrete-time Fourier transform of numerator polynomial
b	multiplicative parameter,
	average time delay in inhibitory postsynaptic potentials
$b(n)$	signal envelope
b_i	feedforward coefficient of a linear, time-invariant system
β	waveform duration parameter,
	exponential weighting factor
β_l	discretized waveform duration parameter,
	l^{th} ECG measurement on the observed signal
$\tilde{\beta}_l$	l^{th} ECG measurement on the reconstructed signal
$\beta(n)$	time-varying waveform duration parameter
C_i	interaction between neuron subpopulations
$\mathbf{C}_x$	covariance matrix of $\mathbf{x}$
C_ψ	normalization factor in the inverse wavelet transform
$C_x(t, \Omega)$	general Cohen's class time–frequency distribution
c	constant
c_i	partial fraction expansion coefficient
$c_j(k)$	dyadic scaling expansion coefficient
$c_j^u(k)$	$c_j(k)$ with zeros inserted
c_w	constant
D	decimation/interpolation factor,
	signal duration,
	interval preceding EP stimulus
D_0, D_1	QRS detection threshold parameters (refractory period)
$D_{\text{HT}}(\Omega)$	Fourier transform of $d_{\text{HT}}(t)$
$D_{\text{HT}_k}(\Omega)$	Fourier transform of $d_{\text{HT}}(t)$ at $k/\overline{T}_0$
$D_{\text{E}}^u(\Omega)$	Fourier transform of $d_{\text{E}}^u(t)$
$D^u(\Omega)$	Fourier transform of $d^u(t)$
$D(\Omega)$	Fourier transform of $d(t)$
$\mathbf{d}$	vector of unevenly distributed signal samples at $d(t_i)$
$d^u(t)$	unevenly sampled signal (continuous-time)
$d^i(t)$	interpolated signal (continuous-time)
$d^e(t)$	evenly sampled signal (continuous-time)
$d_{\text{HT}}(t)$	heart timing signal
$d_{\text{IT}}(k)$	interval tachogram signal
$d_{\text{IIT}}(k)$	inverse interval tachogram signal

$d_{\mathrm{IF}}(t)$	interval function
$d_{\mathrm{IIF}}(t)$	inverse interval function
$d^i_{\mathrm{IIF}_s}(t)$	inverse interval function interpolated with RR interval shifting
$d_{\mathrm{E}}(t)$	event series signal
$d_{\mathrm{LE}}(t)$	lowpass filtered event series signal
$d_{\mathrm{SR}}(k)$	interval tachogram from sinus rhythm
$d(n)$	heart rate signal
d_i	pole of a linear, time-invariant system
$d_j(k)$	coefficient of the dyadic wavelet expansion
$d_j^u(k)$	$d_j(k)$ with zeros inserted
$\Delta(n)$	discrete-time segmentation function
$\Delta\mathbf{w}(n)$	time-varying weight error vector
$\Delta\mathbf{w}^b$	bias in the weight vector estimate of the LMS algorithm
$\Delta\omega_i$	3-dB bandwidth of ith spectral component
Δ_t	continuous-time signal duration
Δ_Ω	continuous-time signal bandwidth
Δ_n	discrete-time signal duration
Δ_ω	discrete-time signal bandwidth
Δ_θ	tolerance interval time in QRS detector evaluation
$\Delta\hat{s}_a(n)$	difference signal between two subaverages
Δt_{i_j}	time distance from i^{th} beat to the $(i+j)^{\mathrm{th}}$ beat
$\Delta\beta$	step size for discretized duration parameter β, normalized ECG measurement error
$\delta(t)$	continuous-time unit impulse function (Dirac function)
$\delta(n)$	discrete-time unit impulse function
$e(n)$	discrete-time error signal
$e_p(n)$	prediction error of a pth order AR system
$e_r(n)$	prediction error within a reference window
$e_t(n)$	prediction error within a test window
$e^+(n)$	forward prediction error
$e^-(n)$	backward prediction error
$\mathbf{e}_i$	error signal vector in block LMS algorithm
$E[\cdot]$	expected value
E_s	energy of s(n)
$\mathcal{E}$	mean-square error
$\mathcal{E}(n)$	mean-square error function at time n
$\mathcal{E}_{ex}(n)$	excess mean-square error function at time n
$\mathcal{E}_{min}$	minimum mean-square error
$\mathcal{E}_{\mathbf{w}}$	mean-square error function of the weight vector $\mathbf{w}$
$\mathcal{E}_{\mathbf{w}}(n)$	mean-square error function of the weight vector $\mathbf{w}$ at time n
ϵ	fraction of a number

ε	tolerance in data compression
ϵ^2	least-squares error
F	continuous-time frequency
F_i	constant frequency values for series development
F_s	sampling rate
F_c	cut-off frequency
F_I	mean repetition frequency
$\mathbf{F}_m(n)$	matrix used in LMS filtering
f	normalized discrete-time frequency
f_c	normalized cut-off frequency
$f(\cdot)$	nonlinear function, e.g., a sigmoid
$\boldsymbol{\Phi}$	matrix defining a set of basis functions
$\Phi(\Omega)$	Fourier transform of $\varphi(t)$
ϕ	phase
$\boldsymbol{\varphi}$	vector basis function
$\varphi(n)$	discrete-time basis function
$\varphi(t)$	continuous-time phase function, scaling function in wavelet representation
$\varphi_{j,k}(t)$	dyadically sampled scaling function
$G(t,\Omega)$	two-dimensional Fourier transform of the kernel function $g(\tau,\nu)$
$g(\tau,\nu)$	two-dimensional continuous-time kernel function
$g(l,n)$	sum across basis functions of the product at samples l and n
g_M	update factor in recursive weighted averaging
$g(n)$	QRS detector threshold (refractory period)
$g(\cdot)$	interpolation function
$g(n,\varepsilon)$	slope function with tolerance ε for SAPA data compression
$|\Gamma_{xy}(e^{\jmath\omega})|^2$	magnitude squared coherence of $x(n)$ and $y(n)$
$\Gamma_{\mathrm{SPI}}(n)$	spectral purity index
$\Gamma(\nu)$	Gamma function
γ	reflection coefficient of a lattice filter, trimming factor in the trimmed mean estimator
$\mathbf{H}$	matrix of sine/cosine basis functions
$H(e^{\jmath\omega})$	frequency response of $h(n)$
$H(e^{\jmath\omega},n)$	time-varying frequency response of $h(k,n)$
$H^c(e^{\jmath\omega})$	clipped frequency response of $h(n)$
$H(z)$	transfer function of $h(n)$
$H_{ij}(z)$	cross-transfer function relating $x_i(n)$ and $x_j(n)$
$H_p(z)$	transfer function of order p AR system
$\mathcal{H}_i$	Hjorth descriptor
h	complex-valued parameter
$\mathbf{h}$	vector of discrete-time impulse response, vector of unevenly spaced samples

$h(k)$	impulse response of a discrete-time, linear, time-invariant filter
$h_\varphi(n)$	sequence of scaling coefficients
$h_\psi(n)$	sequence of wavelet coefficients
$h(k,n)$	impulse response of a discrete-time, linear, time-varying filter
$h(t)$	continuous-time impulse response
$h_e(t)$	impulse response (excitatory postsynaptical potentials)
$h_i(t)$	impulse response (inhibitory postsynaptical potentials)
$\mathbf{I}$	identity matrix
$\mathbf{i}$	column vector whose top element is one and the remaining zero
J	finest scale in wavelet decomposition,
	total number of leads
$\Im(\cdot)$	imaginary part
$\mathcal{J}(\cdot)$	mean-square error for continuous functions
$\jmath$	$\sqrt{-1}$
K	size of truncated set of basis functions,
	number of occurrence times,
	periodogram segmentation parameter,
	subset used in trimmed mean
K_ψ	number of vanishing moment for a certain wavelet
$K_{mc}(x)$	kurtosis of signal $x(n)$
$\kappa(t)$	continuous-valued indexing function
$\kappa_x(k_1, k_2)$	third-order moment (cumulant) of discrete-time process $x(n)$
L	filter length,
	segmented signal length
$L(\mathbf{x})$	likelihood ratio
$\mathcal{L}$	Lagrangian function
$\mathbf{\Lambda}$	diagonal matrix of eigenvalues
λ	eigenvalue,
	Lagrange multiplier
λ_r	average firing rate
M	number of channels/leads/events
$M(\Omega)$	Fourier transform of $m(t)$,
	number of realizations in an ensemble
$\mathcal{M}(p)$	function for model order determination
$m(t)$	continuous-time modulation function
m_i	constant amplitude factors in function series development
m_x	mean value of stationary process $x(n)$
$\mathbf{m}_x$	vector of mean values for $\mathbf{x}$
$m_x(n)$	mean value of the process $x(n)$
μ	adaptation parameter in the LMS algorithm,
	fraction of peak amplitude for QRS detection threshold selection
N	signal length,

	negative amplitude peak of an EP
N_φ	number of scaling coefficients $h_\varphi(n)$
N_ψ	number of wavelet coefficients $h_\psi(n)$
N_{FN}	number of false negative detections
N_{FP}	number of false positive detections
N_{TN}	number of true negative detections
N_{TP}	number of true positive detections
N_D	number of true detections
N_M	number of missed detections
N_F	number of false alarms
n	sequence index
∇_x	gradient with respect to the x
$\nabla_{\mathbf{w}}$	gradient with respect to the vector $\mathbf{w}$
η	threshold value
$\eta(n)$	influence function for robust, recursive averaging
$\eta_I(n)$	interval-dependent threshold
η_T	threshold parameter for wavelet denoising
Ω	continuous-time radian frequency
$\overline{\Omega}$	center of gravity of $X(\Omega)$
ω	discrete-time radian frequency
$o(k)$	binary variable
P	power,
	positive amplitude peak in evoked potentials
P_D	probability of detection
P_F	probability of false detection
P_M	probability of missed detection
P_i	power of the ith component of a rational power spectrum
$P_x(\Omega)$	characteristic function of x
$\mathcal{P}_{\text{CR}}$	data compression ratio
$\mathcal{P}_{\text{PRD}}$	percentage root mean-square difference
$\mathcal{P}_{\text{RMS}}$	root mean-square parameter
$\mathcal{P}_{\text{WDD}}$	weighted diagnostic distortion
p	model order
$p_{1,2}$	pair of complex-conjugate poles
$p_x(x)$	probability density function of x
$p(\mathbf{x};\theta)$	probability density function of $\mathbf{x}$ with θ as a parameter
$p(\mathbf{x};\boldsymbol{\theta})$	probability density function of $\mathbf{x}$ with $\boldsymbol{\theta}$ as a parameter vector
$\Psi(\Omega)$	continuous-time Fourier transform of the mother wavelet
$\psi(\cdot)$	influence function for robust averaging
$\psi(t)$	mother wavelet
$\psi_{s,\tau}(t)$	continuous-time family of wavelet functions
$\psi_{j,k}(t)$	dyadic discretized family of wavelet functions

q	model order
$q(n)$	discrete-time signal
R	threshold value in the IPFM model
$\mathcal{R}$	eigenvalue-based performance index
r	radius in complex plane,
	steepness of a sigmoid function between the two levels
r_i	RR interval preceding the ith beat
$\mathbf{r}_j$	autocorrelation vector of j lags of $x(n)$
$r(n)$	instantaneous RR interval estimate
$r_x(k)$	autocorrelation function of $x(n)$
$r_x(n_1, n_2)$	autocorrelation function of $x(n)$ between the samples n_1 and n_2
$r_x(k; n)$	time-varying autocorrelation function of $x(n)$
$r_x(\tau)$	autocorrelation function of $x(t)$
$r_{xy}(k)$	cross-correlation function of $x(n)$ and $y(n)$
ρ	cross-correlation coefficient,
	exponential damping factor
ρ_{ij}	cross-correlation coefficient between $x_i(n)$ and $x_j(n)$
$\rho_q(l)$	energy-normalized autocorrelation function of $q(n)$
$\mathbf{R}_x$	autocorrelation matrix of $\mathbf{x}$
$\mathbf{R}_v(n)$	spatial correlation matrix between different channels at time n
$\mathbf{R}_V$	$M \times M$ correlation matrix between the M different EPs in $\mathbf{V}$
$\mathbf{r}_{xy}$	cross-correlation vector of $x(n)$ and $\mathbf{y}(n)$
$\text{Res}[\cdot, \star]$	residue of a complex-valued function $(\cdot)$ at pole $(\star)$
$\Re(\cdot)$	real part
$S_x(e^{\jmath\omega})$	power spectrum of $x(n)$
$S_A(e^{\jmath\omega})$	discrete-time Fourier transform of the analytic signal $s_A(n)$
$S_x^r(e^{\jmath\omega})$	rhythmic activity of the EEG power spectrum
$S_x^a(e^{\jmath\omega})$	unstructured activity of the EEG power spectrum
$S_x(z)$	complex power spectrum of $x(n)$
$S_x(\Omega)$	energy spectrum of $x(t)$ in time–frequency representations
$S_m(\Omega)$	power spectrum of $m(t)$
$S_d^u(\Omega)$	power density spectrum of $d^u(t)$
$S_x(t, \Omega)$	spectrogram of $x(t)$
$S_{mc}(x)$	skewness of $x(n)$
s	IPFM model parameter related to ectopic beats
$\mathbf{s}$	signal vector
$s(n)$	discrete-time signal
$\check{s}(n)$	Hilbert transform of $s(n)$
$s_A(n)$	the analytic signal of $s(n)$
$s(t)$	lowpass envelope signal
$s_c(t/\beta)$	lowpass envelope signal with varying duration β
$\hat{\mathbf{s}}_a$	vector ensemble average estimator

$\hat{\mathbf{s}}_{a_l}$	vector ensemble subaverage estimator
$\hat{\mathbf{s}}_{e,M}$	vector exponential average estimator based on M EPs
$\hat{\mathbf{s}}_{r,M}$	recursive, robust average estimator
$\hat{\mathbf{s}}_w$	vector weighted average estimator
$\hat{s}_a(n)$	ensemble average estimator
$\hat{s}_{a,M}(n)$	ensemble average estimator based on M EPs
$\hat{s}_{a_l}(n)$	ensemble subaverage estimator
$\hat{s}_{e,M}(n)$	exponential average estimator based on M EPs
$\hat{s}_{\mathrm{med}}(n)$	ensemble median estimator
$\hat{s}_{\mathrm{tri}}(n)$	trimmed mean estimator
ξ	signal energy
$\xi(n)$	running signal energy
σ_v^2	variance of $v(n)$
σ_v	standard deviation of $v(n)$
σ	width parameter,
	variance of a Gaussian PDF
T	sampling interval
$\mathbf{T}$	synthesis matrix for ECG lead transformation
T_I	mean interval length
t	continuous-time
$\bar{t}$	"center of gravity" of $x(t)$
t_i	occurrence time of the ith beat
t_{k_e}	occurrence time of a sinus beat preceding an ectopic beat
$\hat{t}^f_{k_e+1}$	forward extended time (sinus rhythm replacing the ectopic)
$\hat{t}^b_{k_e}$	backward extended time (sinus rhythm preceding the ectopic)
t_e	occurrence time of an ectopic beat
τ	time delay/latency (continuous-time),
	convergence time for the LMS algorithm
$\boldsymbol{\theta}$	unknown vector parameter
θ	unknown scalar parameter
θ_i	discrete-time occurrence time
θ'_i	discrete-time segmentation onset of ith PQRST complex
U	window signal power normalization factor
$u(t)$	continuous-time unit step function
$\mathbf{V}$	$N \times M$ matrix modeling the noise of M different EPs
$V(z)$	transfer function of the discrete-time noise $v(n)$
$V[\cdot]$	variance of a estimate
$\mathcal{V}_j$	space spanned by the translated scaling function at scale j
$\mathbf{v}$	vector of discrete-time noise samples
$\mathbf{v}_i$	vector of noise signal samples from i^{th} evoked potential
$v(t)$	continuous-time noise

$v(n)$	discrete-time noise
$\mathbf{v}(n)$	$M \times 1$ vector of noise samples at time n of M channels
ν	continuous-time radian frequency, conduction velocity
W	window length
$W_B(e^{j\omega})$	discrete-time Fourier transform of the Bartlett window
$W_x(t, \Omega)$	continuous-time Wigner–Ville distribution
$W_{x_1, x_2}(t, \Omega)$	cross Wigner–Ville distribution between $x_1(t)$ and $x_2(t)$
$\mathcal{W}_j$	space spanned by the translated wavelet function at scale j
$\mathbf{w}$	weight vector
$\mathbf{w}^o$	optimal weight vector
$\mathbf{w}(n)$	weight vector at time n
w_i	scalar weight
$w_{j,k}$	coefficient expansion of the discrete wavelet transform
$w(n)$	discrete-time window or weighting function
$w(s, \tau)$	continuous wavelet transform
$w_B(n)$	Bartlett window
$\overline{\omega}_i$	i^{th} order spectral moment
ω_i	peak frequency of the ith spectral component
ω_c	normalized cut-off frequency
$\mathbf{X}$	data matrix of an ensemble of signal
$\mathbf{X}_M$	data matrix of an ensemble of M signals
$X(z)$	transfer function of $x(n)$
$X(e^{j\omega})$	discrete-time Fourier transform of $x(n)$
$X(\Omega)$	Fourier transform of $x(t)$
$X_A(\Omega)$	Fourier transform of the analytic signal $x_A(t)$
$X(t, \Omega)$	STFT of $x(t)$
$x(n)$	observed discrete-time signal
$x_{i,l}(n)$	observed discrete-time (i^{th} beat and l^{th} lead)
$\tilde{x}(n)$	reconstructed signal
$\hat{x}_p(n)$	p^{th} order linear prediction of $x(n)$
$x^{(i)}(n)$	i^{th} order discrete-time "derivative" of $x(n)$
$x(t)$	continuous-time signal
$x_c(t)$	continuous-time signal
$x_A(t)$	analytic signal of $x(t)$
$x(n)$	observed signal
$x_i(t)$	i^{th} continuous-time signal in an ensemble, wavelet approximation signal at scale i of $x(t)$
$\mathbf{x}$	vector of signal samples
$\tilde{\mathbf{x}}$	reconstructed/decompressed signal vector
$\mathbf{x}_i$	vector of signal samples from i^{th} evoked potential

$\mathbf{x}'_i$	piled lead vector of signals $\mathbf{x}_{i,j}$ from i^{th} beat
$\mathbf{x}_{i,j}$	vector of signal samples from i^{th} QRS complex at j^{th} lead
$\mathbf{x}_p$	vector of p signal samples preceding n
$\mathbf{x}(n)$	$M \times 1$ vector of samples at time n of M channels
$\mathcal{X}$	space expanded by a set of basis functions
$y(n)$	discrete-time filter output
$y_\infty(t)$	continuous-time ECG baseline wander signal
$y(t)$	cubic spline approximation of baseline wander
$y_j(n)$	wavelet decomposition detail signal at scale j
$\mathbf{Z}$	complex variable matrix
z	complex variable
$\tilde{z}_{e,i}$	exponentially updated peak amplitude at i^{th} beat
$z_{1,2}$	pair of complex-conjugate zeros
$z(n)$	discrete-time signal
$z_d(n)$	discrete-time signal decimated from $z(n)$
$z_u(n)$	discrete-time signal with zeros inserted

B.2   Abbreviations

A/D	analog-to-digital (conversion)
AEP	auditory evoked potentials
AIC	Akaike information criterion
ANS	autonomic nervous system
AR	autoregressive
AR(p)	autoregressive process of order p
ARMA	autoregressive moving average
ARV	average rectified value
AV	atrioventricular
AZTEC	amplitude zone time epoch coding
BAEP	brainstem auditory evoked potentials
BCI	brain–computer interface
BPM	beats per minute
BLMS	block least mean-square
CCU	coronary care unit
CNS	central nervous system
CPU	central processing unit
CSA	compressed spectral array
CWD	Choi–Williams distribution
CWT	continuous wavelet transform
dB	decibel

DC	direct current
DFT	discrete Fourier transform
DSP	digital signal processor
DTFT	discrete-time Fourier transform
DWT	discrete wavelet transform
DWPT	discrete wavelet packet transform
ECG	electrocardiogram
ECoG	electrocorticogram
EEG	electroencephalogram
EG	electrogram
EGG	electrogastrogram
EMG	electromyogram
ENG	electroneurogram
EOG	electrooculogram
EP	evoked potential
ERG	electroretinogram
FT	Fourier transform (continuous-time)
FFT	Fast Fourier Transform (discrete-time)
FIR	finite impulse response
GAL	gradient adaptive lattice
HRV	heart rate variability
HT	heart timing
Hz	Hertz
IBIS	integrate body mind information system database
ICU	intensive care unit
IF	interval function
IIF	inverse interval function
IIR	infinite impulse response
IIT	inverse interval tachogram
IT	interval tachogram
IPFM	integral pulse frequency modulation
KL	Karhunen–Loeve
KLT	KL transform
kHz	kilohertz
LMS	least mean-square
LTST	long-term ST database
MA	moving average
MDL	minimum description length
MEG	magnetoencephalogram
ML	maximum likelihood
MSE	mean-square error
MMSE	minimum mean-square error

MRI   magnetic resonance imaging
MUAP   motor unit action potential
MVC   maximal voluntary contraction
NN   normal-to-normal RR interval
PC   personal computer
PDF   probability density function
PET   positron emission tomography
pNN50   pairs of NN RR intervals differing by more than 50 ms
PNS   peripheral nervous system
PRD   percentage root mean-square difference
PWVD   pseudo Wigner–Ville distribution
REM   rapid eye movement
RMS   root mean-square
Res   residue
rMSSD   root mean-square of successive differences
ROC   receiver operating characteristic
SA   sinoatrial (node)
SAPA   scan-along polygonal approximation
SDNN   standard deviation of NN intervals
SEM   spectral error measure
SEP   somatosensory evoked potentials
SNR   signal-to-noise ratio
SPA   spectral parameter analysis
SPECT   single photon emission computed tomography
SPI   spectral purity index
SQUID   superconducting quantum interference device
SSW   spikes and sharp waves
STFT   short-time Fourier transform
SVD   singular value decomposition
SVPB   supraventricular premature beat
TINN   triangular interpolation index
VCG   vectorcardiogram
VEP   visual evoked potentials
VPB   ventricular premature beat
VT   ventricular tachycardia
WCT   Wilson central terminal
WVD   Wigner–Ville distribution
WDD   weighted diagnostic distortion

Index

10/20 electrode system, 37, 186
12-lead vectorcardiogram, 426

50/60 Hz powerline interference, 76, 190, 202, 441, 473–483

abbreviations, 658
abrupt changes, 62
action potential, 7, 28
activity (Hjorth descriptor), 100
adaptive filter (LMS), 83, 91, 279, 482
adaptive signal whitening, 355
afferent, 27
Akaike information criterion (AIC), 118
all-or-nothing principle, 7
all-pole modeling, *see* autoregressive
        modeling
alpha rhythm, 34, 78
alternate ensemble average, 244
ambiguity function, 142–147
ambulatory monitoring
    ECG, 439, 441, 444
    EEG, 43
analysis filter bank, 300
analytic signal, 145
angina pectoris, 438
arrhythmia, 430
    atrial, 434
    bigeminy, 434
    bradycardia, 431
    paroxysmal, 431
    persistent/permanent, 431
    respiratory sinus, 432
    tachycardia, 431
    trigeminy, 434
    ventricular, 436
artifact cancellation, 78–91
artifacts in EEG, 73–91
    cardiac, 75, 83
    electrode, 76
    equipment, 76
    eye movement, 73

    muscle, 74
asphyxia, 125
atrioventricular node, 414
atrium, 413
auditory EP, 185, 234, 252
autocorrelation/covariance methods, 106
automaticity, 414
autoregressive modeling, 648
    autocorrelation method, 106
    Burg's method, 112
    covariance method, 106
    impulse input, 67
    model order, 118
    modified covariance method, 110
    multivariate, 67
    sampling rate, 119
    segmentation, 131
    spectral parameters, 119–125
    stability, 116
    time-invariant, 65, 104
    time-variant, 66
autoregressive moving average modeling, 64, 122, 647
average beat subtraction, 534
average firing rate, 342, 373
average rectified value, 355
averaging, *see* ensemble averaging
axon, 28, 338
a posteriori filter, 241, 249

baroreceptor reflex, 615
Bartlett window, 94
baseline wander, 441, 457–473
basis functions, 260, 500
    cal, 263
    Karhunen–Loève, 264
    sal, 263
    sine, cosine, 260, 481
    Walsh, 263
beta rhythm, 34, 75, 78
bias, 645
bigeminy, 434

bioelectricity, 6
bispectrum, 63
blinks, 73, 78, 82
block LMS algorithm, 279
bradyarrhythmia, 431
bradycardia, 431
brain–computer interface (BCI), 47
brainstem auditory EP, 185, 214
Burg's method, 112

cascade algorithm, 308
central sulcus, 30
cerebral cortex, 30
characteristic function, 225, 380
Choi–Williams distribution, 154
circadian rhythm disorders, 45
circulant matrix, 482
clustering, 387
    MUAPs, 387
    QRS morphologies, 456
cognitive EP, 189
Cohen's class, 153–158
coherence function, 616
Coiflet wavelets, 310
comb filter, 464
compensatory pause, 433
complexity (Hjorth descriptor), 101
compressed spectral array, 139
compression ratio, 541
conduction blocks, 438
conduction velocity estimation, 365–371
    multichannel, 369
    two-channel, 367
consistency, 645
continuous wavelet transform (CWT), 289
correlation
    ergodicity, 645
    estimator, 93
    function, 643
    matrix, 59, 643
cortex, 30, 68, 185, 187
covariance matrix, 59, 643
cross Wigner–Ville distribution, 150
cross-correlation coefficient, 200, 603
cross-correlation matrix, 643
cross-power spectrum, 615
cubic spline baseline estimation, 470

damped sinusoids, 272
data acquisition, 14
data compression, 3
data compression (ECG), 456, 514–544

AZTEC, 520
    compression ratio, 541
    direct methods, 519–526
    fan method, 525
    long-term predictor, 536
    lossless, 514, 517–518
    lossy, 514
    PRD, 541
    quantization and coding, 540
    rate distortion curve, 542
    RMS, 542
    SAPA, 522
    transform-based methods, 526–533
    wavelet packets, 531
    wavelets, 531
    WDD, 543
databases, 17
    AHA, 18
    European ST–T, 18
    IMPROVE, 18
    LTST, 18
    MIMIC, 18
    MIT–BIH arrhythmia, 18, 542
Daubechies wavelets, 310
decimation (sampling rate), 465
decomposition (EMG), 383
deconvolution, 229
delta rhythm, 34, 73
dendrites, 28, 70
denoising, 312–318
    hard thresholding, 313
    soft thresholding, 313
depolarization, 7, 414–419
direct decomposition, 257
direct sum, 257
discrete wavelet transform (DWT), 291
dyadic sampling, 290

ECG, *see* electrocardiogram
ectopic beat correction, 605–614
    correlation-based, 608
    heart timing, 611
    interpolation-based, 609
EEG, *see* electroencephalogram
efferent, 27
eigenvalue, 638
eigenvector, 638
electrocardiogram (ECG), 11, 411
    ambulatory monitoring, 444
    beat morphology, 438
    filtering, 454

generation, 415–419

high-resolution, 447

intensive care monitoring, 444

late potentials, 447

noise and artifacts, 440–443

recording techniques, 419–426

resting, 443

rhythms, 430

stress test, 445

waves and time intervals, 426–430

electrocorticogram (ECoG), 11

electrode motion artifacts, 344, 441

electrodes, 9

ECG, 419

EEG, 37

EMG, 343

electroencephalogram (EEG), 11, 25

10/20 electrode system, 37, 186

alpha rhythm, 34

amplitude, 32

artifacts, 73–91

beta rhythm, 34

delta rhythm, 34

frequency, 32

gamma rhythm, 34

ictal, 35

mental tasks, 47

mu rhythm, 49

recording techniques, 37

sampling rate, 39

sleep rhythms, 35

spikes and sharp waves, 34, 62

theta rhythm, 34

video recording, 42

electrogastrogram, 14

electrogram (EG), 11, 412

electromyogram (EMG), 12, 75, 77, 337

amplitude estimation, 347–360

diagnostic, 345

ergonomics, 345

kinesiology, 345

prosthesis control, 346

recording techniques, 343

spectral analysis, 361–364

electromyographic noise, 441, 484

electroneurogram (ENG), 12

electrooculogram, 73

electrooculogram (EOG), 14

electroretinogram (ERG), 13

EMG, *see* electromyogram

endogenous response, 189

ensemble averaging, 83, 181, 192

alternate, 244

as a linear filter, 200–202

exponential, 202–205

homogeneous, 193–200

inhomogeneous, 207–218

latency correction, 230

robust, 219

SNR, 200

weighted, 207–218

ensemble correlation, 236–241

ensemble median, 221

ensemble variance, 198

envelope, 501

EOG, *see* electrooculogram

EP, *see* evoked potentials

epilepsy, 40, 68, 75, 76, 141

partial seizures, 41

primary generalized seizures, 41

ergodicity, 645

estimation–subtraction filter, 479

Euclidean norm, 637

event series, 373, 576, 582–585

evoked potentials (EPs), 11, 181

auditory, 185, 234, 252

brainstem auditory, 185, 214

cognitive, 189

latency, 181

noise and artifacts, 190

somatosensory, 187, 214, 252

visual, 188, 197, 214, 223, 234, 252

wave definitions, 182

excess mean-square error, 87

exogenous response, 189

exponential averaging, 202–205, 224, 254, 285

eye movement, 73, 78–91, 191

eye-closing period, 505

feature extraction, 2, 48, 386, 387, 456

fiducial point, 568

filter

a posteriori, 241, 249

comb, 464

FIR, 88, 105, 110, 250, 460

forward-backward IIR, 462

IIR, 245

inverse, 229

lattice, 108

least mean-square, 83, 91, 279

lowpass differentiator, 498

noncausal, 245
nonlinear, 476
notch, 473
prediction error, 105
time-varying, 252, 467, 484
time-varying Gaussian, 484
filtered-impulse signal model, 374
firing pattern, 342
firing rate, 29, 70
forward-backward IIR, 462
Fourier series, 261
Fourier transform
    continuous-time, 227
    discrete-time, 60, 92
    fast (FFT), 92
    short-time, 61
Frank lead system, 424
French–Holden algorithm, 585
Frobenius norm, 281, 637

gamma rhythm, 34
Gaussian PDF
    definition, 59
    EEG analysis, 57, 197
    EMG analysis, 349
generalized eigenvalue problem, 209
generalized Gaussian PDF, 219
gradient adaptive lattice algorithm, 160

Haar wavelet, 290, 297
heart rate variability (HRV), 431, 567
    ectopic beats, 605–614
    generalized IPFM model, 577, 587, 611
    heart rhythm representations, 573
    IPFM model, 574–578
    pNN50, 570
    rMSSD, 570
    SDANN, 570
    SDNN, 570
    spectral analysis, 589–603
    time domain measures, 570–573
    TINN, 572
    triangular index, 572
heart rhythm representation, 578–589
    event series, 582–585
    heart timing, 585–589
    interval function, 579
    interval tachogram, 578
    inverse interval function, 579
    inverse interval tachogram, 578
    lowpass filtered event series, 583
heart surgery, 140

heart timing, 585–589, 611
hemispheres, 30
Hessian matrix, 396
higher-order moments, 62, 363
Hilbert transform, 502
Hjorth descriptors, 100–102
hypersomnia, 44
hyperventilation, 42
hypoxia, 125

indexing function, 577
influence function, 223
inner product, 255, 636
insomnia, 44
instantaneous LMS algorithm, 278
integral pulse frequency modulation model,
    574–578
interfiring intervals, 342
interpolation (sampling rate), 465
interval function, 579
interval tachogram, 578
inverse z-transform, 123
inverse filtering, 229
inverse interval function, 579
inverse interval tachogram, 578
isoelectric line, 417
isometric force, 343
isopotential map, 183

Karhunen–Loève expansion, 264
Karhunen–Loève transform (KLT), 528
knee-jerk reflex, 30
kurtosis, 62
K complexes, 35, 67

Lagrange multipliers, 107, 209, 266, 640
Laplacian PDF, 219, 353
late potentials, 476
latency
    definition, 181
    estimation, 229
    shifts, 225
lattice filter, 108, 160
lead
    augmented unipolar limb, 420
    bipolar, 419
    bipolar limb, 420
    unipolar, 419
lead system
    ECG, 419
        bipolar, 420
        orthogonal, 423

precordial, 420
standard, 419
synthesized, 425
Frank, 424
lead vector, 416
leader–follower clustering, 387
leakage, 94
least mean-square (LMS), 83
algorithm, 85, 159, 279, 482
block, 279
block algorithm, 285
convergence, 85
excess mean-square error, 87
instantaneous, 278
least-squares error, 106
least-squares solution, 392, 641
Levinson–Durbin recursion, 108, 113
likelihood function, 217
limiter
hard, 224
sign, 224
linear equations, 641
linear models, 63
list of symbols, 649
lobe
frontal, 30
occipital, 30
parietal, 30
temporal, 30
locked-in syndrome, 47
Lomb's periodogram, 597–603
look-back detection mode, 507
lossless compression (ECG), 514, 517–518
lossy compression (ECG), 514
lowpass differentiator, 498
lowpass modeling, 261
lumped-parameter model, 70

magnetoencephalogram, 26
magnitude squared coherence, 616
Mahalanobi distance, 388
marginal condition
frequency, 149
time, 149
matched filter, 232, 490
matrix definitions, 633–639
matrix inversion lemma, 389, 637
matrix optimization, 640
maximal voluntary contraction, 342
maximum likelihood estimation
amplitude, 347

ensemble correlation, 239
latency, 230
occurrence time, 490
occurrence time and amplitude, 491
occurrence time, amplitude and
duration, 494
signal waveform, 216, 220
time delay, 367
mean frequency (MNF), 361
mean instantaneous frequency, 152
mean-square error (MSE), 79, 256, 264, 279
with constraint, 91
median, 221
median frequency (MDF), 361
method of steepest descent, 84
Mexican hat wavelet, 290
minimum description length, 118
mobility (Hjorth descriptor), 101
model order, 118
Akaike information criterion, 118
minimum description length, 118
modified covariance method, 110
mother wavelet, 288
motor imagery, 48
motor nerves, 27
motor unit, 338–343
motor unit action potential (MUAP), 338
motor unit recruitment, 339
moving average modeling, 648
mu rhythm, 49
MUAP resolution, 391
MUAP train, 342, 372
amplitude, 377
model, 372
power spectrum, 378
multiresolution signal analysis, 292–300, 513
myocardial infarction, 439
myocardial ischemia, 438
myocardium, 413
myopathy, 340

needle EMG, 343
negative predictive value, 17
nervous system, 27
autonomic, 27
central, 27
parasympathetic, 27, 414, 430, 431, 570,
590
peripheral, 27
somatic, 27
sympathetic, 27, 414, 430, 431, 570,
590

neuromuscular junction, 338, 342
neuron
      inter-, 28, 68
      motor, 28
      postsynaptic, 28
      sensory, 28
neuropathy, 340
Newton's method, 395
noise reduction
      ensemble averaging, 192
      linear filtering, 241–253
nonlinear transformation, 68, 486, 492, 500
nonparametric spectral analysis, 91–97
nonstationarity, 61
norm
      Euclidean, 637
      Frobenius, 637
normal equations, 108

oddball task, 190
ordinary eigenvalue problem, 266
orthogonal expansions, 254–272
      Karhunen–Loève, 264
      sine,cosine, 260
      SNR, 258
      truncation, 257
      Walsh, 263
orthogonal lead system, 423
orthogonal matrix, 635
outer product, 636

P wave, 427
pacemaker, 412
parasomnia, 45
Parseval's theorem, 130
partial fraction expansion, 121
pattern reversal, 188
peak-and-valley picking strategy, 497
performance evaluation, 16, 507
periodogram
      definition, 93
      Lomb's, 597–603
      mean, 93
      variance, 95
photic stimulation, 42, 139
Physionet, 19
polynomial fitting, 470–473
polyphasic MUAPs, 341
polysomnography, 46
positive definite, 636
positive predictive value, 17
positive semidefinite, 636

postsynaptic potential
      excitatory, 29, 31, 69
      inhibitory, 29, 69
power spectrum, 60, 92, 646
powerline interference, 76, 190, 202, 344,
      441, 473–483
PQ interval, 429
PQRST delineation, 510–513
prediction error filter, 105
premature beat, 433
      supraventricular, 433
      ventricular, 433
probability density function (PDF)
      Gaussian, 59
      generalized Gaussian, 219
      Laplacian, 219, 353
      uniform, 219, 227
Prony's method
      least-squares, 277
      original, 274
pseudo Wigner–Ville distribution, 151
Purkinje fibers, 414

Q wave, 428
QRS complex, 428
QRS detection, 455, 485–509
      decision rule, 486, 504–507
      nonlinear transformation, 486, 500
      performance evaluation, 507
      preprocessing, 497
      signal and noise problems, 487
      signal modeling, 488
      threshold, 504
QT interval, 429

R wave, 428
rank of a matrix, 634
rapid eye movement (REM), 35, 58, 74
rate distortion curve, 542
receiver operating characteristic (ROC), 509
recording techniques
      auditory EP, 185
      ECG, 419
      EEG, 37
      EMG, 343
      somatosensory EP, 187
      visual EP, 188
recruitment, 339
reentry, 430
refinement equation, 294
reflection coefficients, 113
refractory period, 8, 30, 415, 490, 505

repolarization, 7, 414–419
respiratory sinus arrhythmia, 432
rhythms
    brain, 31
    heart, 430, 457
RR interval, 429
running average, 102

S wave, 428
sampling jitter, 227
sampling rate alteration, 464
scaling function, 292
scalogram, 291
segmentation, 62, 125–135
    dissimilarity measure, 127
    periodogram-based, 128–131
    reference window, 127
    test window, 127
    whitening approach, 131–134
seizure, 34, 68, 141, 162
sensitivity, 17
sensory nerves, 27, 183, 187
septum, 413
short-time Fourier transform (STFT), 137–142
sigmoid function, 70
signal acquisition, 14
signal decomposition (EMG), 383
similarity measure, 388, 603
simulation, 21
single-trial analysis, 182, 197, 253–278
singular value decomposition, 639
sinoatrial node, 414
sinus rhythm, 431
skewness, 62
sleep rhythms, 35, 67
sleep spindles, 35
sliding window, 352
smearing, 94
soma, 28
somatosensory EP, 187, 214, 252
spatial correlation, 80, 356
specificity, 17
spectral analysis
    EMG, 361
    model-based, 103–119
    moments, 100
    nonparametric, 91–97
    segmentation, 128
spectral averaging, 95
spectral compression, 361

spectral parameters
    AR-based, 119–125
    Hjorth descriptors, 100
    peak frequency, 99
    power in bands, 98
    spectral purity index, 102
    spectral slope, 99
spectral purity index, 102
spectrogram, 138
spectrum of counts, 592–596
spike and sharp waves, 34, 62
spike-wave complexes, 35
split trial assessment, 198
ST segment, 429
standard 12-lead ECG, 419
stationarity, 60
steepest descent, 84
stochastic process, 642
subaveraging, 197
surface EMG, 343
synapse, 28
synthesis filter bank, 303
synthesized leads, 425

T wave, 429
T wave alternans, 439
tachyarrhythmia, 431
tachycardia, 431
theta rhythm, 34, 73
time delay estimation
    multichannel, 369
    two-channel, 367
time–frequency analysis, 135–162
    ambiguity function, 142–147
    Choi–Williams distribution, 154
    Cohen's class, 153–158
    cross Wigner–Ville distribution, 150
    GAL algorithm, 160
    LMS algorithm, 159
    mean instantaneous frequency, 152
    pseudo Wigner–Ville distribution, 151
    short-time Fourier transform, 137–142
    Wigner–Ville distribution, 147–152
Toeplitz matrix, 60, 81, 108, 110, 350, 635
trace of a matrix, 636
triangular window, 94
trigeminy, 434
trimmed means, 222

uncertainty principle, 139
uniform PDF, 219, 227

vagus nerve stimulator, 44
vectorcardiogram (VCG), 419
ventricle, 413
vertex waves, 35
visual EP, 188, 197, 214, 223, 234
visualEP, 252
volume conductor, 9

Walsh functions, 263
wave delineation, 455, 510–513
waveform resolution (MUAPs), 391
wavelet data compression, 531
wavelet denoising, 312–318
wavelet equation, 296
wavelet function, 295
wavelet packets, 531
wavelet series expansion, 296
wavelet transform, 288–292, 513
wavelets, 286–318
    Coiflet, 310
    Daubechies, 310
    dyadic sampling, 290
    Haar, 290, 297
    Mexican hat, 290
    sinc, 306
weighted averaging
    Gaussian noise, 216
    MSE, 208
    signal-to-noise ratio, 209
    varying noise variance, 210
    varying signal amplitude, 214
Welch's method, 97
white noise, 63, 230, 488, 646
wide-sense stationary, *see* stationary, 644
Wiener filtering, 90, 246
Wiener–Hopf equations, 246, 251
Wigner–Ville distribution, 147–152
window
    Bartlett, 94
    Hamming, 143, 461
    rectangular, 460
    triangular, 94
windowing, 95
Woody's method, 229–236

z-transform, 245, 646

Mexican road sign of particular importance in biomedical signal processing: *Do not maltreat the signals*! (Photo by LS)